ENCYCLOPEDIA OF PHYSICS

EDITED BY

S. FLÜGGE

VOLUME XLV

NUCLEAR INSTRUMENTATION II

CO-EDITOR

E. CREUTZ

WITH 293 FIGURES

SPRINGER-VERLAG

BERLIN · GÖTTINGEN · HEIDELBERG

1958

HANDBUCH DER PHYSIK

HERAUSGEGEBEN VON

S. FLÜGGE

BAND XLV

INSTRUMENTELLE HILFSMITTEL DER KERNPHYSIK II

MITHERAUSGEBER

E. CREUTZ

MIT 293 FIGUREN

SPRINGER-VERLAG
BERLIN · GÖTTINGEN · HEIDELBERG
1958

ISBN 978-3-642-45905-4   ISBN 978-3-642-45903-0 (eBook)
DOI 10.1007/978-3-642-45903-0

© by Springer-Verlag oHG. Berlin · Göttingen · Heidelberg 1958
Softcover reprint of hardcover 1st edition 1958

Contents.

Page

Ionization Chambers in Nuclear Physics.

By

H. W. FULBRIGHT.

With 29 Figures.

I. Introduction.

Since the discovery of x-rays ionization devices have been widely used for the detection of radiation and for the measurement of its intensity and quality. Cosmic radiation was discovered in a series of ionization measurements made by HESS and KOLHÖRSTER[1], between 1911 and 1914, the existence of neutrons was demonstrated by CHADWICK[2] in 1932 using an ionization chamber to detect recoiling protons, and the large amount of energy released in fission was shown by FRISCH[3] in 1939 using an ionization chamber. In addition to these outstanding examples, there is a long list of less sensational applications which could be written down even in the field of x-rays alone. In recent years ionization chambers have been developed especially for studying single ionizing events due to fast particles from radioactive decay, nuclear disintegrations, and in a few experiments, cosmic radiation.

The state of knowledge about the physical processes taking place in ionization chambers can be summed up simply: we know enough to be able to use ionization chamber methods with confidence in making physical measurements of many kinds, and we have a very good conception of most of the phenomena occurring in ionization chambers, but there remain several points on which the picture is slightly hazy. One of them concerns the stopping of charged particles moving with velocities comparable with, or less than, those of the electrons in the molecules through which the particles are moving. Another is the matter of the velocity distribution of slow electrons moving through a gas under the action of an electric field. This distribution could in principle be calculated if a number of types of interactions between the electrons and the gas molecules were to be taken into account in detail, but the conditions have so far proved too complicated to permit a complete analysis except in the case of the noble gases. Nevertheless the behavior of the electrons is well enough known for a qualitative understanding of ionization chamber characteristics in terms of the various interactions of the ionization electrons with the gas.

In subdivision II of this article a brief survey of the processes by which ion pairs are created in the chamber will be given, along with references to the extensive literature on that subject. In subdivision III the behavior of slow, charged particles in the gas will be described. Subdivision IV is concerned with the utilization of the signal. Finally, in Subdivision V, a description of a variety of typical

---

[1] HESS: Phys. Z. **12**, 998 (1911); **13**, 1084 (1912); **14**, 610 (1913). — KOLHÖRSTER: Verh. dtsch. phys. Ges. **15**, 1111 (1913); **16**, 719 (1914).
[2] CHADWICK: Proc. Roy. Soc. Lond. **136**, 692 (1932).
[3] O. R. FRISCH: Nature, Lond. **143**, 276 (1939).

experiments will be given, along with references to others. For discussions of methods of construction and many other practical matters the reader is referred to ROSSI and STAUB [7], and to the references cited.

II. General principles of ionization chamber operation.

1. Definitions and remarks. Although the ionization apparatus used by the early experimenters consisted of nothing but an electroscope in a container, in modern times the two essential parts, the container of gas in which the ionization occurs and the indicating device, are almost invariably separated. The change has come about partly because the electronic amplifier has become so useful in the indicator. Therefore we will speak of an ionization chamber as an enclosure containing a substance in which the ionization is produced, and having two or more electrodes between which electrical currents can flow when the charges separated from each other in the ionization process move about inside

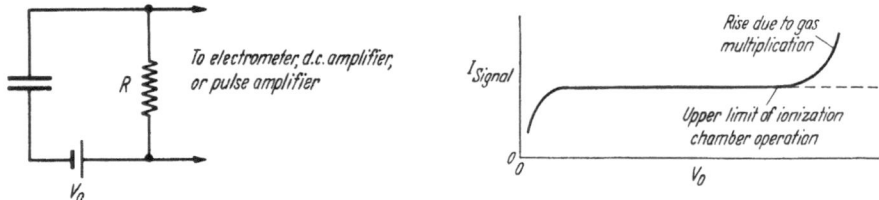

Fig. 1a and b. a) Ionization chamber connections. b) Output signal dependence on applied voltage for constant intensity of radiation.

the chamber under the influence of an applied electric field. The induced currents constitute the signal. Frequently the chamber contains only two electrodes. The electric field may be supplied by an external circuit of high impedance which maintains a constant average potential difference (bias) between the electrodes (see Fig. 1). Charges of the two electrical signs migrate toward their respective electrodes, generating voltage pulses by electrical induction. The proportional counter and the Geiger counter, which both developed from the ionization chamber, are excluded from the definition because their operation is based upon "gas multiplication", an avalanching process initiated when the original ionization electrons have gained sufficiently great kinetic energies from their motion through the strong electrical fields characteristically applied in those devices to create ion pairs by collisions with gas molecules.

Our definition implicitly includes solid and liquid as well as gas fillings for the chamber, but in practice only gas fillings have been of much importance[1]. The discussion which follows will be entirely in terms of gas-filled chambers.

In using an ionization chamber for measurements of intensities of radiation an experimenter needs to know the relationship between the true intensity and the response of his instrument, a relationship which he can usually determine by means of a calibration experiment. Once he has obtained a calibration he can proceed to use the instrument with confidence, providing that the calibration

---

[1] The "crystal counter" is a solid ionization chamber. P. J. VAN HEERDEN: The Crystal Counter, a New Instrument in Nuclear Physics. Amsterdam: N. V. Noord Hollandsche Uitgevers Maatschappij 1945. — R. HOFSTADTER: Phys. Rev. **72**, 1120 (1947). — Nucleonics **4** No. 4, 2 (1949); **4**, No. 5, 29 (1949), and CHAMPION: Progr. Nucl. Phys. **3**. Liquid-filled chambers are mentioned by JAFFE [8] and have been proposed by a number of people, but there appear to have been few applications of importance. See, however, J. H. MARSHALL: Phys. Rev. **91**, 905—909 (1953) and Rev. Sci. Instrum. **26**, 521 (1955) for a liquid argon counter. The performance of a decane-filled chamber is described by M. A. VAN DILLA: Phys. Rev. **85**, 705 (1952).

remains constant. It turns out that the problem of calibrating can usually be reduced to that of finding the effective average number of ion pairs produced per electron volt of energy lost by the primary ionizing particles in the chamber. This quantity may vary slightly with the type of particle, and it certainly varies with the type of gas, but in many gases it is for most practical purposes a constant.

One might therefore argue that a detailed knowledge of the physical processes by which the ionization is produced is not necessary. However that is not entirely true, because questions involving such matters as the straggling in ionization frequently arise, therefore a certain familiarity with the processes of ionization is highly desirable. A brief account of those processes will now be given, along with references to the extensive literature on the subject.

2. The primary ionizing process. Any atom, ion, or charged particle moving through the gas of an ionization chamber will lose energy by collisions with the gas molecules, and will tend toward thermal equilibrium with them. The slowing down of fast charged particles in matter is due largely to ionization and atomic excitation losses until the velocity of the particle has fallen near or below that of the outermost atomic electrons, then other loss mechanisms become important, especially nuclear elastic scattering [1]. At relativistic velocities primary electrons lose appreciable energy in the form of bremsstrahlen and Cerenkov radiation. However, for present purposes losses due to collisions with atomic electrons are by far the most important. We will call a fast charged particle, ion, or atom which passes through the gas of an ionization chamber initiating a series of individual ionizing events a *primary ionizing particle*. Typically it may be an alpha-particle from the radioactive decay of a sample inside the chamber, or an electron ejected from the walls or gas by an incoming photon, or it may be a charged particle or ion released in a nuclear reaction between a nucleus in the gas and an incoming neutron. We will refer to the passage of the primary ionizing particle as a *primary ionizing event*.

The conservation laws set a lower limit on the energy which a primary particle must have in order to be able to produce ionization. If a particle of mass M collides with a stationary atom of mass M_1, the maximum excitation energy which the atom can receive is

$$E' = \frac{M}{M_1 + M} E,\tag{2.1}$$

where E is the energy of the incident particle. Therefore a primary electron must have an energy slightly in excess of I_0, the first ionization potential of the atom, before ionization can occur. The corresponding velocity threshold is roughly the velocity of the outermost electrons in the atom. On the other hand, according to (2.4) if $m = M$ the threshold energy is $2I_0$, so atoms moving through a gas of their own type may create ions in single collisions even when moving very much slower than the outermost atomic electrons, providing, of course, that a physical mechanism is available through which the ionization can occur. Thresholds at the limit imposed by the conservation laws may have been observed for collisions between neutral gas atoms [5]. Ordinarily the primary ionizing particles observed in experiments in nuclear physics are charged and have at the start energies 10^2 to 10^7 times the absolute limit for ionization.

If, following J. J. Thomson[1], one views ionization as a process in which an essentially free electron receives energy from the incident particle during a collision, escaping if the energy acquired exceeds its particular binding energy in the atom, one finds another energy limit. The maximum energy which the

---

[1] J. J. Thomson: Phil. Mag. **23**, 449 (1912).

electron can acquire is

$$W_{max} = \frac{4\,m\,M}{(m+M)^2}\,E\,, \tag{2.2}$$

non-relativistically, where $m =$ mass of electron, and $M =$ mass of the incident particle. For heavy particles this reduces to

$$W_{max} = \frac{4\,m\,E}{M}\,. \tag{2.3}$$

Setting this equal to I_0 gives the energy threshold for a heavy primary of mass M,

$$E_{min} = \frac{M\,I_0}{4\,m}\,. \tag{2.4}$$

For oxygen $I_0 = 12.2$ ev, so according to this simple model a proton would need a minimum energy of 5600 ev to be able to create an ion pair in oxygen. For ionizing the K shell it would need at least 723 kev ($I_k = 530$ kev). However, when the full effects of the binding of the electrons in the atoms are taken into account this minimum energy requirement is no longer strictly true. For example, Henneberg[1] showed quantum mechanically that if screening is neglected the probability of K-shell ionization by heavy particles rises as E^4 at low energies, eventually reaches a maximum at $E = 13.5\,\dfrac{M\,Z^2}{m}$ ev, where Z is the atomic number of the target atom, and thereafter falls slowly. When screening is taken into account the maximum comes at lower energies. For protons in oxygen Henneberg's formulas give unscreened and screened values of 1.6 and 0.5 Mev, respectively, the latter being rather close to the value 0.723 Mev found above using the simple model. Henneberg's calculations have been verified fairly well by a number of experiments[2].

On the free electron assumption, classical mechanics gives for the upper energy limit of the electrons knocked out of atoms by heavy ionizing particles

$$W_{max} = \frac{2\,m\,v^2}{(1-\beta^2)}\,, \tag{2.5}$$

where $\beta = v/c$, and $v =$ the velocity of the primary particle. In the non-relativistic limit $W_{max} = 2\,m\,v^2 = 4\,m\,E/M$, so the maximum energy of the secondary electrons from bombardment by, say, 1 Mev protons is 2.17 kev. Additional, secondary, electrons will in turn be knocked out of other atoms by the more energetic electrons (delta rays) from the primary ionization process. The delta ray energy spectrum can be calculated approximately from the Rutherford scattering formula. This gives the following non-relativistic result for the number of delta rays per unit track length having energies between W and $W + dW$:

$$\frac{dn}{dx} = \frac{e^4\,z^2\,N Z\,dW}{E\,W^2}\quad \left(I_0 \ll W < \frac{4\,m\,E}{M}\right) \tag{2.6}$$

where M and E are the mass and energy of the primary and z its charge, and NZ is the number of electrons per cm^3. According to quantum mechanical calculations of the Henneberg type the electron spectrum does not cut off sharply, but instead persists to very much higher energies than the limit (2.5) indicates, but at very low intensity. Continuous backgrounds of electrons emitted by targets under intense proton bombardment in the course of Coulomb excitation experiments have been attributed to the primary ionization process[3].

[1] W. Henneberg: Z. Physik **86**, 592 (1933).

[2] For example H. W. Lewis, B. E. Simmons and E. Merzbacher: Phys. Rev. **91**, 943 (1953).

[3] T. Huus, J. H. Bjerregaard and B. Elbek: Dan. Vid. Selskab. Mat.-Fys. Medd. **30**, No. 17 (1956).

The relative numbers of primary and secondary ions formed can be found by experiment. Primary ionization can be measured by means of a cloud chamber[1], or can be calculated from the measured efficiency of Geiger counters[2]. Total ionization can be found by use of cloud chambers, ionization chambers, or proportional counters. Experimental results of TERROUX and WILLIAMS obtained using beta particles having energies up to about 1 Mev in a cloud chamber indicate that with beta particles as primaries $\frac{1}{2}$ to $\frac{2}{3}$ of the total is secondary ionization. With electrons as primaries many of the delta rays are quite energetic, the upper energy limit of their spectrum being very nearly the full energy of the primary electron. The fact that so much energy goes into producing so few delta rays has important statistical consequences (Sect. 5). With heavy primaries a much smaller fraction of the total ionization is produced by fast delta rays than is the case with electrons of the same energy, as can be estimated from (2.6). There does not seem to be much direct experimental information on this point. It is found that the average energy, w, required to produce an ion pair is of the order of twice the ionization potential of the molecules, that is, about half the energy lost by collisions eventually is wasted insofar as ionization chamber operation is concerned (see Table 6).

The problem of calculating the stopping power of matter for charged particles by the process of collision is an old one which has been solved formally in a good approximation. To arrive at a solution it was necessary to find the probability of ionization or excitation in each of the various atomic shells. The calculation has proved too messy to carry out in detail for any but the lightest elements. However, the practical difficulty was reduced greatly by the derivation of an approximate expression for collision stopping power in which the remaining problems in calculation are buried in a single parameter, the mean excitation potential I of the atoms after the collision, which can be determined for any element by a single experiment.

The well known formula for the stopping power for heavy particles, derived by BETHE[3] and MØLLER[4] is

$$- \frac{dE}{dx} = \frac{4\pi e^4 z^2}{m v^2} N Z \left[\log \frac{2 m v^2}{I} - \log (1 - \beta^2) - \beta^2 \right] \qquad (2.7)$$

where

$e =$ electronic charge,
$m =$ electronic mass,
$z e =$ charge of the heavy particle,
$N Z =$ number of electrons per unit volume of stopping material,
$\beta = v/c$, where $c =$ velocity of light,
$v =$ velocity of the heavy particle, and
$I =$ the mean excitation potential.

Notice that (2.7) is a function of the velocity, but not the mass, of the particle.

The use of the Born approximation in the derivation suggests that (2.7) should be valid only for values of $v \gg v_k$, the velocity of the K-shell electrons. If that were so it would have only a limited usefulness, because the requirement could not be satisfied by most particles involved in ordinary nuclear reactions, even in as light a gas as oxygen, as we have seen. Fortunately the approximation

[1] C. T. R. WILSON: Proc. Roy. Soc. Lond. **104**, 192 (1923). — E. J. WILLIAMS: Proc. Roy. Soc. Lond., Ser. A **135**, 108 (1935). — E. J. WILLIAMS and F. R. TERROUX: Proc. Roy. Soc. Lond., Ser. A **126**, 289 (1930).
[2] F. L. HEREFORD: Phys. Rev. **74**, 574 (1948).
[3] H. A. BETHE: Z. Physik **76**, 293 (1938). — Ann. Physik **5**, 325 (1930).
[4] C. MØLLER: Ann. Physik **14**, 531 (1932).

was shown by Mott and Henneberg[1] to be very much better than had been thought. It turns out that (2.7) is valid even for $v < v_k$, if corrections are made for the binding energies of the electrons, especially in the heavier elements. Correction terms C_K, C_L, etc. corresponding to binding energies of electrons in the K, L, etc. atomic shells then appear inside the brackets in (2.7). For a discussion of this and other phases of the primary ionizing process, see Ashkin and Bethe [1] and Bohr [3]. According to a theory due to Bloch[2] I should vary approximately linearly with the atomic number of the atoms, that is $I = KZ$ with K a constant. Actually K is not constant, having a value of about 11.5 for Al and showing a slow variation from 9.3 to 8.8 as Z goes from 26 (Fe) to 92 (U) (see Table 1).

Table 1. *I, K for different elements* [1].

Z	Element	I	$K = I/Z$	Z	Element	I	$K = I/Z$
1	H	15.6	15.6	29	Cu	276	9.5
3	Li	34.0	11.3	47	Ag	418	8.9
4	Be	60.4	15.1	50	Sn	463	9.2
6	C	76.4	12.7	74	W	655	9.2
13	Al	150	11.5	82	Pb	705	8.6
26	Fe	241	9.3	92	U	811	8.8

Values of I can be determined most directly by a measurement of stopping power, followed by a calculation using (2.3). Such experiments have been carried out with low energy protons ($20 E < 2000$ kev) at Copenhagen, Chicago, Los Alamos and Pasadena, and with high energy protons at Berkeley (180, 270 and 340 Mev) and at Harvard (around 100 Mev), among others. See Allison and Warshaw [4] and Ashkin and Bethe [1] for discussion. Table 1 shows values of I for various stopping materials based on the 340 Mev measurements of Bakker and Segrè[3]. The value of 150 ev for I_{Al} obtained by Wilson[4] using protons of 2 to 4 Mev was used as a standard.

A different stopping formula analogous to (2.7) was derived for primary electrons by Bethe[5] using Møller's formula[6] for the scattering of electrons by electrons. For electrons moving with relativistic velocities

$$- \frac{dE}{dx} = \frac{2\pi e^4}{m v^2} NZ \left\{ \log \frac{m v^2 E}{2 I^2 (1 - \beta^2)} - (2\sqrt{1 - \beta^2} - 1 + \beta^2) \log 2 + \\ + 1 - \beta^2 + \frac{1}{8}(1 - \sqrt{1 - \beta^2})^2 \right\}. \tag{2.8}$$

This expression gives the same results as (2.7) within ten percent for protons and electrons of the same velocity up to a proton energy of 10 Bev [1]. For small velocities (2.8) becomes

$$- \frac{dE}{dx} = \frac{4\pi e^4 NZ}{m v^2} \log \frac{m v^2}{2 I} \sqrt{\frac{e}{2}}. \tag{2.9}$$

This is different from (2.7) because of exchange effects due to the identical nature of the primary and atomic electrons.

---

[1] N. F. Mott: Proc. Cambridge Phil. Soc. **27**, 553 (1931). — W. Henneberg: Z. Physik **86**, 592 (1933).
[2] F. Bloch: Z. Physik **81**, 363 (1933).
[3] C. J. Bakker and E. Segrè: Phys. Rev. **81**, 489 (1951).
[4] R. R. Wilson: Phys. Rev. **60**, 749 (1941).
[5] H. A. Bethe: Handbuch der Physik, Vol. 24, p. 273. Berlin: Springer 1933.
[6] C. Møller: Ann. Physik **14**, 531 (1932).

3. Secondary ionizing events. *α) Delta rays.* Since the primary ionizing process accounts directly for fewer than half the ions ultimately formed, second ary events are clearly important, particularly ionization by delta rays. When the primaries are heavy particles, δ-ray energies are usually no greater than a few tens of kilovolts, hence their energy loss by bremsstrahlung is negligible and they lose most of their energy in a series of ionizing collisions frequently involving cascading. Since they are electrons they can readily produce ionization when their energy is quite low; furthermore their average energy loss by elastic and inelastic collisions with molecules is very small. Therefore all but a small fraction of the kinetic energy of the delta rays is spent in producing ions or optically excited atoms or molecules.

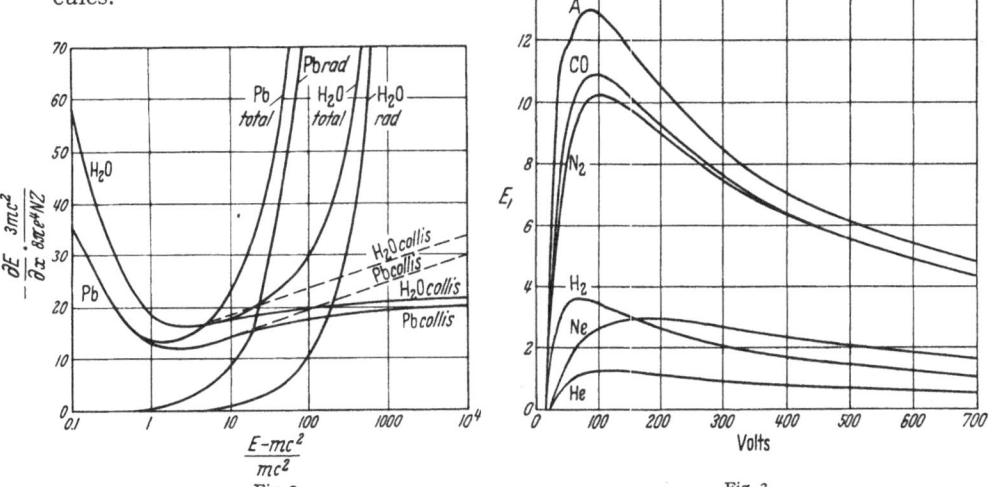

Fig. 2.   Fig. 3.

Fig. 2. Shape of stopping power curve. Average energy loss plotted against kinetic energy in mc^2. Curves given for collision, radiation, and total loss. Dotted curves calculated without, solid curves with, polarization of the medium by the electric field of the particle taken into account. HEITLER [24].

Fig. 3. Dependence of efficiency of ionization in helium, neon, argon, carbon monoxide, and hydrogen on energy of bombarding electrons (SMITH, quoted by BRODE [25]).

β) Optical transitions. Roughly half the energy spent by the primary particles in the gas will be used in processes other than the production of ions. This is immediately clear from a comparison of the average energy required to create an ion pair with the ionization potential of the gas atoms or molecules. Most of the energy lost is radiated. Electromagnetic radiation from excited atoms, ions, or molecules may be absorbed or transmitted by the walls, it may produce ion pairs in the gas, or it may liberate electrons from solid material in the chamber through the photoelectric effect[1,2]. The relative amount of energy used in freeing charges depends upon the physical conditions, which will vary from one

---

[1] Other processes, such as molecular dissociation may also be involved, but those listed are the important ones for present purposes.

[2] There are roughly 10^5 photons emitted per Mev of energy lost in the gas. Much of the radiant energy is in the ultraviolet, but can be converted to visible light by means of a "shifter" such as a sheet of polystyrene containing tetraphenyl butadiene. The visible light can be detected by use of a photomultiplier. Gaseous scintillation counters have already been used. For reviews see C. EGGLER and C. M. HUDDLESTON: Nucleonics **14**, No. 4, 34 (1956), and J. A. NORTHROP and R. NOBLES: Nucleonics **14**, No. 4, 36 (1956). The light signal is fast, $\sim 10^{-8}$ sec, so applications combining a fast photomultiplier signal for coincidence detection and a slower ion chamber pulse for energy resolution are possible.

experiment to another. In most cases the current of photoelectrons from the surfaces of solid matter is negligible, but the photo-production of ions in the gas may be appreciable.

Photoionization by radiation from optical (outer electron) transitions cannot occur as a simple process in pure gases, but may occur in mixtures of gases. The amount of available experimental information on this subject applicable to ionization chambers appears to be small. The recovery of x-radiation can be illustrated by the following example.

Let there be n_K ions with vacancies in the K shells of one kind of atom in the gas. The fraction which will emit K radiation is defined as w_K, the fluorescence yield.

$$w_K = \frac{\sum\limits_i n_{K\alpha_1} + n_{K\alpha_2} + \cdots + n_{K_i} + \cdots}{n_K} \tag{3.1}$$

where $n_{Ki} = $ the number of transitions of the type Ki. The K series x-ray energy radiated will be

$$E_{n_K} = \sum_i P_{Ki} N_{Ki} (h\nu)_{Ki}, \tag{3.2}$$

where $P_{Ki} = $ relative probability of emission of quanta of energy $(h\nu)_{Ki}$. Of this, the energy reaching the walls will be

$$E'_{n_K} = \sum_i P_{Ki} N_{Ki} (h\nu)_{Ki} e^{-\varrho \mu_{Ki} \cdot d} \tag{3.3}$$

where $\mu_{Ki} = $ mass absorption coefficient for the radiation Ki, $\varrho = $ density of the gas, and $d = $ mean effective path length to the solid matter, a function of the geometry. The $K\alpha$ term in (3.3) will be the largest. The term due to $K\alpha$ radiation is

$$\left.\begin{aligned} E'_{n_{K\alpha}} &= P_{K\alpha} n_{K\alpha} h\nu_{K\alpha} e^{-\varrho \mu_{K\alpha} \cdot d} \\ &= (\text{radiated } K\alpha \text{ energy}) \times e^{-\varrho \mu_{K\alpha} \cdot d} . \end{aligned}\right\} \tag{3.4}$$

For argon, with pressure $= 1$ atm, $d = 4$ cm, and $\mu_{K\alpha} = 200$ cm^2/g this gives about 30% loss. However the absolute loss will be appreciable only if the number of quanta emitted is sufficiently large. For argon, w_K is only 0.15, and n_K will usually be very small. For example, on the basis of Henneberg's theory, Hansteen[1] estimates that the cross section per argon atom for K-shell ionization by protons rises from roughly 2×10^{-21} cm^2 at 1 Mev to a peak of about 6×10^{-21} cm^2 at 4 Mev, from which it is estimated that on the average only 1 to 10 K-shell vacancies are produced when 5 Mev protons are stopped in argon.

On the average w_K varies smoothly and monotonically through the periodic table from practically zero at $Z = 10$ (neon) to 0.65 at $Z = 36$ (krypton) and 0.90 at $Z = 82$ (lead). However, with heavy particles the velocity requirement essentially rules out the excitation of K x-rays in most cases. In fact, among the gases commonly used in ionization chambers argon seems to be the only one in which much energy can ordinarily be lost in the form of K x-radiation when the ionizing particles are heavy. With electrons as primaries, especially when the gas pressure is low and the chamber volume small, noticeable x-radiation losses can conceivably occur in various gases.

Tellez-Plasencia[2], and Tellez-Plasencia and Theron[3] have worked out more complete expressions for the relationship between incident x-ray energy

[1] J. M. Hansteen: Arch. Mat. Naturvidenskab B **53**, No. 6 (1956), and private communication.

[2] H. Tellez-Plasencia: J. Phys. Radium **9**, 230 (1948); **12**, 89 (1951).

[3] H. Tellez-Plasencia and P. Theron: J. Phys. Radium **11**, 93 (1950).

and ionization, taking into account the effect of fluorescent radiation, for various geometrical conditions.

γ) *The Auger effect.* Excited atoms and ions may lose their energy through radiationless transitions in which the excess energy is carried away by Auger electrons. The process can only occur if there is a vacancy in one of the inner shells outside of which there are at least two electrons. This is the chief competitor to the emission of x-radiation. In argon it accounts for 85 % of the de-excitations of the K-shell, and in neon virtually all. It is obviously a very effective mechanism for the production of secondary ions when it occurs.

δ) *Metastable atoms.* The processes mentioned above all occur in times of the order of 10^{-7} sec or less. In pure noble gases metastable atoms formed in the primary process may remain excited for times as long as a second. They eventually lose their energy in collisions either with impurity molecules or with a surface of solid material. In either case electrons may be emitted in the process. If in the former case the quenching molecules are polyatomic, molecular excitation and dissociation are competing processes.

A very considerable fraction of the excitation energy may be involved in such cases, as JESSE and SADAUSKIS have recently demonstrated quite clearly. They measured the average energy loss w per ion pair formed in He and Ne when 5.3 Mev alpha particles were stopped, first using spectroscopically pure gases, then repeating after 0.12% of argon had been added. The results (Table 2) show a 30% reduction in w, corresponding to an increase of about 40% in the number of ions formed. This is consistent with observations of BIONDI and others working with electrical discharges in gases [5]. The explanation is that the impurity molecules quench the metastable states during collisions in which electrons emitted carry away the excess energy.

Table 2. *Energy per ion pair for α-particles in helium, neon, and argon, with and without impurities.*

Gas	w (ev)	Reference
Pure helium	41.3	1
Pure helium	42.7	2
Helium plus 0.13% argon	29.7	1
Pure neon	36.3	1
	36.8	2
Neon plus 0.12% argon .	26.1	1
Pure argon	26.4	2
Argon plus 0.2% acetylene	21.0	2
Argon plus 2% acetylene[4]	20.5	3

The observation of this effect has accounted for the excessive discrepancy between values of w_{He} found earlier by various experimenters. It also brought the experimental results into better agreement with theoretical results. Recent Born approximation calculations by ERSKINE [6] taking higher order ionizing events into account have yielded a value $w_{He} = 41.1$ ev.

In chambers filled with pure noble gases metastable atoms diffusing to the cathode will produce electrons at the surface which will contribute to the total signal. In pulse chambers fluctuations in this current will be a source of noise.

ε) *Liberation of electrons at the cathode by positive ion impact.* How important this process may be under ion chamber conditions is not clear. The experimental

[1] W. P. JESSE and J. SADAUSKIS: Phys. Rev. **88**, 417 (1952).

[2] W. P. JESSE and J. SADAUSKIS: Phys. Rev. **90**, 1120 (1953).

[3] C. E. MELTON, G. S. HURST and T. E. BORTNER: Phys. Rev. **96**, 643 (1954).

[4] Lowest w was found with 2% acetylene. Other impurities, even those with ionization potentials higher than the metastable states of argon, also reduce w. See reference 3 above, and J. SHARPE: Proc. Phys. Soc. Lond. A **65**, 859 (1952), and G. BERTOLINI, M. BETTONI and A. BISI: Nuovo Cim. **11**, 458 (1954).

[5] BIONDI: Rev. Sci. Instrum. **22**, 500 (1951). — BIONDI and BROWN: Phys. Rev. **75**, 1700 (1949). — KRUITHOF and DRUYVESTEYN: Physica, Haag **4**, 450 (1937).

[6] G. A. ERSKINE: Proc. Roy. Soc. Lond., Ser. A **224**, 362—373 (1954).

evidence all seems to have been obtained with ions having energies of 10 ev or more, and extrapolation to energies under 1 ev may not be justified. Furthermore the yield of electrons per ion, γ_i, is a sensitive function of the purity of the surface. HAGSTRUM[1] has recently made very careful experiments on this effect, using extreme care to ensure that no more than a small fraction of a monomolecular impurity film could have collected on the cathode before the determination of γ_i was made. He used singly and doubly charged noble gas ions impinging on tungsten and molybdenum plates. Under these conditions γ_i was found to be very nearly constant both for singly and doubly charged ions as the energy varied from 10 to 1000 ev. HAGSTRUM also measured the energy spectrum of the electrons and established that the upper limit was just below E_i—twice the work function of the metal. These results seem to indicate

Table 3. *Electrons yielded, γ_i, per 10 ev ion striking clean Mo and W surfaces.*

Ion	γ_i for Mo	γ_i for W	γ_{i_2} for Mo (doubly charged ions)
He	0.300	0.290	0.80
Ne	0.254	0.213	0.67
A	0.122	0.095	0.42
Kr	0.069	0.050	0.38
Xe	0.022	0.013	0.29

that an Auger effect is responsible for the emission. Table 3 gives γ_i for 10 ev singly and doubly charged ions. The yield from impure surfaces is smaller. From the relative consistency of ionization chamber results on total ionization found with the same type of gas in different chambers it seems likely that the effect is not very important under typical ionization chamber conditions, but no study seems to have been made.

4. Total ionization vs. energy relationship. It is customary to discuss the ionization I (number of ion pairs) yielded when a particle of energy E is stopped in a gas in terms of the average energy lost, w, per ion pair produced, a quantity which must be determined by experiment for each gas. Until very recently rather large variations existed in the values of w found by various experimenters, but the latest results have come into better agreement, partly because of the recognition of the importance of metastable states in He and Ne (see above) and of the necessity of using very high electric fields in order to minimize recombination effects. Unfortunately the absolute values are still somewhat rough. One

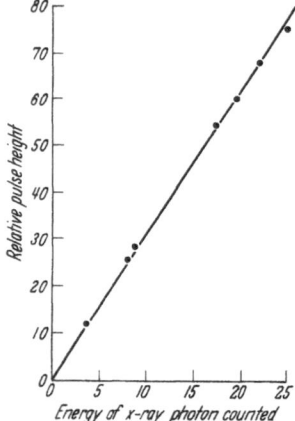

Fig. 4. Total ionization in nitrogen as a function of the energy of soft x-rays entering a proportional counter. CURRAN, ANGUS and COCKROFT: Phil. Mag. 40, 36 (1949).

must remember that the experimental values for w depend not only upon the nature of the gas, but also upon the shape, dimensions and cathode material of the chamber, as well as the gas pressure, all of which may influence the results slightly.

The shapes of the I vs. E curves for *electrons as primaries* in various gases have been investigated by a number of workers using proportional counters. In these experiments all instrumental conditions could be kept constant while the energy of the primary electron groups was altered. The shapes of the curves are therefore obtained under favorable conditions. The best experimental results indicate that I is proportional to E up to 300 kev or more within experimental error in A, Kr, and N_2. Points have been taken as low as 200 to 300 ev. See

[1] H. D. HAGSTRUM: Phys. Rev. **104**, 672 (1956).

Fig. 4 for N_2 results. For these gases then

$$I = \frac{E}{w} \quad \text{(primary particles electrons)} \quad (4.1)$$

where w is a constant. Less complete information on other gases suggests that (4.1) is probably valid generally. There appears to be no clear evidence to the contrary.

The determination of absolute values for w is more difficult. Table 4 gives some recent results obtained under widely differing conditions. Values of w for a particular gas fluctuate about the mean by a few percent from experiment to experiment.

In the columns labeled R are given values of w_{gas}/w_{O_2} for the various experiments. These ratios are remarkably nearly equal for N_2, CO_2, air, Kr, Xe, CH_4, and C_2H_4. The large fluctuations for He and Ne can be understood in terms of the metastable states and varying amounts of impurities. It seems possible, therefore, that the fluctuations in the w's of N_2, CO_2, etc. are due to the difficulty in getting an absolute instrumental calibration. On this assumption we have the result that (4.1) is valid to better than 1 Mev, and possibly to many Mev.

With *protons and heavier primary particles* the results are not quite so simple. Much of the information available was obtained by observing pulses produced by 5 to 9 Mev alpha particles. Absolute values could then be calculated from the absolute pulse height and the measured capacitance of the ionization chamber

Table 4. *Absolute values, w, and values, R, relative to oxygen, for the average energy required to produce one ion pair in different gases with electrons as primaries.*

Energy, method and authority	0.2 to 46 kev P. counter and I.C.		0 to 17 and 0 to 60 kev I.C.[2]		0 to 1.7 Mev Extrapol. ch.[3]		0 to 1.7 Mev Extrapol. ch.[5]		1 to 34 Mev I.C.[6]		Averages[8]	
	w (ev)[1]	R	w (ev)	R	w (ev)[4]	R	w (ev)[4]	R	w (ev)	R[7]	w (ev)	R
Hydrogen .	38.0	1.18	36.3	1.14			36.3	1.17	37.8	1.22	36.9	1.16
Helium . .	32.5[9]		42.3	1.37			40.3	1.29	44.5	1.43	41.3	
Nitrogen .	35.8	1.12	34.7	1.12	34.8	1.12	34.6	1.11	34.8	1.12	34.9	1.12
Oxygen . .	32.2	(1.00)	30.9	(1.00)	30.9	(1.00)	31.2	(1.00)			31.3	(1.00)
Neon . . .			36.6	1.19			35.3	1.13			35.9	1.16
Argon . .	27.0	0.86	(26.4)	0.85	25.5	0.82	25.8	0.83			26.3	0.84
Krypton .			24.2	0.78			24.7	0.79			24.4	0.78
Xenon . .			22.2	0.72			22.0	0.71			22.1	0.72
Air[10] . . .	35.0	1.09	33.9	1.10	33.9	1.10	33.9	1.09			34.2	1.09
Carbon dioxide .			32.8	1.06	32.6	1.06					32.7	1.06
Methane .	30.2	0.94	27.3	0.89	26.8	0.87					28.1	0.90
Ethylene .					26.3	0.85	26.4	0.85			26.4	0.85
Acetylene .			26.1	0.84								
Pentane . .			24.6									

[1] J. M. VALENTINE: Proc. Roy. Soc. Lond., Ser. A **211**, 75 (1952).

[2] W. P. JESSE and J. SADAUSKIS: Phys. Rev. **97**, 1668 (1955).

[3] J. WEISS and W. BERNSTEIN: Phys. Rev. **98**, 1828 (1955).

[4] Stopping power calculations [Eq. (2.3)] are involved in the determination of these numbers.

[5] J. WEISS and W. BERNSTEIN: Phys. Rev. **103**, 1253 (1956).

[6] W. C. BARBER: Phys. Rev. **97**, 1071 (1955).

[7] Calculated from the average value for nitrogen.

[8] Not including the 1—34 Mev results.

[9] Impure gas.

[10] Additional values of w are: 32.5 (rough average) found by EISL, Ann. d. Phys. **3**, 277 (1929), and GERBES, Ann. d. Phys. **23**, 648 (1935), for 10 to 60 kev electrons, and 33.5, GROSS, WINGATE and FAILLA, Radiology **66**, 101 (1956), for S^{35} beta rays.

and amplifier input circuit. In such an experiment Jesse, Forstat and Sadaus-kis[1] found an I vs. E curve for argon which was a straight line which, extrapolated, passed through the origin. In a similar experiment performed with a reported relative accuracy of 0.1% in the determination of pulse amplitudes, Cranshaw and Harvey[2] also found linearity in the 5 to 9 Mev region, but their straight line, extrapolated, intercepted the E axis at about $+85$ kev, reproducible within about 10 kev. This result is qualitatively consistent with the idea that the heavy particles cease to ionize efficiently when they have been slowed to velocities comparable with those of the outer electrons of the gas molecules.

There is experimental evidence supporting this picture. Careful investigations have revealed that for heavy particles at low energies w is indeed not

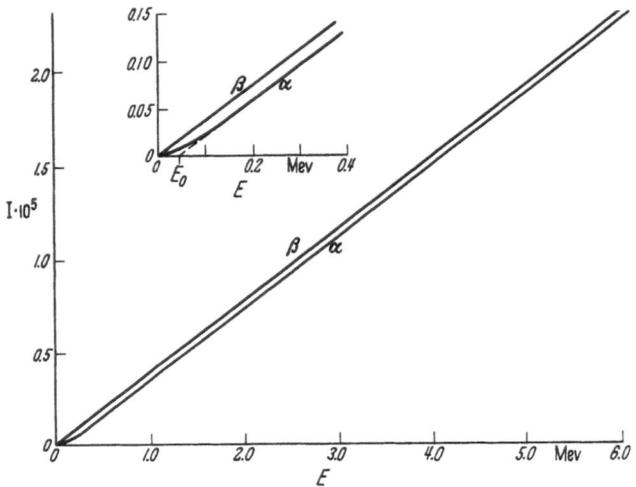

Fig. 5. Qualitative comparison of total ionization vs. energy loss relationships for beta particles (β) and heavier particles (α), as indicated by the best available information. The exact shape of the low energy end of curve α is uncertain. The spacing between the lines is also uncertain. For primaries no heavier than alpha particles the spacing shown is probably too large.

entirely independent of the type of particle. In working with the B^{10} (n, α) Li^7 reaction Rhodes, Franzen, and Stephens[3] found that the ratio of the ionization produced in argon by the alpha particles to that produced by the recoiling lithium fragments is not exactly equal to the value of 1.7529 expected from their known masses and the requirement of momentum conservation, but is actually 1.878 ± 0.014. Stebler, Huber, and Bichsel[4] had previously found a ratio of 1.89 ± 0.02. Hanna[5] reported a value of 1.838 ± 0.023. Finally, de Juren and Rosenwasser[6] using a parallel plate chamber with fast electron collection, but without a grid, obtained a ratio of 1.7859 in disagreement with the others. Milton, Rutledge, and Lennox[7] have recently analyzed de Juren and Rosenwasser's results, giving very careful attention to questions of resolving power and have corrected them upward to a value of 1.854 ± 0.020 which is in satisfactory agreement. Milton et al. point out that this ratio would be consistent

[1] W. P. Jesse, H. Forstat and J. Sadauskis: Phys. Rev. 77, 782 (1950).
[2] T. E. Cranshaw and J. A. Harvey: Canad. Res. 26, 243 (1948).
[3] Rhodes, W. Franzen and W. Stephens: Phys. Rev. 87, 141 (1952).
[4] A. Stebler, P. Huber and H. Bichsel: Helv. phys. Acta 22, 362 (1949).
[5] G. C. Hanna: Phys. Rev. 80, 530 (1950).
[6] J. A. de Juren and H. Rosenwasser: Phys. Rev. 93, 831 (1954).
[7] J. C. D. Milton, A. R. Rutledge and P. I. Lennox: Chalk River (Canada) Report CRP-632. 1956.

with an oversimplifying assumption that both particles continue to ionize the argon with full effect until they have been slowed to the speed of the outermost electrons (when the alpha particle still has 48 kev of kinetic energy and the Li^7 fragment 84 kev) then cease to ionize.

Additional information on the weakness of the ionizing process in stopping very slow particles comes from the work of MADSEN[1] who examined the ionization produced by recoil fragments from the decay of Po^{210}, ThC, and ThC'. The fragments had energies of 104, 118, and 170 kev, respectively. Argon plus 5% air was used, and a value of $w = 67$ ev per ion pair was found. When the results were plotted they were found to lie on a straight line with slope $= 67$ ev per ion pair and with energy axis intercept about 40 kev. So it seems that ionization is not a negligible process for very heavy ions, even when they have speeds much lower than those of the atomic electrons, although it is considerably weaker than for higher velocities. MADSEN's results are supported by the recent experiments of JESSE and SADAUSKIS[2] who have found that the recoils from the decay of Po^{210} give w values in He and A which are respectively 4.1 and 4.5 times the average value for alpha particles[3].

Fig. 5 shows a plot representing qualitatively the suspected true shape of the I vs. E curve for protons, deuterons, alpha particles, and the lighter nuclear recoils of initial energy E brought to rest in a gas. The energy at which the curvature becomes appreciable is expected to be greater for the heavier particles.

Table 5 gives w and R values for heavy particles under a variety of conditions. Absolute values of w show smaller fluctuations generally than was the case with electrons, except for the 340 Mev protons, for which the absolute values are in doubt. Again the R values for N_2, CO_2 and air are quite consistent. In addition those for A, H_2 and CH_4 are also quite close. The agreement of the 340 Mev R's with the averages from the lower energy work, usually within a few percent, is striking. So is the agreement of the average R values with the corresponding ones from the electron experiments, especially for N_2, air, CO_2 and CH_4. For noble gases the averages of w's are generally in fairly good agreement with those found with electrons, although the data are somewhat scanty except in the case of argon. For other gases the w's are generally a few percent larger. The best direct comparison between the w's found for various gases with heavy particles and with electrons as primaries appears to be that of JESSE and SADAUSKIS[4], who report tentatively that $w_\alpha = w_\beta$ for the noble gases and hydrogen, but $w_\alpha > w_\beta$ for N_2, O_2, CO_2, air, CH_4, C_2H_2, and C_2H_6. In making the comparison they used earlier results[5] obtained with 5.3 Mev α's and new ones for 1.2 Mev α's, and compared them with new w_β results obtained with different equipment, using argon as a standard. In all cases the noble gases (and presumably not H_2) were circulated through a purifier. In the 5.3 Mev case the average discrepancy was about 4%; in the 1.2 Mev case, 8%. They remark that the differences may be due to columnar recombination, which is reported to be rather bad in O_2, CO_2 and CH_4[6]. A more direct experimental comparison of w_α with w_β is needed. One would expect w_α to be slightly greater than w_β on the average, because of the relative ineffectiveness of the ionization process for very slow particles, and

[1] B. S. MADSEN: Dan. Mat.-Fys. Medd. **23**, No. 8 (1945).

[2] W. P. JESSE and J. SADAUSKIS: Phys. Rev. **102**, 389 (1956).

[3] However, this leaves unexplained the result of GERTHSEN and GRIMM [Z. Physik **120**, 476 (1943)], who found $w = 36$ for air, compared with the value $34-35$ accepted for lighter particles.

[4] W. P. JESSE and J. SADAUSKIS: Phys. Rev. **97**, 1668 (1955).

[5] W. P. JESSE and J. SADAUSKIS: Phys. Rev. **90**, 1120 (1953).

[6] See Sect. 15.

Table 5. *Absolute values, w, and values, R, relative to oxygen, for the average energy loss required to produce one ion pair in different gases with heavy particles as primaries.*

	α's 5 Mev[1]		α's 5 Mev[2]		α's 5 Mev[3]	
	w (ev)	R	w (ev)	R	w (ev)	R
Hydrogen . . .	35.96 ± 0.15	1.12	37.0 ± 0.4	1.15		
Helium.			46.0 ± 0.5	1.43	29.6 ± 0.3[9]	
Nitrogen[10] . . .	36.5 ± 0.15	1.14	36.3 ± 0.4	1.13	36.3 ± 0.15	1.13
Oxygen.			32.2 ± 0.3	(1.00)	32.17 ± 0.15	(1.00)
Neon.						
Argon			26.4 ± 0.3	0.82	26.25 ± 0.12	0.82
Krypton						
Xenon						
Air[10]	34.95 ± 0.18	1.09	35.0 ± 0.3	1.09		
Carbon dioxide .	34.3 ± 0.3	1.07	34.3 ± 0.3	1.06	33.5 ± 0.3	1.04
Methane	29.0 ± 0.15	0.91	29.4 ± 0.3	0.91		

	5.3 Mev α's[4]		5 Mev α's[5]		340 Mev p's[6]		Average	Values
	w (ev)	R	w (ev)	R	w (ev)[7]	R	w (ev)[8]	R
Hydrogen . . .	36.3	1.12	37.0	1.19	35.3	1.12	36.6	1.14
Helium.	42.7	1.31	31.7[9]		29.9[9]		44.4	1.38
Nitrogen[10] . . .	36.6	1.13	36.0	1.16	33.6	1.06	36.3	1.13
Oxygen . . .	32.5	(1.00)	32.2	(1.00)	31.5	(1.00)	32.1	(1.00)
Neon.	36.8	1.13			28.6[9]		36.8	1.15
Argon	26.4	0.82	25.9	0.81	25.5	0.81	26.25	0.85
Krypton	24.1	0.74					24.1	0.74
Xenon	21.9	0.67					21.9	0.67
Air[10]	35.5	1.09	35.2	1.13	33.3	1.06	35.2	1.10
Carbon dioxide .	34.5	1.06					34.15	1.06
Methane	29.2	0.88	29.0	0.94			29.1	0.92

w_{prot} should be intermediate. On the assumption that the true I vs. E curve is given by Fig. 5, the slope of the straight part is

$$\frac{\partial I}{\partial E} = \frac{1}{w_{\alpha\,ave}} \frac{E}{E - E_0}. \tag{4.2}$$

If the reciprocal of the slope is set equal to w_β, as seems the best guess, then for 5 Mev α's in argon w_α should be about 1.5% larger than w_β if $E_0 = 80$ kev. The overall average w_α in Table 5 is about 2.5% greater than the overall average w_β in Table 3, but this may be fortuitous.

Fission fragments are a special case. As they begin passing through the gas they have an average charge of about 14 units[11], but this becomes smaller toward the end of the track through electron capture, with the result that the relative

---

[1] C. Biber, P. Huber and A. Müller: Helv. phys. Acta **28**, 503 (1955).

[2] T. E. Bortner and G. S. Hurst: Phys. Rev. **93**, 1236 (1954).

[3] W. Haeberli, P. Huber and E. Baldinger: Helv. phys. Acta **26**, 145 (1953).

[4] W. P. Jesse and J. Sadauskis: Phys. Rev. **90**, 1120 (1953).

[5] J. M. Valentine and S. C. Curran: Phil. Mag. **43**, 964 (1952).

[6] C. J. Bakker and E. Segre: Phys. Rev. **81**, 489 (1951).

[7] Bakker and Segre estimate that these values are a few percent low because of the effect of star formation.

[8] Computed without the 340 Mev and impure gas results.

[9] Impure gas.

[10] In addition, Alder, Huber and Metzger found (36.3 ± 0.14) and (34.7 ± 0.5) ev for N_2 and air, respectively.

[11] N. O. Lassen: Dan. Vid. Selskab. Mat.-Fys. Medd. **26**, No. 5 (1951).

Table 6. *Single ionization potentials, I_0, and recommended energies per ion pair for different gases with electrons as primaries.*

Gas	I_0 (ev)[1]	Energy of metastable state (ev)	w (ev)	Remarks and references
H_2	15.4		36.8	Table 4
He, pure[2]	24.6	19.8	41.3	Table 4
N_2	15.5		34.9	Table 4
O_2	12.2		31.3	Table 4
Ne, pure[2]	21.6	16.53, 16.62	35.9	Table 4
A, pure[2]	15.8	11.49, 11.66	26.4	Table 4
Kr, pure	14.0	9.86, 10.50	24.4	Table 4
Xe, pure	12.1	9.28, 9.4	22.1	Table 4
H_2O, vapor	12.6		30.0	3
CO_2	13.7		32.7	Table 4
Air			34.2	Table 4
CH_4	13.1		28.1	Table 4
C_2H_2	11.35		26.1	4
C_2H_4	10.5 − 10.8		26.4	Table 4
C_2H_6	11.8		24.6	4
C_4H_{10}.	10.8		23.0 ± 0.3	5
C_2H_5OH	10.7		32.6 ± 0.7	5
Cl_2	13.2		23.5 ± 1.2	6
CCl_4	11.0 ± 1.0		25.9 ± 0.7	5
BF_3	17		35.3 ± 0.4	5
H_2S			23.4 ± 0.6	5
NH_3			30.5 ± 0.4	5
SO_2			32.5 ± 0.7	5

amount of energy lost in nuclear collisions increases. One can write for the energy of a fission fragment $E = w_\alpha I_t + E_0$. For light and heavy fission fragments $E_0 \approx 5.7$ and 6.7, respectively[7].

5. Statistical fluctuations in ionization. *α) Fluctuations in the total number of ions formed in an element of track length.* The straggling in energy loss by ionization in a very short length of the track of a primary particle cannot be calculated correctly by methods based upon the assumption of random fluctuations in the total number of ion pairs formed in that element. Such a calculation would give too small a result, basically because so much of the energy loss is concentrated in the kinetic energy of a relatively small number of delta rays from the primary ionization process. This is especially true when the primary ionizing particles are electrons. An analytic theory of the fluctuations was first given by LANDAU[8]. Refinements have been made by BLUNCK and LEISEGANG[9], SYMON[10], and MOYAL[11]. MOYAL shows that LANDAU's distribution function correctly gives the fluctuations in stopping power and ionization, even for very small values of primary ionization, and that atomic structure and quantum

---

[1] LANDOLT-BÖRNSTEIN Tables, Vols. 1 and 3. Berlin: Springer 1951.

[2] See also Table 2.

[3] Based on $W_{H_2O}/W_{air} = 0.878 \pm 0.02$ for 5 Mev α's. R. K. APPLEYARD: Nature, Lond. **164**, 838 (1949).

[4] W. P. JESSE and J. SADAUSKIS: Phys. Rev. **97**, 1668 (1955).

[5] These values are those found by C. BIBER, P. HUBER and A. MÜLLER: Helv. phys. Acta **28**, 503 (1955) using 5 Mev α particles.

[6] A. GIBERT, F. ROGGEN and J. ROSSEL: Helv. phys. Acta **17**, 97 (1944). Value for 1−2 Mev protons from (n, p) reaction.

[7] H. W. SCHMITT and R. B. LEACHMAN: Phys. Rev. **102**, 183 (1956).

[8] W. LANDAU: J. Phys. USSR. **8**, 201 (1944).

[9] O. BLUNCK and S. LEISEGANG: Z. Physik **128**, 500 (1950).

[10] K. SYMON: Thesis. Harvard University 1948.

[11] J. E. MOYAL: Phil. Mag. **46**, 263 (1955).

resonance effects cannot change it very much. The latter result seems to be in contradiction with the conclusions of Blunck and Leisegang. For small losses the distribution-in-energy-loss curve is asymmetric, with a tail on the high loss side. For this case Rossi [6] gives useful curves to aid in making calculations. For larger losses the curve becomes approximately Gaussian in shape.

The validity of these theories has been shown experimentally for electrons by Chen and Warshaw[1] and for protons by Igo, Clark, and Eisberg[2]. However, some experiments with electrons[3] have shown most probable energy losses

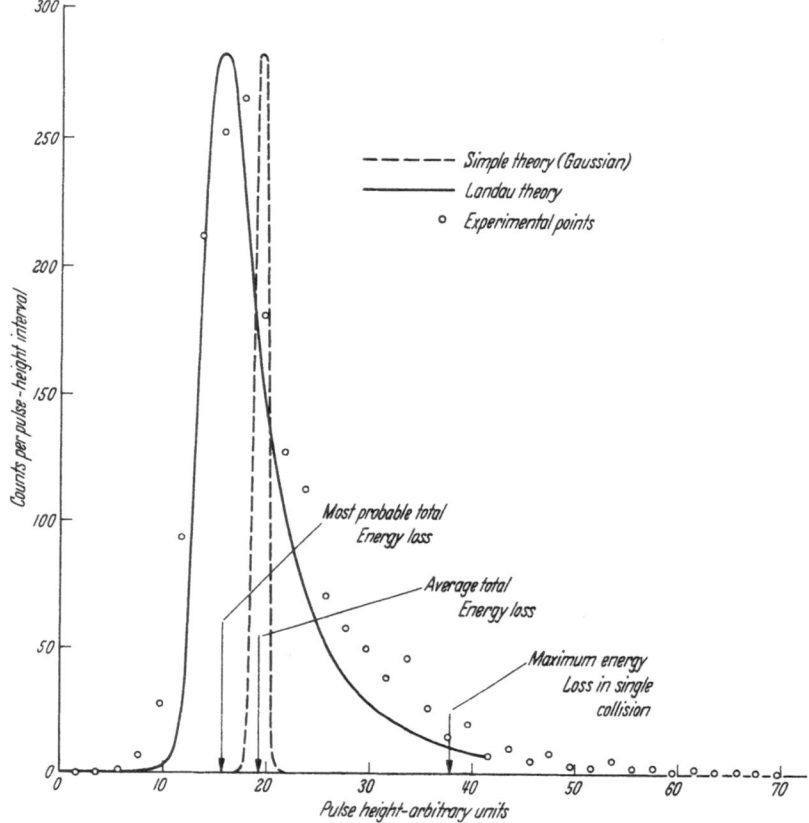

Fig. 6. Frequency distribution of energy losses of 31.5 Mev protons in a proportional counter. The theoretical distribution was calculated from the theory of Symon, as given in [6]. The energy scale of the experimental points has been adjusted arbitrarily to give the best average fit with the theoretical curve (Igo, Clark, and Eisberg).

roughly 10% lower than predicted, with widths of straggling curves somewhat larger than predicted. Fig. 6 shows the experimental results of Igo, Clark, and Eisberg plotted along with the prediction of the Landau-Symon theory as well as the prediction which is obtained from the incorrect simple statistical method.

The Landau distribution is approximately symmetrical when the following condition is satisfied.

$$\frac{2\pi e^4 z^2 NZ}{m v^2} x > W_{\max} \tag{5.1}$$

[1] J. J. L. Chen and S. Warshaw: Phys. Rev. **84**, 355 (1951).
[2] C. J. Igo, D. D. Clark and R. M. Eisberg: Phys. Rev. **89**, 879 (1953).
[3] For example, see West: Proc. Phys. Soc. Lond. A **66**, 306 (1953).

x is the length of the element of track. When the primary particles are electrons the asymmetric distribution always applies. With protons condition (5.1) is frequently satisfied. The most probable energy loss is then equal to the average energy loss, and the mean square deviation from the average is given approximately by Bohr's equation

$$\Omega^2 = 4\pi z^2 e^4 N Z x,$$ (5.2)

a particularly simple expression independent of v.

β) *Fluctuations in total ionization along a complete track.* Here, again, the simple statistical approach is invalid; in fact Fano[1] has shown that in the case of hydrogen it gives mean square fluctuations roughly three times too large. The reduction is basically a consequence of the physical fact that particles of energy E always lose exactly that much energy in being brought to rest in the gas. In principle his arguments could be extended to more complex gases with somewhat similar results, but the details would be difficult to work out. The experimental information on this subject is scanty, but experiments of Kirkwood, Ponte-corvo, and Hanna[2] give qualitative support to Fano's theory.

Fluctuations in total ionization set an ultimate limit on the resolving power of ionization pulse type spectrometers. It appears that in most cases amplifier noise fluctuations will be worse.

III. Events following ionization.

6. General. Within about 10^{-7} second after the fast primary has passed, most of the ionization electrons except those due to the quenching of metastable states can be expected to have been reduced to energies of the order of 10 ev or less. Those which remain free will begin to drift across the chamber under the action of the applied electric field, moving in very irregular paths because of collisions with gas molecules. A local drift velocity is quickly established which depends in an extremely complicated way upon the constitution of the gas. The temperature, pressure, and field strength also influence the drift rate.

Electrons may be captured by positive ions in a process called recombination, which can be minimized by the application of a strong collecting field to separate the charges rapidly. Electrons may also be captured by neutral atoms, in a process called attachment. The negative ions formed are ordinarily quite stable, but they may give up an electron to a positive ion in a subsequent collision. This neutralizing process is also referred to as recombination.

The positive ions will drift along the electric field, also following irregular paths because of molecular collisions.

7. Diffusion of ions. From kinetic theory it can be shown[3] that if N particles are created at the origin at a time $t=0$, and if they drift through the gas with a velocity v in the z-direction, the density of particles at a later time will be

$$n = \frac{N}{[\sqrt{2\pi} l(t)]^3} e^{-[x^2+y^2+(z-vt)^2]/2l^2(t)}$$ (7.1)

where $l^2(t) = 2Dt$, and D is the proper diffusion constant. This equation can be used to describe the spreading out of groups of ions and electrons as they drift through the gas along the electric field lines, given the initial distribution.

---

[1] U. Fano: Phys. Rev. **72**, 26 (1947).
[2] D. H. W. Kirkwood, B. Pontecorvo and G. C. Hanna: Phys. Rev. **74**, 497 (1948).
[3] See Jaffé [8] or Rossi and Staub [7].

8. Drift velocity, mobilities of ions. It is an experimental fact that the drift velocity of a positive or negative ion (electrons are excluded) in a gas under the action of an electric field is given by

$$v_\pm = \mu_\pm \frac{X}{p} \qquad (8.1)$$

where $\mu_\pm$ is a constant called the mobility of the ion, p is the pressure and X is the electric field strength.

Eq. (8.1) holds accurately over a wide range of X and p, including those values characteristic of ionization chambers. With X in volt/cm and p in atmospheres, μ_+ and μ_- both have the order of magnitude unity, but μ_+ is generally slightly smaller than μ_- (see Table 7). However, this experimental difference may be due to the presence of an indeterminate number of free electrons along with the negative ions.

Table 7. *Diffusion constants and ion mobilities for different gases[1] at 300° K.*

Gas	μ_+	μ_-	D_+	D_-
H_2	6.0	8.0	0.12	0.19
D_2	7.5[2]			
He	5.9[3]	6.3		
N_2	1.3	1.8	0.029	0.041
O_2	1.3	1.8	0.025	0.04
CO_2	0.82	0.85	0.23	0.26
A	1.3	1.7		
Air	1.4	1.9	0.029	0.042
	(1.25)[4]	(1.61)[4]		

For A_+ ions in argon gas at $p=1$ atm a field of 1300 volt/cm gives a drift velocity of 10^3 cm/sec. The collection time in a chamber 1 cm across would therefore be about 10^{-3} sec, which therefore can be taken as the order of magnitude characteristic of signal rise times in ion collection chambers.

9. Drift velocities of electrons. No expression analogous to Eq. (8.1) can be written down for the drift velocity of electrons, which is generally a very sensitive function of the constitution of the gas and in some cases does not even increase monotonically with the strength of the applied field. In principle the drift velocity could be calculated, given the velocity distribution of the electrons in the "swarm" and the angular distributions of the scattering of electrons of the various energies from the gas molecules. Except for low energy electrons moving in pure noble gases no reliable velocity distribution seems available [9]. Experimentally a determination of the velocity distribution is also extremely difficult. However, drift velocities can readily be measured directly for gas mixtures of interest. Under ordinary conditions drift velocities of electrons are roughly 1000 times those of the positive ions (see Figs. 7 to 11).

Although the behavior of the electrons cannot be predicted in detail, a listing of the main physical processes whose complex interaction determines the velocity distribution of the electrons can be given, along with a qualitative description of the effect of each. In moving along the electric field the electrons gain kinetic energy until sufficient energy loss processes come into play to establish a gain-loss energy balance. In pure monatomic (noble) gases the only mechanism for energy loss by electrons with kinetic energies under the first excitation potential is elastic scattering, which operates weakly because of the large ion-to-electron mass ratio, the fractional loss per collision being of the order of $2m/M$. In pure argon the first excitation level is at 11.5 volt, and the mean agitation energy may rise to 8 or 10 volt before equilibrium is established. The electron

---

[1] Except as noted values are rough averages from tables by K. Przibram: Handbuch der Physik, Second Edit., Vol. 22, Chap. 4. Berlin: Springer 1933.

[2] Bennett and Thomas: Phys. Rev. **62**, 41 (1942).

[3] May be too small. L. M. Chanin and M. A. Biondi: Phys. Rev. **106**, 473 (1957) give 10.3 and 20.8 for He^+ and He_2^+, respectively.

[4] E. Montel and T. Ouang: J. Phys. Radium **15**, 586—587 (1954).

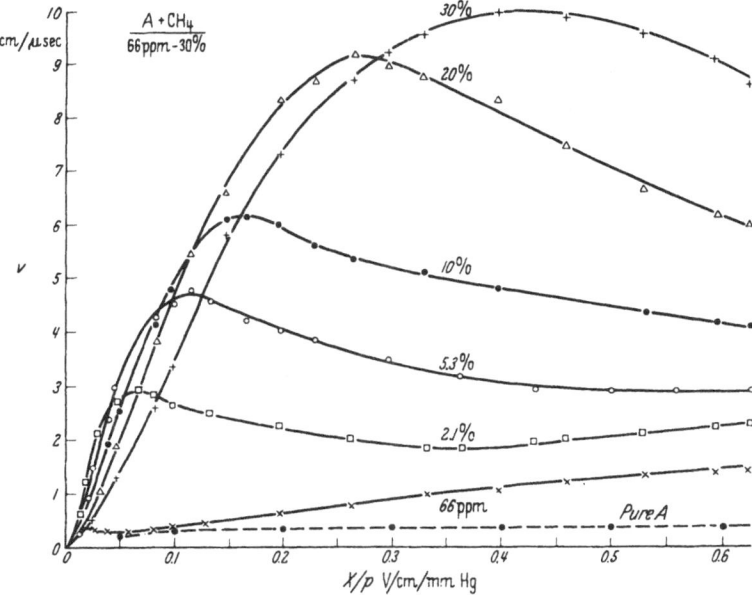

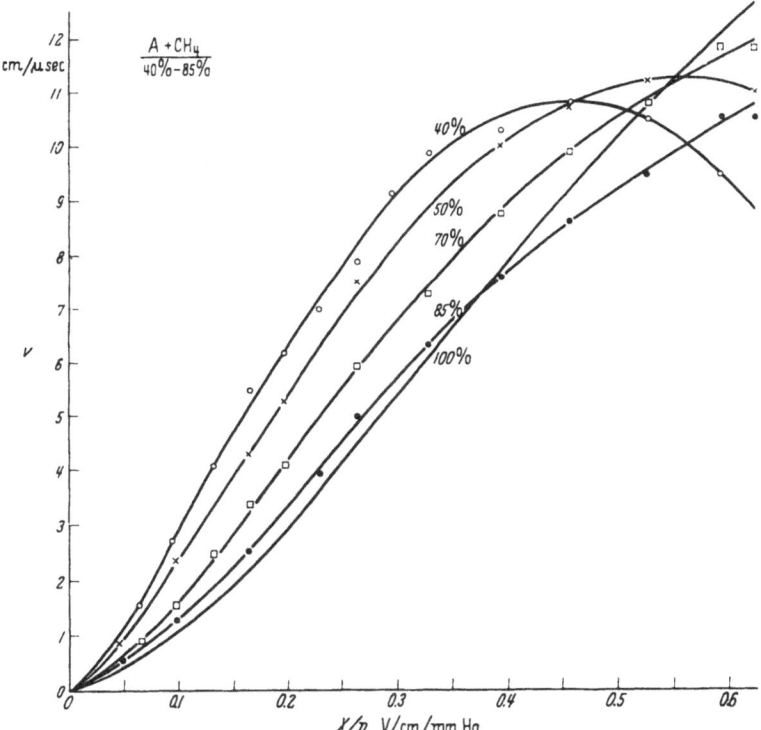

Fig. 7. Drift velocity of electrons in argon-plus-methane mixtures as a function of the reduced pressure, X/p. ENGLISH and HANNA [26]. The curve for pure argon is from BORTNER, HURST and STONE: Rev. Sci. Instrum. **28**, 103 (1957).

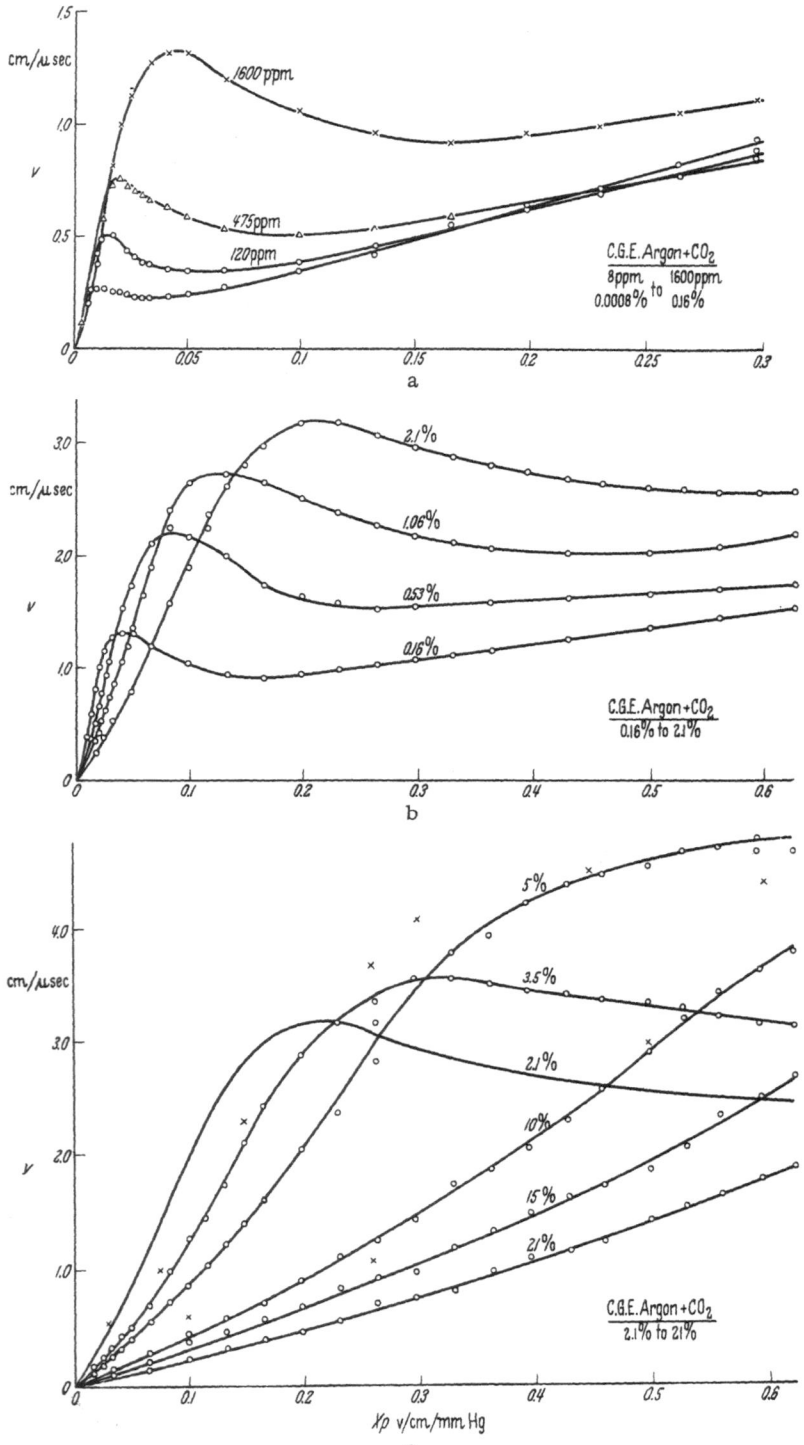

Fig. 8 a—c. Drift velocity of electrons in argon-plus-carbon dioxide mixtures. ENGLISH and HANNA [26].

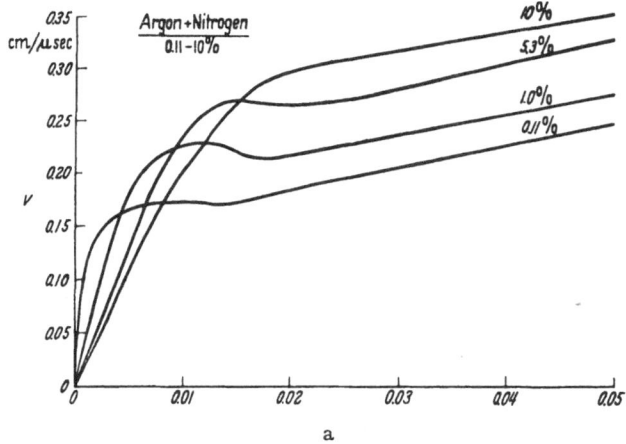

a

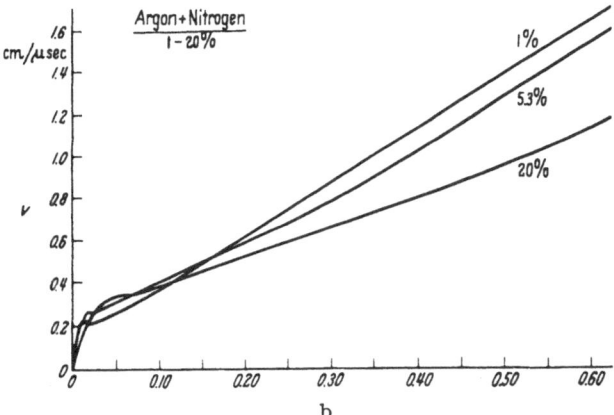

b

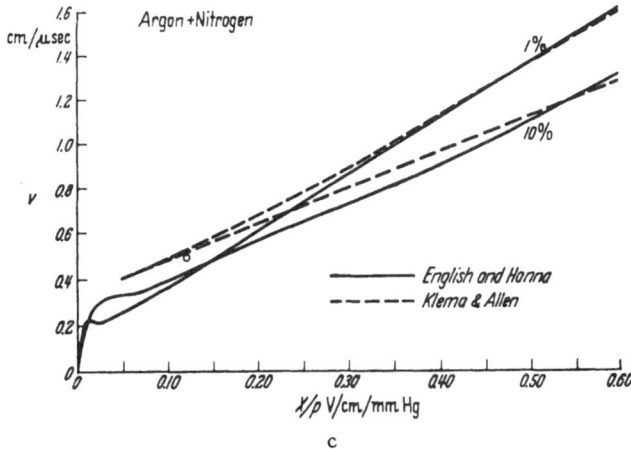

c

Fig. 9 a—c Drift velocity of electrons in argon-plus-nitrogen mixtures. ENGLISH and HANNA [26].

temperature will then be about 200 times the thermal equilibrium value. The result will be a low drift velocity w, as can be seen from the following argument.

From kinetic theory it can be shown [7] that the drift velocity can be expressed by the equation

$$v = \frac{D}{\varepsilon kT} eX \qquad (9.1)$$

where εkT is $\frac{2}{3}$ times the mean agitation energy and D is the diffusion coefficient of electrons in the gas. However, in this case D itself is a function of X. On

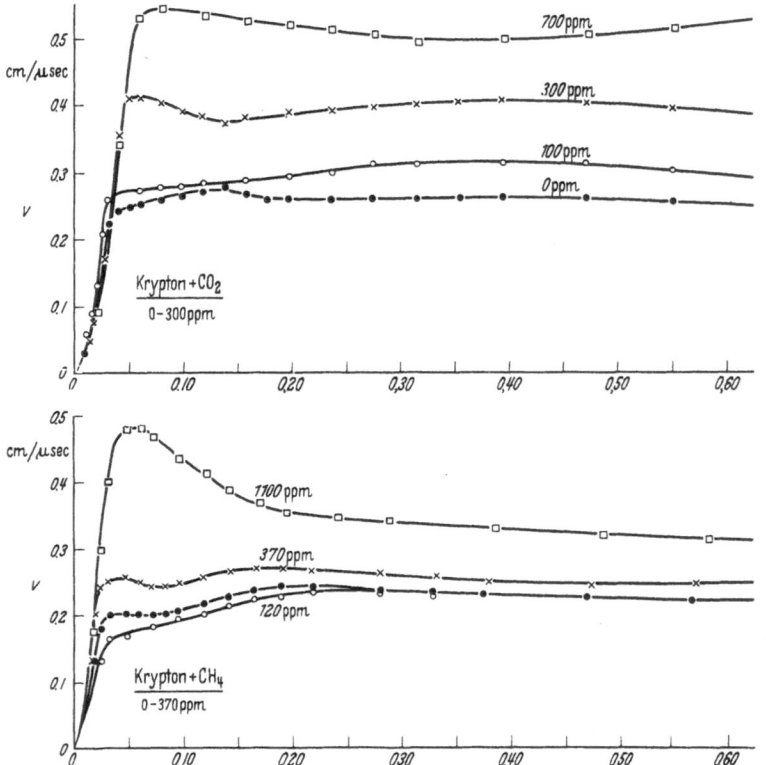

Fig. 10a. Drift velocity of electrons in krypton-plus-carbon dioxide and krypton-plus-methane mixtures.

the crude assumptions (a) that all electrons have the same agitation energy and (b) that their scattering is isotropic, Eq. (9.1) leads to $v = \dfrac{e}{m} \dfrac{\lambda}{u} \dfrac{X}{p}$, where λ is the mean free path of the electrons and u is the rms agitation velocity. Although this expression is not exact it serves to illustrate the fact that the drift velocity is roughly proportional to the mean free path and inversely proportional to the square root of the mean agitation energy, a result which can immediately be used to explain qualitatively the increase in drift velocities in noble gases when polyatomic impurities are added.

Polyatomic impurity molecules have low-lying vibrational states which can be excited by collisions with the free electrons through the Franck-Condon mechanism. Furthermore they have rotational states which can be excited by slow electrons. Gerjuoy and Stein have shown by detailed calculation that the rate of energy loss by slow electrons in exciting rotational states in N_2 is much greater than the elastic collision loss rate[1]. As a result of molecular excitation the mean agitation energy of the electrons is very much lower than it would be in a pure noble gas. Therefore a fraction of a percent of N_2 (first vibrational state at 0.29 ev), CO_2, or CH_4 added to argon will reduce the mean agitation energy drastically, with a corresponding increase in v.

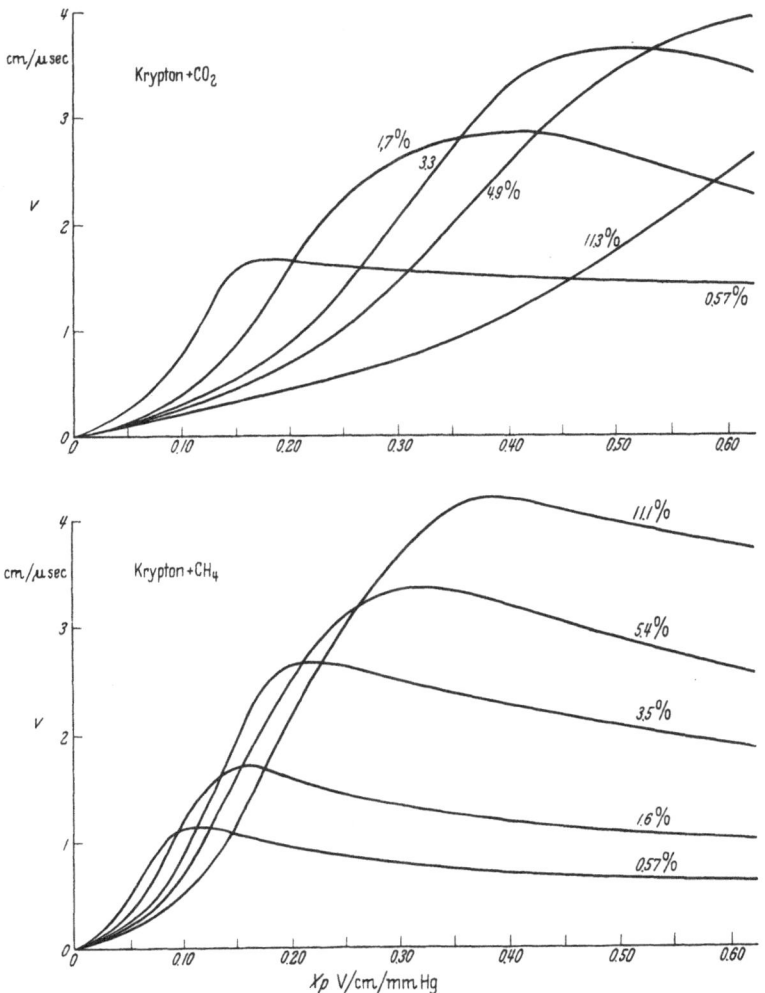

Fig. 10b. Drift velocity of electrons in krypton-plus-carbon dioxide and krypton-plus-methane mixtures.

Actually the increase observed would be even greater than expected, because another phenomenon also acts to increase w. The Ramsauer effect is the appearance of unusually small cross sections for the scattering of electrons from certain gas atoms at energies of about 1 volt[2] (see Fig. 12). Because of this phenomenon the lowering of the agitation energy has the effect of increasing the mean free path with a corresponding additional increase in w (see Figs. 7 to 11).

[1] E. Gerjuoy and S. Stein: Phys. Rev. **97**, 1671 (1955).
[2] Ramsauer: Ann. Phys. **64**, 513 (1921); **66**, 546 (1921).

The peaks in the drift velocity curves for the noble gases are interpreted as a manifestation of the Ramsauer effect[1]. While only the noble gases A, Kr and Xe show the strong Ramsauer effect, the elastic scattering cross sections of most gases vary considerably with the energy of the electrons [25].

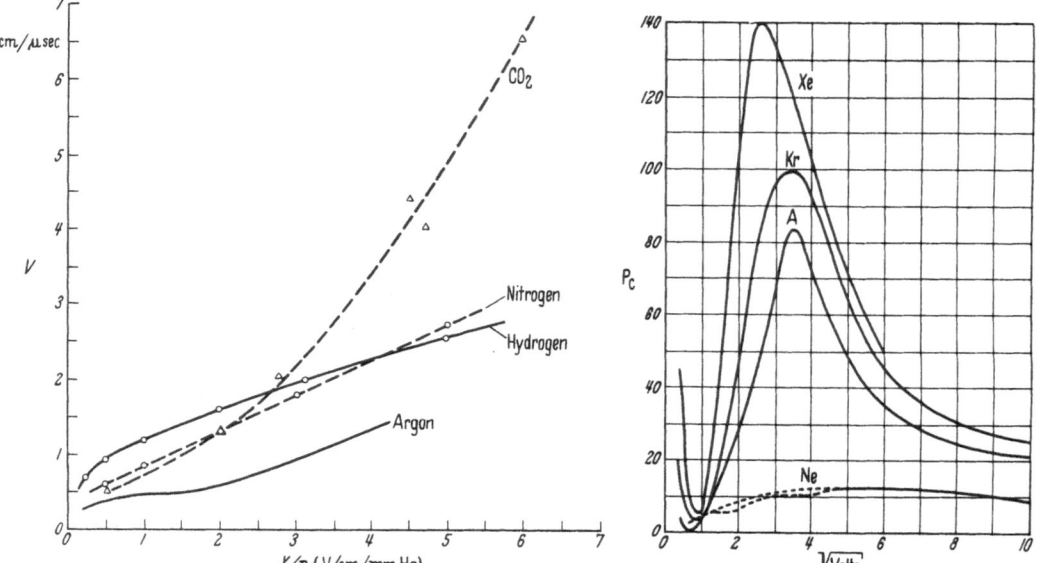

Fig. 11. Drift velocity of electrons in argon, nitrogen, hydrogen, and carbon dioxide at reduced field strengths up to 6 V/cm/mm Hg [11]. The curve for argon is from R. A. NIELSEN: Phys. Rev. 50, 950 (1936).

Fig. 12. The probability of collision of an electron with molecules of neon, argon, krypton, and xenon as a function of the energy of the electron. From BRODE [25].

10. Electron attachment. The capture of a slow electron by a molecule has a cross section which depends strongly upon the nature of the molecule involved. Those with similar chemical properties tend to behave similarly. In particular molecules of the strongly electro-negative elements such as oxygen and the halogens capture electrons readily. On the other hand the noble gases, N_2, CH_4, H_2, and D_2 capture no electrons.

The cross section for attachment is strongly energy-dependent when the electron is captured into an energy state with a finite lifetime. Various mechanisms of capture are in principle possible for removing the excess energy. Radiative capture, in which a gamma quantum is emitted, apparently has too small a probability to contribute significantly [10]. Another possibility is a three body collision in which a second molecule, atom, or electron acquires energy. This will be especially favored if the third body is a molecule in which vibrational motion can absorb energy. A third possibility in the case of polyatomic molecules is capture with the excitation of a vibrational state of energy E_v in the ion formed. Then, unless a third body is involved,

$$E_e + E_v = E_e^0 + E + E_v^0 \tag{10.1}$$

where E_e^0 and E_v^0 are the energies of initial electronic and vibrational states, E_e is the energy of that electronic state into which capture occurs, and E is the energy released in the capture. This kind of process may result in the formation of negative ions of the initial molecules, or it may lead to molecular dissociation

---

[1] ENGLISH and HANNA: [26].

into a negative ion and a neutral atom. The former is not likely to occur at low pressures where third body collisions are infrequent because the energy balance equation is so rarely satisfied.

Measurements of the probability of capture of electrons of low energies by O_2 have been made by BRADBURY[1] and HEALEY and KIRKPATRICK[2], with results of the general nature shown in Fig. 13. Two peaks are indicated, one near $E = 0$, the other near $E = 2$ ev. The explanation of these two peaks is not known. They may both be due to the formation of O_2^- in a process involving a third body[3]. Dissociative attachment apparently does not occur below about 3 volt. In any case the results are important here because oxygen is a common impurity in ionization chambers. McCUTCHEN[4] studied the effect of admixtures of 0.1% of O_2 to pure argon and found that attachment losses became appreciable. He then added 2% of CO_2 and found that the losses became very much less, presumably because the average agitation energy of the electrons was reduced to correspond to the minimum in the O_2 attachment curve at about 1 volt. A similar effect has been observed by FACCHINI and MALVICINI[5] using CH_4 and N_2 instead of CO_2. For discussions of both the theory and the experimental results with other gases, see MASSEY [10] and MASSEY and BURHOP [9].

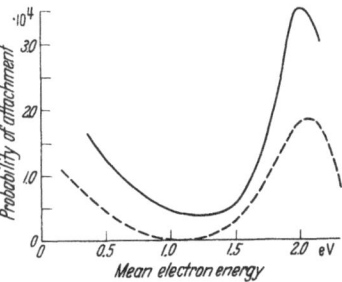

Fig. 13. The probability of electron attachment in oxygen per collision. Solid line: experiments of BRADBURY, Phys. Rev. 44, 883 (1933). Broken line: experiments of HEALEY and KIRKPATRICK [11].

11. Recombination. Positive ions may combine with negative ions or electrons with which they collide. The rate at which the process occurs is

$$\frac{d n_+}{d t} = - \alpha\, n_+\, n_- \tag{11.1}$$

where n_+ is the number of positive ions per cm³, n_- is the corresponding number of negative ions or electrons and α is the appropriate coefficient of recombination. The value of α depends upon the type, pressure, and temperature of the gas.

In ionization chamber work it has become customary to speak of three types of recombination depending upon geometrical conditions:

$\alpha)$ *Volume recombination.* The ions are generally distributed throughout the volume. Eq. (11.1) applies.

$\beta)$ *Columnar recombination.* The ions all lie initially along the track of the particle, e.g. an alpha particle, which produced them. JAFFÉ's theory of columnar ionization of heavy ions [8] arbitrarily assumes that ion pairs are initially distributed symmetrically around the line of motion of the primary particle, the ion density being a Gaussian function of r, then gives the saturation curves expected as a function of the angle between X and the direction of the track. The results are in good agreement with experiments showing that saturation is obtained with the lowest fields when X is normal to the track direction[6]. The theory is given only for heavy negative ions, not electrons.

[1] N. BRADBURY: Phys. Rev. 44, 952 (1933). — J. Chem. Phys. 2, 827 (1934).

[2] Quoted in HEALEY and REED [11].

[3] F. BLOCH and M. BRADBURY, MASSEY and BURHOP: [9], p. 262.

[4] C. W. CUTCHEN: Thesis. Princeton University 1950.

[5] U. FACCHINI and A. MALVICINI: Nuovo Cim. 1, 1255 (1955). — Nucleonics 13, No. 4. 36 (1955).

[6] G. JAFFÉ: Phys. Z. 30, 849 (1929). — W. R. KANNE and J. A. BEARDEN: Phys. Rev. 50, 935 (1936).

γ) Preferential recombination. Ionization electrons which are reduced to a low velocity at a small distance from a positive ion may recombine with it "preferentially", because the strong electric field of the ion may prevent the escape of the electron until it can be captured. This effect is found to be most pronounced at the higher pressures, as would be expected (see Sect. 34). It appears to be a fundamental limitation virtually equivalent to an increase in w with p.

12. Recombination of positive ions and electrons. In recombination, as in attachment, excess energy must be removed. Four possible processes are:

α) Radiative capture.

β) Dielectronic recombination. In this case an electron is captured, and at the same time a second electron already in the positive ion is simultaneously raised to a more energetic orbit and holds the energy of capture until that energy can be lost by collision or radiation.

Table 8. *Electron-ion recombination coefficients in different gases* [5], [10].

Gas	α cm^3/sec
H_2	$< 3 \times 10^{-8}$ [2]
H_2	$\sim 5.9 \times 10^{-11}$ [3]
He	1.7×10^{-8}
N_2	1.4×10^{-6}
O_2	2.7×10^{-7}
Ne	2.1×10^{-7}
A	8.8×10^{-7}
A	1.1×10^{-6}

γ) Dissociative recombination. This can occur if a polyatomic ion captures an electron and then breaks up into two or more neutral atoms.

δ) Three body recombination. Here a third body involved in the collision carries away the excess energy.

Of the four, apparently only (γ), where physically possible, and (δ) can be important in ionization chambers. However, the effects of collisions were not taken into account in the calculations giving the low probability for radiative capture which therefore cannot be ruled out entirely on the basis of present knowledge [10].

It has been estimated roughly [10] by means of a theory due to J. J. Thomson[1] that the three body recombination coefficient has values $(\alpha_{3e})_{He} \approx 10^{-11} p$ cm^3 per sec, and ($p < 30$ atm) and $(\alpha_{3e})_{Air} \approx 2 \times 10^{-10} p$ cm^3/sec, ($p < 20$ atm), where $p =$ pressure in mm of Hg. Thus for $p < 0.1$ mm three body recombination is negligibly small, of the same order of magnitude as the radiative and dielectronic types. For $p \approx 15$ atm, α_{3e} becomes equal to α_{Dis}.

Experimental values for α derived mostly from observations on the afterglow of microwave discharges are given in Table 8. These values are in the order of 10^{-6} cm^3/sec in contrast with values of the order of 10^{-10} expected for dielectronic or radiative recombination.

At the pressures prevailing in the experiments three body recombination should be negligible and, in fact, α proved independent of p. It was established that recombination in He and Ne under these conditions is *dissociative*. In He it turned out that about 70% of the ions were He_2^+. The dissociative process is

$$He_2^+ + e \rightarrow He' + He''.$$

The situation with respect to argon is reported to be more complex.

An excellent discussion of electron-ion recombination is given by Massey [10][4]. Unfortunately the experimental conditions under which the available data were found are so different from those prevailing in most ionization chamber work

---

[1] J. J. Thomson: Phil. Mag. **47**, 337 (1924).
[2] K. B. Persson and S. C. Brown: Phys. Rev. **100**, 729 (1955).
[3] R. F. Whitmer: Phys. Rev. **104**, 572 (1956).
[4] Also see M. Bayet and D. Quemada: J. Phys. Radium **16**, 334 (1955).

that the results may have little relevance. For example, if no molecular ions are formed there will be no dissociative recombination, and α will probably be of the order of 10^{-11} cm³/sec, except, possibly, at high pressures.

13. Recombination of positive and negative ions[1]. In this case there are three possible processes:

α) *Radiative.* This is too weak to be important.

β) *Three-body ionic recombination.* This can be calculated from J. J. THOMSON's theory which gives

$$\alpha_3 = \frac{64 \sqrt{2\pi}}{81} \frac{e^6}{(kT)^{\frac{5}{2}}} \left(\frac{M_1 + M_2}{M_1 M_2}\right)^{\frac{1}{2}} \left(\frac{1}{\lambda_1} + \frac{1}{\lambda_2}\right) \tag{13.1}$$

where M_1 and M_2 are the masses of the two ions, λ_1 and λ_2 their mean free paths, and

$$r_0 = \frac{2e^2}{3kT}. \tag{13.2}$$

The Thomson formula gives saturation with increasing pressures, the saturation value being

$$\alpha_3 = \frac{16 \sqrt{2\pi}}{9} \frac{e^4}{(kT)^{\frac{3}{2}}} \left(\frac{M_1 + M_2}{M_1 M_2}\right)^{\frac{1}{2}}. \tag{13.3}$$

At higher pressures LANGEVIN's theory must be used, and α_{3i} falls off as $1/p$. Fig. 14 shows experimental results for air.

γ) *Mutual neutralization.* In this process positive and negative ions brush against each other and the extra electron

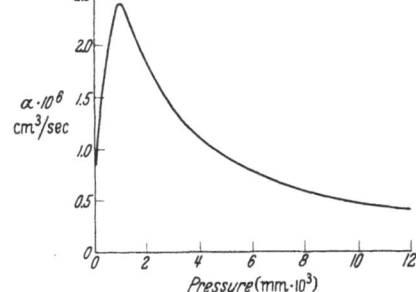

Fig. 14. Variation of recombination coefficient for positive and negative ions in air as function of the pressure. J. SAYERS: Proc. Roy. Soc. Lond., Ser. A 169, 83 (1939).

is transferred to the positive ion, the excess energy being carried away by the neutral atoms. There is apparently very little experimental information on this matter. Estimates for $O^- + O^+ \rightarrow O' + O''$ give $\alpha \sim 10^{-8}$ cm³/sec.

14. Generation of the signal. Even though in practice the cooperative signals from several hundred ion pairs are required to give detectable signals (see Sect. 16), the mechanism of signal induction is most readily discussed with reference to an individual ion pair. Let a positive ion and an electron lie close to each other initially somewhere near the middle of an ionization chamber having two electrodes, and let the electrodes be connected to a hypothetical battery having voltage V_0 and internal impedance zero. Furthermore, let the density of charge in the gas be so small that space charge effects can be ignored, as is usually the case in practice. This means that the potential at any point inside the chamber is determined by V_0 and the electrode geometry. Under the action of the electric field the two charges separate and move toward their respective collectors, the electron moving much faster. As soon as motion commences, a current begins to flow through the external circuit. The strength of the current as a function of time can in principle be calculated by a solution of the electrostatic boundary value problem if the position-vs-time relationship of the charges is known. However there is an easier approach: the result can be found in terms of the potential distribution in the chamber.

Let the potential at the initial position of the charges be V_1, and let the electron move to a new position where the potential is $V_1 + \Delta V$. Then the energy gained

---

[1] For a full discussion see MASSEY and BURHOP [9].

by the electron from the field will be $e\,\Delta V$, and the effective charge, q', moving between the electrodes can be calculated from energy conservation.

$$q' = e\,\Delta V/V_0, \tag{14.1}$$

a result valid only in the absence of space charge effects. The same result can be reached in various other ways, in the same approximation [7], [23]. Since the potential distribution within the chamber due to the applied field can almost always be determined either by calculation or by analogue measurements, (14.1) is a very convenient relationship to use.

Clearly the electron will make its entire contribution to the current and will have been collected before the positive ion motion has had much effect, since the electron will drift about 1000 times as fast. Eventually, however, a total effective charge $q' = e$ will be collected for the ion pair, the fraction due to the electron being given by (14.1) as eV_1/V_0, the fraction due to the ion being $q'_+ = e(V_0 - V_1)/V_0$. If the electron is replaced by a negative ion the analysis is the same, but of course the signal then has no fast component (see Fig. 15).

Any signal large enough to be observable will be composed of contributions from numerous ion pairs, which in many cases will have been created throughout an extended region of the chamber.

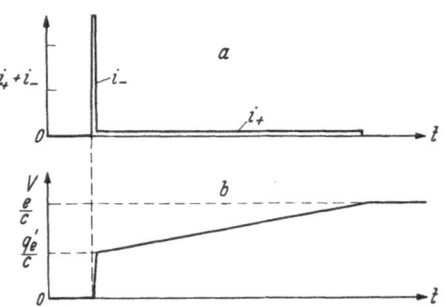

Fig. 15a and b. a) Idealized fast and slow components of signal current generated by the motion of a single electron and a single positive ion in a parallel-plate ionization chamber. b) The corresponding voltage signal.

15. Saturation. As a general rule the signal of an ionization chamber exposed to a constant flux of radiation (integrating operation) or to particles all of which lose a fixed amount of energy in the gas (pulse operation) shows a rapid initial rise, followed by a transition to a constant value as the collecting field X is raised from zero. See Fig. 16 for typical saturation curves. Saturation is obtained when X is made high enough so that the charges are collected before appreciable losses due to recombination and diffusion to the walls occur. When the gas pressure is less than about 10 atm one is usually safe in assuming that essentially all the ions are being collected when X exceeds the saturation value. At higher pressures that assumption is likely to be erroneous because of the onset of preferential recombination, which is a sensitive function of p, but not of X (see Sect. 34 for examples). Chambers filled with noble gases, particularly helium, argon, or neon, containing impurities capable of de-exciting metastable states may give anomalous saturation curves in pulse operation, especially if the collection time is long, because of the creation of additional ionization during the collection time.

Fast electron collection chambers give saturation curves which are influenced by attachment as well as by recombination and diffusion losses at the walls. Since attachment losses may increase with X, depending upon the electronegative impurities present, the "saturation" curve may be atypical in such cases. In certain rather special cases the situation is further complicated by the fact that v decreases with increasing X over some ranges of field strength. The signal from fast collection chambers may actually be largest for intermediate values of X, or may have a maximum and a minimum value before a higher saturation plateau, not necessarily completely flat, is reached. Experimenters working with demountable electron collection chambers have frequently used continuously operating purifiers through which the counter gas circulates by convection, or is pumped.

The active agent is a hot metal such as calcium, uranium or copper[1], which removes electronegative impurities by chemical action (see Table 9).

The saturation curve for ion collection integrating chambers has been studied both theoretically and experimentally. Typical curves for N_2 and CO_2 are shown in Fig. 16, obtained by DICK et al.[2] using a Po alpha source. It is found that the

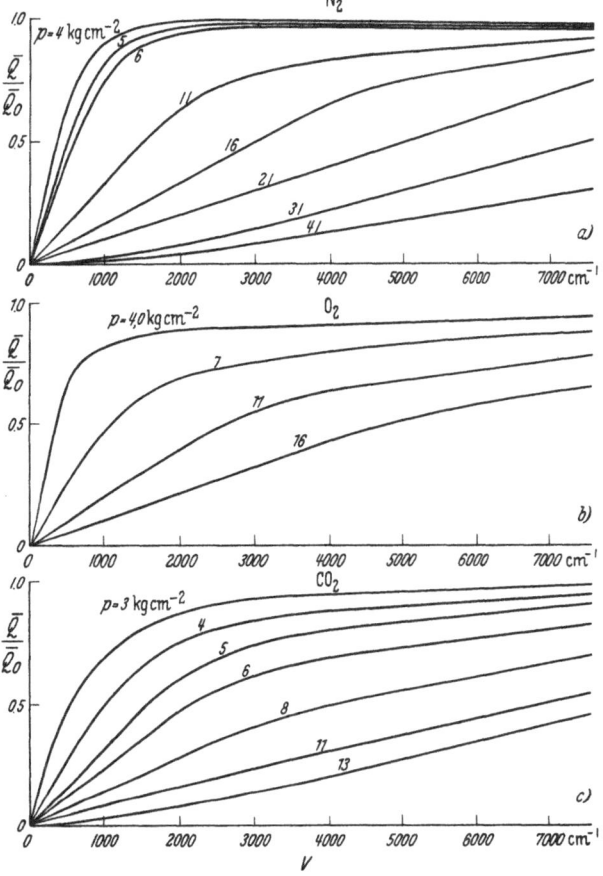

Fig. 16a—c. Saturation curves for nitrogen, oxygen, and carbon dioxide obtained by DICK, FALK-VAIRANT, and ROSSEL using polonium alpha particles in a parallel-plate ionization chamber.

saturation current and voltage are related by[3]

$$I_{sat} \approx K V_{sat}^n \qquad (15.1)$$

where K is a constant and $n \approx 2$.

Experiments with intense radiation[4] have shown that in parallel plate chambers space charge effects are easily made negligible, even with rather intense

[1] C. CERNIGOI and G. POIANI: Nuovo Cim. 2, 677—678 (1955) report that incandescent copper strips are superior to calcium for purifying argon.

[2] L. DICK, P. FALK-VAIRANT and J. ROSSEL: Helv. phys. Acta 20, 357 (1947).

[3] M. BLAU and J. CARLIN: Rev. Sci. Instrum. 18, 715—721 (1947).

[4] J. W. BOAG: Brit. J. Radiol. 23, 601—611 (1950). — J. W. BOAG and T. WILSON: Brit. J. Appl. Phys. 3, 222—227 (1952).

Table 9. *Some methods of purifying different gases.*

Gas	Method	Reference
Hydrogen, deuterium	Hot Pd filter; decomposition of UH_3	G. H. Stafford: Nature, Lond. **162**, 771 (1948). — Franzen [16]
Noble gases, hydrogen, deuterium, nitrogen	circulation over calcium at 300° C after outgassing at 400° C. Ca removes N_2 above 300° C	Klema and Barschall: Phys. Rev. **63**, 18 (1943); Rossi and Staub [7]; Klema and Allen: Phys. Rev. **77**, 661 (1950)
Noble gases	circulation over uranium metal at 750 to 800° C. Spongy U made by thermal decomposition of UH_3 is very reactive	Franzen [16]; Allen and Ferguson: AERE Report NP/R 1720. (1955)
Noble gases	circulation over incandescent Cu strips	Cernigoi and Poiani: Nuovo Cim. **2**, 677 (1955)
Helium	liquid nitrogen cold trap.	Bromley: Phys. Rev. **88**, 565 (1952)
Argon	liquid nitrogen, oxygen, cold trap methods	Sentfle and Farley: Rev. Sci. Instrum. **27**, 238 (1956); Fulbright [12], p. 184
Carbon dioxide	circulation over Ca at 140°C	Rossi and Staub [7]
Methane	fractionating column in liquid oxygen. Metallic Na deposit on container walls	Beghian and Halban: Proc. Phys. Soc. Lond. A **63**, 395 (1949); Allen and Ferguson: AERE Report NP/R 1720. 1955
Boron trifluoride	chemical decomposition of $CaF_2 \cdot BF_3$ thermal decomposition of $C_6H_5BF_4$	Bistline: Rev. Sci. Instrum. **19**, 842 (1948) Rossi and Staub [7]

radiation, by keeping X high. Lapsley[1], who used a cylindrical chamber, reports that a higher voltage was required to produce saturation when the central electrode was positive. He attributed the effect to positive ion space charge. Boag and Wilson[2] discuss empirical theories of saturation curves for chambers with the three simple symmetries.

The importance of columnar recombination in air, O_2, CO_2, CCl_4, etc. has been shown by the careful saturation curve determinations of Dick *et al.*[2], Kimura *et al.*[3], and Biber *et al.*[4], using alpha particle sources.

IV. Utilization of the signal.

16. Detection. The signal is never observed undistorted. There is always an inertial element somewhere between the observer and the chamber. It may be a simple galvanometer or electrometer, it may be an RC network in an electronic amplifier, or something else. The indicator as well as the ionization chamber conditions must be chosen to fit the particular application.

The discussion of electrometers and D.C. amplifiers for integrating applications is outside the scope of this paper. Many excellent references on the subject of electronic pulse amplifiers are available. See Elmore [14], Elmore and

---

[1] A. C. Lapsley: Rev. Sci. Instrum. **24**, 602—605 (1953).
[2] J. W. Boag and T. Wilson: Brit. J. Appl. Phys. **3**, 222 (1952).
[3] K. Kimura, R. Ischiwari, K. Yuasa, S. Yamashita, K. Mikaye and S. Kimura: J. Phys. Soc. Japan **7**, 111 (1952).
[4] C. Biber, P. Huber and A. Müller: Helv. phys. Acta **28**, 503 (1955).

SANDS [15], VALLEY and WALLMAN[1], and the papers cited in Subdivision V for detailed discussions of principles and practice. See the recent review by HIGIN-BOTHAM[2] for pulse height analyser references. A number of excellent pulse amplifiers and pulse height discriminators are sold commercially. However it seems worth while to mention here some of the basic requirements of pulse amplifiers.

α) *Amplification.* Ion chamber signals from single alpha particles are of the order of a few millivolts, e.g. for Po alpha particles stopped in argon $\frac{5.3 \times 10^6}{26.3} = 2 \times 10^5$ ion pairs are formed, and if the input capacitance of the amplifier together with the ion chamber and connectors is 20 $\mu\mu$f the pulse height will be $\frac{2 \times 10^5 \times 1.6 \times 10^{-19}}{20 \times 10^{-12}} = 1.6$ millivolts. Output signals of the order of 100 volt are conveniently handled, so an overall gain in excess of 60000 is required, 600000 being a suitable value. Of course gain controls are needed.

β) *Stability and linearity.* The gain should be stabilized by use of inverse feedback. Many amplifier designs call for the use of three feedback loops, one in a preamplifier closely attached to the ionization chamber, and two more in a main amplifier unit. Inverse feedback reduces distortion of pulse amplitudes by the amplifier to a negligible level.

γ) *Pulse shaping and signal-to-noise-ratio.* Perhaps the most difficult part of the problem of choosing design constants for a pulse amplifier is connected with pulse shaping. The gain characteristics of the amplifier as a function of frequency must be chosen to give the maximum signal-to-noise ratio consistent with experimental requirements on counting rates, pulse rise times, etc. If all special experimental requirements are ignored the problem is simplified somewhat, but remains fairly complicated.

Fundamentally there are two major sources of noise outside the ion chamber: the input circuit, and the amplifier itself. Noise generated in the amplifier consists mainly of shot noise due to fluctuations in the plate current of tubes; flicker effect noise due to variations in the emissivity of cathodes, giving appreciable contributions up to 50000 cycles or so; and microphonic and pickup noise due to extraneous sources. The last can be minimized by obvious means. The flicker effect can be minimized by choosing an amplifier frequency pass band which excludes low frequencies, if possible, and by choosing for use in the preamplifier particularly good tubes from a large assortment. The shot effect noise is a fundamental limitation which is most important in the first amplifier tube, where in effect it contributes part of the input circuit noise. The other major source of noise in the input circuit is the load resistor, which, even though perfect mechanically, always acts as a noise generator producing an effective voltage V_n given by

$$\overline{V_n^2} = 4\,\mathrm{k}\,T\,R \tag{16.1}$$

per unit frequency band. For a discussion of the implications of this, see EL-MORE [14], WILSON[3], and VALLADAS[4], and VAN DER ZIEL[5].

---

[1] G. E. VALLEY and H. WALLMAN: Vacuum Tube Amplifiers. New York: McGraw-Hill 1948.
[2] W. A. HIGINBOTHAM: Nucleonics **14**, No. 4, 61 (1956).
[3] R. WILSON: Phil. Mag. **41**, 66—76 (1950).
[4] G. VALLADAS: J. Phys. Radium **13**, 521—526 (1952). This paper includes a discussion of the limitations imposed by noise on the method of determining track orientations by pulse shape analysis used by R. SHERR and R. PETERSON: Rev. Sci. Instrum. **18**, 567 (1947).
[5] A. VAN DER ZIEL: Noise. London: Chapman & Hall 1955.

As a result of the compromises which must be made in connection with the problem of noise, it has become customary to use in amplifiers for amplifying fast electron signals a large value for the load resistance R, a practice which actually reduces the effective resistor noise, because the input capacitance is

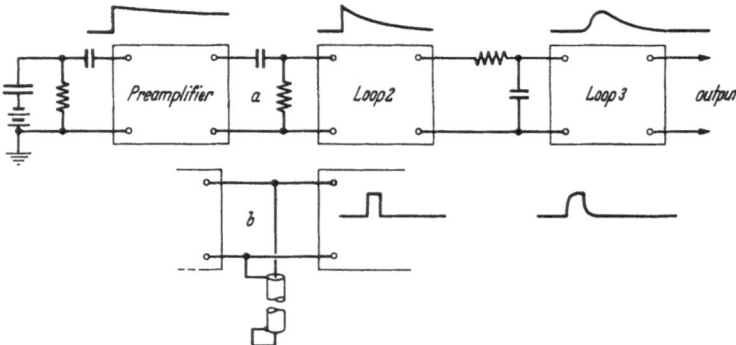

Fig. 17 a and b. Linear pulse amplifier block diagram and idealized wave forms. a) With $R - C$, and b) with shorted delay line pulse forming.

shunted across the resistor[1]. In this case the mean square noise at the output terminals of the amplifier is

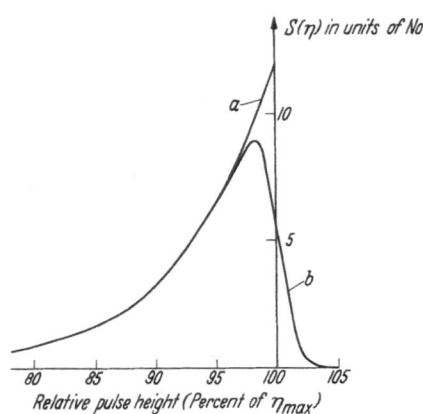

Fig. 18. The effect of noise upon a pulse-height spectrum. Curve (a): distribution expected ideally from monoenergetic short range protons randomly distributed throughout a cylindrical ionization chamber having radii a and b (see text). Curve (b): the same curve after amplifier noise having an rms value ($g/\sqrt{2}$) of 1.28% has been folded in.

$$V_n^2 = 4kTR \int_{f_1}^{f_2} \frac{g(f)\,df}{1 + (2\pi RCf)^2} \qquad (16.2)$$

where f_1 and f_2 are the frequency pass band limits of the amplifier, and $g(f)$ is its gain characteristic. In most applications the signal is first amplified, without special shaping, by a preamplifier, which has a voltage gain of roughly 30, then it passes through a high pass filter before reaching the first loop of the main amplifier. That first filter may be an R-C circuit, or preferably (with fast signals) a shorted delay line arrangement.

Between the two feedback loops of the main amplifier, in the feedback branch of the first loop, is a second filtering circuit having low pass characteristics. Fig. 17 gives a block diagram with idealized typical waveforms.

δ) *Noise and resolving power.* If electrically generated pulses of constant amplitude η_0 are fed into a pulse amplifier the output pulse height spectrum will not be an infinitely sharp line, but will have a Gaussian shape corresponding to the noise generated in the amplifier, $S_n(\eta) = \dfrac{N_0}{g\sqrt{\pi}} \exp\left(\dfrac{-(\eta - \eta_0)^2}{g^2}\right)$. If the pulses are fed into the normal input circuit through a very small capacitor, the output

---

[1] The total input capacitance should be kept small; the input tube of the amplifier should have a high mutual conductance. The input tube can even be mounted inside the ionization chamber to reduce the input capacitance: G. Bertolini and A. Bisi: Nuovo Cim. **9**, 1022 to 1024 (1952).

noise will include the noise of the normal input circuit, therefore the overall noise can be determined by a measurement of the output spectrum. The amplifier noise will distort any pulse height spectrum amplified. The "experimental" spectrum is given by

$$S_{\exp}(\eta) = \frac{1}{g\sqrt{\pi}} \int_0^{\eta'_{\max}} S(\eta')\, e^{-\frac{(\eta'-\eta)^2}{g^2}}\, d\eta',\tag{16.3}$$

where

$\eta =$ the pulse height,

$S(\eta') =$ the true spectrum,

$\eta'_{\max} =$ the maximum pulse height in $S(\eta')$,

$g =$ a constant $=$ rms noise $\times \sqrt{2}$.

Fig. 18 shows the effect of noise on a particular spectrum. A measured slowly and smoothly varying spectrum can be corrected approximately for the effect of amplifier noise by use of the method of OWEN and PRIMAKOFF[1], which gives

$$S(\eta) \approx S_{\exp}(\eta) - \frac{g^2}{4}\, \frac{d^2\, S_{\exp}(\eta)}{d\eta^2}.\tag{16.4}$$

17. Average current (or integrating) operation. In this case large numbers of primary particles are involved. The indicator smooths out statistical fluctuations in the signal and presents a reading of the current averaged over a time long compared with the average interval between primary events. The averaging time in many cases is of the order of a second. An electrometer or a DC amplifier indicator with RC smoothing may be used. Since fast components of the signal would not be seen anyway, electron attachment is not a factor, so almost any gas may be used, but operation on the saturation plateau is desirable, especially to maintain a constant sensitivity.

18. Slow pulse operation (response to heavy ion motion). Here it may be desired simply to count relatively infrequent pulses (say of the order of 10 per second, or fewer), or it may be desired to measure the pulse height spectrum. In either case it is advantageous to amplify the signal with an inverse-feedback-stabilized linear pulse amplifier. The amplifier must have a pass band including audio frequencies corresponding to the lower frequency Fourier components of the slow signal, so in practice there is often a troublesome amount of noise in the output due to microphonics, hum from power line pickup, and to the flicker effect in the amplifier tubes. The first and second can be minimized by careful design and construction, but the last is an inherent noise generator which seriously limits the resolving power of slow pulse spectrometers unless the pass band is made very narrow. Vibrating reed electrometers[2] have been used very successfully in slow pulse work (Sect. 29α).

In slow pulse operation attachment *per se* causes no difficulty, but, recombination as always, should be minimized.

19. Fast pulse operation. In this case the fast components of the signal due to electron motion are all-important; the currents due to heavy ion motion are discriminated against by the amplifier, which has a pass-band that effectively cuts off well above the audio range. Having the pass-band at high frequencies helps enormously to reduce the noise from the three sources mentioned above.

[1] G. E. OWEN and H. PRIMAKOFF: Phys. Rev. **74**, 1406 (1948).

[2] H. PALEVSKY, R. K. SWANK and R. GRENCHIK: Rev. Sci. Instrum. **18**, 298 (1947).

The biggest advantage of fast pulse operation is that counting rates can be quite high so that a lot of information can be obtained in a short time. Also, high-resolving-power coincidence methods can be employed. The chief disadvantage is that the pulse heights obtained will usually depend upon the spatial distribution of the initial ionization because of the effective charge reduction mentioned in Sect. 14. However several ways around this problem have been devised.

Let us suppose that we have the problem of designing a fast pulse chamber for α particle spectrometry, that the source is to be placed inside the chamber, and that the gas pressure is so high that the particles will stop before reaching the walls. Then, if the range of the particles is of the order of the chamber dimensions, we will have to face the geometrical problem of choosing a design such that all electrons carry to the collector approximately the same effective charge, preferably e, regardless of differences in their positions of liberation. The necessity for special provision is illustrated by a calculation of the alpha

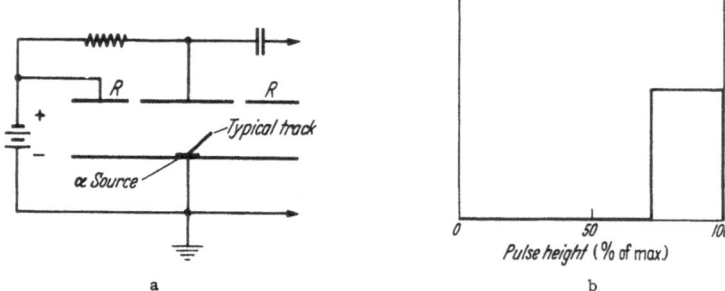

Fig. 19a and b. a) Parallel-plate ionization chamber with guard ring. b) Idealized spectrum expected from a source of mono-energetic alpha particles in such a chamber when only the fast component of the signal is observed.

particle pulse height spectrum for the parallel plate geometry. The pulse spectrum can easily be shown to be flat-topped (if edge effects are ignored), with effective charge ranging from

$$Q^-_{max} = Ne, \quad \text{to} \quad Q^-_{min} = Ne\left(1 - \frac{\bar{\varrho}}{d}\right), \tag{19.1}$$

where $\bar{\varrho}$ is the distance from the origin of the track to the center of gravity of the ionization along the track[1].

The problem then is: How can the chamber be built so that the signal generated while the electrons are moving throughout the entire region in which the ionization occurs is negligible? Two ways will be described.

α) *Shielding method.* One way is to introduce a third electrode, a "Frisch grid", separating the chamber into two parts shielded electrostatically from each other, the electron collector being on one side, the volume in which the ions are formed being on the other[2] (see Fig. 20). If the potentials of the electrodes are properly chosen electrons can be drawn through the grid without serious loss to the grid wires and reach the collector. The grid thus serves as the virtual source of electrons, and all electrons carry over almost exactly the same effective charge, e. The effectiveness of the grid as a shield has been calculated for the

---

[1] It should be observed that $\bar{\varrho}$ appears in (19.1) because the uniform field distribution characteristic of parallel plate geometry leads to the possibility of using such an average.

[2] Named for O. R. FRISCH, who suggested its use. British Atomic Energy Report BR 49. 1945.

parallel plate case by BUNEMANN, CRANSHAW, and HARVEY[1]. The results of their calculations, and confirming experiments, show that the spread in the pulse height distribution due to departure from perfect shielding is easily made less than the spread due to other causes, especially amplifier noise.

As would be expected, the calculations showed that the shielding action of the grid is best when the wires are large and the openings small, but, of course, the number of lines of force from the ionization volume terminating on the grid wires will then tend to be large and there will tend to be a correspondingly large loss of electrons. The inefficiency of shielding is given by

$$\sigma \equiv \frac{dX_B}{dX_Q} = \frac{l}{b+l} \approx \frac{d}{2\pi b} \log \frac{d}{2\pi r}, \quad (19.2)$$

where

$$l = \frac{d}{2\pi}\left(\frac{1}{4}\varrho^2 - \log \varrho\right),$$

$$t = \frac{d}{2\pi}\varrho^2 = \varrho r \quad \text{and} \quad \varrho = \frac{2\pi r}{d}.$$

See Fig. 20. The conditions for all of the field lines from Q to terminate on the collector is given by

$$\frac{V_B - V_G}{V_G - V_A} \geq \frac{b + b\varrho + 2l\varrho}{a - a\varrho - 2l\varrho}. \quad (19.3)$$

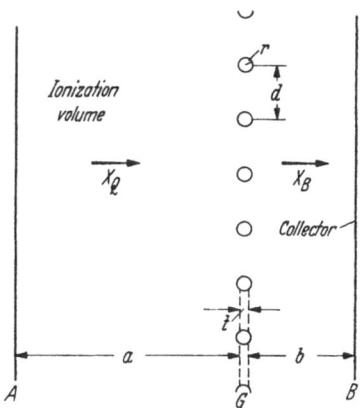

Fig. 20. The geometry of an ionization chamber with a Frisch grid.

One should not conclude that in practice no electrons will be lost to the grid if Eq. (19.3) is satisfied. Actually diffusion will cause some losses, which, however, can be made small. HERWIG, MILLER and UTTERBACK[2] have made experiments to study the loss of electrons to grids and have found a correlation between field conditions required for saturation and the lateral diffusion in various gases. They also studied gas multiplication effects caused by the strong fields near grid wires as well as saturation effects characteristic of He, Ne, A, Kr, N_2, He + 3% CO_2, A + 3% CO_2 and A + 5% N_2. CO_2 was the most favorable from the grid-loss view point.

Many other shielding geometries are also possible.

$\beta)$ *The extreme geometry method.* In this method a two-electrode design is used, the anode being so small that very little charge is induced on it by any particle not quite nearby. A cylindrical geometry with a fine wire anode is an example (see Fig. 21). Suppose that such a chamber is filled with N_2 gas and is irradiated uniformly with slow neutrons (as in the experiments of FRANZEN, HALPERN, and STEPHENS[3]). Then short range monoenergetic protons from the exothermic N^{14} (n, p) C^{14} reaction $(Q = 0.6$ Mev) will produce pulses. The pulse height spectrum to be expected can be calculated from Eq. (14.1) if wall and end effects are ignored. If the inner and outer radii are a and b, respectively, and r a radial position in between the cylinders

$$V_r = \frac{V_0 \log\left(\frac{r}{a}\right)}{\log\left(\frac{b}{a}\right)} \quad (a < r < b) \quad (19.4)$$

[1] O. BUNEMANN, T. E. CRANSHAW and J. A. HARVEY: Canad. J. Res. A **27**, 191 (1949).
[2] I. O. HERWIG, G. H. MILLER and N. G. UTTERBACK: Rev. Sci. Instrum. **26**, 929—936 (1955).
[3] W. FRANZEN, J. HALPERN and W. E. STEPHENS: Phys. Rev. **77**, 641 (1950).

and the corresponding effective charge of an electron produced at r is

$$q' = \frac{e \log \left(\dfrac{r}{a}\right)}{\log \left(\dfrac{b}{a}\right)}. \tag{19.5}$$

The relative probability that an electron will start at a radius between r and $r + dr$ is $2\pi r \, dr$, or in terms of q'

$$S = 2\pi a^2 \left(\frac{b}{a}\right)^{2\frac{q'}{e}} \log \left(\frac{b}{a}\right) \tag{19.6}$$

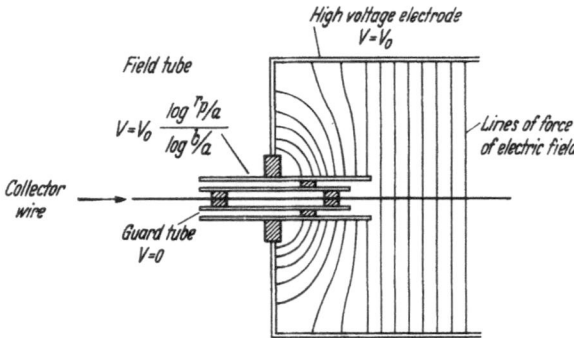

Fig. 21. Geometry of a cylindrical ionization chamber. One end only shown, illustrating the use of an auxilliary electrode to define the active volume and to eliminate end-effect distortion of the symmetry of the field in the active volume. COCKROFT and CURRAN: Rev. Sci. Instrum. 22, 37 (1951).

which, when normalized to correspond to N_0 total counts becomes the spectrum

$$S_c(\eta) = \frac{2 N_0}{\left(\dfrac{b}{a}\right)^2 - 1} \left(\frac{b}{a}\right)^{2\eta}, \tag{19.7}$$

where $\eta = \dfrac{q'}{e} =$ relative pulse height.

This method can readily be applied to the spherical case, with the result[1]

$$S_s(\eta) = \frac{N_0}{\left(\dfrac{b}{a}\right)^3 - 1} \left[\frac{1}{1 - \eta\left(1 - \dfrac{a}{b}\right)^3} - 1\right]. \tag{19.8}$$

From Eq. (19.7) it is clear that S_c will be sharpest when b/a has the largest possible value. On the other hand there is a conflicting requirement: the electric field must be high enough everywhere to produce saturation. Therefore b/a ratios of the order of several hundred are the largest ordinarily used, with the result that the resolving power is no better than about 5% (see Fig. 18).

If the pulses observed arise from the uniform irradiation of the chamber with short bursts of x- or γ-rays, the effective electron component of charge in the cylindrical case will be [7]

$$Q^- = e N_0 \left[\frac{b^2}{b^2 - a^2} - \frac{1}{2 \log \left(\dfrac{b}{a}\right)}\right] \tag{19.9}$$

---

[1] W. FRANZEN [16].

or approximately

$$Q^- = e N_0 \left[1 - \frac{1}{2 \log \left(\dfrac{b}{a} \right)} \right] \qquad (b \gg a) \qquad (19.10)$$

where N_0 is the total number of ion pairs formed.

If the polarity is reversed, the total effective electron charge will be reduced to

$$Q^-_{\text{rev}} = e N_0 \left(\frac{1}{2 \log \left(\dfrac{b}{a} \right)} - \frac{a^2}{b^2 - a^2} \right), \qquad (19.11)$$

which, of course, is equal to the total effective positive ion charge in the normal case.

If the range of the primary particles is not negligibly small compared with b, wall and end effects cannot be neglected. The calculation of the geometrical effect is then generally much more tedious, especially since the distribution of ionization density along the track (Bragg curve[1]) must be taken into account. The spectrum may have to be obtained by graphical analysis, especially in case the chamber has no simple symmetry, and $V(x, y, z)$ has been determined by an analogue measurement.

V. Applications.

A number of typical applications of ionization chamber techniques will now be given. They have been chosen as representative uses from various fields and they do not by any means constitute a complete survey.

20. The older applications. Before about 1930, when vacuum tube amplifiers came into practical use, virtually all of ionization chamber work was done with the help of an electrometer indicator. Often the chamber and the electroscope were combined in one unit. In other cases the two were separated in space, but connected with a conductor. In still others the two were quite separate, and the experiment consisted of giving the chamber a known initial charge, exposing it to radiation, then connecting it to the electroscope to determine the amount of charge lost. Modern versions of each of these integrating methods are cited below.

21. Radiation survey instruments: health protection instruments. Ionization chambers, notably the Victoreen R-Meter, have for a long time been used as standard devices for the measurement of x-ray doses and fluxes. It was natural to extend the method to the measurement of intensities and doses of nuclear radiations. The basic integrating ionization chamber used in health monitoring measurements is the cavity ionization chamber, which is usually an air chamber with rather thick walls of material similar to biological tissue with respect to ionization energy loss of radiation, for example polystyrene. Then the dose in the walls can be obtained from the Bragg-Gray equation[1]

$$D_m = J_m w P_m, \qquad (21.1)$$

where D_m is the dose in the walls (ergs/gm), J_m is the number of ion pairs detected per unit mass of gas, w is the average energy loss per ion pair in the gas, and P_m is the ratio of the mass stopping power of the wall to that of the gas for the type

---

[1] RUTHERFORD, CHADWICK and ELLIS: Radiations from Radioactive Substances. Cambridge: University Press 1930. — G. A. W. RUTGERS and J. M. W. MILATZ: Physica, Haag **7**, 508—514 (1940).

of primary ionizing particle involved. Eq. (21.1) holds rather well under many conditions (see GRAY[1], FANO[2], and MARINELLI[3]). It is not exactly correct however[4]. In particular errors are noticeable when the atomic numbers of the gas and the walls are very different, especially if the cavity is large and the pressure is high. Calculations taking into account the finite range of delta rays give results in agreement with observed deviations[5].

The original radiation dose unit, the *r (Roentgen) unit*, was defined as the quantity of radiation producing 1 esu of charge of both signs in 1.293 mg of dry air[6]. The equivalent amount of energy absorbed is about 84 ergs per gram in air or 93 ergs per gram in soft tissue. A later unit, the *rep* (Roentgen equivalent, physical), was defined as the quantity of radiation yielding 93 ergs per gram of energy absorption at the place under consideration inside the tissue. A still newer unit, the *rad*[7], is based upon the absorption of *100 ergs per gram of tissue*. It seems possible that the rad may become the permanent unit. An additional unit, derived from the rad, is the *rem* (Roentgen equivalent, man).

Table 10. *Values of relative biological effectiveness for different kinds of radiation* [13].

Radiation	RBE
X-rays, γ-rays, β-rays	1
Slow neutrons	3 to 4
Fast neutrons, protons, and α-particles .	10
Heavy recoil nuclei	20

$$\text{Dose (rem)} = \text{Dose (rad)} \times \text{RBE}. \qquad (21.2)$$

See Table 10. According to present standards people can safely tolerate an average weekly dose of *0.3 rem* for an indefinitely long time.

α) *The extrapolation chamber.* In connection with the problem of determining the back scatter contributions to skin doses in x-ray therapy FAILLA[8] suggested the procedure of plotting the ionization measured in a flat ionization chamber as a function of plate spacing. Then, if ion current is plotted against spacing, and the curve is extrapolated to zero spacing, the slope at the origin has a value which can be used with the Bragg-Gray relationship [Eq. (21.1)] to calculate the ionization energy loss in the solid material with the effect of the gap eliminated. Chambers of this sort have found applications in other fields[9].

β) *The pocket ionization chamber.* In many laboratories persons engaged in work with radiation are required to carry in their pockets one or more pocket ionization chambers in order to establish fairly accurate records of the doses of radiation which various parts of their bodies receive. These are pencil-like units about $1\frac{1}{2}$ cm in diameter and 15 cm long (see Fig. 22). They are charged before

[1] L. H. GRAY: Proc. Roy. Soc. Lond. **156**, 578 (1936).
[2] U. FANO: Radiation Res. **1**, 237—240 (1954).
[3] L. D. MARINELLI: Ann. Rev. Nucl. Sci. **3**, 249—270 (1953).
[4] See, for example, F. H. ATTIX and L. DE LA VERGNE: J. Res. Nat. Bur. Stand. **53**, No. 6, 393—402 (1954).
[5] L. V. SPENCER and F. H. ATTIX: Radiation Res. **3**, 239—254 (1955), and P. R. J. BURCH: Radiation Res. **3**, 361—378 (1955). For other discussions of the limitations see: [3]. WHITE, MARINELLI and FAILLA: Amer. J. Roentgenol. **44**, 889 (1940) and T. I. WANG: Nucleonics **7**, 55—71 (1950).
[6] Report on International Congress on Radiation Units. London 1953. Nucleonics **8**, No. 1, 28 (1951).
[7] A similar report for the 1954 meeting: Nucleonics **12**, No. 1, 11 (1954).
[8] G. FAILLA: Radiology **20**, 202—215 (1937).
[9] See for example J. WEISS, W. BERNSTEIN and J. B. H. KUPER: J. Chem. Phys. **22**, 1593 (1954).

work begins with the help of a special instrument sometimes called a minometer, and their loss of charge is read directly in R units on the scale of the same instrument at the end of the work period. The charging potential is about 150 volt. The main practical problem here is to provide such good insulation that the loss of charge through leakage is negligible. Teflon (C_nF_{2n+2}) is highly recommended[1,2]. A common scale for the reading instrument is 200 milliroentgens. SUTHERLAND[3] describes the constructional details of a practical portable dynamic electrometer for reading the doses. The original minometer was an electroscope. These devices have proved quite useful. The accuracy of dose determination by this means is about 10%.

Workers sometimes need to be able to plan their activities according to the amount of dosage that they have received, and frequent excursions to a master

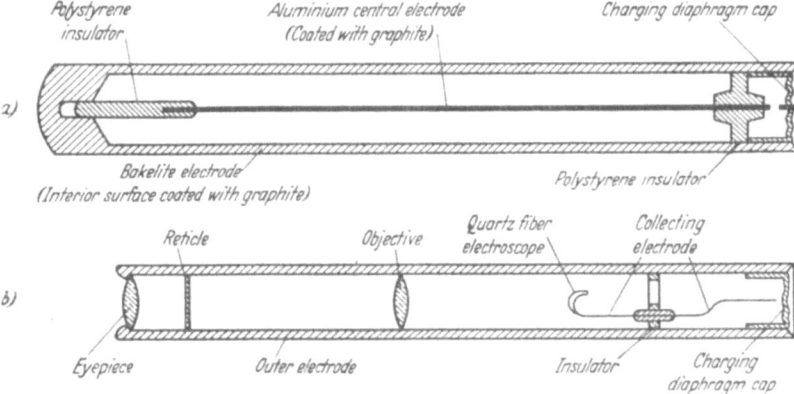

Fig. 22a and b. a) A pocket ionization chamber. b) A self-reading pocket ionization chamber.

reading device may be impractical. In such cases pocket chambers with built-in electroscopes are often used.

γ) *The Lauritsen electroscope.* A somewhat bulkier, but still portable unit containing both ionizing chamber and built-in electroscope is the well-known Lauritsen electroscope[4]. It is given a full charge at the start by means of an auxiliary supply, and the scale reading then gives a running account of the quantity of radiation which has passed through the chamber. According to GLASSTONE [13] a Lauritsen electroscope is accurate to about 1%, and the version with boron-lined walls is the most accurate instrument available for the monitoring of slow neutron fluxes. In slow neutron monitoring it may be desirable to use a second electroscope identical with the first except for having no boron lining so that the effect of the gamma ray component of the radiation will be cancelled when the difference of the two readings is taken. (Special compensating chambers will be described under Sect. 23, below.)

δ) *Portable instruments giving continuous readings of radiation intensity.* For a great many purposes, such as monitoring the interior surfaces of a laboratory

[1] H. LYSONS: J. Sci. Instrum. 27, 334—335 (1950).
[2] H. H. Ross and G. FAILLA: Nucleonics 14, No. 2, 32—37 (1956). This article gives details of the construction of tissue-equivalent chambers for gamma and neutron dosimetry.
[3] W. H. SUTHERLAND: Brit. J. Radiol. 22, 255—260 (1949).
[4] C. C. LAURITSEN and T. LAURITSEN: Rev. Sci. Instrum. 8, 438 (1937).

in which radiochemistry is done, portable instruments which read the momentary intensity of the radiation are needed. Many different models based upon open air ionization chambers have been built, and a number of them are sold commercially. Typically they consist of a small box weighing a few pounds containing the chamber, a DC amplifier, batteries, and a meter on which the output of the amplifier is indicated. The most sensitive scale of the meter is usually about 0 to 0.1 milliroentgens per hour, and a switch is provided for reducing the sensitivity in several steps. Often one side of the ionization chamber has a thin window or wire screen through which alpha particles, X-rays, soft γ-rays, and beta particles can pass, and one or more removable cover plates are provided to make possible discrimination between radiations of different penetrabilities[1]. A variation of this unit has the ionization chamber mounted on the end of a long pole so that it can be used as a probe. For surveying weaker fields of radiation portable counters or counting rate meters based upon proportional or Geiger counter tubes are frequently used.

ε) *Slow neutron monitors.* It was stated above that the boron-lined Lauritsen electroscope is a good slow neutron measuring device. Clearly any other chamber containing in its gas or on its walls nuclei in which slow neutron-induced reactions can occur with the emission of charged particles will be sensitive to neutrons. He^3 and BF_3 are typical gases used in fillings, and boron or uranium (in which fission can occur) are most often used in linings of slow neutron counters. Either counting or integrating methods may be used, since the signals from the individual reactions are quite large, especially, of course, in the case of fission. Even so, the boron activated counters are frequently used as proportional counters.

The design of chambers with multiple detecting layers is described by Lowde[2], who gives formulas for correct layer thickness and neutron counting efficiency. He describes a miniature instrument with a sensitive volume of 0.4 cm³ which is said to be 24% efficient in counting slow neutrons. Veall[3] describes another small chamber. Many other examples of borontype counters are described in the literature[4]. For the design and operation of a fission-type chamber see Aves *et al.*[5] and Holmes *et al.*[6]. For further discussion of slow neutron detectors, see Sect. 23, below, and Rossi and Staub [7].

ζ) *Fast neutron monitors.* Ionization chambers have not been used very much in fast neutron counting for health monitoring purposes. Two reasons contribute to the explanation. First, the total cross sections for nuclear reactions tend to be small, of the order of a barn or less for most reactions in the Mev range. The total cross section for the n-p collision for example is 4.5 barn at 1 Mev and is falling off approximately as $E_n^{-\frac{1}{2}}$. Second, fast neutrons are very effective in damaging biological tissues so that the number of fast neutrons which constitute an average tolerance dose is small, only about 50/cm²/sec according to present standards. The tolerance dose as measured by air ionization chambers is of the order of $\frac{1}{10}$ of that for x-rays or γ-rays.

[1] For example D. G. Wyatt: J. Sci. Inst. Phys. Ind. **26**, 13—16 (1949), and M. S. Freedman: U.S. AEC Document AECD-2549-A, p. 28. 1948. M. Briere, A. Rogozinski and J. Well: J. Phys. Radium **12**, 697—699 (1951), describe a unit with logarithmic response.
[2] R. D. Lowde: Rev. Sci. Instrum. **21**, 835 (1950).
[3] N. Veall: J. Sci. Instrum. **24**, 331 (1947).
[4] For example K. E. Larsson and C. Taylor: Ark. Fysik 3 (9), 131 (1952). — C. C. Jonker and J. Blok: Physica, Haag **15**, 1032 (1949). — H. v. Ubisch: Ark. Fysik **35** (11), 1 (1948).
[5] R. Aves, D. Barnes and R. B. Mackenzie: J. Nucl. Energy **1**, No. 2, 110 (1954).
[6] J. E. R. Holmes, D. D. McVicar, L. R. Shepherd and R. D. Smith: J. Nucl. Energy **1**, No. 2, 117 (1954).

For other fast neutron monitoring considerations see the division by H. BAR-SCHALL in this volume and MARINELLI [18]. HURST et al.[1] give a description of proportional counter methods. For very high energy monitors ($E_n > 50$ Mev) see WIEGAND[2] and DE JUREN[3].

22. General radiation monitors for cyclotron rooms. Because of a number of factors, including the great range of the particles in the beam, it is difficult to measure the beam current of the larger nuclear accelerators, so some other continuous indication of level of output is needed. Ionization chambers have been useful in this connection because of their reliability of operation and their general simplicity.

23. Ionization chambers as the sensing elements in atomic pile control systems. The control of the power level in an atomic pile is a very serious matter because, as has happened several times, when the power level is allowed to get much too

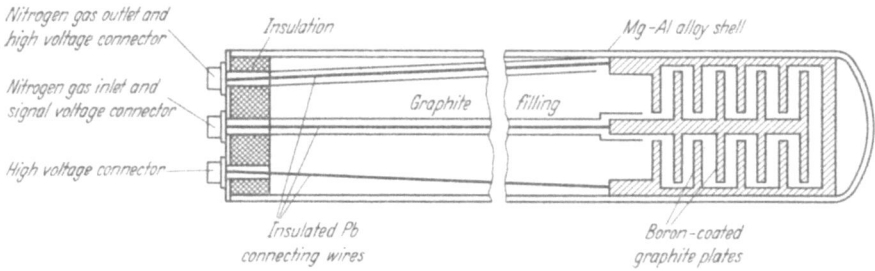

Fig. 23. A parallel circular-plate (PCP) ionization chamber for use inside an atomic pile.

high even for a short time, the result is that the pile must be rebuilt, and the entire laboratory and part of the countryside are likely to have been contaminated. The reliability and stability of the ionization chamber have made it the accepted high level neutron flux detector in power pile control systems. The rest of the system often consists in part of an electronic amplifier and servo arrangement driving control rods in such a way that the power is held at a present level. Cox[4] has described an additional refinement. Electrical differentiation of the ionization chamber signal gives a voltage proportional to the rate of change of ionization, which is a measure of the rate at which the power level is rising. That signal can be fed into the feedback mechanism in such a way that the power level automatically rises at a preset rate, then levels off and remains constant.

The most common ionization chambers for pile control purposes are those having enriched B^{10} linings or $B^{10}F_3$ gas fillings. U^{235} linings can also be used, but the fission products which are formed accumulate and produce a variable background which interferes unless a continuous flow of gas is directed through the chamber. With enriched boron a maximum ion current of 10^{-16} amp/cm² per neutron per cm² can be obtained [13]. To obtain sufficient sensitivity with chambers of reasonable size it sometimes turns out to be advantageous to build large surfaces into the chambers, for example electrodes out of nestling sets of boron-coated cylinders. An alternating disc geometry can also be used (see Fig. 23).

[1] G. S. HURST, R. H. RITCHIE and W. A. MILLS: Proceedings of international conference on peaceful uses of atomic energy, Vol. 14, p. 220. New York: United Nations. Also see B. THOMPSON: Nucleonics **13**, No. 3, 44 (1955).

[2] C. WIEGAND: Rev. Sci. Instrum. **19**, 790 (1948): Bismuth fission.

[3] J. DE JUREN: U.S. AEC Report UCRL 1090. 1951.

[4] R. J. COX, A.E.R.E.: Quoted by DENIS TAYLOR, Nucleonics **12**, No. 10, 12 (1954).

The chambers used in pile control systems are sometimes placed outside the pile at a considerable distance so that an average neutron yield of a sort is detected. In other cases they are placed directly inside the pile, a number of them being distributed throughout its volume so that an average level can be measured. In such cases the physical design of the chamber becomes very much more difficult, because attention must be paid to such matters as radiation damage to insulators and intense radioactivity induced in the parts of the chamber leading to a serious amount of ionization, especially from beta rays. A number of successful designs have been developed (see Fig. 23). Some of them are "compensation" chambers. They consist of two ionization chambers built into one envelope, one containing no boron,

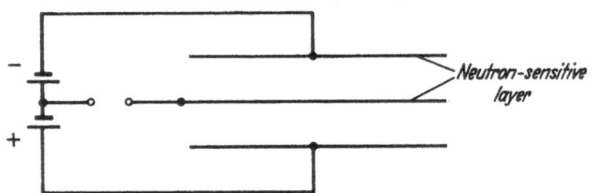

Fig. 24. A compensation chamber. Only one side has a special neutron-sensitive deposit on its walls. The net current due to gamma ray ionization in the two sides is zero.

but otherwise having the same sensitivity as the second. The difference between the currents flowing through the two sides is a measure of the neutron induced ionization (see Fig. 24).

24. The routine measurement of strengths of sources. Two air-filled ionization chambers used at the United States Bureau of Standards for the comparison of radioactive samples with standard sources are described by Seliger and Schwebel[1]. One, having a solid angle of acceptance of $\sim 2\pi$ is used for β-ray observations, the other, having a larger acceptance angle, is used for γ-ray measurements (see Figs. 25a and b). A null circuit based on a Lindemann-Ryerson electrometer is used with each. The apparatus is very stable: readings are reproducible to within better than 0.5% after a year has passed. Individual readings are reproducible to within 0.3%.

An ionization chamber of annular shape suitable for the continuous monitoring of γ-radiation from radioactive liquids and gases flowing through a pipe along its axis is described by Persiani[2].

Ionization chambers have been used for the routine analysis of gases containing radioactive nuclei. For example, Janney and Moyer[3] describe the use of an ionization chamber setup with Lindeman electrometer indicator in the routine analysis of CO_2 gas containing some C^{14}. They were able to measure samples in which the specific activity was as low as 10^{-9} curie/gm. They found that some of the sample remained behind after the measurement, apparently in chemical combination with the walls of the chamber, thus producing a "memory" effect, which, however, disappeared after new chambers with walls previously coated with carbonate or plated with gold were used. They found it desirable to feed the gas into the chamber through a filter in order to do away with a transient effect in the ionization current associated with filling for which they surmise that charged dust particles were responsible. They describe other refinements, including a bridge circuit designed to remove troublesome effects due to fluctuations in the biasing battery voltage. Brownell and Lockhart[4] describe another application to C^{14} assay.

[1] H. H. Seliger and A. Schwebel: Nucleonics **12**, No. 7 (1954). Also C. C. Smith and H. M. Seliger: Rev. Sci. Instrum. **24**, 474 (1953).
[2] P. J. Persiani: U.S. AEC Document ANL-4702. 1951.
[3] C. D. Janney and B. J. Moyer: Rev. Sci. Instrum. **19**, 667 (1948).
[4] G. L. Brownell and H. S. Lockhart: Nucleonics **10**, (2) 26 (1952).

A flow type chamber for measuring low energy beta radiation in gases is described by QUINN[1]. Gas flowing at the rate of 32 liter per minute and containing between 10^{-10} and 10^{-11} curie per liter of S^{35} in the form of SO_2 was monitored continuously.

WILZBACH et al.[2] describe an ionization chamber apparatus with a vibrating reed electrometer indicator for use in the routine analysis of the H^3 content of gas samples.

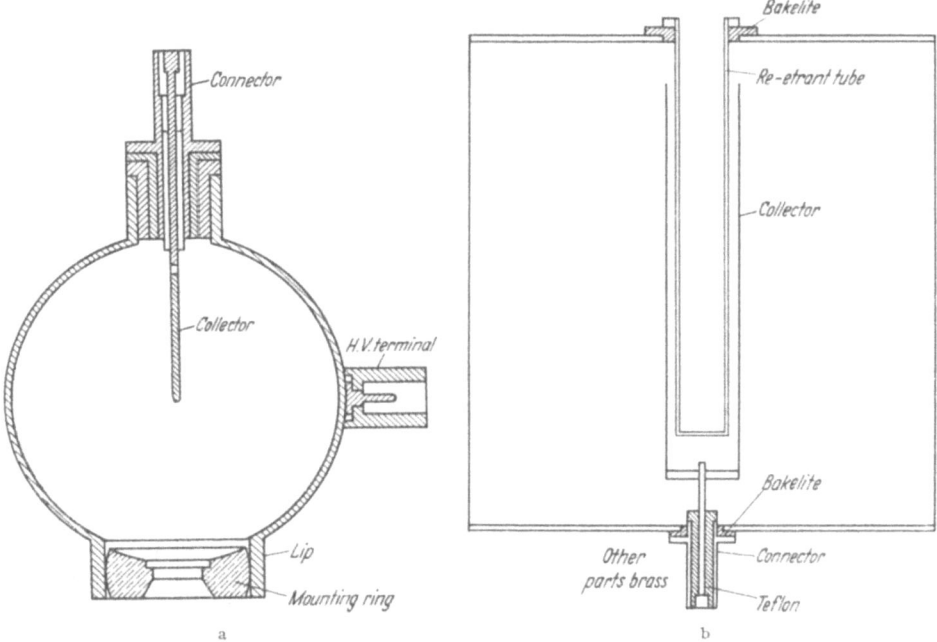

Fig. 25a and b. Ionization chambers for secondary standard monitoring at the U.S. National Bureau of Standards. a) Beta-ray chamber. b) Gamma-ray chamber.

25. Balanced chamber method. BAINBRIDGE and GOLDHABER[3] used a balanced pair of chambers in an experiment which demonstrated the effect of chemical binding on the lifetime of the decay of metastable technecium 99. Samples with different chemical forms, having initial activities closely matched, were placed in the two chambers. The difference in currents in the two chambers was measured as a function of time. The results showed the small nuclear effect quite conclusively. The use of such balanced chambers was suggested by RUTHERFORD in 1911. This method is also adaptable to the measurement of radiations of low intensities relative to background.

26. Cosmic ray applications. COOR and SNOW[4] sent an ionization chamber aloft in a balloon to observe bursts of ionization. Their apparatus consisted of a spherical ionization chamber filled with argon gas and connected to the laboratory on the ground by a radio felemetering system.

[1] R. L. QUINN: U.S. AEC Document Y-791. 1951.
[2] K. E. WILZBACH, A. R. VAN DYKEN and L. KAPLAN: U.S. AEC Document ANL-5134. 1953.
[3] K. T. BAINBRIDGE and M. GOLDHABER: Phys. Rev. **84**, 1260—1261 (1951).
[4] T. A. COOR and G. SNOW: Phys. Rev. **73**, 1252 (1948).

Neher[1] and Neher and Johnston[2] describe an automatic ionization chamber for balloon flight measurements of the total cosmic ray energy at high altitudes. This apparatus consists of a chamber with auxiliary electrometer and circuits which recharge the chamber after a fixed amount of charge has flowed across it. The recharging operation generates a signal which is recorded.

A cloud chamber filled with a gas which permits the rapid collection of electrons by a biased electrode, thus generating a pulse which can then be used to trigger the expansion of the gas, is described by Cohen[3]. He found that a mixture of argon and iso-amyl alcohol was a satisfactory filling gas. A similar cloud-ion chamber is described by Lewis et al.[4]

Since the initial discovery, ionization chambers have contributed little to the study of cosmic radiation because they cannot be applied to the accumulation of detailed information on high energy events which can so readily be obtained, for example, with photographic plates.

27. The measurement of variations in X-ray or γ-ray intensities. Rossi and Staub [7] describe the use of ionization chambers operating with fast electron collection in observing rapid changes in the intensity of x-rays and γ-rays. They obtained time resolutions of the order of half a microsecond. There seems to be nothing to prevent the extension of the method to the observation of even faster changes in the intensity of ionizing radiation.

28. Absolute determination of the strength of α-particle sources. The parallel-plate ionization chamber can be used for these measurements. If an infinitesimally thin sample is deposited on one of the flat surfaces every α-particle entering the gas should be counted, except the extremely rare ones scattered back into the electrode by a nuclear collision occurring quite near the source. However, in practice the sources have appreciable thickness, so a fraction of the counts will be lost because of loss of energy by particles which start out beneath the surface of the source, moving almost parallel with the surface. Furthermore, an appreciable number of particles starting out heading into the source backing will be scattered into the gas and counted. This means that the effective solid angle for counting cannot be assumed to be exactly 2π steradians. Rossi and Staub [7] discuss both points. The first effect can often be made negligible by increasing the specific activity of the samples, but the second represents an intrinsic limitation of the geometry. If a very thin sample holder is made the common wall of a two-sided chamber the second difficulty can be avoided (see Fig. 29). This geometry is also suitable for absolute β-particle counting, but in that case proportional or Geiger counter operation is necessary because of the small energies of many of the β-particles.

29. Alpha particle spectroscopy. In recent years pulse height spectrometry of alpha particles has been extremely helpful, especially in work with transuranic elements[5]. Its convenience, stability, low background counting rate, adequate resolution (1 or 2%), large and predictable solid angle for counting, and acceptance of particles of all energies in the emission spectrum have all contributed

[1] H. V. Neher: Rev. Sci. Instrum. **24**, 99—102 (1953).

[2] H. V. Neher and A. R. Johnston: Rev. Sci. Instrum. **27**, 173 (1956).

[3] M. J. Cohen: Phys. Rev. **75**, 1329 (1949). — Rev. Sci. Instrum. **22**, 966—977 (1951).

[4] H. W. Lewis, W. W. Brown, D. O. Seevers and E. W. Hones: Rev. Sci. Instrum. **22**, 259—263 (1951).

[5] For references see I. Perlman and F. Asaro: Ann. Rev. Nucl. Sci. **4**, 157 (1954); G. T. Seaborg and I. Perlman: Rev. Mod. Phys. **20**, 585 (1948), and F. Asaro and I. Perlman: Rev. Mod. Phys. **26**, 456 (1954).

to the usefulness of this type of spectrometer. Both slow and fast operation have been employed. Examples of each type will be described.

α) Slow pulse spectrometers. In this case the chamber has only two electrodes and the signal received full contributions from the slowly moving ions. JESSE and his collaborators[1] have used such a chamber in conjunction with a quite slowly responding indicator consisting of a vibrating reed electrometer and a moving pen recorder which records individual pulses. The response time is about one second, and the reproducibility and resolving power are said to be excellent, in spite of the mechanical friction of pen against paper in the recorder. BAL-DINGER and HUBER[2] achieved a resolving power of about 1.4% at 4.5 Mev with a nitrogen-filled slow pulse chamber with parallel plates spaced 1 cm apart having a potential difference of 9 kV. They present a spectrum of the alpha particles from natural uranium clearly showing the U^{235} peak between the much higher peaks from U^{234} and U^{238} (see Fig. 26).

β) Fast pulse alpha spectrometers with gridded chambers. The energies of a number of α-spectra including $_{84}Po^{213}$, $_{85}At^{217}$, $_{87}Fr^{221}$, $_{89}Ac^{225}$, $_{92}U^{233}$, and $_{94}Pu^{239}$, were measured by CRANSHAW and HARVEY[3] using a gridded chamber. For a description of the general operation of such chambers see Sect. 19α. The other main pieces of the spectrometer included a fast linear pulse amplifier with stabilized gain and multichannel pulse height analyzer. They found a linear pulse height-energy re-

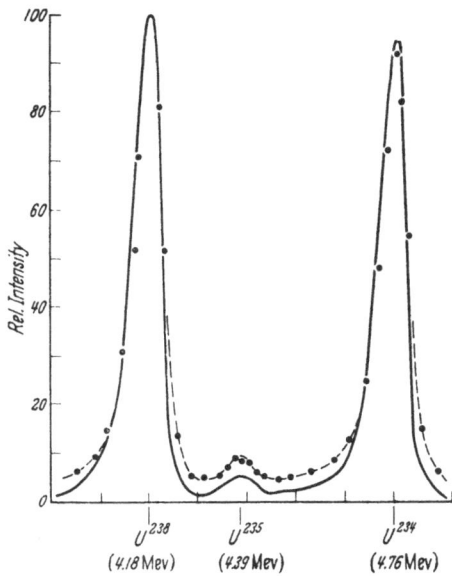

Fig. 26. Pulse height spectrum of natural uranium. BALDINGER and HUBER: Helv. phys. Acta 22, 365 (1949).

lationship for argon between 5 and 9 Mev. At 5.3 Mev the energy corresponding to the full width at half maximum of the pulse height spectrum was 80 kev, and at 4.79 Mev it was 100 kev. In one set of measurements the width at 5.3 Mev was only 50 kev. It was estimated that the corresponding standard deviation 22 kev, was made up from 17 kev from amplifier noise, 14 kev from straggling of ionization, source thickness, etc., and only 2 kev from imperfect shielding.

30. Fast, gridded chambers for the analysis of charged particles from nuclear reactions. In several laboratories, gridded ionization chambers are part of the standard equipment for analyzing charged particles from nuclear reactions occurring outside of the chamber. A thin window is provided through which the particles can enter the ionization volume without excessive energy loss. The geometry part of the chamber design problem is thus made slightly harder. Fig. 27 shows the construction of such a chamber used at the University of Rochester. The thin window is mounted on an extension of the inner wall of the chamber. The end of the grid, which is slightly curved, can fit beneath it. The

[1] W. P. JESSE, H. FORSTAT and J. SADAUSKIS: Phys. Rev. **77**, 782 (1950).
[2] E. BALDINGER and P. HUBER: Helv. phys. Acta **22**, 365−368 (1949).
[3] T. E. CRANSHAW and J. A. HARVEY: Canad. J. Res. **26**, 243 (1948).

collector shown is a metal rod bent to match the shape of the grid. Ski-shaped metal plates have sometimes been used. With such relatively unsymmetrical

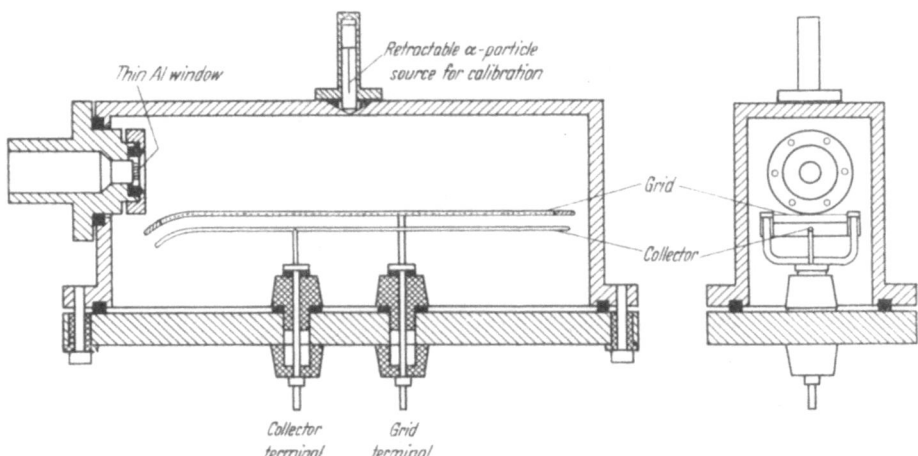

Fig. 27. Gridded chamber used for the analysis of charged particle groups from nuclear reactions.

electrode geometries one cannot easily predict the best ratios of electrode potentials, which must be determined by experiment. The chamber shown was designed

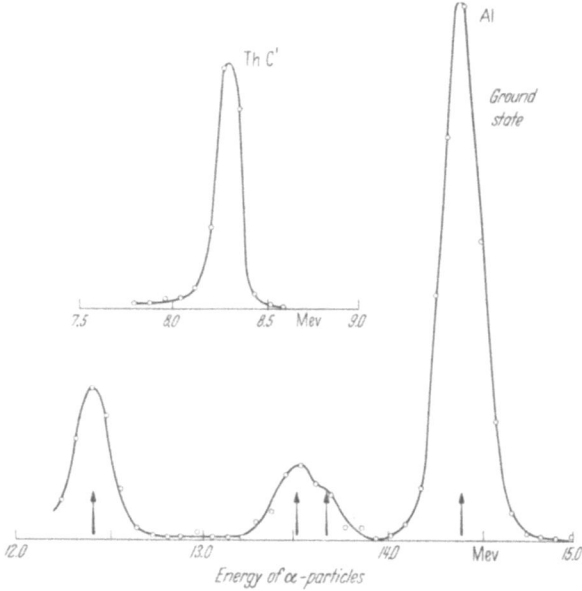

Fig. 28. Below: pulse height spectrum of 20 Mev alpha particles scattered from aluminum, observed at 90° with an ionization chamber of the type shown in Fig. 27. Arrows mark the expected positions of groups due to known levels. Above: line due to ThC′ alpha particles shot into the chamber through the window for calibration. Results obtained by Fulbright, Lassen and Poulsen, using the Copenhagen cyclotron.

for use inside the vacuum of a scattering chamber. The entire top section of the shell was operated at a negative potential so that the collector could be kept at grond potential. It is preferable to use a simpler construction where possible.

An electromagnet (not shown) was used for remote control of the alpha particle source. A gas mixture of argon plus 5 to 20% of CH_4 gives good results. It is advisable to use a continuously operating hot calcium purifier.

The particles necessarily suffer energy loss in passing through the window material. The statistical variations in loss produce a spread in the pulse height distribution (Sect. 5α) which can be minimized by the use of as thin a window as is consistent with the mechanical strength requirement. An extraordinarily strong plastic material called mylar, available in thin sheets, is admirably suited for this purpose. The inner surface of a mylar window should be coated with a transparent, electrically conducting layer of aluminum, grounded to avoid electrostatic charging.

Since the signal voltage of the ionization chamber is only in the order of a millivolt, care must be exercised in shielding and grounding the chamber and the various parts of the amplifier system, especially in cyclotron applications where there is always at least one source of radio frequency power in the vicinity. In practice it turns out to be possible to make the shielding so effective that noise from all external sources is negligible. Apparently little work, if any, has been done using this type of spectrometer for protons having energies above 20 Mev.

A chamber of this sort, properly operated in conjunction with an amplifier stabilized by means of inverse feedback, is remarkably stable and will accurately reproducible results for an indefinite length of time, even if taken apart and reassembled, providing only that the same gas mixture is always used, and that the electrical capacitance is not altered. Its insensitivity to gamma radiation makes it especially useful in cyclotron experiments.

31. Study of nuclear reactions occurring inside the chamber. In these studies the particle inducing the reaction inside the chamber must be uncharged, otherwise the background ionization would be far too great. Most work of this kind has been done with neutrons, slow or fast, although experiments with gamma rays have also been made.

α) *Measurement of the energy released in slow neutron induced reactions.* A good example of such work is the study of the reaction N^{14} (n, p) C^{14}, made by FRAN-ZEN, HALPERN, and STEPHENS[1]. They used a cylindrical chamber filled with N_2, and observed the pulse height distribution of fast pulses due to the low energy protons released when slow neutrons were captured by N^{14} nuclei in the filling gas. The shape of the observed distribution fits excellently the spectrum expected from the geometrical effect (see Sect. 19β). The absolute value of the energy released was obtained by comparing the position of maximum with the corresponding peak for polonium alpha particles from a source introduced on a probe. Using a gas mixture containing 2 atmospheres of argon, 0.1 atmospheres of N_2, and 0.5 atmospheres of enriched helium having an He^3 content of $5 \times 10^{-3}\%$ they also observed the proton group from the $He^3(n, p)$ H^3 and obtained a Q value for that reaction.

RHODES *et al.*[2] have examined the B^{10} (n, α) He^4, Li^7 reaction using apparatus very similar to that described above, but this time the source was a thin deposit of boron on the inner surface of the counter wall. The pulse height spectrum o`bserved had a peak due to the outgoing α particles and another due to the Li^7* recoils. The positions of the peaks showed that the ratio of the ionization produced by the α's to that produced by the Li^7* recoils in argon was 1.878 ± 0.014 instead of the value 1.7529 expected from momentum conservation and the assumption

[1] W. FRANZEN, J. HALPERN and W. E. STEPHENS: Phys. Rev. **77**, 641 (1950).

[2] RHODES, FRANZEN and STEPHENS: Phys. Rev. **87**, 141 (1952).

of the same exact proportionality relationship between total ionization and energy for both particles. The implications of this and similar measurements are discussed in Sect. 4.

The reaction $Cl^{35}(n, \alpha)$ has been investigated by Metzger, Huber, and Alder[1] by a similar method using a heated chamber containing CCl_4 vapor.

β) Elastic scattering of fast neutrons. When monoenergetic neutrons are scattered elastically from nuclei the energy of each recoiling nucleus is simply related to the angle ϑ of recoil.

$$E_r = \frac{4A}{(1 + A)^2} E_n \cos^2 \vartheta \tag{31.1}$$

where E_n means the energy of the incident neutron, and A the mass of the scattering nucleus.

The center of mass differential cross section $\sigma(\vartheta_0)$ is related to the recoil energy spectrum through the equation

$$S(E_r) = \frac{\pi}{E_n} \frac{(A + 1)^2}{A} \frac{\sigma(\vartheta_0)}{\sigma} \tag{31.2}$$

where ϑ_0 is the center-of-mass scattering angle, and σ the total cross section for elastic scattering[2].

If the scattering nuclei are in the ion chamber gas, the resulting pulse height spectrum can therefore be transformed into the angular distribution of the scattered neutrons, providing that the experimental conditions are such that the observed spectrum, corrected if necessary for known distorting influences such as wall effects and amplifier noise, can be considered a true spectrum of the nuclear recoils. Experiments of this sort can be carried out easily only in case the first excited state of the scattering nucleus lies at an energy higher than the energy of the neutron beam, otherwise recoils occurring because of inelastic scattering will interfere; furthermore thresholds for other nuclear reactions should be higher than the energy of the neutron beam.

The elastic scattering of neutrons from helium has been studied by a number of workers using various ionization chamber techniques. A good example is the early work of Baldinger, Huber, and Staub[3] who used parallel plate geometry, a slow pulse amplifier, and chamber filling of argon (4 atm) and He (7 atm). Barschall and Kanner[4] also used a slow pulse apparatus having a set of grids through which the ions were drawn to the collectors by the electric field. They used 7 atm of He in the chamber. Further work was done by Huber and Baldinger[5] using 3 to 4.14 Mev neutrons and slow pulse detection of recoils in a parallel plate chamber filled with 16 atm of He.

γ) A gamma ray-induced reaction. Wilson, Collie and Halban[6] studied the photodisintegration of the deuteron by 2.76 Mev γ-rays by means of a spherical ionization chamber filled with deuterium gas at a pressure of 6 atm. The deuterium was purified partly by being admitted to the chamber through a palla-

---

[1] F. Metzger, P. Huber and F. Alder: Helv. phys. Acta **20**, 236 (1947).

[2] If the scattering is isotopic in the center of mass system $[\sigma(\vartheta_0) = \text{const}]$ the recoil spectrum corresponding to a given E will be flat. For $n - p$ collisions $\sigma(\vartheta_0) = \text{const}$ up to 15 Mev.

[3] E. Baldinger, P. Huber and H. Staub: Helv. phys. Acta **11**, 245 (1938).

[4] H. H. Barschall and M. H. Kanner: Phys. Rev. **58**, 590 (1940).

[5] P. Huber and E. Baldinger: Helv. phys. Acta **25**, 435 (1952).

[6] R. Wilson, C. H. Collie and H. Halban: Nature, Lond. **162**, 185 (1948); **163**, 245 (1949).

dium leak. (This method of purification has also been used with hydrogen.) Pulses due to disintegration protons with energies down to 150 kev were observable when the chamber was irradiated with gamma rays. The results were analyzed to yield a value for the cross section.

32. The study of fission. The fact that a large amount of energy is released in nuclear fission was first proved by observations of the very large ionization pulses produced[1]. Subsequently the study of fission pulses from ionization chambers under various experimental conditions has yielded considerable information about the fission process itself. For example, the energy distribution of fragments from the fission of U^{233}, U^{235}, and Pu^{239} has been measured[2,3] carefully in a double ionization chamber with the sample mounted on a thin foil in between the two sections. In any single event the heavy particle entered one side, the light particle the other. The mass spectrum derived from the results was in good agreement with that expected from chemical analyses. The angular distribution of fission fragments has also been studied by an ion chamber thod[4,5].

33. Fast neutron spectroscopy. Here the ionization pulses from nuclei recoiling after being struck by neutrons are observed. In some designs a "radiator", usually a thin foil containing a hydrogeneous compound such as paraffin or polyethylene, provides nuclei which, upon being struck by neutrons, enter the gas of the ionization chamber where they are brought to rest. Some form of directional discrimination must be employed, otherwise monochromatic neutrons would give pulses of all energies from zero up to the energy of the beam. This can be accomplished in a variety of ways, some involving collimators of solid material with channels through which recoils from the radiator can pass only if they have approximately the desired direction with respect to the neutron beam (usually 0°), others involving coincidence or anticoincidence methods. For a discussion and references the reader is referred to BARSCHALL's article in this volume. Also see HOLT and LITHERLAND[6] for a practical application involving "residual range" determination of the recoils by measurement of the collection time of the ionization electrons from the end of the track, and SCHMIDT-ROHR[7] for a total ionization measuring device. Both of these can be used with neutrons having energies from several Mev to 25 Mev. Also see BROMLEY[8] for a helium recoil counter used to discriminate between ground state and excited state neutrons in studying the angular distribution of neutrons from the $C^{13}(d, n)N^{14}$ reaction.

34. High pressure operation. The use of an electron collection chamber filled with quite pure hydrogen at pressures up to 90 atm has been investigated by STAFFORD[9]. He found that pulse height saturation was obtained even at 60 pounds per square inch, but the pulses had only 0.32 times the height expected from the known value of w_{H_2}. At 88 atm a similar effect was observed, but the saturation pulse height was only 0.18 times the maximum value. He attributed this result to preferential recombination (Sect. 11). Counters filled with very pure methane and

[1] O. R. FRISCH: Nature, Lond. **143**, 276 (1939).
[2] D. C. BRUNTON and G. C. HANNA: Canad. J. Res. **28**, 190—227 (1950).
[3] D. C. BRUNTON and W. B. THOMPSON: Canad. J. Res. **28**, 498—508 (1950).
[4] J. E. BROLLEY jr. and W. C. DICKINSON: Phys. Rev. **94**, 640—642 (1954).
[5] R. L. HENKEL and J. E. BROLLEY jr.: Phys. Rev. **103**, 1292 (1956).
[6] J. R. HOLT and A. E. LITHERLAND: Rev. Sci. Instrum. **25**, 298 (1954).
[7] U. SCHMIDT-ROHR: Z. Naturforsch. **8a**, 470 (1953).
[8] D. A. BROMLEY: Phys. Rev. **88**, 565 (1952).
[9] G. N. STAFFORD: Nature, Lond. **162**, 771 (1948).

D_2 have been investigated up to 35 atm[1,2]. Loss of only 3% of the electrons in H_2 at 20 atm is reported[2].

Fulbright and Milton[3] have used an argon-filled chamber[4] with electron collection at pressures up to 135 atm for the investigation of the shapes of forbidden beta spectra including those of Be^{10}, K^{40}, RaE, and Cl^{36} (see Fig. 29). They report pulse height saturation and a linear pulse-height-vs-energy curve at least up to 1.2 Mev. The apparent W_A at about 40 atm pressure was (36 ± 3) ev (for 662 kev electrons), compared with the value of 26.3 ev found at pressures of a few atmospheres. This result may also be due to preferential recombination. Their success in reproducing accurately the shapes of a wide variety of spectral distributions shows the reliability of the method up to 135 atm. Attempts made to check the P^{32} spectrum at 165 atm were unsuccessful; the F-K plots were slightly concave downward.

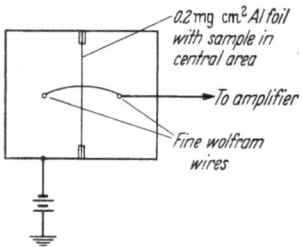

Fig. 29. Geometry used by Fulbright and Milton for beta-ray spectroscopy. Also suitable for absolute alpha-particle counting and (proportional counter operation) absolute beta-particle counting.

The construction and performance of an air-filled chamber operated at pressures up to 78 atmospheres for measuring weak gamma sources is described by Spiers[5]. Raising the pressure (from atmospheric) increased the gamma ray response by a factor of 20, while the background increased only by a factor of 4. As little as 5×10^{-3} μgm of radium content is said to be detectable with this chamber.

Generally speaking, raising the pressure in an ionization chamber invites recombination troubles, which limit the usefulness of high pressure chambers. With fast electrons as primaries the difficulties are relatively light, since preferential recombination is usually the only type involved, but with heavier primaries such as alpha particles, columnar recombination becomes very important at pressures over about 10 atmospheres. Recombination cannot be overcome simply by raising the applied voltage because of the rapidity of the increase of the difficulty with pressure.

Bibliography.

[1] Ashkin, J., and H. Bethe: In Experimental Nuclear Physics, Vol. 1, edit. by E. Segrè, pp. 166—357. New York: Wiley 1953. — A very good general reference on the passage of ionizing radiation through matter.

[2] Mott, N. F., and H. S. W. Massey: The Theory of Atomic Collisions. Oxford: Clarendon Press 1950.

[3] Bohr, N.: The Penetration of Atomic Particles Through Matter. Dan. Vid. Selskab. Mat.-Fys. Medd. **18**, No. 8 (1948). — Gives good physical descriptions of the processes involved.

[4] Allison, S. K., and S. Warshaw: Rev. Mod. Phys. **25**, 779 (1953). — Thorough review of experimental data on the stopping of charged particles by matter.

[5] Engel, A. v.: Ionized Gases. Oxford: Clarendon Press 1955. — General reference on electrical conduction in gases.

---

[1] L. E. Beghian and H. H. Halban: Proc. Phys. Soc. Lond. A **62**, 395—396 (1949).

[2] R. Wilson, L. Beghian, C. H. Collie, H. Halban and G. R. Bishop: Rev. Sci. Instrum. **21**, 699—705 (1950).

[3] H. Fulbright: In Beta and Gamma Ray Spectroscopy, Kai Seigbahn, Editor. Amsterdam: North Holland Publishing Co. 1955, and J. C. D. Milton: Thesis. Princeton University 1950.

[4] Sometimes used as a proportional counter with low gas gains.

[5] F. W. Spiers: Brit. J. Radiol. **22**, 169 (1949). Another high pressure chamber (30 atm) for γ-ray detection is described by J. Sharpe and F. Wade: Atomics (London) **4**, 32 (1953).

[6] Rossi, B.: High-energy Particles. Englewood Cliffs, N. J.: Prentice-Hall 1952. — Good for the interaction of relativistic particles with matter. Gives useful curves for Landau-Symon calculations of straggling in energy loss by ionization in a short length of the track of a charged particle.

[7] Rossi, B., and H. Staub: Ionization Chambers and Counters. New York: McGraw Hill 1949. — This is the best single general reference on ionization chambers. It discusses fundamental principles and practice, gives constructional details, and contains much information of practical value.

[8] Jaffé, G.: Ann Physik **42**, 303 (1913). — The classic paper on columnar recombination.

[9] Massey, H. S. W., and E. H. S. Burhop: Electronic and Ionic Impact Phenomena. Oxford: Clarendon Press 1952. — Probably the best and most complete general reference on its subject.

[10] Massey, H. S. W.: In Adv. Physics **1**, 395 (1952). — Excellent review of electron-ion recombination including a discussion of the implications of the results of afterglow experiments.

[11] Healey, R. H., and J. W. Reed: The Behavior of Slow Electrons in Gases. Sidney: Amalgamated Wireless, Ltd. 1941. — Contains much conveniently tabulated data from a variety of experiments, including drift velocities, agitation energies, etc.

[12] Siegbahn, K.: Beta and Gamma Ray Spectroscopy. Amsterdam: North-Holland Publishing Co. 1955. — A general reference which includes, among many others, discussions of the interaction of fast electrons with matter, the Auger effect, and application of ionization chamber techniques to beta-ray spectroscopy and the study of gamma-ray induced reactions.

[13] Glasstone, S.: Principles of Nuclear Reactor Engineering. New York: Van Nostrand 1955. — A major general reference in its field. Contains useful descriptions of various types of ionization chambers used in pile control and health monitoring.

[14] Elmore, W.: Nucleonics **2** (3) 16; **2** (4) 43 (1948). — Probably the best short reference on fast pulse amplifier fundamentals, including signal-to-noise considerations.

[15] Elmore, W., and M. Sands: Electronics. New York: McGraw Hill 1949. — A good general reference for all sorts of electronic circuits (except the recently-developed distributed amplifiers, etc.) of interest in nuclear physics. Inclues some constructional details and many practical circuit diagrams.

[16] Franzen, W.: Theory and Use of Pulse Ionization Chambers. Report AT (30—31). Princeton University 1951. — Unpublished report on fast electron collection chambers.

[17] Rutherford, E., J. Chadwick and C. D. Ellis: Radiations from Radioactive Substances. Cambridge 1930.

[18] Marinelli, L. D.: Ann. Rev. Nucl. Sci. **3**, 249 (1953). — Good review article on radiation monitoring for health protection. Gives many references.

[19] Gray, L. H.: Proc. Roy. Soc. Lond., Ser. A **156**, 578 (1936); **159**, 263 (1937). — On the measurement of x and γ-radiation by use of ionization methods.

[20] Meeks, J. M., and J. D. Craggs: Electrical Breakdown of Gases. Oxford: Clarendon Press 1953. — Good general reference on conduction in gases.

[21] Curran, S. C., and J. D. Craggs: Counting Tubes and Their Applications. London: Butterworth Scientific Publications 1949.

[22] Wilkinson, D. H.: Ionization Chambers and Counters. Cambridge 1950. — Considerable discussion of fundamental processes.

[23] Staub, H. H.: In Vol. 1 of Segrè, Experimental Nuclear Physics, pp. 1—165. New York: Wiley 1953. — Principles of nuclear particle detection. General reference on fundamentals of ionization chambers, proportional Counters, and Geiger counters. Gives data, from Healey and Reed [11], on behavior of slow electrons in various gases.

[24] Heitler, W.: The Quantum Theory of Radiation, Third Edit. Oxford: Clarendon Press 1954.

[25] Brode, R. B.: Rev. Mod. Phys. **5**, 257 (1933). — Gives a collection of information, including many graphs, on the cross-sections for collisions of electrons of various energies with gas molecules.

[26] English, W. N., and G. C. Hanna: Canad. J. Res. **31**, 768 (1953). — Experiments on the drift velocities of electrons in various gases under the influence of electric fields.

Geiger Counters.

By
SERGE A. KORFF*.

With 11 Figures.

I. Introduction.

1. History. G. M. Counters, or Geiger-Müller Counters, or, as they are now becoming known, Geiger 'Counters, are today approaching the halfcentury anniversary of their first discovery. In 1908 a paper appeared in England in which RUTHERFORD and GEIGER [26] described a device they had found suitable for detecting the pulses produced by alpha particles. They had suspended a wire along the axis of a cylinder, applied a potential difference between the wire and cylinder, projected the alpha particles axially and had found on the wire pulses such that "the current through the gas due to the entrance of an alpha particle into the detecting vessel was magnified ... sufficiently to give a marked deflection to the needle of a galvanometer of moderate sensibility". They has produced a counter, and making use of gas amplification near the wire of the pulses, had increased the pulse size by a factor of "several thousand".

In this experiment they had employed cylindrical symmetry. Within a few years, spherical geometry had also been introduced. GEIGER realized that what was required was a high field, and that since such fields existed around points, a sharp point would also serve. Devices using sharp points [8] were at first called "Geiger point counters". Later the word "point" was frequently omitted in the literature, and during one time interval the words "Geiger counter" were understood to mean a point counter, while a cylindrical counter was called a Geiger-Müller counter. However, in recent years, point counters have become much less used, and the demands for the simplification of terminology have been insistent, so that today the words "Geiger counter" are taken as referring to the cylindrical arrangement. The terminology went through still further changes, for in 1928 GEIGER and KLEMPERER [9] constructed a successful proportional counter, and for a while such proportional counters were called "Geiger-Klemperer counters". In this case the two mens names were dropped and today the device is known as a "proportional counter". Indeed it should be remarked that the first counter in 1908 was in fact operating as a proportional counter, for the voltage used was not high enough to enable it to count beta particles as well. Yet, while this situation was understood by the early observers, the term "proportional counter" did not enter the literature until much later.

The next important advance came with the development of the coincidence technique by BOTHE and KOLHÖRSTER [1], ROSSI [27] and TUVE [35]. The simultaneity of pulses in two or more counters was made the criterion of detection. This advance made possible directional studies through the alignment of several units. It enabled the study of complex events involving, for example, scattering or secondary production, to be carried out by out-of-line arrangements, absorption studies by interposition of material, and many other types of investigation.

* This work was in part supported by the joint program of the Office of Naval Research and the Atomic Energy Commission and in part by the Air Force.

Indeed it inaugurated the modern era of cosmic ray research and is also today a widely used technique of investigation in nuclear physics applications. Thus the modern era dates from about twenty five years ago.

Another important milestone in counter operation came with the discovery of organic vapors as quenching agents. It was found by TROST [36] that by introducing a vapor of a polyatomic substance, the action of the counter could be speeded up, and it could be made to recover faster to receive the next count. We shall discuss the mechanism in detail later. We know today that perhaps the action is not speeded up as much as was at one time thought, but in fact the necessary electronic circuits are what is simplified. Nevertheless, looking back upon the development, it was this discovery which in fact materially increased the utility of counters and was the last necessary step before the rapid advance of the last twenty years could be undertaken.

2. Description of ion collection. In order to lay the foundation for the later detailed description and for the definitions of terms to be introduced, we must briefly review the events which take place when a counter discharges and the ions are collected. First we define ionization as any process whereby a previously neutral atom or molecule becomes charged, either positively or negatively, i.e., either by the loss or by the acquisition of one or more electrons. We shall next assume that an ionizing event has taken place within the counter. By this we imply that some radiation, i.e., one or more chaged particles, or photons, or neutral particles have by some process produced ionization in the sensitive volume of the counter. For simplicity we shall assume the counter to be a cylinder with a wire along its axis, although we recognize many other forms are possible. We further assume that the wire is connected to an electronic circuit capable of measuring its changes in potential, and that an electric field is produced by connecting a source of potential across the counter, the cylinder being negative and the wire positive. The entire arrangement is placed in some sort of container, in which the kind and pressure of gas can be known and controlled.

After the ionizing event has occurred, the positive ions will drift toward the cylinder and the negative ions and electrons toward the wire. As the electrons approach the wire, they enter a region where the field in a short distance becomes quite high, and therefore the electrons acquire high velocities. They make impacts upon other atoms in their way, and may produce additional ionization. The additional electrons thus produced add to the original ones and all proceed toward the wire. Each electron may initiate an avalanche, since one electron after an ionizing impact is accompanied by a second, and the two after their next impacts yield four, and so on. It is obvious that such a process is rapidly divergent, and only a small number of collisions is needed to produce a large avalanche. Since the signal on the wire depends on the number of electrons arriving there, a signal due to one single electron may be amplified in this manner to much larger size. The process is called "gas amplification".

While this process is taking place the positive ions, which are much heavier than the electrons, move slowly out toward the cylinder. Upon arrival at the cylinder they are neutralized. In this simple event, the discharge process is finished, and the counter is ready to receive a new count. But it can happen that in the process of being neutralized the positives may cause the liberation of electrons. If this should occur, the new electrons will move inward and initiate new avalanches and cause the processes to repeat. In such a case we have the essential ingredients for a self-sustaining discharge, and the counter will not return to its previous state of sensitivity. Therefore, it is essential somehow to prevent the formation of secondary or subsequent electrons. The term "quenching" has been applied to the process or processes whereby the liberation of subsequent electrons is inhibited, the discharge is terminated, and the conditions for reinitiation are removed.

Further complications are introduced if any of the electrons are captured to from negative ions. Since negative ions move much more slowly than do electrons, it can happen that two signals may arrive on the central wire corresponding to one single ionizing event, one signal being produced by the electrons and the second by the ions. This process we shall also consider, since it unfortunately may occur in practice unless certain precautions are taken.

The size of the avalanche which we have discussed above depends on the field. Clearly if the field is low enough, no avalanche will occur, and the electron or electrons formed in the initial ionizing event will simply be collected. This situation takes place in the ionization chamber, and its operation has been discussed in other chapters. Further, if the field is of intermediate values, such that the gas amplification is relatively small, constant and not a function of the number of initial ions, the counter is said to operate in the proportional region. This case too has been discussed elsewhere, and we shall not consider it.

3. Definitions of terms to be used. With this background, we may now proceed to a definition of terms to be used. These terms gone through complex evolutions, and have been used in differing ways by many observers in many lands.

In order to bring some semblance of order into this confused situation, the present author ten years ago proposed a set of definitions. About seven years after that, the American Institute of Electrical Engineers and the Institute of Radio Engineers (I.R.E.) appointed a committee to study the problem of terminology and of standardized testing procedures. The report of the Radiation Counter Tube Subcommittee, of which the present author was privileged to be a member, has been published in the *Prodeedings of the I.R.E.*, Vol. 40, p. 924 to 930, No. 8, August, 1952. Reprints of the standards, designated 52 I.R.E. 7.S2 (testing procedures) and 52 I.R.E. 7.S3 (definitions), may be purchased for $ 0.75 and $ 0.50 each respectively, from the I.R.E., 1 East 79th Street, New York 21, New York, USA. The definitions listed below are those of the Committee, and full credit is hereby extended to them.

In order to save space, in this chapter, the device which is defined as a "radiation counter tube" is called in this present work "counter". For the engineering profession the three words are required to (a) identify the nature of the entities being counter, (b) to distinguish the devices from mechanical counters, and (c) to suggest that these devices are in the category of gas-discharge tubes. Since we in this chapter shall only consider such tubes we shall shorten the name. Further, in this chapter we shall generally omit the name of Müller, sometimes rendered in English spelling as Mueller, and speak of "Geiger counters".

In what follows we now are giving the I.R.E. Standard Definitions.

Avalanche. A cascade multiplication of ions.

Background Counts. Counts caused by radiation coming from sources other than that measured.

Count (in a Radiation Counter). A single response of the counting system. Note— See also Tube Count.

Counting-Rate Versus Voltage Characteristic. Counting rate as a function of applied voltage for a given constant average intensity of radiation.

Dead Time. The time from the start of a counted pulse until an observable succeeding pulse can occur. Note— See also Recovery Time.

Efficiency (of a Radiation Counter Tube): The probability that a tube count will take place with a specified particle or quantum incident in a specified manner.

Externally Quenched Counter Tube. A radiation counter tube that requires the use of an external quenching circuit to inhibit reignition.

Gas Amplification. The ratio of the charge collected to the charge liberated by the initial ionizing event. Note— See also Standards on Gas-Filled Radiation Counter Tubes: Methods of Testing, Sect. 12.

Gas-Filled Radiation Counter Tube. A gas tube, in a radiation counter, used for the detection of radiation by means of gas ionization.

Geiger-Mueller Counter Tube. A radiation counter tube designed to operate in the Geiger-Mueller region.

Geiger-Mueller Region (of a Radiation Counter Tube). The range of applied voltage in which the charge collected per isolated count is independent of the charge liberated by the initial ionizing event.

Geiger-Mueller Threshold. The lowest applied voltage at which the charge collected per isolated tube count is substantially independent of the nature of the initial ionizing event.

Hysteresis (of a Radiation Counter Tube). The temporary change in the counting rate versus voltage characteristic caused by previous operation.

Initial Ionizing Event. An ionizing event that initiates a tube count.

Ionizing Event. Any interaction by which one or more ions are produced.

Multiple Tube Counts (in Radiation Counter Tubes). Spurious counts induced by previous tube counts.

Normalized Plateau Slope. The slope of the substantially straight portion of the counting rate versus voltage characteristic divided by the quotient of the counting rate by the voltage at the Geiger-Mueller threshold.

Overvoltage. The amount by which the applied voltage exceeds the Geiger-Mueller threshold.

Plateau. The portion of the counting rate versus voltage characteristic in which the counting rate is substantially independent of the applied voltage.

Plateau Length. The range of applied voltage over which the plateau of a radiation counter tube extends.

Predissociation. A process by which a molecule that has absorbed energy dissociates before it has had an opportunity to lose energy by radiation.

Proportional Counter Tube. A radiation counter tube designed to operate in the proportional region.

Proportional Region. The range of applied voltage in which the gas amplification is greater than unity and is independent of the charge liberated by the initial ionizing event. Note — The proportional region depends on the type and energy of the radiation.

Quenching (in a Gas-Filled Radiation Counter Tube). The process of terminating a discharge in a radiation counter tube by inhibiting reignition.

Radiation. In nuclear work, the term is extended beyond its usual meaning to include moving nuclear particles, charged or uncharged, and electrons moving with sufficient speed to enter into nuclear processes.

Radiation Counter. An instrument used for detecting or measuring radiation by counting action.

Recovery Time (of a Radiation Counter). The minimum time from the start of a counted pulse to the instant a succeeding pulse can attain a specific percentage of the maximum value of the counted pulse.

Region of Limited Proportionality. The range of applied voltage below the Geiger-Mueller threshold, in which the gas amplification depends upon the charge liberated by the initial ionizing event.

Reignition (of a Radiation Counter Tube). A process by which multiple counts are generated within a counter tube by atoms or molecules excited or ionized in the discharge accompanying a tube count.

Relative Plateau Slope. The average percentage change in the counting rate near the midpoint of the plateau per increment of applied voltage. Note — Relative plateau slope is usually expressed as the percentage change in counting rate per 100-volt change in applied voltage.

Resolving Time (of a Radiation Counter). The time from the start of a counted pulse to the instant a succeeding pulse can assume the minimum strength to be detected by the counting circuit. (This quantity pertains to the combination of tube and recording circuit.)

Self-Quenched Counter Tube. A radiation counter tube in which reignition of the discharge is inhibited by internal processes.

Sensitive Volume (of a Radiation Counter Tube). That portion of the tube responding to specific radiation.

Spurious Tube Counts (in Radiation Counter Tubes). Counts in radiation counter tubes other than background counts and those caused by the source measured. Note — Spurious counts are caused by failure of the quenching process, electrical leakage, and the like. Spurious counts may seriously affect measurement of background counts.

Tube Count. A terminated discharge produced by an ionizing event in a radiation counter tube.

II. The gas discharge mechanism.

4. Non-selfquenching counters. We shall first discuss the operation of non-selfquenching counters. We shall assume that the ionizing event has taken place, and shall consider subsequent operation.

In the initial ionizing event, one or more positive ions and electrons were separated from each other, or, in other words, ionization was produced. Consider first the electrons. These find themselves in a field in which they are accelerated toward the central wire. The field is described by the relation:

$$E = \frac{V}{r \log (r_2/r_1)} \tag{4.1}$$

where E is the field in volts per cm, V is the applied potential in volts, r is the radius at the point under discussion, and r_2 and r_1 are the radii of the cylinder and the wire respectively. Evidently r must lie between r_2 and r_1. This field is a rapidly varying function of r as r approaches r_1, and reaches high values near the wire. For example, in a typical counter, with cylinder radius 1 cm, potential V of 1000 volts across the counter, and wire radius 5×10^{-3} cm, corresponding closely to a 4 mil central wire, the field at the wire surface is 37400 volts per cm, and drops to 187 volts per cm at the cylinder. Indeed 1 mm out from the wire the field will be 1870 volts per cm. Thus it is clear that the entire high field volume is a region of small radius around the central wire. The average field, over the entire counter is 1000 volts per cm, so that throughout most of the volume the actual field differs from the average value by les than a factor of five, but immediately adjacent to the wire it rapidly mounts to high values.

An electron, or an ion, will dirft in a field with a velicity v cm/sec given by:

$$v = k E/p \tag{4.2}$$

where E is the field in volts/cm, P is the pressure, and k is a constant called the mobility. Clearly the mobility is the drift velocity in unit field at unit pressure. The pressure p may be measured in atmospheres, or in mm Hg. In computing velocities, it is important to be careful which units are used, for some authors use one and some the other. Values of k have been measured and are to be found in the literature for both electrons an for various ions and various gases commonly used in counters. We shall here give a few typical values, together with some references to other values in the literature.

Table 1. *Mobilities of positive and negative ions in various gases, units of $cm^2/volt\ sec$.*
References: International Critical Tables; L. B. LOEB, Kinetic Theory of gases; Smithsonian Physical Tables.

Gas	H_2	He	A	N_2	O_2	Ethane (neg. ions)	Ethyl (pos.)	Alcohol (neg.)
Mobility, k in cm/sec at unit field, and at pressures of one atm	5 to 6	5.1	1.3	1.3	1.3	1.07	0.34	0.27

From these figures it is seen that the mobilities of most ions in most gases are of the order of unity, slightly larger for hydrogen and helium, slightly less for large molecules like ethyl alcohol. It should be added that the figures above differ between observers and are probably not accurate to the second decimal place.

For electrons the situation is less simple. The drift velocities if plotted against E/p in some cases show approximately straight lines over limited ranges of E/p, but in some cases level off. For example, observers report that electrons in argon increase in drift velocity with E/p until a value of about 10^6 cm/sec is attained at $E/p = 0.4$ volt/cm/mm Hg, and then no further increase takes place up to values of E/p of 2.5. On the other hand, roughly linear plots are given for the following cases:

Table 2. *Electronic drift velocities in gases; units of E/p are volt/cm/mm Hg.*
References: R. H. HEALEY and J. W. REED, The Behaviour of Slow Electrons in Gases, Amalgamated Wireless, Ltd., Sydney, 1941; B. ROSSI and H. STAUB, Ionization Chambers and counters, McGraw-Hill, New York 1949.

Gas	E/p interval	velocity interval, $\times 10^6$ cm/sec
CO_2	0 to 8	0 to 9
BF_3	0 to 6	0 to 70
N_2 and H_2	0 to 50	0 to 20
Ne and He	0 to 6	0 to 7 for Ne, 3 for He

Here again, there appears to be considerable variation in the data, and the disagreements between several authors seem to be far greater than the curves themselves would suggest. Therefore it seems one should not consider these data too precise, and should only use them as an approximate guide. Further, these data are in units of E/p of volt/cm/mm Hg. For example a velocity of 9×10^6 cm/sec at E/p of 8 corresponds to a velocity of 1.48×10^3 cm/sec at unit field and STP. We see therefore that electron velocities may be taken as between 1000 and 10000 times those of ions in like gases, in terms of fields of 1 volt/cm and 760 mm Hg pressure.

In a typical counter having an average field of 1000 volt per cm and a pressure of 10 cm Hg, E/p would be 10 volts/cm/mm Hg. Its excursions would run from an E/p of 370 immediately adjacent to the wire down to 2 near the cylinder.

The high field region is quite outside the range of most of the available experiments and little data exists here. We do, however, have the data with which to discuss the motion of electrons in the main body of the counter.

Let us use a numerical illustration. Consider an electron moving in toward the central wire. It will move with increasing velocity as it nears the wire, finally attaining speeds at which the energy gained per mean free path of forward motion is sufficient to enable it to make ionizing collisions with the atoms or molecules of the gas in the counter.

The mean free path of an ion in a gas is given by approximately:

$$L = \frac{1}{n \pi r^2} \tag{4.3 a}$$

where L is the mean free path in cm, n is the number of atoms or molecules per cm³, and r is the average molecular radius for the collision under discussion, a quantity which depends on (a) the kind of gas, (b) the kind of ion, and (c) the velocity of the average ion's drift. In this equation no account is taken of the Maxwellian distribution. If this distribution is considered the equation becomes:

$$L = \frac{1}{\sqrt{2}\, n \pi r^2} , \tag{4.3 b}$$

the factor $\sqrt{2}$ being the essential difference. In this, it is assumed that the radius of the ion is the same order as that of the gas atom. Should the radii differ

greatly, as in the case of the electron moving through a gas, where the radius of
the electron is small compared to that of the gas atoms, and where the velocity
of the electron is large compared to the mean velocity with which the atoms are
moving about in accord with our familiar kinetic theory of a gas, the expression
for the mean free path becomes:

$$L = \frac{4}{n \pi r^2}.$$ (4.3c)

To give some illustrative and typical values, it will be recalled that atomic
radii are of the order of one angstrom for processes of this sort. The Loschmidt
number is 2.7×10^{19} atoms or molecules of any kind per cm^3 at .STP, and since
counters usually run at pressures below one atmosphere, it is at once evident
that in counters the free paths will be of the order of 1.2×10^{-3} cm for ions and
perhaps 5×10^{-3} cm for electrons. Thus an electron will make an appreciable
number of collisions per cm of advance, and will make some 20 collisions in the
last mm advance toward the central wire. It is precisely here that the electron
is in the high field region. At a distance of half a millimeter from the wire the
field is 3740 volts/cm, the free path 5×10^{-3} cm, and an electron will gain 18 volts
per free path, an amount sufficient to ionize most atoms which it might strike.
Thus each time the electron makes a collision it will produce another electron,
and this electron too will produce still further electrons. In this manner an
avalanche of electrons is formed. The magnitude of such an avalanche is given by:

$$N = N_0 e^{ax}$$ (4.4)

where N is the number of electrons formed by an initial N_0 electrons, which have
advanced a distance x through the gas, a is a constant called the First Townsend
coefficient and e is the base of natural logarithms. Such an avalanche can readily
be of appreciable size. It is evident that if the collision ionization process starts
a mere 20 free paths out from the wire the size of the avalanche already will
number millions per initial electron.

In this avalanche, many excited atoms are also formed. Some recombinations
will take place, and some ions will return from excited states by radiating. In
general at the low pressures in counters the mean free time between collisions
is sufficiently long to permit at least some radiative de-excitation, so that the
avalanche becomes a source of photons. These photons may, if they reach the
cylinder and if their energy is above the photoelectric threshold of the cylinder
material, cause the ejection of photoelectrons from the cylinder. These new
photoelectrons will again drift across the counter, enter the high field region,
and there cause still further avalanches. Moreover, especially if the counter
contains a mixture of gases rather than just one pure gas, the photons emitted
by one kind of atom may have sufficient energy to ionize the other kind, so that
additional ionization of the gas itself may also occur. If this process occurrs,
we shall have two sources of additional electrons.

In the end, we have the requirements for a self sustaining discharge if at each
avalanche enough photons are formed to produce at least one new electron. The
probability of this happening depends on (a) the number of photons formed in
the avalanche, (b) the probability of the photon reaching the cylinder, i.e., the
chance that it may be absorbed while passing through the gas, and (c) the work
function and photoelectric efficiency of the material of which the cylinder is
made. If for example a million photons are formed in the avalanche, then even
with poor photoelectric efficiencies we may have the necessary one electron
formed, and it will therefore not astonish us to learn that a counter will in such
cases usually go into a self sustaining discharge.

It is clear that at this point we must provide some mechanism for terminating the discharge in order to restore the counter to its state of being sensitive to the next ionizing event. This termination may be accomplished in two ways, (a) by doing something externally such as lowering the potential below that value necessary to sustain the discharge, or (b) internally, by providing such a situation that the supply of secondary electrons is cut off. In this section we shall consider that situation (a) obtains, and that we are providing some mechanism to terminate the discharge from the outside. Later we shall discuss possible mechanisms, and in the following section we shall discuss internally or self-quenched counters.

To complete the discussion of what happens in the original avalanche, we must next realize that the discharge spreads down the length of the wire, and will eventually result in covering the entire wire with a thin sheath in which ionizing events are occurring. This discharge may be seen by looking at a counter filled for example with neon or a neon-hydrogen mixture, and supplied with a transparent window. The discharge is often fairly faint, so that a dark room and a dark-adapted eye are needed, but the flashes of light around the wire can be made visible to the unaided eye without much difficulty. The spectrum of such discharges has been photographed and shows the recombination radiation lines which one might expect.

Unto the present point we have discussed only a part of the problem, namely the formation of the electron avalanche. The formation of each electron always leaves behind a positive ion, which being much heavier and moving with slower speeds, (see table of values of k) will be left behind and will not move much during the time taken by the electrons to build up the electron avalanche. The presence of the positive charges will then produce marked effects. First, we shall have a positive ion space charge sheath formed around the wire. Inside this sheath, owing to the effective electrostatic shielding produced by such a space charge, the field will be lower. Indeed the MONTGOMERYS [20] have shown that the field will fall to such a value that the discharge can no longer be sustained as avalanche formation no longer occurrs. Hence the discharge actually terminates itself and the counter discharge will stop then and there if no subsequent additional electrons are formed.

Another important effect which follows is that because of the positive charges in the space charge sheath, electrons are held in the wire by electrostatic image forces. Consideration of the geometry of the image in a cylinder of a point charge one or more radii distant shows that most but not necessarily all the electrons are held on the wire, and do not begin to move off the wire until after the space charge has moved outward an appreciable distance. Thus the signal reaching the detector is controlled by the fact that most of the electrons are at first bound. However, as the space charge begins to move outward, the image forces become progressively less important, and more electrons are freed to move off the wire to the detector. Hence the signal received from the wire does not reach full amplitude at the instant of the end of the electron avalanche, but at a somewhat later time.

The electron avalanche is over very quickly. The electron is in a high field region, where the field is far above the average value. We recall that an electron with an energy of 10 electron volts will travel at a speed of 1.8×10^8 cm/sec, and will require 3×10^{-10} sec to go a distance of half a millimeter. Hence we see that the time needed to complete the electron avalanche is short compared to the other times involved.

We shall next consider the outward movement of the positive ion space charge sheath. This space charge moves toward the cylinder in a field which is at first strong but which rapidly falls to average and below-average values. Hence the positive ions can not be expected to produce much new ionization by collision processes on their way out. Further, since the drift velocities are of the order of 10^4 cm/sec in the average conditions in a counter, i.e., say E is 1000 volt/cm, p is 0.1 atm and k is 1 cm/sec in unit field and pressure, it will require about 100 microseconds for the positive ions to cross the dimensions of the counter and reach the cylinder. This is about 100 times longer than the times required to complete the electron avalanche and the spread of the discharge along the wire.

When the positive ions reach the cylinder, they will be neutralized. This process of neutralization may be thought of as one in which an electron is drawn out of the metal over or through the potential barrier. Since the photoelectric work function of most metals is perhaps 4 or 5 volts, and the ionization potential of many gases commonly used in counters is perhaps 15 volts (for example, argon, 15.68 volts) there will be an excess of energy equal to the difference between the two to be dissipated. Further, when an ion is neutralized, expecially it it is a simple atom or even in most cases a diatomic molecule, a photon of recombination radiation will be emitted. This recombination radiation generally has more energy than that corresponding to the work function of the metal surface, and hence we may expect that electrons will be emitted by surfaces bombarded by such radiation.

By the time the positive ion space charge cloud has reached the cylinder, the field near the wire has recovered to normal, and any electrons formed in such processes and coming in to the field will initiate a new avalanche. Thus we have here the mechanism of the reinitiation of the discharge which had actually been in the first instance quenched by the lowered field inside the space charge sheath when that sheath was near the wire.

A further complication will occur in counters using a single pure gas. If the atoms or molecules of this gas have any metastable states, it may be expected that some metastables will be formed in the discharge. If the gas is pure, these metastables will not be de-excited by collisions with other gas atoms, but only by collisions with the walls of the container, in this case the cylinder of the counter. It has been known for thirty years that collisions of the metastable atoms or molecules upon surfaces results in the liberation of electrons from these surfaces. Since the majority of metastable energies are greater than most photoelectric work functions, the conditions required for liberation of additional electrons are present. Moreover, since metastables in some cases have long lives, often hundreds or more microseconds, the metastable may reach the wall long after the discharge has been terminated and the counter returned to sensitivity for the next pulse. Therefore counters filled with pure gases are found to show spurious pulses. Indeed it is virtually impossible to quench such a counter as metastables may reach the wall at any time and reinitiate the avalanche. The procedure of using mixed gases fo deexcite the metastables by collision has been recognized for over twenty years. For example, in a non-selfquenching counter, a mixture of argon plus ten percent of hydrogen has been found to work well and to do the necessary suppression of the metastables. Similarly, in typical halogen-quenched counters, using a mixture of argon, neon and chlorine, each kind of gas can de excite the metastable atoms of the other or others.

It is clear therefore that in counters containing simple monatomic or diatomic gases such as nitrogen or hydrogen, conditions are such that the discharge once started will continue, owing to the continuous supply of new electrons. To terminate the discharge it is necessary to alter these conditions sufficiently so that the supply of electrons becomes converging instead of diverging. The simple way to do this is to lower the field around the central wire sufficiently so that the electron avalanches are smaller. It is not necessary to reduce the field to zero, for the photoelectric efficiencies of surfaces are so poor that even if many thousand photons reach the surface, on the average less than one electron will be liberated. The potential across the counter need only be dropped below the starting potential, and does not need to go to zero. Indeed it is important that it should not go to zero, as that would eliminate the cleanup of ions, and leave them drifting about in a fieldfree space. Reduction of the voltage below the starting potential can be accomplished either (a) by using a high resistance, the RC time of the counter wire system being longer than the time needed for the action to terminate, or (b) by using an electronic circuit to reduce the potential for the desired time. In modern times, the resistance systems are seldom used, since the RC values usually

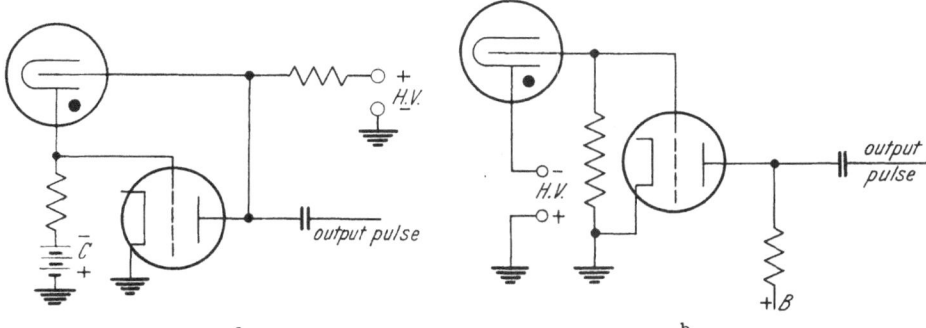

Fig. 1 a and b. (a) Neher-Harper quenching circuit. (b) Neher-Pickering quenching circuit.

must be long enough to give the counter poor resolution times. Electronic circuits, on the other hand, can be set to operate at any desired speed, and are much more flexible and practical.

There are many quenching circuits which have been described in the literature. A detailed discussion of such circuits is out of place here, and we shall treat the matter very briefly. Most of the quenching circuits are modifications of the original NEHER-HARPER [23] and NEHER-PICKERING [24] circuits, shown schematically in Fig. 1 a and b. We shall not attempt to give the analysis of the operation of these circuits here, but shall merely remark that in each case the net effect of the arrival of electrons upon the central wire, is to lower the potential across the counter for a time just long enough to permit the counter to recover, for the positive ions and stray electrons formed in the neutralization process to be collected without having an opportunity to start new avalanches. Thus the counter recovers and is ready to receive the next count.

Since the main factor in the length of time the counter requires to operate is the travel time of the positive ions, it is clear that if these could be collected sooner, the counter would operate faster. A circuit which accomplishes this purpose was developed by SIMPSON [30] as a part of his thesis research at New York University. The circuit reverses the potential across the counter as soon as the signal produced by the arrival of the first electrons on the central wire is communicated to the grid of the first tube. The electron avalanche being already complete after the first two microseconds or so (allowing for the time needed for the discharge to spread down along the wire: see next section) the circuit then reverses the potential so that the wire is negative. The positive ions have only a small distance to go, and that distance is wholly in a high field region. Hence the ions rapidly travel to the wire, where they are neutralized. The circuit then again reverses the potential, and returns the counter to its original state. SIMPSON found that this circuit permitted an marked speedup of the operation of the counter to be achieved. However, this gain is achieved at the price of appreciable added electronic complexity, for the requirements of a circuit which will do this are quite severe. The circuit must be able to produce an almost exactly square wave, of amplitude perhaps 2000 volts (in the case of a counter normally operating at 1500 volts, with a 500 volt potential across it when reversed) for a period of a few microseconds, and the pulse must not "overshoot" or have long tails.

5. Self quenching counters. As we have pointed out above, it is the phenomenon of the production of additional electrons at the cylinder upon the neutralization of the positive ions which causes the reinitiation of the discharge. Without this reinitiation, the discharge terminates and the counter is ready for the next event. It was found by TROST [36] in 1937 that by adding a small amount of an organic vapor the phenomenon of reinitiation was suppressed. In such a counter the phenomena followed the same pattern as in the non self quenching types up to the point at which the act of secondary electron production took place, and after that the action differed in that no additional electrons were produced. Thus we may begin the discussion by considering that an initial electron has been formed, that it has moved in toward the wire, and in the high field region near the wire has produced an avalanche. The discharge has spread along the length of the wire, and the electrons have been collected. Then the positive ions have moved out to the cylinder and there been neutralized. Here the process ends. As no secondary electrons were formed, there was no reinitiation of the discharge and the counter was then ready for the next count.

In discussing the difference between this type of counting action and the other, we must explain two things. First we must show that photons formed in the discharge do not reach the cylinder; second, we must explain the absence of electron production upon the neutralization of the ions. An explanation for this effect was given by KORFF and PRESENT [13] in 1944 and may be summarized as follows. First, what happens to the photons formed in the discharge? The majority of organic compunds have complex absorption spectra in the far ultraviolet. Since the work functions of the usual counter cylinders are perhaps 4 or 4.5 volts, an energy corresponding to some 3000 or 2500 Å, we need not be concerned with the photons of less than this energy. They can produce no electrons in any case. But if the vapor has strong absorption bands in that part of the ultraviolet where the recombination photons are to be found, the photons will be absorbed in the gas and will not reach the cylinder. Examination of the data on absorption spectra, studied by SPONER and TELLER [32], suggests that the absorption should be strong and usually continuous in this region. The reason for this is that 4 volts is already an energy comparable to the strength of the molecular bond, so that the electronic transitions and the rotation-vibration bands generally are heavy structureless systems, converging on dissociation. The larger and the more complex the molecule, the larger will in general be the parts of the ultraviolet which are blotted out and the stronger will be the bands. Hence we have in this strong absorption the reason why the photons formed in the discharge, with energy high enough to cause photoelectric emission, do not reach the cylinder.

We come next to the second part of the discussion, namely what happens when the positive ions reach the cylinder and are neutralized. This process again has two parts, both necessary to the operation. The first process is electron transfer, insuring that only positive ions of the complex molecule arrive at the cylinder. The second is the process of predissociation, which leads to the neutralization of the ion without the production of electrons. Let us consider these two in some detail.

In the discharge around the central wire, it is evident that many positive ions will be formed from all the constituent molecules and atoms in the counter. Since it is usual to employ a vehicular gas, such as for example argon, to which a relatively small amount, perhaps of the order of ten percent of the organic vapor is added, it is clear that there will be many argon ions formed in the avalanche. The actual number of argon ions N_A will be given by

$$\frac{N_A}{N_v} = \frac{P_A\,\sigma_A}{P_v\,\sigma_v}\,, \tag{5.1}$$

where N_v is the number of vapor ions, P is the pressure and σ the ionization cross section of the argon or the vapor as indicated by the subscript. It will suffice to say that most of the ions will be those of the vehicular gas, simply because there is so much more of that present. These ions on their way out to the cylinder will make collisions with neutral molecules of the vapor. Since in general the ionization potential of the vehicular gas is greater than that of the vapor, (for example, argon, 15.68 volts, alcohol, 11.3 volts) conditions are favorable for electron transfer to take place. Thus the reaction:

$$A^+ + M \to M^+ + A + E \tag{5.2}$$

takes place, where A is the argon atom, M the organic molecule, these symbols with the superscript $(+)$ the corresponding ions, and E is energy. The difference in energy, in the illustration 4.4 volts, must of course be dissipate d,but this can occur as radiation of a photon or as kinetic energy of the two entities involved in the collision. Further, since the free paths between collisions are of the order 10^{-3} to 10^{-4} cm, the ion will make some thousands of collision on its way out to the cylinder, of which collisions this number multiplied by the partial pressure of the vapor will be with vapor molecules. This is a sufficiently large number of collisions to ensure that transfer will in all probability take place. Indeed KALL-MANN and ROSEN [15] have measured the probability of such transfer and have found it to be of the order of gas-kinetic cross sections if the ion is moving through a gas of its own kind, and somewhat smaller if the gas is a different kind. It is a sufficiently probable process so that after some thousands of collisions a negligible fraction of the argon ions have not partaken. The result of this process is that all the ions arriving at the cylinder are of the vapor, and not of the vehicular gas.

It may at this point be mentioned that the transfer process is favored if the ionization potentials are not far apart, and is less probable if they are very different. Thus if one desires to construct a counter with a minimum amount of quenching vapor, one should seek a combination of gas and vapor such that the ionization potential of the gas is just slightly above that of the vapor. Further, it is possible to find combinations in which the reverse situation obtains. For example, a counter filled with a xenon-methane mixture, will show very poor self-quenching properties since the ionization potential of the xenon is less than that of the methane, and electron transfer is energetically impossible.

We are now ready to discuss the second part of the process. We shall show why the neutralization of the vapor ion at the cylinder does not result in electron production. The reason for this is that when the polyatomic ion is neutralized, predissociation takes place. It is a general property of complex molecules that because of the crossing over of the potential energy curves, the energy excess in the molecule upon neutralization does not manifest itself as radiation but gives rise to radiationless transitions, the energy being expended in breaking molecular bonds. Usually the act of predissociation takes place in times of the order of 10^{-13} sec, which is short compared to the time required for radiative deexcitation of the excited states. Because of this mechanism, no photons are formed, and therefore no new electrons. In the absence of a supply of new electrons to maintain the discharge the discharge terminates upon completion of the process at the wire.

In general the avalanche of which we have spoken takes place in a small volume. It also takes place in a short time. Another phenomenon follows. Photons from the initial avalanche will ionize the gas at other places, and still new avalanches will form. This action continues, and the discharge propagates itself down the length of the wire until the entire wire, or the entire high field region is involved. Thus when the electron avalanche is finished, a positive ion space charge cloud will surround the wire along its length. Two questions come immediately to mind in considering this process. The first is about the speed of propagation of the discharge. The second is the detailed mechanism of its propagation. The present author suggested the experiment to one of his students [38] and simultaneously an independent group in Europe [11] was making the study. In summary, the results agreed, and it was found that the discharge propagates down the wire at a velocity of about 10^7 cm/sec, the exact value depending somewhat on the nature of the gas and other factors.

When we come to a detailed consideration of the mechanism, we see at once one of the important differences in operation between the self-quenched counters and the nonself-quenched types. Consider first the nonself-quenched case. In this case the photons formed in the first avalanche, as a result of the collisions, will spread outward, often reaching the cylinder where they will free photoelectrons. These photoelectrons will travel inward to the wire and initiate new avalanches. The photons will be formed in the recombination and deexcitation processes, and many of them in such gases as argon will lie in the far ultraviolet, well above the photoelectric thresholds of most cathode materials. The time required for such a process is mainly the time needed for the electrons to traverse the counter from the cylinder to the wire. The photon travel time is much less than this. So also is the deexcitation time, the lifetimes of most of the excited states being some 10^{-8} sec for radiative deexcitative transitions. The electrons, in average E/p of more than 10 volts/cm/mm Hg, will move at speeds of around 10^7 cm/sec, and will require times of the order of tenths of microseconds to cross the counter. Beautiful oscilloscopic records of the successive groups of electrons were obtained by the Montgomerys [20]. Moreover, since the photons can travel down the length of the counter, the discharge can proceed by jumps and the new avalanches can occur at any place next.

A further proof of this behavior is obtained in the studies of beads on the central wire. When beads or disks are placed on the central wire, it is found that a nonself-quenching counter discharges as before, thus indicating that some agency, in this case photons, can travel from one part of the counter to the other, jumping past the beads. But when the experiment is done on self-quenching counters, the discharge usually is confined to one section between two beads. In this case therefore, the discharge is not propagated by something which can move freely throughout the counter. The organic vapor usually is the determining factor, since the vapor can absorb the photons, so that the free path of the photons is quite small. It is found that beads of the order of 1 mm in radius are sufficient to stop the discharge, the critical size depending on the overvoltage as well as on the nature and pressure of the gas. Thus we can conclude that the free path for photons has been shown by experiment to be of the order of 1 mm in many typical self-quenched counters. In this case, since the deexcitation time is of the order of 10^{-8} sec, we shall have a series of jumps of one mm, followed by a wait of 10^{-8} sec, followed by another photon jump. Hence the observed propagation velocity of 10^7 cm/sec appears to have a reasonable explanation.

Thus to summarize, we have the following action: first the initial electron or electrons move in, and start an avalanche a few free paths from the wire. The photons formed in the process ionize further atoms of the gas near the wire, and new avalanches are set up. This process spreads along the wire at some 10^7 cm/sec. The positive ions which require of the order of a hundred microseconds to cross the counter, will all still have moved only a negligible distance while the avalanches are building up.

The positive ions will move outwards to the cylinder. During their outward movement, the potential of the wire changes, and the field near the wire starts to return to its normal value. The positive ions, by the electron-transfer process, tend to become entirely ions of the organic molecules. These, upon reaching the cylinder are neutralized, and predissociate, resulting in the liberation of no new electrons. The process is terminated, and the counter has returned to its original state, and is ready for the next count.

In this brief account we have cited only a part of the available experimental evidence. Another part, which bears upon the process of predissociation, is also of interest from the point of view of practical considerations. This is the progressive decomposition of the filling gas. It is clear that the theory requires that such an effect take place. The progressive decomposition was first demonstrated by W. Spatz [31] as part of his thesis work at New York University. He showed that (a) the pressure in the counter increased as a function of its use, (b) that the slope of the plateau became less flat, and (c) that the starting potential increased. Later, the work was continued by S. Friedland, Krumbein, Soberman and Korff [33], [16] who made mass spectroscopic analyses of the gas in a counter after various amounts of use. These analyses showed that the amount of the original quenching vapor decreased, and that various stable gases into which it decomposed increased. This progressive change has several important practical consequences which we shall discuss in appropriate places to follow.

III. Special topics.

6. Efficiency. The efficiency of a G.M. counter is, according to the accepted definition, the probability that a count will take place when the specified particle or quantum is incident in a specified manner. We shall briefly review the factors determining this quantity.

First we shall assume that the particle or quantum which we desire to detect has reached the interior of the detector. In other words we shall not discuss window or wall effects here. Then we assume that we have a G.M. counter, operating in the Geiger region with the necessary circuits, potentials and presentation devices attached. If the counter is actually operating in the Geiger region, then the condition necessary to produce a count is that at least one electron shall be produced in the sensitive volume of the counter. If such a single electron is produced, it will initiate a count in the manner we have already discussed. The probability of one electron being produced can in various circumstances be computed.

Consider first the detection of charged particles. If a particle produces x electrons per cm advance through the gas in the counter at unit pressure, then the probability of this particle passing through the counter and not leaving a single electron behing it is e^{-x}, and the efficiency G is evidently:

$$G = 1 - e^{-x}. \tag{6.1}$$

The quantity x is clearly the specific ionization of the particular particle under discussion, in the appropriate units. If we take the units for the specific ionization, s, as ions per cm at one atmosphere, then we have:

$$x = s l p \tag{6.2}$$

where s is the specific ionization, p the pressure in atmospheres, and l the length of the path which this particle follows through the counter. Tables of s will be found in the literature and we will only cite a few here. They are listed in Table 3.

Table 3. *Specific ionization, ions per cm, produced in gas shown, at pressure of 1 atm, by particles moving at minimum ionization (for example, μ mesons at energies of 400 Mev or so).*

Gas	Symbol	s (ions per cm at 1 atm)	Reference
Hydrogen . . .	H_2	6	1, 2
Helium	He	5.9—6.5	1, 3
Neon	Ne	12	4
Argon.	A	29.4	1
Methane. . . .	CH_4	16	5
Air		21	2

To give a specific example, suppose we have a counter containing argon at a pressure of 15 cm Hg, (0.2 atm), and suppose the particle describes a path length of 2 cm through the counter, and further suppose that the particle is a meson or other particle moving with an energy such as to produce minimum ionization, so that we may take s as 40 ionpairs per cm, then in this case the particle will leave 16 ions behind while traversing this counter, and hence its efficiency G is better than 99.99%. Similarly if we fill the same counter with hydrogen at 7.6 cm pressure, s being 6, G will be 80%. It is clear therefore that it is possible to build counters with virtually 100% efficiency. The disadvantage of low efficiency counters is especially emphasized if we use such counters in coincidence circuits, for the efficiencies are multiplicative for such systems. In anticoincidence circuits, an inefficient counter may in fact fail to exclude the type of event which such a system is designed not to detect, and thus lead to entirely wrong conclusions. On the other hand it must also be pointed out [14] that low efficiency counters have possible uses, such as being devices which permit measurement of s. If for example we were studying a mixed radiation, low efficiency counters will preferentially respond to particles with high specific ionization, and in the case of cosmic rays have been used to find average values for s.

---

[1] M. COSYNS: Bull. Tech. Ing. Ecole Polytech. Brux. **1936**.

[2] W. E. DANFORTH and W. E. RAMSEY: Phys. Rev. **49**, 854 (1936).

[3] W. E. HAZEN: Phys. Rev. **63**, 107 (1943).

[4] H. K. SKRAMSTAD and D. H. LOUGHRIDGE: Phys. Rev. **50**, 677 (1936).

[5] W. D. SPATZ: Private communication.

For the detection of photons, different considerations obtain. Here the chance that a photon will produce an electron in the gas, by either photoelectric absorption, or by the Compton effect, is quite small, compared to the probability that a photon will eject one or more electrons from the walls. For photons of visible or ultraviolet light, special photosensitive surfaces may be used, the surface being chosen so as to have a photoelectric work function exceeded in energy by the $h\nu$ of the quantum which it is desired to detect. Photoelectric efficiencies are in general poor, but special surfaces can be prepared for special spectral regions which have relatively good efficiencies. For X-rays of energy sufficient to penetrate through the walls of the counter, it is customary to use material of high atomic number for the walls, as in this case the probability of electron ejection is increased. Thus for example, bismuth cylinders have been used, as has lead.

7. Lifetime and rejuvenation. The main factor determining the life of a self-quenching counter is the decomposition of the quenching constituent during the discharge. The life, or lifetime, of a counter is defined as the number of counts which that counter can detect before its operating characteristics change so much as to make it useless in the particular circuit used. Naturally, since the changes are progressive, and since for some uses the tolerances are much wider than for others, exact values can only be defined for special cases. However, we can discuss the governing considerations.

The main factor determining the life of a self quenching counter is the decomposition of the quenching constituent in the discharge. Some numerical examples can be considered here. Consider a counter containing argon with an alcohol quenching agent. Suppose that there is 15 cm argon plus 10% alcohol, a usual admixture. Then there will be 5.4×10^{17} alcohol molecules per cm^3 in the original mixture. Let the volume of such a counter be 200 cm^3, and we shall have about 10^{20} alcohol molecules in total. Further suppose it is operating as a Geiger counter and is delivering pulses of ten volt amplitude. Then, if the capacitance of the central wire system is supposed to be $20\mu\mu F$, the charge per count is 2×10^{-10} Coulombs, corresponding to 1.4×10^9 electronic charges. Hence there will be this number of positive ions being neutralized at the cylinder per count. Since each positive ion upon being neutralized predissociates, we may expect that after about 8×10^{10} counts all the alcohol molecules will have been decomposed. Actually, it is observed that the counter changes its characteristics markedly long before this happens, so that the effective life of such a counter may be of the order of 10^{10} counts. In general we often find useful lives of this order, although variations by as much as a factor of ten in either direction are not uncommon.

It is immediately obvious that two procedures will lengthen the effective life of counters. The first is to put in more of the quenching agent. This has the disadvantage that the vapor pressure of most organic quenching agents is only a few cm at room temperature. Hence if we put in an amount equal to the vapor pressure, if the operating temperature of the counter drops a few degrees, the quenching agent will start to condense out and the presence of liquids will give usually produce undesirable effects by allowing leakage paths across insulating surfaces inside the counter. The second procedure is to employ vapors, the molecules of which can decompose more than once, still leaving a residual capable of quenching (i.e., predissociating). An example of such a vapor is ethyl acetate.

An apparent step, which in reality is not advantageous, is to fill the counter entirely with a gas which is a quenching agent. For example, methane, propane or butane at room temperature all can be used for filling counters. However, in this case another trouble arises. Methane and a number of other compounds of this general type tend to polymerize in the discharge, forming compounds which are solid or liquid at room temperature. There being a high field in the counter, these compounds will often preferentially precipitate out on the central wire and the cylinder. Unfortunately, even the presence of a part of a mono-molecular layer on the central wire will interfere badly with the buildup of the electron avalanche and the spread of the discharge along the wire. It has been found that a counter filled with pure methane will often become quite useless after as few as 10^8 counts. That this is the right explanation was proven by WEISZ [37], who transferred the gas out a of a methane counter after it had become useless, and put it into a new counter, which then operated properly; new gas put into the old counter failed to do or show any improvement; but flashing the wire, which drove off the accumulated polymerization products "rejuvenated" the counter and brought it back to usefulness. Rejuvenation is thus accomplished by cleaning out the old counter, and refilling it with fresh gas if the gas has also started to show signs of decomposition.

Lifetimes of nonself-quenching counters are controlled by quite a few other factors such as the pitting of the central wire, and are generally much longer.

8. Flatness of plateau; effects of impurities. The flatness of the plateau of a counter is an important feature, and it is worth while considering what factors determine this quantity. It turns out that several factors are operative, and we shall discuss the situation quantitatively where possible.

We recall that the counting rate curve, counts per unit time as a function of voltage for a constant source of radiation, is in fact an integral curve, as we are measuring all pulses greater than the size determined by the minimum which the circuit will accept. In terms of a pulse size distribution, a flat plateau means that we have a group of pulses of a certain size and none larger as long as we remain on the plateau. The derivative with respect to voltage of the counting rate curve is a measure of the pulse size distribution. Thus any factors in the discharge which give rise to irregularities in the pulse height will cause a non-flat plateau. One process operating in this manner is negative ion formation. If some of the electrons are captured to form negative ions, then these negative ions will drift toward the central wire but much more slowly than did the electrons. Entering the high field region, the negative ion may lose its electron near the wire after the initial avalanche is over, and thus initiate a new count, not due to a new entity to be counted but related in time to the previous count. If the negative ion is formed very close to the wire, it may drift in and not take much part in avalanche formation; thus again leading to the formation of a pulse not exactly the same size as the pulse formed without it. Thus negative ions may lead to the pulse size distribution not being one single group but containing a few of varying sizes. Therefore it is to be expected that the gases which form negative ions will exhibit a property of diminishing the flatness of the plateaus of the counters.

Such gases, tending to form negative ions readily, include (a) the halogens, and (b) the familiar O_2 and H_2O. To test this hypothesis, SPATZ [31] built a series of counters filled with carefully purified argon and alcohol vapor, and then introduced specific quantities of ordinary air. It was found that the introduction of quite a small addition of air made an observable difference. Thus for example, a counter with 95% argon and 5% alcohol, might have a plateau which showed a rise of 1% in the counting rate for a change of 100 volt. Admitting as little as 2% air changed the slope to 15% per 100 volt. Vapors showing strong negative ion formation have long been known to give steep slopes to plateaus. In 1937 the present author tried CCl_4 as a quenching agent and found that this vapor produced a plateau so non-flat that it was hard to say just where the plateau began or ended. Later, BROWN and MARONI [4] made a systematic study of flatness of plateau as a function of the tendency to form negative ions. They found an excellent correlation of plateau slope with electron attachment coefficient. They further showed that the slope was not a function of the noble gas used.

The study of the effects of the progressive decomposition of the quenching constituent carried on over a long period at New York University also bears upon this point, for it was found that some quenching constituents decomposed into fragments which tend to attach electrons. For example ethyl alcohol, C_2H_5OH apparently in the discharge tends sometimes to lose the OH group, which is known to have a quite appreciable electron attachment coefficient. As an argon-alcohol counter was used, it was found that the slope of its plateau got progressively steeper, which was attributed to the progressive release of OH. Further, if the O atom should be detached from the alcohol molecule, or if the OH group should pick up a hydrogen atom and become H_2O, in either case a constituent with is formed which has a tendency to favor the formation of negative ions.

We may summarize this experience by saying that impurities which tend to form negative ions should particularly be avoided.

9. Low voltage counters. For certain applications it is desirable to have counters which operate at as low voltages as possible. For example, in the case of balloon borne equipment the saving in battery weight is important. Hence some attention has been devoted to the problems of reducing the operating potential.

It will be recalled from the above discussion that the operating potential of a counter is determined, in practical terms, by two factors; the first is the need to have a field around the central wire high enough to produce gas amplification. Therefore it is evident that the two factors determining the voltage required are (a) the diameter of the central wire and the cylinder, and (b) the kind and amount of gas used. As for the diameter of the central wire, it is immediately obvious that reducing the wire diameter will increase the field around it, since the field approaches infinity when the wire diameter approaches zero. But after the wire diameter has been reduced until it is 2.5×10^{-3} cm in diameter, (corresponding to 1 mil wire, in the American notation) the radius is already about equal to about one free path, in the gas at the reduced pressures generally used in counters. Therefore further reduction in size would not increase the gas multiplication for there would not be enough additional travel for the electron to increase the avalanche size appreciably. Moreover, a wire of 1 mil size is already quite fragile, and represents about the practical minimum. Further, if the counter is to have a size adequate for detection of small amounts of radiation, it is not possible to reduce the cylinder diameter too far. In any event, reduction of this dimension affects the potential only logarithmically, according to Eq. (4.1). It is clear, therefore that to reduce the potential of counter we shall have to study carefully the dependence of this factor upon the nature and pressure of the gas.

For a simple case, the dependence of starting or operating potential upon pressure has the form shown in Fig. 2. It will be noted that the curve has a minimum which in many cases is at around a few mm pressure, and that it rises rather slowly on the high pressure side of this minimum. Unfortunately, it is usually impossible to operate a counter at the minimum, for the reason that the pressure is so low here that the efficiency, see Eq. (6.1), would be unusably low. What we must do, then, is to find a gas or combination of gases for which the minimum occurs at a high enough pressure to be useful, or is broad enough to give us a low voltage at a pressure permitting an efficient filling. In this connection, the work of Druyvestyn and Penning [6] is of importance. They studied the breakdown phenomena between coaxial cylinders for a variety of gas mixtures. They employed mixtures which in many cases consisted of a noble gas with a small added amount of some other gas.

Before discussing this situation in detail, we shall merely recall that some gases exhibit much higher starting potentials than do others. Assuming that a counter is filled first with argon, and then to the same pressure with nitrogen, it will be found that the nitrogen-filled counter will have a much higher starting

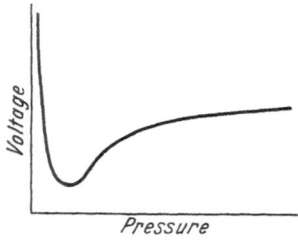

Fig. 2. Usual form for relation between starting potential of counter and gas pressure in the counter.

potential then will the aron-filled counter. The gases showing the high starting potential are such as N_2, O_2, H_2, CH_4, BF_3, and many others, while those showing low starting potentials are generally the noble gases such as He, A, Ne, and Kr. In general, He is avoided owing to its low specific ionization; even Ne is sufficiently low in specific ionization to make it less good gas then argon. The heavier noble gases are excellent for filling counters, but not much used on account of high cost. Argon is therefore the most commonly used gas, since it combines high specific ionization, low cost and low operating potential.

However, it is not possible to use just one gas in a counter if this gas has metastable states. Unfortunately, all the noble gases have such metastable levels. Since the metastable states have relatively long lifetimes, compared to the time required for the positive ions to drift outward to the cylinder, and since the result of a metastable atom diffusing into the wall of the counter is the possible liberation of an electron, the presence of atoms in metastable states is likely to produce spurious counts. Hence it is usual, when using a counter with primarily an argon filling, to add some other gas to deexcite the metastable states by collision. In the usual case in which some quenching constituent is added, this requirement is automatically fulfilled. Further, when the metastable level of one gas is close to the ionization potential of the other, collision may result in ionization. For example, Neon has a metastable level at 16.6 volts; while the ionization potential of argon is 15.69 volts. Thus a collision of a neon metastable with an argon atom can result in a neon atom in its ground-state plus an argon ion. This kind of energy-transfer is thought to play an important role in low-voltage operation.

We are now ready to consider specific cases. For example, consider a counter filled with neon, at pressures between 6 and 15 cm Hg, and with 0.002% of argon added. Such a mixture exhibits a broad minimum in the curve of starting potential as a function of pressure, such as Fig. 2, and an actual minimum starting potential of a little over 100 volts. Increasing the argon pressure to 0.012% makes little difference; but if we increase the argon pressure to 0.4%, the minimum will move up to about 160 volts at around 4 cm pressure. This pressure is already fairly low. Hence it is seen that low voltage operation can be achieved, but that it requires careful control of the amounts of the gases. Further, since the amounts are so small, purities are important, and contaminants must not be large compared to the figures cited. Mercury vapor, which as we know is not of any importance in ordinary counters, is of importance in this case, and must also be avoided.

Low voltage counters must be filled on systems not using mercury in traps or pumps, or having these continually isolated by cold traps.

With low voltage counters of this type, several other unusual features manifest themselves. First, the wire diameter is not at all critical, and can be quite coarse. Second, in some cases the counter will work just as well with the potentials reversed, and even better in some cases. Thus in the case of a counter filled with neon plus 0.002% argon, the curves for cylinder positive and negative cross at a pressure of 9 cm, and at 30 cm the counter will operate at 130 volts if the wire is negative, whereas it operates at 230 volts if the wire is positive. The radii in this case are 2.3 cm for the cylinder and 0.087 cm for the wire. At pressures below 9 cm, the operation is at the lower voltage when the wire is positive.

In the case of these low voltage counters, no quenching constituent is used, and it is clear that the counter must be used with a quenching circuit. Further, we repeat, careful control of gas purity is required, for a few parts in a hundred thousand of contaminants will nullify the low voltage properties mentioned above.

10. Halogen fillings. Whereas halogens have long been considered undesirable as constituents for filling counters, because of their tendency to form negative ions, nevertheless these gases do exhibit quenching characteristics. PRESENT [25] has pointed out that also in the case of halogens, since the band structure in the ultraviolet shows continuous absorption, predissociation will probably take place, and therefore this gas will serve as a quenching agent. LIEBSON and FRIEDMAN [18] have applied this to counters, and found that a mixture of four parts of argon to one of chlorine, using neon as the main body of the gas, made a satisfactory counter. Thus for example, in a counter they used 0.25 mm of the chlorine-argon mixture, with the neon pressure anywhere between 5 and 60 cm. Further, this mixture gave quite low starting potentials, and at a pressure of 5 cm for the neon, started at 300 volts.

It will at once be asked why such counters do not exhibit marked effects due to negative ions. The reason for this lies in the small amount of the halogen. An electron traversing the counter will make some 10^4 collisions; but of these only six will on the average be with chlorine atoms. Now it requires perhaps 2000 collisions on the average to produce capture, and hence few negative ions are formed. Certainly at higher halogen pressures, trouble would arise from negative ion formation.

Several other unusual properties characterize counters using halogen quenching constituents. First, a quite coarse central wire may be used. Wires up to 30 mils, ten times the usual diameter, have been found to work well. Second, due to the chemical activity of the filling, corrosion effects may be marked. It is usual to minimize these effects by using stainless steel cylinders, or tantalum, and glass walls or insulators. Copper, brass or silver, so often used in ordinary counters, give poor results. Further, the plateaus of such counters are usually not very flat. The optimum amount of halogen is failry critical. Also, not all halogens will work. For example, using a neon counter with an argon-flourine quenching mixture would be unsatisfactory since the ionization potential of the flourine is high enough so that electron transfer from the halogen to the noble gas would be energetically impossible. Bromine and chlorine are the two gases generally used.

Perhaps the most important single factor in the halogen counters is that the lifetime is long. The halogen may dissociate in the discharge, but when the diatomic molecule dissociates, the monatomic constituents will have some tendency to reunite; thus an equilibrium is soon established, and after this, further us of the counter does not alter the amount of quenching constituent present. Thus, lifetimes much greater then those found with the organic vapors can be achieved, at the price of non-flat plateaus.

Counters of this type have very small temperature coefficients in the ordinary range of temperatures. Dead times are found to be of the order of 350 microseconds. Time lags have been observed, of the order of one to several microseconds. The exact reason for these time lags is at present being studied by SHERMAN [34] and in a preliminary report is indicated as being due to the effect of the metastable states in the neon, which soak up part of the energy of the avalanche, and cause the buildup to be somewhat slower than usual. The time lags, in many cases are of the order of a fraction of a microsecond to several microseconds. They are found to have a time distribution, and to decrease as the overvoltage is increased. Further, the efficiency of these counters is not uniform. The efficiency was, the cases tested, found to be a function of geometry, and depended on where in the sensitive volume the primary ionizing event occurred.

11. External cathode types. It has been known for many years that certain kinds of glass are slightly conducting, in the sense that ions can migrate through them. Making use of this property, MAZE [22] developed a counter with the cathode on the outside. He painted a cathode surface, using aquadag, on the outside of a glass tube which served as the envelope of the counter. The kind of glass used for these counters is generally what is termed "soft glass", soda-glass or lime-glass. The harder glasses such as pyrex have very much lower conductivities and are not suitable for this construction.

The advantages of this type of construction are several. First, when the positive ions reach the glass interior surface, they encounter a substance with a high photoelectric work function, and therefore the number of secondary electrons formed is minimized. Hence these counters have good quenching characteristics and exceptionally long and flat plateaus. Second, the counters are easy to construct. They require one less seal, no cylindrical internal electrode, and if operated with the cylinder grounded have as good safety features as other types.

The disadvantages arise mainly from the long time required for the ions to migrate through the glass. The counters tend to polarize at high counting rates. For example, BERETTA and ROSTAGNI [5] have measured such resolving times and found in the case of those counters they used that the time was around 0.1 sec. Naturally this limits the counter to applications in which the counting rate is slow. In certain applications, such as the cosmic ray case, sizes of counters can be selected such that the counting rates, which are known in advance, will not be in excess of the resolving ability. The second disadvantage is that a great many laboratories employ pyrex systems, so that graded seals must be used to attach the non-pyrex counters.

IV. Design and use considerations.

12. Constructional features. α) *Envelopes.* Counters may be made with either glass or metal envelopes, depending on the intended use. Each type has certain advantages and disadvantages. Which design should be chosen depends on the relative importance of these factors in the particular use contemplated for the counter.

Consider first the metal types. Almost invariably, the cylinder or envelope also serves as the cathode. Since metals with moderate work functions are available, it is possible to make the cylinder of such substances as copper or brass.

Of these two, copper is preferred since the zinc component of the brass tends (a) to reduce the effective work function, and (b) to boil out if the counter is heated during the outgassing process, and will then settle on end surfaces which may have their good insulating characteristics impaired by this effect. Aluminum is generally to be avoided because (a) of the difficulty of making good seals to it, (b) aluminum is often porous and continues to give off gas for long periods, and (c) aluminum is chemically active, reacting with filling constituents, and with cleaning agents. Stainless steel has a good work function characteristic, but is (a) expensive, (b) hard to seal to insulators and leads.

With the metal envelope types, thought must be devoted to the end design. Usually the metal itself has ends which are perforated to permit the electrodes to enter. In this case it is important to see that no high field regions are developed inside the active part of the counter. Glass sleeves should surround the central wire system where it passes through the envelope (see Fig. 3).

Metal envelope types are almost invariably operated with the cylinder at ground potential. Since the metal almost completely surrounds the wire system this design provides (a) good electrostatic shielding and (b) prevents or minimizes photons from outside reaching possibly photosensitive surfaces. Again it must

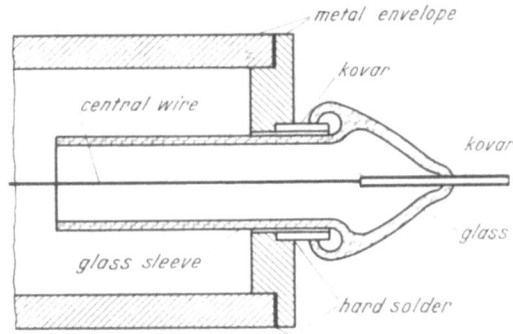

Fig. 3. End arrangements in metal-envelope counter.

be emphasized that it is important to make an actual ground connection, to minimize the possibility of picking up stray pulses.

Many types of metal-to-glass seals exist. Thin copper can be sealed to some kinds of glass (the so-called "housekeeper seals"), tungsten leads can be sealed through glass "squeezes", such as are used in introducing electrodes in conventional vacuum tubes, and certain alloys such as Kovar can be directly sealed to special glasses. All these designs are sufficiently heat-resistant to permit the seal to be baked out during the evacuation preliminary to filling. Clearly, the old-fashioned sealing-wax seals, waxes generally, hard rubber or plastic insulators should be avoided since with such seals, outgassing is impossible. Modern techniques for vacuum-tube manufacture should be followed in general, since in the case of vacuum tubes it is well known that good vacuum technique, good seals and gas-free metals are of importance.

The author has long been of the opinion that counters could be made by modern vacuum-tube making machinery, on a semi-automatic basis, with almost 100% reproducibility. Metal envelope techniques have in this field been carried to very satisfactory developments. For example, a counter could be made inside a tube such as a 6L6, using the shell as the cathode, and mounting the central wire on the usual glass lead-in press.

Turning next to the glass envelope types, these usually employ a separate metal electrode inside for the cathode. The cathode may be a thin layer of metal evaporated or chemically deposited on the inside (or in the case of the Maze counters, on the outside) of the cylinder, or a metal cylinder of some metal such as copper (or brass, see above). The inside of the cylinder should be free from sharp points or irregularities, which can be the source of spurious counts. Also, the point at which the conductor which leads to the outside, makes contact with the cylinder, should be on the outer side of the cylinder, and the conducting lead should at no point be nearer to the central wire then the radius of the cylinder. The ends of the cylinder should be carefully rounded so that no sharp points or burrs are left (Fig. 4). If the cylinder is made by rolling a sheet of metal, special care should be devoted to the manner of folding the metal so that the metal does not terminate on the inside (see Fig. 5). Naturally, if it is possible to use seamless tubing, it is better to do so.

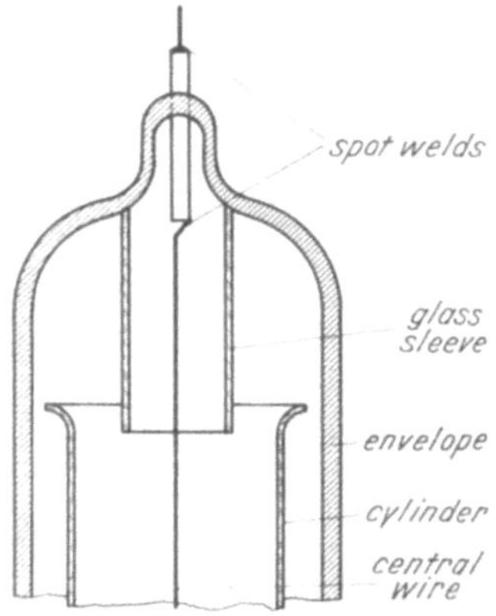

Glass envelope types can be operated with either the cylinder of the wire at ground potential. This feature gives some added flexibility for certain purposes. Glass envelopes are obviously somewhat more fragile than metal cylinders. The metal cylinder should not be thicker than necessary, (a) because of absorption effects in the cylinder and (b) since a heavy cylinder puts an undesirable mechanical strain on the glass envelope.

β) The central wire. It is usual to make the central wire out of tungsten (wolfram). This material is readily available in wire form, spotwelds and seals satisfactorily, and has a high enough melting point so that heating for outgassing is possible. Wire sizes of 2, 3 or 4 mil (radii of 2.5, 3.7 or 5×10^{-3} cm) are often used. Wire sizes in excess of 4 mil is coarse enough to make appreciable increases in the operating potential for all counters except certain types as we have discussed above; while wire sizes of 1 mil or less are extremely fragile and may be broken in ordinary use unless special care is taken. Other metals may be used if for any reason they are desirable, since the element of which the wire is made seems to make little difference in the operation. The problem is mainly one of finding a metal with desirable mechanical properties.

Fig. 4. Arrangement at end of cylinder.

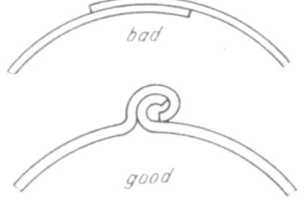

Fig. 5. Preferred arrangement in attaching edges when cylinder is made of a sheet of metal rolled into cylindrical shape.

Especial care should be taken at the point at which the central wire is attached ot the heavy leads. At this point, a spot-weld is generally employed, and the possibility of leaving sharp points exists. Such sharp points should not be present anywhere in the high field region. Usually it can be arranged that the place where this attachment occurs is outside the cylinder and out of the high field. If not, then arrangements should be made to cover it.

With regard to the position of the central wire, this is usually shown on diagrams as being along the axis of the cylinder. Actually, it has been found by test that as long as the wire is parellel with the axis, it does not need to be in the exact geometrical center of the counter. The present author conducted an extensive set of tests [*14*] some years ago to establish this point. On the other hand it is important that the wire should be (a) tight, and (b) parallel with the axis. It is evident that if these conditions are not met, the field will vary along the wire, and be greater at the places where the wire is nearer the cylinder. Such a situation generally results in loss in flatness of plateaus.

If the central wire ends inside or near the end of the cylinder, it is well to cover the end with a small sphere, since sharp points are often produced at the end of a wire in the cutting process. Further, it is very important that no dust or abrasive grains be left on the wire to form sharp points.

γ) *Gases, kinds and purities.* The main considerations about gases have been outlined above. Whereas almost any gas will work, certain gases are much better than others. In general, argon is usually used as the main constituent, because of (a) its low starting potential, (b) high specific ionization, and (c) its low price. Neon is similar in the first respect, somewhat less good for the second and third. The normal stable gases of the atmosphere, nitrogen and oxygen have higher starting potentials; so does hydrogen, and the latter also has a low specific ionization. Hydrogen is not used except in certain special applications, such as recoil counters, where its unique properties are of importance. Electronegative gases should be avoided because of their tendency to form negative ions. Except for certain special applications discussed above, oxygen, the halogens, and water vapor should be avoided.

For quenching constituents, many organic vapors serve. Alcohol has been widely used. Ethyl acetate has been found to give good flat plateaus [*14*], and may be somewhat better than alcohol in that its decomposition products tend even less to form negative ions. Ethyl ether, petroleum ether and other organic compounds have also been used. Organic compounds such as methane, ethane and propane tend to polymerize and thus shorten the life of the counter. General considerations are: (a) vapors should be pure, and such possible contaminants as water vapor should be avoided. (b) many of the liquids such as alcohol, when stored in air, absorb air, which should be removed before using them in counters. (c) fillings should never be made with substances to such pressures that the vapor will start to condense out at any temperature at which counter is to be used. If such condensation occurs, undesirable changes are often experienced Thus for example, if ethyl alcohol is used, since its vapor pressure at room temperature is around 2 cm, the pressure put into the counters should not exceed. 1 cm, so that condensation will not take place if the room temperature should be lowered somewhat.

Purity of the permanent gases should also be considered. Certain contaminants are much worse than others. Small amounts of a oxygen as a contaminant in argon for example, will tend to spoil the flatness of the plateau, while considerable additions of neon would not produce undesirable effects.

δ) *Vacuum technique; pumping and cleaning.* If it is desired to produce numbers of counters with as nearly identical characteristics as possible, and with as long useful life as possible, then chemical stability is a requisite. It is self evident that if the composition of amount of the gas inside the counter changes with time, the operating characteristics will likewise change. These considerations have been known for years in the vacuum tube industry, and yet are not always recognized in the making of counters. Hence the counter should first be thoroughly cleaned and pumped to secure a good vacuum. The procedures for this are well enough known so that only a brief resume is indicated. The metals of which the surfaces are made should also be chemically stable, and have as high photoelectric work functions as possible. All dirt, grease and soldering or welding flux or debris should be removed. This can generally be accomplished by first washing out the counter with a dilute acid, then with acetone to remove greases, and then cleaning out the remains of the acid or other liquids with distilled water. The counters are next attached to the vacuum system, pumped and baked. They should be baked at as high a temperature as is possible without softening the glass or soldered joints. If possible, the central wire should be glowed. This glowing serves two purposes. First it removes such tiny irregularities or even dust-grains as may be present on the surface, and second it drives occluded gases out. To glow the wire, either it must be electrically available on both ends, or if it is not, then induction heating is needed. Finally, thought should be devoted to whether there are any chemical reactions that might occur between the filling gases and the electrodes.

After the counter has been thoroughly baked and pumped, and found to be free from leaks, it may be filled. In filling, if a vapor is used, the vapor should be admitted first, and then the gas such as argon, for the vapor will take a very long time to diffuse to a uniform concentration against the argon pressure. After the argon has been admitted, time enough for diffusion should be allowed, for the argon will tend to push the vapor ahead of it and may result in concentrating more in some counters and less in others. Often it is desirable to test the counters while they are still attached to the filling system, so that errors can be corrected and refillings can be carried out if necessary. Dipping the ends of glass covered counters into a wax with good electrical characteristics is good technique, for the wax covering will minimize the formation of surface films of moisture which are troublesome when the humidity is high. Wax-soluble dyes can also be added to reduce the photoelectric effect, due to photons from the outside reaching the interior of the cylinder.

Perhaps the most serious single thing which should be considered in making counters is the avoidance of sharp points anywhere in the sensitive volume. These sharp points may produce fields so high that field emission of electrons will take place, and thus may give rise to spurious counts, or in extreme cases, to useless counters. Small grains of dust, in the counter, will tend to precipitate in the field in the very worst places. Hence the need for thorough cleaning mentioned above. Similarly, sharp points at junctions or welds should be avoided, or if inevitable, should be so located as to be outside the high field regions, or if they must be in the high field region, they should be covered so as to reduce the field in their vicinity.

13. Windows. α) *Thin end or side.* When the radiation to be detected has very little penetrating power, it is necessary to devise thin windows and to attach them to counters in such a manner that the radiation can reach the sensitive volume. There are essentially two types of windows, the first being a thin window placed at one end (or side) of an ordinary counter. The second type is a counter on which the walls are thin, so that radiation from any direction can enter. It is clear that the first type of window can be made much thinner than the second because of the smaller area subject to the mechanical strain due to the pressure difference. Since end windows can be made thin enough to admit alpha particles from ordinary radioactive substances, these are often called "alpha particle counters", while the thin side walled counters will usually not permit alpha particles to enter but will allow beta particles to be counted and are hence often referred to as "beta particle counters".

Let us consider first the thin end window types, or alpha particle counters. Since the ranges of alpha particles are of the order of three to six cm of air, it is clear that the windows must be thin compared to the equivalent of this. A substance of density one gram per cm³, if rolled out into a sheet one thousandth of a cm in thickness, will have a stopping power approximately equivalent to 1 cm air, provided its atomic number Z does not differ very much from that of air.

Thus for example, a 1 mil cellophane foil corresponds to about 2.5 cm air. Such a foil would be transparent to alpha particles. Mica can also be produced in very thin sheets, which will pass alpha particles. Since the density of air is about 1.2 mg/cm^3, we can talk in terms of foils 1 to 2 mg/cm^2 as being suitable for alpha particle work. Thin metal foils can also be made, but aluminum in thin layers tends to have pin-holes, and copper or silver has the drawback that its Z is high enough so that the foil has to be still thinner to correspond to air equivalent. Thus, a copper foil with 1 mg/cm^2 would be about 1.4×10^{-4} cm thick. So thin a foil is hard to make, and has little mechanical strength.

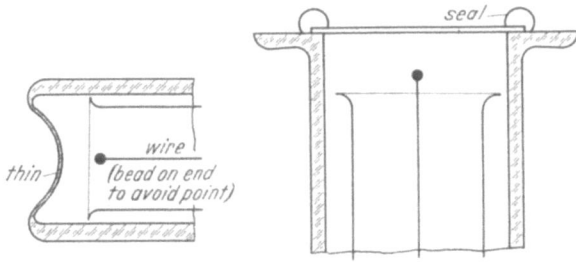

Fig. 6. Typical thin end alpha-particle types. The illustration at the left shows an all-glass construction, while on the right a window of mica or other thin material is sealed on. Note: bead at end of wire; and note termination of cylinder, which is flared outward.

On the other hand glass can be blown out into extremely thin bubbles. Hence the usual materials for such foils are glass, mica or cellophane. If a large surface is required, mechanical support must be provided. Such support can be in the form of a perforated grid, with the thin foil supporting the pressure difference between the atmosphere and the gas inside of the counter only at the holes. Fig. 6 shows some typical thin window constructions.

One of the problem is that of sealing the window. In the case of glass windows, the problem has a simple solution by merely making the counter and the window out of one piece. A skillful glassblower can blow an extremely thin bubble, slightly

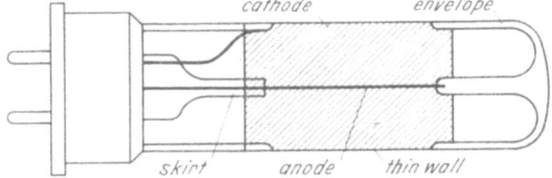

Fig. 7. Thin side-wall beta counter, available commercially.

curved inward, which will support a full atmosphere over a square cm or so. The purpose of the slight curve is that the glass should always be under tension, in which position it is especially mechanically strong. In the case of the thin mica windows, the procedure often used has been to place the mica on the metal flange of a metal-envelope counter, and then put some powdered glass or other sealing compound over the junction, and heat the entire assembly until the glass fuses to both the metal and the mica (see Fig. 6). Cellophane may be attached by the use of various plastic solvents or cements. However, cellophane will not permit nearly as high temperatures as to be used in baking, pumping and evacuating the counter as either the glass or the mica construction.

The thin side windows can also be constructed by a skillful glassblower, by pulling out the central section of a counter while it is hot. The glass can then be chemically silvered or coppered, or aquadag painted on, or metal evaporated on it, for a cathode. Fig. 7 shows some typical constructions of thin-walled counters. Copper or aluminum can be machined down to thin dimensions by skillful lathe operators or a thin metal foil can be stretched over a framework and brazed along the edges.

Thin windows:

β) *Sample inside counter.* In some cases the radiation to be detected has so little penetrating power that it is desirable to have the radiation originate inside the counter so that it does not have to pass through any window at all. This situation obtains for example when counting the beta rays from C^{14}, which are extremely soft. In this case it is possible to put the sample inside the counter by

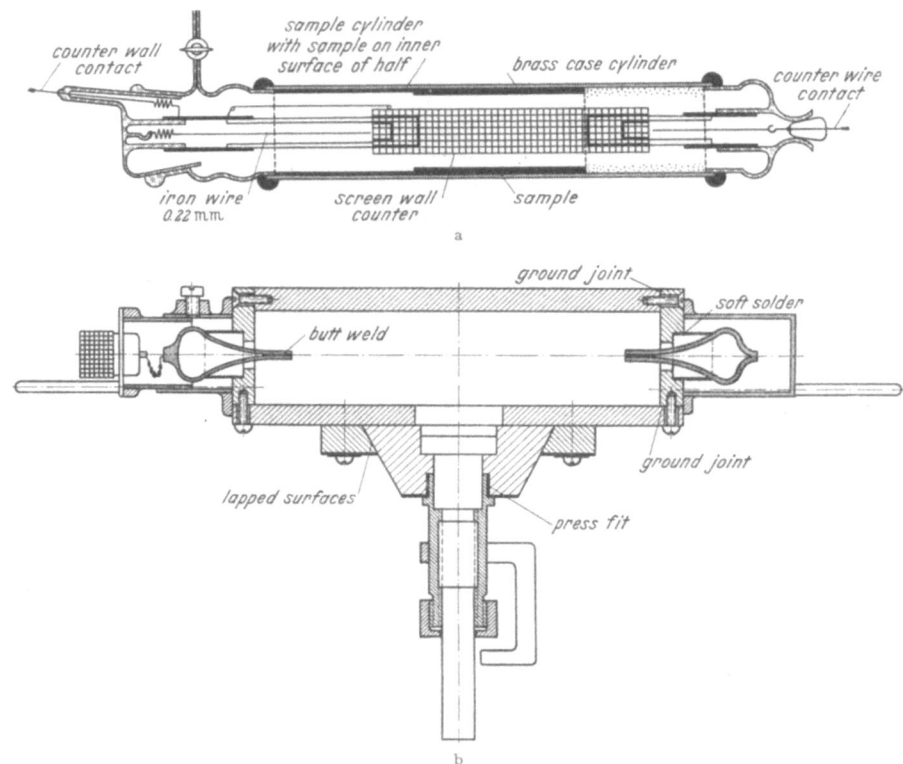

Fig. 8a and b. (a) Counter developed by LIBBY *et al.* for Radiocarbon dating. [Ref.: W. F. LIBBY and D. D. LEE: Phys. Rev. **55**, 245 (1939). — W. F. LIBBY *et al.*; Rev. Sci. Instrum. **22**, 225 (1951).] (b) Counter developed by SIMPSON for detecting soft radiation. Gas flow type. Sample internally exposed. [Ref.: J. A. SIMPSON: Rev. Sci. Instrum. **18**, 884 (1947); **19**, 733 (1948).]

one of several measures. First, the sample can be chemically made into a gas, in the case of carbon as CO_2 or methane, and the gas introduced as the main constituent of the gas in the counter. Another possibility is to mix powdered carbon (or other radioactive substance) with a chemical binder and paint it on the inside of the cylinder. A third technique is to have a hole at the side of the cylinder, and to place the sample on a metal slab which will just fit the hole, thus presenting one side of the slab directly to the gas inside the counter. In this case, some sort of seal or joint has to be provided, through which the sample can be introduced. Further, in this case it is often convenient to operate the counter at a pressure of one atmosphere of filling gas, so that there is no pressure difference between the inside and the surrounding air. This can be accomplished by allowing gas from a container to flow through the counter while the run is in progress. Such counters are called "flow counters" and several have already appeared as commercial models. Fig. 8 shows some possible arrangements.

14. Internal radiators. Another technique which has been developed in an attempt to have the radiation to be detected originate inside the counter, is to place "radiators" at suitable points inside the active volume itself. In each case, the radiation to be detected passes into the counter, where for various reasons the detection efficiency is low. Then it encounters a radiator, from which it ejects secondary particles, which secondary particles make the detection easier or more efficient. We shall cite three quite different illustrations. First, consider a counter which is to detect X radiation. Such a counter usually operates by counting the production of secondary electrons from the walls of the counter as the X radiation traverses the device. The production of secondary rays can this case be in augmented by placing more material inside the counter, such as foils or sheets of some substance from which secondaries might emanate. Whether these radiators should be conductors or dielectrics depends on the geometry, on the shape and position of each. In each case the whole arrangement, cylinder, radiators and collecting wires, is designed to give the maximum surface area from which secondaries can be ejected into the active volume. A second type of radiator is the paraffin or other hydrogenous radiator placed inside a counter to serve as a source of recoil protons in fast neutron counting. Third, in slow neutron counting, radiators placed inside the counter may be coated with boron, with lithium or with uranium, so that the fragments produced by the slow neutron capture reactions will in turn

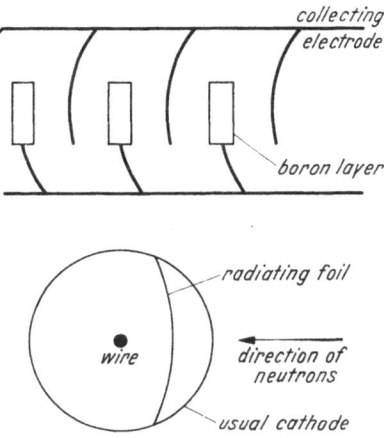

Fig. 9. Typical arrangements of internal radiators.

emerge and produce ionization in the active volume of the counter. In each case, the main point is that the radiation to be detected is not very ionizing, and therefore the efficiency can be improved by providing more substance from which secondaries can be ejected. Fig. 9 shows several arrangements of such internal radiators.

15. Gamma ray counters. The basic problem in building counters which are to detect gamma rays is to develop a design in which the probability is maximized that the gamma ray, in passing through the counter shall produce at least one electron. Since gamma rays passing through a gas will ionize only very slightly, a problem exists which differs materially from that involved in counting fast charged particles. For the case of fast charged particles, when these are moving at that speed which causes a minimum of ionization, we have the measurements already cited on specific ionization. Any charged particle moving at other speeds will ionize more than the figures cited in Sect. 4 and the efficiency of its detection will be larger than Eq. (6.1) indicates. But for gamma rays, the probability of producing ionization is much less. Indeed the probability is so low that serious thought must be given to alternative mechanisms. The most usual single mechanism is that of the ejection of electrons from the walls by the gamma ray. The usual mechanism then, of gamma ray detection depends on this probability of electron ejection from the walls. This probability depends both on the material of the wall and on the energy of the gamma ray.

In this discussion we shall consider a photon, or quantum of electromagnetic radiation such as for example a gamma ray or an X-ray, impinging upon a metal surface. It is known that such quanta cause the ejection of secondary electrons. Further, the efficiency of this process is known to be low, and other things being equal, the efficiency will increase with the atomic number Z of the wall and with the energy of the photon. Two procedures are at once obvious for improving the efficiency of counters for gamma-ray detection. The first is to increase the surface area of the cathode, and the second is to make the cathode of material with as high Z as possible. Metals of high Z are lead and bismuth. Both have been used successfully in counters. Further, of the cathode is made in the form of a grid of narrow mesh, the surface area of metal exposed to the interior can be increased somewhat. Similarly, by using a corrugated or other design for the cathode, the area can be slightly augmented. Finally, actual radiators (see Fig. 9) can be placed inside the counter itself to give additional surfaces from which the gamma rays can eject secondary electrons.

No simple analyses other than empirical data on electron emission in counters over a wide range of frequencies have as yet been made, but experience on ways to produce good counters for various wavelength intervalls does exist. Thus for example, X-ray counters with bismuth cylinders have been used by HART [10] and his colleagues, who reports efficiencies of of 0.7% at 0.7 Mev, increasing to 2% at 3 Mev. Lead, brass and aluminum cathodes for gamma rays have been used by BRADT [3] and colleagues. MAIER-LEIBNITZ [21] reports that lead cylinders showed uniform sensitivity in the range 0.15 to 1.5 Mev, while with brass the sensitivity was proportional to the energy in the interval from 0.1 to 3 Mev. At low energies, below 0.1 Mev, he found a thin cylinder to tin to be satisfactory. Dependance on wall thickness was studied by SUFFCZYNSKI [29] who found that, as the gamma ray energy increases, greater sensitivity is obtained by increasing the wall thickness. He used lead and brass cylinders, and gamma rays from natural radioactive substances.

16. Directional counters. The usual method of obtaining directional properties with counters is to make use of the technique of coincident discharges. This method was introduced by ROSSI [27] and TUVE [35] about twenty five years ago. Two or more counters are arranged in a line. The electronic circuits receiving the pulse are so adjusted that no pulse is registered unless all the counters discharge. If the counters are in a line, then the line defines the path of the particle being counted. Certain generalizations are at once apparent. The counters do not need to determine a straight line if information about a curved or bent path is sought; they merely define the path. In the case of complex events, a set of counters may include one group in line and another out-of-line. It has long been realized that great flexibility exists, and the possible arrangements are limited more by the ingenuity of the experimenter than by the characteristics of the counters themselves.

The electronic procedure in arranging a coincidence circuit is very simple. A number of possible circuits have been devised, which are discussed elsewhere, and here we shall mention only one, the Rossi type, which illustrates the point. Consider two, or indeed n vacuum tubes, the plates of which are connected together. Suppose further that these tubes are all operating with the grid at or close to cathode potential, so that the tubes are conducting. Let the grids receive signals from the wires of their respective counters. Then since the signal results from the arrival of electrons, it will be a negative pulse. The tubes will cease to conduct. Before the signal arrives, the potential of the common point at the plates of all the tubes will be determined by the value of the B voltage divided in the ratio of the resistor R to the parallel combined resistance in each of the tubes when conducting. The net (reciprocal of the) resistance of a group of tubes in parallel each of resistance R_i being given by $\sum 1/R_i$, it is evident that if one tube becomes non-conducting its resistance goes to infinity and the term corresponding to it goes to zero. Should the resistances already be fairly high, say of the order of the load resistor R, the potential of the plate undergoes only a small change when only one of the tubes becomes nonconducting, but undergoes a very large change if all the tubes cease to conduct. The user of such a circuit should compute the actual changes which he may expect to find, in order to adjust the following circuits so that they will not respond when a signal is produced by any other than the desired number of tubes produce signals.

Further, it is evident that such coincidence arrangements may be used with many modifications. Coincidence can be arranged between Geiger counters and proportional counters or other devices, such as scintillation counters. In the latter case it must be kept in mind that the pulse lengths may be different, and pulse-lengthening precedures may need to be used. The counter arrangement may be used to trigger other devices such as cloud chambers.

Another useful modification consists in the so-called "anticoincidence" arrangement, in which the condition is imposed that an output pulse is obtained only when one or more "anti-counters" do not discharge. It is at once evident that to accomplish such operation, we need only reverse the sign of the pulse, as for example by inserting one more stage in the vacuum tube amplifier, so that when pulsed a nonconducting tube conducts momentarily. If the plate of such a tube is connected to the common point, then the output signal will not appear if the anti-tube receives a pulse. Such anti-coincidence operation is especially useful in shielding, when for example it is desired to exclude cosmic radiation from a counting operation, as in the case of C^{14} counting. The anti-counters in effect desensitize the circuit during the time that a cosmic ray passing through might produce a count. Fig. 10a shows the arrangement of a conventional coincidence arrangement determining the line from which a particle may have come. Fig. 10b shows an out-of-line arrangement for studying a complex cosmic ray event. Fig. 10c shows the combined use of a set of counters, a proportional counter, and a trigger arrangement; Fig. 10d shows an anticoincidence arrangement which in effect determines that a particle enters an experimental arrangement but does not emerge, Fig. 10f shows an anticoincidence guard arrangement for eliminating cosmic ray effects. These arrangements are typical, and many other modifications have been used or will suggest themselves for particular experiments.

A further modification of the coincidence technique (see Fig. 10e) comes from the use of delay-lines. A pulse can be delayed, and fed into a circuit at a desired time after (or before, by reversing the position of the delay-line) a given event. Such arrangements have been used in studying meson-decay, where for example it was desired to establish that a particle entered an arrangement and that a certain time afterwards a meson decayed into an electron which could be countered in the delayed counters.

One condition usually present is that if a particle is to take part in coincidence counting the particle must survive. A neutron will be used up if it is captured, and hence a slow neutron counter making use of a capture reaction will not be suitable for more than a single component. The same general considerations apply to photons in the ultraviolet, where the energy is sufficient to produce one electron but then the photon does not survive to produce others. Similarly, coincidence counters for X-ray photons have a problem, in that since efficiencies are much less than unity, coincidence arrangements will have efficiencies which

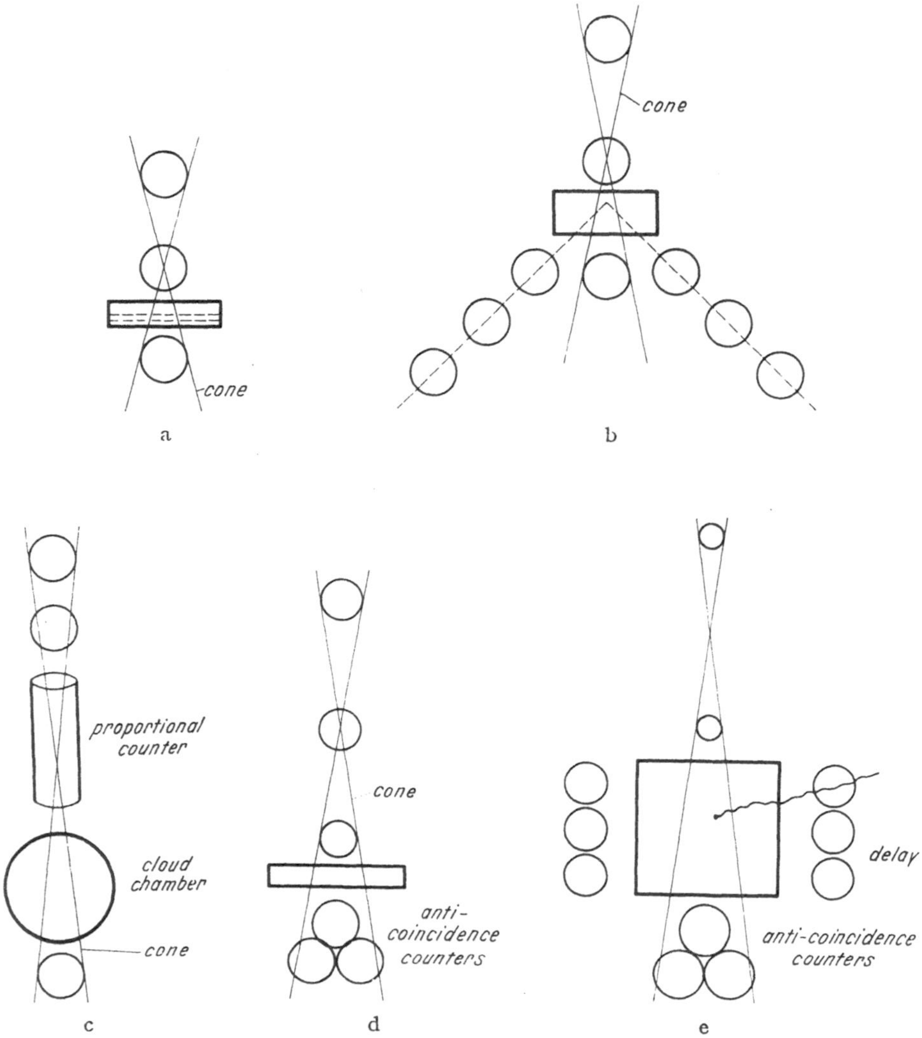

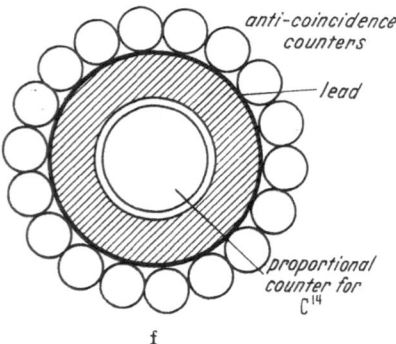

Fig. 10a—f. (a) Conventional directional coincidence arrangement; cone defined by extreme counters passes through sample of variable thickness for absorption studies. (b) Typical out-of-line arrangement to study complex events and secondary production. (c) Typical arrangement to use counters as triggers and also for measuring specific ionization in proportional counters and or cloud chambers. (d) Use of anti-coincidence units to establish that entity enters experimental arrangement but does not emerge. (e) System used in meson-decay study; top counters determine direction of meson; anticoincidence counters determine that meson stops in absorber. Counters at side are connected through variable delay line and measure time before decay-electron emerges. (f) Arrangement for shielding radiocarbon counter. Proportional counter is in center. It is surrounded by a lead shield to reduce local radioactive background. Outside this is a ring of anticoincidence counters which eliminate most of the cosmic ray background.

depend on multiplying the two or more already small quantities. If the efficiency of a gamma ray counter is 1%, then a threefold coincidence arrangement for such gamma rays will have an efficiency of 10^{-6}.

It is possible to build single counters with directional characteristics. Aside from the obvious directional properties due to windows or outside shields, another method exists. It is known that electron emission can be asymmetric, and is a function of Z. High energy photons have been found by BRAGG and MADSEN [2] to produce more electrons when entering than when leaving material of high Z, while the opposite is true for material of low Z. Thus if a counter cylinder is made by cutting it in half longitudinally, and having half lead and half aluminum, such an arrangement will be more efficient when the aluminum half faces the direction of arrival of the gamma rays. Fig. 11 shows such a device which was used by RAJEWSKI [28].

17. Procedures to shorten receiving times. We shall recapitulate points we have discussed above, and summarized those which tend to shorten resolving times. These are (a) the use of gases in which mobilities are larger, (b) operation at higher voltages, (c) the use of potential reversing circuits, (d) these use of counters in the proportional region with the consequently higher gain in the amplifiers, (e) the use of beads on the wire to limit the lateral spread of the discharge.

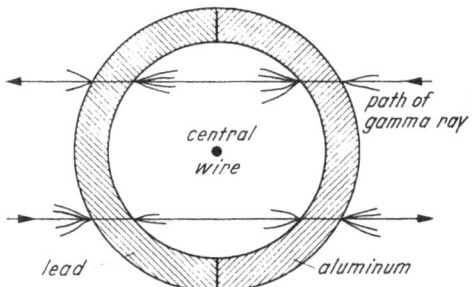

Fig. 11. Directional counter, using differential emission of secondary electrons from lead or aluminum by gamma rays. Detection efficiency is greater if electron enters from right side, as shown by upper arrow. Short irregular tracks represent secondary electron emission.

Actually the first two factors do not permit resolving times to shorten by much. The various possible gases do have a few percent different mobilities, but none differ by large factors. Similarly, one must operate the counter on its plateau and as this is only at most a few hundred volts in length, no great improvement is obtainable here either. Further, operation at high overvoltage uses up the quenching constituent faster. We may therefore disregard the first factors for all practical purposes.

The third factor we have discussed above. It has been shown that this does in fact permit speeding up the counter by a factor of ten, and perhaps by more, if care is taken. The action is however at the expense of greater electronic complexity. Hence, if the electronic complexity is not too servere an argument against the use of potential reversing circuits, considerable gain may be made with their aid.

Another procedure in speeding up the counting action is to operate the counter in the proportional region. In this voltage interval, the initial electrons produce an avalanche, but the avalanche does not spread the full length of the wire. Consequently, other parts of the counter remain sensitive, and a proportional counter can count at rates much faster than a Geiger counter. In this case, a necessary condition after a count is that the next ionizing event takes place somewhere else inside the sensitive volume of the counter. The first count and the first avalanche have rendered some part of the counter "dead". The size of the part involved depends on the gas amplification. It will vary all the way from being negligibly small when the gas amplification is close to unity, to becoming quite appreciable in the case of high gas amplification. Indeed as one enters the region of limited proportionality more and more of the counter is affected by each count. Hence, if one wants to gain much in shortening the resolving time, it is necessary to operate the counter with as low gas amplification as feasible. Naturally, a high gain amplifier is required for this purpose.

Finally, speeding up the operation has also been achieved by placing a lot of beads on the central wire, operating the counter as a self-quenching counter and accepting the discharge of one section only, between any two adjacent beads. It should however be pointed out that the main factor determining the resolving time is the migration time of the positive ions. Therefore, shortening the effective length of the counter again does not speed up the operation by any large factor.

18. Time lags. An important feature of counter operation is the time lag. This is defined as the interval of time between the instant at which the initial ionizing event takes place and the instant at which the pulse upon the central wire system first becomes observable. This time lag is not to be confused with the recovery time we have discussed before.

Clearly, if the primary ionizing event results in the production of one or more electrons, and these move in to the wire, and in its neighborhood initiate an avalanche, the factors involved in the time lag will be (1) the length of time taken for the electron or electrons to traverse the counter, (2) the length of time needed for the avalanche to build up, and (3) any electrical factors determining the determining the amplitude of the pulse which has to appear on the central wire before it is observed.

The electron transit times have been measured by J. Heirtzler [12] and others. The time were found to be of the order of less than microseconds. Further, the buildup time for the avalanche is also very fast, and hence this is not usually the determining factor. However, if an electron should be captured to form a negative ion while it is on its way in to the wire, it will of course move at the velocity characteristic of ions, which is slower than that of electrons by factors of thousands. Hence if this happens at all, notable lags are found. Such lags, due to negative ion formation have been observed by the Montgomerys [20] and by Laufer [19] who used oxygen in counters and measured the time lags due to the electron capture effects. Lags of several, and up to hundreds of microseconds can be observed, depending upon circumstances. This effect has been discussed above and is repeated here only for completeness.

19. Factors increasing the useful life of the counter. In recapitulating the factors which will tend to increase the useful life of counters, we shall divide them first into (a) manufacturing procedures, (b) design considerations, and (c) procedures to be followed when using the counters.

First, with regard to manufacture procedures, we should point out that the problem has some similarities to that of producing good radio tubes. A vast amount of manufacturing experience in this domain already exists. In summary we should state that radio tubes are very thoroughly pumped and baked, such that the vacuum in them may be expected to be both good and permanent. The filling gases should be pure, the insides of the counter as clean as possible. All these procedures will tend to minimize subsequent outgassing and other changes in the counter. Materials which are chemically active should be avoided, as should materials which are hard to clean or outgas. Glowing the wire is recommended where feasible.

In considering the design, perhaps the first problem to be considered is the quenching procedure to be used. If self-quenching is indicated and poor plateau flatness can be tolerated, halogen quenching may be used. In general such gases as tend to polymerize readily, such as for example methane, should be avoided as quenching or as main constituents, and also such gases as tend to form negative ions. Relatively heavy vapors at concentrations well below that at which condensation at room temperature takes place are desirable, for example ethyl acetate.

In using counters, several considerations may be kept in mind. For self-quenched types, operation at low overvoltages is recommended. The high voltage should be disconnected from the counter when the counter is not in use. Protective resistors should be used whenever possible, to limit the flow of current if the voltage is inadvertently made too high. If the counter should go into continuous discharge, the high voltage should be disconnected as fast as possible.

References.

[1] BOTHE, W., and H. KOLHÖRSTER: Z. Physik 56, 571 (1929).
[2] BRAGG, W., and MADSEN: Phil. Mag. 16, 918 (1908).
[3] BRADT, H. et al.: Helv. phys. Acta 19, 77 (1946).
[4] BROWN, S. C., and C. MARONI: Rev. Sci. Instrum. 21, 241 (1950).
[5] BERETTA, E., and A. ROSTAGNI: Nuovo Cim. 6, 391 (1949).
[6] DRUYVESTYN, M. J., and F. M. PENNING: Rev. Mod. Phys. 12, 87 (1940).
[7] FRIEDLAND, S. S.: Phys. Rev. 74, 898 (1948). — FRIEDLAND and KATZENSTEIN: Phys. Rev. 84, 591 (1954).
[8] GEIGER, H., and E. RUTHERFORD: Phil. Mag. 24, 618 (1912).
[9] GEIGER, G., and O. KLEMPERER: Z. Physik 49, 753 (1928).
[10] HART, R. J., K. RUSSELL and M. STEFFEN: Phys. Rev. 81, 460 (1951).
[11] HUBER, P., F. ALDER and E. BALDINGER: Helv. Phys. Acta 19, 207 (1946); 20, 73 (1947).
[12] HEIRTZLER, J.: Rev. Sci. Instrum. 25, 243 (1954).
[13] KORFF, S. A., and R. D. PRESENT: Phys. Rev. 65, 274 (1944).
[14] KORFF, S. A.: Electron and Nuclear Counters, 2nd ed. New York: Van Nostrand 1955.
[15] KALLMANN, H., and B. ROSEN: Z. Physik 61, 61 (1930).
[16] KORFF, S. A., and A. D. KRUMBEIN: Phys. Rev. 76, 1412 (1949). — S. S. FRIEDLAND: Phys. Rev. 71, 377 (1947).
[17] KRUMBEIN, A. D.: Rev. Sci. Instrum. 22, 821 (1951).
[18] LIEBSON, S. H, and H. FRIEDMAN: Rev. Sci. Instrum. 19, 303 (1948).
[19] LAUFER, A.: Rev. Sci. Instrum. 21, 244 (1950).
[20] MONTGOMERY, C. G., and D. D. MONTGOMERY: Phys. Rev. 57, 1030 (1940).
[21] MAIER-LEIBNITZ, M.: Z. Naturforsch. 1, 243 (1946).
[22] MAZE, R.: J. Phys. Radium 7, 164 (1946).
[23] NEHER, H. V., and W. W. HARPER: Phys. Rev. 49, 940 (1936).
[24] NEHER, H. V., and W. H. PICKFRING: Phys. Rev. 53, 316 (1938).
[25] PRESENT, R. D.: Phys. Rev. 72, 243 (1947).
[26] RUTHERFORD, E., and H. GEIGER: Proc. Roy. Soc. Lond., Ser. A 81, 141 (1908).
[27] ROSSI, B.: Nuovo Cim. 8, 49, 85 (1931). — Z. Physik 68, 64 (1931).
[28] RAJEWSKY, B.: Z. Physik 120, 627 (1943).
[29] SUFFCZYNSKI, M.: Acta Phys. Polon. 10, 270 (1951).
[30] SIMPSON, J. A.: Phys. Rev. 66, 39 (1944).
[31] SPATZ, W. D.: Phys. Rev. 64, 236 (1944).
[32] SPONER, H., and E. TELLER: Rev. Mod. Phys. 13, 75 (1941).
[33] SOBERMAN, R., S. A. KORFF, S. S. FRIEDLAND and H. S. KATZENSTEIN: Rev. Sci. Instrum. 24, 1058 (1953).
[34] SHERMAN, H.: In press.
[35] TUVE, M. A.: Phys. Rev. 35, 651 (1930).
[36] TROST, A.: Z. Physik 105, 399 (1937).
[37] WEISZ, P., and W. P. KERN: Phys. Rev. 75, 899 (1949).
[38] WANTUCH, E.: Phys. Rev. 71, 646 (1947).

Scintillation and Čerenkov Counters.

By

WILLIAM E. MOTT and ROGER B. SUTTON.

With 87 Figures.

1. Introduction. A scintillation or Čerenkov counter generally consists of three parts: the medium in which the light is produced by moving charged particles; a photomultiplier tube in which the light is converted to electrons which are then multiplied up to a detectable signal; and often, but not always, a light pipe to lead the light from the scintillant or Čerenkov medium to the photomultiplier. Fig. 57 shows a typical scintillation counter. Figs. 75 and 86 show Čerenkov counters. These assemblies must be enclosed in light tight containers so that the relatively weak light flashes from the media will not be swamped by stray light. Part I contains a discussion of the properties of many of the commercially available photomultipliers and a brief treatment of light pipes. In Parts II and III the characteristics of scintillation materials are discussed, the former dealing with inorganic compounds and the latter with organics. Čerenkov counters and their uses are reviewed in Part IV, and the properties of large scintillation and Čerenkov counters for high energy β- and γ-ray spectroscopy are considered in Part V. The important characteristics of these counters which have led to such rapid widespread use include: resolving times down to the region of 10^{-9} sec; large sensitive area (counters with dimensions of the order of one meter have been constructed); proportionality; simplicity of construction; and versatility.

For spectrometry, where a large scintillation efficiency is required, inorganic scintillants, e.g., NaI(Tl) are generally used; for γ-ray work the relatively large photoelectric absorption coefficients of these phosphors are also very desirable. However, they usually have long decay times of the order of 10^{-6} sec. Thus, the organic scintillators, which as a rule have shorter decay times, down to 10^{-9} sec, are most useful for fast coincidence work with charged particles. Čerenkov counters give considerably smaller pulses than scintillation counters and are used only when advantage is to be taken of the special properties of the radiation as is discussed in Part IV.

No discussion is given of counters using gases as scintillants. These generally consist of a noble gas at several atmospheres in a container, the walls of which are coated with a wave length shifter such as 1,1,4,4-tetraphenyl-1,3-butadiene (TPB) so that the radiation will match the sensitivity curve of the photomultiplier. The decay time of the scintillation is of the order of 10^{-9} sec; the pulse height is linear with energy loss but is rather small ($\sim 10^5$ photons/Mev). These counters have been discussed in papers by EGGLER and HUDDLESTON[1], by NOBLES[2] and by FORTE[3].

To obtain an idea of the general behavior of scintillation counters, let us consider a specific example. A high energy proton traversing a plastic scintillator

[1] C. EGGLER and C. M. HUDDLESTON: I.R.E. Trans. Nuclear Sci., N.S. **3**, No. 4, 36 (1956).
[2] R. A. NOBLES: Rev. Sci. Instrum. **27**, 280 (1956).
[3] M. FORTE: Nuovo Cim. **3**, 1443 (1956).

2 cm thick will lose about 5 Mev energy. If the scintillant has an efficiency of 3%, 150000 ev will be converted to useful light; this will correspond roughly to 60000 photons. If these are collected at the photocathode of a multiplier with 5% cathode efficiency, there will be 3000 electrons produced. With a photo-multiplier gain of 10^6, the pulse at the output will consist of 3×10^9 electrons or 5×10^{-10} coulombs. The resultant pulse height will depend on the particular arrangement of phosphor, multiplier, output circuit, and efficiency of light collection; however, it is clear that generally little amplification is required after the multiplier.

I. Photomultiplier tubes.

2. Photomultiplier types and characteristics. During the ten years or so that scintillation and Čerenkov counters have been in use, considerable improvement

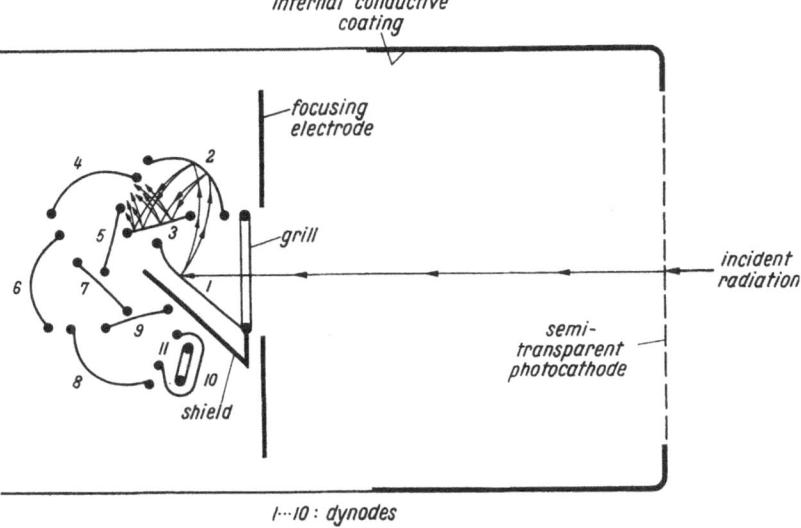

Fig. 1. RCA circular array type of dynode structure.

in photomultiplier construction has been achieved. Three companies, the Radio Corporation of America (Harrison, New Jersey, U.S.A.), Allen B. DuMont Laboratories, Inc. (Clifton, New Jersey, U.S.A.) and E. M. I. Electronics, Ltd. (Hayes, Middlesex, England), produce by far the largest number of tubes used. Tables 1 to 3, which give the characteristics of the multipliers available from these companies, thus show the tubes which are easily available for counter construction. The values listed are those given by the manufacturers and are, for the most part, average values. Considerable variation is found in such characteristics as sensitivity, spectral response, dark current, and maximum voltage before break-down from one tube to another of the same model. For critical applications it is necessary to select from a batch the one which best fits the requirements.

Besides the above companies there are others in Europe producing tubes[1]. These are: the Laboratoires d'Electronique et de Physique Appliqués and the Compagnie de Télégraphie Sans Fils, both in Paris, France; the Eidgenössische Technische Hochschule in Zürich, Switzerland. Several types are produced in

---

[1] G. A. Morton: I.R.E. Trans. Nuclear Sci., N.S. **3**, No. 4, 141 (1956).

Moscow, U.S.S.R.[2]. Although these companies produce fewer numbers, tubes of outstanding quality can sometimes be obtained, such as the 17-stage high-gain, low noise, high-vacuum tube and others developed by SCHAETTI[3] at the Eidgenössische Technische Hochschule.

The multipliers listed in Tables 1 to 3 can be divided into four groups as far as dynode structure is concerned. These structures are shown in Figs. 1 to 4. The circular array (Fig. 1) has been used by RCA for most of its models and has the property of providing high electric fields to collect electrons from each dynode. Thus, the transit time is short and the transit time spread small. Fig. 2 shows the "in-line" structure used in the 14-stage RCA-6810 which also is characterized by small transit time spread. The box structure (Fig. 3) is used by DuMont in all their tubes. The transit time spread is greater than for the previous structures. However, there is in general a preference for these tubes where energy resolution is important and time resolution is not. The "venetian-blind" structure of most of the EMI tubes (Fig. 4) has a moderately fast response. The pulse height spread of these seems to be as small as that of the DuMont tubes (i.e., the tube width is about 4% for a light pulser simulating Cs^{137} γ-rays on NaI(Tl)].

Most of the photomultipliers have a spectral response, designated S-11 in the tables, which is obtained with a CsSb cathode inside a glass envelope. It should be noted that all three companies now manufacture types with ultraviolet transmitting windows which extend the response down to about 2000 Å rather than the 3500 Å for the glass window. Fig. 5 shows two typical response curves as given by RCA. The S-13 response is for the RCA-6903, in which CsSb is deposited on a fused silica window.

Stray magnetic fields can have quite large effects on the output current of most photomultipliers. Shielding to reduce the earth's or other fields to about 0.1 gauss should be provided, particularly if related readings are to be taken with the tube in various positions. In most cases μ-metal shields can be obtained from the manufacturer. These shields reduce the external fields to about 4%

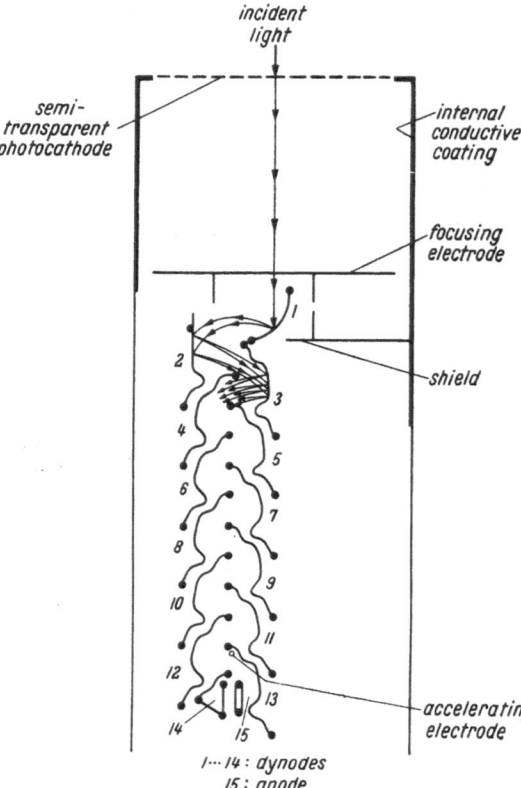

Fig. 2. RCA linear type of dynode structure.

incident light

semi-transparent photocathode

internal conductive coating

focusing electrode

shield

accelerating electrode

1···14: dynodes
15: anode

---

[2] Y. A. NEMILOV, V. M. OVCHINNIKOV, A. N. PISAREVSKY and E. D. TETERIN: J. Nuclear Energy **4**, 358 (1957).

[3] N. SCHAETTI and W. BAUMGARTNER: Z. angew. Math. u. Phys. **1**, 268 (1950). — Helv. phys. Acta **23**, 524, 869 (1950). — Le Vide **1951**, No. 34—35, 1041.

for fields up to 300 oersted, thus it is often necessary to provide additional shielding when the photomultipliers are to be used in the vicinity of magnets.

The voltage values to be applied to the various dynodes are usually indicated by the manufacturer. However, for optimum performance these should be individually adjusted for each tube. The tube can be operated with the cathode below ground or with the anode above ground. If any object at anode potential touches the outside of the glass envelope, spurious pulses may appear. This effect can be prevented by wrapping the assembly with a metallic shield held at cathode potential or by physically isolating the tube and phosphor system. STUMP and TALLEY[4] suggest that by providing separate, high current, voltage

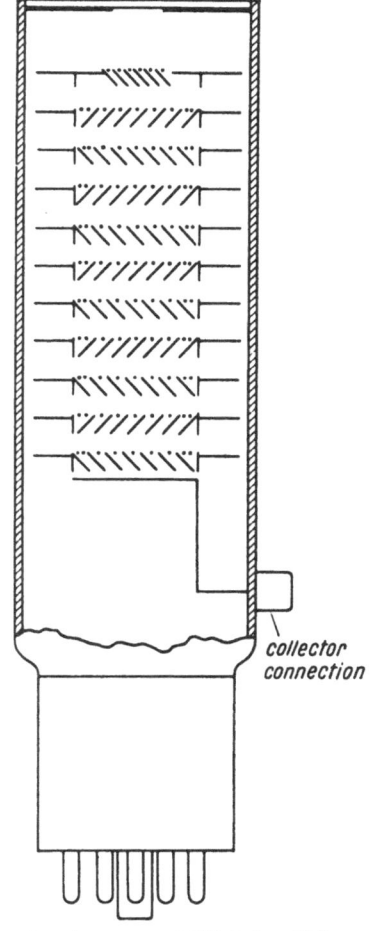

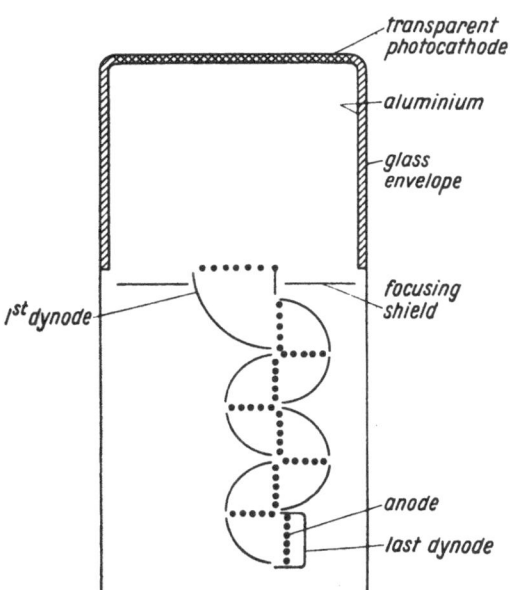

Fig. 3. Dynode structure of a DuMont photomultiplier.

Fig. 4. Dynode structure of EMI photomultiplier.

supplies for the last two or three dynodes, the breakdown voltage of a multiplier can be considerably increased. In one case they were able to increase the voltage, by this means, from 1200 to 1700 volt.

Although the DuMont tubes are favored for spectroscopic work, it appears that for some of these the gain depends on counting rate; this effect seems to be smaller for the RCA tubes tested. A fairly slow change with counting rate was first reported by CALDWELL and TURNER[5]. With a DuMont-6292 and a NaI(Tl) crystal, the pulse height of the pair peak for 4.5 Mev γ-rays increased at a rate of about 15% per hour after the radiation on the crystal was increased from

[4] R. STUMP and H. E. TALLEY: Rev. Sci. Instrum. **25**, 1132 (1954).

[5] R. L. CALDWELL and S. E. TURNER: Nucleonics **12**, No. 12, 47 (1954).

Table 1. *Characteristics of RCA photomultipliers.*

Type	1P21 (a)	1P28 (a)	2020 (b,d)	5819 (c)	6199 (d)	6342 (d)	6372 (e)	6655 (d)	6810 (d)	6903 (d)
Minimum photocathode size (in.)	15/16 × 5/16	15/16 × 3/16	1 1/2	1 11/16	1 1/4	1 11/16	4 1/8 × 2 1/4	1 11/16	1 11/16	1 5/8
Photocathode window	—	Corning No. 9741	—	—	—	—	—	—	—	Fused silica
Photocathode material	Cs–Sb	Cs–Sb	Cs–Sb	Cs–Sb	Cs–Sb	Cs–Sb	Cs–Sb	Cs–Sb	Cs–Sb	Cs–Sb
Spectral response	S-4	S-5	S-11	S-11	S-11	S-11	S-11	S-11	S-11	S-13
Wave length of max. response (Å)	4000±500	3400±500	4400±500	4400±500	4400±500	4400±500	4400±500	4400±500	4400±500	4400±500
Median cathode efficiency * (μa/lumen)	40	40	60	50	45	60	35	50	60	60
Avg. peak quantum efficiency (%)	12.5	18.2	11.2	11.2	10.9	13.5	7.4	11.2	13.5	13.2
Cathode dark current at 25°C (10^{-15} A)	0.02	0.03	0.35	0.35	0.18	0.42	3.5	0.35	0.36	0.40
Dynode material	Cs–Sb	Cs–Sb	Cs–Sb	Cs–Sb	Cs–Sb	Ag–Mg	Cs–Sb	Cs–Sb	Ag–Mg	Ag–Mg
Number of stages	9	9	10	10	10	10	10	10	14	10
Max. volts per stage	~105	105	125	105	105	125	100	~105	~125	~125
Avg. multiplier gain	2×10^6(f)	1.25×10^6(f)	5.6×10^5(g)	5×10^5(f)	6×10^5(f)	6×10^4(g)	6×10^5(f)	5×10^5(f)	12.5×10^6(h)	4×10^6(f)
Max. peak anode current (ma)	0.1	0.5	2.0	0.75	0.75	2.0	0.75	0.75	—	—
Transit time (mμsec)	13(?)	13 (?)	25	30	25	27	—	30	47	—
Transit time variation over photocathode (mμsec)	—	—	4	4	3.5	4	—	4	5–6	—
Transit time spread (mμsec) (center of cathode)	—	—	—	—	—	—	—	—	—	—
Seated bulb length (in.)	3 1/8	3 1/8	4 1/4 ± 3/16	4 7/8 ± 3/16	3 7/8 ± 3/16	4 7/8 ± 3/16	7 1/4	4 7/8 ± 3/16	6 11/16 ± 3/16	5 5/8 ± 3/16
Max. bulb diameter (in.)	1 3/16	1 3/16	2 1/8 ± 1/16	2 1/8 ± 1/16	1 9/16	2 1/8 ± 1/16	2 9/16	2 1/8 ± 1/16	2 1/8 ± 1/16	2 1/8 ± 3/32

* Tungsten filament lamp at color temperature of 2870° K. The light flux is 0.01 lumen and 200 V are applied between cathode and all other electrodes connected together as one anode. Load resistor has a value of 0.01 megohm.

(a) Interior, non-transparent photocathode. Side-on type.
(b) Similar to RCA-6342 but with low-resistivity photocathode.
(c) Head-on tube; not flat-faced.
(d) Head-on tube; flat-faced.
(e) Photocathode on side of bulb wall.
(f) Supply voltage 1000 V.
(g) Supply voltage 1500 V.
(h) Supply voltage 2000 V.

Table 2. *Characteristics of DuMont photomultipliers.*

Type	6292	6363	6364	6365*	6467	6291*	K1328*	K1384	6935
Minimum photocathode size (in.)	1 1/2	2 1/8	4 9/16	1/2	1	1 1/4	14	11	1/4
Photocathode window	lime glass	lime glass	lime glass	lime glass	lime glass	lime glass	lime glass	lime glass	lime glass
Photocathode material	Cs–Sb	Cs–Sb	Cs–Sb	Cs–Sb	Cs–Sb	Cs–Sb	Cs–Sb	Cs–Sb	Cs–Sb
Spectral response	S-11	S-11	S-11	S-11	S-11	S-11	S-11	S-11	S-11
Wave length of max. response (Å)	4400±500	4400±500	4400±500	4400±500	4400±500	4400±500	4400±500	4400±500	4400±500
Avg. cathode efficiency (μa/lumen)	60	60	60	50	60	60	45	45	45
Avg. peak quantum efficiency (%)	2	4.5	8	~11	~11	~11	~9	~9	~9
Cathode dark current at 22°C (10^{-15} A)	~11	~11	~11	0.2	0.8	1.4	100	70	0.16
Dynode material	Ag–Mg	Ag–Mg	Ag–Mg	Ag–Mg	Ag–Mg	Ag–Mg	Ag–Mg	Ag–Mg	Cs–Sb
Number of stages	10	10	10	6	10	10	12	12	10
Max. volts per stage	145	150	150	170	150	150	105	105	105
Avg. multiplier gain (10^5 V per stage)	2×10^6	2×10^6	2×10^6	3×10^5	2×10^6	2×10^6	5×10^3	5×10^3	3×10^5
Max. peak anode current (ma)	~25	25	25	5	25	25	25	25	5
Transit time (mμsec)	~60	~60	~20	—	~20	~20	~60	—	—
Transit time variation over photocathode (mμsec)	3	~20	—	—	—	—	—	—	—
Transit time spread (mμsec) (center of cathode)	~20	—	—	—	—	—	—	—	—
Seated bulb length (in.)	4 7/8 ± 3/16	5 7/8 ± 3/16	6 7/8 ± 3/16	2 9/16 ± 1/16	4 ± 3/16	4 1/8 ± 1/4	13 1/4 ± 1/4	11 1/2 ± 1/4	4 3/4 ± 1/8
Max. bulb diameter (in.)	2 1/8 ± 1/16	3 ± 3/32	5 1/4 ± 3/32	3/4 ± 1/16	1 1/4 ± 1/16	1 1/2 ± 1/16	15 7/8 ± 1/4	12 1/2 ± 1/4	3/4 ± 1/16

* Denotes tubes are in the last stages of development at the time of completion of this table.

Table 2. (Continued.)

Type	*K* 1306	*K* 1386*	*K* 1209	*K* 1213	*K* 1295	*K* 1161	*K* 1162	*K* 1382	*K* 1193
Min. photocathode size (in.)	$1^1/_2$	21	$4^1/_2$	$2^3/_4$	$1^1/_2$	$1^1/_2$	$1^1/_4$	$^1/_2$	$^1/_2$
Photocathode window	quartz	lime glass	lime glass	lime glass	lime glass	lime glass	lime glass	—	—
Photocathode material	Cs—Sb	Cs—Sb	Cs—Sb	Cs—Sb	Cs—Sb	Cs—Sb	Cs—Sb	Cs—Sb	Cs—Sb
Spectral response	—	*S*-11	*S*-11	*S*-11	*S*-11	*S*-11	*S*-11	*S*-11	*S*-11
Wave length of max. response (Å)	4400 ± 500	4400 ± 500	4400 ± 500	4400 ± 500	4400 ± 500	4400 ± 500	4400 ± 500	4400 ± 500	4400 ± 500
Avg. cathode efficiency (μa/lumen)	60	—	60	60	60	60	60	50	50
Avg. peak quantum efficiency (%)	~15	—	~11	~11	~11	~11	~11	~9	~9
Cathode dark current at 22° C (10^{-15} A)	2	200	8	4.5	2	2	1.4	—	—
Dynode material	Ag—Mg	Ag—Mg	Ag—Mg	Ag—Mg	Ag—Mg	Cs—Sb	Cs—Sb	Cs—Sb	Ag—Mg
Number of stages	10	12	12	12	12	10	10	10	10
Max. volts per stage	150	105	95 (avg.)	95 (avg.)	95 (avg.)	105	105	105	105
Avg. multiplier gain (105 V per stage)	2×10^5	7×10^5	2×10^5	2×10^5	2×10^5	9×10^5	9×10^5	3×10^5	1.5×10^5
Max. peak anode current (ma)	25	25	25	25	25	25	25	1.0	5
Transit time (mμsec)	~60	—	—	—	~60	~60	~60	—	—
Transit time variation over photocathode (mμsec)	3	—	—	—	3	3	—	—	—
Transit time spread (mμsec) (center of cathode)	~20	—	—	~20	~20	~20	~20	—	—
Seated bulb length (in.)	$4^7/_8 \pm ^3/_{16}$	$17 \pm ^1/_4$	$8^1/_2 \pm ^1/_4$	$6^3/_8 \pm ^1/_8$	$6 \pm ^3/_{16}$	$4^7/_8 \pm ^3/_{16}$	$4^1/_4 \pm ^1/_4$	$5^1/_{16} \pm ^1/_8$	$5^1/_{16} \pm ^1/_8$
Max. bulb diameter (in.)	$2 \pm ^1/_{16}$	$21 \pm ^1/_4$	$5^1/_4 \pm ^3/_{32}$	$3 \pm ^1/_8$	$2 \pm ^1/_{16}$	$2 \pm ^1/_{16}$	$1^1/_2 \pm ^1/_{16}$	$^3/_4 \pm ^1/_{32}$	$^3/_4 \pm ^1/_{32}$

* Denotes tubes are in the last stages of development at the time of completion of this table.

Table 3. *Characteristics of EMI photomultipliers.*

Type	6095 *B*	6097 *B*	6097 *F*	6097 *S*	9514 *B*	9514 *S*	9502 *B*	6256 *B*	6255 *B*	6094 *B*	6099 *B*	6099 *F*	6098 *B*(a)	9526(a,d)
Photocathode size (in.)	$1^3/_4$	$1^3/_4$	$1^3/_4$	$1^3/_4$	$1^3/_4$	$1^3/_4$	0.4	0.4	$1^3/_4$	0.4	$4^1/_2$	$4^1/_2$	$^7/_8$	$^7/_8$
Photocathode window (b)	pyrex	pyrex	pyrex	pyrex	pyrex	pyrex	kodial	quartz	quartz	kodial	pyrex	pyrex	soda	quartz
Photocathode material	Cs—Be—Ag	Cs—Sb	Cs—Sb	Cs—Sb	Cs—Sb	Cs—Sb	Cs—Sb	Cs—Sb	Cs—Sb	Cs—Sb	Cs—Sb	Cs—Sb	Cs—Sb	Cs—Sb
Spectral response	*S*-11 (?)	*S*-11 (?)	*S*-11 (?)	*S*-4 (?)	*S*-11 (?)	*S*-4 (?)	*S*-11 (?)	*S*-11 (?)	*S*-11 (?)	*S*-11 (?)	*S*-11 (?)	*S*-11 (?)	*S*-11 (?)	—
Wave length of max. response (Å)	4600	4300	4300	4100	4300	4100	4300	4300	4300	4300	4300	4300	4300	4300
Avg. Cathode Efficiency (μ a/lumen)	30	55	55	30	55	30	45	45	55	45	40	40	55	55
Avg. peak quantum efficiency (%)	5	12	12	8	12	8	10	10	12	10	9	9	12	12
Cathode dark current (c) (10^{-15} A)	10	10	40	10	10	10	11	1	10	1	100	400	10	10
Dynode material	Cs—Sb	Cs—Sb	Cs—Sb	Cs—Sb	Cs—Sb	Cs—Sb	Cs—Sb	Cs—Sb	Cs—Sb	Cs—Sb	Cs—Sb	Cs—Sb	Cs—Sb	Cs—Sb
Number of stages	11	11	11	11	13	13	13	13	13	11	11	11	11	11
Max. volts per stage	180	180	180	180	180	180	180	180	180	180	180	180	180	180
Avg. multiplier gain (160 V per stage)	2×10^7	1.1×10^7	0.9×10^6	1.3×10^7	1.1×10^8	1.3×10^8	1.3×10^8	1.3×10^8	1.1×10^8	1.3×10^7	1×10^7	0.6×10^6	1.1×10^7	1.1×10^7
Max. peak anode current (ma)	10	10	10	10	10	10	10	10	10	10	10	10	4	4
Transit time at 180 V per stage (mμsec)	30	30	30	30	36	36	36	36	36	30	—	—	—	—
Transit time variation over photocathode (mμsec)	—	—	—	—	—	—	—	—	—	—	—	—	—	—
Transit time spread for single electron at 180 V per stage (mμsec)	10	10	10	10	~10	~10	~10	~10	~10	10	10	10	—	—
Seated bulb length (in.)	$4^1/_2$	$4^1/_2$	$4^1/_2$	$4^1/_2$	4.8	4.8	4.8	4.8	4.8	$4^1/_2$	9	9	$4^1/_2$	$4^1/_2$
Max. bulb diameter (in.)	2	2	2	2	2	2	2	2.04	2.04	2	$5^1/_8$	$5^1/_8$	$1^1/_8$	$1^1/_8$

(a) Box and grid type dynode structure.
(b) All windows flat.
(c) Maximum dark current at minimum overall gain.
(d) Tentative data.

13 mr per hour to 45 mr per hour. A more rapid effect has been reported by BELL, DAVIS and BERNSTEIN[6]. Fig. 6 shows, for four DuMont-6292 multipliers, the manner in which the height of the Cs[137] peak changed when the counting rate was changed by moving the source position. It is interesting that the slope can have either sign. Fig. 7 gives the product of particle flux times energy vs. pulse

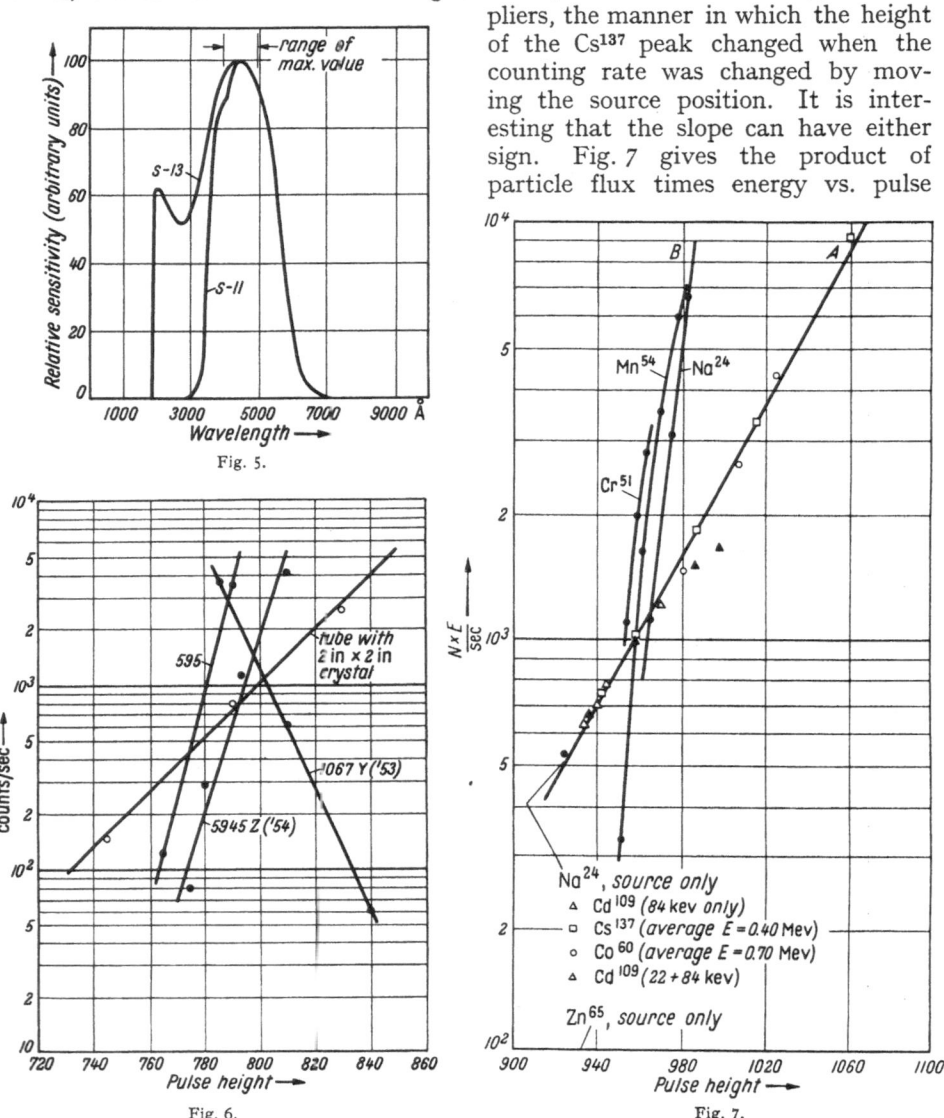

Fig. 5.

Fig. 6.

Fig. 7.

Fig. 5. Typical photocathode response curves for equal values of radiant power at all wavelengths.

Fig. 6. Shift in pulse height with counting rate for four DuMont-6292 photomultipliers (BELL, DAVIS and BERNSTEIN[6]).

Fig. 7. Plot of $N \times E$ av/sec vs. pulse height; curve A was obtained with a DuMont-6292, curves B with a DuMont-6363 photomultiplier. In curve A the 2.75-Mev line of Na[24] was observed, as additional counts of Cd[109], Cs[137], and Co[60] were added individually. In curves B the 1.12-Mev line of Zn[65] was observed as additional counts of Cr[51], Mn[54], and Na[24] were individually added (BELL, DAVIS and BERSTEIN[6]).

height for a DuMont-6292 and -6363. For a given tube the determining quantity is the flux of photons on the cathode. In the worst cases, a 20% change

---

[6] P. R. BELL, R. C. DAVIS and W. BERNSTEIN: Rev. Sci. Instrum. 26, 726 (1955).

in pulse height occurred for a factor of several hundred in counting rate. The effect described here was rapid, the full change occurring in 30 sec or less; however, a smaller, long-time shift was also observed. Illumination of the photocathode with weak light gave rise to similar effects, and the resolution was also affected. RCA-6342 and -5819 tubes showed the effect to a much smaller or zero degree. The experiments indicated that the gain-change is a dynode effect.

Measurements have indicated that response of photomultipliers with semi-transparent photocathodes may be limited by resistance of the cathode. ALLEN[12] has investigated the effect for an RCA-5819 and for several RCA-C7164 tubes. He illuminated the photocathode with a rectangular light pulse of variable intensity and with duration of 120 μsec. The current to the first dynode increased with light intensity up to 0.5 μa; at higher intensities the current pulse showed an initial peak, which increased with intensity of light, followed by a flat plateau which did not increase. The peak presumably represented the initial flow of current from the cathode whereas the lower plateau occurred due to a voltage drop across the cathode which caused the central portions to be ineffective for high currents. The peak appeared for the RCA-5819 at 20 μa, and generally at a few μa for the RCA-C7164 tubes. The above results were obtained with 135 volt between cathode and first dynode.

The resistance of these photocathodes depends on the temperature and past history of the tubes. A multiplier stored in darkness for a number of hours shows a resistance which decreases with time as current is drawn from the cathode; the rate of decrease is greater the higher the current and the higher the temperature. These effects have been investigated by WIDMAIER and ENGSTROM[13]. At low light levels, such as obtained in scintillation counters and at room temperature, this effect may not cause loss of sensitivity. However, a loss of sensitivity for multipliers cooled to temperatures below 150° K has been reported by MURRAY[14] and by HAHN and ROSSEL[15]. The former refers to DuMont-6292's while the latter to RCA-5819's. This effect is perhaps related to an increase of cathode resistance at low temperatures*.

Conflicting results on the variation of multiplier gain with temperature, around room temperature, have been reported by several investigators. SELIGER and ZIEGLER[16] reported an increase of response with temperature of 0.13% per degree centigrade for three RCA-5819's and an increase of 0.2% per degree centigrade for two DuMont-6292's, while a third 6292 gave results somewhat in between. KINARD[17] found a decrease in sensitivity of from about 0.1 to 0.5% per degree centigrade for DuMont-6292's, RCA-5819's and RCA-6655's; he further concluded that the gain change was associated with the dynode structure rather than with the cathode. BALL, BOOTH and MACGREGOR[18] have likewise measured

[12] J. S. ALLEN: Los Alamos Scientific Laboratory (U.S.) Report No. LA-1613 (Unpublished) (1953).

[13] W. WIDMAIER and R. W. ENGSTROM: RCA-Review 16, No. 1, 109 (1955).

[14] R. B. MURRAY: Oak Ridge National Laboratory (U.S.) Report No. 56-11-5 (Unpublished) (1956).

[15] B. HAHN and J. ROSSEL: Helv. phys. Acta 26, 803 (1953).

* EMI has produced a variant of a standard tube (e.g. 6097) which is reported to give standard and unchanged characteristics with the photocathode at − 180° C. Also, RCA-6342's and DuMont-6292's with a thin metallic film over the photocathode have been obtained by MURRAY (private communication). He found that these tubes operate well at liquid nitrogen temperature.

[16] H. H. SELIGER and C. A. ZIEGLER: I.R.E. Trans. Nuclear Sci., N.S. 3, No. 1, 62 (1956).

[17] F. E. KINARD: Nucleonics 15, No. 4, 92 (1957).

[18] W. P. BALL, R. BOOTH and M. H. MACGREGOR: Bull. Amer. Phys. Soc., Ser. II 1, No. 4, 183 (1956).

a decrease in response with increasing temperature for RCA-6199's which varied from 0.2 to 0.5% per degree centrigrade. These three investigations were undertaken in the temperature range about 250 to 300 °K. The discrepancies in some of the above results have not been explained.

The appearance of satellite or secondary pulses, following the main pulse, has been observed in many photomultipliers. These afterpulses can have amplitudes from those corresponding to one cathode electron up to values near that of the main pulse; however, they generally correspond to less than 10 cathode electrons. They appear a few millimicroseconds to several microseconds after the initiating pulse. The nature and origin of these pulses have been studied by Godfrey, Harrison and Keuffel[19], by Mueller, Best, Jackson and Singletary[20], by Lanter and Corwin[21], by Allen[22], by Meyer and Maier[23], and by Wells[24]. There appear to be two sources of these pulses. One is due to photons produced by the electron avalanche near the anode in the residual gas of the tube. Some of these photons may strike the cathode and produce a small pulse. These may be reduced by optical shielding between the last stages and the cathode and by the use of dynode materials such as AgMg or BeCu which can be well outgassed and which do not tend to give rise to vapor such as do CsSb surfaces. This type of afterpulse occurs at times after the main pulse corresponding to the electron transit time of the multiplier. The second source of afterpulses is due to production of positive ions near the first dynode. This effect has been most carefully studied by Allen[22]. Apparently gases trapped on the surfaces at the region of the first dynode become ionized by the cathode electrons. These ions drift to the cathode and produce electrons when they strike. The transit time for the ions depends on the charge and atomic weight but is in the vicinity of several microseconds. By measuring these times Allen has identified ions of H_2^+, O^+ or N^+, O_2^+ or N_2^+ or CO^+, mass 80 to 90 and Hg^+.

Multipliers with AgMg dynodes, which can be baked at higher temperatures than those with CsSb dynodes, are less likely to show these afterpulses. The amount of afterpulsing is a function of the particular tube as well as the type, and therefore should be investigated for any tube used in experiments where the effect can cause trouble.

At present, work aimed at the improvement of all vital elements of the tubes is in progress. New types of photocathode materials[7], such as the multialkali Sb—K—Na and Sb—K—Na—Cs surfaces, show promise of higher sensitivity; the absence of Cs in the former perhaps will lead to lower dark current and may permit the application of higher tube voltages. In order to reduce transit time spread, better design of the dynode structure and particularly of the cathode-first dynode geometry is being investigated. The new RCA-6810 has brought some improvement, however, as for most tubes, the greatest contribution to the time spread occurs in the region between cathode and first dynode; work on improvement in this direction is being carried out*. The new RCA-C7170,

[19] T. N. K. Godfrey, F. B. Harrison and J. W. Keuffel: Phys. Rev. **84**, 1248 (1951).

[20] D. W. Mueller, G. Best, J. Jackson and J. Singletary: Nucleonics **10**, No. 6, 53 (1952).

[21] R. J. Lanter and R. W. Corwin: Rev. Sci. Instrum. **23**, 507 (1952).

[22] J. S. Allen: Los Alamos Scientific Laboratory (U.S.) Reports No. LA-1459 (1952) and No. LA-1613 (1953) (Unpublished).

[23] K. P. Meyer and A. Maier: Helv. phys. Acta **27**, 57 (1954).

[24] F. H. Wells: Nucleonics **10**, No. 4, 28 (1952).

[7] A. H. Sommer: Rev. Sci. Instrum. **26**, 725 (1955). — I.R.E. Trans. Nuclear Sci., N.S. **3**, No. 4, 8 (1956).

* The RCA-6810A, which recently superseded the RCA-6810, has a curved photocathode and as a result has less spread than its predecessor.

similar to the 6810 but with a 5-inch diameter curved cathode, apparently is considerably faster than most other tubes. Investigations are also being made of new secondary emission surfaces and development work is in progress on tubes with larger photocathodes[8, 9]. The use of accelerator grids, in the in-line dynode structure, will permit higher currents to be drawn before space charge limiting occurs[9, 10]. The transmission secondary electron emission type of dynode structure described by STERNGLASS and WATCHEL[11] seems to offer the best hope for considerably better time resolution, perhaps for time interval measurements down to 10^{-10} sec.

3. Light pipes and reflectors. A part of many counters using photomultipliers is the medium which transmits the light from the scintillator or Čerenkov medium to the photocathode. In general, if good optical contact between the light source and tube can be made, it is better to avoid any light pipe*, since some intensity is always lost in the transmission. However, in many cases, it is impractical to use direct coupling; for example, if large magnetic fields are present in the region of detection or if space is a problem. For the case of large scintillators, better uniformity, and for some pulses, higher intensity will be obtained by use of a light pipe which tapers

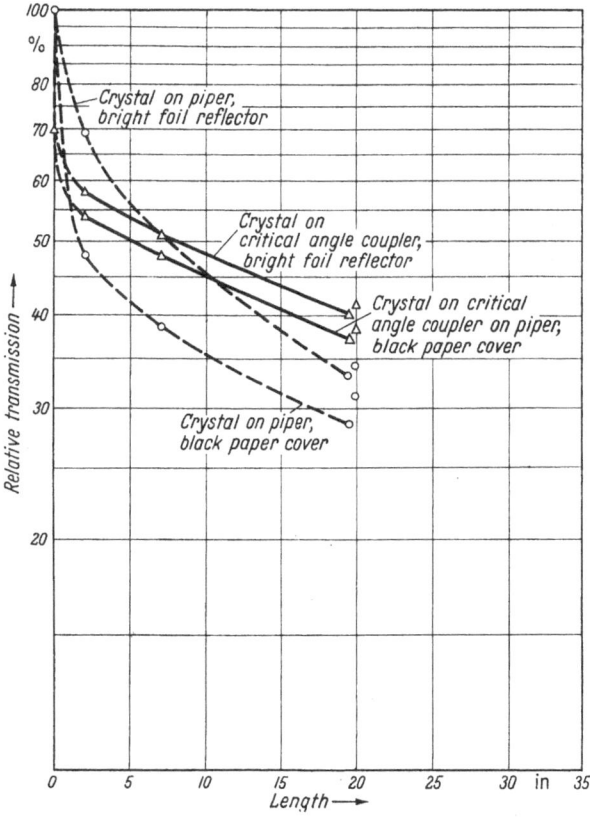

Fig. 8. Transmission characteristics of 3-in. Lucite light pipe for Cs[137] γ-rays on a ½-in thick × 1½ in diameter NaI(Tl) crystal. A DuMont 6363 photomultiplier was used (HARRIS and BELL[28]).

from the size of the viewing surface to that of the photocathode. Discussions of the properties of such systems are given by GARWIN[25], and by BRINI, PELI, RIMONDI and VERONISI[26].

Use of long Lucite light pipes and couplers to optically connect a crystal to a photocathode of equal or larger size has been studied by TOVE[27] and by

[8] B. R. LINDEN: I.R.E. Trans. Nuclear Sci. N.S. **3**, No. 4, 33 (1956).
[9] G. A. MORTON: I.R.E. Trans. Nuclear Sci., N.S. **3**, No. 4, 112 (1956).
[10] J. S. ALLEN and L. R. MEGILL: I.R.E. Trans. Nuclear Sci., N.S. **3**, No. 4, 112 (1956).
[11] E. J. STERNGLASS and M. M. WATCHEL: I.R.E. Trans. Nuclear Sci. N.S. **3**, No. 4, 29 (1956).
* Except perhaps for very thin crystals.
[25] R. L. GARWIN: Rev. Sci. Instrum. **23**, 143 (1952).
[26] D. BRINI, L. PELI, O. RIMONDI and P. VERONISI: N. 4 del Suppl. al Vol. 2, Ser. X, del Nuovo Cim. **1955**, 1048.
[27] P. A. TOVE: Rev. Sci. Instrum. **27**, 143 (1956).
[28] C. C. HARRIS and P. R. BELL: I.R.E. Trans. Nuclear Sci., N.S. **3**, No. 4, 87 (1956).

Harris and Bell[28]. In both of these studies the largest pulse height and best resolution were obtained by mounting the crystals directly on the tubes, and coupling with an oil film so as to match indices of refraction as closely as possible. Any Lucite coupler used decreased the light collection efficiency. In each of these experiments two couplers were tried. Tove used a logarithmic spiral type and a more complicated shape, both designed to collect by internal reflection

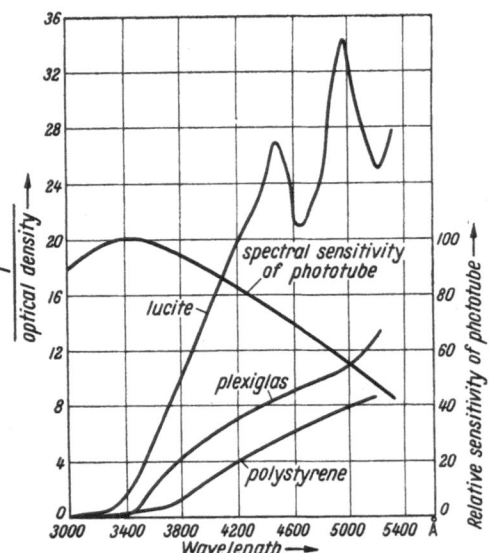

Fig. 9. Relative transmission of 9-in.long samples of Lucite, Plexiglas and polystyrene, and relative sensitivity of an RCA 1 P28 photomultiplier (Timmerhaus et al.[29]).

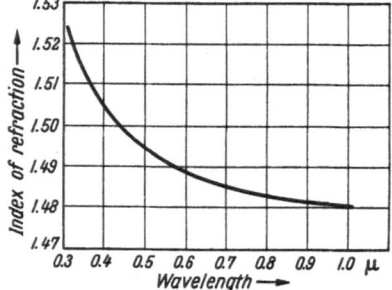

Fig. 10. Index of refraction of Lucite as a function of wavelength.

100% of the light crossing the input surface. Harris and Bell used a logarithmic spiral type and one with straight walls tapered so that the walls were less steep than the critical angle. When light pipes of several inch diameter are used with comparable sized crystals and photomultipliers, it is definitely better to use such a coupler as described. Fig. 8 shows the transmission vs. length for a 3-inch diameter pipe, from which it can be seen that for any but a short pipe a coupler is important. Similar results were obtained by Tove for a 1-inch pipe. It is also apparent from the studies that it is important to depend as much as possible on internal reflection. A good polish should be applied to the surfaces of coupler and pipe, and if a reflecting foil is used, it should be only very loosely in contact with the pipe and coupler. The resolution seems to increase less rapidly than one would expect from the assumption that the width of the pulse height peak is proportional to $1/\sqrt{N}$, where N is the number of photoelectrons obtained.

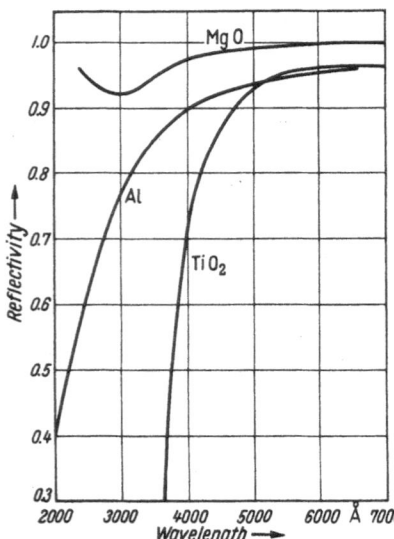

Fig. 11. Reflectivity versus wavelength of Al, TiO₂, and MgO.

Since Lucite (Perspex, Plexiglas) is so widely used for light pipes and couplers,

[29] K. D. Timmerhaus, E. B. Giller, R. B. Duffield and H. G. Drickamer: Nucleonics 6, No. 6, 37 (1950).

the transmission of a 9 in. thick sample is shown in Fig. 9, and the index of refraction in Fig. 10. The transmission of the commercial products can vary somewhat, so care should be taken to obtain selected optical grade material.

Reflectors, poorly coupled around a crystal and light pipe, generally increase the light reaching the photomultiplier in any assembly. The most satisfactory specular reflector is Al foil. The reflection coefficient, shown in Fig. 11, holds up well down to short wave length. MgO and TiO_2 are the most commonly used diffuse reflectors. The reflection coefficients are also shown in Fig. 11. It should be noted that TiO_2 becomes a poor reflector below 4000 Å.

II. Inorganic phosphors.

The following discussion is concerned primarily with the properties and behavior of the most important inorganic phosphors and is intended as a guide for those who desire to utilize or study these scintillators. Considerable emphasis is placed on sodium iodide because of its widespread use in fundamental and applied research and in routine industrial applications. Of the other inorganic scintillators, only the cesium halides and the lithium iodide phosphors are considered in any detail. For further information, the reader is referred to the several reviews[1-5] of the development, theory, and application of inorganic crystals which have appeared in recent years.

a) Sodium iodide.

4. Scintillation spectra and decay times. It is the purpose of the ensuing to illustrate how the emission properties of a NaI crystal vary with the activator concentration, the temperature, and the specific energy loss (dE/dx) of the particle exciting the crystal. To this end, the data of VAN SCIVER[6-9] have been selected as representative and will be described in some detail. Additional information on the subject can be obtained by referring to the work of BONANOMI and ROSSEL[10] who originally observed the emission properties of activated and unactivated sodium iodide at low temperatures; and of EBY and JENTSCHKE[11] who studied the spectra and decay times of sodium iodide crystals at room temperature as a function of the thallium concentration (0.00006 to 0.008 mol fraction Tl) and type of exciting radiation (4.7 and 23 Mev α-particles, 11.5 Mev deuterons, and Co^{60} γ-rays).

VAN SCIVER has measured the emission spectra from unactivated and thallium-activated crystals in the temperature range -190 to $+20°$ C using ultraviolet light, γ-rays, and α-particles as sources of excitation. The emission curves for NaI (pure), NaI (10^{-6} mol fraction Tl), and NaI (10^{-3} mol fraction Tl) crystals

    [1] G. F. J. GARLICK: Progress in Nuclear Physics, Vol. 2. London: Pergamon Press; New York: Academic Press 1952.
    [2] J. B. BIRKS: Scintillation Counters. New York: McGraw-Hill Book Co.; London: Pergamon Press 1953.
    [3] S. C. CURRAN: Luminescence and the Scintillation Counter. New York: Academic Press; London: Butterworth Scientific Publ. 1953.
    [4] R. K. SWANK: Annual Rev. Nucl. Sci. **4**, 111 (1954).
    [5] P. R. BELL: The Scintillation Method in Beta- and Gamma-Ray Spectroscopy. New York: Interscience Publ. 1955.
    [6] W. VAN SCIVER and R. HOFSTADTER: Phys. Rev. **97**, 1181 (1955).
    [7] W. VAN SCIVER: Stanford University (U.S.) Report No. HEPL-38 (Unpublished) (1955).
    [8] W. VAN SCIVER: Nucleonics **14**, No. 4, 50 (1956).
    [9] W. VAN SCIVER: I.R.E. Trans. Nuclear Sci., N.S. **3**, No. 4, 39 (1956).
    [10] J. BONANOMI and J. ROSSEL: Helv. phys. Acta **24**, 310 (1951); **25**, 725 (1952).
    [11] F. S. EBY and W. K. JENTSCHKE: Phys. Rev. **96**, 911 (1954).

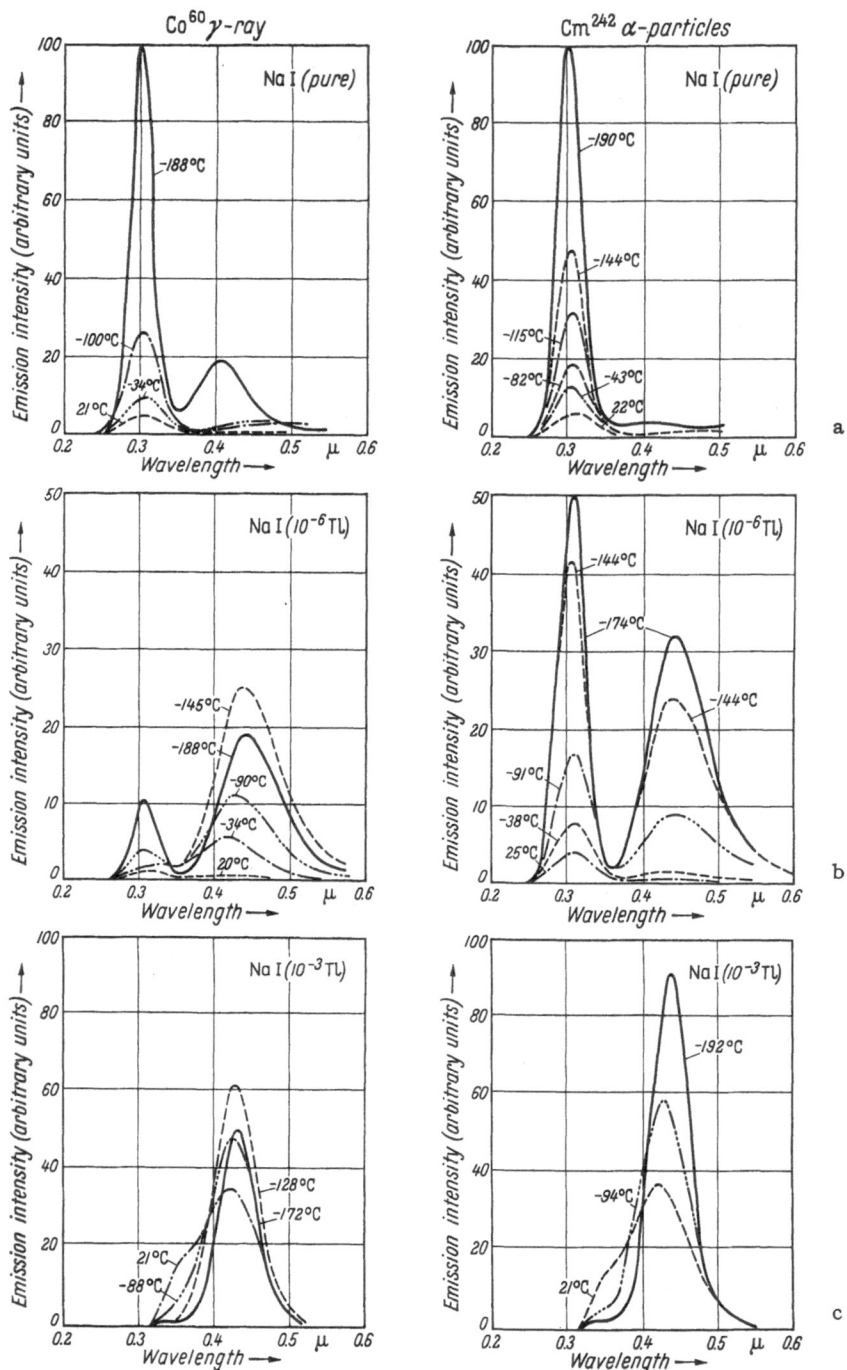

Fig. 12a—c. Emission spectra of unactivated and thallium-activated NaI crystals excited by Co⁶⁰ γ-rays and Cm²⁴² α-particles The ordinates are in arbitrary units with the normalization between the different families of curves only approximate
(van Sciver [7⁻⁹]).

excited by Co^{60} γ-rays and Cm^{242} α-particles (6.1 Mev) are shown in Fig. 12a – c and are discussed below.

a) NaI (pure). The characteristic emission from NaI (pure) is in a strongly temperature-dependent band which peaks at approximately 0.31 μ at room temperature and at 0.303 μ at $-188°$ C. The intensity of this emission is very weak at room temperature but increases with decreasing temperature. γ-ray and α-particle excitations differ in that, compared with the latter, the former produces considerable emission in the vicinity of 0.41 μ at $-188°$ C (Fig. 12a).

Concerning the decay time of the scintillations from unactivated NaI, VAN SCIVER found that the light in the 0.3 μ band has a decay which varies between 10^{-8} sec at $20°$ C and 3×10^{-8} sec at $-190°$ C (Fig. 13). At $-188°$ C, the 0.41 μ band was observed to have a very slow complex decay involving characteristic times of the order of hours.

*b) NaI (10^{-6} Tl)***. When a crystal containing a trace of thallium ($\sim 10^{-6}$ mol fraction) is excited by γ-rays, the intensity of the emission in the 0.31 μ band is reduced to approximately 10% of the value obtained with NaI (pure). Emission in the α-particle-excited 0.31 μ band is also reduced but only by a factor of two. The presence of thallium is further characterized by an emission band centered at 0.43 μ (Fig. 12b).

*c) NaI (10^{-3} Tl)****. For the commonly used NaI (10^{-3} Tl) crystal, the emission curves at room temperature are of the same shape whether excited by γ-rays

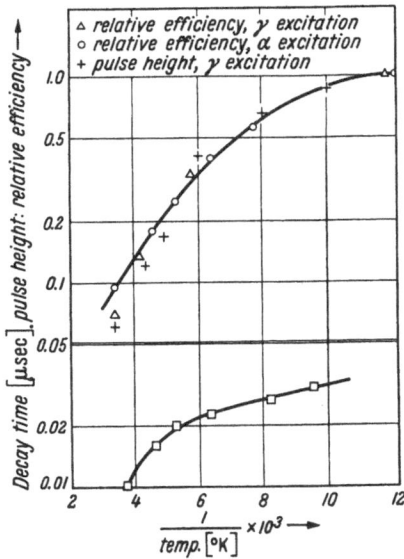

Fig. 13. Relative luminous efficiency, relative pulse height, and decay time of the 0.3 μ band in a NaI (pure) crystal as a function of the temperature (VAN SCIVER[8]).

or α-particles and the emission is essentially all in a band centered at approximately 0.42 μ. With decreasing temperature, the intensity of the γ-ray induced peak increases, reaches a maximum at an intermediate temperature ($\sim -130°$ C), and then decreases, whereas the peak in the α-particle spectrum rises continuously. In both cases the position of the peak shifts toward the longer wavelength region as the temperature decreases (Fig. 12c). The characteristic decay time for a crystal activated with 10^{-3} mol fraction of thallium is approximately 25×10^{-8} sec [11, 12].

5. Scintillation response. The fluorescent response of sodium iodide crystals to electrons, protons, deuterons, and α-particles has been the subject of numerous experimental investigations in recent years and in general a fairly clear picture of the relative response to different particles has emerged. Of the incongruities

---

* Grown from Mallinckrodt analytical reagent in an evacuated quartz tube using the Stockbarger-Bridgman method. Particular care was used to prevent contamination by thallium.

** Grown from analytical reagent sodium iodide in a platinum crucible without addition of thallium using a furnace which had previously been used to grow thallium-activated crystals. The Tl concentration was proved to be between 10^{-6} and 10^{-7} mol fraction.

*** Obtained from the Harshaw Chemical Company, Cleveland, Ohio.

[12] R. HOFSTADTER: Phys. Rev. **79**, 796 (1949).

which persist, probably the most perplexing are in the work on the response of NaI(Tl) to photons. In fact, the results have suggested such strikingly different conclusions that they warrant careful consideration.

Pringle and Standil[13] initially observed that the pulse height versus energy curve for photons was nonlinear below 150 kev with a curvature corresponding to higher scintillation efficiencies at lower photon energy. This point, however, was almost immediately disputed by several research groups. Notable are the investigations by West, Meyerhof and Hofstadter[14], Taylor et al.[15], Eriksen and Jenssen[16], and Bannerman, Lewis and Curran[17], which show that the pulse height increases linearly with photon and electron energy from 1 kev to at least 650 kev. In one case (Taylor et al.), there are indications of a deviation from linearity below 1 kev, but the curvature is in the opposite direction to that noted by Pringle and Standil.

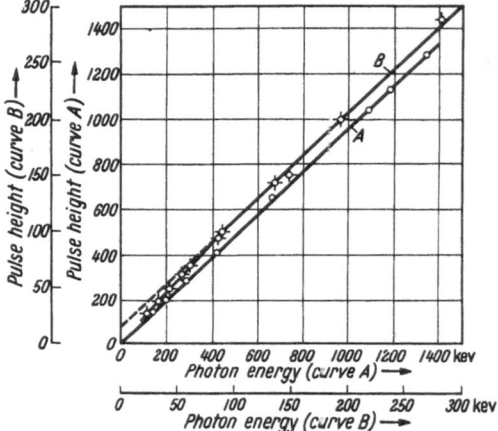

On the basis of the above, most writers on the subject of the scintillation response of NaI(Tl) to photons have concluded that the average pulse height is a linear function of the photon energy. There are now, however, additional data which are not only incompatible with this conclusion but which substantiate the view of Pringle and Standil.

Freedman et al.[18] have reported on a response curve for NaI(Tl) which exhibits a distinct curvature over the energy range 24 to 146 kev. The direction of the curvature indicates that the pulse height per unit energy is greater at low than at high energies. A similar nonlinearity has been observed by Bernstein[19] for

Fig. 14. Response of 1.25-in. diameter × 0.50-in. long NaI(Tl) crystal to internally produced electrons. Curve A shows the response from 100 to 1330 kev; curve B, the region below 300 kev. The straight line portion of B intercepts the energy axis at −16 kev (Engelkemeir[20]).

photon energies less than 200 kev. In this case an extrapolation of the linear portion of the curve gives an intercept of −25 kev. And recently, Engelkemeir[20] published a detailed account of his studies on the photon response of sodium iodide over the energy range 10 to 1500 kev. He not only found a pronounced nonlinearity below 100 kev (Fig. 14) but also evidence of deviation from linearity over the entire region investigated (Fig. 15).

Since a consistent physical explanation of the effect observed by the above authors and by Pringle and Standil has not been advanced, it is worthwhile to examine some of the possible sources of nonlinearity in a NaI(Tl) scintillation spectrometer. These are: (a) Instrumental effects. (b) Photomultiplier effects. (c) Variation of light collection efficiency within the crystal. (d) Escape from the

[13] R. W. Pringle and S. Standil: Phys. Rev. 80, 762 (1950).

[14] H. I. West, W. E. Meyerhof and R. Hofstadter: Phys. Rev. 81, 141 (1951).

[15] C. J. Taylor, W. K. Jentschke, M. E. Remley, F. S. Eby and P. G. Kruger: Phys. Rev. 84, 1034 (1951).

[16] V. O. Eriksen and G. Jenssen: Phys. Rev. 85, 150 (1952).

[17] R. C. Bannerman, G. M. Lewis and S. C. Curran: Phil. Mag. 42, 1097 (1951).

[18] M. S. Freedman, A. H. Jaffey, F. Wagner jr. and J. May: Phys. Rev. 89, 302 (1953).

[19] W. Bernstein: Nucleonics 14, No. 4, 46 (1956).

[20] D. Engelkemeir: Rev. Sci. Instrum. 27, 589 (1956).

crystal of some of the initial photon energy in the form of photoelectrons, x-rays, or Compton scattered electrons or quanta. (e) Variation of fluorescence efficiency with electron energy.

ENGELKEMEIR has considered each of these in detail and on the bases of several tests has hopefully ruled out (a) and (b) as sources for the nonlinearity appearing in his work. Furthermore, it does not seem reasonable to attribute the effect to (c) since a nonuniform light collection efficiency normally causes

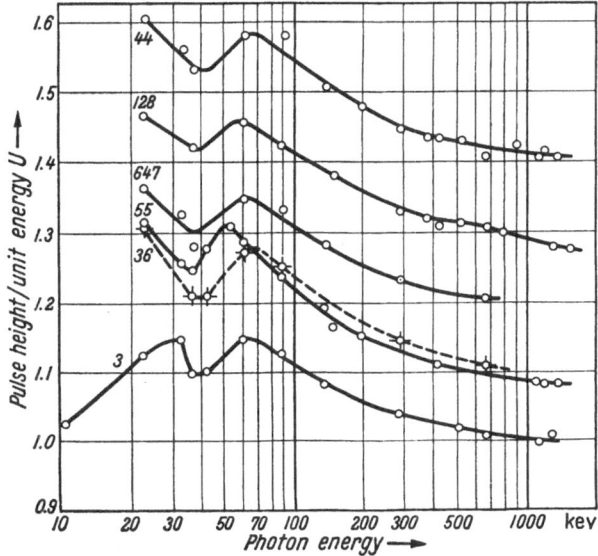

Fig. 15. Pulse height per unit energy versus energy for NaI(Tl) crystals. The pulse height per unit energy at 662 kev has been normalized to a value of unity and the successive curves for different crystals have been displaced upward 0.1 unit from the preceding one (ENGELKEMEIR [20]).

Crystal No	diameter (in.)	height (in.)	Reflector	NaI Surface	Photomultiplier
3	1.22	0.12	Al	ground	Du Mont-6292
55	1.25	0.50	MgO	ground	RCA-5819
36	1.25	0.50	Al	smooth	RCA-5819
647	1.5	1.0	MgO	ground	RCA-5819
128	2.5	2.12	MgO	ground	Du Mont-6363
44	4.0	4.0	MgO	ground	Du Mont-6364

the pulse height per unit energy to increase with increasing energy rather than to decrease. This occurs because low energy photons are principally absorbed near the top face of the crystal, whereas higher energy photons, which penetrate deeper, are more uniformly absorbed. Thus, with an inefficient optical reflector around the crystal, on the average disproportionally more light is collected for the higher energy photon events.

Experimentally one can determine if optical effects are producing a nonlinear response in a spectrometer unit by studying the response curves from crystals of widely different thicknesses. Both BERNSTEIN and ENGELKEMEIR (Fig. 15) have done this and in all cases the nonlinearity was present.

Although the importance of (d) was not thoroughly evaluated, conceivably the photopeak pulse height distributions could be sufficiently distorted by the escape of K x-rays and primary and secondary photoelectrons from the crystal to produce an observable nonlinearity. Moreover, since it is known that the frequency of escape events (excluding K x-ray escape) increases with increasing energy[21] for a given crystal and geometry, one might expect the ratio of the average pulse height observed for a photopeak to the incident photon energy to decrease as the photon energy increases. Again, it might be possible experimentally to recognize the effect by varying the crystal size and the source-crystal geometry.

In the absence of another explanation it is always tempting to associate a nonlinear response with the scintillation efficiency of the crystal. In this case, however, one should proceed with caution until additional data are acquired even though there is some evidence that the fluorescence efficiency may increase at very low energies*.

Turning now to the response of NaI(Tl) to other types of ionizing particles, it is necessary to note that an exhaustive study of the subject has yet to be made. However, sufficient data have accumulated to outline the general characteristics of the response curves and to establish some approximate values for the relative scintillation efficiencies. Of particular importance to the discussion are the observations of EBY and JENTSCHKE[11] which indicate that the pulse height resulting from excitation by a particular particle and energy is a function of the Tl concentration. For example, for deuteron and α-particle excitation, the pulse heights increase sharply with Tl concentration for mol fractions smaller than about 0.0013 and decrease for higher concentrations (Figs. 12 and 16). This dependence of the scintillation response on the Tl concentration makes it rather difficult to compare the work of the different investigators since in most cases the activator concentrations are not known.

The scintillation response of commercially available NaI(Tl) crystals (approx. 10^{-3} mol fraction Tl) to protons has been studied for bombarding energies in the regions 60 to 400 kev[23] and 1 to 18 Mev[11,15,24,25]. The results show that the proton pulse height versus energy curve does not deviate significantly from a straight line in the energy region 1 to 18 Mev, but leave considerable doubt whether or not the response is linear below 1 Mev. To a large extent the latter point of view is based on the data of BROLLEY and RIBE[25] which indicate that the response curve is linear over the region investigated (2 to 5 Mev) but extrapolates to an intercept on the energy axis of 0.29 Mev. In addition, it is felt that even though the low energy points of ALLISON and CASSON[23] (60 to 400 kev) fall on a straight line passing through the origin, they are not conclusive evidence for linearity because of the significant gap which remains in the measurements below 1 Mev.

A very similar situation exists in the case of deuterons where the response has been studied for $E_d = 60$ to 630 kev[23] and 1 to 11 Mev[11,15,25]. Again, the two portions of the experimental response curve are indistinguishable from straight lines with the measurements of BROLLEY and RIBE suggesting a saturation type nonlinearity below 1 or 2 Mev.

Although only a relatively small amount of information has been reported on the relative scintillation efficiencies for electrons, protons, and deuterons, it is probable that for a given energy loss in the crystal the responses do not differ

---

* WRIGHT [Phys. Rev. 96, 569 (1954)] has shown that at low particle energies statistical processes in the photomultiplier tube result in a pulse-height distribution which is markedly asymmetrical with the position of the maximum always lying below that of the mean. He applied this correction to the data of WEST, MEYERHOF and HOFSTADTER[14] and found a response curve for NaI(Tl) which suggested that the fluorescence efficiency increases at energies below approximately 2 kev.

[21] K. LIDÉN and N. STARFELT: Ark. Fysik 7, 427 (1954).
[23] S. K. ALLISON and H. CASSON: Phys. Rev. 90, 880 (1953).
[24] J. K. LIKELY and W. FRANZEN: Phys. Rev. 87, 666 (1952).
[25] J. E. BROLLEY and F. L. RIBE: Phys. Rev. 98, 1112 (1955).

greatly, at least for energies less than 10 Mev. For example, ALLISON and CASSON[23] directly observed a $p:d$ ratio of $1:0.96$ and by an indirect method determined that the response of protons was approximately the same as for electrons of the same energy.

For α-particles excitation of commercial NaI(Tl) crystals at room temperature, it has been clearly demonstrated by several investigators[11,15,23,26] that the pulse height per unit energy decreases with increasing specific energy loss causing a nonlinear response below 20 Mev. This effect of the specific energy loss on the scintillation efficiency is such that at 5 Mev the ratio of the pulse height to energy for α-particles is approximately 0.6 that for electrons. A typical α-particle response curve is plotted in Fig. 17; for comparison, the response of the same crystal to protons is also shown.

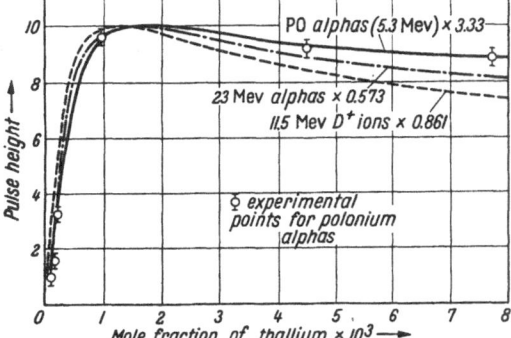

Fig. 16. Variation in pulse height resulting from α-particle and deuteron excitation of NaI(Tl) as a function of the Tl concentration (EBY and JENTSCHKE[11]).

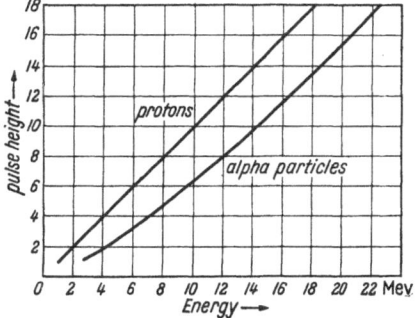

Fig. 17. Scintillation response of NaI(Tl) crystal containing 0.0013 mole fraction Tl to α-particles and protons (EBY and JENTSCHKE[11]).

The temperature dependence of the scintillation efficiency of α-particles relative to electrons has been investigated by DER MATEOSIAN, McKEOWN and MUELHAUSE[27] in both "pure" and thallium-activated NaI crystals grown by the Bridgman[28] and Kyropoulos[29] methods. For the purposes of this discussion two of their results are significant:

1. The scintillation response of NaI(Tl) crystals to electrons (662 kev) and α-particles (5.3 Mev) decreased with decreasing temperature, whereas that of unactivated NaI increased as the temperature was lowered. For example, at room temperature the pulse heights from the unactivated crystals were less than 0.02 times the pulse heights from the thallium-activated crystals; but below −115° C for crystals grown by the Kyropoulos method and −180° C for those grown by the Bridgman method, the pulse heights were indistinguishable. The pulse height versus temperature curves for excitation by polonium α-particles are shown in Fig. 18.

2. The ratio of the pulse height per unit energy of the α-particle to that of the electron was dependent on both the method of preparing the crystal and the temperature of its operation. As shown in Fig. 19, this ratio had a value of 1.5 at approximately −150° C for unactivated crystals grown by the Kyropoulos method, indicating that the α-particle produced light more efficiently than the electron under these conditions. For the unactivated and the thallium-activated

[26] R. H. LOVBERG: Phys. Rev. **84**, 852 (1951).
[27] E. DER MATEOSIAN, M. McKEOWN and C. O. MUEHLHAUSE: Phys. Rev. **101**, 967 (1956).
[28] P. W. BRIDGMAN: Proc. Amer. Acad. Arts Sci. **60**, 305 (1925).
[29] S. KYROPOULOS: Z. anorg. allg. Chem. **154**, 308 (1926).

crystals grown by the Bridgman method, the ratios appeared to be temperature independent and had a value of approximately 0.7 (cf. Fig. 13).

In evaluating the above results, careful consideration should be given to the fact that there is presently no strong evidence that the differences between the crystals grown by the Kyropoulos and the Bridgman methods will always appear. Furthermore, in view of the observations of van Sciver[6-9] on the effect of a trace of thallium on the intensity, wavelength, and decay time

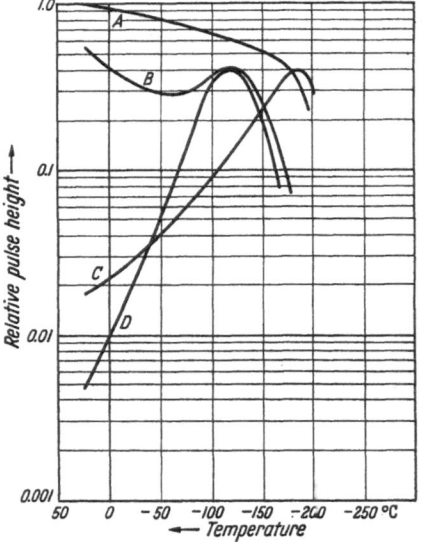

Fig. 18. Temperature dependence of the response of sodium iodide crystals to α-particles. Curve A: NaI(Tl) grown by the Bridgman method. Curve B: NaI(Tl) grown by the Kyropoulos method. Curve C: Unactivated NaI grown by the Bridgman method. Curve D: Unactivated NaI grown by the Kyropoulos method (der Mateosian, McKeown and Muehlhause[27]).

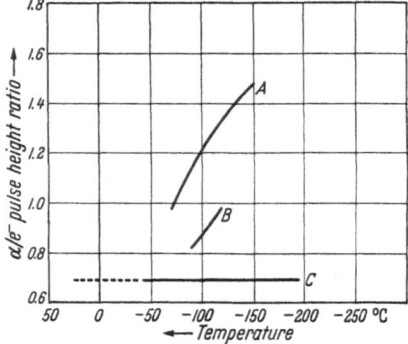

Fig. 19. Ratio of pulse heights per unit energy for α-particles and electrons as a function of temperature. Curve A: Unactivated NaI grown by the Kyropoulos method. Curve B: NaI(Tl) grown by the Kyropoulos method. Curve C: Both activated and unactivated NaI grown by the Bridgman method (der Mateosian, McKeown and Muehlhause[27]).

of α- or γ-produced scintillations in "pure" sodium iodide crystals, one should exercise care in using the data on "pure" crystals presented in Figs. 18 and 19 until additional data are accumulated.

Even though NaI(Tl) is exhibiting signs of saturation at low α-particle energies and the mean range of light fission fragments is approximately three-fourths that of 4.8 Mev α-particles, Milton and Fraser[30] have shown that the scintillations produced in NaI(Tl) by fission fragments are about twice as large as those from the α-particles. The relative pulse heights for the light and heavy fragment groups and for U233 α-particles from NaI(Tl), KI(Tl), and stilbene crystals are summarized in Table 4. As the two fission fragment groups were well resolved by the NaI(Tl) and the KI(Tl) crystals, the larger pulse heights are associated with the more energetic light-fragment group.

Table 4. *Relative pulse heights for fission fragments and U233 α-particles in NaI(Tl), KI(Tl), and stilbene crystals from Milton and Fraser[30].*

Phosphor	Light/α	Heavy/α	α
NaI (Tl) . .	2.1	1.6	17.8
KI (Tl) . .	3.5	2.3	3.8
Stilbene . .	2		1

6. Absolute scintillation efficiency. Very little information of any import has appeared in the literature on the absolute scintillation efficiency of thallium-activated sodium iodide. In fact, to our knowledge, there has been no direct

---

[30] J. C. D. Milton and J. S. Fraser: Phys. Rev. 96, 1508 (1954).

measurement of the ratio of the energy converted into light to the energy dissipated in a NaI(Tl) crystal by charged particles or by photons *. Estimates based on the number of photoelectrons released at the photocathode of a particular tube per Mev of electron energy absorbed in the crystal[31-33] are considered highly unreliable since values for the photoelectron production efficiency of the photocathode, the light collection efficiency, etc., were rather arbitrarily selected and could almost as reasonably have been chosen to give efficiencies differing by a factor of 2 or 3.

Of course, it is possible in principle to obtain the absolute efficiency from relative measurements and some work of this type has been done using anthracene as the reference material. For example, SANGSTER and IRVINE[34] have reported that for Co^{60} γ-ray excitation of NaI(Tl) the average pulse height relative to anthracene = 100 is 120 and on the same basis SWANK[35] has indicated a pulse height of 210 for β-particle excitation. Unfortunately, the range of these relative efficiency values is not representative of the range of uncertainty on the absolute efficiency because of the large discrepancies which appear in the measurements of the absolute scintillation efficiency of anthracene (see Table 12).

These considerations suggest that the generally accepted value[2, 3, 36] of 8% for the efficiency of photon production by electrons in NaI(Tl) at room temperature is a rough approximation rather than an established experimental fact. Although a priori one cannot exclude the possibility that this value is correct, there is good reason to suspect its reliability*.

7. Energy loss and range of particles in NaI. There have been no direct measurements of the specific energy loss (dE/dx) of electrons, protons, and α-particles in sodium iodide; consequently, the values must be calculated from available data on the constituent elements. In making these calculations, it is assumed that the stopping power of an atom is unaltered by the presence of atoms of other types since the chemical binding energies are usually negligible compared with the mean ionization and excitation potential I. Thus, for all practical purposes the stopping power of NaI can be defined as simply the sum of the stopping powers of the sodium and iodine.

NELMS[37] has recently calculated mean energy loss values due to ionization and excitation for electrons and positrons with energies below 1.2 Mev in Na and I and has derived the values of dE/dx for NaI on the basis of this additivity rule. These energy loss values and the ranges obtained by integrating the reciprocal stopping power from 0 to E are given in Table 5.

[31] J. A. McINTYRE and R. HOFSTADTER: Phys. Rev. **78**, 617 (1950).

[32] R. HOFSTADTER and J. A. McINTYRE: Phys. Rev. **80**, 631 (1950).

[33] C. J. BORKOWSKI and R. L. CLARK: Rev. Sci. Instrum. **24**, 1046 (1953).

[34] R. C. SANGSTER and J. W. IRVINE jr.: J. Chem. Phys. **24**, 670 (1956).

[35] R. K. SWANK: Nucleonics **12**, No. 3, 14 (1954).

[36] G. A. MORTON: Proc. of the Internat. Conference on the Peaceful Uses of Atomic Energy **14**, 246 (1956).

* W. J. VAN SCIVER and L. BOGART [Bull. Amer. Phys. Soc., Ser. II **2**, 142 (1957)] are presently measuring the absolute efficiencies of scintillation phosphors by a direct method which gives accurately repeatable results. Their procedure is to substitute an RCA C-7387 light pulser tube for the phosphor in a typical scintillation counter setup and adjust its intensity to give a pulse height equivalent to that of the Cs^{137} photopeak. An electrometer is used to measure photomultiplier cathode current which when divided by pulse rate gives photocathode electronic charge per pulse. The combination of this value with the photomultiplier cathode sensitivity (which is measured with a calibrated thermopile and monochromator) and primary particle energy yields the phosphor's scintillation efficiency. For a 1 in. × 1 in. Harshaw NaI(Tl) crystal with MgO smoked reflector and DC-200 silicone oil coupling, the value of (14.1 ± 0.7)% was obtained.

[37] ANN T. NELMS: Nat. Bur. Stand., Circ. No. 577 (Unpublished) (1956).

Table 5. *Energy loss and range of electrons and positrons in NaI according to Nelms[37].*

E (Mev)	$-\dfrac{dE^-}{dx}\left(\dfrac{\text{Mev cm}^2}{\text{gm}}\right)$	R^- (gm/cm²)	$-\dfrac{dE^+}{dx}\left(\dfrac{\text{Mev cm}^2}{\text{gm}}\right)$	R^+ (gm/cm²)
0.01	11.6	0.000513	13.3	0.000433
0.02	7.15	0.00162	8.00	0.00142
0.03	5.38	0.00328	5.93	0.00292
0.04	4.41	0.00532	4.80	0.00478
0.05	3.79	0.00781	4.09	0.00707
0.06	3.35	0.0106	3.60	0.00966
0.07	3.03	0.0138	3.23	0.0126
0.08	2.79	0.0172	2.96	0.0158
0.09	2.59	0.0243	2.74	0.0194
0.10	2.43	0.0249	2.56	0.0231
0.15	1.95	0.0482	2.01	0.0455
0.20	1.69	0.0759	1.73	0.0725
0.25	1.55	0.107	1.56	0.103
0.30	1.45	0.140	1.45	0.136
0.35	1.38	0.176	1.38	0.172
0.40	1.33	0.213	1.32	0.209
0.45	1.30	0.251	1.29	0.247
0.50	1.27	0.290	1.26	0.286
0.55	1.25	0.329	1.23	0.327
0.60	1.23	0.370	1.21	0.368
0.65	1.22	0.410	1.20	0.409
0.70	1.21	0.451	1.19	0.451
0.75	1.21	0.493	1.18	0.493
0.80	1.20	0.534	1.18	0.535
0.85	1.20	0.576	1.17	0.578
0.90	1.20	0.618	1.17	0.621
0.95	1.19	0.660	1.17	0.664
1.0	1.19	0.701	1.16	0.707
1.2	1.19	0.869	1.16	0.879

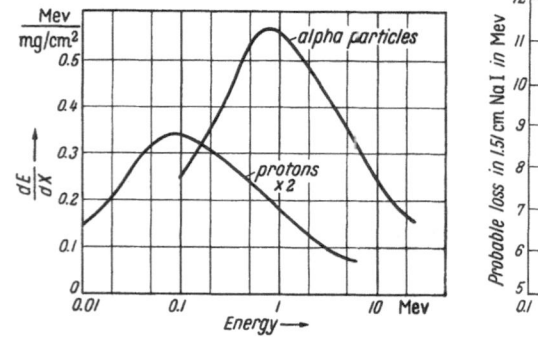

Fig. 20.

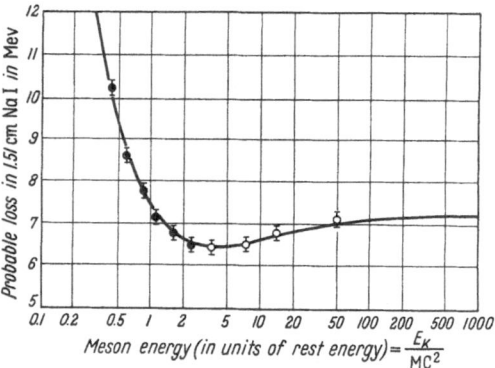

Fig. 21.

Fig. 20. Specific energy loss of protons and α-particles in NaI (Eby and Jentschke[11]).

Fig. 21. Probable energy loss of mesons in 1.51 cm of NaI(Tl). The solid circles represent cyclotron data using pions, except for the highest point at $E_K/Mc^2 = 2.3$ using muons. The open circles show the cosmic-ray muon results. The curve is calculated from the theory of Sternheimer (Bowen[38]).

Fig. 20 shows the dE/dx versus energy curves for protons and α-particles which were calculated by Eby and Jentschke[11] from the stopping power and range-energy data of several investigators. The dE/dx values for protons and for

α-particles with energies greater than 2 Mev are thought to be reliable to $\pm 8\%$. For α-particles with energies less than 2 Mev, the uncertainty is not less than $\pm 10\%$.

In contrast to the above, a reasonably detailed experimental study has been made of the energy loss of mesons in thallium-activated sodium iodide crystals (BOWEN [38]). The results are shown in Fig. 21 with a curve calculated from STERN-HEIMER'S [39] theory. It is noted that there is good agreement between theory and experiment, particularly in the energy region below the minimum of ionization.

In conclusion, we note that a complete investigation of the energy loss and range of charged particles in sodium iodide would be a highly desirable undertaking. From both the experimental and the theoretical points of view, an accurate knowledge of these properties is essential if maximum results are to be obtained from experiments based on scintillation phenomena in sodium iodide crystals.

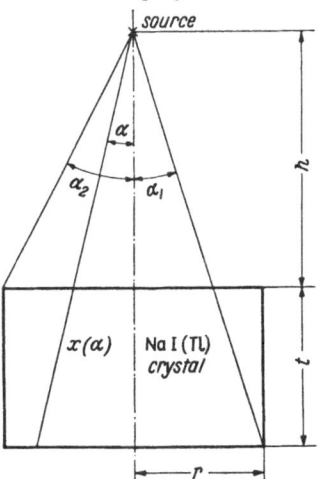

Fig. 22. Geometry for calculating efficiency of NaI(Tl) crystal.

8. Absolute γ-ray intensity measurements with NaI(Tl) crystals. α) *Total detection efficiency.* The total detection efficiency ε_t of a crystal for γ-rays of energy E_γ is

$$\varepsilon_t(E_\gamma) = \frac{n_t}{n_0 \, \Omega_c} \qquad (8.1)$$

where n_0 is the number of γ-rays emitted per second by the source; n_t, the total number of γ-rays of energy E_γ detected per second; and $4\pi\Omega_c$, the solid angle subtended by the crystal at the source. Thus, for the case of a point source of monoenergetic radiation located on the axis of a cylindrical crystal (Fig. 22) and no loss of radiation between the source and the crystal, $\varepsilon_t(E_\gamma)$ is given by

$$\varepsilon_t(E_\gamma) = \frac{\frac{1}{2}\int_0^{\alpha_2}(1 - e^{-\tau x(\alpha)}) \sin \alpha \, d\alpha}{\frac{1}{2}\int_0^{\alpha_2} \sin \alpha \, d\alpha} \qquad (8.2)$$

where

$$\tau = \mu + \sigma + \varkappa \qquad (8.3)$$

and α is the angle between the axis of the crystal of radius r and length t and the direction of an incident γ-ray; $x(\alpha)$, the distance a γ-ray making an angle α travels in the crystal

$$\left.\begin{array}{ll} x = t \sec \alpha, & \text{for } 0 \leq \alpha \leq \alpha_1 = \arctan \dfrac{r}{h+t}, \\[2ex] x = r \operatorname{cosec} \alpha - h \sec \alpha, & \text{for } \alpha_1 \leq \alpha \leq \alpha_2 = \arctan \dfrac{r}{h}; \end{array}\right\} \qquad (8.4)$$

and τ, the linear absorption coefficient [photoelectric (μ) + Compton (σ) + pair production $(\varkappa)$] of sodium iodide.

[38] T. BOWEN: Phys. Rev. **96**, 754 (1954).
[39] R. M. STERNHEIMER: ·Phys. Rev. **88**, 851 (1952); **91**, 256 (1953).

In order to compute the efficiency of a detector as a function of energy from Eq. (8.2), the values of τ must be determined either experimentally[40-42] or theoretically[43,44]. Since absorption processes which do not result in a transfer of energy to the detecting medium must be excluded from efficiency computations, total absorption coefficients should not be used unless the contributions due to coherent (Rayleigh) scattering are subtracted. White[44] has calculated total absorption coefficients for sodium iodide for photon energies from 10 kev to 100 Mev and these have been corrected for coherent scattering by Bell[5]. The values of τ, μ, σ, and $\varkappa$ for the energy region 10 kev to 10 Mev are given in Fig. 23.

It is apparent from Fig. 22 that ε_t depends not only on the γ-ray energy and the dimensions of the crystal, but also on the source-crystal distance h. Because the average length of path in the crystal is a function of α_2 (being the longest for very large and very small source-crystal spacings), the efficiency decreases, passing through a minimum at a distance h which is dependent on the crystal size, and then increases as $h \to 0$ from infinity.

The detection efficiencies for $1^1/_2$ in. diameter $\times 1$ in. high and 3 in. $\times 3$ in. right circular cylinders of sodium iodide and for a 3 in. $\times 3$ in. cylinder with the top beveled at $45°$ to give a 2 in. diameter face have been computed by the Mathematics Panel, Oak Ridge National Laboratory[45] for various values of the parameter τ and h by numerically integrating Eq. (8.2). Recently, these calculations were extended by Wolicki, Jastrow and Brooks[46] to include all standard sizes of cylindrical NaI crystals. Curves of ε_t versus E_γ for the three above-mentioned crystals are given in Figs. 24 to 32 and of $\varepsilon_t \Omega_c$ versus E_γ for several other crystal sizes ranging from 0.50 in. diameter $\times 0.50$ in. high to 5.0 in. $\times 4.0$ in. in Figs. 33 to 49. Efficiencies for many crystal sizes which were not included in the calculations can be obtained directly from these curves since the interaction probability calculated for a set of parameters t, h, r, and τ is equal to that for nt, nh, nr, and τ/n, where n is any number[46]. In cases where the r/t for a crystal is different from that for any of the crystal sizes given, the efficiencies can be obtained[46] by interpolating between the known results for two crystals 1 and 2 where

$$\frac{r_1}{t_1} < \frac{r}{t} < \frac{r_2}{t_2}. \tag{8.5}$$

A plot of the pulse spectrum for monoenergetic γ-rays shows a distribution of pulses from externally scattered radiation (Compton γ-rays scattered into the crystal by the shielding, photomultiplier, crystal housing, etc.) in addition to the spectrum of pulse heights resulting from primary photon interactions in the crystal. The peak of this distribution appears at an energy corresponding to 180° Compton scattering of the primary γ-rays and hence is primarily due to backscattered photons emitted from the photomultiplier tube window. In the pulse height distribution for the 662 kev γ-ray of Cs^{137} shown in Fig. 50, the backscatter peak is seen at approximately 185 kev. In practice, therefore, a measurement of a gamma emission rate (n_0) requires the use of a scintillation spectrometer

---

[40] P. R. Bell, R. C. Davis, D. S. Hughes, W. H. Jordan and C. A. Randall: Oak Ridge National Laboratory (U.S.) Report No. 1415 (Unpublished) (1953).
[41] P. R. Howland and W. E. Krieger: Phys. Rev. 95, 407 (1954).
[42] L. H. Th. Rietjens, G. J. Arkenbaut, G. F. Walters and J. C. Kluguer: Physica, Haag 21, 110 (1955).
[43] C. M. Davisson and R. D. Evans: Rev. Mod. Phys. 24, 79 (1952).
[44] G. R. White: Nat. Bur. Stand. Report No. 1003 (Unpublished) (1952).
[45] Mathematics Panel, Oak Ridge National Laboratory, Oak Ridge, Tennessee.
[46] E. A. Wolicki, R. Jastrow and F. Brooks: Naval Research Laboratory (U.S.) Report No. 4833 (Unpublished) (1956).

in order that the shape of this undesirable contribution to the spectrum may be determined and the total pulse spectrum corrected accordingly. Once the corrections have been made, an integration under the counting rate versus pulse height curve gives n_t. Generally, it is not difficult to obtain n_t for monoenergetic γ-ray sources; the task, however, becomes extremely complex for sources emitting two or more γ-rays.

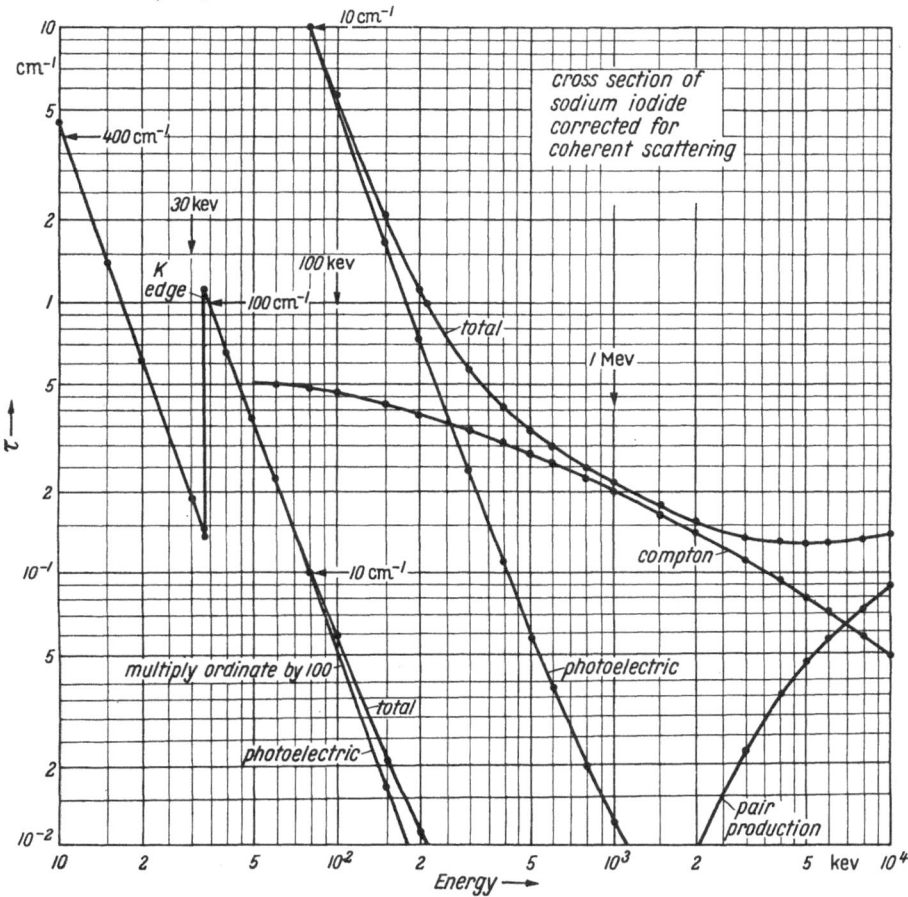

Fig. 23. Gamma-ray absorption coefficients for NaI (BELL[8]).

β) *Photopeak efficiency.* The problems encountered in relating an observed counting rate to the intensity of γ-rays emitted by a given source are greatly simplified if the total number of counts per second in the full-energy peak (photopeak), rather than n_t, is used as the standard. The shapes of the photopeaks are fairly insensitive to radiation scattered into the crystal and background corrections are normally small. Also, since the shape of the photoelectric peak in the pulse height distribution from a monoenergetic γ-ray source is nearly Gaussian, the total counts per second in the peak (area under photopeak) can be directly related to the peak counting rate. Thus, if the photopeak efficiency $\varepsilon_p(E_\gamma)$ is defined as the ratio of the number of counts per second in the photopeak (n_p) to the number of γ-rays of energy E_γ incident per second on the surface of the crystal [or expressed alternatively, $\varepsilon_p(E_\gamma)$ is the probability that an incident

γ-ray of energy E_γ will produce a pulse in the full-energy peak] then

$$n_0(E_\gamma) = \frac{n_p(E_\gamma)}{\varepsilon_p(E_\gamma)\,\Omega_c}$$ (8.6)

where

$$\varepsilon_p(E)_\gamma = \frac{\frac{1}{2}\int_0^{\alpha_2}\sin\alpha\,d\alpha\int_0^x\mu_p\,e^{-\tau\,x(\alpha)}\,dx}{\frac{1}{2}\int_0^{\alpha_2}\sin\alpha\,d\alpha}$$ (8.7)

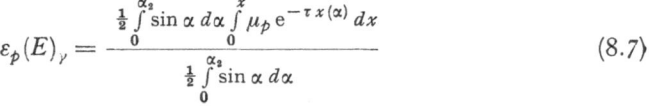

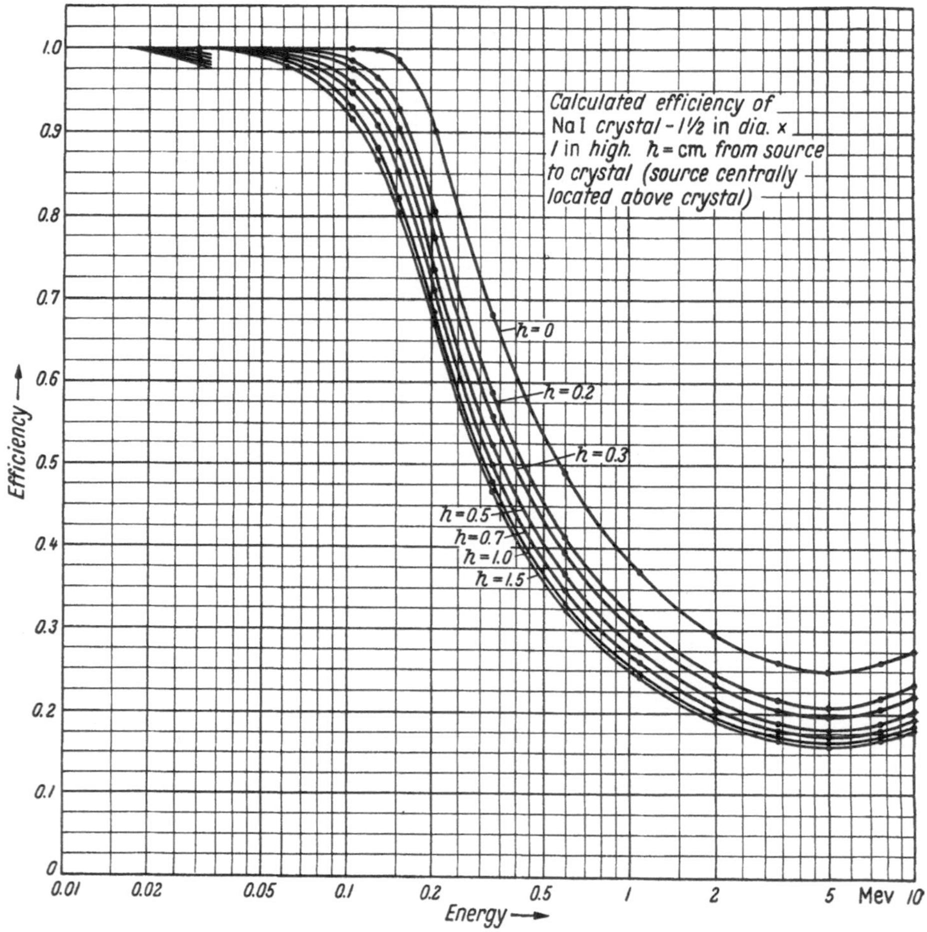

Fig. 24. Detection efficiency of $1^{1}/_2$ in. diameter × 1 in. high NaI(Tl) crystal (Mathematics Panel, ORNL[45]).

and μ_p is the effective photo-absorption coefficient. It is important to note that μ_p refers not only to absorption by the photoelectric process but also to absorption by all multiple processes which produce a pulse in the full-energy peak, i.e., where the incident photon transfers all its energy to the crystal. For example, if a Compton scattering event is followed by the photoelectric absorption of the scattered photon in the crystal, the two processes will cause one single pulse corresponding to the full energy E_γ.

In the region below 100 kev, where the photoelectric absorption coefficient is equal to the total absorption coefficient, $\mu_p = \mu$, and the photopeak efficiency is identical to the total detection efficiency of the crystal. For $E_\gamma > 100$ kev, however, it is difficult to calculate ε_p directly from Eq. (8.7) because the large number of multiple processes in the crystal prevents μ_p from being expressed explicitly as a function of energy.

To determine $\varepsilon_p(E_\gamma)$ experimentally (for a monoenergetic source), it is convenient to use the calculated value of $\varepsilon_t(E_\gamma)$ and the measured value of the ratio (R)

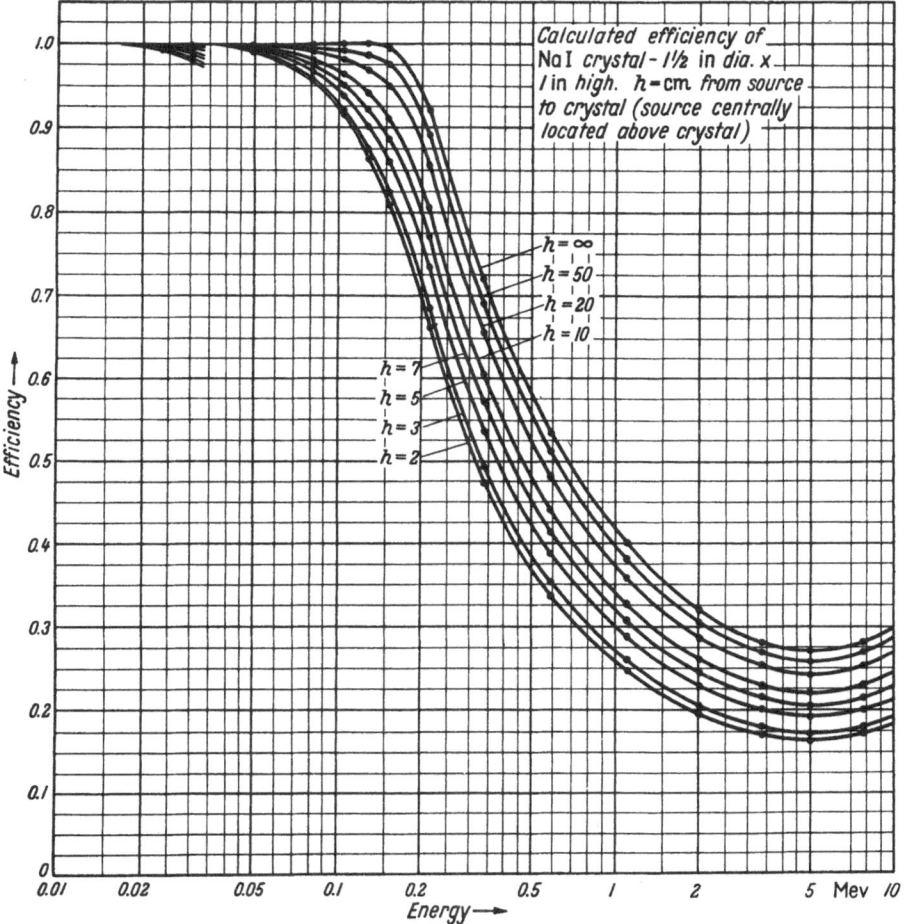

Fig. 25. Detection efficiency of $1^1/_2$ in. diameter $\times$ 1 in. high NaI(Tl) crystal (Mathematics Panel, ORNL[43]).

of the integral under the photopeak section of the pulse spectrum (area under photopeak) to the integral under the entire pulse spectrum. Since

$$R(E_\gamma) = \frac{\frac{1}{2}\int\limits_0^{\alpha_2} \sin \alpha \, d\alpha \int\limits_0^x \mu_p \, e^{-\tau \, x(\alpha)} \, dx}{\frac{1}{2}\int\limits_0^{\alpha_2}(1 - e^{-\tau \, x(\alpha)}) \sin \alpha \, d\alpha} = \frac{n_p}{n_t}, \qquad (8.8)$$

it follows from Eqs. (8.2) and (8.7) that

$$\varepsilon_p(E_\gamma) = R \, \varepsilon_t(E_\gamma). \qquad (8.9)$$

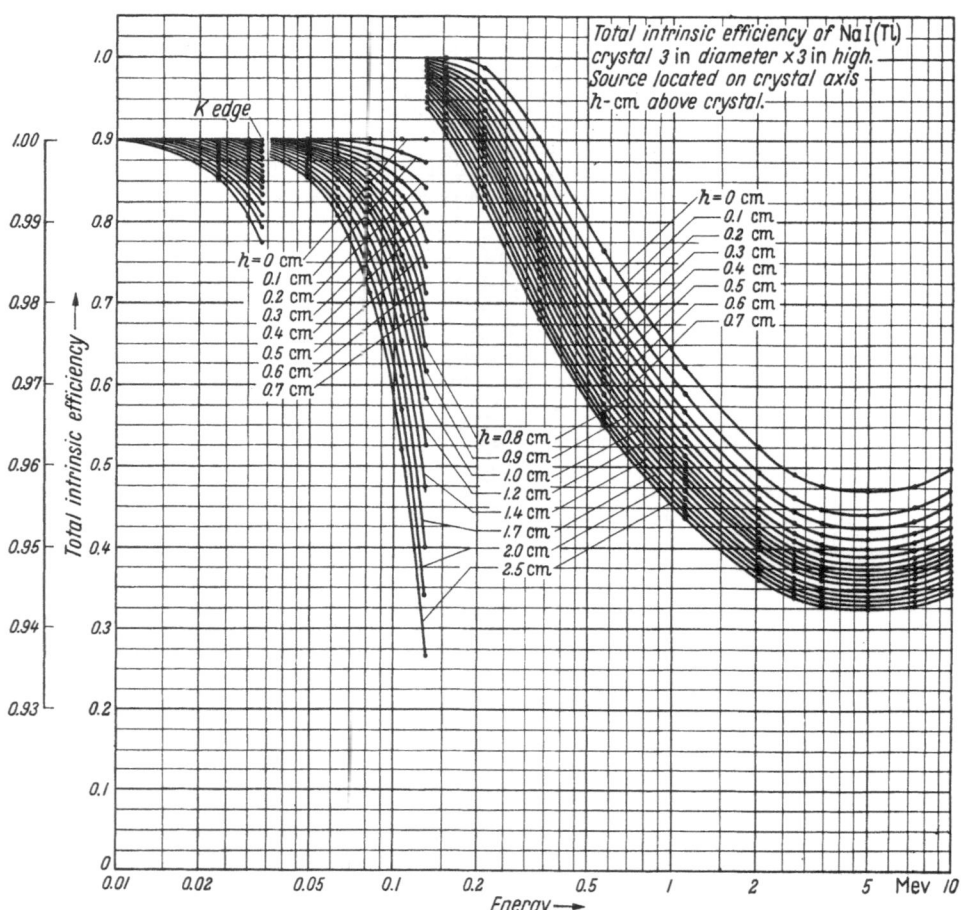

Fig. 26. Detection efficiency of 3 in. diameter × 3 in. high NaI(Tl) crystal (Mathematics Panel, ORNL[45]).

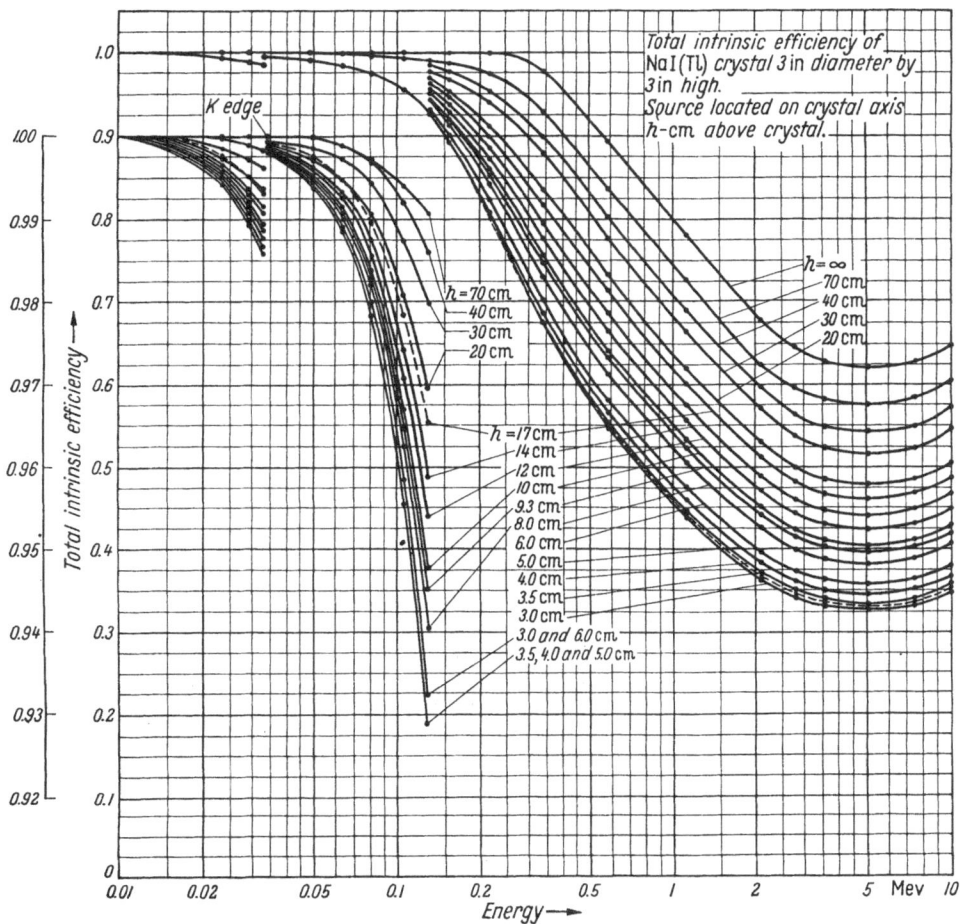

Fig. 27. Detection efficiency of 3 in. diameter × 3 in. high NaI(Tl) crystal (Mathematics Panel, ORNL[45]).

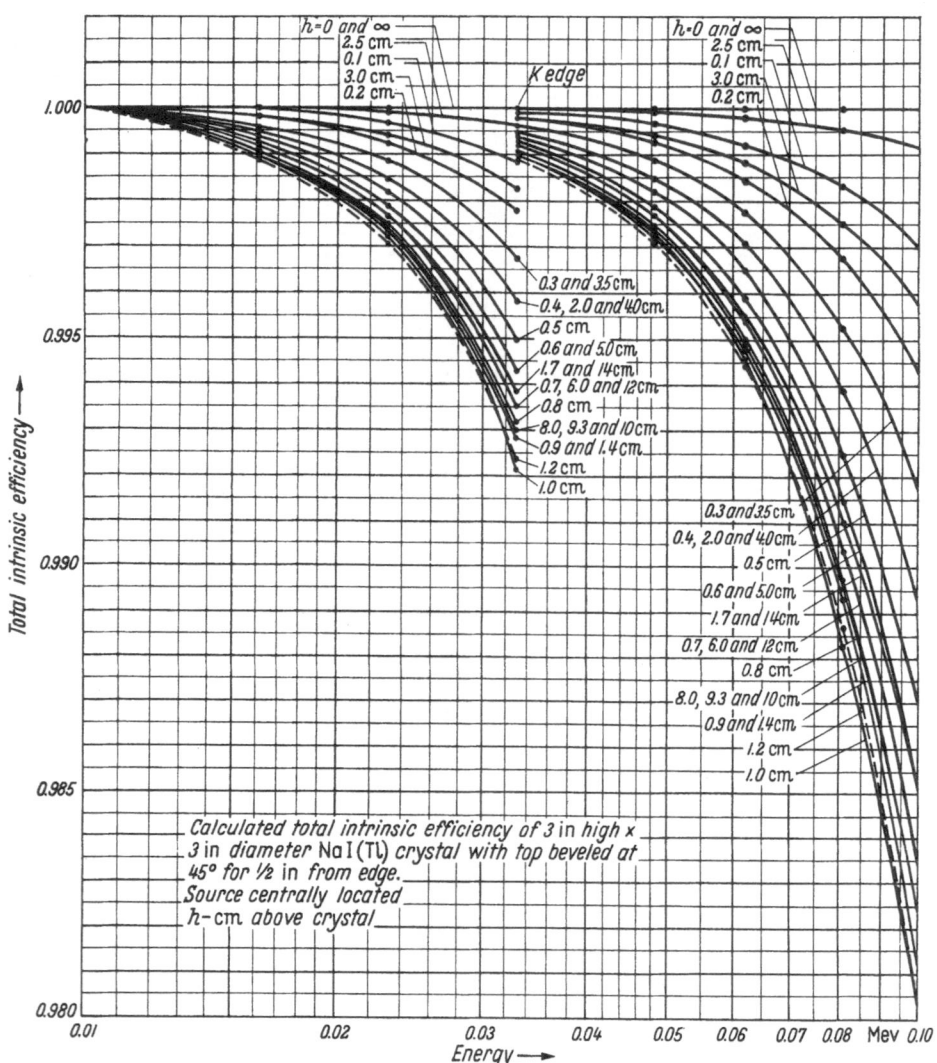

Fig. 28. Detection efficiency of 3 in. diameter × 3 in. high NaI(Tl) crystal (Mathematics Panel, ORNL[48]).

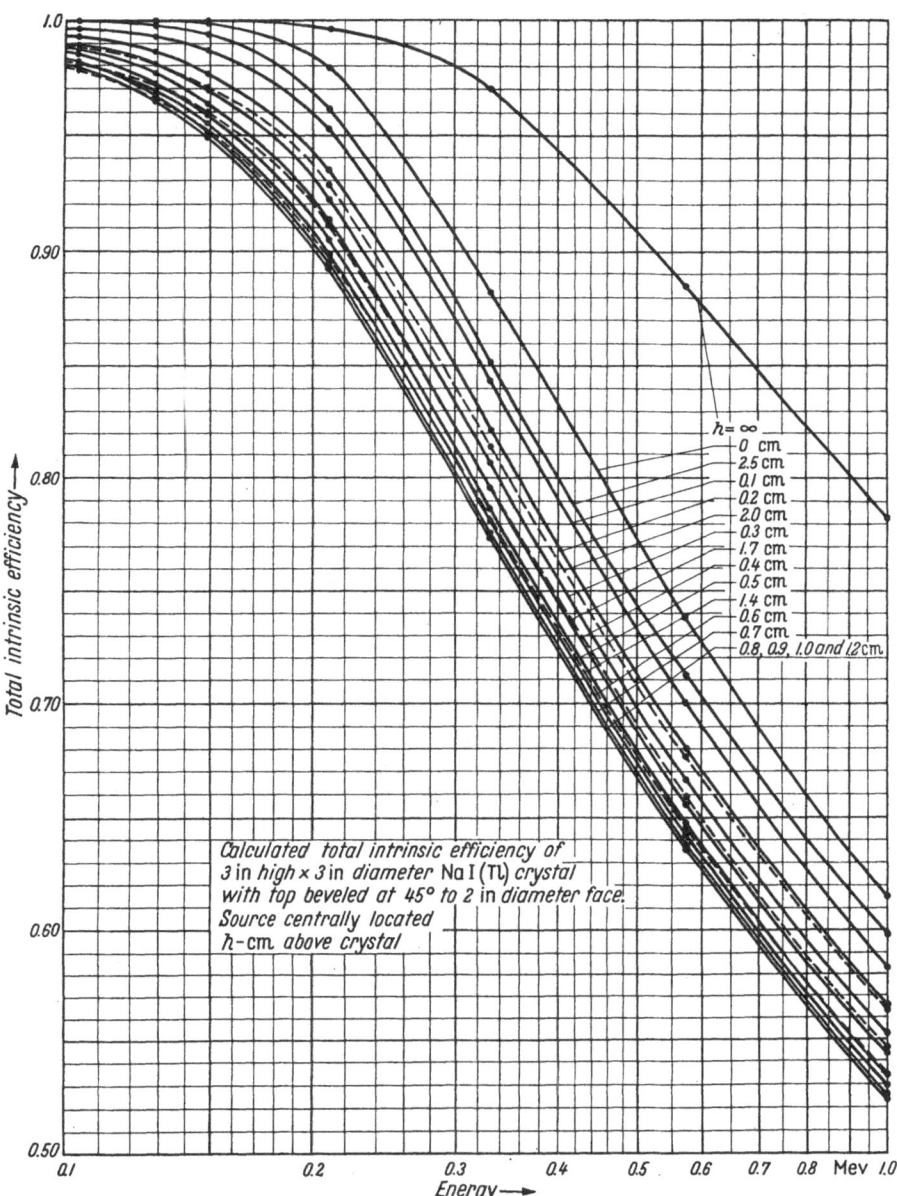

Fig. 29. Detection efficiency of 3 in. diameter × 3 in. high NaI(Tl) crystal (Mathematics Panel, ORNL[45]).

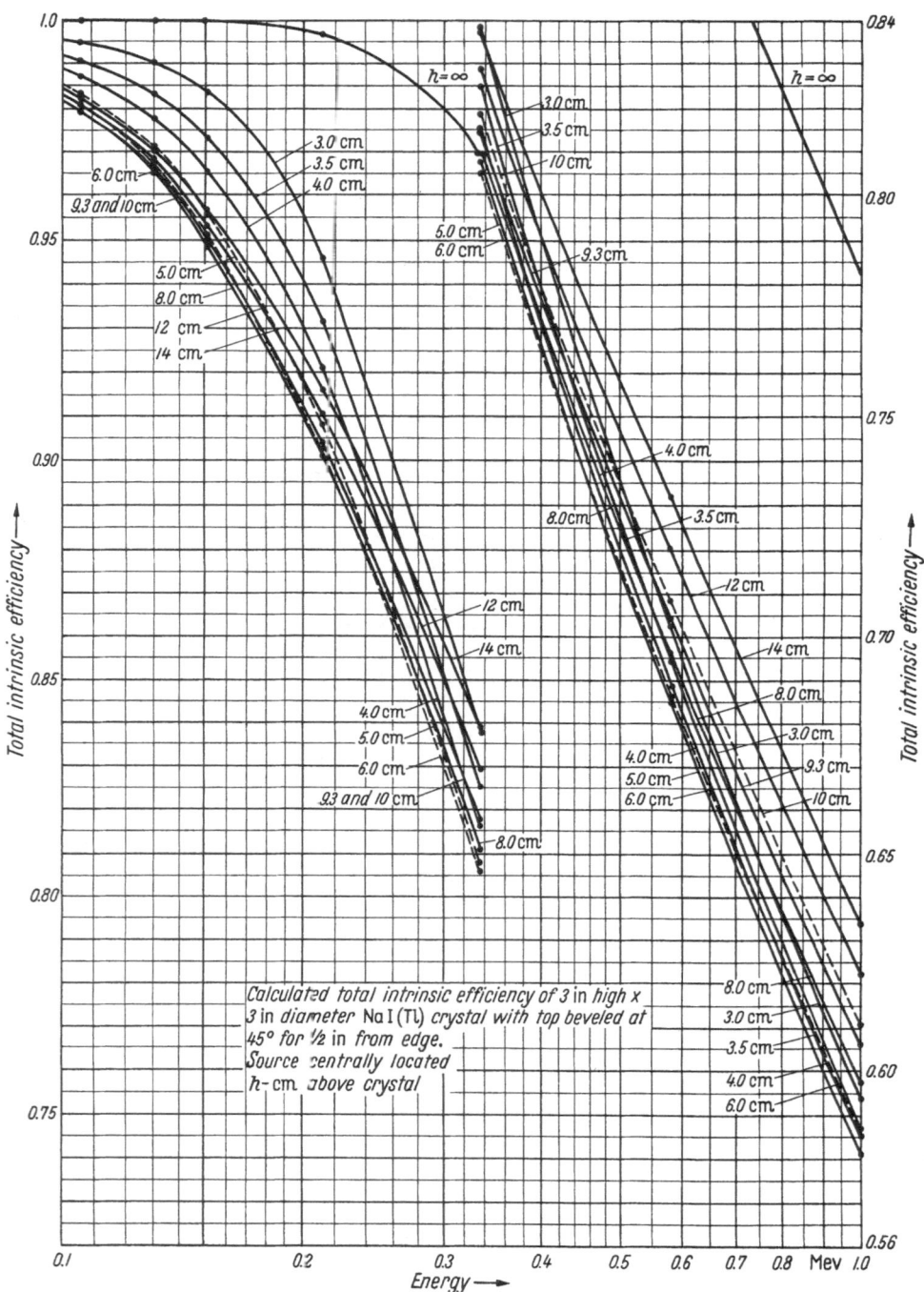

Fig. 30. Detection efficiency of 3 in. diameter × 3 in. high NaI(Tl) crystal (Mathematics Panel ORNL[45]).

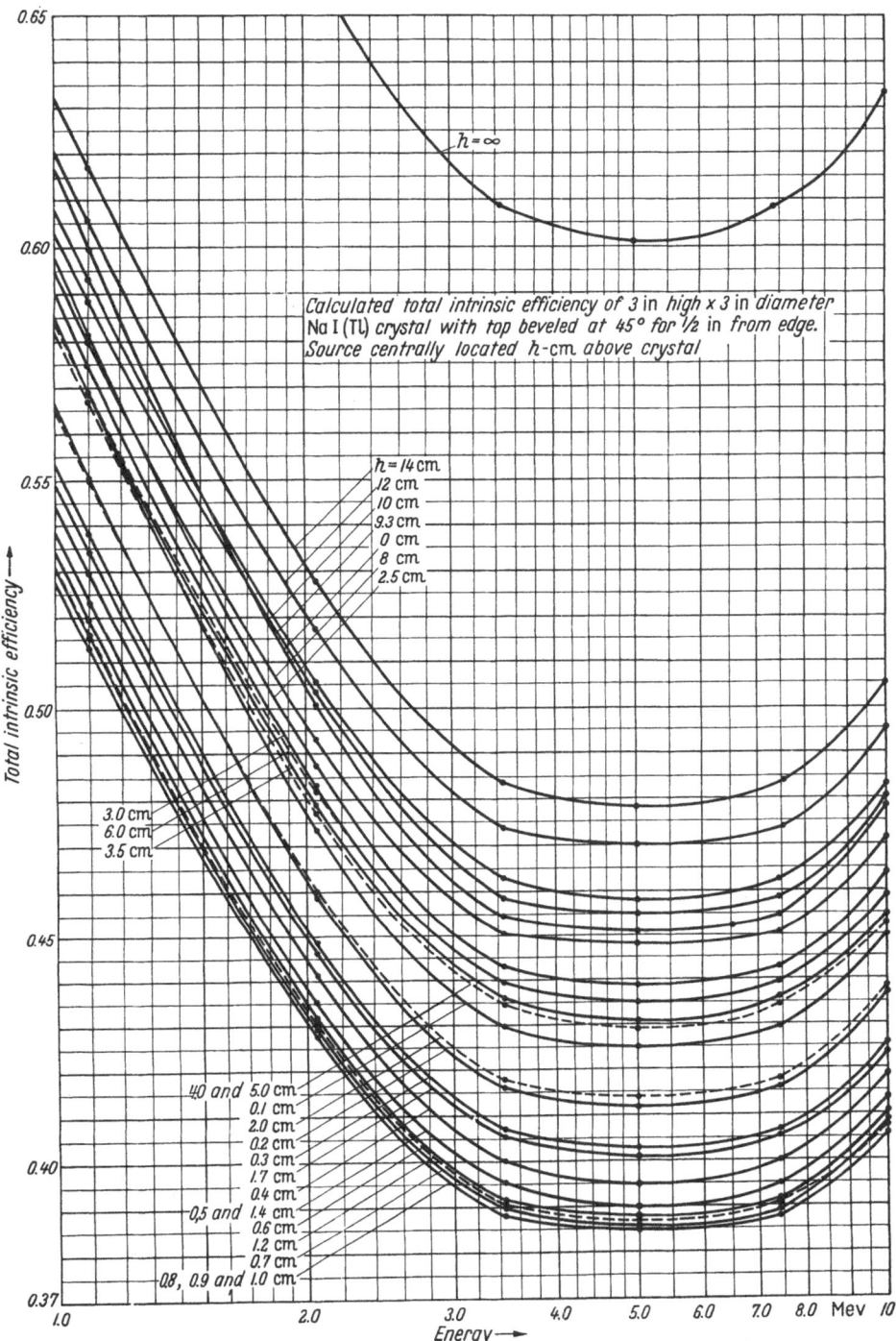

Fig. 31. Detection efficiency of 3 in. diameter × 3 in. high NaI(Tl) crystal (Mathematics Panel, ORNL[45]).

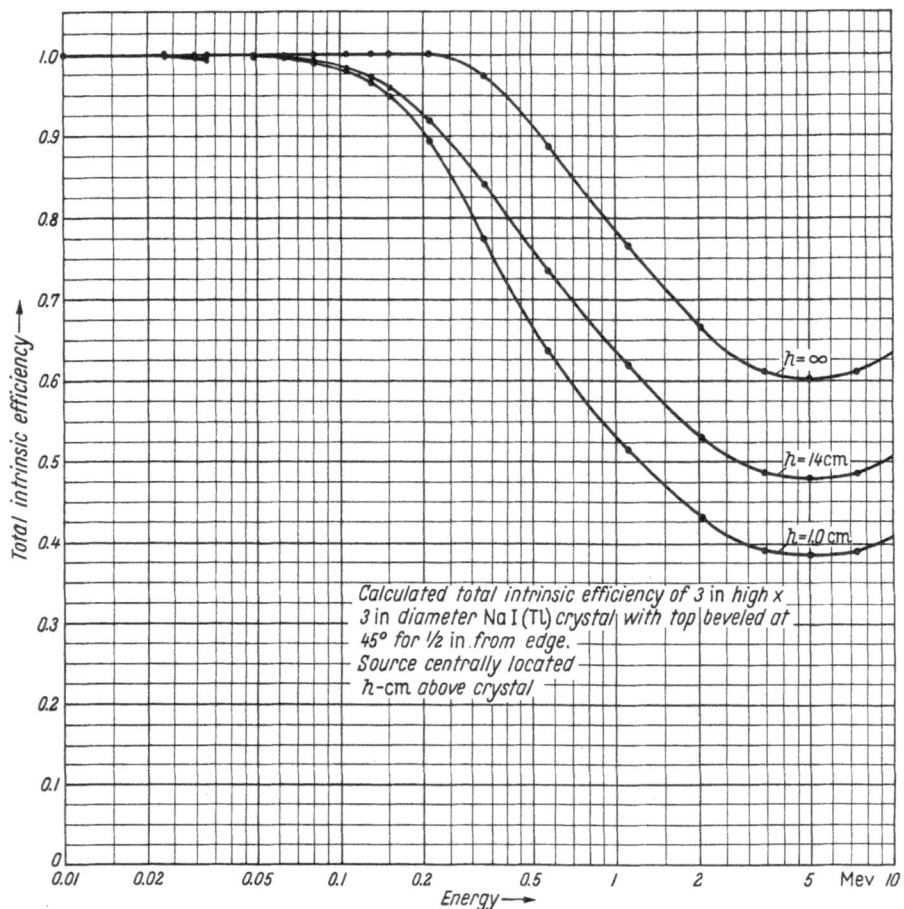

Fig. 32. Detection efficiency of 3 in. diameter × 3 in. high NaI(Tl) crystal (Mathematics Panel, ORNL[45]).

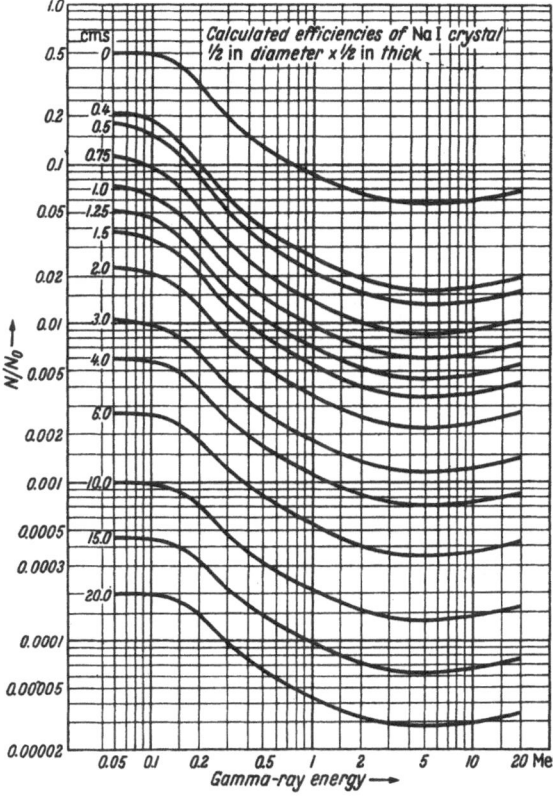

Fig. 33 Calculated efficiency $(\varepsilon_t \Omega_c)$ of $^1/_2$ in. diameter $\times$ $^1/_2$ in. high NaI crystal for various source-to-crystal distances (WoLICKI, JASTROW and BROOKS[46]).

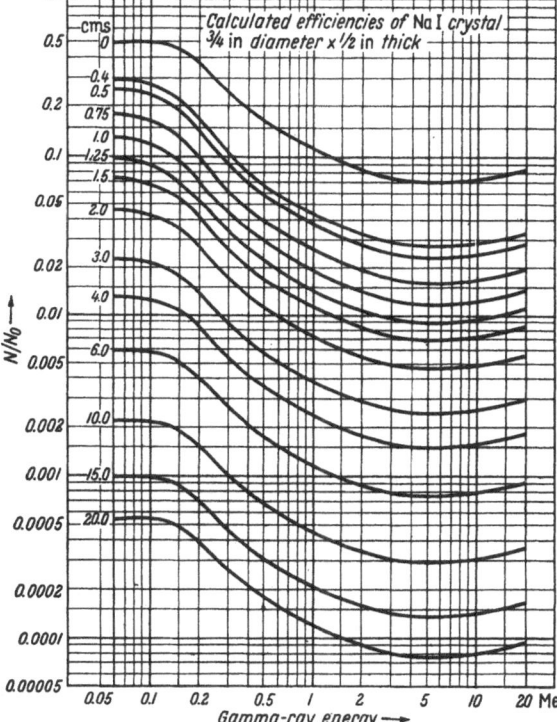

Fig. 34. Calculated efficiency $(\varepsilon_t \Omega_c)$ of $^3/_4$ in. diameter $\times$ $^1/_2$ in. high NaI crystal for various source-to-crystal distances (WoLICKI, JASTROV and BROOKS[46]).

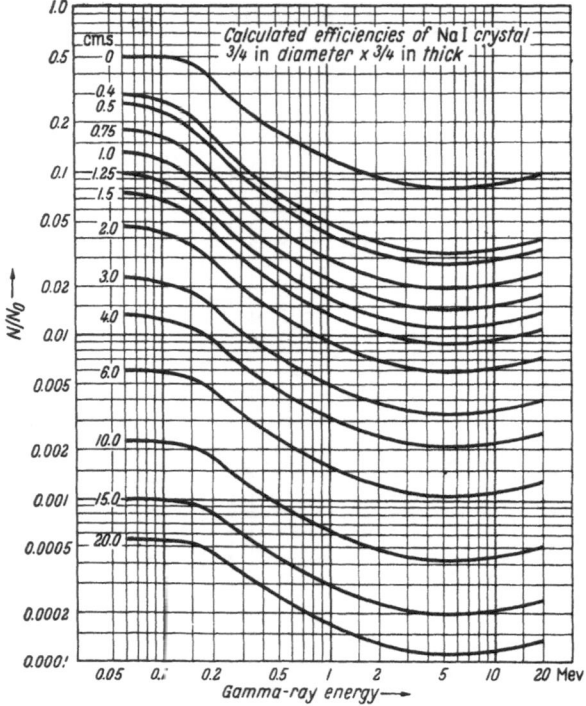

Fig. 35. Calculated efficiency ($\varepsilon_t \Omega_c$) of ³/₄ in. diameter × ³/₄ in. high NaI crystal for various source-to-crystal distances (Wolicki, Jastrow and Brooks[46]).

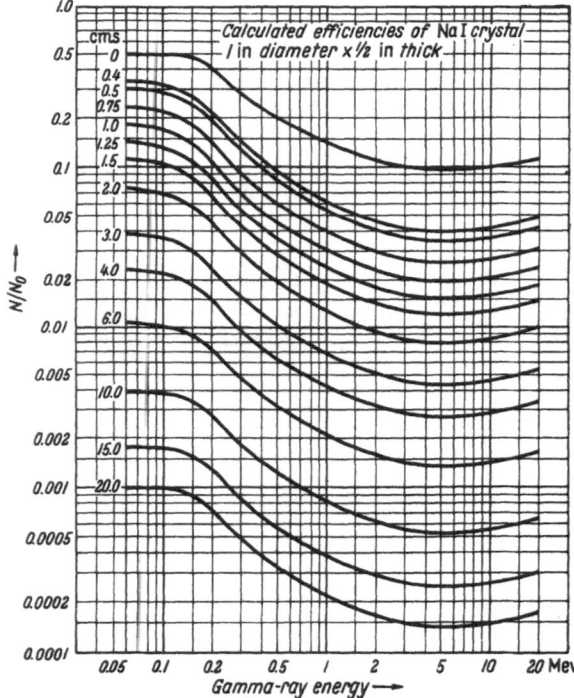

Fig. 36. Calculated efficiency ($\varepsilon_t \Omega_c$) of 1 in. diameter × ¹/₂ in. high NaI crystal for various source-to-crystal distances (Wolicki, Jastrow and Brooks[46]).

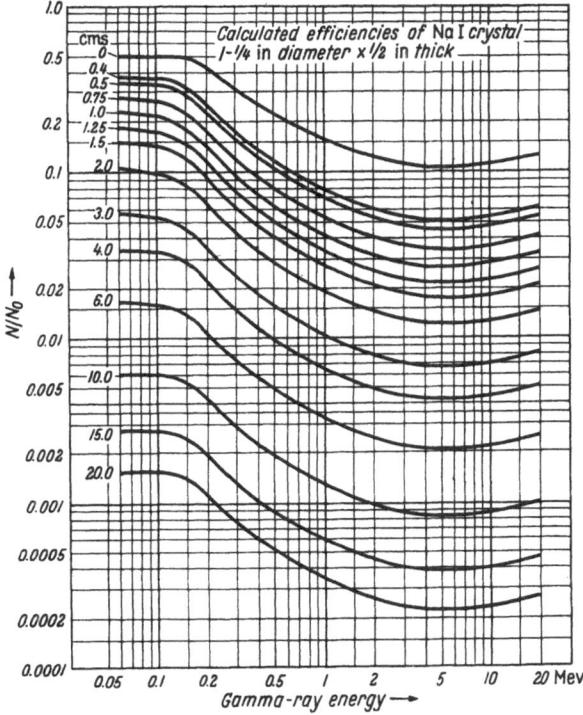

Fig. 37. Calculated efficiency ($\varepsilon_t\Omega_c$) of $1\frac{1}{4}$ in. diameter $\times \frac{1}{2}$ in. high NaI crystal for various source-to-crystal distances. (WOLICKI, JASTROW and BROOKS[46]).

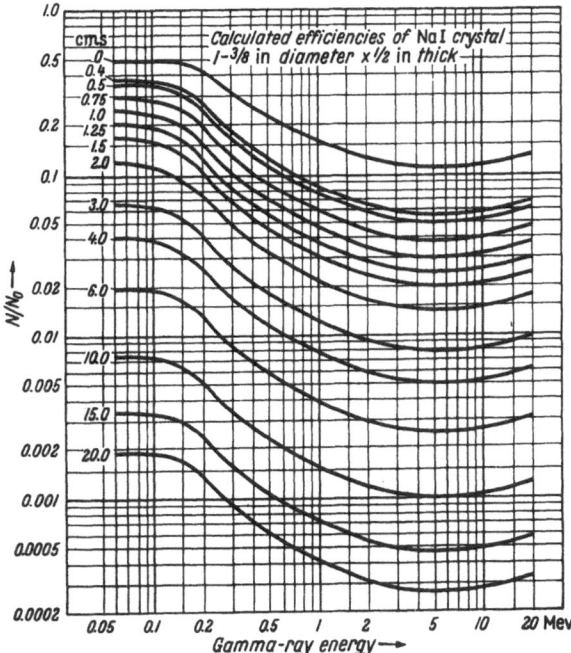

Fig. 38. Calculated efficiency ($\varepsilon_t\Omega_c$) of $1\frac{3}{8}$ in. diameter $\times \frac{1}{2}$ in. high NaI crystal for various source-to-crystal distances. (WOLICKI, JASTROW and BROOKS[46]).

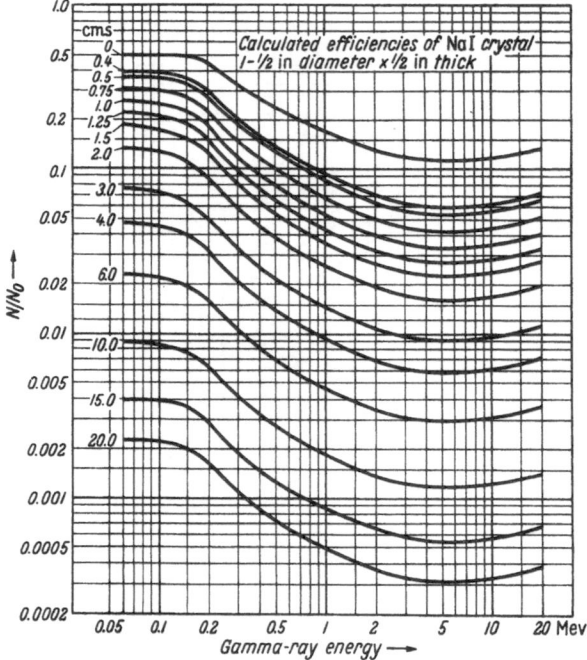

Fig. 39. Calculated efficiency $(\varepsilon_t \Omega_o)$ of $1\frac{1}{2}$ in. diameter $\times \frac{1}{2}$ in. high NaI crystal for various source-to-crystal distances (Wolicki, Jastrow and Brooks[46]).

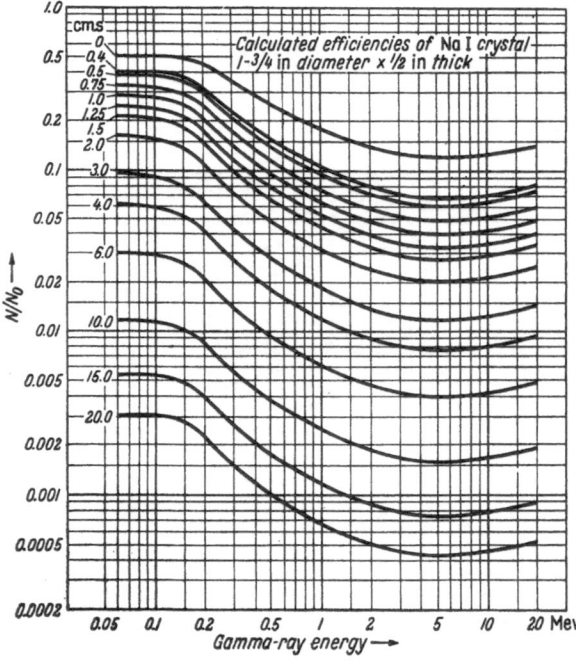

Fig. 40. Calculated efficiency $(\varepsilon_t \Omega_o)$ of $1\frac{3}{4}$ in. diameter $\times \frac{1}{2}$ in. high NaI crystal for various source-to-crystal distances (Wolicki, Jastrow and Brooks[46]).

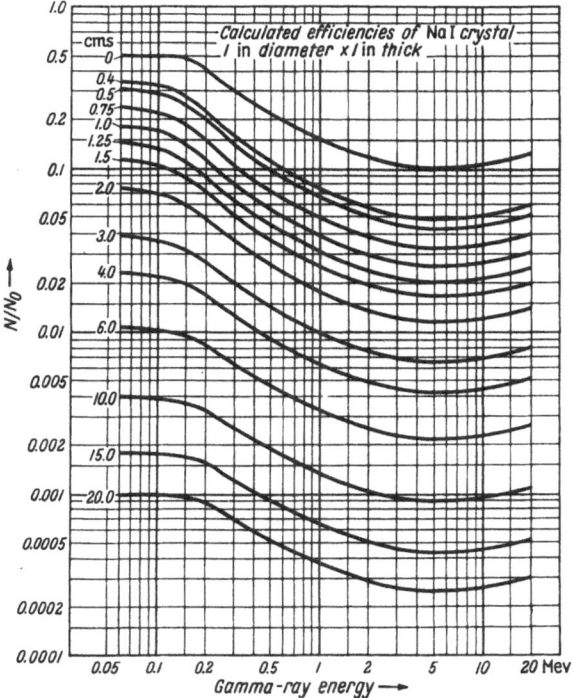

Fig. 41. Calculated efficiency ($\varepsilon_t \Omega_c$) of 1 in. diameter × 1 in. high NaI crystal for various source-to-crystal distances (Wolicki, Jastrow and Brooks[46]).

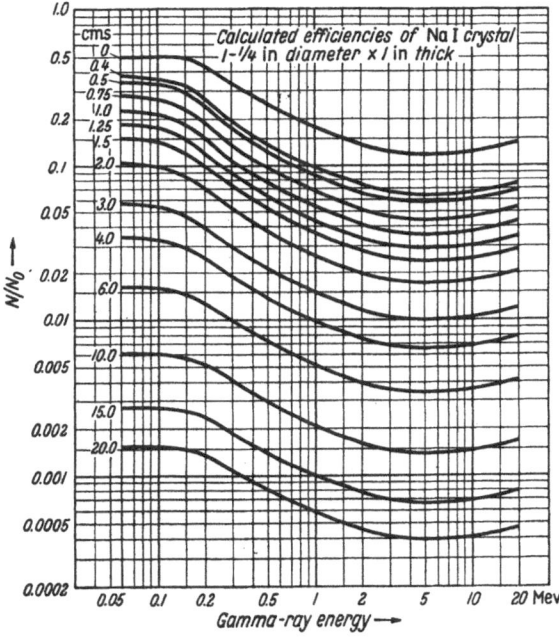

Fig. 42. Calculated efficiency ($\varepsilon_t \Omega_c$) of $1^1/_4$ in. diameter × 1 in. high NaI crystal for various source-to-crystal distances Wolicki, Jastrow and Brooks[46]).

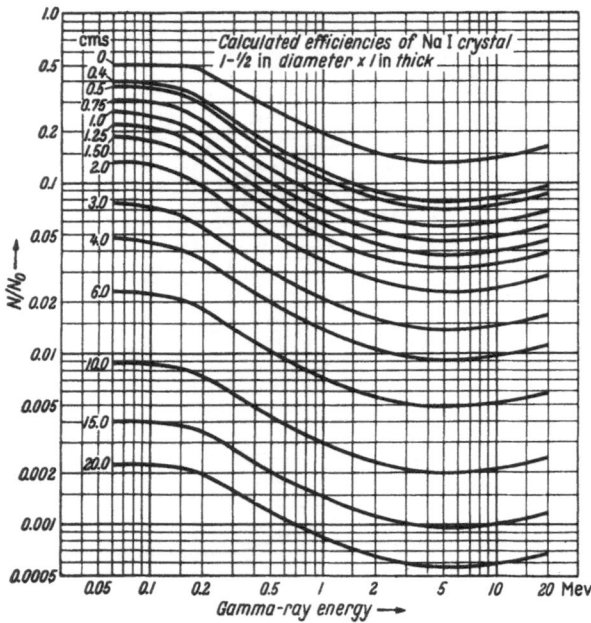

Fig. 43. Calculated efficiency $(\varepsilon_t \Omega_c)$ of $1^1/_2$ in. diameter $\times 1$ in. high NaI crystal for various source-to-crystal distances (WOLICKI, JASTROW and BROOKS[46]).

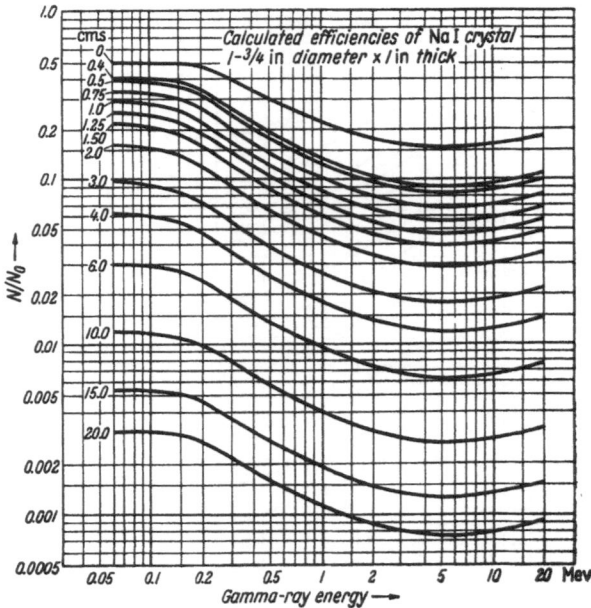

Fig. 44. Calculated efficiency $(\varepsilon_t \Omega_c)$ of $1^3/_4$ in. diameter $\times 1$ in. high NaI crystal for various source-to-crystal distances (WOLICKI, JASTROW and BROOKS[46]).

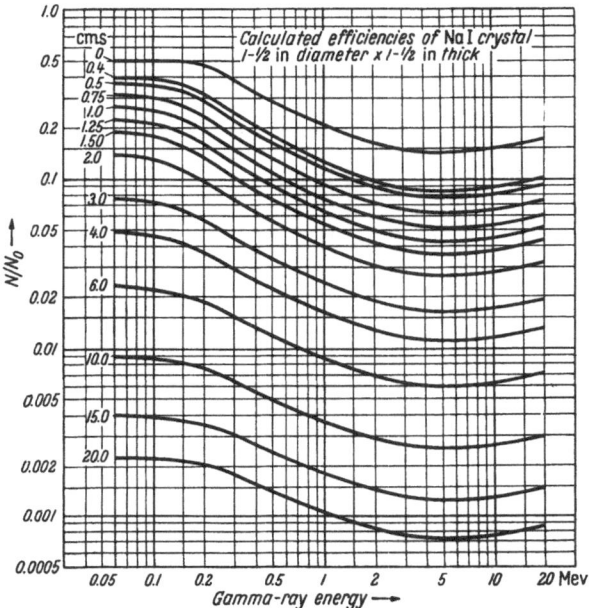

Fig. 45. Calculated efficiency $(\varepsilon_t \Omega_c)$ of $1^{1}/_{2}$ in. diameter $\times 1^{1}/_{2}$ in. high NaI crystal for various source-to-crystal distances (WOLICKI, JASTROW and BROOKS[46]).

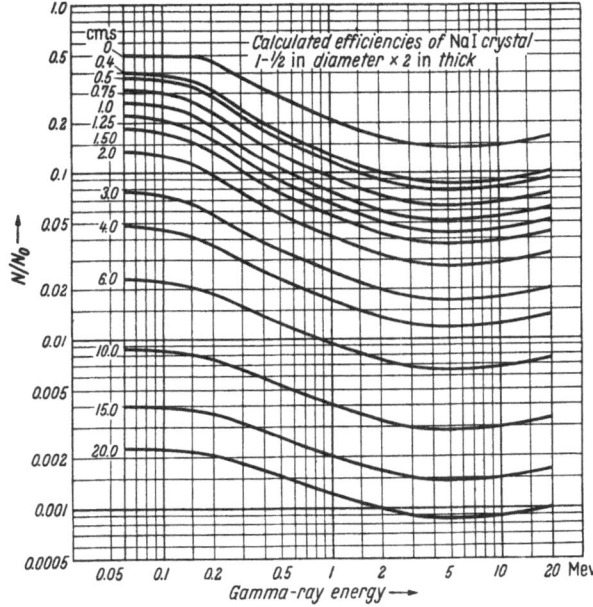

Fig. 46. Calculated efficiency $(\varepsilon_t \Omega_c)$ of $1^{1}/_{2}$ in. diameter $\times 2$ in. high NaI crystal for various source-to-crystal distances (WOLICKI, JASTROW and BROOKS[46]).

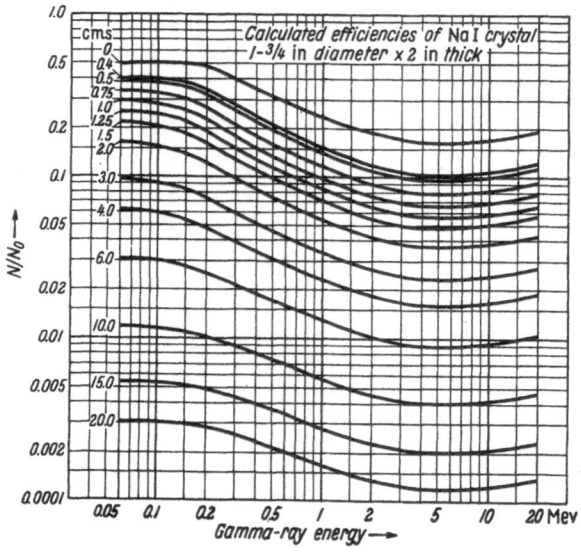

Fig. 47. Calculated efficiency $(\varepsilon_t \Omega_c)$ of $1^3/_4$ in. diameter $\times 2$ in. high NaI crystal for various source-to-crystal distances (WOLICKI, JASTROW and BROOKS[46]).

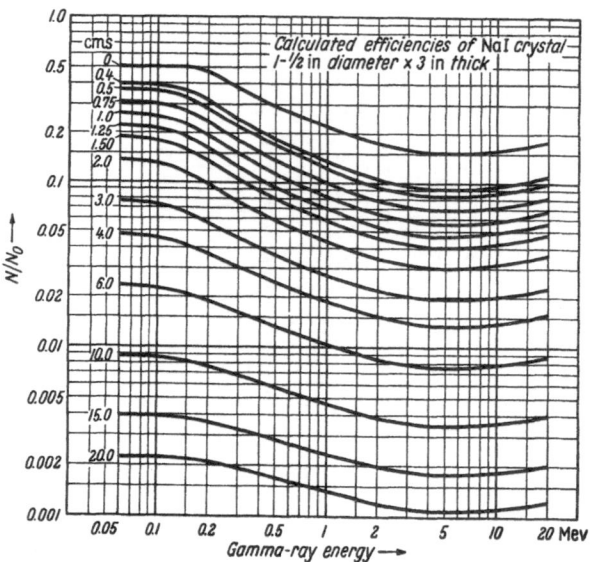

Fig. 48. Calculated efficiency $(\varepsilon_t \Omega_c)$ of $1^1/_2$ in. diameter $\times 3$ in. high NaI crystal for various source-to-crystal distances (WOLICKI, JASTROW and BROOKS[46]).

In addition to being a function of energy, the peak-to-total ratio, R, is also a function of crystal size and source-detector spacing. This may be attributed to the fact that the probability of a photon transferring all its energy to the crystal varies as the source-detector configuration is altered.

LAZAR et al.[47] have measured R versus energy at several distances for various sodium iodide crystals using sources of Ce^{141} (0.145 Mev), Cr^{51} (0.320 Mev), Be^7 (0.479 Mev), Sr^{85} (0.513 Mev), Cs^{137} (0.661 Mev), Nb^{95} (0.759 Mev), Mn^{54} (0.820 Mev), Zn^{65} (1.114 Mev), Na^{22} (1.275 and 0.511 Mev), Y^{88} (1.85 and 0.908 Mev), and Na^{24} (2.76 and 1.38 Mev). For the energy region above 1.114 Mev, where only sources that emitted two γ-rays were available, a correction was made

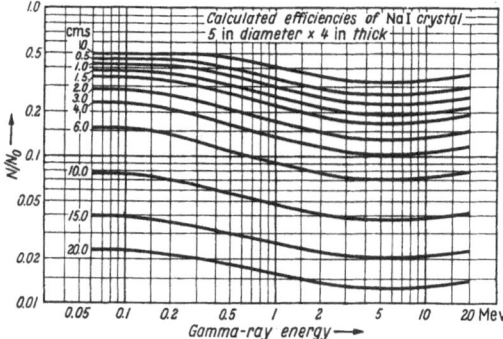

Fig. 49. Calculated efficiency ($\varepsilon_t \Omega_c$) of 5 in. diameter × 4 in. high NaI crystal for various source-to-crystal distances (WOLICKI, JASTROW and BROOKS[46]).

for the contribution to the total area from the lower-energy γ-rays by using the value of R already determined at the lower energy with a monoenergetic source. At 1.78, 2.14, and above 2.76 Mev, γ-rays were observed from (p, p'), $(p, γ)$, and $(p, α)$ reactions. To minimize scattering effects, the crystals were suspended in an empty

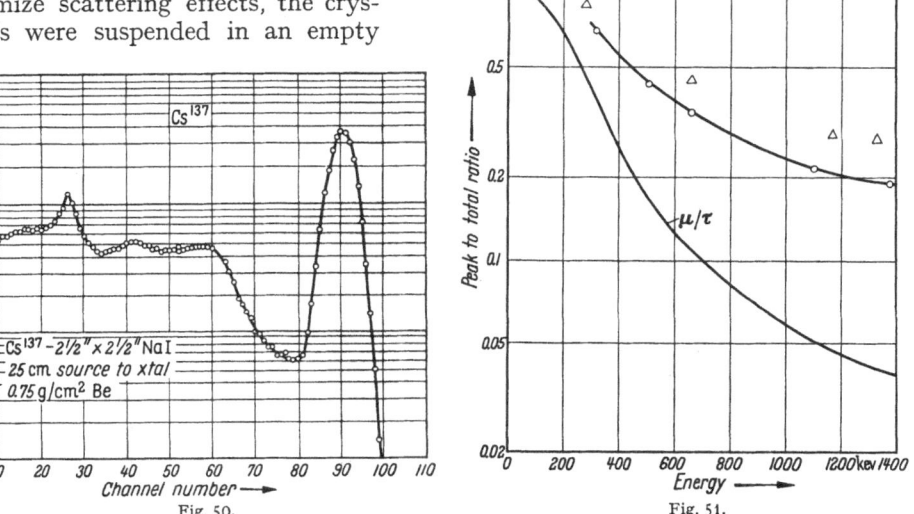

Fig. 50.

Fig. 50. Pulse height distribution for Cs^{137} γ-rays on a $2\frac{1}{2}$ in. diameter × $2\frac{1}{2}$ in. high NaI(Tl) crystal (BORKOWSKI[57]).

Fig. 51. Ratio of photopeak pulses to total pulses for a $1\frac{1}{2}$ in. diameter × 1 in. high NaI(Tl) crystal. For comparison, μ/τ is also given. O, Experimental, source at 2.5 cm (from LAZAR et al.[47]). Δ, Theoretical, collimated radiation (BERGER and DOGGETT[49]).

room in which the nearest walls were ≥ 4 feet away. A typical curve of R versus energy is shown in Fig. 51. It is to be noted that R differs considerably

[47] N. H. LAZAR, R. C. DAVIS and P. R. BELL: Nucleonics 14, No. 4, 52 (1956).

[57] C. J. BORKOWSKI: I.R.E. Trans. Nuclear Sci., N.S. 3, No. 4, 71 (1956).

[49] M. J. BERGER and J. DOGGETT: Rev. Sci. Instrum. 27, 269 (1956). — J. Res. Nat. Bur. Stand. 56, 355 (1956).

from μ/τ, the value it would have if multiple processes and electron escape could be neglected. The curves of photopeak efficiency versus energy resulting from these experiments* are given in Fig. 52. Efficienciss were checked at several energies by a β-γ coincidence experiment using sources where the β-rays and γ-rays were known to be in coincidence. In every case, the agreement was within

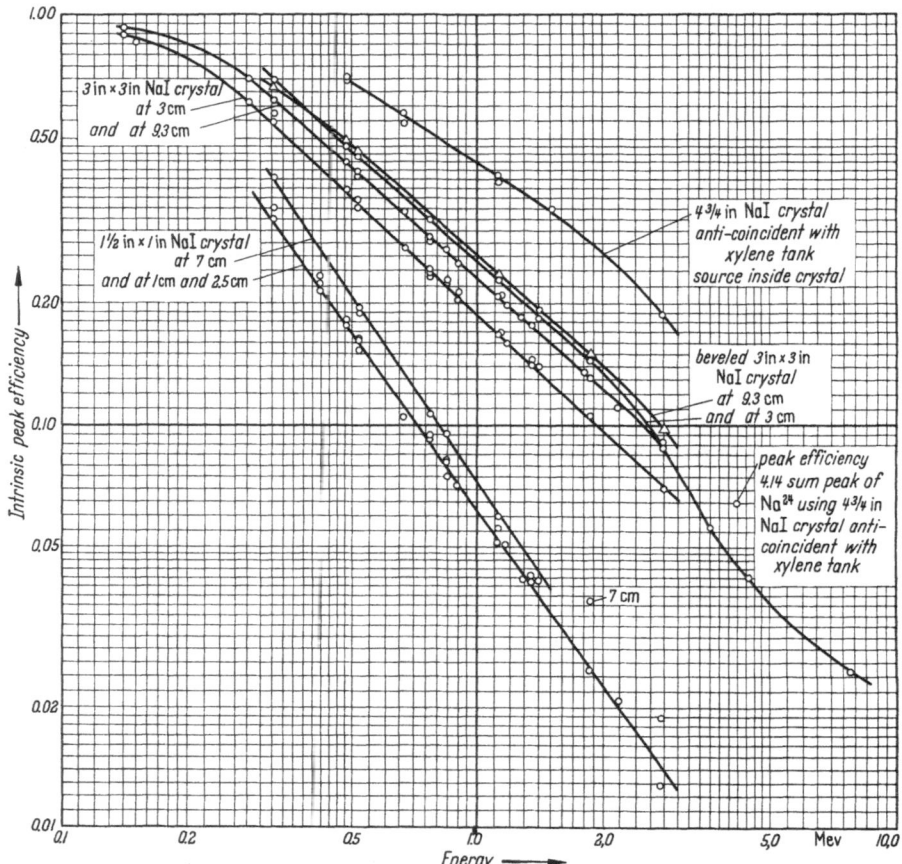

Fig. 52. Photopeak efficiency versus energy for NaI(Tl) crystals (LAZAR et al.[47]).

3%. Intrinsic peak efficiency curves[48] for several other NaI(Tl) crystals, including $1^3/_4$ in. diameter×2 in. high and 3 in.×3 in. well-type crystals, are shown in Fig. 53.

A computation of $R(E_\gamma)$ by the Monte Carlo method has recently been completed by BERGER and DOGGETT[49] using the National Bureau of Standards (U.S.) automatic computer (SEAC). Results were obtained for collimated monoenergetic γ-rays (with energies from 0.279 to 4.45 Mev) centrally incident on the endface of cylindrical NaI(Tl) crystals ranging in size from 0.5 in. diameter×0.5 in. high

---

* $\varepsilon_p(E)$ was calculated with the aid of Eq. (8.9) using experimental values of $R(E)$ and the calculated values of $\varepsilon_t(E)$ given in Figs. 24 to 32.

[48] J. E. FRANCIS, C. C. HARRIS and J. I. TROMBKA: Oak Ridge National Laboratory (U.S.) Report No. 2204 (Unpublished) (1956).

to 5 in. × 9 in. Although these values (Table 6) are in good agreement with those calculated analytically by MAEDER et al.[50], they are consistently higher than experimental values[5, 51-53]. BERGER and DOGGETT note that the agreement might have been better if the escape of primary photoelectrons and photo-electrons produced by the photo-electric absorption of Compton photons had been considered in the Monte Carlo calculations. There is also the possibility that background radiation (scattered radiation from the surroundings, etc.) was counted in the experimental work. This would depress the experimental values of R.

It was implied earlier that once the photopeak efficiency has been determined for a standard source-crystal configuration, absolute γ-ray intensity determinations could be made by measuring only the area under the photopeak portion of the spectrum. Attention is now drawn to a correction that is required when the intensity of low energy γ- and x-radiation is being measured.

For monoenergetic radiation below 100 kev, where $\mu = \tau$, the ratio of the number of pulses in the photopeak to the number in the total spectrum would be unity if all of the K x-rays following photoelectric absorption in iodine were absorbed in the crystal, i.e., every event would contribute to the photopeak.

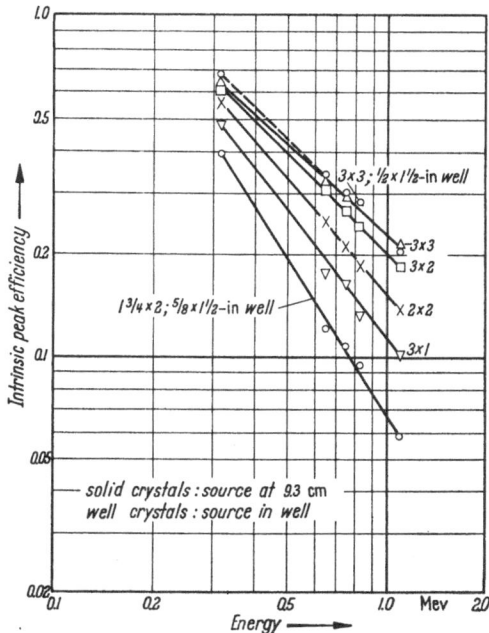

Fig. 53. Photopeak efficiency versus energy for NaI(Tl) crystals (FRANCIS et al.[48]).

Table 6. *Peak-to-total ratio for collimated radiation incident centrally on the endface of a cylindrical NaI (Tl) crystal.* (BERGER and DOGGETT[49]).

Crystal radius: length:	2.5 in. 9.0 in.	2.5 in. 8.0 in.	2.5 in. 4.0 in.	0.875 in. 2.0 in.	0.75 in. 2.0 in.	0.75 in. 1.5 in.	0.75 in. 1.0 in.	0.5 in. 1.0 in.	0.25 in. 0.5 in.	No. of Monte Carlo Histories
Energy (Mev)										
0.279	0.973 ±0.002	0.973 ±0.002	0.971 ±0.002	0.915 ±0.004	0.900 ±0.004	0.882 ±0.004	0.843 ±0.005	0.814 ±0.006	0.711 ±0.008	6000
0.661	0.887 ±0.003	0.884 ±0.003	0.821 ±0.004	0.542 ±0.005	0.508 ±0.006	0.481 ±0.006	0.442 ±0.007	0.377 ±0.007	0.243 ±0.008	10000
1.17	0.778 ±0.006	0.775 ±0.006	0.682 ±0.007	0.368 ±0.008	0.343 ±0.008	0.315 ±0.009	0.282 ±0.010	0.235 ±0.010	0.140 ±0.011	5000
1.33	0.767 ±0.006	0.758 ±0.006	0.667 ±0.007	0.351 ±0.009	0.325 ±0.009	0.303 ±0.009	0.272 ±0.010	0.224 ±0.010	0.120 ±0.011	5000
2.62	0.653 ±0.008	0.643 ±0.009	0.531 ±0.010	0.243 ±0.014	0.226 ±0.015	0.201 ±0.016	0.175 ±0.017	0.144 ±0.017	0.0950 ±0.018	6000
4.45	0.621 ±0.015	0.608 ±0.015	0.491 ±0.020	0.169 ±0.033	0.157 ±0.033	0.132 ±0.035	0.119 ±0.036	0.0826 ±0.037	0.0890 ±0.039	5000

[50] D. MAEDER, R. MÜLLER and V. WINTERSTEIGER: Helv. phys. Acta **27**, 3 (1954).
[51] W. E. KREGER: Phys. Rev. **96**, 1554 (1954).
[52] K. LIDÉN and N. STARFELT: Ark. Fysik **7**, 427 (1954).
[53] R. S. FOOTE and H. W. KOCH: Rev. Sci. Instrum. **25**, 746 (1954).

Actually, a number of the K x-rays escape from the crystal (primarily from the front face) and an "escape peak" is produced in the pulse spectrum at energy $E - E_K$, where E_K is the mean value of the K_α and K_β lines weighted according to intensities and is 29 kev for iodine. The recorded pulse spectrum therefore consists of two peaks, namely, a photoelectric peak of energy E and an escape peak at $(E - E_K)$. This is illustrated in Fig. 54 where the response of a small NaI(Tl) crystal to 44 kev x-rays is shown. For full energy peak which appears at approximately 78 volts both the iodine photoelectron and the K x-ray were absorbed, whereas the peak at 27 volts (15 kev) was produced by the occasional escape of the K x-ray from the front face of the crystal.

The ratio (P) of the number of pulses in the escape peak (n_{esc}) to the number in the photopeak (n_p) depends upon the energy of the incident photons (E), the

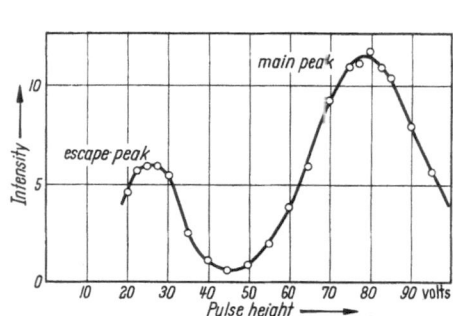

Fig. 54. The response of a NaI(Tl) crystal to 44 kev x-rays (WEST et al.[14]).

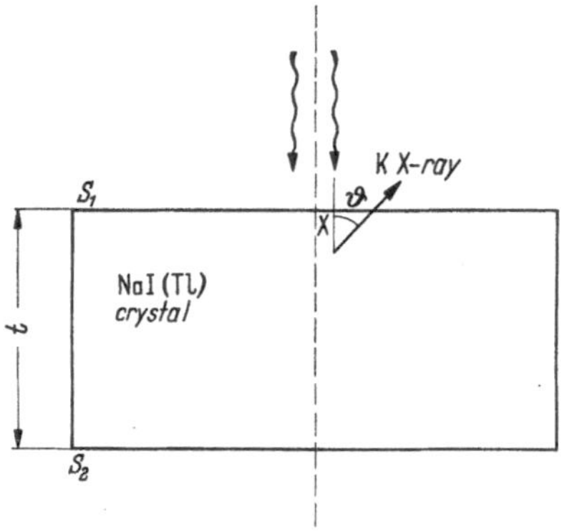

Fig. 55. Geometry for calculating the iodine K x-ray escape factor for the front face of a NaI(Tl) crystal. Case of normally incident collimated radiation.

dimensions of the crystal, and the experimental geometry. For the purposes of most calculations over the energy range where the escape peak is prominent (33 kev to $\sim$150 kev), $P(E)$ can be expressed as

$$P(E) = \frac{p(E)}{1 - p(E)} \tag{8.10}$$

where

$$p(E) = \frac{n_{esc}(E)}{N_0(E)} \tag{8.11}$$

is the probability that an incident photon will undergo a photoelectric effect with the iodine K x-ray escaping from the crystal, and $N_0(E)$ is the number of incident photons.

For the special case of a collimated beam of photons normally incident on a surface S_1 (Fig. 55) whose linear dimensions are large compared with the mean free path of the K x-rays, the escape probability is

$$p(E_\gamma) = \tfrac{1}{2} \omega_{\bar{K}} \, \delta_K \, \mu \int\limits_0^t \int\limits_0^{\pi/2} e^{-\left(\tau + \frac{\tau_K}{\cos\vartheta}\right)x} \sin\vartheta \, d\vartheta \, dx \tag{8.12}$$

where
   $\tau =$ linear absorption coefficient,
   $\tau_K =$ linear absorption coefficient of the iodine K x-rays,
   $\mu =$ photoelectric absorption coefficient,

$\delta_K =$ the fraction of photoelectric events taking place in the K-shell,
$\omega_K =$ K-fluorescence yield of iodine,
$x =$ depth of penetration of incident photon into crystal of thickness t,
$\vartheta =$ angle between the direction of the normal to the surface and the K x-ray.

Because of the high absorption coefficient of the iodine K x-rays in sodium iodide, escape from $x > 2$ mm is small and t may be replaced by ∞. Upon integrating,

$$p(E) = \frac{1}{2}\,\omega_K\,\delta_K\,\frac{\mu}{\tau}\left[1 - \frac{\tau_K}{\tau}\ln\left(1 + \frac{\tau}{\tau_K}\right)\right]. \tag{8.13}$$

It is important to note that Eq. (8.13) gives the escape factor for the front surface of the crystal only. However, since most practical crystals can be considered infinitely thick for photons with energies in the range under consideration, escape from the back surface is negligible compared with that from S_1. Detailed calculations on the probability of escape of K x-rays from the back and side surface of a sodium iodide crystal for normally incident radiation have been carried out by LIDÉN and STARFELT[52]; AXEL[54] has derived formulas for geometries in which the crystal subtends any conical shell of radiation from a source.

Calculated curves of $P(E)$ versus energy for several different geometries are given in Fig. 56. In each case Eq. (8.10) has been used to relate $p(E)$ to $P(E)$. For comparison, some experimental values obtained by MEYERHOF and WEST[55] and by McGOWAN[56] are also shown.

A quantitative measurement of the rate of emission of photons of energy E from a source with a scintillation spectrometer therefore requires a knowledge

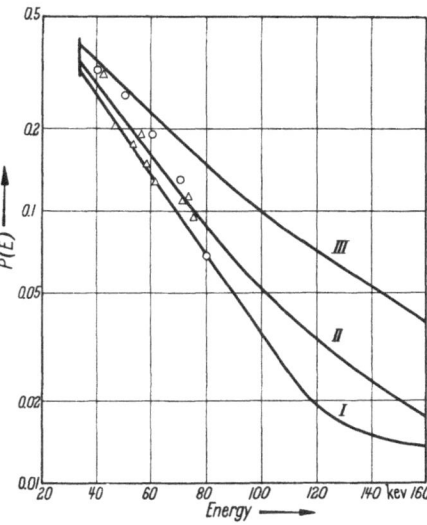

Fig. 56. Ratio $P(E)$ of the number of pulses in the escape peak to the number in the photopeak for a sodium iodide crystal with dimensions large compared with the mean free path of the incident photons (no escape from back and side faces of the crystal). The calculated curves I, II, and III are for normally incident radiation [Eq. (8.13)], incident radiation in cone of half-angle 60° (AXEL[54]), and incident radiation in cone of half-angle 90° (AXEL[54]), respectively. Experimental points: O, Collimated radiation (MEYERHOF and WEST[55]). △, Geometry unknown (McGOWAN[56]).

of the total detection efficiency of the crystal $\varepsilon_t(E)$, the peak to total ratio $R(E)$, the escape peak to photopeak ratio $P(E)$, and an integration of the pulse spectrum in the full-energy peak. Of considerable import to the user of the scintillation spectrometer is the fact that once $\varepsilon_t(E)$, $R(E)$, and $P(E)$ have been determined for a given crystal, another crystal of the same size and for the same source-crystal configuration will have identical values.

b) Energy resolution with scintillation spectrometer.

One of the most useful tools available on a routine basis to the experimental nuclear physicist and chemist today is the scintillation spectrometer. For this reason, many workers have been and will be faced with the problem of developing quickly and easily an instrument with adequate energy resolution to allow the investigation of a broad range of problems. Particularly at the small laboratory

[54] PETER AXEL: Rev. Sci. Instrum. **25**, 391 (1954).
[55] W. MEYERHOF and H. WEST jr.: Rev. Sci. Instrum. **25**, 1025 (1954).
[56] F. K. McGOWAN: Phys. Rev. **93**, 163 (1954).

where the worker is unable to carefully select photomultiplier tubes and crystals (mainly for economic reasons), the desired result can often be difficult and tedious to achieve. In the following, some of the techniques which are used to obtain good γ-ray energy resolution with a NaI(Tl) spectrometer are reviewed. The remarks are applicable to most of the other inorganic phosphors except that some do not need to be protected from water. Factors affecting energy resolution and intrinsic crystal resolution are also discussed.

9. Crystal preparation and mounting. Even though the quality of photomultiplier tubes and NaI(Tl) crystals has improved considerably in recent years, the outcome of attempts to obtain good energy resolution with a NaI(Tl) scintillation spectrometer still depends largely upon the method of preparing the phosphor and mounting it on the phototube. Most of the difficulties encountered are directly associated with two properties of the material. In the first place, sodium iodide is chemically unstable in the presence of water vapor and due to the release of free iodine, a yellow discoloration occurs which damages the optical properties of the crystal. This means that every precaution must be taken to prevent water from contacting the surfaces of the mounted crystal if the spectrometer is to remain free of drifts in pulse height and energy resolution for long periods. Secondly, sodium iodide has a very high index of refraction ($\sim$1.8) in the wavelength region of its fluorescence. Because of the small critical angle with respect to air, most of the light generated by the phosphor will be totally reflected at the crystal-air interface between the crystal and the photomultiplier tube unless a good optical coupling material is used. However, even with an effective crystal-phototube glass (refractive index $\sim$1.5) interface, the light loss due to reflection is still severe and methods must be devised for redirecting the light and returning it to the exit face of the crystal within the critical angle. Or in other words, the objective is to prevent the light from undergoing successive total internal reflections and being permanently trapped within the crystal. Techniques for increasing the light collection efficiency of the crystal-phototube combination and for protecting the crystal against moisture are described below [5, 33].

Starting with a crystal which has been cut to shape with a string saw or other appropriate tool, all surfaces should be rough ground with dry emery or carborundum paper (No. 120 to 320 grit) in a dry box having a relative humidity less than 10%. In effect, the grinding produces a surface which is free of water and free iodine and which efficaciously diffuses the light. In general, no improvement in light output or energy resolution is obtained by finer grinding or by polishing of any kind.

The exit face of the crystal should next be optically coupled to the phototube envelope (or light pipe*) with a material such as Dow-Corning DC-200 silicone oil (viscosity: 10^6 centistokes) or white opthalmological petrolatum. A thin uniform layer of coupling medium is preferable because there is then less tendency for flow when the photomultiplier tube is used in a horizontal position. The excess of material which is squeezed from the joint should be completely removed; grinding paper should be used to remove fluid from the exposed surfaces of the crystal. Since a good optical joint reduces the amount of light which is ultimately lost, its importance cannot be overemphasized.

In order to effectively return the light striking the top and lateral surfaces of the crystal to the exit face, the crystal should be coated with a highly efficient diffuse reflector, e.g., magnesium oxide or α-aluminum oxide [58]. Either of these

---

* Light pipes are frequently used with small- to moderate-sized crystals to minimize the effect of photocathode nonuniformity on the energy resolution.

[58] Linde "A" Abrasive, Linde Air Products Company, Chicago, Illinois.

reflectors will give a larger light output and better energy resolution than available specular surfaces. Although the reflecting power of α-alumina is somewhat less than that of MgO, it is preferred by some workers because its efficiency is greater when traces of oil are present. Also, the alumina does not react with NaI, whereas the MgO appears to react slightly[5].

A rather ingenious method of mounting large and irregularly shaped hygroscopic phosphors has been described by BELL[5]. The mounting arrangement for a crystal having the same diameter as the photomultiplier tube is shown in Fig. 57. In present usage the reflector, a mixture of α-alumina and sodium silicate, is sprayed on the inside of a 0.005 in. thick aluminum foil can which has been sealed with R-313[59]. The can is carefully lowered over the crystal and sealed to the photomultiplier tube with Apiezon Q vacuum wax and black Scotch electrical tape No. 33. A hypodermic needle is inserted through the seal and a vacuum applied until the thin-walled can is collapsed around the photomultiplier and crystal. The needle is withdrawn while pressing on the Apiezon seal. Sufficient mechanical support is provided by the assembly to allow a 3 in. diameter photomultiplier with a 3 in. diameter × 3 in. long NaI(Tl) crystal to be placed in any position for extended periods without danger. One of the chief advantages of

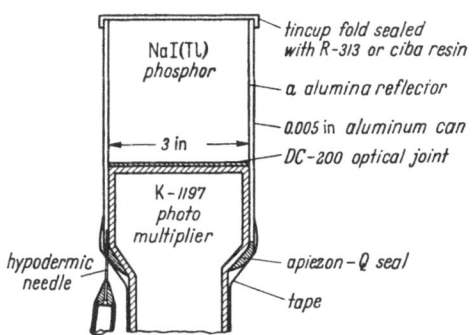

Fig. 57. Mounting technique for crystal having same diameter as the photomultiplier tube (BELL[5]).

this type of mount is that a minimum of scattering material is employed around the crystal. Thus, the loss of performance which results from counting Compton electrons and γ-rays produced outside of the crystal is minimized. Using this mounting arrangement, a width (full width of peak at half-maximum counting rate) of about 8% can readily be obtained for the 0.661 Mev γ-rays from Cs[137] with a 3 in. × 3 in. NaI(Tl) crystal and a DuMont No. 6363 photomultiplier tube.

10. Factors affecting energy resolution. Statistical fluctuations in the various processes by which the energy dissipated in a phosphor is converted into a voltage pulse at the output of a scintillation counter are responsible for the variations in magnitude of the output pulses produced by monoenergetic charged particles (photoelectrons, recoil protons, incident α-particles, etc.). Experimental and theoretical studies[60-71] of the width of the generated pulse height distribution

[59] Carl H. Biggs Company, Los Angeles, California.
[60] G. A. MORTON and J. A. MITCHELL: RCA-Review **9**, 632 (1948). — Nucleonics **4**, No. 1 16 (1949).
[61] G. A. MORTON: RCA-Review **10**, 525 (1949).
[62] R. C. HOYT: Rev. Sci. Instrum. **20**, 178 (1949).
[63] F. SEITZ and D. W. MUELLER: Phys. Rev. **78**, 605 (1949).
[64] G. F. J. GARLICK and G. T. WRIGHT: Proc. Phys. Soc. Lond. B **65**, 415 (1952).
[65] P. W. ROBERTS: Proc. Phys. Soc. Lond. A **66**, 192 (1953).
[66] G. T. WRIGHT and G. F. J. GARLICK: Brit. J. Appl. Phys. **5**, 13 (1954).
[67] G. T. WRIGHT: J. Sci. Instrum. **31**, 377 (1954).
[68] E. BREITENBERGER: Progress in Nuclear Physics, Vol. 4, p. 56. London: Pergamon Press 1955.
[69] G. F. J. GARLICK: J. Sci. Instrum. **32**, 369 (1955).
[70] G. G. KELLEY, P. R. BELL, R. C. DAVIS and N. H. LAZAR: Nucleonics **14**, No. 4, 53 (1956). — I.R.E. Trans. Nuclear Sci., N.S. **3**, No. 4, 57 (1956).
[71] J. J. HOPKINS: Rev. Sci. Instrum. **22**, 29 (1951).

have shown that although the statistical processes occurring in the photomultiplier tube certainly limit the pulse height resolution capabilities of the scintillation spectrometer, there are a number of other very important factors. In general, then, for a given phosphor-photomultiplier arrangement, the following processes contribute to the broadening of the pulse height distribution for monoenergetic particles:

α) *Emission of photons by the phosphor.* In addition to the normal statistical fluctuations in the number of photons per scintillation (n_i) emitted in the wave length region where the photocathode surface is sensitive, there may be variations in n_i due to local variations in the luminescence efficiency of the phosphor, e.g., in an impurity-activated inorganic crystal the luminescence efficiency would vary from point to point if the activating ions were nonuniformly distributed. Also, successive particles may lose different amounts of energy to the phosphor due to interaction, edge, and scattering effects, and possibly the nature of the luminescence process itself is such that the conversion of absorbed energy to photon energy fluctuates for successive particles.

β) *Collection of the emitted photons by the photocathode.* Since successive scintillations never occur at exactly the same points inside the phosphor, the optical geometry of the system, and hence the photon collection efficiency of the photocathode will, in general, be different for each photon of each scintillation even for optically perfect and geometrically regular phosphors. Moreover, the fraction of emitted photons collected by the photocathode may vary if the crystal contains optical flaws or is not optically transparent.

γ) *Emission of photoelectrons by the photocathode.* Of prime importance here are the normal statistical fluctuations in the number of photoelectrons per scintillation released from the photocathode. However, point-to-point variations of the response of the photocathode and the random emission of thermal electrons by the photocathode may also add to the variance of the pulse height distribution.

δ) *Collection of photoelectrons by the first dynode and multiplication by successive stages.* Although the variance from the multiplying section of the photomultiplier tube is fundamentally due to the normal statistical fluctuations in the multiplication process, contributions may arise from variations in the fraction of photoelectrons collected by the first dynode, in the collection efficiencies of subsequent dynodes, and in dynode response.

The analysis leading to the derivation of a general expression for the fractional variance of the output pulse height distribution in terms of the variations described above has been given by WRIGHT[67] and by BREITENBERGER[68]. For the special case of a counter in which there is no thermionic emission of electrons from the photocathode and the dynodes are uniform and identical in response, the equations for the mean number of electrons arriving at the anode following each scintillation and the fractional variance are respectively[67]

$$\overline{N} = \overline{n}\,\overline{p}\,\overline{s}^t \tag{10.1}$$

and

$$\frac{\mathrm{Var}\,(N)}{\overline{N}^2} = \left(1 + \frac{\sigma_p^2}{\overline{p}^2}\right)\frac{\sigma_n^2}{\overline{n}^2} + \frac{\sigma_p^2}{\overline{p}^2} + \frac{1}{\overline{n}\,\overline{p}}\left(\frac{\overline{s}}{\overline{s}-1} - \overline{p}\right) \tag{10.2}$$

where $\overline{n}$ is the mean number of photons produced per scintillation [in Eq. (10.2) it is assumed that $\overline{n} \gg 1$]; $\overline{p}$, a mean probability that a photon produces a photoelectron which reaches the first dynode, i.e., if p_{ij} is the probability that the j-th photon of the i-th scintillation (particle of energy E_0 gives n_i photons) produces a useful photoelectron, then

$$\bar{p}_i = \sum_{j=1}^{n_i} p_{ij}/n_i \quad \text{and} \quad \bar{\bar{p}} = \sum_{i=1}^{\varkappa} \bar{p}_i/\varkappa; \tag{10.3}$$

$\bar{s}$, the mean gain per stage; t, the number of stages; and σ_n^2 and σ_p^2, the variances (squared standard deviations) of n_i and $\bar{p}_i$, respectively, i.e.,

$$\sigma_n^2 = \sum_{i=1}^{\varkappa} (n_i - \bar{n})^2/\varkappa \quad \text{where} \quad \bar{n} = \sum_{i=1}^{\varkappa} n_i/\varkappa \tag{10.4}$$

and

$$\sigma_p^2 = \sum_{i=1}^{\varkappa} (\bar{p}_i - \bar{\bar{p}})^2/\varkappa. \tag{10.5}$$

If the pulse distribution described by Eqs. (10.1) and (10.2) is Gaussian, the fractional half-width η (full width at half-maximum) is related to the fractional variance by the equation

$$\eta^2 = 5.56 \frac{\text{Var}(N)}{\bar{N}^2}. \tag{10.6}$$

Under the further assumptions that there are no intrinsic effects in the phosphor and only normal statistical fluctuations occur in the rest of the system ($\sigma_n^2/\bar{n}^2 = 1/\bar{n}$; $\sigma_p^2 = 0$), Eqs. (10.2) and (10.6) reduce to

$$\eta^2 = \frac{5.56}{\bar{n}\bar{\bar{p}}} \left(\frac{\bar{s}}{\bar{s} - 1} \right). \tag{10.7}$$

Thus it is seen that for an ideal counter a plot of the square of peak width against $1/\bar{n}\bar{\bar{p}}$ yields a straight line which passes through the origin.

In the case where all the variations existing in Processes α and β and only the normal statistical fluctuations in Processes γ and δ contribute,

$$\eta^2(E_0) = \eta_c^2(E_0) + \frac{5.56}{\bar{n}\bar{\bar{p}}} \left(\frac{\bar{s}}{\bar{s} - 1} \right) \tag{10.8}$$

where $\eta_c(E_0)$ represents the contribution to the width due to intrinsic effects in the phosphor for monoenergetic particles of energy E_0. Under these conditions, a plot of $\eta^2(E_0)$ versus $1/\bar{n}\bar{\bar{p}}$ gives an intercept on the ordinate equal to $\eta_c^2(E_0)$.

The variation of η^2 with $1/\bar{n}\bar{\bar{p}}$ has been experimentally investigated by WRIGHT and GARLICK[66] for several phosphors, including anthracene, KI(Tl), and CaWO$_4$, and by KELLEY et al.[70] for NaI(Tl). The former authors used a neon lamp pulse source with neutral filters (to vary $\bar{n}\bar{\bar{p}}$) to check the effect of photomultiplier characteristics on the pulse spread. As predicted by Eq. (10.7), they found that the square of the width for the uniform light flashes from the neon lamp was proportional to the inverse of the mean number of photons striking the photocathode. Data on intrinsic crystal width $\eta_c(E_0)$ were obtained from Eq. (10.8) by exciting a phosphor with 5.3 Mev α-particles and inserting neutral filters between the phosphor and photomultiplier tube to vary $1/\bar{n}\bar{\bar{p}}$. Although a linear relation between η^2 and $1/\bar{n}\bar{\bar{p}}$ was obtained for the crystals tested, the line slopes were not all equal to the slope of the line for the neon lamp pulses. These differences in slope possibly show the influence of photon distribution over a nonuniform photocathode on the tube width (η_{tube}) and suggest that Eq. (10.8) is not applicable to these curves. Hence, the intercepts may not give η_c^2 alone but η_c^2 plus an additive constant due to photomultiplier effects.

In the experiments of Kelley et al., the value of $\eta(E_0)$ for the Cs^{137} photo-peak (661 kev) from a NaI(Tl) crystal was compared with the width of the line produced by the artificial scintillations of an RCA C-73687 cathode-ray-tube light flasher which was substituted for the phosphor and adjusted to give the same size output pulse. On the assumption that the flasher line width (η_{flasher}) repre-tented the tube width, η_c was calculated from the equation

$$\eta_c^2 = \eta^2 - \eta_{\text{flasher}}^2. \qquad (10.9)$$

Widths at other values of $1/\bar{n}\,\bar{p}$ were obtained by inserting (1), a bare light

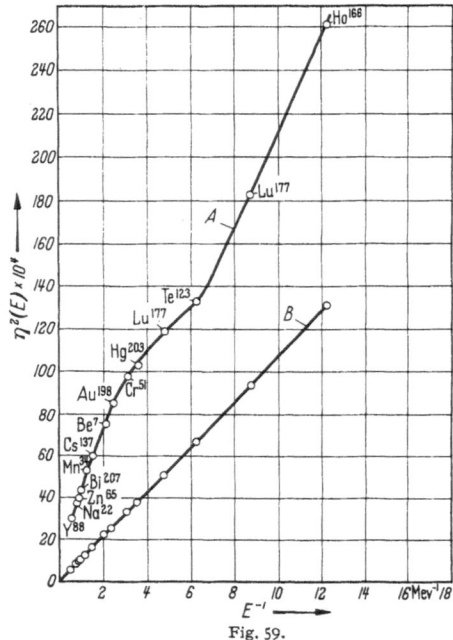

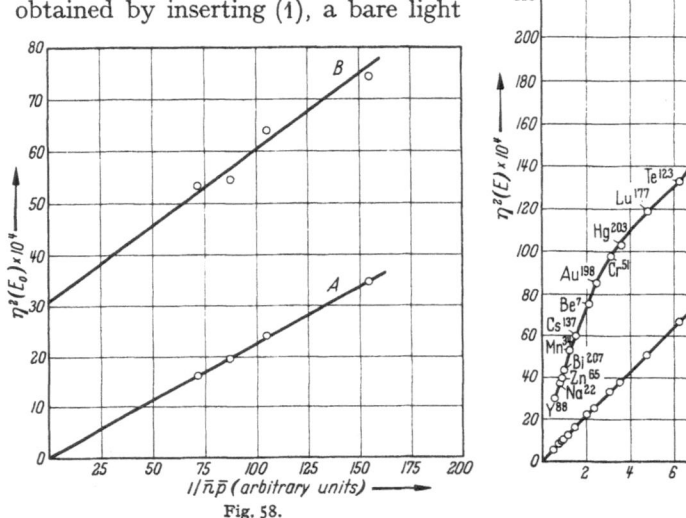

Fig. 58.

Fig. 59.

Fig. 58. Variation of η^2 with $1/\bar{n}\bar{p}$. Curve A: RCA-C73687 cathode-ray-tube light flasher. Curve B: NaI(Tl) crystal and Cs^{137} source (data from Kelley et al.[49]).

Fig. 59. Variation o $^2\eta^2$ as a function of energy. Curve A: 3 in. × 3 in. NaI(Tl) crystal and DuMont-6363 photomultiplier. Cs^{137} half-width is 7.7%. Curve B: calculated for tube on the basis of an assumed line width of 4% at 661 kev (data from Kelley et al[49]).

piper; (2), a light piper with aluminum foil; and (3), a light piper wrapped with black Scotch tape between the source of photons and the photomultiplier. The original data are given in Table 7 and plots of η^2 versus $1/\bar{n}\bar{p}$ for the NaI(Tl) crystal and the flasher in Fig. 58. While these certainly demonstrate the general existence of an intrinsic-loss of resolution in NaI(Tl), there are insufficient data to establish whether or not the differences of Eq. (10.9) accurately represent $\eta_c^2(E_0)$.

The above discussion on intrinsic crystal widths has been restricted to cases where the phosphor was excited by particles of a single energy and $\bar{n}\bar{p}$ was varied by artificial means, i.e., by varying $\bar{p}$. It is now of interest to consider the depend-ence of η_c on the energy of the particles incident on the phosphor; before discussing NaI(Tl) brief reference will be made to the data on anthracene.

For α-particle excitation of anthracene, Wright and Garlick found that as the energy of the incident particles was decreased, and consequently $\bar{n}\bar{p}$ decreased ($\bar{n} \propto E$ if the response of the phosphor is linear with energy), there was a marked increase in η^2, and, on the basis of their earlier data from the neon lamp pulser, concluded that η_c^2 varied as $(1/\bar{n}\bar{p})^2$ after a certain value of $1/\bar{n}\bar{p}$ was reached but was zero for lower values. Since the results indicated that this effect was associated with events occurring near the end of the α-particle path, it was attributed to a variation in number of knock-on protons produced by successive α-particles.

In the only other study on anthracene, BREITENBERGER calculated $\eta^2(E)$ for monoenergetic electrons incident on a 13 mm thick cylindrical crystal using the data published by HOPKINS[71]. The plot of $\eta^2(E)$ versus $1/\overline{E}$ is a straight line which extrapolates to approximately 0.0083 at infinite energy.

Table 7. *Light attenuation experiment* (KELLEY et al.[70]).

Condition	Pulse height	η	η_{Flasher}	$\sqrt{\eta^2 - \eta^2_{\text{Flasher}}}$
No light piper.	1400	7.3	4.0	6.1
Light piper plus Al foil.	1150	7.4	4.4	6.0
Bare light piper	950	8.0	4.9	6.4
Light piper plus black tape . . .	650	8.65	5.9	6.4

Measurements of line widths as a function of γ-ray energy for NaI(Tl) have been reported by KELLEY et al.[70] for a spectrometer consisting of a 3 in. × 3 in. crystal and a DuMont 6363 photomultiplier. As shown in Fig. 59 the observed values of $\eta^2(E)$ do not lie on a straight line when plotted against $1/E$, indicating that η_c is energy dependent. In order to analyze the data, KELLEY et al. assumed a tube width of 4% at 661 kev and calculated $\eta_c^2(E)$ from Eq. (10.9). The results given in Table 8 led them to conclude that $\eta_c(E)$ was very nearly inversely proportional to the fourth root of the energy.

Table 8. *Total line width and intrinsic crystal width as a function of γ-ray energy for NaI (Tl)* (KELLEY et al.[70]).

Source	γ-ray energy (kev)	η	η_c	η_{Tube}	η_c^2	$\eta_c^2 E^{\frac{1}{2}}$
Ho^{166}	81	16.19	11.5	11.4	132.0	119.0
Lu^{177}	113	13.5	9.41	9.67	88.5	94.1
Te^{123}	159	11.5	8.1	8.16	65.43	82.5
Lu^{177}	208	10.9	8.27	7.13	68.4	98.6
Hg^{203}	279	10.14	8.06	6.15	65.0	109.0
Cr^{51}	320	9.89	8.1	5.75	64.8	116.0
Au^{198}	411	9.21	7.69	5.07	59.1	120.0
Be^7	478	8.62	7.22	4.70	52.2	114.0
Cs^{137}	661	7.7	6.6	4.0*	43.6	112.0
Mn^{54}	835	7.26	6.33	3.56	40.0	116.0
Bi^{207}	1067	6.56	5.75	3.15	33.1	108.0
Zn^{65}	1114	6.29	5.48	3.08	30.1	100.0
Na^{22}	1277	6.07	5.35	2.87	28.6	102.0
Y^{88}	1850	5.45	4.9	2.39	24.0	103.0

Similar measurements have been made by BERNSTEIN[19] with a $1\frac{1}{2}$ in. diameter × 1 in. long NaI(Tl) crystal and a DuMont 6292 photomultiplier but with slightly different results. His plot of $\eta^2(E)$ versus $1/E$ is linear at high energies with an intercept on the ordinate of $\eta^2 = 0.0017$. The low energy portion of the curve (below 200 to 300 kev) is also approximately linear but extrapolates to a much larger intercept.

* Assumed.
[70] G. G. KELLEY, P. R. BELL, R. C. DAVIS and N. H. LAZAR: Nucleonics **14**, No. 4, 53 (1956). — I.R.E. Trans. Nuclear Sci. N.S. **3**, No. 4, 57 (1956).
[71] J. J. HOPKINS: Rev. Sci. Instrum. **22**, 29 (1951).

Although a number of experiments were performed by the above authors in an attempt to define the physical effects causing the non-linear variation of η^2 with $1/E$, the source of intrinsic width in NaI(Tl) crystals essentially remains unknown. It appears that non-normal fluctuations in the number of photons per scintillation emitted by the phosphor definitely contribute to the pulse height spread, but there is not yet sufficient information to establish the nature of the effect or whether or not it is removable.

c) Cesium halide phosphors.

11. With the exception of NaI, cesium iodide [72-78] is probably the most interesting of the inorganic phosphors studied to date. This material is nonhygroscopic and is suitable for scintillation counting in both the activated and unactivated forms (see below). Large colorless crystals are readily grown although in some cases color centers are formed in CsI(Tl) upon exposure to weak ultraviolet radiation. The high atomic number constituents (55 and 53) and the high density (4.51 gm/cm³) result in relatively large absorption coefficients [73] for x-rays and γ-rays.

VAN SCIVER and HOFSTADTER [72] have reported that relative to NaI(Tl), CsI(Tl) has a scintillation efficiency of about 0.28. Thus, the nonhygroscopic property of CsI(Tl) makes it an attractive replacement for NaI(Tl) in applications where the shorter fluorescence decay time (0.25 µsec versus 0.55 µsec) and larger fluorescence conversion efficiency of sodium iodide are not required*. Unfortunately, there is only meager information available on the scintillation response of this phosphor to photons and charged particles, the only published work being on the response to protons in the energy range 0.88 to 4.33 Mev (GALONSKY et al. [78]). These measurements show that the pulse height versus proton energy curve is a straight line which extrapolates to (0.07 ± 0.02) Mev at zero pulse height.

Unactivated cesium iodide has been investigated by HAHN [75] and HAHN and ROSSEL [76,77]. The crystals were cooled with liquid nitrogen and used in combination with an RCA-5819 photomultiplier in the manner shown in Fig. 60. The principal results from these studies were as follows:

(a) The energy conversion efficiency increased with decreasing temperature. This is illustrated in Fig. 61 where pulse height per unit energy for Cs¹³⁷ γ-rays and Po α-particles are plotted as an inverse function of the temperature.

(b) The pulse height at 100° K increased linearly with γ-ray energy from 40 kev to 1.3 Mev, and, with the same optical geometry, was nearly twice that from NaI(Tl).

(c) The scintillation efficiency decreased with increasing specific ionization causing a non-linear response to α-particles. However, the response of CsI to α-particles was more nearly proportional to energy than either the NaI(Tl) or the KI(Tl) response. At 77° K the pulse height per unit energy for Po²¹⁰ α-particles (5.3 Mev) was 85 % of the value for γ-rays.

[72] W. VAN SCIVER and R. HOFSTADTER: Phys. Rev. **84**, 1062 (1951).

[73] G. J. BRUCKER: Nucleonics **10**, No. 11, 72 (1952).

[74] J. BONANOMI and J. ROSSEL: Helv. phys. Acta **25**, 725 (1952).

[75] B. HAHN: Phys. Rev. **91**, 772 (1953).

[76] B. HAHN and J. ROSSEL: Helv. phys. Acta **26**, 271 (1953).

[77] B. HAHN and J. ROSSEL: Helv. phys. Acta **26**, 803 (1953).

[78] A. GALONSKY, C. H. JOHNSON and C. D. MOAK: Rev. Sci. Instrum. **27**, 58 (1956).

* The use of CsI(Tl) may be limited however by a long-lived phosphorescence that has been observed by FRANCIS and BELL: [Oak Ridge National Laboratory (U.S.) Report No. 1975, (Unpublished)] (1955). Their preliminary measurements indicate a decay time for this phosphorescence of approximately 220 seconds.

(d) At 100° K the full width at half maximum of the 145 kev photopeak of Ce[141] was 11%. This small width indicated an absolute scintillation efficiency of about 40%.

Although the value of 40% for the absolute efficiency of CsI (without activator) is not well established, it appears certain that the true value is larger than that for NaI(Tl). Further work is needed on this phosphor before its properties can be thoroughly evaluated.

Small anhydrous cesium fluoride (unactivated) crystals [79,80] have been grown which scintillate predominantly in the blue ultraviolet. The decay time of

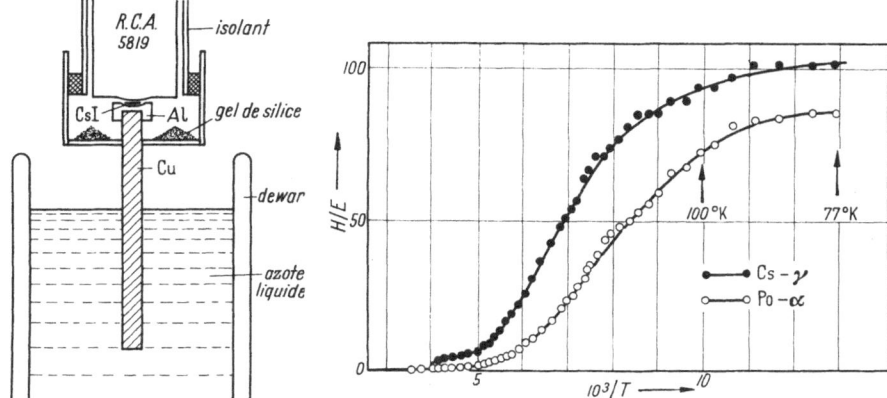

Fig. 60. Experimental arrangement used by HAHN and ROSSEL[77] for measurements on CsI (unactivated).

Fig. 61. Effect of temperature on the energy conversion efficiency of CsI (unactivated). The pulse height per unit energy or Cs[137] γ-rays at 77° K has arbitrarily been given the value 100 (HAHN and ROSSEL[77]).

Table 9. *The cesium halide phosphors.*

Phosphor	CsI(Tl)	CsI (100° K)	CsF	CsF(Tl)
Density (gm/cm³)	4.51	4.51	3.59	3.59
Melting point (°C)	621	621	684	684
Fluorescence decay time (μsec)*	0.55	0.5	0.005 (β) 0.02 (α)	0.15
Phosphorescence decay time (sec)*	220	—	—	—
Wavelength of maximum emission (Å)	—	Blue	3900	3400, 3900
Appearance	Colorless	Colorless	Corlorless	—
Relative pulse height** NaI (Tl) = 100	28	200	5	—
Response to heavy particles	$\alpha/\beta = 0.5$	$\alpha/\beta = 0.85$	$\alpha/\beta = 0.2$	—
Miscellaneous	nonhygroscopic	nonhygroscopic	hygroscopic	hygroscopic
Reference	4, 72, 81, 103	74—77	79, 80	80

* Time for light pulse to fall to 1/e of initial value.
** β-particle excitation.

4 R. K. SWANK: Annual Rev. Nucl. Sci. **4**, 111 (1954).

72 W. VAN SCIVER and R. HOFSTADTER: Phys. Rev. **84**, 1062 (1951).

74 J. BONANOMI and J. ROSSEL: Helv. phys. Acta **25**, 725 (1952).

75 B. HAHN: Phys. Rev. **91**, 772 (1953).

76 B. HAHN and J. ROSSEL: Helv. phys. Acta **26**, 271 (1953).

77 B. HAHN and J. ROSSEL: Helv. phys. Acta **26**, 803 (1953).

79 W. VAN SCIVER and R. HOFSTADTER: Phys. Rev. **87**, 522 (1952).

80 W. VAN SCIVER: Stanford University (U.S.) Report No. HEPL-38 (Unpublished) (1955).

81 J. E. FRANCIS and P. R. BELL: Oak Ridge National Laboratory (U. S.) Report No. 1975, (Unpublished) (1955).

103 H. KNOEPFEL, E. LOEPFE and P. STOLL: Helv. phys. Acta **29**, 241 (1956).

5×10^{-9} sec for γ-ray produced pulses is the fastest known for an inorganic scintillator and approaches the speed of some of the best solid organic phosphors. α-particle pulses have a decay time of 2×10^{-7} sec which suggests that the luminescence process differs in the two cases. Relative to NaI(Tl), the scintillation efficiency of CsF (excited by 1.6 Mev electrons) is 0.05 and is independent of temperature in the range $+125°$ C to $-188°$ C. Since the material is deliquescent and a very inefficient phosphor, CsF crystals are not likely to be used in routine scintillation counter applications. There may, however, be applications in the field of high energy nuclear research where there is a need for high density, rapid decay time phosphors.

The main properties of the cesium iodide and cesium fluoride phosphors are listed in Table 9.

d) Lithium iodide phosphors.

12. Lithium iodide crystals are given special attention as slow neutron detectors because the large cross section of the $Li^6(n, \alpha) H^3$ reaction at low neutron

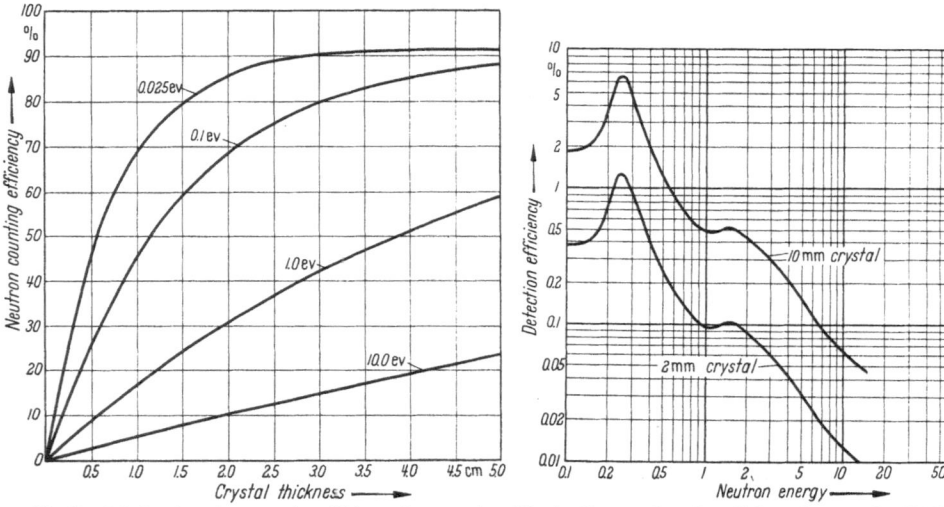

Fig. 62. Calculated neutron counting efficiency for normal lithium iodide as a function of thickness for energies between 0.025 and 10 ev. Effects due to neutron capture by the iodine have been included.

Fig. 63. Neutron detection efficiency for crystals of Li⁶I as a function of neutron energy. The shape of the curve is determined entirely by the energy dependence of the Li⁶(n, α)H³ cross section (Murray[90]).

energies suggests a high neutron counting efficiency (Fig. 62). Furthermore, since the α-particle and the triton have a combined energy of 4.78 Mev, relatively large, uniform light pulses are produced when neutrons are absorbed by the Li^6 in a uniformly activated phosphor. Single crystals have been grown and studied with a number of different activators, including Tl[82-84,86], Sn[85,89], Eu[87-89], Sm[89], In[84], and Ag[84].

[82] R. Hofstadter: Nucleonics **6**, No. 5, 70 (1950).

[83] R. Hofstadter, J. A. McIntyre, H. Roderick and H. I. West jr.: Phys. Rev. **82**, 749 (1951).

[84] W. Bernstein and A. W. Schardt: Phys. Rev. **85**, 919 (1952).

[85] J. Schenck and R. L. Heath: Phys. Rev. **85**, 923 (1952).

[86] J. Schenck: Nucleonics **10**, No. 8, 54 (1952).

[87] J. Schenck: Nature, Lond. **171**, 518 (1953).

[88] J. Schenck and J. H. Neiler: Scintillation Counter Symposium, Washington, D.C., 1954. Nucleonics **11**, No. 3, 28 (1954).

[89] K. P. Nicholson and G. F. Snelling: Brit. J. Appl. Phys. **6**, 104 (1955).

Although thallium was the first activator tried[83], it was later shown[84] that crystals grown with an excess of TlI activator (0.5% by weight) exhibited variations in pulse height over the crystal volume as a result of large nonuniformities in Tl concentration. Considerable success has been experienced, however, in growing crystals uniformly activated with Sn and Eu. On the basis of the work reported by SCHENCK[87] and by NICHOLSON and SNELLING[89], one concludes that LiI(Eu) is the most useful scintillation detector for low energy neutrons. The europium-activated phosphor is fairly easy to grow and has a fluorescent conversion efficiency at least five times greater than that of LiI(Sn). In general it appears that the efficiencies of all the lithium iodide phosphors are reduced for heavily ionizing particles. Data for LiI scintillators are summarized in Table 10.

Table 10. *The lithium iodide phosphors**.

Phosphor	LiI(Tl)	LiI(Sn)	LiI(Eu)		LiI(Sm)	LiI(In)	LiI(Ag)
Activator	TlI	SnI_2	$EuCl_2$	$EuCl_2$	SmI_3	InI	AgI
Activator concentration (mole %)	~0.01	0.1	0.05	0.03	0.02	—	0.2
Fluorescence decay time** (μsec)	1.2	0.7	2.0	1.4	0.25	—	—
Wavelength of maximum emission (Å)	Blue-Green	5300	~4400	~4400	Blue	Whitish-Orange	Green-Yellow
Appearance	Colorless	Yellow-Green	Almost Colorless	Almost Colorless	Colorless	—	Colorless
Relative pulse height*** NaI (Tl) = 100	10	~4	36	35	3.3	1—2	~4
Energy resolution for products of Li^6 (n, α) H^3 reaction (%)	—	15.1	6	11.5	13.8	—	—
Measured energy released by products of Li^6 (n, α) H^3 reaction (Mev)	4.1	4.5	4.1	4.55	3.6	—	—
Reference	83, 84	4, 85	4, 87	4, 89	89	84	84

As part of a program directed toward the development of a more efficient (see Fig. 63) and versatile fast-neutron spectrometer, MURRAY[90] has investigated the scintillation properties of LiI(Eu) crystals with various concentrations of activator and Li^6 isotope to neutrons in the energy region 1 to 15 Mev. Thin crystals (2 mm thick) were normally used in this work in order to minimize the high energy γ-ray contribution to an observed fast-neutron pulse height spectrum; however, the pulse height spectra obtained with 10 mm crystals were quite similar to those from the 2 mm crystals. The spectrum from 5.79 Mev neutrons incident on a 2 mm thick crystal of Li^6I(Eu) at room temperature is shown in Fig. 64; the large thermal peak was produced by neutrons which were degraded in energy by scattering from the walls, floor, etc., and consequently were readily absorbed by the crystal because of the large (n, α) cross section at thermal energies. MURRAY ascribed the difference in shapes of the fast-neutron peak†

---

* Lithium iodide has a density of 4.06 gm/cm³, a melting point of 446° C, and is very hygroscopic.

** Time for light pulse to fall to $1/e$ of initial value.

*** β-particle excitation.

4 R. K. SWANK: Annual Rev. Nucl. Sci. **4**, 111 (1954).

90 R. B. MURRAY: Oak Ridge National Laboratory (U.S.) Report No. 56-11-5 (Unpublished) (1956). — R. B. MURRAY and J. SCHENCK: Bull. Amer. Phys. Soc., Ser. II **1**, 296 (1956).

† In a study using 5.79 Mev neutrons on an 8 mm thick Li^7I(Eu) crystal, no evidence of a fast-neutron peak was observed, although a small thermal peak appeared due to a trace of Li^6 impurity.

and the thermal peak to a difference in the scintillation response of LiI(Eu) to α-particles and tritons. That is, since the kinetics of the (n, α) reaction are such that for the fast-neutron induced reaction both the α-particle and the triton have a range of energies available to them, whereas for the thermal-neutron reaction, the energy distribution between the two particles is uniquely specified, the pulse height distribution from fast monoenergetic neutrons

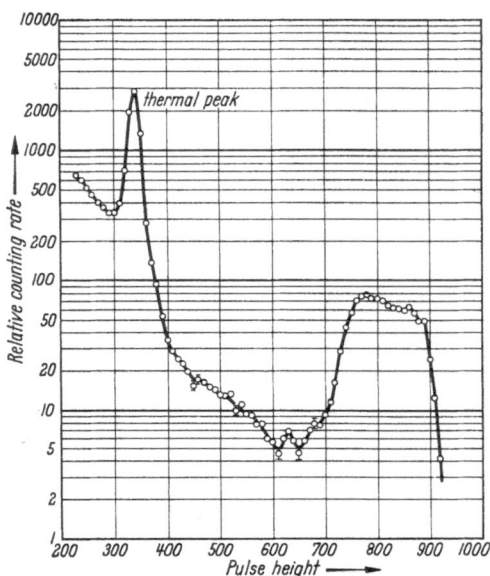

Fig. 64. Pulse height spectrum from 5.79 Mev neutrons incident on a 24 mm diameter × 2 mm thick Li⁶I (0.025 mol-% Eu) crystal at room temperature (Murray[90]).

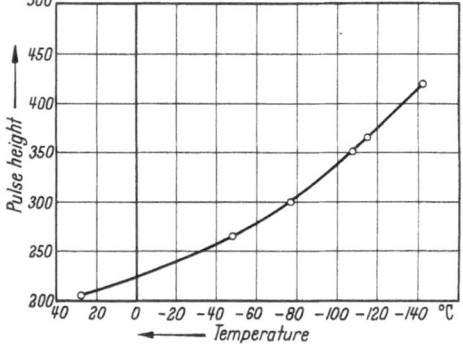

Fig. 65. Thermal-neutron pulse height as a function of temperature for a normal LiI (0.1 mol-% Eu) crystal (Murray[90]).

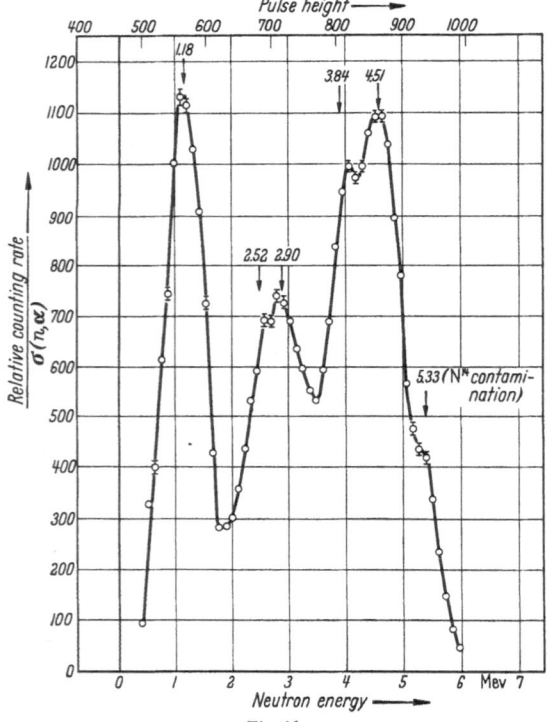

Fig. 66.

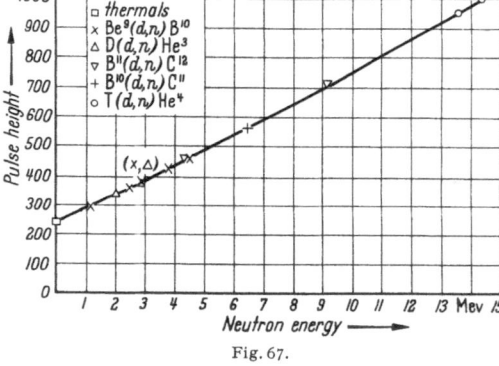

Fig. 67.

Fig. 66. Pulse height spectrum from Be⁹(d, n) B¹⁰ neutrons on a 2 mm Li⁶I(Eu) crystal at −196° C. The counting rates have been corrected for the Li⁶(n, α)H³ cross section dependence on energy and the thermal neutron peak has been subtracted. Arrows indicate expected neutron groups (Murray[91]).

Fig. 67. Pulse height versus neutron energy for a 30 mm diameter × 2 mm thick Li⁶I(Eu) crystal at −196° C (Murray[91]).

incident on LiI(Eu) would be much broader than that from thermal neutrons if the scintillation efficiency was a function of dE/dx.

Upon cooling lithium iodide crystals to $-142°$ C, MURRAY observed two effects. First, the scintillation efficiency increased with decreasing temperature as shown in Fig. 65 where thermal-neutron pulse height is plotted as a function of temperature for a normal LiI (0.1 mol % Eu) crystal. And second, the fast-neutron peaks assumed a nearly gaussian shape at about $-142°$ C and showed a marked improvement in resolution. For example, the half-width of the peak from 5.3 Mev neutrons on a Li^6I (0.06% Eu) crystal decreased from 18% at room temperature to 10% at $-142°$ C.

In a recent extension of this work, MURRAY[91] examined fast neutron spectra with Li^6I(Eu) crystals operating at $-196°$ C with very encouraging results. The spectrum obtained from Be^9 (d, n) B^{10} neutrons on a 2 mm crystal is given in Fig. 66. It is seen that the curve shows resolved neutron groups between 0 and 5 Mev neutron energy, corresponding to the first five levels in B^{10}. Fig. 67 indicates that the relationship between pulse height and neutron energy is essentially linear over the range of interest.

e) Other inorganic phosphors.

13. Of the other inorganic materials which have been investigated as scintillators, only potassium iodide[10,92-94], zinc sulfide[95-99], and calcium and cadmium tungstate[100-102] are of any practical importance. Some of the properties of these phosphors are listed in Table 11.

Table 11. *Miscellaneous inorganic scintillators.*

Phosphor	KI(Tl)	ZnS(Ag)	ZnS(Cu)	CaWO₄	CdWO₄	NaI(Tl)
Density (gm/cm³)	3.13	4.10	4.10	6.06	7.90	3.67
Melting point (°C)	686	1850	1850	1535	1325	651
Fluorescence decay time (μsec)*	>1	~10	~10	~4	~6	0.25
Wavelength of maximum emission (Å)	4100	4500	5200	4300	5200	4100
Relative pulse height**	25	~100 (β) ~180 (α)	~100 (β) ~160 (α)	~50	~65	100
Miscellaneous	Phosphorescent	Multi-crystalline powders	Multi-crystalline powders	Small crystals	Small crystals	Hygroscopic
Reference	10, 92	4, 97	4, 97	2, 101	2, 101	—

[91] R. B. MURRAY: Private Communication (to be published) (1958).
[92] J. C. D. MILTON and R. HOFSTADTER: Phys. Rev. 75, 1289 (1949).
[93] W. FRANZEN, R. W. PELLE and R. SHERR: Phys. Rev. 79, 742 (1950).
[94] L. M. BELYAEV, M. D. GALANIN, Z. L. MORGENSHTERN and Z. A. CHIZHIKOVA: Dokl. Akad. Nauk. USSR. 99, 691 (1954); 105, 57 (1955).
[95] J. CHARITON and C. A. LEA: Proc. Roy. Soc. Lond. 122, 304 (1929).
[96] I. BROSER, H. KALLMANN and U. M. MARTIUS: Z. Naturforsch. 4a, 204 (1949).
[97] H. KALLMANN: Phys. Rev. 75, 623 (1949).
[98] W. HOOGENSTRAATEN: Nederl. Tijdschr. Natuurk. 21, 150 (1955).
[99] J. GOLDEMBERG, E. SILVA and S. S. VILLACA: Anais Acad. brasil. Ci. 27, No. 2, 141 (1955).
* Time for pulse to fall to 1/e of initial value.
** β-particle excitation. NaI(Tl) = 100.
[10] J. BONAMONI and J. ROSSEL: Helv. phys. Acta 24, 310 (1951); 25, 725 (1952).
[4] R. K. SWANK: Annual Rev. Nucl. Sci. 4, 111 (1954).
[2] J. B. BIRKS: Scintillation Counters. New York: McGraw-Hill Book Co.; London: Pergamon Press 1953.
[101] R. H. GILLETTE: Rev. Sci. Instrum. 21, 294 (1950).
[100] R. J. MOON: Phys. Rev. 73, 1210 (1948).
[102] J. A. McINTYRE and R. HOFSTADTER: Phys. Rev. 78, 617 (1950).

III. Organic phosphors.

KALLMANN and FURST[1,2,3] made the first general study of organic materials which could be used in scintillation counters. Their investigation included both organic crystals such as anthracene and naphthalene, and organic liquid solutions such as p-terphenyl in toluene. The most recent review of the subject has been made by BROOKS[43].

14. Organic crystals. An extensive study of organic crystals has recently been made by SANGSTER and IRVINE[4]. The most commonly used of these phosphors are anthracene and stilbene. The efficiency of anthracene is about 30% greater than that of stilbene and twice that of the best liquid and plastic scintilants (Tables 12 to 14). However, anthracene suffers from having a considerably longer decay time than do the other materials discussed here. The absolute efficiency of anthracene is of interest, apart from its intrinsic value, in that many phosphors are compared to it as a standard. It should be kept in mind that the value of this quantity, as measured by different workers, varies from 1 to 10%[4], the variation arising from a number of possible sources, e.g., variations in crystal purity and surface conditions, differences in crystal size, variation in photomultiplier response, or variation in the fraction of light collected. (These efficiency values are given as the percentage of the energy loss in the crystal that appears as useful light. The value of 2.7 ev may be used for photon energy.)

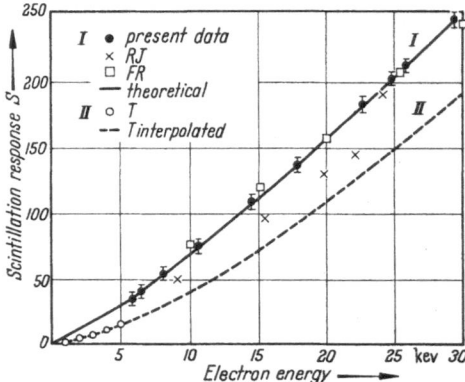

Fig. 68. Response of anthracene to γ- and β-rays (BIRKS and BROOKS[11]). "Present data": BIRKS and BROOKS[11]. RJ: ROBINSON and JENTSCHKE[52]. FR: FOWLER and ROOS[12]. T: TAYLOR et al.[13].

The scintillation properties of anthracene have been more extensively studied than those of any other of the organic materials. Values of the efficiency reported in the literature are given in Table 12. These are all for excitation by β- or γ-rays for which organic crystal phosphors appear to have a light output which varies linearly with energy loss except perhaps for energies below 10 kev. Fig. 68 (curve I), a compilation by BIRKS and BROOKS[11] of their data together with that of FOWLER and ROOS[12] and of ROBINSON and JENTSCHKE[52], shows the response of anthracene for γ-rays of 6 to 20 kev; there appears to be a slight curvature. That some nonlinearity exists seems to be in agreement with the observations[13,8] for higher energies, for which linearity is observed, but for which the curve extrapolates to zero pulse height at about 24 kev.

[1] H. KALLMANN and M. FURST: Phys. Rev. **79**, 857 (1950).
[2] H. KALLMANN and M. FURST: Phys. Rev. **81**, 853 (1951).
[3] M. FURST and H. KALLMANN: Phys. Rev. **85**, 816 (1952).
[43] F. D. BROOKS: Progr. Nucl. Physics **5**, 252 (1956).
[4] R. C. SANGSTER and J. W. IRVINE: J. Chem. Phys. **24**, No. 4, 670 (1956).
[11] J. B. BIRKS and F. D. BROOKS: Proc. Phys. Soc. Lond. B **69**, 721 (1956).
[12] J. M. FOWLER and C. E. ROOS: Phys. Rev. **98**, 996 (1955).
[52] W. H. ROBINSON and W. JENTSCHKE: Phys. Rev. **95**, 1412 (1954).
[13] C. J. TAYLOR, W. K. JENTSCHKE, M. E. REMLEY, F. S. EBY and P. G. KRUGER: **Phys.** Rev. **84**, 1034 (1951).
[8] J. I. HOPKINS: Rev. Sci. Instrum. **22**, 29 (1951).

Table 12. *Organic crystal phosphors.*

Phosphor	Density (gm/cm³)	Melting point (°C)	Scintillation efficiency						Wave length of maximum emission (Å)	Scintillation decay time		
			absolute efficiency (a) (%)	relative pulse height (b)	reflector	phototube type	crystal size	reference		decay time (e) (mμsec)	crystal size	reference
Anthracene ($C_{14}H_{10}$)	1.25	217	3.5 5.0 3.3 (c) 10 4.2 1.1 1.9 3.8 (d)	100	Al none none — Al Al —	RCA-C7140 RCA-5819 EMI-5060 — RCA-5819 RCA-C7140 EMI-6266	1/2″ dia.×1/4″ 1 cm³ 1 cm²×2 mm 5 mm flake 1.3 to 2.5 cm thick >1.5 mm thick	4 10 7 9 5 8 8 6	4450	23 to 38	≥1 mm thick	43
p-Terphenyl ($C_{18}H_{14}$)	1.23	213	—	30 40	Al —	RCA-C7140 RCA-5819 or DuMont-6292	1/2″ dia.×1/4″ —	4 44	3900, 4100	4.5	1/2″ dia.×1/4″	32
p, p′-Quaterphenyl ($C_{24}H_{18}$)	—	318	—	94 85	Al —	RCA-C7140 RCA-5819 or DuMont-6292	1/2″ dia.×1/4″ —	4 44	4350	4.2	1/2″ dia.×1/4″	32
trans-Stilbene ($C_{14}H_{12}$)	1.16	124	—	46 60	Al —	RCA-C7140 RCA-5819 or DuMont-6292	1/2″ dia.×1/4″ —	4 44	3850	<3 to 8.2	≥1 mm thick	43
Diphenylacetylene ($C_{14}H_{10}$)	1.18	62.5	—	32 45	Al —	RCA-C7140 RCA-5819 or DuMont-6292	1/2″ dia.×1/4″ —	4 44	4000	3.5 5.4	1/2″ dia.×1/4″ 13/16″dia.×1/2″	32 32

(a) For high energy β-particle excitation.
(b) Relative to anthracene = 100 for high energy β-particle excitation.
(c) For external 8 kev electrons.
(d) For 5.3 Mev α-particle excitation; data of HOPKINS[8] used to obtain electron efficiency.
(e) Time for intensity to fall to 1/e of initial value.

[4] R. C. SANGSTER and J. W. IRVINE: J. Chem. Phys. 24, No. 4, 670 (1956).
[43] F. D. BROOKS: Progr. Nucl. Physics. 5, 252 (1956).
[10] G. T. WRIGHT: Proc. Phys. Soc. Lond. B 68, 929 (1955).
[7] D. K. BUTT: Proc. Phys. Soc. Lond. A 66, 940 (1953).
[9] M. FURST, H. KALLMANN and B. KRAMER: Phys. Rev. 89, 416 (1953).
[5] F. B. HARRISON: Nucleonics 10, No. 6, 40 (1952).
[8] J. I. HOPKINS: Rev. Sci. Instrum. 22, 29 (1951).
[6] J. B. BIRKS and M. E. SZENDREI: Phys. Rev. 91, 197 (1953).
[32] R. K. SWANK and W. L. BUCK: Rev. Sci. Instrum. 26, 15 (1955).
[44] R. K. SWANK: Annual Rev. Nucl. Sci. 4, 111 (1954).

Table 13. *Organic liquid phosphors.*

Solvent	Density (g/cm³)	Primary solute (g/l)	Secondary solute (g/l)	Scintillation efficiency					Wave length of maximum emission (Å)	Short wave length limit of emission (Å)	Scintillation decay time		
				relative pulse height* (anthracene=100)	reflector	phototube type	sample size	reference			decay time(a) (mµsec)	sample size (in.)	reference
Xylene (C₈H₁₀)	0.86	TP(5)	αNPO	48	none	RCA-1 P28	—	3	(3250 to 4000)**				
Xylene		TP(5)		40	—	DuMont-6292	10 cm³ in 20 ml beaker	28					
Xylene		TP(5)	BBO	49	—	DuMont-6292	"	28					
Xylene		TP(5)	POPOP	44	—	DuMont-6292	"	28					
Xylene		PPO(3)		57	Al	DuMont-6292	1 ml	24, 25					
o-xylene	0.88	TP(6)		26	porcelain	RCA-1 P28	—	1					
m-xylene	0.86	TP(5.5)		44	porcelain	RCA-1 P28	—	24, 25					
m-xylene		PPO(3)		58	Al	DuMont-6292	1 ml	1					
p-xylene	0.86	TP(5)		16	porcelain	RCA-1 P28	—	24, 25					
p-xylene		PPO(3)		60	Al	DuMont-6292	1 ml	1					
p-xylene		PBD(10)(b)		70	Al	DuMont-6292	1 ml	25	(3400 to 3700)**				
Toluene (C₇H₈)	0.87	TP(9)(c)		36	MgO	RCA-6342	—	27	3530	3200	2.2 ± 0.3		45
Toluene		TP(9)(b)	αNPO (0.1)(b)	48	MgO	RCA-6342	—	27	3530	3200			
Toluene		TP(9)(d)	DPH (0.1)(b)	19	MgO	RCA-6342	—	27	3530	3200			
Toluene		TP(4)	POPOP (0.1)(b)	—	—	—	—	—	3530	3200			
Toluene		TP(4)		61	Al	DuMont-6292	1 ml	25, 34	4320	3800	< 2.9	5/8 dia.×3/8	32
Toluene		TP(4)		54	Al	DuMont-6292	1 ml	23, 25	4150	3700	< 3.2	5/8 dia.×3/8	32
Toluene		TP(4)		40	Al	DuMont-6292	1 ml	25					
Toluene		PBD(8)(b)		62	Al	DuMont-6292	1 ml	23, 25	4220	3800	< 2.8	5/8 dia.×3/8	32
Toluene		PPO(4)(b)		63	MgO	RCA-6342	—	27	3700		≦ 3.0	5/8 dia.×3/8	32
Toluene		TP(3)		54	none	RCA-6342	—	27	3820	3400	< 2.9	5/8 dia.×3/8	32
Phenylcyclohexane (C₁₂H₁₄)	0.94	TP(5)	αNPO	46	—	DuMont-6292	10 cm³ in 20 ml beaker	28	(3200 to 4100)**				
Phenylcyclohexane		TP(5)	BBO	53	—	DuMont-6292	"	28					
Phenylcyclohexane		TP(5)	POPOP	54	—	DuMont-6292	"	28					
Phenylcyclohexane		TP(5)	PBD	48	Al	DuMont-6292	1 ml	24, 25					
Phenylcyclohexane		PPO(3)		55	Al	DuMont-6292	1 ml	24, 25					
Triethylbenzene (C₁₂H₁₈)	0.86	PPO(3)		52									

PPO = 2,5-diphenyloxazole.   αNPO = 2-(1-naphthyl)-5-phenyloxazole.   POPOP = 1,4-di-[2-(5-phenyloxazolyl)]-benzene.   PBD = 2-phenyl-5-(4-biphenylyl)-1,3,4-oxadiazole.
BBO = 2-5-di-(4-biphenylyl)-oxazole.   TP = p-terphenyl.   ** Spectral range.   (a) Time for intensity to fall to 1/e of initial value.   (b) Air saturated.   (c) Oxygen saturated.   (d) Oxygen saturated.

* For high energy β-particle excitation.

²² M. FURST and H. KALLMANN: Phys. Rev. **85**, 816 (1952).
²⁸ R. C. DAVIS: Oak Ridge National Laboratory (U.S.) Report No. 1975 (Unpublished).
²³ F. N. HAYES, B. S. ROGERS and P. C. SANDERS: Nucleonics **13**, No. 1, 46 (1955).
³⁴ F. N. HAYES, D. G. OTT and V. N. KERR: Nucleonics **14**, No. 1, 42 (1956).
¹ H. KALLMANN and M. FURST: Phys. Rev. **79**, 857 (1950).
²⁷ R. K. SWANK and W. L. BUCK: Argonne National Laboratory (U.S.) Report No. ANL-5554 (Unpublished) (1956).
⁴⁵ S. SINGER, L. K. NEHER and R. A. RUEHLE: Rev. Sci. Instrum. **27**, 40 (1956).
²⁴ R. K. SWANK and W. L. BUCK: Rev. Sci. Instrum. **26**, 15 (1955).
²⁵ F. N. HAYES and W. L. ROGERS: Atomic Energy Commission (U.S.) Report No. AECU-3073 (Unpublished) (1954).
³ F. N. HAYES: Los Alamos Scientific Laboratory (U.S.) Report No. 1639 (Unpublished) (1953).
² H. KALLMANN and M. FURST: Phys. Rev. **81**, 853 (1951).

Table 14. *Organic plastic phosphors.*

Solvent	Density (g/cm³)	Primary solute (% by wt.)	Secondary solute (% by wt.)	Scintillation efficiency					Wave length of maximum emission (Å)	Short wave length limit of emission (Å)	Scintillation decay time		
				relative pulse height* (anthracene=100)	reflector	phototube type	sample size (in.)	reference			decay time(b) (mμsec)	sample size (in.)	reference
Polystyrene (C₈H₈)ₙ	1.06	TPB (1.7)(a)		36	Al	RCA-5819	¹³/₁₆ dia.×¹/₂	33	4450	4250	4.6	—	44
Polystyrene		p-terphenyl (2.5)(a)		28	Al	RCA-5819	¹³/₁₆ dia.×¹/₂	33	3550	3450	≤3.0	—	44
Polystyrene		PPO(1)(a)	αNPO (0.05)	24	Al	RCA-5819	¹³/₁₆ dia.×¹/₂	33	3800	3600	—	—	—
Polystyrene		p-terphenyl (0.9)		36	none	DuMont-6292	0.6 dia.×0.2	46	4150	3750	2.2±0.3	—	45
Polystyrene		p-terphenyl (3.4)	TPB (0.02)	39	—	RCA-5819 or DuMont-6292	—	44	4450	—	4.0	—	44
Polyvinyltoluene (C₉H₁₀)ₙ	—	TPB (1.4)(a)		37	Al	RCA-5819	¹³/₁₆ dia.×¹/₂	33	4450	—	4.6	—	44
Polyvinyltoluene		p-terphenyl (4.0)(a)		32	Al	RCA-5819	¹³/₁₆ dia.×¹/₂	33	3550	—	≤3.0	¹³/₁₆ dia.×¹/₂	32
Polyvinyltoluene		p-terphenyl (4.0)	TPB (0.02)(a)	45	Al	RCA-5819	¹³/₁₆ dia.×¹/₂	33	4600(c)	—	4.0	¹³/₁₆ dia.×¹/₂	32
Polyvinyltoluene		p-terphenyl (3.4)	DPS (0.09)(a)	52	Al	DuMont-6292	¹/₂ dia.×0.3	25	~3800	—	≤3.0	¹³/₁₆ dia.×¹/₂	32
Polyvinyltoluene		p-terphenyl (3.4)	POPOP(0.1)	51	Al	DuMont-6292	¹/₂ dia.×0.3	25	~4300	—	—	—	—
Polyvinyltoluene		PBD (1 to 3)	POPOP	47	—	—	—	48	~4300	—	—	—	—

TPB = 1,1,4,4,-tetraphenyl-1,3-butadiene
PPO = 2,5-diphenyloxazole
αNPO = 2-(1-naphthyl)-5-phenyloxazole
DPS = p,p'-diphenylstilbene
PBD = 2-phenyl-5-(4-biphenylyl)-1,3,4-oxadiazole
POPOP = 1,4-di-[2-(5-phenyloxazolyl)]-benzene

* For high energy β-particle excitation.
(a) Dissolved gases removed.
(b) Time for intensity to fall to $1/e$ of initial value.
(c) See Ref. 47.

[33] R. K. SWANK and W. L. BUCK: Phys. Rev. 91, 927 (1953).
[44] R. K. SWANK: Annual Rev. Nucl. Sci. 4, 111 (1954).
[46] M. M. HOFFMAN, R. W. PETERSON and M. JANCO: Los Alamos Scientific Laboratory (U.S.) Report No. LA-2069 (Unpublished) (1956).
[45] S. SINGER, L. K. NEHER and R. A. RUEHLE: Rev. Sci. Instrum. 27, 40 (1956).
[32] R. K. SWANK and W. L. BUCK: Rev. Sci. Instrum. 26, 15 (1955).
[25] F. N. HAYES, D. G. OTT and V. N. KERR: Nucleonics 14, No. 1, 42 (1956).
[48] L. J. BASIK: Argonne National Laboratory (U.S.) Report No. ANL-5554 (Unpublished) (1956).

The above discussion on linearity of the pulse height curve refers to excitation by electrons produced inside the crystal. Additional nonlinearity at low energies is observed when external particles are used. This effect is apparently due to surface effects such as loss of radiation from the bombarded face and quenching due to surface impurities or dislocations[14,15]. The response for external electrons is also shown in Fig. 68 (curve II). It appears that loss of scintillation efficiency occurs for a region a few hundredths of a mm in depth.

For particles heavier than electrons considerably more nonlinearity in the pulse height vs. energy curve is observed. Fig. 69 gives the response curves for α-particles, deuterons, protons, mesons and electrons. Such curves as these and that in Fig. 68 for internal electrons can be fitted by an expression of the form

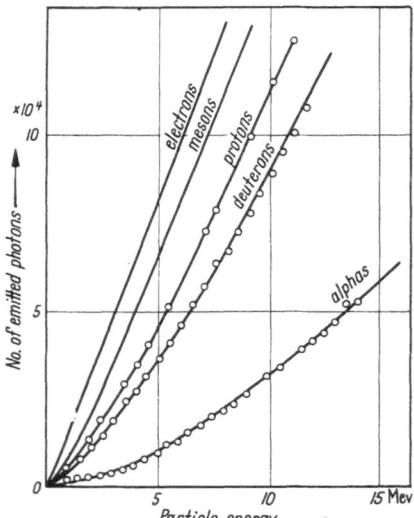

Fig. 69. Scintillation response of anthracene to ioniz-ing particles. The solid lines are the theoretical curves. Points are experimental (Wright[17]).

$$\frac{ds}{dr} = \frac{A\,\dfrac{dE}{dr}}{1 + C\,\dfrac{dE}{dr}}$$

first proposed by Birks[15,16], and which ap-proximates (for dE/dr small) an expression later suggested by Wright[17]. Here ds/dr, the light output per cm path of the particle, is given in terms of dE/dr, the energy loss per cm. A is simply a pro-portionality factor between useful ener-gy loss and light output. The energy-loss dependent term in the denominator, which is included to explain the above-mentioned nonlinearities in the pulse height curves for interior energy loss, is related to quenching mechanisms arising from highly excited or damaged mole-cules along the particle path. All the solid curves in Fig. 69 have been fitted with the same values of A and C[17]. The data of Zimmerman[18], for protons and α-particles of energy below 400 kev, are rather interesting. Over the range of energies studies, dE/dx for protons is decreasing while for α-particles it is in-creasing except for the higher energies. The pulse height vs. energy curve for protons is convex towards the energy axis, whereas that for α-particles is generally concave, except at the higher energies where an apparent inflection occurs. From the data of Zimmerman it is not clear that the value of the constant C, in the above formula, can be chosen so as to fit exactly the results for all par-ticles.

The decay time of a phosphor is as interesting as the response. Again many values are found in the literature and considerable spread in these values occurs (Table 12). The fluorescence decay time of a crystal may depend considerably on

[14] G. T. Wright: Phys. Rev. **100**, 588 (1955).
[15] J. B. Birks: Scintillation Counters. London: Pergamon Press; New York: McGraw-Hill Book Co. 1953.
[16] J. B. Birks: Proc. Phys. Soc. Lond. A **64**, 874 (1951).
[17] G. T. Wright: Phys. Rev. **91**, 1282 (1953).
[18] E. J. Zimmerman: Phys. Rev. **99**, 1199 (1955).

crystal size. BIRKS and LITTLE[19] have measured photofluorescence lifetimes for various samples of anthracene, stilbene, diphenylacetylene and terphenyl. For microcrystals of anthracene and stilbene these were about 3.5 and 1.7 mμsec respectively, whereas for a 1 cm cube of anthracene the decay time was 14 mμsec and for a 2 mm sample of stilbene, 3 mμsec. The values for diphenylacetylene and terphenyl were not very dependent on size and were 3.0 and 3.5 mμsec respectively for 2 mm thick crystals. The dependence on size can perhaps be related to the amount of overlap of the absorption and emission spectra of the materials which, as BIRKS and WRIGHT[20] showed, is very considerable for anthracene and rather small for the other materials.

The lifetimes of the decay for γ-ray and charged particle excitation are considerably greater than for ultraviolet excitation. WRIGHT[21] gave 31 mμsec for a 1 mm sample of anthracene excited by electrons; PHILIPS and SWANK[22] found 8.2 mμsec for stilbene and SWANK and BUCK[32] measured about 5 mμsec for diphenylacetylene. Further, for the same crystal used for the electron measurement, WRIGHT[21] found a decay time of 53 mμsec for α-particle bombardment of anthracene. WRIGHT explained the extended lifetime under charged particle irradiation, and its apparent dependence on ionization density, as due to ion recombination times in the highly ionized columns left by the incident particles.

15. Liquid and plastic phosphors. The properties of liquid solutions have been, and still are, undergoing extensive investigation by HAYES et al.[23-26], by SWANK and BUCK[27], by DAVIS[28], and by other groups. A development after the early work of KALLMANN and FÜRST in the field of organic solutions has been the introduction of solid solutions, e.g., terphenyl in polystyrene. These were first investigated by SCHORR and TORNEY[29] and by KOSKI[30] and have been further studied, particularly by BUCK and SWANK[31-33]. The important characteristics of the most useful of the presently known of these phosphors are listed in Tables 13 and 14. We cannot emphasize too strongly, however, that the values of efficiency may depend greatly on the exact geometry used in the measurements (sample size and reflector), on the photomultiplier cathode spectral characteristics (these may vary widely from one tube to the next of the same type) and on the purity of the materials used (e.g., the amount of absorbed gases). Decay times may also be dependent on sample size and purity. The later investigations of those mentioned above have been undertaken with these problems in mind. These effects

[19] J. B. BIRKS and W. A. LITTLE: Proc. Phys. Soc. Lond. A **66**, 921 (1953).

[20] J. B. BIRKS and G. T. WRIGHT: Proc. Phys. Soc. Lond. B **67**, 657 (1954).

[21] G. T. WRIGHT: Proc. Phys. Soc. Lond. B **69**, 358 (1956).

[22] H. B. PHILLIPS and R. K. SWANK: Rev. Sci. Instrum. **24**, 611 (1953).

[23] F. N. HAYES: Los Alamos Scientific Laboratory (U.S.) Report No. 1639 (Unpublished) (1953).

[24] F. N. HAYES, B. S. ROGERS and P. C. SANDERS: Nucleonics **13**, No. 1, 46 (1955).

[25] F. N. HAYES, D. G. OTT and V. N. KERR: Nucleonics **14**, No. 1, 42 (1956).

[26] F. N. HAYES, D. G. OTT, V. N. KERR and B. S. ROGERS: Nucleonics **13**, No. 12, 38 (1956).

[27] R. K. SWANK and W. L. BUCK: Argonne National Laboratory (U.S.) Report No. ANL-5554 (Unpublished) (1956).

[28] R. C. DAVIS: Oak Ridge National Laboratory (U.S.) Report No. 1975 (Unpublished) (1956).

[29] M. SCHORR and F. TORNEY: Phys. Rev. **80**, 474 (1950).

[30] W. S. KOSKI: Phys. Rev. **82**, 230 (1951).

[31] W. L. BUCK and R. K. SWANK: Nucleonics **11**, No. 11, 48 (1953).

[32] R. K. SWANK and W. L. BUCK: Rev. Sci. Instrum. **26**, 15 (1955).

[33] R. K. SWANK and W. L. BUCK: Phys. Rev. **91**, 927 (1953).

were demonstrated vividly by Hayes and Rogers[34] who showed that the pulse-height ratio for two different liquid phosphors may vary from 1.25 to 2.26 depending on the particular multiplier (even of a given cathode type) and reflector used; this range could have been greater if different samples sizes were used. These variations were due to the fact that different scintillants have different emission spectra and to the spectral dependence of the absorption, reflection and photosensitivity characteristics of the materials used. For example, a TiO_2 (tygon) reflector has a low reflection coefficient below 4000 Å, whereas Al or MgO are fairly uniform down to 2500 Å so that the pulse height for a given sample and multiplier depends on reflector. For 8 g/l of p-terphenyl in toluene the ratio of pulse height obtained when using TiO_2 to that with Al is 0.86. Included in Table 13 are data of Swank and Buck[27] showing the effect of dissolved air and oxygen on pulse height. The oxygen quenching factor is of the order of about two. Thus, where possible we have included in Tables 13 and 14 the available pertinent information on the particular conditions under which the data were obtained.

In general, the organic phosphors have a lower efficiency than the inorganics. However, for fast coincidence experiments, where the photomultiplier pulses are clipped after a few mμsec, larger pulses can often be obtained with the organics, because of their shorter decay time, than with the inorganics. Further, the liquid or solid solutions can be prepared in sizes much greater than any single crystals can be; plastic scintillant discs 42 in. in diameter and 3.5 in. thick[35,36] and liquid scintillants of 420 liter volume[37] have been used.

Liquid and plastic organic solution scintillators have similar characteristics. They are mixtures of two or three hydrocarbons. The bulk of the mixture, the solvent, is a liquid, e.g., xylene or toluene, or a plastic, e.g., polystyrene or polyvinyltoluene. Part of the energy lost by a charged particle moving through these materials will appear as electromagnetic radiation. However, this radiation is generally of too short a wave length to fall within the sensitive region of any photomultiplier, and, more important, is too strongly absorbed by the medium to be able to emerge from any finite-sized sample. If there is a solute present whose absorption band overlaps the emission band of the solvent, the excitation of the solvent molecules is quickly transferred to the solute, even though the latter is present only in a concentration of a few percent. The emission band of the solute, by appropriate choice, will fall in a region where solvent absorption is small and where the photomultipliers are sensitive. For example, the luminescence spectrum of a solution of a few percent of p-terphenyl in toluene is essentially that of p-terphenyl, modified only slightly by solvent absorption and by solute absorption and re-emission. The transfer of excitation from solvent to solute may be partly due to the emission of ultraviolet radiation by the solvent and absorption by the solute. However, it has been shown by Swank and Buck[33] that the most important mechanism is by radiationless transfer from an excited solvent molecule to a nearby solute molecule.

It is often advantageous to add a second solute to the solution. The absorption band of this second solute should overlap the emission band of the primary solute. The purpose is to further shift the final emission to longer wave lengths to match better the photomultiplier and particularly to avoid solution absorption of the

[34] F. N. Hayes and B. S. Rogers: Atomic Energy Commission (U.S.) Report No. AECU-3073 (Unpublished) (1954).

[35] G. W. Clark, F. Sherb and W. B. Smith: Massachusetts Institute of Technology Laboratory for Nuclear Studies (U.S.) Technical Report No. 69 (Unpublished) (1956).

[36] B. Rossi: Bull. Amer. Phys. Soc., Ser. II 2, No. 1, 53 (1957).

[37] E. C. Anderson, R. L. Schuch, J. D. Perrings and W. H. Langham: Nucleonics 14, No. 1, 26 (1956).

radiation when very large samples are used. In favorable cases up to 80% as many photons may be emitted by the second solute as would have been emitted by the first. The secondary solute needs only to be present in a concentration a few percent of that of the primary. If the amount is increased, the efficiency will finally begin to fall. The values in Tables 13 and 14 are for the optimum concentrations. That the final emission is that of the secondary solute, after the latter reaches sufficient concentration, is shown vividly in Fig. 70 from HAYES[23]. Here is plotted the wave length of the radiation emitted vs. concentration of α NPO in a solution of 6 g/l of p-terphenyl in toluene; at 1 g/l of α NPO the emission spectrum is essentially that of α NPO, whereas at 10^{-3} g/l the spectrum is that of p-terphenyl. The advantage of the second solute for large solutions

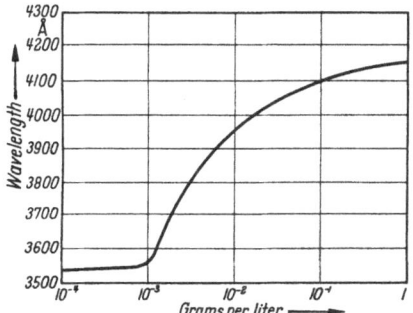

Fig. 70. Wavelength of fluorescent radiation as a function of the concentration of αNPO (g/l) in a solution of 6 g/l of p-terphenyl in toluene (HAYES[23]).

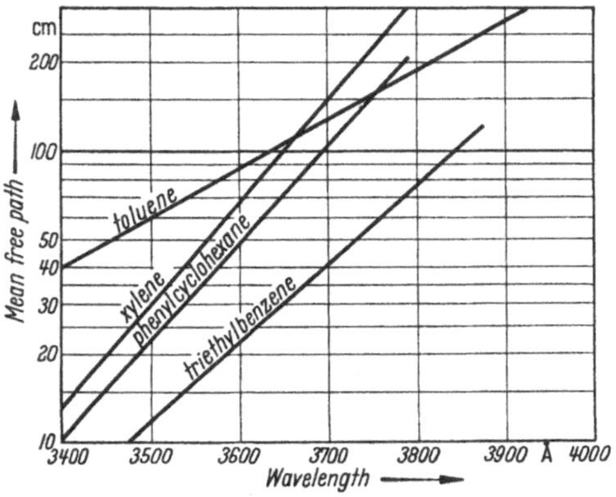

Fig. 71. Absorption mean free path versus wave length for several solvents (HAYES et al.[24]).

can be seen from measurements of HAYES, OTT and KERR[25]. They find that for sample sizes 0.3 in. high $\times$ 0.5 in. in diameter, the pulse-height ratio of toluene plus 4 g/l p-terphenyl to toluene plus 4 g/l p-terphenyl plus 0.1 g/l 1,4-di-[2-(5-phenyloxazolyl)]-benzene (POPOP) is 0.52, whereas for 60-liter samples the ratio drops to 0.18. The reasons for not using the secondary as the only solute may include low saturation concentration, expense, poor transfer of excitation from the solvent, and too large absorption of radiation. The presence of foreign materials such as oxygen, as mentioned above, gives rise to a quenching effect. Presumably the mechanism is the transfer of excitation to the molecules after which they undergo radiationless transition to the ground state, the energy going to thermal excitation. Thus, precautions must be taken to use pure ingredients for these scintillators. Further, due to this quenching effect, difficulty is encountered in introducing neutron sensitive materials to increase the neutron detection efficiency.

Liquid scintillators most commonly used employ toluene as a solvent rather than xylene. Although the latter often gives greater efficiencies for small samples such as with p-terphenyl as primary solute, it does not in larger samples. Also, xylene tends to react to a greater extent than does toluene with some reflector materials such as TiO_2 and MgO. Triethylbenzene may be used as a solvent if fire hazard is to be borne in mind.

Apparently most of the attenuation of the radiation emitted by the final solute arises from absorption by the solvent. Absorption mean free paths vs. wave length are given in Fig. 71. The reason why solutions in toluene are more efficient than those in xylene for large volumes, whereas the reverse is true for small volumes, is clear from these curves. Also, the value of using secondary solutes in toluene solutions, particularly for large samples, is apparent.

As for the organic crystals, the efficiency of liquid and plastic scintillants depends on the ionization density of the bombarding particle. The pulse height obtained with 5 Mev α-particles is about 10% of that obtained with 5 Mev electrons for most of the materials listed. This ratio depends on energy. Fig. 72, from Hoffman, Peterson and Janco[46], shows the relative pulse size from anthracene, stilbene and plastic (polystyrene plus 0.9% p-terphenyl plus 0.05% αNPO)

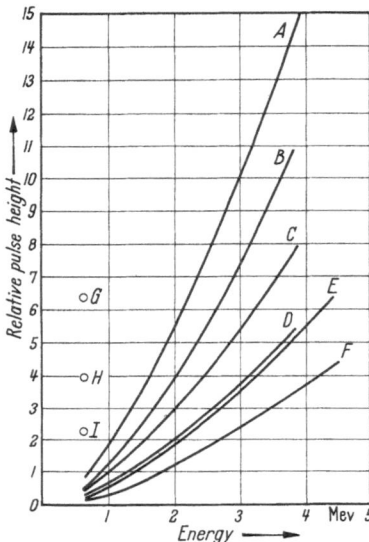

for electrons, protons and α-particles. Only one energy value is given for electrons since, as for the crystals, the pulse height vs. energy curve is linear except perhaps at very low energy.

16. Temperature dependence of organic phosphors. The temperature variation of the efficiency of various organic scintillants has been investigated by several groups. All results indicate that the efficiency decreases with temperature, at least in the region above 0° C. Kelley and Goodrich[38] used anthracene on a 1 P 28 multiplier and the conversion electrons from Cs^{137} as a source. The measured pulse height was constant over the region − 160 to 0° C, but fell by about 10% at 25° C. In their experiment the multiplier was at the same temperature as the crystal, so that only the overall characteristics of the system were measured. Using a 1 P 21 they found the pulse height increased by 30% in going from 25 to 60° C, which variation they suggested was due to a lowering of the short wave length limit of the multiplier. Minarik and Drickamer[39], with a 1 P 28 held at dry ice temperature, measured the counting rate from anthracene irradiated by radium. The rate dropped by a factor of 10 from 0 to 200° C

Fig. 72. Relative pulse height from anthracene, stilbene and plastic scintillant. A Anthracene to protons. B Anthracene to deuterons. C Stilbene to protons. D Stilbene to deuterons. E Plastic to protons. F Plastic to deuterons. G Anthracene to Cs¹³⁷ conversion electrons. H Stilbene to Cs¹³⁷ conversion electrons. I Plastic to Cs¹³⁷ conversion electrons. (Hoffman et al. [46].)

(not linearly) and thereafter fell very rapidly (the melting point of anthracene is about 217° C). Ball, Booth and MacGregor[40] have measured the pulse height variation of anthracene, stilbene and plastic, irradiated with Cs^{137} conversion electrons, over the temperature region 5 to 40° C. The resulting temperature coefficients (percent change in pulse height per degree centigrade, corrected for multiplier response) were: anthracene, − 0.27 ± 0.03; plastic, − 0.29 ± 0.03; stilbene, − 0.19 ± 0.03. Liebson[41] reported a light output (essentially counting rate) for anthracene which fell continuously with temperature over the range − 170 to 40° C, the rate of fall increasing with temperature; the total change

[46] M. M. Hoffman, R. W. Peterson and M. Janco: Los Alamos Scientific Laboratory (U.S.) Report No. LA-2069 (Unpublished) (1956).

[38] G. G. Kelley and M. Goodrich: Phys. Rev. 77, 138 (1950).

[39] W. L. Minarik and H. G. Drickamer: Rev. Sci. Instrum. 22, 704 (1951).

[40] W. P. Ball, R. Booth and M. MacGregor: Bull. Amer. Phys. Soc., Ser. II 1, No. 4, 183 (1956).

[41] S. H. Liebson: Nucleonics 10, No. 7, 41 (1952).

was almost a factor of two. For organic liquid scintillators SELIGER and ZIEGLER [42] have made measurements in the region -33 to $30°$ C. The temperature coefficients (percent change in pulse height per degree centigrade) for the scintillants investigated (solutions in m-xylene) are DPO (4 g/l), 0.5; PBD (8 g/l), 0.33; αNPO (3.2 g/l), 0.3. On the basis of their measurements they suggested that DPO in toluene at $-35°$ C with POPOP as a secondary solute might have a higher efficiency than anthracene.

The decay times of organic crystals have been investigated to some extent and seem to increase with temperature. KELLEY and GOODRICH [38] found a linear increase for anthracene; at $-180°$ C the decay constant was 0.012 μsec and at $+40°$ C, 0.038 μsec. LIEBSON [41] reported a nonlinear increase from 0.012 to 0.03 μsec for anthracene as the temperature was increased from 3 to 270° K. The temperature coefficient of the stilbene decay constant was much smaller.

17. Preparation of plastic scintillators. The following is a method for producing scintillating plastics for fast counters as described by SIEGEL and BLOCK [49]. The technique is similar to one originated by WOUTERS [50], and has been used to make excellent plastics as large as 10 in. in diameter by 10 in. long. The method involves heating styrene monomer to the boiling point, with the scintillating material (terphenyl) as a solute, and maintaining the mixture at that temperature until polymerization is complete.

The process is carried on in a glass reaction vessel immersed in a heated oil bath. Pyrex is recommended because of its resistance to thermal shock, but laboratory brown bottles up to 3 kg in capacity have been used successfully. The oil bath, heated by an electric hot plate, is maintained at about 155° C throughout the entire process. On the top of the reaction vessel a reflux column is attached through a ball joint to condense vapors; for making large plastics two reflux columns in series are necessary. Joints may be attached to the vessels with ceramic cement (Saureisen Insa-Lute).

The substances and quantities used during the process are:

Styrene	97.46% (by weight)
Terphenyl	2.5%
Tetraphenylbutadiene	0.03%
Zinc stearate	0.01%.

Plastics containing other concentrations of terphenyl and diphenylbutadiene instead of tetraphenylbutadiene [51] were found to be inferior. Also, two commercial brands of styrene were tried and were found equally good. The cost of raw materials is about $ 2.00 per kg of finished plastic.

When working with these materials it should be remembered that styrene is flammable and toxic, and that terphenyl and tetraphenylbutadiene are possibly carcinogenic.

At the beginning of the process only $1/4$ to $1/3$ the total amount of styrene + solutes is heated in the reaction vessel. This is done to avoid dangerously violent boiling and overloading of the reflux column. While this first charge is being initially heated, the reflux column should be kept off as long as possible so that any water vapor can be driven away. (Trapped water contributes to cloudiness in the final plastic.) When the reaction temperature is reached, about 150 to 155° C, the styrene will boil for about 20 minutes until it becomes slightly viscous.

[42] H. H. SELIGER and C. A. ZIEGLER: I.R.E. Trans. Nuclear Sci., N.S. **3**, No. 4, 62 (1956).

[49] R. SIEGEL and B. BLOCK: Carnegie Institute of Technology. Private Communication.

[50] L. F. WOUTERS: University of California Radiation Laboratory (U.S.) Report No. UCRL-4516 (Unpublished) (1955).

[51] Pilot Chemicals, Inc., 47 Felton St., Waltham 54, Mass. (U.S.).

Then violent foaming will occur intermittently, after which a foam will remain on the plastic while the polymerization is being completed. About an hour after all the foam and all the bubbles in the plastic have disappeared, the next charge is added. The overall time between charges is 3 to 4 hours.

After the first charge has been polymerized, the temperature of the bath should be maintained at the reaction temperature during all successive charging times. This is done to avoid a difference in index of refraction between the different charges. Such an interface was found when the temperature was lowered while charges were being added; the interface was less noticable when the temperature was maintained at the normal point. It is further suggested that part of each styrene charge be added before any of the solid matter.

When the final charge is free of all bubbles, the temperature is kept at 155° C for about 3 more hours. Then it may be gradually lowered to about 80° C, where it is kept for a few hours until the plastic is solid. The glass vessel should be broken while still at 80° C and the plastic placed in an insulated box to anneal for a few days to a week depending on the size of the plastic.

Pulse heights measured on a Tektronix 517 oscilloscope were 60 to 70% as large as those from stilbene.

At various stages during machining of the plastic into counter elements some relieving of internal strains may be desirable. Two hours at 75 to 80° C will anneal surface strains. Little or no annealing takes place below 75° C and above 80° C softening and deformation occurs. For thorough annealing of rough shapes, temperatures of 110 to 120° C have been maintained for several hours.

IV. Čerenkov counters.

18. General properties of Čerenkov radiation. Čerenkov radiation is emitted by a charged particle when it travels through a dispersive medium at a velocity greater than the velocity of light in the medium. The first quantitative information about this effect was given by the experimental investigations of Čerenkov[1-5] and the classical theory of Frank and Tamm[6,7]. The theory, which agreed with the experimental results, gave for the intensity and angular distribution of the radiation the following relations:

$$I\,d\nu = \left(\frac{2\pi e Z}{c}\right)^2 \left(1 - \frac{1}{\beta^2 n^2}\right)\nu\,d\nu \tag{18.1}$$

and

$$\cos\vartheta = \frac{1}{n\beta} \tag{18.2}$$

where β is v/c; I, the energy (ergs) radiated in the frequency interval ν to $\nu+d\nu$ per cm of path of the particle; eZ, the charge of the particle (esu); c, the velocity of light (cm/sec); v the particle velocity; and n, the index of refraction of the medium for radiation of frequency ν. The wave front of the radiation is conical with the axis of the cone along the path of the particle and apex at the particle; the normal to the wave front makes an angle ϑ with respect to the particle direction. The radiation is polarized with the magnetic vector tangent to the conical wave front and perpendicular to the direction of motion of the particle.

[1] P. A. Čerenkov: C. R. Acad. Sci. USSR. **8**, 451 (1934).
[2] P. A. Čerenkov: C. R. Acad. Sci. USSR. **12**, 413 (1936).
[3] P. A. Čerenkov: C. R. Acad. Sci. USSR. **14**, 102 (1937).
[4] P. A. Čerenkov: C. R. Acad. Sci. USSR. **14**, 105 (1937).
[5] P. A. Čerenkov: Phys. Rev. **52**, 378 (1937).
[6] I. Frank and I. Tamm: C. R. Acad. Sci. USSR. **14**, 109 (1937).
[7] I. Tamm: J. Phys. USSR. **1**, 439 (1939).

Later, more precise experiments confirmed the above properties of the radiation. COLLINS and REILING[8], using a monoenergetic electron beam as a source of particles, instead of electrons with a continuous energy spectrum as used by ČERENKOV, checked both Eqs. (18.1) and (18.2) and also examined the spectrum of the radiation, which was found to have the proper continuous distribution. WYCOFF and HENDERSON[9], using an electron beam, the energy of which could be varied from 240 to 815 kev, carefully tested Eq. (18.2) over the region where β is changing rapidly. An even more refined test of the theory was made by MATHER[10]. Here the particles were protons, and relation (18.2) was checked by measuring ϑ for energies of 225, 270 and 340 Mev. Thus, incidentally, the effect was shown to be independent of particle mass. With a precise experimental arrangement in which the first order chromatic dispersion was compensated by an appropriate prism, MATHER found that the remaining angular spread (about 0.5° full width at $1/_2$ maximum intensity) of the light could be explained by second order chromatic effects and beam effects, such as energy and angular spread, scattering, slowing down in the medium and diffraction due to finite path length. Hence, the angular distribution is apparently the delta-function predicted by the theory. Also, the light was shown to be completely plane polarized as predicted. WINCKLER, MITCHELL, ANDERSON and PETERSON[11], using charged π-mesons traveling in Lucite, have measured the intensity of the radiation as a function of energy. They confirmed the variation predicted by Eq. (18.1) in the energy region near threshold, $n\beta = 1$.

Quantum electrodynamical as well as further classical theoretical treatments of the phenomenon have been given. These do not alter the predictions of Eqs. (18.1) and (18.2) insofar as we are here concerned with the application to counters. A recent article by TIDMAN[12], which presents a quantum theory treatment, also gives references which cover the theoretical work. Expression (18.2) for the angular distribution of the radiation can be derived in a simple way using a HUYGENS' construction[10] or by an argument based on conservation of energy and momentum for the particle-photon system[13].

The first applications of the Čerenkov effect to the field of particle counters were made by GETTING[14] and DICKE[15] in 1947 and by JELLEY[16] in 1951. The radiation produced in these counters was detected by photomultiplier tubes, as has been the case in all Čerenkov counters used since. Following these demonstrations of their practicability, Čerenkov counters have been used in numerous experiments and many are described in review articles by MARSHALL[17] and by JELLEY[18-20].

Before describing particular counters, it is profitable to consider the properties of the radiation in a little more detail. Čerenkov radiation is emitted in the spectral region where the index of refraction is such that $n > 1/\beta$. However,

[8] G. B. COLLINS and V. G. REILING: Phys. Rev. 54, 499 (1938).
[9] H. O. WYCOFF and J. E. HENDERSON: Phys. Rev. 64, 1 (1943).
[10] R. L. MATHER: Phys. Rev. 84, 181 (1951).
[11] J. R. WINCKLER, E. N. MITCHELL, K. A. ANDERSON and L. PETERSON: Phys. Rev. 98, 1411 (1955).
[12] D. A. TIDMAN: Nuclear Phys. 2, No. 4, 289 (1956).
[13] S. M. NEAMTAM: Phys. Rev. 92, 1362 (1953).
[14] I. A. GETTING: Phys. Rev. 71, 123 (1947).
[15] R. H. DICKE: Phys. Rev. 71, 737 (1947).
[16] J. V. JELLEY: Proc. Roy. Soc. Lond. A 64, 82 (1951).
[17] J. MARSHALL: Annual Rev. Nucl. Sci. 4, 141 (1954).
[18] J. V. JELLEY: Atomics 4, 81 (1953).
[19] J. V. JELLEY: Brit. J. Appl. Phys. 6, 227 (1955).
[20] J. V. JELLEY: Progr. Nucl. Phys. 3, 131 (1953).

for a counter we are interested only in the part of the spectrum for which the detector, a photomultiplier, is sensitive and the medium is not very absorbent. For most photomultipliers the long wavelength cut-off is about 6000 Å and the short wavelength cut-off about 3500 Å. This latter limit may be extended to about 2000 Å by use of photomultipliers with ultraviolet transmitting windows, such as the RCA 6903, the EMI 6255 B or 6256 B, and the DuMont K 1306, provided the medium also transmits radiation in the ultraviolet region.

From Eq. (18.1) the intensity of the radiation per cm path of the particle is given by

$$\left. \begin{aligned} I\,dv &= 1.0 \times 10^{-38} Z^2 \left(1 - \frac{1}{\beta^2 n^2}\right) v\,dv \text{ erg} \\ &= 1.53 \times 10^{-12} Z^2 \left(1 - \frac{1}{\beta^2 n^2}\right) dv \text{ photons} \end{aligned} \right\} \tag{18.1a}$$

where dv; is sec^{-1}. A plot of $\left(1 - \frac{1}{\beta^2 n^2}\right)$ vs. $n\beta$ isshown in Fig. 73. From this curve and the assumption that n is constant over the region, one can obtain the number of photons per cm by multiplying the ordinate by $1.53 \times 10^{-12} Z^2 \Delta v$. Fig. 73 thus essentially shows the variation of pulse height with velocity for a given index of refraction. As an example of a pulse height calculation let us consider the spectral region 3500 to 5500 Å ($\Delta v = 3.1 \times 10^{14}$ sec^{-1}) as being of interest (which is roughly the case for photocathodes with S-11 response). In the limit of $1/n\beta$ approaching zero, the upper limit of 475 photons per cm path would be emitted; for a particle with $\beta = 1$ traveling in water ($n = 1.33$), there would be about 205 photons per cm. Assuming 100% collection of this light and 6% conversion efficiency for the photocathode, a maximum of 28 photoelectrons would be liberated from the photocathode per cm of path of the particle, whereas for the case of the relativistic particle in water, the number would be 12. In some cases these numbers may be increased. Heiberg and Marshall[21] have achieved an increase in pulse height of 20 to 30% for a water radiator by adding an appropriate fluorescent material which produced a shift of some of the ultraviolet radiation to longer and hence useful wave lengths. Two materials were the sodium salts of 2-amino-6,8-naphthalene-disulfonic acid and 2-naphthol-3,6-disulfonic acid. Likewise, Jones, Kratz and Rouvina[22] by saturating the CCl_4 in a large tank with 1,4-di-[2-(5-phenyl-oxazyolol)]-benzene (POPOP) have increased the Čerenkov pulse output, for shower electrons, by about a factor of two. However, the action of these shifters tends to make the radiation isotropic and hence the directional property is lost. A second way to increase the useful yield is to make use of a multiplier with an ultraviolet transmitting window, as has been mentioned above. By this means, the number of useful photons from a water radiator can be about doubled; no Lucite light pipe should be used or all the advantage will be lost.

In most cases some light will be lost in reflections and by absorption. Thus the pulses from a Čerenkov counter will generally be small and show large fluctuations. For this reason Čerenkov counters are not used unless advantage is taken of one of the several unique characteristics of the radiation, such as the angular distribution, existence of a velocity threshold, or short-time duration of the light pulse. Further, in order to obtain a particle detection efficiency approaching 100%, it is customary to use scintillation counters in coincidence with the Čerenkov counter so that signal pulses may be separated completely from noise pulses.

[21] E. Heiberg and J. Marshall: Rev. Sci. Instrum. **27**, 618 (1956).
[22] W. B. Jones, H. R. Kratz and J. Rouvina: General Electric Research Laboratory Report No. 57-RL-1675 (Unpublished) (1957).

The angular distribution of the radiation is of importance in many Čerenkov counters. Fig. 74 shows the value of ϑ as a function of $n\beta$. For example, since n for water decreases from 1.3489 at 3500 Å to 1.3344 at 5500 Å, for $\beta = 1$ the radiation between these limits will be emitted in a forward cone of $\vartheta = 42°$ half angle and with an angular width, $\varDelta\vartheta$ of about 0.6°. It is important to note that the radiation will have an angle of incidence equal to or greater than the critical angle at a surface parallel to the path of the particle, provided the medium outside has $n = 1$. Hence, at such surfaces internal reflection will occur. It should also be noted that for certain combinations of β, n, and geometry, internal reflection will occur at all surfaces. Consequently, in a rectangular cylinder of $n = \sqrt{2}$, there will be entrapment of the light from particles of $\beta = 1$ traversing the medium parallel to the cylinder axis. For higher indices entrapment will occur

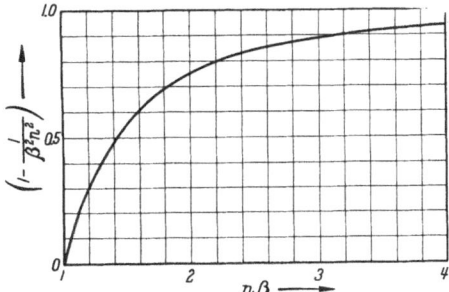

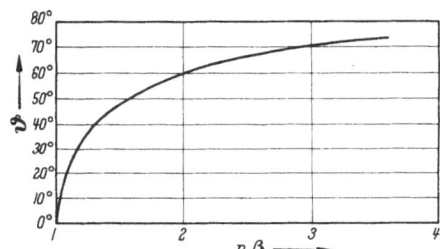

Fig. 73. Variation of intensity of Čerenkov radiation with $n\beta$ where the number of photons per cm of path is given by

$$1.53 \times 10^{-12} Z^2 \left(1 - \frac{1}{\beta^2 n^2} \right) d\nu.$$

Fig. 74. Angle of emission of Čerenkov radiation as a function of $n\beta$.

for smaller β. Thus, poor optical contact between the photomultiplier and medium could result in the loss of much light.

A quite thorough discussion of the optical properties of Čerenkov counters has been given by MANDO [23]. The dependence of the directional response, uniformity of response, time of response, and the efficiency, on the nature of the reflecting surfaces (specular or diffuse) is considered in detail. This investigation is related mainly to counters which have a sensitive area large compared to the photocathode area. As might be expected, counters constructed with diffusing walls show more uniform response, lower maximum efficiency, less directional behavior, and longer time response than those with specularly reflecting walls.

The remainder of this section is devoted to descriptions of counters which have been designed to make use of the specific properties of Čerenkov radiation.

19. Threshold counters. Most Čerenkov counters are used to take advantage of the threshold effect*. For instance, by the use of such a counter much of the

---

[23] M. MANDO: Nuovo Cim. **12**, (1), 5 (1954).

* It is sometimes important to remember that a heavy particle can transfer sufficient energy to an electron so that the latter has even higher β than the incident particle. The maximum kinetic energy an electron can obtain is given by (BRUNO ROSSI: High-Energy Particles. New York: Prentice Hall 1952.)

$$E_m = 2 m_e c^2 \frac{\beta^2}{1 - \beta^2}$$

for incident particles with $p \ll \dfrac{m^2 c}{m_e}$, where m_e and m are the electron and particle masses; p is the incident particle momentum; and β refers to the incident particle. Thus a 1.36 Bev proton or a 200 Mev π-meson ($\beta = 0.91$) can give rise to a knock-on electron of energy about 5 Mev ($\beta = 0.992$). Likewise, through nuclear reactions, fast secondaries can be produced by a slower primary; for example, a proton of $\beta = 0.88$ can produce a π-meson of β up to about 0.99.

background existing in the vicinity of a high-energy accelerator can be eliminated when counting electrons. A typical example is described by Hildebrand[24] and shown in Fig. 75. The assembly can be adapted easily to use a solid rather than liquid medium. With water, this counter was 100% efficient for 145 Mev π^--mesons when used in coincidence with scintillation counters. Hildebrand[25] used such counters to detect electrons from the conversion in lead of the γ-rays from π^0 decay, the π^0-mesons coming from the reaction $n + p \rightarrow \pi^0 + d$.

Because the light is emitted in the forward direction, it is usually preferable to construct a counter in a manner such as above so as to minimize the loss of intensity due to reflections. Furthermore, when possible, internal

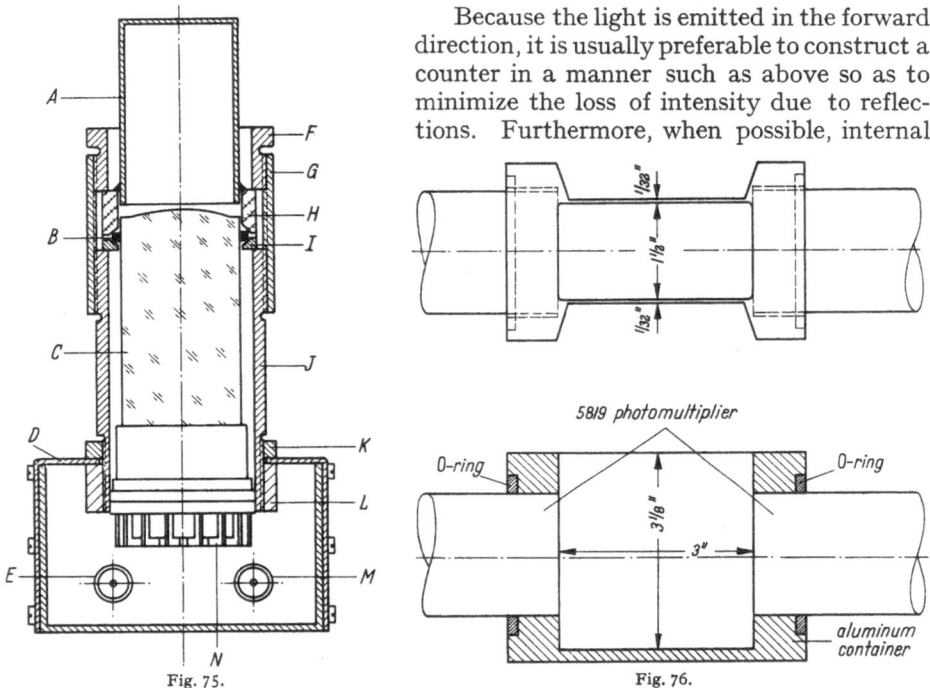

Fig. 75.

Fig. 76.

Fig. 75. Counter assembly. (A) Liquid container. (B) 2 in.×2³/₈ in. "○" ring. (C) RCA-5819 multiplier phototube. (D) 3 in.× 4 in.× 5 in. I.C.A. Fleximount Aluminum case. (E) High voltage connector set into case. (F) 2¹/₂ in. insulating conduit bushingmale. (G) 2¹/₂ in. conduit coupling. (H) Flange 2.660 in. o.d., 2.070 in. i.d., ³/₄ in. long, 60° bevel. (I) Washer 2.660 in. o.d., 2.070 in. i.d., 60° bevel. (J) 2¹/₂ in. pipe nipple 4 in. long. (K) 2¹/₂ in. conduit lock nut. (L) 2¹/₂ in. insulating conduit bushingfemale-relieve thread. (M) Signal connector set into case. (N) Amphenol-medium diheptal tube base No. 59-417. (Hildebrand[24].)

Fig. 76. Transmission type Čerenkov counter (Lindenbaum and Pevsner[27]).

reflection should be used to direct the light to the multiplier, and reflectors should make poor optical contact with the medium; the counter shown in Fig. 75 might suffer less light loss if the container for the liquid was a clear plastic loosely wrapped with Al foil. However, if a transmission counter is required, it is desirable to avoid having the particle traverse the multiplier and a design such as that described by Lindenbaum and Pevsner[27] can be followed (Fig. 76). The cell, to contain a liquid, was 3 in.×3 in.×1.5 in.; particles traversed the 1.5 in. dimension. The cell was lined with aluminum foil and viewed by two RCA 5819 photomultipliers. A large amount (up to 80%) of the light was lost, but with sufficient amplification an efficiency of almost 100% was obtained using the counter in a coincidence telescope with scintillators.

[24] R. H. Hildebrand: Rev. Sci. Instrum. 24, 463 (1953).
[25] R. H. Hildebrand: Phys. Rev. 89, 1090 (1953).
[27] S. J. Lindenbaum and A. Pevsner: Rev. Sci. Instrum. 25, 285 (1954).

For cosmic ray investigations of slow, heavy particles, where for example the particles are to be photographed in a cloud chamber, it is undesirable to trigger the chamber on relativistic particles which occur in large numbers. Here a Čerenkov counter may be used in anticoincidence with a Geiger counter telescope as described by YORK[28]. The medium was Lucite for which the critical value of β is about 0.66. Due to the inefficiency of the Čerenkov detector, the type of event on which the chamber triggered most often was still a single fast particle, but the frequency of such events was considerably reduced compared to that of the desirable events.

For a somewhat similar purpose, DEURDEN and HYAMS[29] have used a water Čerenkov counter in anticoincidence with Geiger counters, so as to count only cosmic-ray protons in the momentum range 700 to 1100 Mev/c. The arrangement was such that the particles with sufficient range to traverse the absorber in a Geiger counter telescope, but with insufficient velocity to give a Čerenkov pulse, were necessarily protons and not mesons. Hence, practically all Geiger counts for which there were no Čerenkov counts corresponded to proton events. A counting rate of 3.5 per hour, at sea level, was observed for protons in the above momentum range. The Čerenkov counter was constructed of a cubical box (24 cm on a side) to contain the water and was coated inside with $MgCO_3$ for diffuse reflection. The photocathode of an EMI $VX\,5045$ photomultiplier projected slightly into the water at the middle of one side; particles traversed the counter perpendicular to the multiplier axis. The box and multiplier assembly was tested with a light source; a 15% light collection efficiency was found. When tested with cosmic ray μ-mesons traversing the counter, an average of 30 photo-cathode electrons was measured, a number in rough agreement with the collection efficiency.

20. Directional counters. The first of the counters described above will give a larger pulse height when the particle travels along the counter axis towards the multiplier than in the opposite direction. If it travels away from the photomultiplier, more reflections are necessary in order for the light to reach the photocathode; the amount of light reaching the photocathode in this case may be reduced to practically zero by making the end of the medium away from the multiplier absorbing, e.g., by painting with black paint in a manner so as to prevent internal reflection. WINCKLER and ANDERSON[30,31] have described the behavior of a counter with a Lucite radiator of dimensions 4.25 in. × 2.25 in. × 8 in. They used an RCA C-7157 tube with a 4.25 in. × 3.125 in. cathode which was against the 4.25 in. × 2.25 in. surface of the Lucite; the opposite surface was blackened. Internal reflection was used for light collection. Tested in a Geiger counter telescope (containing 6 in. of lead) with cosmic ray μ-mesons, an efficiency of 100% was obtained with the mesons coming towards the multiplier and zero when they traveled in the opposite direction. The total width at one-half maximum of the number vs. pulse height curve was between 25 to 50% of the pulse height at maximum, depending on the particular tube. This counter was sent up in balloon flights to measure the ratio of downward to upward flux in cosmic radiation; the experiment indicated that not more than 5% of the particles with $\beta > 0.7$ were traveling in the upward direction.

The directional property of Čerenkov counters has also been used for the detection of K-particles by a measurement of their decays. ROBINSON[32] described

[28] C. M. YORK jr.: Phys. Rev. **96**, 1635 (1954).
[29] T. DEURDEN and B. D. HYAMS: Phil. Mag. (7) **43**, 717 (1952).
[30] J. WINCKLER and K. ANDERSON: Rev. Sci. Instrum. **23**, 765 (1952).
[31] J. WINCKLER and K. ANDERSON: Phys. Rev. **93**, 596 (1954).
[32] K. W. ROBINSON: Phys. Rev. **99**, 1606 (1955).

an arrangement in which K-mesons, produced in lead by cosmic rays, went through a Geiger counter array and a liquid scintillation counter. Below the scintillation counter was a bank of water radiators, blackened on the bottom (by black velvet in the water) and with the photomultipliers above. Thus, a Čerenkov pulse (delayed with respect to the scintillation pulse) was detected only when a particle stopped in the region of the Čerenkov counters and its decay particle passed upward through the water. The mean lifetime of such particles was given by measurement of these delays; after a correction for $\pi \to \mu \to e$ counts, the value obtained was $(8.05 \pm 0.66) \times 10^{-9}$ sec.

21. Counters for measurement of the charge of a particle. Since the intensity of Čerenkov light depends on the factor $Z^2 e^2$ [Eq. (18.1)] it would appear to be easy to separate the pulses arising from particles with different Z but with the same β. For this reason Čerenkov counters have been applied to the measurement of the intensity of the components of primary cosmic radiation. The great advantage of Čerenkov counters over ionization counters in this application is that the former will not respond, for example, to a slow proton which might give ionization equal to that of a fast α-particle. The first such investigation was made by WINCKLER and ANDERSON[31] as a byproduct of their experiment described earlier. Their counter was constructed primarily for the asymmetry measurement and not for the purpose of resolving charge components. As a result it was not ideally suited for the latter purpose, since, due to the large particle path in the Lucite, too many large pulses arose from the products of nuclear collisions. However, a rough value for the H:He:Li ratio of 1:0.08:0.02 was obtained.

LINSLEY and HORWITZ[33] have designed Čerenkov counters specifically for the measurement of the primary multiply charged cosmic-ray flux. The best of these consisted of a 4-inch diameter by 1-inch thick Lucite cylinder optically coupled to a DuMont K1198 photomultiplier (5 in. photocathode). This thickness is only a few percent of a geometrical mean-free-path for protons, and hence the separation of proton and α pulses was considerably improved. Aluminum foil was glued to the sides of the Lucite away from the photomultiplier; however, the results probably would have been as good with no reflector. The pulse height distribution obtained with cosmic-ray μ-mesons was fitted to a Poisson distribution of width, at half maximum, of 35 to 40% of the peak pulse height; this width was attributed almost entirely to photomultiplier statistics. The distribution deviated from POISSON at the large pulse height end due to knock-on electrons and at the small pulse height end due to slow μ-mesons. The high-energy tail did not cause the distribution for particles of one charge to overlap that of particles of adjacent charge. LINSLEY and HORWITZ[33] pointed out that the effect on the distribution due to secondary electrons should be somewhat less than the effect for an ionization (or scintillation) device of the same surface density since many of the electrons important in giving the ionization width are too slow to give Čerenkov radiation. However, the Čerenkov counter suffers from fluctuation due to the small number of cathode photoelectrons.

In the experiments[34-37] in which such counters were used, auxiliary devices were necessary to eliminate various backgrounds. LINSLEY[34,35] used the pulse from the Čerenkov counter, in coincidence with Geiger counter pulses, to trigger a cloud chamber. The triggering occurred only when the Čerenkov pulse was

[33] J. LINSLEY and N. HORWITZ: Rev. Sci. Instrum. **26**, 557 (1955).
[34] J. LINSLEY: Phys. Rev. **97**, 1292 (1955).
[35] J. LINSLEY: Phys. Rev. **101**, 826 (1956).
[36] N. HORWITZ: Phys. Rev. **98**, 165 (1955).
[37] W. R. WEBBER and F. B. MCDONALD: Phys. Rev. **100**, 1460 (1955).

about twice that of the mean pulse height for particles of $Z=1, \beta=1$. Examination of the cloud chamber pictures made it possible to separate fast multiply charged particle events from background events such as showers, those in which particles of $Z=1$ produced fast knock-on electrons which traversed the Čerenkov counter, and those in which the Geiger counters were tripped by knock-on electrons. In the experiment of HORWITZ[36], guard counters surrounded the upper Geiger counters and the Čerenkov counter; the tripping of these, coincident with a Čerenkov counter pulse of height corresponding to a fast particle of $Z>1$, indicated a shower rather than a multiply charged particle. Further, an array of Geiger counters below the apparatus indicated an event, arising from a nuclear interaction of a proton, in which more than one particle traversed the equipment. LINSLEY[34] suggested, from the experience of this experiment, that these investigations might be considerably improved by the simultaneous use of an ionization and a Čerenkov counter.

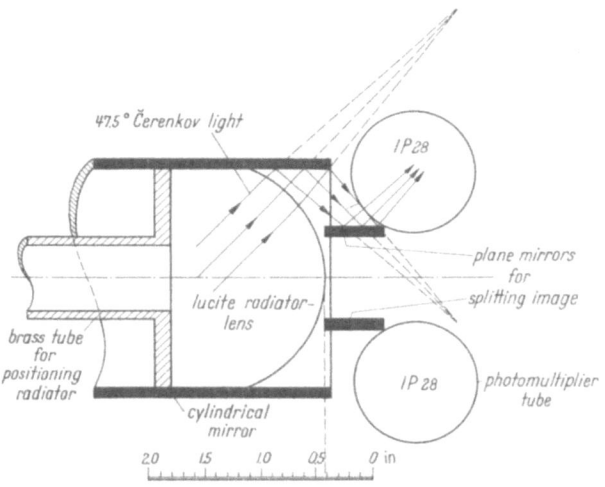

Fig. 77. Focusing Čerenkov counter (MARSHALL[38]).

22. Focusing counters. The first type of Čerenkov counter suggested[14] and tested[15] was a focusing counter, a name applied to one in which the photomultiplier system is sensitive to radiation over a limited angular range and hence the counter is sensitive to a limited velocity range. This type of counter can be used to analyze the energy spectrum of a beam of particles over a limited range of β above $1/n$. For ordinary counting of relativistic particles, these counters have the advantage of being more directionally sensitive than those previously described; for types described below, particles a few degrees off axis are not counted. Hence, they may be useful where random backgrounds of relativistic particles exist. However, they generally have small sensitive areas and a low efficiency for any beam which is not parallel (or which may suffer appreciable multiple scattering in the radiator). As will be seen, focusing counters are considerably more complicated to construct than the nonfocusing type.

MARSHALL[38], and MATHER and MARTINELLI[39] have described somewhat similar types of focusing counters which have been used in beams from accelerators; the former reference also contains a comprehensive discussion of various other models.

The type used by MARSHALL is shown in Fig. 77. The system, except for the photomultipliers and plane mirrors, has symmetry around the axis which is the beam axis. In the absence of any of the mirrors shown, the light from a particle traveling with uniform velocity down the axis would be focused by the spherical surface of the Lucite into a ring at a distance three radii from the center of curvature of this lens (neglecting dispersion). An appropriately placed cylindrical

[38] J. MARSHALL: Phys. Rev. **86**, 685 (1952).
[39] J. W. MATHER and E. A. MARTINELLI: Phys. Rev. **92**, 780 (1953).

mirror of radius one-half the radius of this ring would focus all the light back to the axis. Rather than place one multiplier on the axis to detect the light, it is preferable to split the beam, as was done by the plane mirrors shown in Fig. 77, and to use two photomultipliers (MATHER and MARTINELLI used three mirrors and multipliers) in fast coincidence. The advantage of this arrangement is to reduce background from the following sources: (a) tube noise, (b) particles traversing a multiplier, and (c) particles entering the medium at large angles to the axis.

MARSHALL showed that not all the light from particles which are parallel to but off the axis will be focused to a point; instead it will be spread out over a region between the axis and a distance from the axis given by

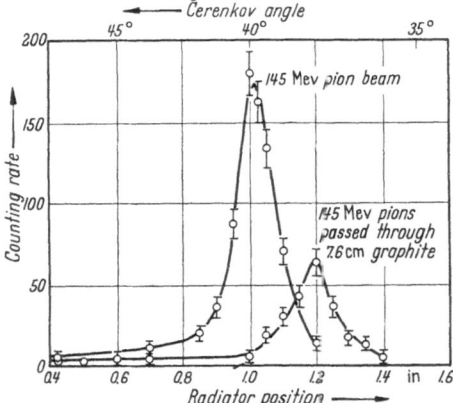

$$D = n\,d\,\frac{\sin\vartheta}{\sin\vartheta'} \qquad (22.1)$$

Fig. 78. Velocity resolution of Čerenkov counter shown in Fig. 77. Radiator position scale corresponds to the scale shown in Fig. 77 (MARSHALL[38]).

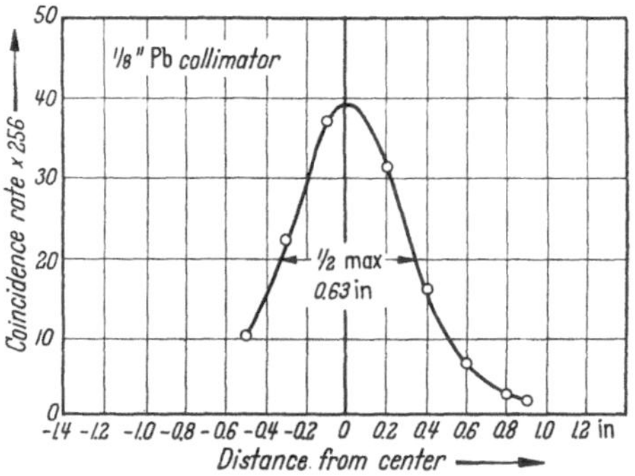

Fig. 79. Variation of counting rate of a focusing Čerenkov counter with distance of beam from axis of counter. Beam parallel to axis (MATHER and MARTINELLI[39]).

where d is the distance of the particle from the axis; n, the index of refraction; ϑ, the Čerenkov angle; and ϑ', the angle of convergence of the light to the focus. Hence, even for a beam of particles which remain parallel through the radiator, the smallest diameter of focal spot will be D for a beam diameter of d. For the construction used by MARSHALL and by MATHER and MARTINELLI, $\vartheta = \vartheta'$, so that $D = nd$.

The resolution of these counters is also broadened somewhat by scattering of the particles, slowing down in the radiator, diffraction, divergence of the beam, dispersion of the medium, aberration of the lens and by the production of fast δ-rays or other fast secondary particles. These effects are discussed by MATHER[10], DEDRICK[40] and MARSHALL[41].

In order to count particles of various velocities, the radiator-lens combination is slid towards or away from the multipliers. An example of the performance of MARSHALL's counter with 145 and 121 Mev π-mesons is shown in Fig. 78. The widths of the curves depended on the coincidence pulse height accepted; with bias just above the level for single counts, the widths were about double those shown. The Čerenkov angles for the two meson energies should be 40.4 and 38.1°, which agree well with the measured angles.

Data on the performance of the counter of MATHER and MARTINELLI were obtained using a collimated 320 Mev x-ray beam of $1/8$ in. diameter. The x-rays

[40] K. G. DEDRICK: Phys. Rev. 87, 891 (1952).
[41] J. MARSHALL: CERN Symposium Proc. 1956, 63.

produced electrons in a thin lead converter in front of the Čerenkov counter. Fig. 79 shows the response of the counter as a function of distance of beam axis from counter axis with these two parallel. The counting rate fell to half value when this distance was 0.3 in. which is compatible [Eq. (18.3)] with the size of the $1P21$ photocathode of about 1 cm (this response curve width of course depends on discriminator bias since there was no plateau). Fig. 80 shows the effect of varying the angle between the beam and counter axes with the beam entering the center of the face of the Lucite radiator. The full width at half maximum of 2.7° demonstrates the directional selectivity of this type of counter. When the electrons traversed the counter in the backward direction no counts were observed. A further test was made in a beam of decay γ-rays (with energy spread from 40 to 120 Mev) from π^0-mesons. With the beam collimated to about

$3/_4$ in. and a $1/_{16}$ in. thick converter, an efficiency of about 20% for counting the γ-rays was measured; this is not far below the efficiency obtained with the same lead thickness but using nonfocusing Čerenkov counters in coincidence with scintillation counters. However, this efficiency is near the peak of the efficiency vs. converter thickness curve, whereas with a nonfocusing counter, which is not so sensitive to the direction of the conversion electrons, the γ-ray efficiency increases almost linearly up to about $3/_{16}$ in. thickness.

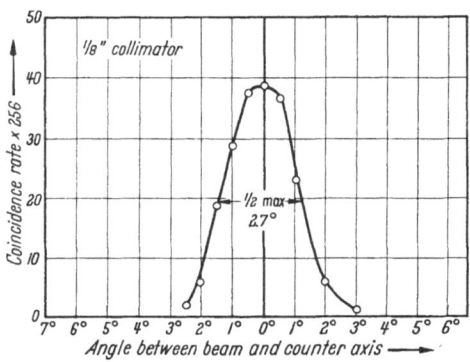

Fig. 80. Variation of counting rate of a focusing Čerenkov counter with angle between beam axis and counter axis (MATHER and MARTINELLI[39]).

One of the most important applications of velocity selecting Čerenkov counters was made by CHAMBERLAIN et al.[42,43] in the experiment in which they detected the antiproton. Here particles of momentum 1.19 Bev/c from a target in the Bevatron were selected by magnets. Since these particles were overwhelmingly π-mesons, it was necessary to make a careful velocity selection to detect the protons. The velocity determinations were made by a time-of-flight measurement (using scintillation counters) and also, more selectively, by a focusing Čerenkov counter somewhat similar to those described above (but without the lens surface). Only particles with β between 0.75 and 0.78 (corresponding to a $\Delta\vartheta$ of about 5°) were counted. About 3% of the mesons ($\beta=0.99$) were also detected by this counter as a result of nuclear collisions. Hence, it was necessary to eliminate mesons as efficiently as possible by inserting before the focusing counter a nonfocusing Čerenkov counter to detect particles with $\beta>0.79$. For the focusing counter, fused quartz was used ($n_D=1.458$) and for the nonfocusing counter, $C_8F_{16}O$ ($n_D=1.276$).

A somewhat different type of velocity selecting counter has been used by FITCH[44] for detecting only K-mesons in a momentum selected beam consisting also of π-mesons and protons. Carbon disulfide was used as the Čerenkov medium. Since $n=1.62$, no radiation was emitted by a particle of $\beta<0.62$; this was the lower limit on the velocity detected. The CS_2 was in a right cylindrical glass

[42] O. CHAMBERLAIN, E. SEGRÈ, C. WIEGAND and T. YPSILANTIS: Phys. Rev. **100**, 947 (1955).
[43] O. CHAMBERLAIN and C. WIEGAND: CERN Symposium Proc. **1956**, 82.
[44] V. L. FITCH: Bull. Amer. Phys. Soc. **1**, 1, 52 (1956).

container. The light originating from a particle which traveled parallel to the axis was internally reflected at the exit end if $\beta > 0.78$. The face of the cylinder through which the particle entered was blackened so that this light would be absorbed. Consequently, the band selected was $0.62 < \beta < 0.78$. The radiation from particles in this velocity range left the exit end of the CS_2 cell and was directed by a diffuse reflector to the photocathode of a multiplier mounted with axis perpendicular to the path of the particle. This method of discrimination against high velocities is particularly sharp since the plane of polarization of the light is the plane of incidence. However, the sharpness of discrimination is affected by the angular divergence of the beam and by multiple scattering of the particles. This type of counter is clearly simpler to construct than the focusing type, and more likely to have high efficiency and large sensitive area.

23. Gas Čerenkov counters. The first experiments using gas as the Čerenkov medium were made by Jelley[19]. The light, produced as cosmic-ray shower particles passed through air, was focussed by a parabolic mirror at a photomultiplier. Such a shower detector has the property of being quite sensitive to the shower direction. From Eq. (18.1 a)

$$I = 900 \, (n\beta - 1) \, Z^2 \text{ photon/cm of path}$$

for $\varDelta\nu = 3 \times 10^{14} \text{ sec}^{-1}$ and $n\beta \gtrsim 1$. Hence, we expect for gases less than 1 useful photon per cm, even for $\beta = 1$. However, gas counters may become quite useful for work at ultra high energy accelerators even though the light intensity is small.

Kinsey[45] has described a gas counter, similar to that of Jelley's, for detecting mesons of several Bev. Nitrogen ($n = 1.000296$ at N.T.P.) was contained in a cylindrical steel tank four feet long. Particles entered at one end; at the other end a parabolic mirror was placed with concave side facing inwards and with its axis on the tank axis. The Čerenkov angle was very small so that the mirror brought the light to the principal focus at which point a photomultiplier was placed (in the beam). When used in a telescope between two scintillators an efficiency of 100% was obtained for counting 3.5 Bev π-mesons; the gas pressure was 3 atmospheres which corresponded to a threshold of 3 Bev.

A somewhat more elegant gas counter has been constructed by Yuan and Lindenbaum[46]. This was built along the lines of the focusing arrangements described by Marshall. The index of refraction could be varied from 1 to 1.1 by varying the gas (CO_2) pressure up to 100 to 200 atmospheres; thus, the velocity of the particles counted was determined by gas pressure.

V. Scintillation and Čerenkov counters for high energy β- and γ-ray spectroscopy.

a) General

24. Lately there has been considerable use of large scintillation and Čerenkov counters for the detection of β- and γ-rays with energies between tens of Mev to several Bev. Such detectors, which can have high efficiency and large sensitive area, are necessary for experiments where the intensity of the radiation is very weak. The principle of operation is simply that the incident β- or γ-ray develops a cascade shower in the counter. If the counter material is a scintillant, a few percent of the energy will appear as luminescence which can be detected with

---

[45] B. B. Kinsey: CERN Symposium Proc. **1956**, 68.
[46] L. C. L. Yuan: CERN Symposium Proc. **1956**, 68.

photomultipliers; if the material is not luminescent but is transparent, a much smaller fraction of the energy will appear as Čerenkov radiation which likewise may be detected. Since the energy appearing as scintillation light is proportional to ionization energy loss of the electrons, the intensity of light produced in the scintillant will be proportional to the energy of the shower. The proportionality between Čerenkov light and shower energy depends on the fact that since the ionization loss by electrons varies only slowly with energy, the total electron path length, which determines the amount of light, is nearly proportional to energy. The counter should be large enough to contain most of the shower, otherwise large fluctuations in pulse height will occur. The problem is to find a dense medium which is either a scintillant or is transparent over a reasonably wide part of the spectral region in which photomultipliers are sensitive. The scintillators which have been used include NaI(Tl), terphenyl in xylene, and discs of plastic scintillant sandwiched between lead discs or discs of lead glass (the use of liquid xenon has been suggested); as Čerenkov radiators, lead glass has been used most widely, but counters employing thallium chloride and CCl_4 have also been built.

The exact performance of these counters can be found only by an experimental determination using monochromatic β- or γ-ray beams of variable energy. However, it is possible to make rough predictions of the pulse height-energy behavior and the width of the pulse height distribution for monoenergetic particles. The simplest procedure is to make use of the data obtained by KANTZ and HOFSTADTER[1,2]. They have measured the average energy loss at various points in a number of media for showers produced by 185 Mev electrons. The materials used were C, Al, Cu, Sn and Pb. These data (Fig. 81) can be used to calculate the fraction of incident energy which will be dissipated in a cylinder of given dimensions for particles axially incident (for γ-rays a correction to the lengths, of the order of several radiation lengths, must be made to account for the position of formation of the first electron-positron pair). For other media, this information can be obtained by the use of the approximation that the behavior of all materials is the same if lengths are expressed in units of radiation lengths and energies in units of the critical energy. At energies higher than those corresponding to the measurements of KANTZ and HOFSTADTER, one may consider that on the average one radiation length is required for two particles to appear in place of the one incident. Thus, roughly, one radiation length or so must be added to the length each time the energy is doubled. The radius of the medium can be increased more slowly with energy. In this manner the energy response of a shower counter can be predicted. Table 15 gives values of the radiation length and critical energy for the materials studied by KANTZ and HOFSTADTER as well as for materials which have been used in counters. This method was used by BROBANT, MOYER and WALLACE[3] to predict the linearity of a lead glass counter 12.25 in. diameter by 14 in. long. The response was expressed in the form

$$E_c = a E (1 - b E)$$

where E is the incident energy; E_c, the energy contained in the glass; and a and b, constants. The predicted value of b over the energy range from 0 to 1.5 Bev was 3.7×10^{-4} Mev^{-1} whereas the measured value was 1.6×10^{-4} Mev^{-1}. It appears that the above approximation may not give very exact results for the energy lost; however, it is adequate for estimates of counter size required for a given energy.

[1] A. KANTZ and R. HOFSTADTER: Nucleonics **12**, No. 3, 36 (1954).
[2] A. KANTZ: Stanford University (U.S.) High Energy Physics Laboratory Report No. 17 1954.
[3] J. M. BROBANT, B. J. MOYER and R. WALLACE: University of California Radiation Laboratory (U.S.) Report No. UCRL-3490 (1956). — Phys. Rev. **101**, 498 (1956).

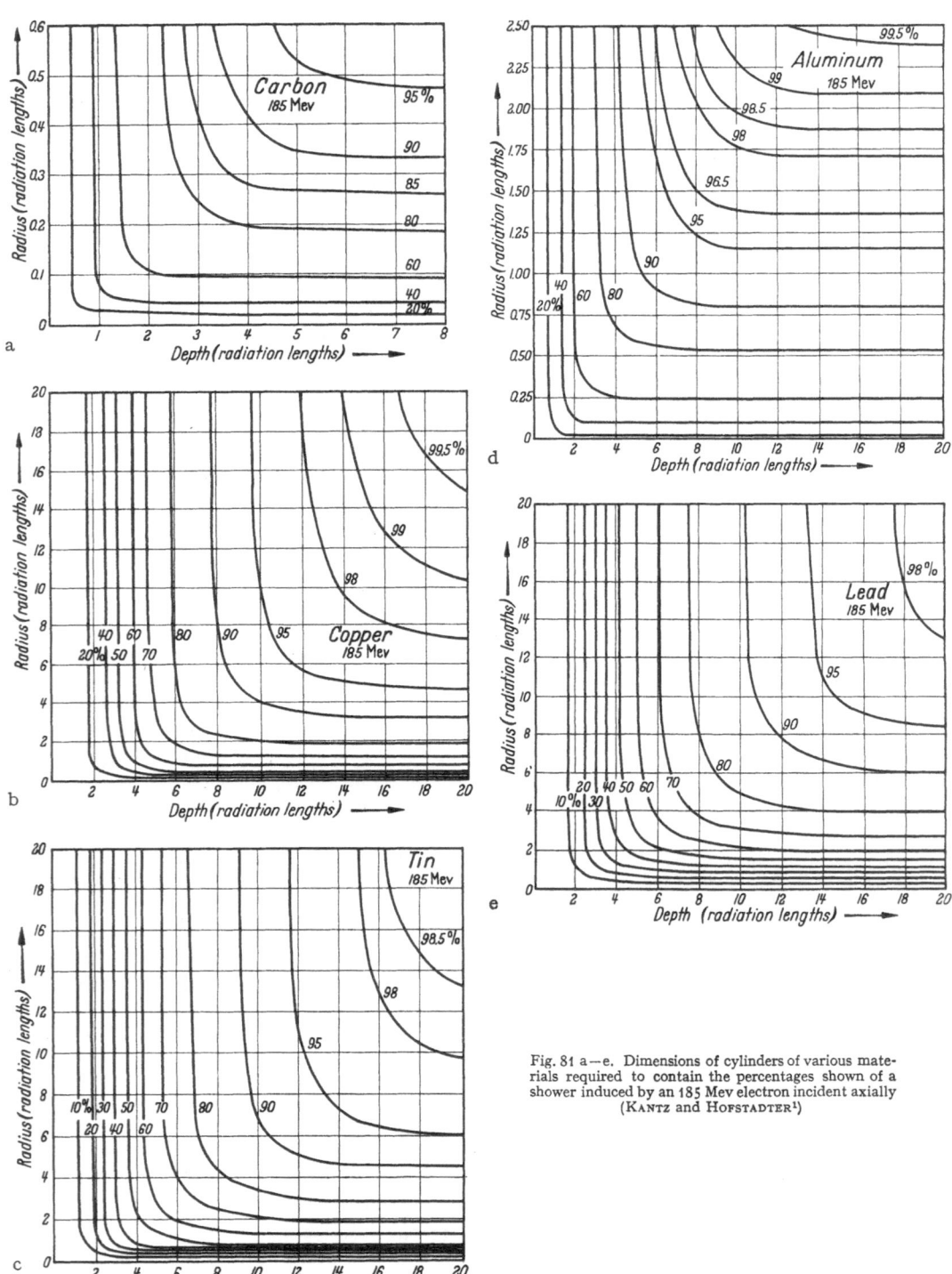

Fig. 81 a—e. Dimensions of cylinders of various materials required to contain the percentages shown of a shower induced by an 185 Mev electron incident axially (Kantz and Hofstadter[1])

The width of the pulse height distribution for a given incident energy cannot be predicted directly from the data of KANTZ and HOFSTADTER[1,2]. However, a rough estimate of the fluctuations due to energy escape from the medium can be made if one assumes that the energy which escapes is in the form of γ-rays of maximum mean-free-path for the material. This is the most important contribution to the total width for scintillation spectrometers. For Čerenkov detectors a comparably large additional contribution is due to photomultiplier statistics since the number of photons produced is small. A smaller part arises from the nonlinear relation between energy lost and Čerenkov light produced; e.g., there are fluctuations in the number of particles falling below the energy at which electrons produce Čerenkov light and from the slow variation in electron ionization energy loss with energy.

Table 15. *Radiation-length and critical energy for various materials.*

Material	Density (g/cm³)	Radiation length (cm)	Critical energy (Mev)	Index of refraction
NaI (Tl)	3.7	2.57	14	1.8
TlCl	7.0	0.83	8.3	2.4
CCl₄	1.6	12.6	38	1.46
Lead glass:				
Chance (a)	3.9	2.56	16.2	1.689 ($\lambda = 4047$ Å)
Corning (b)	3.9	2.77	17.5	1.649 ($\lambda = 5890$ Å)
Schott (c)	4.44	2.01	14.5	—
Lucite	1.18	35	95	1.5
Polystyrene	1.06	50	140	1.6
Toluene	0.86	63	140	1.5
Carbon	1.5	34.7	120	—
Aluminum	2.7	9.8	52	—
Copper	8.94	1.49	22.4	—
Tin	7.28	1.22	12.5	—
Lead	11.3	0.522	8	—

(a) Type EDF 65 3335, Chance Brothers Ltd., Glass Works, Smethwick 40, Birmingham, England.
(b) Code No. 8392, Corning Glass Company, Corning, N.Y.
(c) SF-1, Schott Glass Works, Mainz, Germany.

The amount of light produced if a scintillant is used can be estimated from the energy dissipated in the medium and the fluorescence efficiency of the medium. For example, if 200 Mev were dissipated in NaI(Tl), at least 8% or 16 Mev would appear in the form of light. Using the values of 3 ev per photon and 6% multiplier efficiency, there would be about 5×10^6 photons and 3×10^5 photoelectrons. It is clear that a large pulse would be obtained even if photons were inefficiently collected at the photocathode.

The pulse height expected when a medium is used in which only Čerenkov ligth appears can be estimated by finding a value for the electron track length for the shower, which is the total distance traveled in the medium by all the shower electrons. A rough value for this can be found by assuming a constant collision energy loss for the electrons. For example, if 200 Mev were lost in the Corning glass listed in Table 15, there would be approximately $200/1.47 = 136 \mathrm{g/c^2m}$ or 35 cm of electron path; this is an overestimate since 1.47 is the minimum energy loss per gm per cm². If we assume 300 useful quanta per cm path, we would expect about 10^4 photons. Neglecting loss due to absorption and reflections and assuming 100% collection by the photomultipliers, we would obtain about 600 cathode electrons. In practice, losses by absorption and poor matching of refractive indices at the photomultiplier junction can reduce this number by a factor of 10 to 20 so that the number of electrons would be reduced to between 30 and 60. Fig. 82 gives transmission curves for several materials which have been used in counters; for comparison the S-11 photocathode response is shown.

A more precise prediction of the behavior of these counters can be made by carrying out a Monte Carlo calculation of the production of the showers along the lines of that done by WILSON[4] for lead, and by T. YAMAGATA and M. YOSHIMINE at the University of Illinois for the Corning lead glass. BERGER and DOGGETT[5] and CAMPBELL and BOYLE[6] have carried out similar computations for NaI spectrometers for electrons of energies below 20 Mev. In this type of calculation each of a number of incident photons or electrons, and its secondaries,

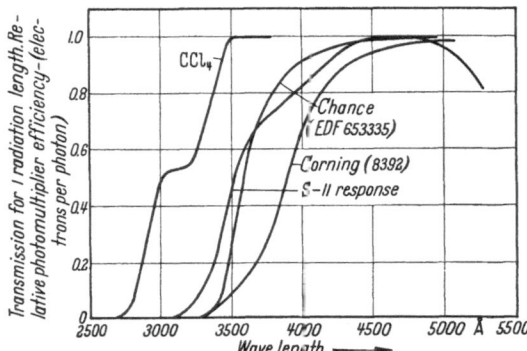

Fig. 82. Transmission versus wavelength for one radiation length of several Čerenkov radiators. For comparison the S-11 photocathode response is also shown.

are followed through the medium until their energy is dissipated or until they escape; the occurrence of the various interactions is determined by random sampling. Such a calculation becomes quite involved if multiple scattering is included in detail so that up to the present none has been exact. The work of YAMAGATA and YOSHIMINE was carried out for photon energies of 50, 100, 150, 200, and 250 Mev and for lengths of lead glass of 2, 7, 14, and 40 radiation lengths. They computed the number of visible quanta versus shower frequency for all combinations of the above energies and lengths. For the two radiation length case, the efficiencies were less than 100%, as would be expected from the pair cross section, and varied from 62% for 50 Mev γ-rays to 72% for 250 Mev γ-rays. For the other sizes the efficiencies were 100%. For γ-ray energies up to 250 Mev, most of the showers were complete in 20 radiation lengths of glass. However, even for 40 radiation lengths, the curves of shower frequency vs. number of visible quanta show a finite width of from 7 to 10% (full width at half maximum) for 250 to 50 Mev γ-rays. The calculations for these curves neglected multiple scattering (hence included no lateral losses); further, they gave quanta produced rather than the experimental quantity which would be observed, i.e., photomultiplier electrons. As a result, considerable additional width will occur due to photomultiplier statistics and to fluctuations arising from the processes of collection of the photons.

A few general remarks contrasting the properties of the scintillation and Čerenkov type counters are in order at this point. The scintillation counter has the advantage of giving a large number of photons so that the pulse height spread is due mainly to energy escape from the medium. However, it suffers from several drawbacks. One is the difficulty of obtaining crystals large enough for the several hundred Mev energy region. NaI(Tl), which has a short radiation length and

---

[4] R. R. WILSON: Phys. Rev. **86**, 261 (1952).

[5] M. J. BERGER and J. DOGGETT: J. Res. Nat. Bur. Stand. **56**, 355 (1956).

[6] J. G. CAMPBELL and A. J. F. BOYLE: Austral. J. Phys. **6**, 171 (1953).

would only have to have dimensions of a foot or so, is too expensive. Liquid and plastic scintillants have a long radiation length and so become very large and cumbersome for energies above about 50 Mev. As will be discussed below, counters have been constructed or alternate lead or lead glass discs and discs of plastic or liquid scintillant. This sampling technique introduces additional pulse height fluctuations so that it may be no better than the Čerenkov type. A further disadvantage of the scintillation type has to do with background. In the vicinity of a high energy accelerator there is generally a large flux of γ-rays and neutrons against which it is almost impossible to shield completely. Large scintillators, particularly NaI(Tl) which has a long decay time, would be swamped with background radiation. Swartz[7] has mentioned the possibility of using liquid xenon as a scintillant (liquid at temperatures below $-20°$ C and pressures above 2.4 atmospheres). The radiation length is about 2.7 cm and it scintillates with a fast decay time. Since it would have the advantage of producing a large amount of light compared to Čerenkov detectors, a few small fast photomultipliers could be used which would help overcome the background problem. However, at present xenon is expensive.

The Čerenkov type of detectors have the advantage of a fast pulse and insensitivity to low energy particles so that the background problem is considerably reduced. Their main disadvantage lies in the small number of photons produced. For a low density solid medium the radiation length is large and the shower extends over a large region yielding a fairly large number of photons but giving rise to long light paths and considerable absorption; such a detector would be somewhat cumbersome. However, for a high atomic number, dense medium of size which can be handled easily and in which not too much light is lost by absorption and at reflections, the electron paths are short and less light is produced. The Čerenkov type of detector, particularly at present, seems to be finding greatest favor.

For γ-ray detection there are two methods of using these spectrometers. First, the γ-rays may be allowed to impinge directly on the medium and therein produce the first electron-positron pair. This method requires several additional radiation lengths of medium in order to obtain resolution comparable to that obtained when electrons are being detected. A second arrangement is to use an auxiliary lead converter in front of the spectrometer. In this case one scintillation counter connected in anticoincidence is placed before the lead, to exclude detection of charged particles, and a second in coincidence is placed between the lead and the spectrometer so as to ensure detection only when γ-ray conversion occurs in the lead. If the medium is not infinite as far as the shower is concerned, better resolution will probably be obtained by this means than by the direct use first described if the lead thickness is less than $1/_2$ radiation length. With such a lead thickness, the counter efficiency is reduced; for $1/_4$ radiation length of lead the efficiency varies from 16 to 21% for γ-rays in the energy range 50 to 250 Mev; for $1/_2$ radiation length the efficiencies are 25 to 32% for the same energies. For greater lead thicknesses, energy loss in the lead gives rise to large pulse height fluctuations. The use of a converter as described provides additonal directional sensitivity and discrimination against background which, in some cases, may overcome the reduced efficiency.

b) Examples of high energy spectrometers.

25. Liquid scintillation counter. Pugh, Frisch and Gomez[8] have used two of the counter types mentioned above. The first was a tank of liquid scintillant,

[7] C. E. Swartz: I.R.E. Trans. Nuclear Sci., N.S. **3**, No. 4, 65 (1956).
[8] G. E. Pugh, D. H. Frisch and R. Gomez: Rev. Sci. Instrum. **25**, 1124 (1954).

16 in. in diameter and 30 in. long. This was viewed by twelve RCA 5819 photomultipliers distributed more or less uniformly over the cylindrical side. The tank was made of clear plastic and wrapped loosely with aluminum foil so as to take advantage of internal reflection whenever possible. It was designed to be used with a lead converter of about $1/2$ radiation length thickness in the manner described above. The pulse height distribution was measured with and without converter by using monochromatic 100 Mev electrons and the width of about 30% was found to be only slightly worse with the converter. For incident γ-rays the resolution would presumably have been worse with no converter. 100 Mev is almost the upper limit of energy for this size of liquid counter; at 150 Mev only about 50% of the energy would be contained.

26. Multiplate counter of plastic scintillant and lead. The second type described by PUGH, FRISCH and GOMEZ was constructed of alternate sheets of plastic

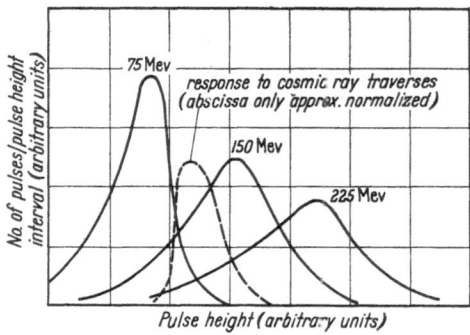

scintillant (0.100 in. thick) and lead (0.047 in. thick). The counter was 8 in. long and 8 in. in diameter. The total lead thickness was 11 radiation lengths. The light collection depended on internal reflection at the faces of the plastic (in poor contact with the lead); over $2/3$ of the light produced should have arrived at the cylindrical edges and passed to photomultipliers through plastic light pipes. They expected 20% of the energy loss to occur in the scintillant so that for a 100 Mev shower over 10^5 photons should have been produced. The pulse height

Fig. 83. Response of a multiplate counter of plastic scintillant and lead to electrons and μ-mesons (PUGH, FRISCH and GOMEZ[8]).

spread arose from the sampling of the shower electrons and from γ-ray loss from the medium. They estimated the fluctuations from the first cause to be about 10%. From a calculation and from the work of KANTZ and HOFSTADTER[1,2] they concluded that 90% of the energy of a 100 Mev electron would be contained. The 10% escaping should have given rise to fluctuations of less than 10%. The pulse height response is shown in Fig. 83 for cosmic ray μ-mesons traversing the counter and for 75, 150 and 225 Mev electrons. The μ-meson pulse distribution was only slightly worse than was expected from the Landau theory, whereas that for electrons was considerably worse than anticipated. The discrepancy was due presumably to an underestimate of either the spread due to the sampling or of that due to energy losses from the spectrometer.

27. NaI(Tl) scintillation counter. KOCH and WYCKHOFF[9] have investigated the response of large NaI(Tl) crystals to electrons of energies up to about 20 Mev. The crystals used were 5 in. in diameter and 4, 5 and 9 in. long. Monoenergetic electron beams with energies ranging from 1.2 to 17.9 Mev and of cross-section about 1 cm × 2 cm were directed at the flat face of the crystals. The energy resolution curves for the three crystal sizes were essentially the same; the widths of the pulse height distribution decreased as the energy increased to about 6 Mev, due to improvement in photomultiplier statistics, and then increased with energy due to energy escape from the sides of the crystal. At 6 Mev the resolution was about 4% and at 18 Mev about 10%.

---

[9] H. W. KOCH and J. M. WYCKOFF: J. Res. Nat. Bur. Stand. **56**, 319 (1956).

28. Thallium chloride Čerenkov counter. Preliminary results obtained using a thallium chloride Čerenkov counter are given by FREGEAU and HOFSTADTER[10]. The crystal was $3^1/_2$ in. diameter and 5 in. long and tapered down at the end opposite to the photomultiplier. The counter was tested with electrons of 75, 125 and 180 Mev. The relative positions of the 75 and 125 Mev peaks agreed with the predictions from the work of KANTZ and HOFSTADTER[1,2]; the 180 Mev peak was at too small a pulse height, presumably as a result of photomultiplier saturation. The full widths at half maximum of the pulse height curves were about 40% of the pulse height with large contributions coming both from statistical fluctuations in the number of photoelectrons produced and from energy escape. Using a $5^1/_2$ in. diameter by $5^1/_2$ in. long crystal of TlCl and one Schaetti $10^1/_2$ in. diameter photomultiplier, HOFSTADTER[11] obtained a width of about 26% (due largely to photomultiplier statistics) for electrons of 200, 250 and 300 Mev.

The position of the peak was proportional to energy. As far as size of counter is concerned, TlCl is very attractive compared to lead glass since the latter must be about three times larger than the former in order to have the same containment factor. A smaller size means less photocathode area and hence less cumbersome magnetic shielding. The electron track length in the glass will be several times that in TlCl and one might expect a larger and narrower response. However, the light path in the glass

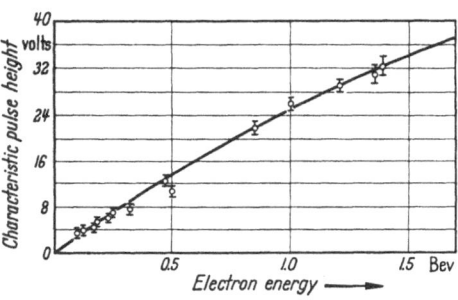

Fig. 84. Pulse height versus electron energy for a lead glass Čerenkov counter (BROBANT, MOYER and WALLACE[3]).

will be longer and so more absorption loss will occur. Further, long absorption paths introduce additional width due to fluctuations in the distribution of photons with wavelength. As a result, it turns out, as will be seen below, that the lead glass counters give results only comparable with the TlCl and suffer from bulkiness.

29. Lead glass counters. The design and characteristics of a lead glass counter used at the University of California Radiation Laboratory have been described by BROBANT, MOYER and WALLACE[3] and by JESTER[12]. It consisted of a cylinder of lead glass 12 in. diameter × 14 in. long (11.02 radiation lengths diameter, 12.85 radiation lengths long) viewed at one end by four 5 in. diameter DuMont 6364 photomultipliers. The counter was calibrated using monoenergetic electrons; the curve of pulse height as a function of electron energy is shown in Fig. 84. The nonlinearity was due to increasing energy loss from the glass as the energy increased. Converting the data of KANTZ and HOFSTADTER[1,2] to the medium and energies used here gave a containment factor which decreased too rapidly with energy as noted earlier. The full width at half maximum of the number-pulse height curve is about 45% of the peak position for electrons above 200 Mev and increases rapidly below this energy to about 100% at 100 Mev.

BERNARDINI[13] has described a counter of the same kind and size of glass but with twelve RCA 5819 photomultipliers to detect the light. Somewhat faster pulses were thereby obtained than for the 5 in. tubes. The pulse height

[10] J. H. FREGEAU and R. HOFSTADTER: Nucleonics **12**, 36 (1954).
[11] R. HOFSTADTER: CERN Symposium Proc. **1956**, 75.
[12] M. H. L. JESTER: University of California Radiation Laboratory (U.S.) Report No. UCRL-2990. 1955.
[13] I. FILOSOFO and T. YAMAGATA: CERN Symposium Proc. **1956**, 85.

distributions for monoenergetic electrons (beam size 2.5×2.5 cm²) and for cosmic ray μ-mesons are shown in Fig. 85.

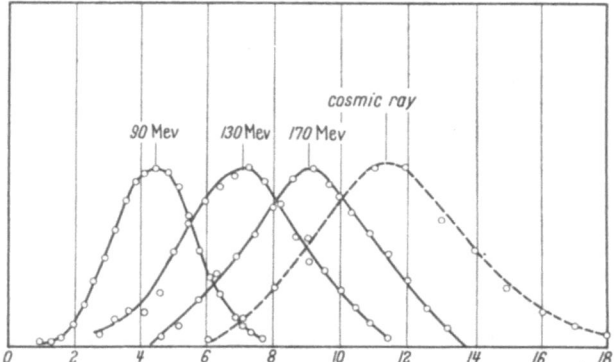

Fig. 85. Response of a lead glass Čerenkov counter to electrons and cosmic-ray μ-mesons. Pulse height is shown on the abscissa and number of pulses/pulse height interval on the ordinate (Bernardini[13]).

It has been suggested that there is an advantage to be gained by tapering such counters as those above and putting the photomultipliers at the large end.

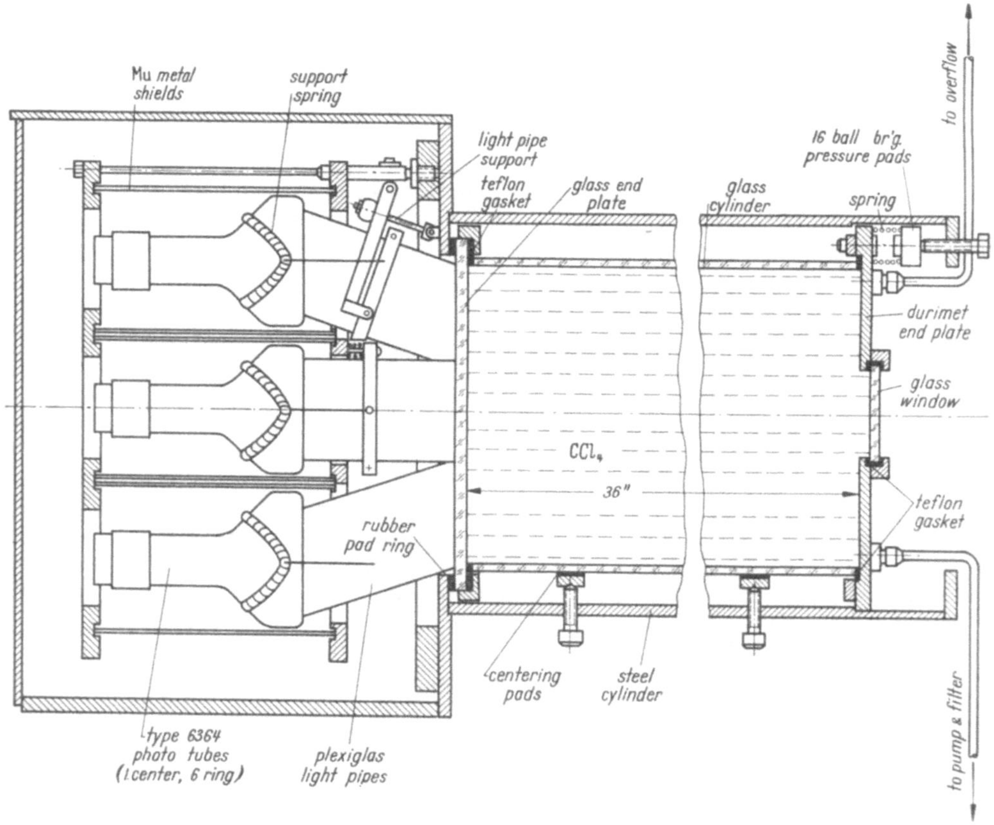

Fig. 86. CCl₄ Čerenkov counter (Jones, Kratz and Rouvina[14]).

The reason for this is that since most of the radiation comes from relativistic electrons, complete internal reflection at the walls of a cylinder is possible only for an electron traveling parallel to the axis. Since the shower electrons generally do not travel exactly axially, tapering will increase the amount of light internally reflected.

30. CCl₄ Čerenkov counter. A Čerenkov counter in which the medium is CCl_4 has been used by JONES, KRATZ and ROUVINA[14] and is shown in Fig. 86. The counter was 12 in. in diameter and 36 in. long (approximately 2.4 radiation lengths diameter and 7.2 long), a size they estimated, from the data of KANTZ and HOFSTADTER[1,2], would contain about 90% of a 300 Mev shower. The small value of the ratio of diameter to length, compared to that for heavier media, is due to the small scattering obtained with low atomic number. Seven DuMont 6364 5 in. diameter multipliers were connected through Plexiglas light pipes to one end of the cylinder. The pulse height from the counter was considerably in-

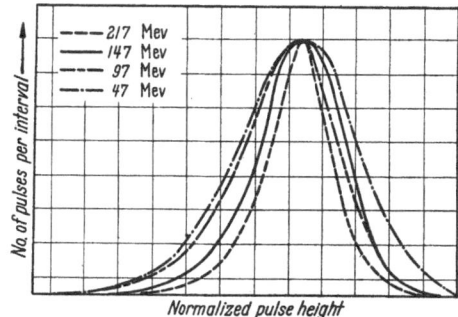

Fig. 87. Response of a CCl₄ Čerenkov counter to electrons (JONES, KRATZ and ROUVINA[14]).

creased, and hence the resolution improved, by saturating the CCl₄ with a frequency shifter 1,4-di-[2-(5-phenyl-oxazyolol)]-benzene (POPOP); for example, the resolution for 217 Mev electrons was reduced from 21 to 14.6% by this means. The pulse height distributions obtained by using a collimated electron beam 0.5 in. in diameter and of energy variable up to 217 Mev are shown in Fig. 87. These distributions, the positions of the maxima of which were proportional to energy, have here been normalized. The resolution decreases as the energy decreases. Curves of pulse height distributions from a pulsed light source gave the contributions to the width due to multiplier statistics; these appear to be half those obtained using electrons.

---

[14] W. B. JONES, H. R. KRATZ and J. ROUVINA: General Electric Research Laboratory (U.S.) Report-No. RL-1675. 1957.

The Proportional Counter as Detector and Spectrometer.

By

S. C. CURRAN.

With 33 Figures.

A. Historical.

1. The proportional counter as we now know it has been developed rapidly since 1948. It is true that prior to this time it was widely appreciated that counting tubes could be used at voltages such that the output pulses differed according to the nature of the ionizing particle passing into or through the vessel [1]. A semi-proportional region of operation of Geiger tubes had received some attention and within this region the tube acted as a counting device for either alphas or protons or other heavily ionizing particles in a background of radiation of the β or γ type. Such semi-proportional devices were not used to measure the energies of the lightly ionizing beta rays for instance and they are indeed now virtually obsolete. We shall neglect them in this account of the proportional counter and consider the term to refer to devices which are really proportional, tubes which may be used for example to analyse a spectrum of alpha radiations or a spectrum of beta particles or electrons of photoelectric or any other kind of origin.

In 1948 the first study of the proportional counter as defined above was made. It was shown by CURRAN, ANGUS and COCKCROFT[1] that the tube could exhibit the complete spectrum of the beta rays of tritium, each beta particle emitted by the source (consisting of a slight trace of tritium gas admixed with the filling gases of the tube) giving rise to a pulse of amplitude proportional to the energy of the primary beta itself. The energy scale was fixed by means of the photo-electrons produced in the gas of the counter by virtue of the photoelectric absorption of standard K X-rays, for example the K X-rays of copper of energy approximately 8 kev ($K\alpha$) and 8.8 kev ($K\beta$). In the same year HANNA, KIRKWOOD and PONTECORVO[2] examined K-capture phenomena in A^{37} by the same method. These developments were highly significant for they demonstrated that in the new type of proportional counter we had a device which was capable of detecting and analyzing in energy the whole range of the ionizing particles, from electrons of minimum specific ionization to those of maximum specific ionization, e.g. fission fragments. If the particle was brought to rest within the counter the total energy expended could be measured while if it passed through the counter its relative specific ionization could be determined. We shall see that the device has been applied in both ways.

Since 1948 many advances in the techniques and understanding of the mechanism of the proportional counter have been made. It is true to say that at present it has made almost obsolete the pulse ion chamber except in the form of the gridded ion chamber which has some special virtues in a more limited field. For much straightforward detection it rivals the Geiger counter and in addition it

[1] S. C. CURRAN, J. ANGUS and A. L. COCKCROFT: Nature, Lond. **162**, 302 (1948).
[2] G. C. HANNA, D. H. W. KIRKWOOD and B. PONTECORVO: Phys. Rev. **75**, 985 (1948).

serves as a spectrometer. With the scintillation spectrometer it shares a very large field indeed. As an indication of its scope we note that it has been operated at pressures in the very wide range of 1 cm of Hg to 50 atmospheres. The temperature range has been pushed upward to 900° C and it has been used in magnetic fields of several thousand gauss. As now conceived it is a most versatile tool.

B. The principle of operation.

I. The multiplication process.

2. The proportional counter is usually cylindrical in form and filled with a gas or mixture of gases or vapours to a pressure of about one atmosphere. The cylinder is generally at a voltage negative with respect to a thin wire stretched

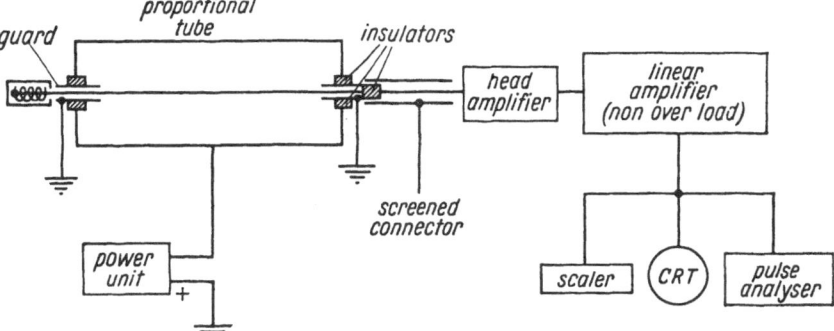

Fig. 1. General arrangement for counting and spectrometry.

along the axis of the cylinder. The wire is taken in this arrangement to ground potential via a resistor which is of the order of one megohm and which serves as the grid leak of the head amplifier. The head amplifier is fed into a linear amplifier (preferably of non-overloading type) which may or may not be designed to handle the extreme range of input signal strength that may be involved. For counting the output pulses are fed directly from amplifier to scaler but for spectrometry the pulses can be fed to some kind of pulse amplitude analyser. The arrangement is shown schematically in Fig. 1.

When an ionizing particle moves through the gas of the counter it gives rise to a number of electrons and an equal number of positive ions along its track. Provided that the gases within the tube do not exhibit appreciable electron affinity the electrons created by the primary ionizing particle can be drawn towards the anode (the positively charged wire in this case). We assume that recombination is not important at the pressures considered or at the specific ionizations involved. If the attractive voltage acting on the electrons is suitably adjusted the electrons can be drawn into the near vicinity of the wire and there enter a region in which the field strength rapidly increases in magnitude. It can readily be shown that the electric field strength X is given by:

$$X = \frac{V_0}{r \ln (b/a)} \tag{2.1}$$

where V_0 is the applied voltage and a and b are the radii of the wire and cylinder respectively. Clearly if the adjustment of V_0 is correct then for small values of r, the radial distance from the axis of the wire, X may be such that ionizing collisions occur, that is the electrons originally produced along the track now acquire between collisions with atoms or molecules sufficient kinetic energy to result in multiplication. The kinetic energy so acquired must be at least equal to the ionization energy of the gas atom or molecule involved. In each such collision an additional electron is released and hence the n original track electrons on the average are replaced by $2n$ electrons. A further stage of multiplication of these $2n$ electrons yields $2^2 n$ electrons and if m ionizing collisions are possible before they arrive at the surface of the wire a total of $2^m n$ electrons are collected instead of the n original electrons. The tube is now said to be multiplying and the gas multiplication factor F is given by:

$$F = 2^m. \tag{2.2}$$

Clearly the value of m is subject to statistical fluctuations around an average value and such variation has been observed experimentally.

We have assumed that each of the track electrons takes part in this straightforward process of multiplication but now it is necessary to consider other aspects of the mechanism. It is obvious that while each electron gives rise to a miniature electron avalanche the positive ions in the gas, since they move relatively slowly, are left as a kind of positive ion cloud in the neighbourhood of the wire. The density of this ion cloud increases rapidly as we move towards the wire surface. The positive ions formed very close to the wire will be accelerated most rapidly towards the cylinder, and the subsequent rapid acceleration of this positive ion cloud away from the wire releases the electrons which have reached the wire and so gives rise to a sharply rising signal voltage across the external resistance. This rapid increase in signal strength is followed by the slower development of the pulse to full amplitude as the positive ions drift more slowly to the cathode under the influence of the decreasing electric field that obtains over most of their path to the cylinder.

II. Quenching of the pulse.

3. Photons produced in the gas. In the case of the Geiger type of discharge it is necessary to pay close attention to the phenomena that arise at the cathode [2]. For the proportional counter conditions on the whole are considerably relaxed. Thus we need consider only two possible causes of malfunction of the tube, namely the production of photons in the gas and the incidence of positive ions at the cathode. As the electrons acquire energy sufficient to yield multiplication of electron charge they can obviously cause excitation of atoms and molecules and the excited atoms or molecules produced will emit photons in various de-excitation processes. Such photons may give rise to photo-electric emission at the cathode and any photoelectrons released in this way will give rise to fresh avalanches at the wire. Provided the photoelectric effect is such that less than n fresh photoelectrons are produced when n electrons are originally in the track the magnitude of any avalanche is on the average less than that of its predecessor and the tube still operates in a proportional manner. In actual fact the gases are chosen so that this marginal condition does not arise. Assuming that excitation is less probable than ionization the number of photons reaching the wall at a multiplication of 10^4 for example will be less than $10^4 n$. Many of the photons produced will moreover be absorbed in non-ionizing processes in the counter gas, particularly in the quenching constituent, e.g. methane. Of those arriving at the wall only a small fraction, probably less than 10^{-4} for most typical wall materials will cause photoemission so the effect rapidly becomes entirely negligible in magnitude. The situation may be much less favourable when the pure rare gases are used. Thus in tubes filled with neon or argon for instance it is not possible to operate at stable multiplication factors in excess of 10^3 and 10^2 is a normal working limit.

4. Positive ions at the cathode. The positive ions produced along the track and more important in the multiplying region at the wire drift relatively slowly to the cylinder. After a time of the order of between 10^{-4} and 10^{-3} sec they arrive at the cathode surface. If they are positive ions of say argon they can give rise to electron emission. However if the positive ion is that of a typical quenching gas or vapour and therefore one of relatively low ionization energy the energy may be spent in the extraction of an electron from the metal of the cylinder and in pre-dissociation. In this case the residual energy is generally less than that required to produce photoemission at the wall and the ions are not responsible for further emission and subsequent multiplication. The processes are exactly the same as those obtaining in the operation of the Geiger-Müller counter except that in the case if the Geiger-Müller tube the conditions are much more severe [2]. In the Geiger regime a single photoelectron can give rise to a spurious discharge but a single photoelectron may be quite negligible in effect in a proportional counter. In practice therefore the gases of a good proportional counter are chosen for the same reasons as apply in the case of the Geiger tube and therefore usually the filling consists of a mixture of a rare gas and a polyatomic gas or vapour. Argon or xenon plus methane are the most popular mixtures but methane alone is frequently used. Carbon dioxide and other gases have been used for special purposes. In work with neutrons boron tri-fluoride is important. In the case of some of these gases high purity is absolutely essential and this is particularly true for operation at high pressure, say in excess of five atmospheres.

III. Electron affinity of gases and vapours [3].

5. In the proportional counter it is essential that the track electrons should not become attached to neutral atoms or molecules during their passage to the wire. Recombination with positive ions in the track itself is in general not an important effect except when heavily ionizing particles are the primary ionizing agents and the gas pressure is relatively high. The electron affinity of certain gases or vapours may be such that during the microsecond or so that the electrons drift into the vicinity of the wire they form negative ions. Such negative ions drift relatively slowly to the wire and two spurious effects follow — the full amplitude of the pulse is not observed since fewer than the proper number of electrons arrive and further the negative ion, arriving late, does not in general give rise to the same multiplication as a free electron. The ion may revert to a free electron and a neutral atom or molecule in the intense field near the wire surface but the probability of occurrence of this process is variable and consequently a cause of large fluctuation in the pulse development. Certain gases, for example the halogens and oxygen, are particularly harmful in this regard and care must be taken to eliminate them as completely as possible.

The phenomenon of electron attachment must be carefully considered in certain applications of the proportional tube. In the study of radioactive materials in gaseous form, an extremely powerful technique to which the proportional counter is well adapted, care must be taken to choose a gas or vapour of the radioelement which does not introduce pulse amplitude variations due to the above mechanism. As an example we may mention that fluorine (say in the study of F^{18}) may be conveniently introduced as boron trifluoride. As free fluorine it would show very markedly the characteristic attachment phenomena. Many other radioelements can likewise be examined when careful choice of the gas is made. The first radioactive gas successfully studied in detail was tritium and this was simply added at very low concentration to the argon plus methane mixture. It has been shown that even water vapour is not particularly harmful. Indeed good proportionality of pulse size has been maintained with water vapour up to partial pressures of 1 cm of mercury[1]. It is relatively seldom that some gas or vapour, especially if those of the organic or organo-metallic form are considered, does not prove suitable for use. We may note here that the use of counters at high temperatures[2], up to more than 800° C, makes the problem of finding a suitable gaseous form less exacting.

IV. Effect of geometrical form.

6. We have so far considered the counter in its ideal form, the field being regarded as that existing between two ideal cylinders of infinite length. In actual practice the shape is always a compromise between conflicting requirements.

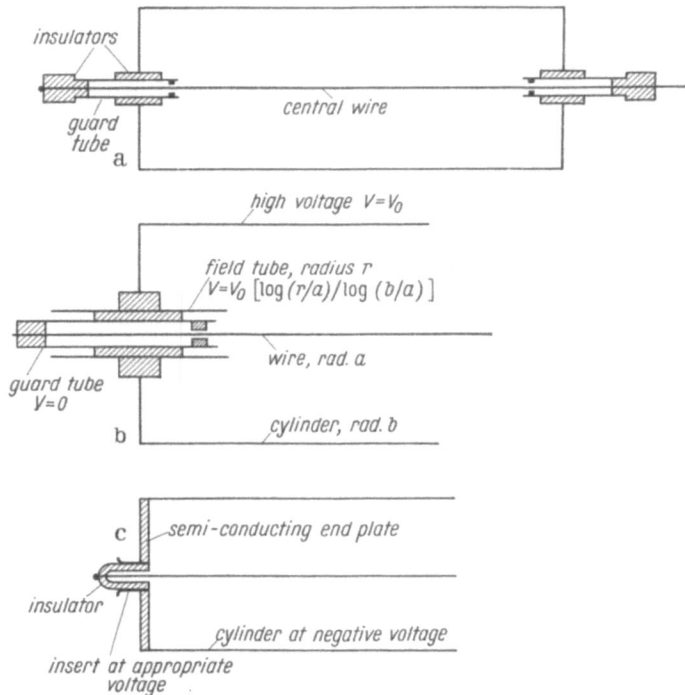

Fig. 2a—c. Arrangements of counter electrodes. (a) Simple counter in schematic form. (b) End of counter with field tube. (c) Schematic form with semiconducting end.

Undoubtedly the main difficulty in producing a counter of practicable dimensions is found in the so-called end effect. The proportional counter in its straightforward form should incorporate some kind of guard arrangement to prevent the arrival of disturbing currents at the wire. These currents usually

[1] R. W. P. Drever and A. Moljk: Rev. Sci. Instrum. 27, 650 (1956).
[2] A. Moljk, R. W. P. Drever and S. C. Curran: Rev. Sci. Instrum. 26, 1034 (1955).

consist of leakage currents across or through the insulating materials used to support the wire and separate it electrically from the applied voltage.

In Fig. 2a we show the arrangement schematically and in this case the guards take the form of small tubes at ground potential. A very considerable variety of design is feasible and there is little to choose between them. Even when the guard electrodes are of small diameter they contribute to the distortion of the electric field and it is extremely desirable that this distortion should be reduced to the minimum.

COCKROFT and CURRAN[1] have given a very practicable way of doing this and it is shown in Fig. 2b. Here the guard tubes are used to support field tubes which are electrically insulated. The field tubes are set at a voltage which is a function of their radius and which is such that the potential at their surface is that at points at the same distance from the counter wire as the surface itself. This can be shown to ensure that the lines of force are radial near the end of the field tube provided this is about equal in length to the radius of the tube. In these circumstances the lines of force within the counting volume can be shown to be radial along the whole length of the operative part of the counter.

With field tubes incorporated in the design a considerable latitude in dimensions is possible without appreciable deterioration of the resolution of the instrument. Assuming that the field tubes together give rise to a "dead" volume of counter

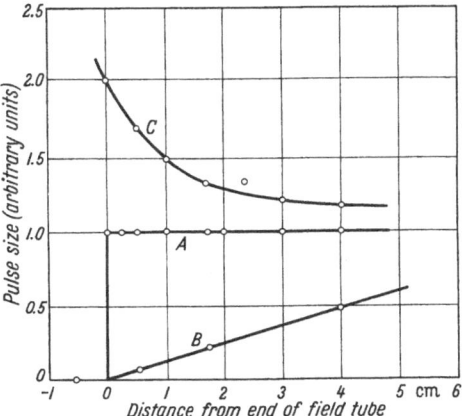

Fig. 3. The variation along the wire of the pulse size for monoenergetic radiation. Curve "A" is for field tube at nearly optimum voltage. Curve "B" is for zero field tube voltage (acting as typical guard tube). Curve "C" is for 20% above optimum voltage.

of total length equal to its diameter we can have counters giving good performance when their cylindrical length exceeds their diameter by a small amount (effectively a pill-box shape) and there is of course no restriction on the maximum length. On the other hand if the field tubes are not used the length must considerably exceed the diameter in order to reduce the effect of the field distortions near the ends of the wire. The principle of field tubes can be extended to field cylinders as shown by COCKROFT and CURRAN. Although the construction is rather more complicated in this case it is possible by such devices as proposed by these authors to cover a very wide range of counter dimensions without introducing undesirable effects. Some results for an end-corrected counter are shown in Fig. 3. A very thin pencil of X-rays was used to measure the gas gain along the length of the counter.

In this same connection it should be noted that a simple method exists in principle. If the central wire is attached at each end to the centre of a circular disc of a semi-conductor the field distribution is correct. ROSSI and STAUB [1] have discussed this method but no extensive use of it in practice has followed. There are interesting developments in the field of semi-conductors and innovations in counter design are likely in the near future. For high sensitivity work with low energy radiation it is very necessary to have a guard arrangement for

---

[1] A. L. COCKROFT and S. C. CURRAN: Rev. Sci. Instrum. **22**, 37 (1951).

the central wire and this condition may always limit the construction with semi-conductors to some extent. Assuming that a suitable semi-conductor can be fashioned the counter may take the form illustrated in Fig. 2c. Here practicaly the whole volume of the vessel is useful as a counter.

C. The construction of proportional tubes.

7. Practical forms. Many practical forms of proportional counter have been described. Most designers adopt the technique of applying a negative voltage

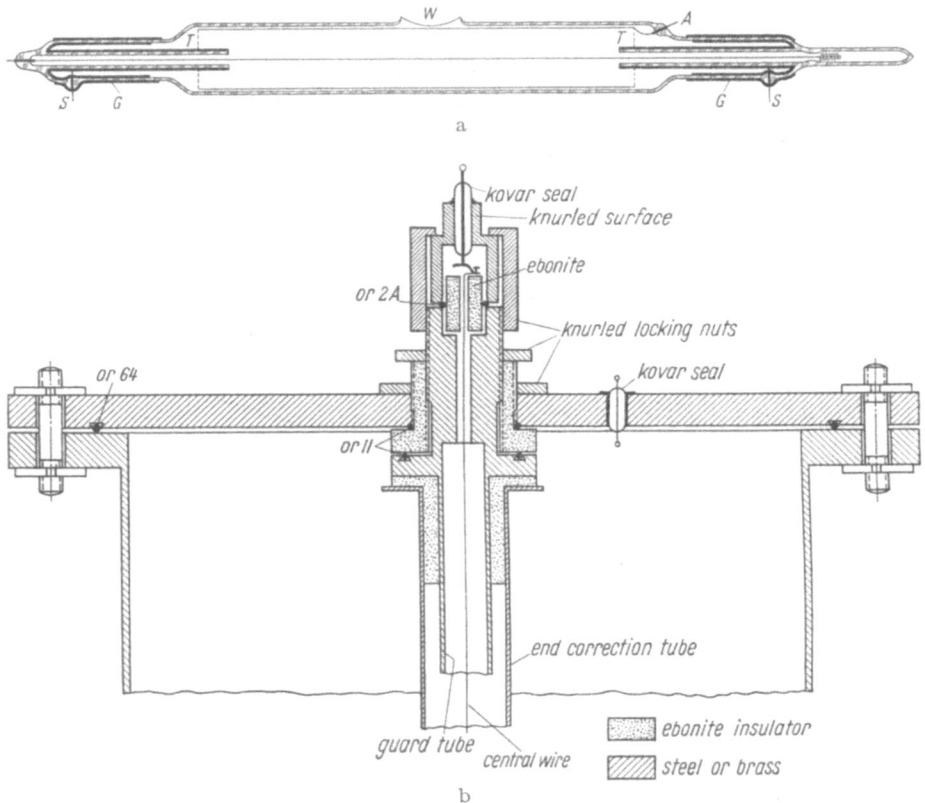

Fig. 4a and b. (a) General purpose glass counter with thin window "*W*". Thin aluminium cathode connected at "*A*". The tubes "*T*" ae coated with graphite and through "*S*" are connected to coatings "*G*" on inside and outside. (b) End of counter employing "O" ring technique.

to the cylinder. This means that the wire as the anode can be connected to earth through a resistor and no complications due to the use of a condenser arise. Except when a condenser is carefully constructed it is likely to act as a source of spurious pulses, and except in rather straightforward applications the circuit of Fig. 1 is much to be preferred.

The amplifier must be a sensitive instrument in many applications; for example when moderate gas gain is employed and the energy expended by the primary ionizing particle is small. Care must therefore be taken in shielding the counter wire from stray electromagnetic fields. The lead from the central wire to the amplifier should be shielded and as short as possible. Indeed the head amplifier should be close to the wire and it may be made a more or less integral part of the counter.

An all-glass type of counter is shown in Fig. 4a[1]. This may serve ase as uful tube in many investigations.

It has the merit that it can be thoroughly outgassed and problems due to gaseous impurities eliminated. For much work an all-metal construction is preferred and here a typical design is illustrated in Fig. 4b. Here the use of rubber rings (○ rings) is involved and the design is such that rapid and easy assembly is secured. Many types of commercial glass-metal seals can be incorporated in

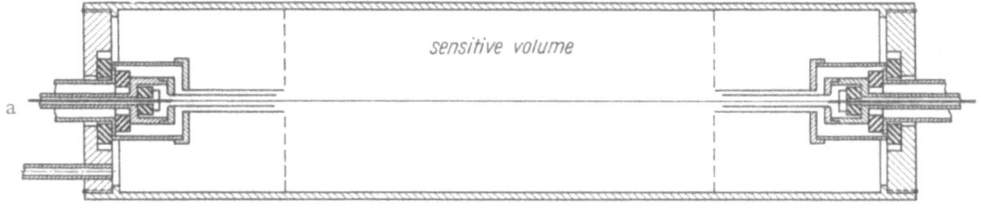

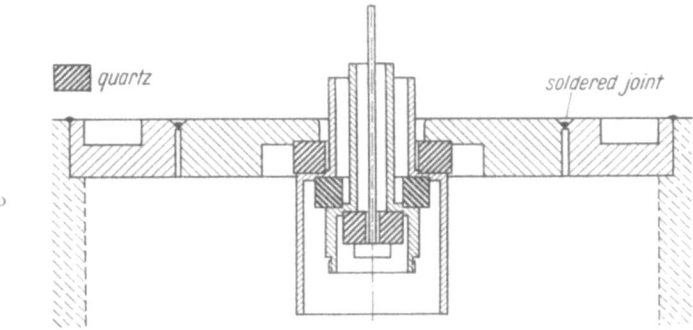

Fig. 5a and b. High pressure counter. (a) A counter for neutron spectroscopy, capable of operating at 40 atmos. (b) The endplate insulator assembly.

counters and the ○ rings thereby avoided. Where great care as to vacuum conditions must be exercised metal gaskets of indium for instance may replace the rubber ones.

A good example of up-to-date technique is shown in Fig. 5. This counter has been employed at pressures up to 40 atmospheres and it illustrates very clearly many of the important aspects of tube design. It was developed and used by BATCHELOR, AVES and SKYRME[2] as a neutron spectrometer with He[3] as the sensitive gas. We shall return to this subject later but meanwhile we consider it as a type of counter assembly.

The body is $1/16$ in. thick, machined from stainless steel and threaded at each end. The end plates are screwed and soft soldered into the ends. An end plate assembly is shown in more detail in Fig. 5b, and it is arranged with high internal pressure in mind. Quartz rings act as insulators and the metal is 36% nickel-iron so that Araldite can be used to make the vacuum-tight joints. The quartz rings are located concentrically by recessing as required. The guard tubes and

[1] S. C. CURRAN, J. ANGUS and A. L. COCKROFT: Phil. Mag. **40**, 36 (1949).
[2] R. BATCHELOR, R. AVES and T. H. R. SKYRME: Rev. Sci. Instrum. **26**, 1037 (1955).

field tubes shown in Fig. 5a are of Dural and they are screwed on to the nickel-iron tubes. The wire is stainless steel of diameter 4 mil. In the particular tube described here the dimensions were: diameter $2^1/_8$ in., total length $9^3/_8$ in., field tube diameter 0.1 in., field tube length $1^1/_{16}$ in., wire diameter 0.004 in., sensitive volume 279.5 cm^3. With these dimensions and a voltage $+V$ applied to the wire and guard tubes, a voltage of $+0.375\ V$ must be applied to the field tubes to eliminate end effects.

The gas system for the above counter was specialised but the authors describe several features of general interest. As regards performance they note that with a filling of 16 cm of He^4 (to stimulate He^3), 160 cm of krypton and 3.2 cm of carbon dioxide the experimentally observed spread was 2.7% as against a pre-

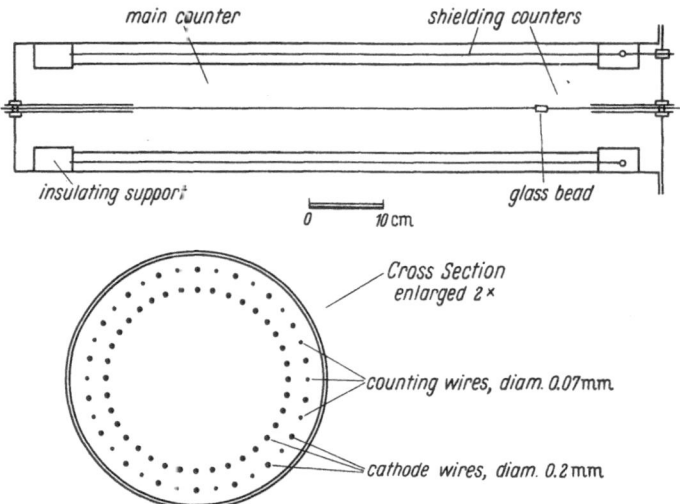

Fig. 6. A proportional counter incorporating its own anti-coincidence system of shielding proportional counters.

dicted intrinsic spread of 2.1%. This result refers to γ-radiation from Am^{241} of energy 60 kev. The pulse height was constant along the length of the counter to within 2%.

A contrasting technique is illustrated in the counter shown in Fig. 6. Here we see the use of $\bigcirc$ rings exploited to the full and for special elaborate equipments they offer many advantages. The counters can be readily assembled, an important feature giving flexibility in research investigations. The counter shown is a very large one used by Moljk, Drever and Curran[1] in achieving extremely low background. The main proportional tube is surrounded by a system of proportional counters operating in the same vessel. We shall say more about this special arrangement below.

Two techniques which are of considerable practical importance should be mentioned. The flow type of proportional counter, see Fig. 19 below, has the advantage that sources of radiation can be rapidly inserted into the vessel and the absorption due to windows avoided. This can be very advantageous in the case of alpha ray sources and the softer beta ray emitters such as C^{14} and S^{35}. In some designs the vessel is opened and the source placed within, generally in the form of an extended foil. It is usual to have the foil just outside an array

[1] A. Moljk, R. W. P. Drever and S. C. Curran: Proc. Roy. Soc. Lond. Ser. A 239, 433 (1957).

of wires which act as the cathode. In some special cases a 4π system is adopted. Here the support for the source as well as the source material itself must be extremely thin to reduce the self-absorption and support absorption to negligible or at least very small value. The technique is particularly valuable in making absolute measurements as the counter is effectively two in parallel and the number of impulses recorded is in most cases equal to the number of disintegrations per second even in those cases where the decay is by no means simple. Thus coincident β-particles and photoelectrons (say from internal conversion of a γ-radiation) appear as single events. The same technique can be applied to

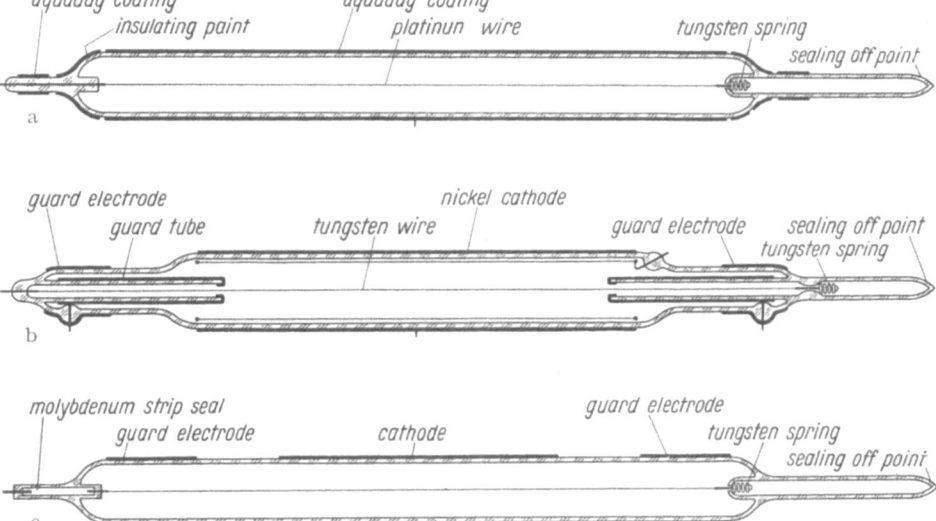

Fig. 7a—c. Maze and high temperature forms of counters. (a) A simple Maze-type tube of soda glass for use at room temperatures. (b) A semi-Maze type of hard glass for use at temperatures up to 450° C. An external conducting coating on the glass is connected to the nickel cathode and the internal coating for guard tube is connected to the external coating for the guard electrode. (c) An external cathode counter of fuzed quartz for operation in the range 200 to 800° C.

Geiger counting. In the flow counter we are not in general concerned with proportionality of pulse amplitude to energy but are satisfied with an effective 100% efficiency of the instrument as a detector. The chief criterion in this case is therefore the existence of a "plateau" and the length and flatness of the "plateau". With methane as the filling gas such "plateaus" on the curve of counting rate as a function of applied voltage may be many hundreds of volts in length and the flatness is generally of the order of 3% per 100 volts or better.

The quality of the plateau with carefully designed proportional counters can be really excellent as regards flatness and various gases and mixtures of gases perform well. Thus any one of the rare gases He, Ne, A, Kr, Xe mixed with some 5 to 10 cm by partial pressure of methane acts as a good mixture. Nitrogen and carbon di-oxide are relatively good. In most instances methane itself is used for the flow type of detector. It is advisable to ensure that the gases are free of electronegative impurities such as oxygen and the halogens. Water vapour may or may not be distinctly harmful. The usual type of purifier is a tube containing calcium turnings heated to 300° C [7]. The purifier should preferably be attached in such a way that circulation of the gas by convection currents is as rapid as possible. An example of the plateau[1] of a good proportional counter

---

[1] J. M. VALENTINE: Proc. Roy. Soc. Lond., Ser. A **211**, 75 (1952).

is given in Fig. 8 below. As is observed the plateau may have distinct steps corresponding to the threshold of detection of various monoenergetic radiations.

The proportional counter can be built in the Maze[1] form. Originally the Maze type of counter was of Geiger-Müller form but later work in the proportional region[2] showed that there was much less difficulty in obtaining satisfactory performance with a greater variety of insulating materials. In the Geiger tube the glass wall is generally of the soda glass variety as with glasses of higher specific resistivity the ions falling on the cathode cause local charging to a serious extent. This is not so true with the Maze construction of a proportional counter. Generally the intensity of the individual ion pulses is less by a factor of 10^4 or more in the proportional operation and even at relatively high rates of counting

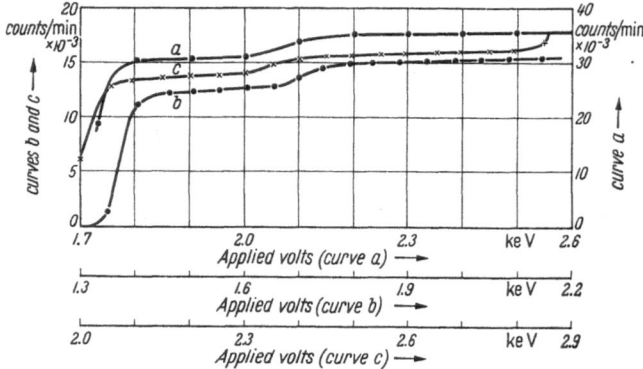

Fig. 8. Variation of count rate with applied voltage. *a* Plateau obtained with A^{37} radiations in 20 cm $CH_4 +$ 1 atm A. *b* Plateau obtained with A^{37} radiations in 1 atm He. *c* Plateau obtained after 19 cm $CH_4 +$ 1 atm A had been added to mixture used for curve *b*. These curves refer to the X-rays and Auger electrons of A^{37} of energy about 2.6 and 0.25 kev. The effect of differing absorption is evident.

the charge can leak away rapidly through the wall. A typical example of such a counter is shown in Fig. 7. Pyrex glass can be used successfully and even quartz is possible at low counting rates and moderate gas multiplication[3].

8. Criteria of performance. We have referred above to the use of the plateau as a criterion of the quality of the counter. In many cases this is not a completely adequate test of the proper functioning of the tube as a detector. A more thorough test depends on the examination of the pulses produced by soft monoenergetic radiation. The X-rays produced by radioactive sources decaying wholly or even partly by K-capture can be used to examine the behaviour of the tube [5]. Thus sources such as Ge^{71}, Fe^{55} or Cr^{51} emit soft X-rays and such a material in the form of a weak source may be placed at a window or in some cases within the tube. The pulse amplitude due to the X-rays can be observed on a cathode ray tube or by means of a single or multi-channel kicksorter. Even a discriminator of variable bias feeding a scaler can be used. If the amplitude of pulses due to radiation of energy E is V volts then if the bias on the scaler is V_0 all radiations of energy greater than EV_0/V are detected. For example if the X-rays from a Ge^{71} source are taken as 10 kev approximately and they give rise to signals of average amplitude 50 volts at the input to the scaler, all radiation of energy greater than a lower limit of 1 kev will be detected when the bias on the scaler is set at 5 volts. The same radioactive sources may of course be used as calibrating agents for the spectrometric analysis of unknown radiation and fluorescence

[1] R. Maze: J. Phys. Radium **7**, 164 (1946).

[2] S. C. Curran, A. L. Cockroft and G. M. Insch: Phil. Mag. **41**, 517 (1950).

[3] A. Moljk, R. W. P. Drever and S. C. Curran: Rev. Sci. Instrum. **26**, 1034 (1955).

X-rays of various elements, for example copper, can be used in similar fashion. Thus a low intensity beam from an X-ray equipment is allowed to strike a thin sheet of say copper, placed close to the window of the tube, and the fluorescence X-rays pass into the counter and give rise to photoelectrons of definite energy when absorption in the gas takes place. In certain limiting cases it is not easy to use this technique. If the gas in the counter is very transparent to the X-rays, as is the case say with methane, the photoelectric absorption is very slight and since at the same time the elements are of low atomic number the Compton effect may make the photoelectric effect relatively less prominent. In these circumstances very soft X-rays are required and the transmission of these through a window into the gas presents a problem. It is possible in such circumstances to add to the gas filling small trace quantities of A^{37} for instance. This element decays by K and L capture and photoelectrons of energy 2.6 kev and about 250 ev are released in the gas (most of these are Auger electrons). If the pulses due to the latter radiation are observed it is easy to guarantee that all radiation of energy above this lower limit is detected. An alternative procedure[1] is to place within the counter an electron source but these are not so readily available. One possible material is Te^{125} (generally derived from Sb^{125} which is a daughter of Sn^{125}, obtained by neutron capture in tin) giving homogeneous electrons. Such procedures may be important when it is vital to be sure that a counter filled with a gas like methane is giving really accurate quantitative data.

Regarding gas filling we have already drawn attention to the fact that care must be exercised. This is especially true at high pressures and for counters filled with carbon dioxide at several atmospheres (as in carbon dating for example) it has been shown that as little as one part per million of oxygen may be harmful not only as regards resolution but as regards plateau[2,3]. The most commonly adopted techniques for purification are usually satisfactory and they need not be discussed here but in certain special cases more refined methods prove profitable. Thus in the case of the He^3 counter discussed above a large glass vessel with a thin layer of sodium on the inner surface was used as a store for the gas before it was used. For continuous maintenance of the gas of the counter in the purified condition over long periods it may often prove essential to have a calcium purifier attached as part of the system so that convection circulation of the gas over the hot calcium turnings is readily achieved by switching on a heating circuit.

Another matter which becomes increasingly important as higher pressure is used is the uniformity of the wire. The actual positioning of the wire within the cylinder is not very critical, as indeed was demonstrated by CURRAN and REID[4] and BECKER et al.[5] with rectangular counters. ROSSI and STAUB [I] calculated that the extreme fractional difference of the field $\Delta E/E$ is given by the formula

$$\Delta E/E = 4a\,d/b^2 \qquad (8.1)$$

where $a =$ wire radius, $b =$ cathode radius and d is the amount of displacement from the axis of the cylinder. But the uniformity of the wire itself may vary and this variation is not too readily detected except of course as a reduction in the resolving power of the instrument as a spectrometer. The electric field

[1] S. C. CURRAN: Physica, Haag 18, 1161 (1952).
[2] HE. DE VRIES and G. W. BARENDSEN: Physica, Haag 19, 987 (1953).
[3] W. H. BURKE and W. G. MEINSCHEIN: Rev. Sci. Instrum. 26, 1137 (1955).
[4] S. C. CURRAN and J. M. REID: Rev. Sci. Instrum. 19, 67 (1948).
[5] J. BECKER, P. CHANSON, E. NAGLOTTE, P. TREILLE, B. T. PRICE and P. ROTHWELL: Proc. Phys. Soc. Lond. A 65, 437 (1952).

strength $X(r)$ at a radial distance r from the axis is given by

$$X = \frac{V_0}{r \ln (b/a)} \tag{8.2}$$

and in spite of the logarithmic variation of X with a it is clear that X and hence the multiplication will depend closely on the value of a. Variations of a may be very localised indeed and could be due to a change of shape. WEST and his colleagues ascribe much of the failure to obtain very good resolving power with energetic radiation at high pressure to this cause. It would be a considerable advantage if wire of very well specified diameter was available for the more specialised counters employed in high resolution work.

In one respect the proportional counter is not difficult to design. Almost any normal metal acts quite satisfactorily as cathode and certainly aluminium, copper, brass, iron, stainless steel, silver, gold and many other materials have been used. It is preferable that the surfaces be smooth and polished but this is probably because they are less likely to be traps for deleterious gases and vapours. We have noted above that many near-insulators can be useful as cathode surfaces e.g. glass. This lack of sensitivity of the cathode surface is important where sources have to be spread over the cathode as in many experiments where the specific activity is very weak and where in addition the radiation may be very soft. Various examples of the use of tubes in this way will be discussed below.

It is not unreasonable to say that the proportional tube is a very flexible instrument so far as design is concerned. It offers many advantages for monitoring in a wide variety of fields.

D. The tube as spectrometer.

I. Avalanches and saturation.

9. The proportional counter may be used with very soft radiations and it can indeed be employed as a pulse photocell as in the work of CURRAN et al.[1]. Although the gas gain can reach very high values for example with single slow photoelectrons as the primary particles, it is more usual to restrict it to a value such that even the most intensely ionizing particles do not show saturation effects. By saturation we mean the mutual interaction of the small avalanches generated by each track electron in the vicinity of the wire. Such interaction interferes with the full development of the pulse. Obviously the spacing of the electrons along the track of the primary particle, as observed from the part of the wire to which they will be drawn, is the important criterion and naturally this depends on a number of factors. The specific ionization of the primary particle is important. For alpha particles and other heavily ionizing radiations it exceeds the value for fast electrons by a factor of the order of 10^2. Hence other things being equal the heavily ionizing particles show saturation effects at lower values of gas gain. Again, the direction of the primary particles has to be taken into account. If they are allowed to fall into the counter in a direction perpendicular to the wire they form avalanches much more closely together than if their tracks are parallel to the wire. Further the pressure influences the spacing of the avalanches as the linear density of electrons along the track is proportional to the density of the gas. In this last instance we note however that the avalanches themselves are less extended than in the gas at lower pressure. It is difficult therefore to express easily the limits of gas multiplication that can be employed

---

[1] S. C. CURRAN, A. L. COCKROFT and J. ANGUS: Phil. Mag. **40**, 929 (1949).

in different applications and it is wise to check the variation of gas gain with voltage in any new arrangement. Certainly for electrons and gases at normal pressures values of gas gain of 10^4 can be employed. Up to such values the gas gain increases exponentially with the applied voltage. An example[1] of the variation is shown in Fig. 9.

If then we are to detect very soft radiation in the presence of harder radiations we have to consider the problem of amplifying an electron charge of between 10^4 and 10^5 electrons arriving at the first grid of the amplifier. We see that it must be a rather sensitive pulse amplifier if the signal has to exceed the noise

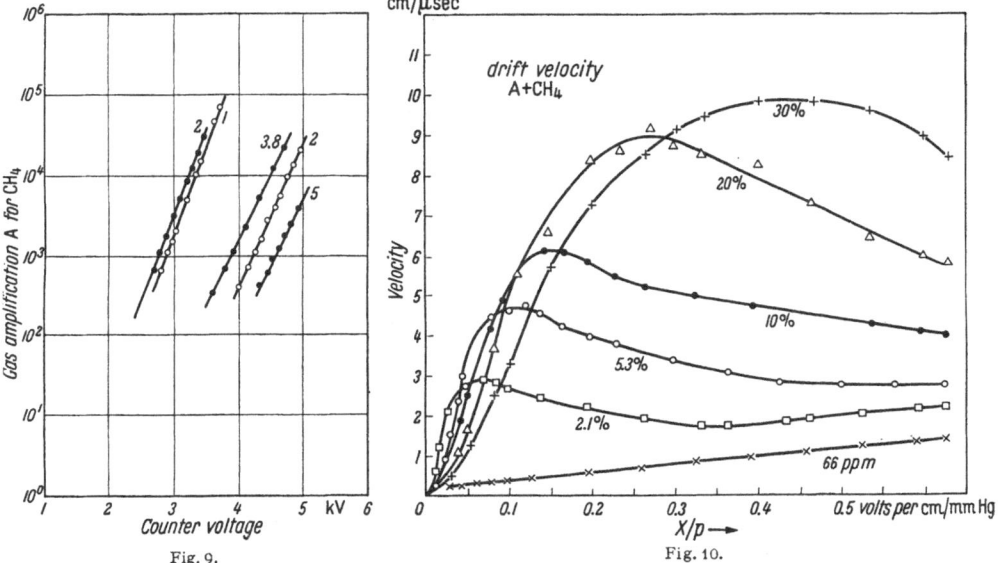

Fig. 9. Gas gain in CH_4 as function of counter voltage for two wire diameters and various pressures (marked in atmospheres on the curves).

Fig. 10. The drift velocity of electrons in argon-methane mixtures for various percentages of methane as marked on the curve.

pulses by a rather large factor: (>10). In theory such noise pulses reach any amplitude but of course we observe a negligible number in practice if the bias level exceeds by a factor of 10 say the r.m.s. value of the noise. A factor of primary importance in this connection is the time-variation of the pulse signal at the wire since this determines the bandwidth of the amplifier itself. We now consider the time-development of a pulse in more detail.

II. Drift velocity and pulse shape.

10. Attention must be paid to the speed of collection of the electrons at the wire. Some studies due to ENGLISH and HANNA[2] of the drift velocity of the electrons in various gases and mixtures are directly related to this question. In Fig. 10 an example of the variation of the velocity for argon-methane mixtures is shown. It is clear that for the 10% admixture of methane to argon and for values of electric fields ($X = $ field strength) and pressures (pressure $= p$) as normally employed the drift velocity is several centimetres per microsecond so that the electrons reach the wire in a typical counter with this type of gas mixture

[1] W. DIETHORN: Doctoral Dissertation 1956, Carnegie Institute of Technology.
[2] W. N. ENGLISH and G. C. HANNA: Canad. J. Phys. **31**, 768 (1953).

in a fraction of one microsecond. The positive ions associated with these electrons are practically wholly confined, at the instant of collection of the electrons, in the near neighbourhood of the wire where the multiplication has taken place. It is only as these positive ions drift outwards from the wire to the cylinder that the electrons are released to develop a signal in the resistor acting as grid leak. Now the field in which the positive ions move is extremely non-linear and the initial stages of development of the pulse correspond to the initial high drift velocity of the ions. The pulse signal rises abruptly therefore and then develops more slowly as the drift velocity decreases. It can be shown easily that the signal reaches half-value amplitude in a very small fraction, $\sim 10^{-3}$ of the total time taken by the positive ions to reach the cathode [7]. Let the drift velocity be denoted by w, the mobility by μ and the gas pressure by p. Then

$$w = \frac{dr}{dt} = \mu \left(\frac{X}{p} \right) \tag{10.1}$$

and

$$x = \frac{V_0}{r \ln (b/a)} \tag{10.2}$$

so that

$$w = \frac{\mu V_0}{r \, p \ln (b/a)} . \tag{10.3}$$

We can then establish that the signal at the wire is

$$S(t) = \frac{-e}{2 C \ln (b/a)} \ln \left\{ 1 + \left(\frac{b^2}{a^2} - 1 \right) \frac{t}{T} \right\}. \tag{10.4}$$

Putting $t = T$, the collection time of the positive ions, the final amplitude of the signal is

$$S_T = - \frac{e}{C} . \tag{10.5}$$

For $t = (a | b) T$ we have

$$S_{a\,T/b} \approx - \frac{e}{2C} . \tag{10.6}$$

With $a/b \sim 10^{-3}$ the signal is shown to reach half-maximum in a time $t \sim 10^{-3} T$. Since $T < 10^{-3}$ sec the signal is found to rise rapidly to about half its final value in $< 10^{-6}$ sec.

A great improvement in sensitivity of the amplifier is possible if the later and much slower development of the pulse from about half amplitude to full amplitude is neglected. The pulses are similar to each other in shape and rejection of the slow stage of pulse development does not interfere appreciably with the proportionality of the signals. We are therefore able to use an amplifier of differentiating time-constant approximately equal to but not less than the integrating time-constant. For most types of counter an integrating value of between 10^{-6} and 10^{-5} sec and a differentiating value of about twice as much prove suitable. Since the amplifier is sensitive it is good practice to have a short screened lead to the input grid and failing that a head amplifier close to the counter with a cathode-follower type of output is desirable.

In covering a whole spectrum of particle energies e.g. beta rays between 1 and 300 kev it is useful if not essential to prevent overloading by big signals. This possibility of overloading must be eliminated at the first or second grid if a flexible amplifier capable of handling a very wide range of input signal strength must be designed. As the signals are fed to a scaling circuit and possibly to a cathode ray tube the 1 kev radiation must correspond to a signal of about 5 volts at the output. In these circumstances the maximum signal strength would be 300×5 volts or some ten times greater than acceptable, illustrating the necessity of eliminating overloading. Several types of non-overloading amplifiers, for example that due to HIGINBOTHAM[1] have been designed.

The design of the amplifier has important bearing on the use of two or more proportional counters in coincidence. The signals take a variable time to reach a predetermined voltage level and we cannot in general define very precisely the time-delay involved. It is therefore necessary to be a little generous in this regard and very short resolving time cannot be used unless care is exercised. The account of the anti-coincidence arrangement due to MOLJK *et al.* illustrates this point.

The counting rate of the device may be rather high. Since the signals have a duration $\sim 10^{-5}$ sec the mean counting rate of 10^3 sec^{-1} is possible without serious loss. Even this value can be exceeded if the counter dimensions are large compared with the average length of track of the primary particles since the ion clouds do not appreciably affect each other as they move towards the cylinder. We must note however that although the signal may be over in 10^{-5} sec or so the ions are still in the tube and they may in some circumstances exert a notable effect on the observed multiplication, even to the extent of introducing so much variation of gain that detection is affected. This effect would occur most readily with localised sources of high specific activity. This circumstance arises with collimated beams of alpha rays and X-rays.

III. The storage of data.

11. Resolving power. Many methods of storing data have been developed [*8*]. For straightforward counting there is virtually everything to be said for some form of electronic scaler and a great variety of these instruments have been described in the literature. Many reliable and versatile units can be purchased. In some types of measurement the simpler counting rate meter can be substituted for the more elaborate scaler but it never offers the statistical accuracy of the absolute instrument.

With the proportional counter operated as a radiation spectrometer it is necessary to provide some form of sorting of pulses in terms of amplitude versus occurrence frequency. Such instruments are generally known as pulse amplitude analysers or kicksorters and considerable ingenuity and skill has been devoted during the last ten years to their design. The subject has been discussed extensively and a number of review articles describing the general features and characteristics are available, for example those due to VAN RENNES and to CHURCHILL and CURRAN [*9*]. The choice of a kicksorter is determined to a large extent by the nature of the work and some general points can be made. As we are dealing essentially with a question involving the energy resolution of the proportional tube we must consider this in some detail.

When an ionizing particle travels through the gas of a counter it gives rise to a number of ion pairs and this number is subject to statistical fluctuation. Moreover the electrons of these ion pairs are the agents for charge multiplication in the neighbourhood of the wire and this multiplication is subject to fluctuation. The overall response of the counter, as measured by the amplitude of the output pulse, is subject to these two component fluctuations and experimentally CURRAN *et al.* showed that they might be separated. The pulse amplitude distribution

---

[1] W. A. HIGINBOTHAM: Brookhaven National Laboratory, Report 234. 1953. See also R. L. CHASE and W. A. HIGINBOTHAM: Rev. Sci. Instrum. **23**, 34 (1952).

due to single slow photoelectrons released at the cathode by light was observed and shown to be of the form indicated in Fig. 11. The curve corresponded rather closely with $y = x^{\frac{1}{2}} e^{-x}$ and of course it represented the contribution due to line width arising from variations in gas gain alone. With the help of this function representing a single electron it was possible to predict the shape of the

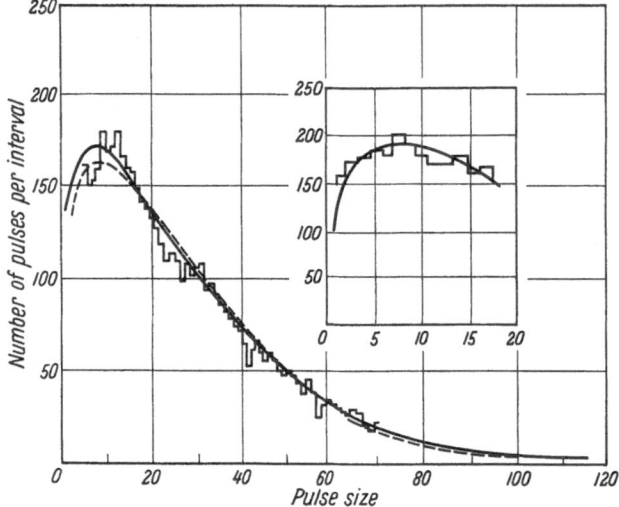

Fig. 11. The pulse size distribution for single slow photo-electrons multiplied in a counter. The full curve fits the experimental data and the broken curve represents $y = x^{\frac{1}{2}} e^{-x}$.

curve that would represent any definite number of slow electrons released simultaneously in the counter and drawn to the wire. Such predicted shapes for various radiations are shown by the broken curves in Fig. 12. For these same

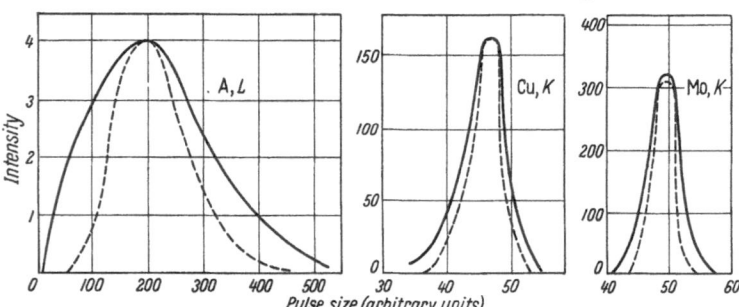

Fig. 12. The width contributed by fluctuation in gas gain (broken curve) shown as part of the total width for different energies of radiation.

radiations (L X-rays emitted by A^{37}, K fluorescence X-rays of Cu and K fluorescence X-rays of Mo) the full curves of Fig. 12 are observed in practice. We must ascribe the difference in the curves, broken and full in each case, to the additional fluctuation in pulse amplitude arising from variations in the number of ion pairs formed in the gas by the radiation. It is possible by examining the curves to deduce that the contributions to line width due to these two causes are roughly equal in magnitude. Moreover for each $\left(\overline{N^2} - \overline{N}^2\right)/\overline{N}^2$ is approximately equal to 0.67. Franzen [7] has pointed out that this conclusion is not in accord with the predictions of an approximate duplication theory of SNYDER[1]. In his treatment of the problem it is assumed that the probability of electron duplication per unit potential traversed by an electron is λ so that for one particle starting at $x=0$ the probability of having N particles at x is given by

$$P_N(x) = e^{-\lambda x} [1 - e^{-\lambda x}]^{N-1}. \tag{11.1}$$

The probability distribution $P_{N\bar{N}}$ about the mean number $\bar{N}(= e^{\lambda x})$ is given by

$$P_{N\bar{N}} = \frac{1}{\bar{N}} \left(1 - \frac{1}{\bar{N}} \right)^{N-1} \left. \right\}$$
or
$$P_{N\bar{N}} \approx (e^{-N/\bar{N}})/\bar{N} \left. \right\} \tag{11.2}$$

for $\bar{N} \gg 1$. For M particles starting from $x=0$ the probability of having a total number $\nu = NM$ at x is given by

$$P_\nu(x) = q_M \left\{ \frac{(\nu - 1)!}{(M - 1)! \, (\nu - M)!} \, e^{-\lambda M x} (1 - e^{-\lambda x})^{\nu - M} \right\} \tag{11.3}$$

where q_M is the probability of having M particles initially.

Hence

$$\frac{\overline{\nu^2} - \bar{\nu}^2}{\bar{\nu}^2} = \frac{\overline{M^2} - \bar{M}^2}{\bar{M}^2} + \frac{1}{\bar{M}} - \frac{1}{\bar{\nu}} \tag{11.4}$$

gives the relative mean square fluctuation or variance of the total number collected. Since ν is generally large the last term is negligible and the variance consists of that of the original number $(\overline{M^2} - \bar{M}^2)/\overline{M^2}$ together with $1/\bar{M}$ the contribution due to the multiplication procress. Eq. (11.2) above gives $(\overline{N^2} - \bar{N}^2)/\overline{N^2} = 1.0$ while $x^{\frac{1}{2}} e^{-x}$ derived from the experimental work leads to $(\overline{x^2} - \bar{x}^2)/\overline{x^2} = 0.67$. Here then we have a disagreement which should be resolved by means of very accurate experimental study and thorough theoretical treatment.

It is clear that in addition to the above uncertainty in the contribution due to multiplication itself there is a definite difference in the nature of the fluctuations in the number of ion pairs created along the complete track of a particle of definite energy as distinct from the fluctuations along a defined path length in the gas. FANO[2] indicated that the former should be less than expected from a simple Poisson distribution about a mean value, and showed that the mean square deviation in the number of ion pairs is given by

$$\overline{(J' - J_0)^2} = \overline{(n - \varepsilon/W)^2} \, J_0/\bar{n} \tag{11.5}$$

where $J' =$ actual number of ion pairs produced, $J_0 =$ average number, $n =$ number of ion pairs liberated per collision, $\varepsilon =$ energy lost per collision and $W =$ average energy of formation of an ion pair.

For a Poisson distribution $\overline{(J' - J_0)^2} = J_0$ and the difference factor $\overline{(n - \varepsilon/W)^2}/\bar{n}$ is less than unity. The method of analysis of CURRAN et al. could be used to determine $\overline{(n - \varepsilon/W)^2}/\bar{n}$ and further work in this field would prove valuable.

Other sources of amplitude fluctuation are present in practice. Amplifier noise and gain change contribute appreciably to the observed line width and

[1] H. S. SNYDER: Phys. Rev. **72**, 181 (1947).
[2] U. FANO: Phys. Rev. **70**, 44 (1946); **72**, 26 (1947).

kicksorters may be potential causes of reduced resolving power although this is not so with modern designs of such units. Likewise variation in the power unit supplying the high voltage can contribute. There are difficulties associated with the uniformity of the wire and purity of the gas (in experiments lasting a considerable time). Most of these difficulties can be effectively eliminated by careful attention to the design of the electronic equipment and the continuous purification of the gases. A curve showing the variation of the resolution with energy, Fig. 13 shows that at high energies the performance is poorer than expected from considerations of fluctuations in the number of electrons collected [5]. Further research may improve this situation.

12. Pulse analysis. The above discussion of resolving power helps us to decide the nature of the kicksorting equipment that should be chosen for use in conjunction with the proportional tube [9]. For all practical purposes however it is probably adequate to consider merely the statistical fluctuations of the number

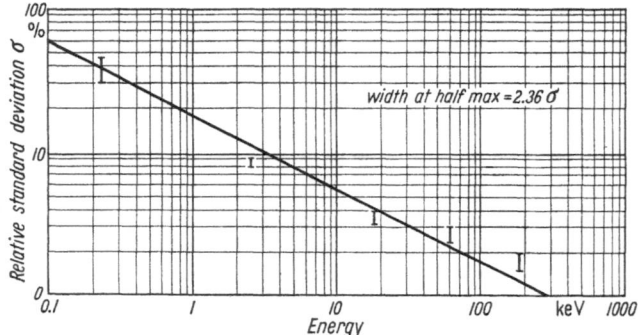

Fig. 13. The resolution as a function of the energy.

of electrons collected at the first grid. At energies E of the order of 100 kev the fluctuation lies between $100\,(0.67E + 0.67E)^{-\frac{1}{2}}/W^{-\frac{1}{2}}$ and $100\,(E + E)^{-\frac{1}{2}}/W^{-\frac{1}{2}}$ where the two major contributions to the fluctuation are indicated by separating the terms in the numerator. For example with $E = 100$ kev and $W = 30$ ev we have on this basis at most $\pm 1.73\%$ for the fluctuation. This shows that a kicksorter of 100 channels is very adequately able to analyse the output of the spectrometer and for more approximate work a 30-channel device may suffice.

Many suitable types of kicksorter have been devised but from many points of view the 100-channel instrument of the Hutchison-Scarrott or recent forms of the Wilkinson type are eminently suitable. The subject has been treated rather fully in the literature. The speed of response is fairly well matched to most types of proportional tube in their more normal applications. The channel width is very stable over long periods of time. Cathode ray display of the pulses is a very useful monitor on the tube performance. In conjunction with such a kicksorter the proportional tube becomes a powerful and speedy spectrometer. Suppose we consider a fairly uniform spectral distribution of pulses. For 1% accuracy in the amplitude of the spectral curve each point should correspond to a total of 10^4 counts. With 100 channels in operation we record 10^6 counts. Such a total can be analysed in about 10 min without difficulty if the avalanches are distributed along the wire (the source is not strictly localised) and the pulse duration is about 10 μsec. This example serves to indicate the speed and accuracy of a suitable combination of counter and kicksorter.

13. Energy per ion pair. The energy expended by a primary particle affects several important aspects of counter performance. Thus we have first to consider whether there is any variation of the energy expended in the production of an ion pair as the energy of a particular primary particle changes. If such an effect was present then the counter would not give pulses linearly proportional to the energy of particles of the type in question. In another respect it is important to consider energy expenditure—the value per ion pair may be different for different primary radiations and this would mean that direct comparison of pulse amplitudes for different radiations could not be made. It seems clear from recent work that the first of these possibilities can be discounted at least for electrons but that the second is in fact definitely established.

Work done prior to 1944 has been reviewed by GRAY[1] and since then a series of accurate studies, which were possible to a large extent by virtue of the introduction of the modern proportional tube, have been made. The earlier work had suggested that there was probably a rise in the value for W for air as the energy of the primary electron fell. The recent much more direct experiments[2] fail to reveal any departure from constancy of W for electrons as primary particles, at least within the range 200 ev to 200 kev and probably as high as 600 kev. Since this range covers effectively the total range of variation of specific ionization of electrons it can be concluded that for nitrogen at least W is constant for electrons to within 2%, a very conservative estimate of the experimental accuracy of the work.

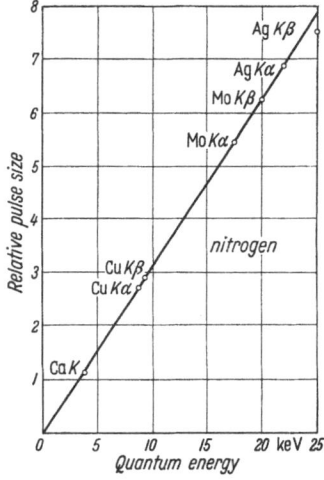

Fig. 14. Linear relationship between pulse size and quantum energy for nitrogen.

A curve[3] which demonstrates the linear relationship between pulse size and electron energy is shown in Fig. 14. It is evident from a curve of this kind that there can be no marked variation of W at different primary particle energies. Most of the curve is built up by observing the amplitude of the pulses due to photoelectrons of definite energy, released within the gas of the counter, as the source of X-radiation giving rise to them is changed. In some cases, no external source of radiation is used but an internal radioactive substance, such as A^{37}, is employed.

The value of W is different for different primary particles in the same gas and care must be exercised when comparing the amplitude of pulses produced by different radiations. In general W for alpha rays and electrons are nearly identical and where they definitely differ the value for alpha rays is usually slightly less than that for electrons. There is not as yet unambiguous evidence of variation of W with energy of the heavy particle in any gas and indeed a slight variation of W with energy is probably present. A good deal of work would seem to lend indirect support to this for alpha rays and protons.

A summary due to VALENTINE of some of the more important points is given in Table 1.

---

[1] L. H. GRAY: Proc. Cambridge Phil. Soc. **40**, 72 (1944).

[2] J. M. VALENTINE: Proc. Roy. Soc. Lond., Ser. A **211**, 75 (1952) and references therein.

[3] S. C. CURRAN, J. ANGUS and A. L. COCKROFT: Phil. Mag. **40**, 36 (1949).

The determination of W by means of the proportional counter is a notable technique. The principle of the method is as follows. Suppose a source of homogeneous radiation, energy E, is detected by means of the proportional counter and gives rise to n counts/sec. If in the same or comparable geometry and counter

Table 1.

Gas	Energy Dissipation W per ion pair and its ratio to that for argon W_A				
	Electrons		Alphas of Po		Protons 340 Mev
	W in ev	W_{gas}/W_A	W in ev	W_{gas}/W_A	W_{gas}/W_A
A	27.0	1.00	25.9	1.00	1.00
He	32.5	1.20	31.7	1.22	1.17
H_2	38.0	1.41	37.0	1.43	1.38
N_2	35.8	1.32	36.0	1.39	1.31
Air	35.0	1.29	35.2	1.36	1.30
O_2	32.2	1.19	32.2	1.24	1.23
CH_4	30.2	1.11	29.0	1.12	—

conditions a saturated ion current through the counter is measured and found to equal I_0 then we have $I_0 =$ number of elementary charges formed per second $= nE/W$.

Hence

$$W = \frac{nE}{I_0}. \tag{13.1}$$

In practice the current I_0 is somewhat difficult to measure accurately if n is chosen to make the effects of counter and scaler dead-time inappreciable. For example with $n = 10^3$/sec and $E = 10^4$ ev we have

$$I_0 = 10^3 \times 10^4/30 = 3.3 \times 10^5/\text{sec}$$

or

$$I_0 = 3.3 \times 10^5/6 \times 10^{18} \text{ amp} < 10^{-13} \text{ amp}.$$

Probably this use of both current and counting rate can be turned to advantage in other investigations in related fields. It is usually advisable to do the counting with a small and known fraction of the source used in measuring the current.

E. Fields of application.

I. The source as gas.

14. Obviously it is most desirable that if a radioactive substance is to be investigated it should if possible be used in the form of a gas within the counter itself. In actual fact a large number of important radioelements have been investigated in this way. Thus the spectra of H^3, A^{37}, C^{14}, S^{35}, Br^{80}, Zn^{65}, Kr^{83} and RaD have been measured, in some cases with an accuracy unobtainable with any other type of source [6].

With the gaseous sources we are completely free of one of the most serious difficulties arising from the use of solid sources on solid supports—the effects of self-absorption and support absorption and reflection. It is true that except in very favourable cases we have to examine with gaseous radioelements the possibility of wall effect. Studies of wall effect arise in other fields, for example in fast neutron detectors of the proton recoil type or in disintegration detectors such as the BF_3 proportional counter or the He^3 detector-spectrometer. In actual practice we must distinguish between various cases; those in which electrons are involved, those in which heavily ionizing particles are encountersed and those in which there are electromagnetic radiations involved e.g. *K*-capture X-rays. With electrons or beta rays the particle track is tortuous and although above 50 or 100 kev it may be approximately straight it is extremely difficult to relate energy expenditure with the distance traversed when the energy of the particle lies below about 100 kev. This makes the analytical problem of allowing for wall effect more difficult than in the case of the heavily ionizing particles where the tracks are approximately straight and of uniform length for monoenergetic rays.

15. Effect of range. In some few instances it is easily possible to make the particle tracks extremely small compared with the dimensions of the counter. Thus when tritium gas is used in the counter the energy of the beta rays is at most 18 kev and with counters of some centimetres diameter at pressures of an atmosphere or more of argon the wall effect is of the order of 1%. On the other hand even with a relatively soft beta radiation such as that of S^{35} the wall effect is very appreciable in the same circumstances. The preponderence of scattering of beta rays makes the calculation at best approximate and if possible empirical procedures should be followed in taking it into account. A fairly sound practice that might serve to illustrate is given by MOLJK and CURRAN[1] in analysing the spectra of C^{14} and S^{35}.

With field tubes in use a precisely defined length of counter of volume V is operative. Now suppose ϱ nuclei decay per sec and per cm^3 and that the fraction of rays with energy between E and $E + dE$ is $N(E) dE$. In the sensitive volume we have $N \varrho V dE$ such electrons produced in each second. If they expend all of their energy within V then they give rise of course to pulses proportional in amplitude to the energy E and they are registered by means of a kicksorter say as occupying the channel E of width dE. But some reach the cylinder walls or pass out through the ends of the sensitive length in each case giving rise to smaller pulses observed in lower-energy channels. The difficulty of defining range arises, particularly for beta rays, but if we can ascribe range R to particles of energy E we can show that for the electrons $N \varrho R/4$ escape per cm^2 per second through the surface S of the sensitive volume. Thus

$$N \varrho V dE \left[1 - \tfrac{1}{4} R (S/V)\right] = N \varrho V dE (1 - \delta_E) \qquad (15.1)$$

gives the number counted correctly in channel E. Here δ_E is the fractional loss to lower channels. At each end of the cylindrical volume V the rate of entry into V of electrons of energy E is $\tfrac{1}{4} N \varrho R$ per cm^2 and these electrons expend part of their energy in V and give rise to pulses in channels lower in energy than E. If we make the rough approximation that the specific ionization of electrons is constant these pulses will fall uniformly into all the energy channels below E. At really low energy the variation of specific ionization with energy is too great to justify such an approximation and the actual distribution of the pulses into the various channels lying below E must be calculated in some way such as by numerical integration using the known range-energy relationship for electrons.

The fraction δ_E not counted correctly in channel E is, as shown above, $\tfrac{1}{4} R (S/V)$ and for a cylinder of radius r and length l this becomes

$$\delta_E = \frac{R}{2l} + \frac{R}{2r}. \qquad (15.2)$$

---

[1] A. MOLJK and S. C. CURRAN: Phys. Rev. **96**, 395 (1954).

This shows that both r and l should be large compared with R if possible. There is a definite advantage to be gained in using gases of high density in order to reduce R. Xenon is a very suitable gas from this point of view but the cost makes argon at higher pressure a more usual choice. Another very practicable method of reducing R is by application of a magnetic field[1]—generally a straight-forward axial field obtained by means of a solenoidal winding. For example with a counter for which $r = 6.8$ cm, $l = 19.15$ cm and which was filled with argon at 5.5 atmos. pressure, R for electrons of energy 100 kev was about 2 cm. With an axial field of strength $B = 3500$ gauss applied the particles describe spirals of maximum curvature $r_m = \dfrac{m v}{e B}$ and the wall effect is reduced in the ratio $R/2r_m$ so that $R/2r$ becomes r_m/r. For this example the component effects are shown in the table. We should note that the magnetic field has a very appreciable influence on the number entering and escaping at the ends of the counter since these move at angles between 0 and 90° to the direction of the field. Experiments indicated that the field produced approximately a reduction by a factor of two in this effect.

Table 2. *Correction factors at 5.5 atmos. of argon and 3500 gauss.*

Electron energy in kev	$R/2l$ %	r_m/r %
25	0.6	2.2
51	1.7	3.3
102	5.7	4.6
153	11.0	5.7

The fraction δ_E of electrons properly belonging to channel E which falls in channels below E decreases with energy. The scattering becomes more prominent with decreasing energy and makes calculation less certain but the corrections as applied by MOLJK and CURRAN to S^{35} yielded very satisfactory agreement between theory and experiment.

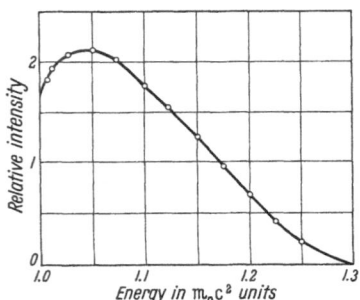

Fig. 15. The full curve represents the theoretical spectrum of C¹⁴, assumed allowed. The circles show the corrected experimental data.

16. Direct method of eliminating wall effect. A more direct method of correcting for wall and end effects can sometimes be applied. We can observe certain spectra such as that for the definitely allowed transition of say S^{35} and deduce from the experimental observations the corrections required to make them fit accurately to the theoretical curve. Such corrections can then be taken over to the correction of experimental observations on other beta ray transitions of similar limiting energy say those of C¹⁴ in this case. We then have an accurate comparative method of study available. The sources themselves are free from particular effects since they are gaseous. The data obtained experimentally for S^{35} was used in this way to give corrected data for C¹⁴ and this corrected data followed the theoretical spectrum very closely indeed according to the results shown in Fig. 15. The full curve in this figure is the theoretical form of the spectrum on the assumption that the transition is allowed. It fits excellently with the adjusted experimental values from less than 3 kev upwards and in this case proves that the spectrum of C¹⁴ is allowed in form.

17. Use of magnetic field. We have seen above that both high pressure and magnetic field may be used to reduce the wall effect and end effect in much work

[1] P. ROTHWELL and D. WEST: Proc. Phys. Soc. Lond. A **63**, 539 (1950).

with beta rays and electrons. We shall consider below the use of still higher pressures in the same connection. As regards the magnetic field ROTHWELL and WEST considered the application of field parallel to the axis of the tube and this of course gives reduction of wall effect but has a less marked influence

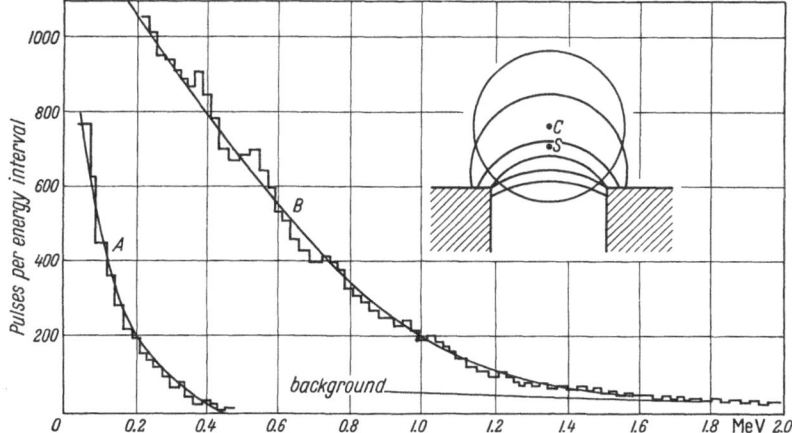

Fig. 16. Histograms of beta rays of P³². Curve "A" no field. Curve "B" with field. Curves can be normalised by making areas under them equal.

on end effect. A restriction of the trajectory of the primary particles in two or in all three dimensions is possible and methods of realising this have been discussed by CURRAN *et al.*[1] and by FRANK[2]. Such methods depend on the use of inhomogeneous magnetic fields and an example of the use of a suitable fringing field is shown in Fig. 16. A good illustration of the use of axial field is shown in Fig. 17, due to ROTHWELL and WEST.

It is probably safe to say that in almost all applications in which the influence

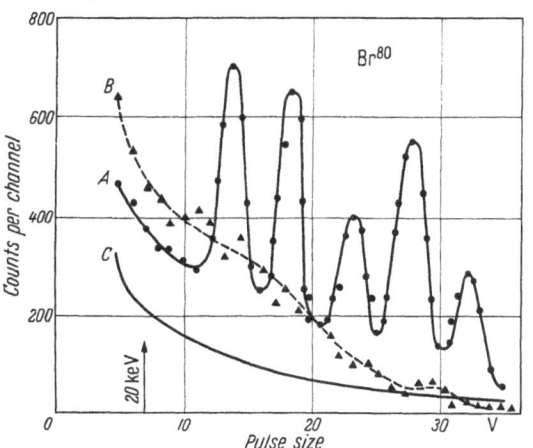

Fig. 17. The effect of a homogeneous magnetic field, parallel to wire, on the spectrum of Br⁸⁰. Curve "A" refers to field on. Curve "B" to field off. Curve "C" is the background due to Br⁸².

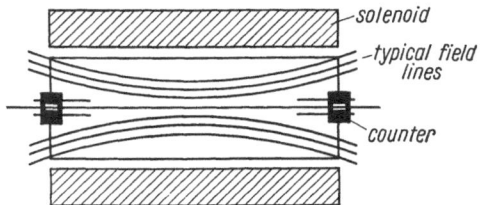

Fig. 18. Application of a solenoidal field counter.

of magnetic field is of value the field produced by a solenoid is suitable. If the solenoid is of a length roughly equal to that of the operating part of the counter then there will be a very appreciable curvature of the lines of the magnetic field away from the axis as indicated in Fig. 18 and the component of the field

---

[1] S. C. CURRAN, A. L. COCKROFT and G. M. INSCH: Proc. Phys. Soc. Lond. A **63**, 845 (1950).

[2] S. G. F. FRANK: Thesis, Cambridge Univ. 1952.

perpendicular to the axis has a strong influence on the passage of electrons across, into or out of, the limiting plane of the counter. The main axial field reduces the wall effect.

It can easily be shown that at normal pressures and temperatures the intrinsic effect of the magnetic field on pulse amplitude should be very small. This has been demonstrated in practice with a variety of orientations of the magnetic field relative to the electric field. The amplitude of pulses produced by particles which do not spend their total energy in the gas may of course be modified for rather obvious reasons. Thus fast electrons leaving the cathode can be deflected back into the surface and so be brought to expend most of their energy in the metal wall. This effect can sometimes be turned to account when we wish to reduce the effect of radiation falling on the wall and exaggerate relatively the effects occurring in the gas itself. Finally we should remark that at high field values $\sim 10^4$ gauss and for a gas such as xenon changes of amplitude of pulse have been noted. These condition are of little moment in practical applications and for most purposes the effect of magnetic field on the processes occurring in the gas multiplication can be discounted.

We have dealt above with gaseous sources of electrons or beta particles. When the ionizing particles are alpha rays or protons, as for example in gas-filled neutron detectors (say boron trifluoride or helium of mass three), the same procedures can be applied to estimation of wall-effect and end-effect and indeed it is simpler on account of the relative freedom from scattering. We must note however that the magnetic field can scarcely be applied advantageously owing to the large $B\varrho$ values of the particles.

18. The absorption of photons. A third and important class of source is the group of gaseous materials emitting X or γ radiation. Here we may be involved in estimating the overall efficiency of the detector and the effect of the walls on the pulse spectrum. The second may be considered in a similar fashion as outlined for electrons above. Firstly the wall material is chosen so that the photo-electric absorption coefficient is as low as possible; this means simply that it should be made of material of low atomic number Z and the ideal would be beryllium, as a coating or inner lining, although good graphite is suitable. Here of course vacuum effects must be considered. In practice much can be done with aluminium as the inner surface of the cathode. The gas will normally be of higher Z value so that the photoelectric effect is as large as possible. In practice it is frequently difficult to employ anything but argon as the main gas. For electromagnetic radiations extending over the band of most practical importance (from ~ 100 ev to 100 kev) the absorption in the gas greatly exceeds that in the wall since we are concerned only with a wall thickness equivalent at most to the range of the photoelectrons in the wall material. Since the photoelectrons released in the gas must have a range in the gas considerably less than the dimensions of the vessel itself we are automatically in the position of having the effective gas absorption dominant over the effective wall absorption. Taken in conjunction with the Z difference this means that photoelectrons released at the wall are not much in evidence in the pulse spectrum. Thus the problem reduces, so far as the pulse spectrum is concerned, to that of a uniform distribution of an electron source through the gas of the counter.

Regarding the overall efficiency the position is not so simple. The X or γ-radiation is emitted uniformly throughout the volume and usually isotropically as regards angular distribution. The path length of the quanta through the gas depends on the position of the emitting atom or molecule and on its direction in space. This problem in integration has been tackled by HAMMERSLEY[1] and we can outline his treatment as follows. He considers the more general case of the uniform distribution of radioactive material through the volume of a cylinder. When the cylinder is such that c is the ratio of its length to its diameter and

$$\mu = \frac{\text{diameter of cylinder} \times \log 2}{\text{half-distance of radiation for absorption frequency } \nu}$$

the integral

$$\varrho(\mu, c) = \frac{1}{V} \int\limits_V dV \int\limits_V dV' \, \frac{\mu \, e^{-\mu r}}{4 \pi r^2}$$

may be evaluated by methods due to HAMMERSLEY and the results tabulated as Table 3. Here the two integrals are taken over the whole of the interior of the cylinder and dV and dV' are two small volume elements and we are considering the amount of radiation emitted from dV and absorbed in dV' which is at a distance r from dV. The integral $\varrho(\mu, c)$ is a transcendental function.

Table 3. *Absorption fraction ϱ as a function of c and μ.*

Values of μ	Values of c									
	1	2	3	4	5	6	7	8	9	10
0.1	0.041	0.048	0.052	0.054	0.055	0.056	0.057	0.057	0.058	0.058
0.2	0.079	0.093	0.099	0.103	0.105	0.107	0.108	0.109	0.109	0.110
0.5	0.181	0.209	0.221	0.227	0.231	0.234	0.236	0.237	0.238	0.239
1.0	0.316	0.357	0.372	0.380	0.385	0.388	0.390	0.392	0.393	0.394
2.0	0.449	0.545	0.562	0.571	0.576	0.579	0.582	0.584	0.585	0.587
5.0	0.739	0.773	0.784	0.790	0.793	0.795	0.797	0.798	0.799	0.800
10.0	0.859	0.880	0.887	0.890	0.892	0.894	0.895	0.896	0.896	0.897
20.0	0.927	0.939	0.942	0.944	0.946	0.946	0.947	0.947	0.948	0.948
100.0	0.985	0.988	0.988	0.989	0.989	0.989	0.989	0.989	0.989	0.990

II. The use of solid sources.

19. Much of the accepted practice of source technique applies directly to studies of solid sources with the proportional counter. In general the solid source may be mounted against a thin window in the cathode provided that the usual limitations of such windows are realised. These limitations are more serious if the form of the energy spectrum of the radiations is involved at least in the case of particles such as β or γ rays. Hence in general external sources are employed when X or γ radiation is under investigation whereas in the analysis of α and β radiation the source is generally mounted internally. As regards windows for electromagnetic radiation it is essential that they should be of low Z material and here the various plastics are very useful. The radiation should not be allowed to strike edges of heavier metals and it should be noted that spurious effects are produced by radiation falling on the inner wall of the tube, as might be the case if the radiation passes across the counter and falls on the part of the cathode opposite the window.

As regards sources of radiation it is of course essential that these should be prepared in as pure a form as possible. Purification by physical and chemalic means is outside the scope of this treatment but the question is of such vital

---

[1] J. M. HAMMERSLEY: J. Math. Phys. **31**, 139 (1952). The values listed in Table 3 are reproduced by permission of the author (the original values are calculated to six significant decimal places).

importance that it must be mentioned here. Associated in many cases with this question of purity is that of specific activity. While the proportional counter is a very efficient spectrometer readily giving an effective 2π geometry, and even a 4π geometry with solid sources, at the same time one of its principal fields of application is the analysis of low energy radiation and this implies the use of very thin sources to reduce to a minimum the effects of energy loss in the source itself and in its support. Such extremely thin sources may make 4π geometry marginal even in respect of counting rate in many cases, for example when the material is one of long lifetime.

III. Support and its effects.

20. We have already discussed the use of sources external to the counter, an arrangement which is satisfactory in the analysis of electromagnetic radiation as windows of high transmission can generally be made.

For soft beta particles the source has generally to be put within the counting vessel and a variety of arrangements can be adopted. To reduce scattering at the support it is necessary to make the supporting foil as thin and light as possible. A foil of some plastic used for support need not be made surface-conducting (as is essential with Geiger counting) as the currents involved are very small indeed but care must be taken to have the electric field in the vicinity of the foil such that electrons go to the wire and positive ions to the cylinder. Thin aluminium foils of about 0.25 mg/cm² are available and these are suitable for much work although not adequately thin for analysis of soft beta radiation in any 4π geometry. In this case it is generally preferable to reduce the problem to one of 2π geometry by having the support a thick and very light material of the hydrocarbon form. Thin plastic films of zapon, formvar, nylon or cellulose acetate are suitable in many studies[1] as they can be made down to 10 μg/cm². As regards the actual mounting of the source this may be at an aperture in the cathode and nearly flush with the surface. It is advisable to design the counter in such cases to keep to a minimum the amount of metal near the source as this reduces scattering of particles which have left the source itself. For the same reason the source can be mounted on the end of a probetype support which is arranged parallel to the wire and possibly quite close to it. The probe and the source are then held at a voltage appropriate to the position in the electric field within the counter. In this way the source is held well away from the walls, perhaps near the middle of the vessel or near the end of the counting volume. The second is useful if we are using a magnetic field parallel to the axis as it gives the beta rays a long path in the gas as measured along the length of the tube while the radial range is confined by the magnetic field itself. In this way beta radiation of higher energy can be analysed without encountering difficulties from absorption at the walls.

In the case of sources of very low intensity the source itself may have to be spread over the whole cathode. The cathode material in this case should be as low in atomic number as possible and aluminium is the most suitable of the common substances. An alternative arrangement which is frequently found useful is to have the operating cathode consisting of a number of wires or light rods. A foil with the source deposited on it can be mounted round these wires so that most of the source is uncovered. The foil may be of a plastic material. The main part of the vessel is a cylinder which surrounds the cage-type counter. Some of the arrangements are sketched in Fig. 19.

It is useful to consider at this stage some of the effects of support, most marked of course in the analysis of soft electron or beta emitting materials. In Fig. 20

---

[1] H. SLÄTIS: P. 259 of "Beta and Gamma Ray Spectroscopy" (Bibliography, ref. [6]).

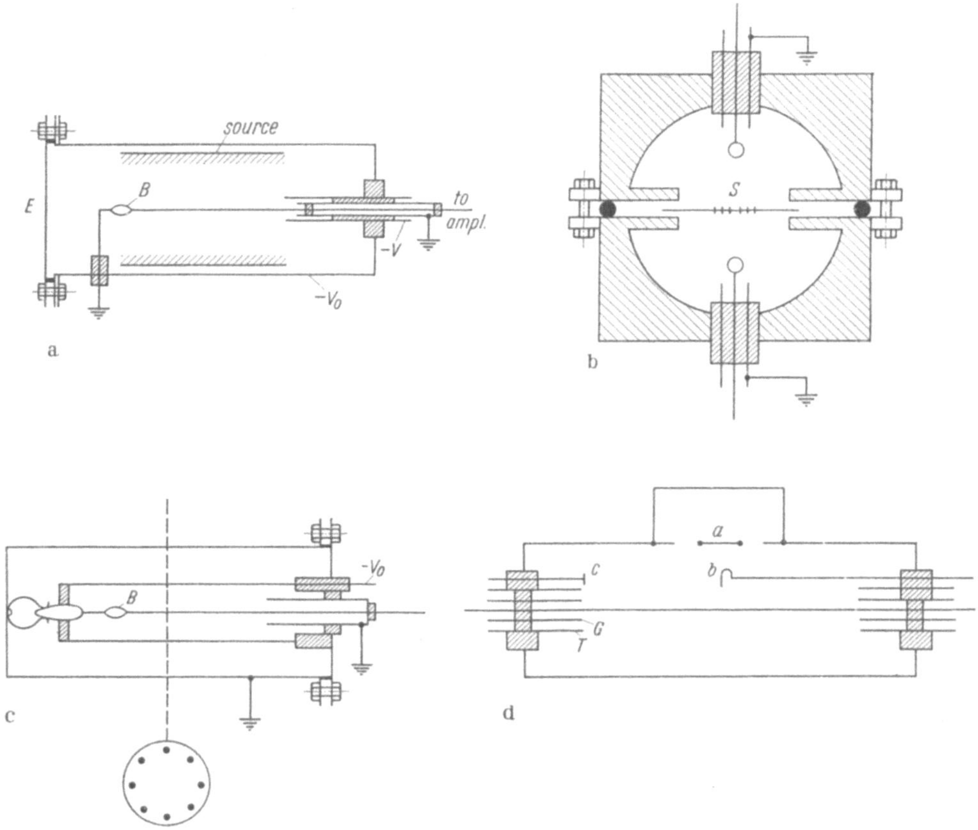

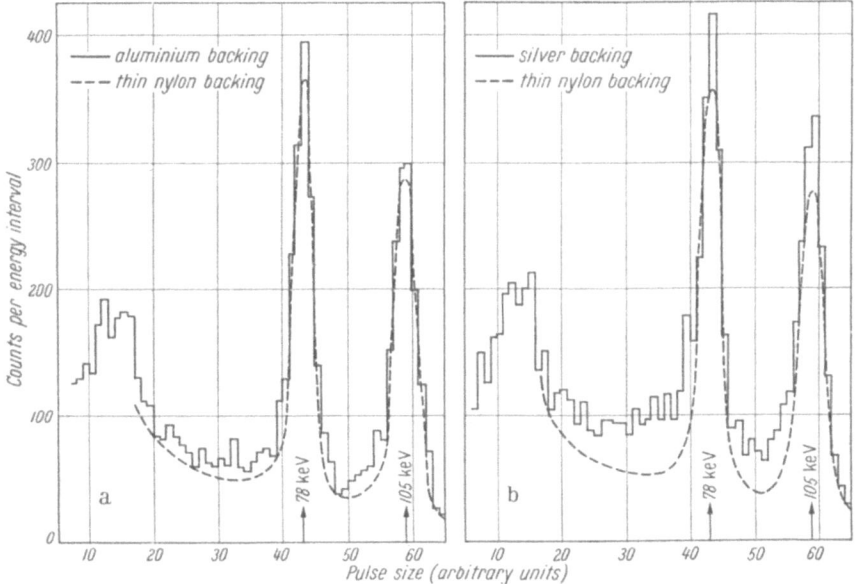

Fig. 19a—d. Various counting arrangements. (a) Large area source spread on thin spilt cylinder which enters at face E. Field tube is at $-V$ volts and B is an insulating bead. (b) 4π loop counter arrangement, usually of gas flow type with source S between complete halves, operated in parallel. (c) Source on foil wrapped round wires. (d) Different source positions are indicated at a, b and c. $T=$ field tube, $G=$ guard.

Fig. 20 a and b. The conversion electron lines of Te$^{125\,m}$ as observed with a source at an aperture in the cathode.

we show the observed lines from a source of isomeric Te[125m]. Here the thin source was mounted on thin nylon at a side aperture and the counter was such that sheets of various materials could be brought immediately behind the source. The effects produced on the conversion electron lines by having aluminium and silver behind the foil are indicated. This type of experiment has been used to study the back-scattering of beta rays and electrons as a function of the nuclear charge of the scattering substance and the energy of the particles. Similar investigations by Burtt and Yaffé[1] and Suzor and Charpak[2] should be mentioned and the more recent results of McNair[3] are pertinent. The use of thick supports of elements of low atomic number does not seriously distort the low energy end of a spectrum and the main effect is to give rise to a "tail" on the low energy side of a line. This remark must be qualified at energies of less than about 10 kev. It would seem to be established by work of Moljk and Drever (private communication) that at about 2 kev it is extremely difficult to realise conditions in which thin sources give rise to line spectra. The adsorption effects and probable chemical effects combined with scattering seem to smooth out lines and a continuum is observed. In this region it is probably essential to use a gaseous source. Another observation is of interest here. Balfour showed that there is no appreciable contribution (certainly less than 2%), to the counting rate arising from slow secondary electrons even when the sources are of the order of 10 μgm/cm². Theoretically extremely thin sources on ideal supports should give quite a large percentage yield of slow secondaries (< 50 eV say).

West et al.[4] noted that for electrons there was no appreciable tail on the low energy side of a peak even when wall effect was relatively large. They explained this advantageous feature in terms of the tortuous nature of the tracks of low energy electrons. Such tracks make large reductions in pulse size due to wall effect more probable. The scattering of beta particles of energy less than 100 kev is very marked indeed as observations in different gases by cloud chamber methods have shown. Thus San-Tsiang et al.[5] showed that the direct distance between start and finish of the track is much less than the total track length; for example in argon it was shown to be 0.55 of the path length.

IV. Sensitivity to electromagnetic radiation.

21. In the application of the proportional tube to the study of electromagnetic radiation we are concerned with both the sensitivity and the resolving power. To obtain good resolving power the main absorption process has to be photo-electric and it must occur mainly in the gas as otherwise the whole energy of the radiation does not appear as equivalent energy of an electron (or electrons) in the gas itself. We must therefore aim at a maximum absorption of the radiation by virtue of photoelectric interaction with the atoms or molecules of the gas. Now the photoelectric absorption coefficient in the energy range of most interest (up to 100 kev) varies approximately as Z^4 where Z is the atomic number of the atoms involved. For high efficiency of detection therefore it is important to have Z as high as possible. In view of the other restrictions on the choice of gas this means that xenon is in many ways to be preferred. We are neglecting

[1] B. P. Burtt: Conference on Absolute Beta Counting, Prelim. Rep. No. 8, N.B.C. 1950; also L. Yaffé, p. 27; L. Yaffé and K. Justus, J. Chem. Soc. 5 (Suppl. 2) 341 (1949).
[2] F. Suzor and G. Charpak: C. R. Acad. Sci., Paris 233, 1356 (1951); 234, 720 (1952).
[3] A. McNair: Thesis, Glasgow University 1956.
[4] D. West, J. K. Dawson and C. J. Mandleberg: Phil. Mag. 43, 875 (1952).
[5] T. San-Tsiang, C. Marty and B. Dreyfus: J. Phys. Radium 8, 269 (1947).

here the relatively insignificant absorption produced by the organic gas or vapour that is usually present. We are also leaving out of account the discontinuities in the value of absorption coefficient that occur in the vicinity of the K and L absorption edges. Our remarks therefore apply really to radiation of energy greater than the K_α of xenon but it is of course mainly for quanta of energy above about 30 kev that there is appreciable difficulty in obtaining high sensitivity. Up to about 30 kev it is possible to use quite successfully argon at reasonable pressures and counters of fairly normal dimensions. These points are brought out by the curves of Fig. 21 due to WEST [5] which show for the gases argon krypton and xenon the absorption of radiation with varying quantum energy. In this connection it is to be noted that the high density of xenon is advantageous in another direction. For the same counter at the same pressure of argon and xenon the range of the photoelectron produced is approximately half in xenon what it is in argon. This means of course that it is much easier to avoid wall and end effects with xenon in the counter.

We see from the curves or from tables of absorption coefficients[1] that up to about 50 kev argon can be used without serious difficulty. For example at an energy of 46 kev about 10% of the incident radiation is absorbed in 8 mg/cm² of gas, largely by photoeffect in the K and L shells of the argon. Now 8 mg/cm² represents a path length of 4 cm approximately

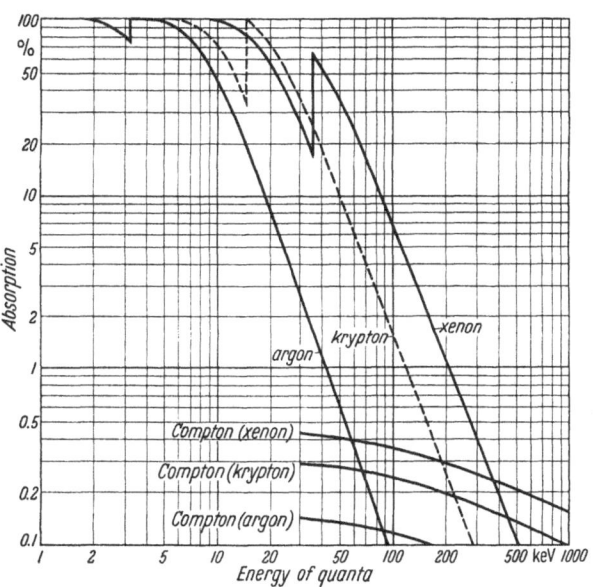

Fig. 21. Photo-electric and Compton absorption in 5 cm of argon, krypton and xenon at N.T.P.

at an argon density of 1.85 mg/cm³ (N.T.P.). This path length is very easily obtained in a counter of moderate size so even atmospheric pressure is adequate to give this efficiency of 10% for detection.

V. Fluorescence yield and escape peak.

22. There is one feature that mitigates against the value of xenon or indeed krypton as compared with argon. This arises from the value of the fluorescence yield[2] and its connection with the so-called escape peak. The photoelectrons released when photoabsorption occurs dissipate their energy in the gas. The fluorescence yield is a measure of the probability that X-radiation will be emitted when the vacancy in the K or L or M shell of the ionized atom is filled. But such X-radiation may not be emitted and Auger electrons may be released so that practically all of the energy $h\nu$ of the incident radiation appears in the form

---

[1] A. H. COMPTON and S. K. ALLISON: X-rays in Theory and Experiment, 2nd ed. (D. VAN NOSTRAND) 1955; also C. M. DAVISSON, Appendix 1 and R. D. HILL, Appendix VI of ref. [6].

[2] E. H. S. BURHOP: The Auger Effect and other Radiationless Transitions. Cambridge: Cambridge University Press 1952.

of electrons which readily produce ionization in the gas. In the light atoms the fluorescence yield is small and for argon it is only 9%. This means that the characteristic X-radiation of argon is not strongly excited and this is frequently a distinct advantage since such X-radiation may in fact escape from the gas itself and so energy is abstracted from the energy $h\nu$ of the incident quanta.

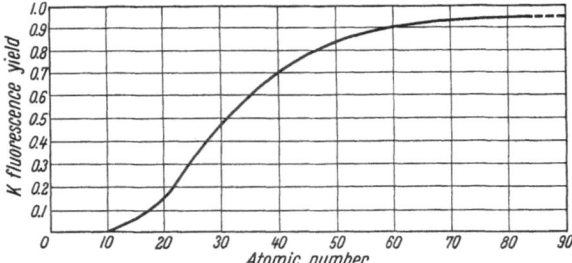

Fig. 22. The K fluorescence yield as a function of atomic number Z (empirical values).

Thus the energy expended in ionization may be not $h\nu$ but possibly $h\nu - h\nu_K$ or $h\nu - h\nu_L$ where $h\nu_K$, $h\nu_L$ are the K, L binding energies of argon. The observed spectrum may therefore show in addition to a peak at an energy of $h\nu$ peaks known as "escape" peaks at energies which are less than $h\nu$ by constant amounts

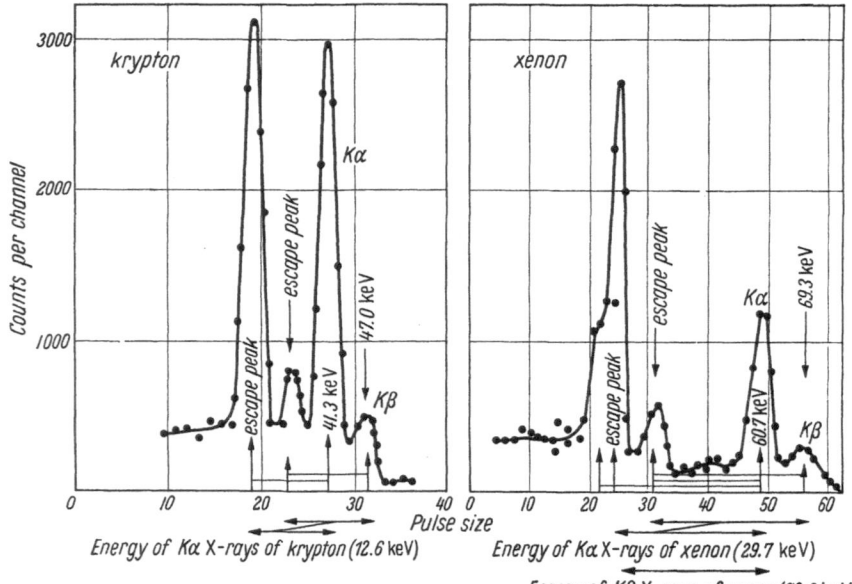

Fig. 23. Examples of escape peaks in krypton and xenon.

characteristic of the absorbing atom in question, say argon or krypton or xenon. The constant separation in energy between the energy of the peak corresponding to that of the incident radiation and that of the escape peak serves as a means of identifying the escape peak if the incident radiation is complex. There are of course refinements to the above argument because of the possible complex transitions in the outer electronic structure of the atoms and in practice it is found that the main escape peak corresponds to the escape of K_α-radiation and a less intense one to escape of K_β. In argon the phenomenon is observable even though the fluorescence yield is 9% but it is usually almost entirely absent if the pressure is fairly high and the counter big. In these circumstances the average probability of escape of the relatively soft K_α or K_β radiation of argon through argon is greatly reduced. On the other hand the larger fluorescence yield of krypton and particularly xenon makes the presence of escape peaks in these gases a notable feature and in many experiments the peaks are a definitely complicating feature.

The fluorescence yield ω is clearly a quantity of critical importance in work with proportional counters. We have seen how its value for the gases used in the tubes influences the nature of the observations and hence its value for such gases must be known. The counter itself has been used in determining the yield for the K-shell and for the gases argon, krypton and xenon, the respective values being (0.11 ± 0.01), (0.67 ± 0.03) and (0.81 ± 0.05). We see that as a function of Z it increases steeply and the form of the curve is shown in Fig. 22, due to BERGSTRÖM[1]. The value for the L and M shells is not so critical but may enter into the detailed consideration of decay schemes of radioactive substances as studied by the counter. Since the tube is of great value in the examination of K-capture and L-capture decay we again require information on the ω values for nuclei of all Z values. Likewise in the observation of internal or external conversion phenomena it is involved.

Examples of escape peaks in krypton and xenon are given in Fig. 23 [5]. In argon the main peak can be observed but generally it is relatively insignificant in intensity. For neon of course it is extremely weak except in rather unusual experimental conditions.

F. Summation effects.

23. In work on many types of radioactive material two or more ionizing particles may give rise to simultaneous pulses in the counter—we have in fact an internal coincidence. This type of coincident event occurs in many practical experimental arrangements because frequently the tube is used at high detection efficiency. The geometry is often 2π or even 4π and in these conditions the summation effects of simultaneous emissions from the sources have to be considered in some detail. The summation may be used to advantage in many cases. The output pulses with a 4π geometry are integral in nature but the phenomenon may become complex, particularly when electromagnetic radiations are involved as it is seldom possible to arrange for nearly 100% absorption of these in the gas of the tube and often a large fraction of such radiation escapes. A good illustration of the phenomenon is found in the study of isomeric $Br^{80\,m}$ by ROTHWELL and WEST [5]. They used the effect to determine the conversion coefficients of the gamma radiation. In many studies much additional information can be secured by systematic changes in the geometry employed, for example small angle, 2π and 4π may be used in turn or again the changes produced by varying the gas pressure may lead to detailed knowledge of the gamma and X-radiation present.

The ideas can be most conveniently discussed in terms of some typical examples. In Figs. 24 and 25 the main results of WILSON and CURRAN [6] on Hg^{203} and of INSCH et al. [6] on RaD respectively are shown. The decay of the isotope Hg^{203} has been extensively studied and it occurs mainly by beta emission to an

---

[1] L. BERGSTRÖM: Auger Electrons emitted from Radioactive Atoms, p. 624 of Bibliography, ref. [6].

excited state at 280 kev above the ground state. The internal conversion of this gamma radiation in both K and L shells occurs with fairly high probability. With a source of Hg²⁰³ mounted within a counter we have various types of event recorded. Thus we can observe

(1) Beta rays only. Here the gamma transition occurs by the emission of a quantum which escapes from the counter. A straightforward beta spectrum results from these events.

(2) Beta rays coincident with K conversion electrons. If the K X-rays arising in this type of decay escape from the tube then a beta spectrum moved along the energy axis by an amount equal to the energy of the conversion electron results. This is the explanation of part II of the spectrum of Fig. 24, neglecting the finer details of the possible variation in energy of the conversion electron.

(3) Beta rays coincident with L conversion electrons. Conversion of the gamma radiation in the L-shell gives rise to a still more energetic conversion electron and a third beta ray spectrum starts at an energy value corresponding to that of this conversion group — hence part III of the curve. Clearly a very effective method of examining the decay scheme is here available and it is relatively easy to deduce the K and L conversion coefficients.

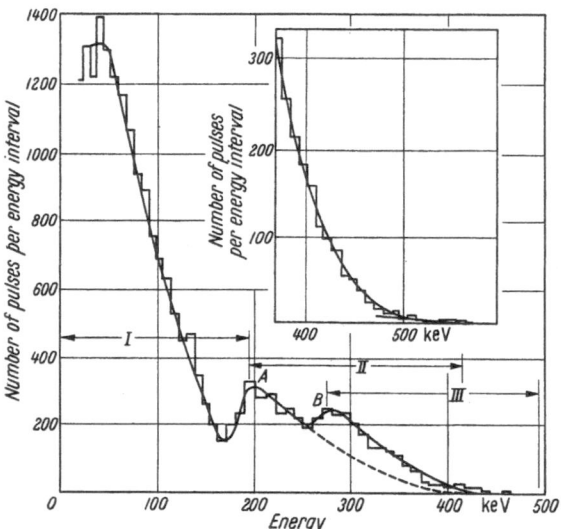

Fig. 24. The L.H. curve shows the spectrum of Hg²⁰³. Part I refers to unaccompanied beta rays, Part II to beta rays coincident with K conversion electrons and Part III to coincidence with L conversion electrons.

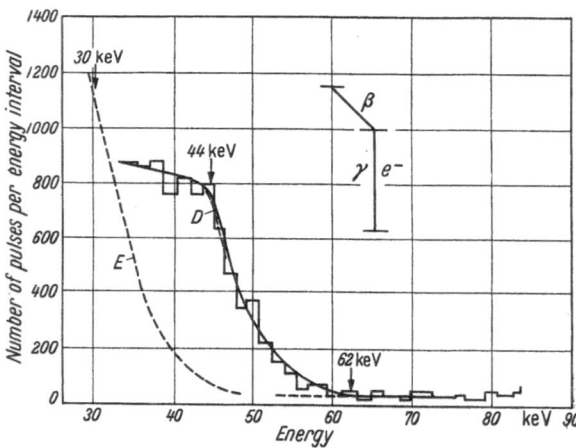

Fig. 25. Spectrum of RaD.

At the same time the example illustrates another of the powerful features of this summation analysis. The primary beta spectrum of Hg²⁰³ is displayed down to very low values in part III without encountering the same difficulties as are experienced in the analysis of the very low energy part of any beta spectrum. The work on Hg²⁰³ has so far been done with a very thin but nevertheless solid source of mercury and it would be interesting to have the same measurements with a source in the form of mercury vapour. The beta rays and conversion electrons are rather energetic and the analysis would be of increased value only if wall and end effects could be made very small.

One of the main objects in the work on Hg^{203} was to study the energy changes in the atom as a whole due to the change in nuclear charge and its division between the beta particle and the neutrino. The experiment seemed to establish that the total energy of beta decay included such redistribution energy and that it was shared between neutrino and beta ray. This result was supported by work on RaD. The total transition energy of the disintegration scheme was found to be 64 kev. The soft beta rays could be detected with comparative ease due to their coincidence with the more energetic photoelectrons arising in the internal conversion of a gamma radiation of energy 46 kev. The M-conversion electrons with an energy of about 44 kev carried the beta particle spectrum along the energy axis and so facilitated examination. These results on RaD showed that the beta spectrum had an allowed shape (within the rather wide limits of the experimental error) and an energy limit of 18 kev. The work was confirmed by the results of JAFFE and COHEN[1] (who used RaD in gaseous form as RaD tetramethyl), and by those of BANNERMAN and CURRAN [6]. The problems of the decay scheme of RaD have been clarified by WU et al.[2] and the main decay scheme appears to be relatively simple.

G. High sensitivity work.

I. Detection within an energy band.

24. The proportional tube is peculiarly well adapted to the examination of low intensity radiations. Most of such radiations have particular energies or they occupy a well-defined energy band. Since the background radiation spreads over a wide energy band it is usually possible to raise the ratio of source counts to background counts by examining radiation within the source band only. As examples we may note that the beta rays of H^3 occupy the energy range from 0 to 18 kev as against a background range of the order of 1 Mev, while the beta particles of C^{14} have a limiting energy of 157 kev. Even more satisfactory performance than these figures would suggest is possible in the examination of K and L capture sources and indeed relatively small branching ratios for K-capture to beta emission have been established, for instance in the recent work of DREVER and MOLJK[3] on F^{18}. Here the positron emitter is shown to have a K-capture ratio of 3%. The same remarks apply of course to weak gamma transitions, especially those of low energy, and to weak conversion lines. Such conversion may be observed by the presence of an electron line as in the case of Cr^{51} say or by the K and L X-radiation arising from the conversion phenomenon itself. This energy factor in reducing background effects is hence very important and it makes the 4π proportional tube, together with the 4π scintillation counter, unique in respect of high sensitivity. It can be maintained that in principle the proportional counter is the most sensitive of the available radiation detectors. We can support this argument by establishing, as has already been shown by DIXON et al.[4], that the main component of the background is due to the action of gamma radiation on the detector (in the case of a well-designed sensitive system). Now a proportional tube filled with the radioactive substance in gaseous

[1] A. A. JAFFE and S. G. COHEN: Phys. Rev. **89**, 454 (1953).

[2] C. S. WU, F. BOEHM and E. NAGEL: Phys. Rev. **90**, 388 (1953); **91**, 319 (1953).

[3] R. W. P. DREVER and A. MOLJK: Phil. Mag. **1**, 1 (1956).

[4] D. DIXON, A. McNAIR and S. C. CURRAN: See p. 212, Proc. 1954 Glasgow Conference on Nuclear and Meson Physics, Pergamon Press, 1955.

or vapour form offers the minimum target of interacting material to the gamma radiation and hence the background pulses are reduced to a minimum. This argument presupposes that the container for the gas, the "wall" of the counter can be made relatively negligible in mass and this is indeed the case. Let us now examine the usual arrangements for reducing background and consider the results in more detail.

II. The nature of the background.

25. The background counting rate of any radiation detector arises from a variety of sources and it has been the subject of many researches, but it has seldom been the main subject; much of the work has been done as an incidental part of other investigations, e.g. the study of cosmic radiation. For an unshielded counter the cosmic radiation sets a lower limit to the rate of counting and this lower limit in the case of Geiger or proportional tubes is of the order of 1 count per min per sq cm of the cross-sectional area of the counter (as viewed from above). There are many other important contributions to background. Clearly the natural radioactive substances, particularly those of the uranium and thorium families, may contribute if they are present in the solid materials of the counter itself as impurities or if they are carried into the counter itself in the filling process—for example radon or thoron present in the atmosphere may contaminate the counter in some circumstances. Again such substances may be in materials in the vicinity of the counter. The lead that is commonly used as a shielding substance is frequently a source of some interfering counts. Other natural radioelements which are distributed widely and therefore must be taken into account are potassium, of which the K^{40} content is responsible, radiocarbon C^{14}, present in fresh wood, graphite etc. and tritium, H^3. There are several rare earths which may cause additional difficulty but the low natural abundance helps here. Perhaps samarium Sm^{147} should be mentioned since it is an alpha source and alpha-emitting substances should be carefully excluded from a sensitive proportional counter.

Some of the most common constructional materials may become contaminated with one or more of the radioelements. Thus aluminium from some sources seems to include a small proportion of potassium and it can prove necessary to specify very high purity material. Some samples of tungsten are more active than others due to potassium impurity. The most notable difficulty occurs with glass containing potassium. Here the counting rate of glass Geiger or proportional tubes is increased and when many of them are used in anti-coincidence shielding of the sensitive detector from external radiation the counters themselves act as sources of spurious radiation and affect the counting rate.

It was thought at one time that thermal emission might be responsible for a component of the background rate but it has been shown to be entirely negligible in any well-designed counting system.

There are several possible ways of reducing and indeed almost completely eliminating the effect of cosmic radiation. The cosmic ray intensity decreases as the amount of absorber is increased. Thus the atmosphere itself reduces the intensity by a factor of about 10 and it has been shown that at depths under the surface of the earth or under water the radiation is considerably attenuated. The most practical way to use such attenuation appears to be to set up the detecting system in a suitable mine or tunnel. Such a step is not always advantageous however. The nature of the surrounding rock must be taken into account. If it consists of shale for instance the gamma radiation from the walls of the chamber may in turn introduce a fresh problem. It can indeed be shown that in a very well designed system the apparent advantage in operating underground almost disappears. Against it there are the usual objections of working in relatively confined space remote from the normal laboratory facilities. It is perhaps fair to say that in practice the advantage of going to depth is mainly one of relative freedom from electromagnetic disturbances. At the very low rates of background counting the presence of spurious signals associated with other equipment can be extremely troublesome. For this reason a rather remote and easily shielded site offers distinct advantages. Shielding here refers to electrical noises.

III. Effective elimination of container.

26. Reduction of gamma interaction. Whatever detector is employed in the analysis of radioactive substances it is true to say that it must contain some material and gamma radiation inter-acting with this material will produce some background counting rate. Clearly therefore we can move in two directions with a view to reducing the background:

(1) The amount of material subject to the gamma bombardment should be as small as possible, consistent of course with the requirement of detection of the radiations involved.

(2) The gamma radiation penetrating to the detector should be made a minimum.

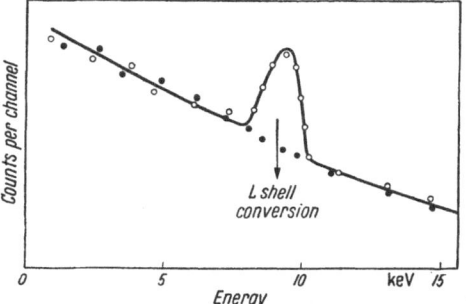

Fig. 26. The solid circles show the form of the background spectrum with a cathode of low Z. The open circles show the excited L radiations in a heavy cathode.

The second requirement is perhaps the most straightforward. Generally speaking rather large masses of lead make the most flexible and suitable screen. Usually the lead is cast in the form of interlocking "bricks" and thicknesses on all sides of the order of three or four inches are employed though much more could be employed with advantage. Since it is difficult to procure lead that is entirely free from natural radioelements it has been demonstrated by KULP and TRYON[1] and confirmed by others that an inner lining to the shield of some other suitable absorber may effect a worthwhile improvement. Mercury in a hollow cylindrical container of steel has been used but it is an expensive substance in pure form in large amounts. Iron itself can be used with rather similar results. The shield is however to some extent a matter of personal judgment and the final choice made by the experimenter will depend largely on the availability of the necessary materials and money.

As regards (1) the situation is much more complex and the answer may possibly differ according to the nature of the particular substance under investigation. We can best discuss the matter by reference to actual studies of various low level activities. During the course of investigations on the activity of rhenium DIXON et al. observed peaks on the spectra of background counts. The most prominent of these was found to correspond to the L X-ray peak of the substance forming the inner wall of the proportional counter. When the material on the inner wall was changed the observed L peak changed likewise. The counter was well shielded and it had an anti-coincidence system of Geiger tubes placed round it. It was therefore likely that the main radiation reaching the inner wall

---

[1] J. L. KULP and L. E. TRYON: Rev. Sci. Instrum. **23**, 296 (1952).

was of the gamma type and that the L (and K) peaks arose in the process of photoelectric absorption of the gamma radiation in the wall itself. The X-rays were detected as homogeneous "lines" on the spectrum while the photoelectrons and Compton electrons gave rise to a continuous background as indicated in Fig. 26. Approximate calculations based on the observed total rate of counting, the rate due to X-rays giving rise to the peaks and the energy spectrum of gamma radiation falling on the wall (as measured in a scintillation spectrometer) seemed to confirm this interpretation of the nature of the greater part of the final background.

27. The self-shielding proportional tube. The complete removal of the gamma radiation presents some formidable problems as we have noted above. The alternative procedure—to make the counter less sensitive to gamma radiation— offered a feasible solution. This method has been pursued by MOLJK, DREVER and CURRAN[1] with considerable success. Their system is primarily intended for application to radiocarbon dating but it can be exploited in other fields of study. The general principle can be stated very briefly—it consists in eliminating as far as possible a solid wall for the counter. We proceed to discuss in more detail the actual form of the detector and its shielding counters.

The cross-sectional diagram of the counter assembly is shown in Fig. 6. We see that an array of peripheral counters of the shielding type are constructed by means of wires. Five wires together with a part of the containing cylinder define the volume of each shielding counter and four of these wires are used in common with the adjacent counters. A single axial wire acts as the central wire of the main proportional tube. Thus the shielded detector has very little solid material forming its cathode—in the example given here the amount of material forming the bounding wires has an effective area of only 4% of that of a solid cylindrical surface of the same diameter. There is therefore a much reduced probability of gamma radiation from outside interacting with the "wall" of the detector. The surrounding counters are proportional in operation and individually they can be made to operate at roughly the same voltage as the main counter since they share the same gas and the diameter of their anode wires may be chosen to suit. In practice the central wires are connected in parallel so that in effect they form one large counter of extremely long wire length. The self capacity is thus high and the pulse amplitude for the total system due to a response in any one of the constituent counters is much reduced. This effect must be taken into account in setting the voltage on the wires of the shielding assembly.

Gamma rays falling on the containing cylinder give rise to ionizing secondary electrons and these must pass into the shielding counter before reaching the inner one. As the shell is in anticoincidence a very large fraction, $\sim 96\%$, of these counts is eliminated. Thus the gamma rays in principle can give rise to recorded events only when they fall on the wires or when they are absorbed in the filling gas itself.

The plan view of the counter shows that it is relatively easy to protect the ends of the inner counter. One end of the counter is protected by a relatively short proportional counter connected like the peripheral one in anti-coincidence with the inner. The wire of this short tube supports the anode wire of the main detector by a short length of glass which insulates the one wire from the other.

The arrangement of voltage supplies to the various counters is relatively simple. The circuits can be fairly standard but it is important to use a good non-over-

---

[1] A. MOLJK, R. W. P. DREVER and S. C. CURRAN: Proc. Roy. Soc. Lond., Ser. A **239**, 433 (1957).

loading type of amplifier particularly on the shielding counters. The pulses may cover an amplitude range of about $10^3:1$. The extremely big pulses are liable to produce very small subsidiary voltages in an amplifier prone to over-load and care must be taken that none of these voltages cause spurious effects. In the system shown in the figure the lead-shielded main counter operates at several hundred counts per minute in the absence of the anti-coincidence counters and it is only too easy to have a small fraction of these passed in error to the register.

The figures that are pertinent are approximately as follows: Total unshielded rate in central counter $= 1000$ to 2000 counts/min. Lead and iron shielded rate $= 500$ to 800 counts/min. Rate with shielding and anti-coincidence counters in operation $= 2.7$ counts/min.

The system can be operated with or without magnetic field applied (by solenoid) in a direction parallel to the length of the counter. The only purpose of using magnetic field in this arrangement is to define the effective boundary between the shielding counters and the inner detector. The magnetic field con-strains the path of the ionizing primary particles very considerably and those starting within the inner counter are confined almost completely to its geometric volume and there is little passage of the particles from the inner to the outer counters (this resulting in their elimination by virtue of the operation of the anti-coincidence circuitry). If the pressure of gas is high and the beta rays involved are of relatively low energy, for example if a pressure of a few atmospheres of gas is used for radiocarbon analyses the boundary is reasonably accurately defined without the field. This means that the system is appreciably simplified although for fairly small or even medium counters the solenoid is not difficult to provide.

28. The ultimate background. We must discuss the nature of the ultimate background achieved in this sensitive counter. Taking the figure of 2.7 counts per minute for the volume of 5.6 litres of fully sensitive counter or about 0.48 counts/min-litre, it is found that 0.1 counts/min-litre are due to the wires forming the boundary and the remainder arises from bremsstrahlung interaction with the gas and neutron interactions. The gas used in obtaining the figure was atmospheric argon which absorbs gamma radiation rather more readily than carbon dioxide or methane but the counting rate for these was somewhat larger, density for density. The boundary wires of stainless steel could be replaced with some advantage by wires of a less dense material, such as aluminium. The final conclusion is that probably 0.2 counts/min-litre is the ulti-mate limit of background rate for a gas such as methane or carbon dioxide at N.T.P. This allows one to date archaeological specimens of ages in the region of 5×10^4 years reasonably well.

It should be noted that the above discussion refers to a system which is set up to detect all beta particles of energy above 3 kev and less than 155 kev so that the counter is close to 100% sensitive to the radiation of C^{14}.

We have been considering above a system which lends itself to the extremely sensitive examination of gaseous sources. This condition it not so difficult to fulfil in view of the extension in the operating range of proportional counters up to more than 800° C as this allows the vapourization to an adequate pressure of many elements or compounds of the elements. It should be remarked that as yet no system like the above has been made for operation up to 800° C but there does not appear to be any real difficulty in so doing. There are instances of course where vapours cannot be used and the nearest approach to the above is then perhaps the most sensitive. A very similar system to that described has

14*

been used by Houtermans and Oeschger[1]. A thin wall is used in place of the bounding wires and the results obtained are very good. Thin sources can be spread on the inner surface of the thin foil forming the cathode of the main counter. If these sources have to be thick then there is of course a real increase in the background due to gamma radiation falling on the source material.

If the specific activity of the material is such that very limited amounts of it will suffice or indeed if only very small amounts are available, say ~100 μgm, then other arrangements may be adopted. Thus the source may be introduced by means of a thin probe into the ultra-sensitive counter of Fig. 6.

29. Remarks on shielding and circuits. Even if the work does not call for limiting sensitivity, and relatively few investigations do, it would appear advisable to replace the more usual Geiger counter shield with a system of proportional detectors. These can be operated readily in the same gas as the central proportional counter and they need not therefore be in a separate vacuum-tight container. A suitable system is shown in cross-section in Fig. 6. Experience in the past has shown that it is definitely advisable to take the wire to earth potential through the grid leak of the head amplifier. It has been shown by Moljk and Drever that a good isolating condenser can be used successfully (without giving rise to any spurious pulses beyond the mixing circuit) in the circuit attached

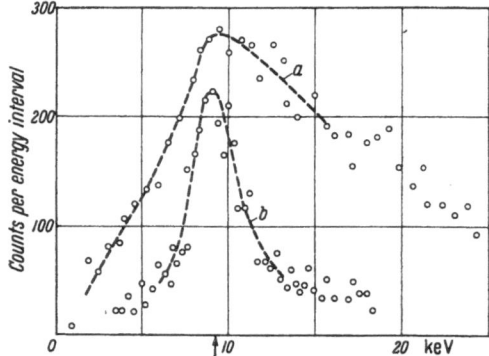

Fig. 27. (a) The energy spectrum of the background pulses in the anticoincidence shield of proportional counters. (b) The energy spectrum of the capture X-rays of Ge⁷¹ observed in the shield. The arrow shows the energy of Ge K X-rays at 9.2 kev.

to the wire or wires of the shielding counters and this facilitates the operation and makes indeed the whole construction of the counter assembly much more practicable.

Three advantages of the self-contained proportional shield should be mentioned. The materials can be chosen for a minimum contamination e.g. they may be free of K^{40} while many glass Geiger tubes are not. Secondly the shield is where it is most effective, as close as possible to the inner counter and thirdly it occupies a minimum volume. This last is important if expensive materials such as mercury form part of the material shield around the assembly. As regards minimum volume a remark on the size of the counters may be appropriate. The radial depth of each shielding counter has usually been about 2 cm. Probably a depth of 1 cm would prove satisfactory and give a useful reduction in the volume of the shield but the reduction itself involves the need for closer tolerances on the various parts of the assembly and the increased self-capacity of the whole system of shielding counters adds to the amplification problem.

The results obtained by Moljk *et al.* with a shield of proportional counters is of value in indicating the feasibility of reducing the radial depth. With a shielding system of 2 cm depth the pulse energy spectrum of the "background" radiation detected in the shield was as shown in Fig. 27. It is evident that there are few very small pulses, energy less than say 1 kev. A flat maximum appears on the pulse energy-frequency curve in the region of 10 kev. For a particle of

[1] F. G. Houtermans and H. Oeschger: Helv. phys. Acta **28**, 464 (1955).

minimum specific ionization we expect an average energy expenditure in one depth of the shield of about $2 \times 2 \times 1.5$ kev, where 2 cm is the path length, 2 kev/mg is the rate of energy dissipation in the gas and 1.5 mg/cm³ is the density of the counter gas at N.T.P. We see therefore that the peak for such particles should occur in the region of 12 kev as the particle generally traverses the shield twice. Clearly a depth of 1 cm would be adequate on this basis and it would represent a very considerable economy of volume and hence of gas or vapour. This is particularly true if the pressure is greater than atmospheric. This may be important where for example the amount of a sample containing C¹⁴ is definitely limited.

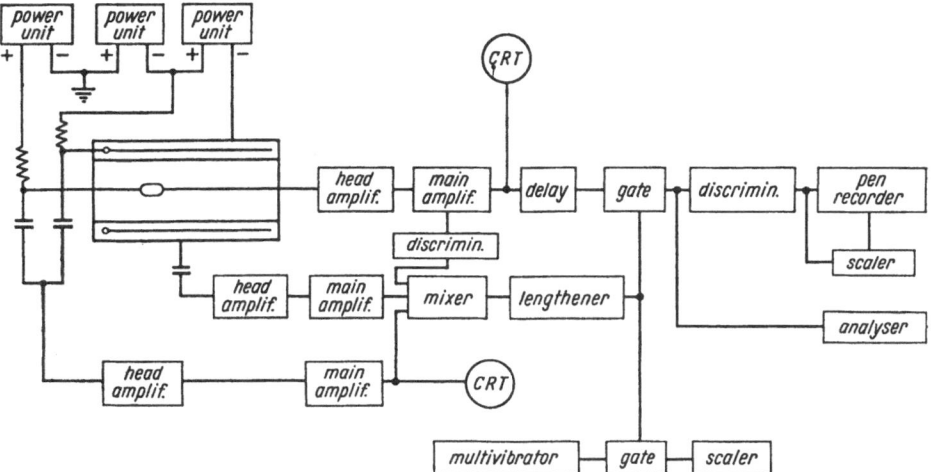

Fig. 28. Block diagram of complete low-level counting system.

The block diagram of Fig. 28 calls for some comment. We have mentioned already the feasibility of condenser coupling to the wire of the anti-coincidence ring of counters. This is perfectly acceptable here because all pulses from the counter ring close the gate circuit. This gate circuit is normally open and such that pulses are passed through in their original form except when it is closed by signals from the anti-coincidence side. Thus if some spurious pulses occur on this side (say from some form of failure of the condenser) the gate is closed by these and no trouble arises. Moreover in the figure a condenser is shown attached to the outer case of the counter. This case at high voltage is frequently a source of spurious signals and the coupling ensures that these signals likewise close the gate. This additional facility has been found very valuable and it makes possible the operation of the system in otherwise rather difficult conditions. Thus computing machines, Tesla coils, and pulsed high energy machines may be mentioned as potential sources of electrical disturbance in a laboratory and this circuit guards very effectively against such sources of electrical noise. For very long runs the time lost by the closure of the gate (due to heavy interference for instance) can be measured accurately by the action of the multivibrator.

The circuits are such that the gate opens after shutting only if there are no signals from the shielding counter. This ensures that after-pulses due to very big signals in the counter (occurring while the gate is closed) are not counted. Finally, there are two discriminators in the circuit, one to set an upper level (155 kev) from the main counter and one to set a suitable lower (~ 3 kev). These figures refer of course to C^{14} in the system. Regarding the power supplies we note that one is used to apply rather higher voltage to the ring of wires. This increase arises mainly because of the high self-capacity of the total wire. The wire could be chosen to have a suitable smaller diameter than the main wire so that it would be possible to operate from a single voltage supply.

A pen-recorder is a valuable adjunct to the system. The frequency-time distribution of the pulses is a good check on the consistency of operation of the complete equipment. This is a matter of importance in long runs and much data and operating time can be saved by the proper rejection of data taken when clearly some fault had developed late in the run. The system as shown is complex as it was used in examining the characteristics of the assembly itself. It is not necessary to include all of it for a practical application.

We have already considered some of the more obvious applications of the proportional counting technique and it remains to amplify to some extent various questions previously discussed and to draw attention to other less obvious uses.

H. Applications of proportional tubes.

I. Form of beta spectra.

30. Soft spectra. The first application of the proportional counter to the study of very low energy radiations was the examination of the form of the beta spectrum of tritium H^3 by Curran et al. [6]. Further work of a similar nature is due to Hanna and Pontecorvo[1]. In addition to giving an accurate spectrum at energies hitherto unexplored the work had a direct bearing on the evaluation of a new upper limit to the rest mass of the neutrino (see Kofoed-Hansen[2]). The excellent agreement between the observations and the theoretical form for an allowed beta transition is illustrated by the straight line plot of Fig. 29 which gives $(N/pWF)^{\frac{1}{2}}$ as a function of the energy of the tritium beta rays. Here N is the number of beta particles per chosen energy interval, p is the momentum of the particles, W the energy and F the Fermi function. A possible discrepancy between theory and experiment at very low energy (<1 kev) was revealed in the earlier experiments but unpublished work of Moljk and Curran shows that even this uncertainty was almost certainly due to the deficiency of the kicksorting equipment in the early work. It is clear that the theoretical and experimental results agree to within about 1% down to less than 500 ev.

This example of the application of the instrument to low energy beta spectra has been multiplied in recent years and in particular the work at very low energies with heavy radioelements has been of value in determining how the energy of rearrangement of the atomic shells was distributed in the act of beta emission. Curran [6] has discussed the data and deduced that the energy is divided between the beta particle and the neutrino and that it must be regarded as part of the total transition energy.

31. High temperature. The very great range of temperature that can now be exploited as a result of the work of Moljk et al.[3] should facilitate the investigation of many elements which cannot be obtained (or which can be obtained only with great difficulty or in some unsuitable form) as a gas or vapour at normal temperatures. The fact that a tube can operate as a proportional counter up to temperatures in excess of 800° C has been demonstrated. As an example of the

---

[1] G. C. Hanna and B. Pontecorvo: Phys. Rev. **75**, 983 (1949).
[2] O. Kofoed-Hansen: Phil. Mag. **42**, 1448 (1951).
[3] A. Moljk, R. W. P. Drever and S. C. Curran: Rev. Sci. Instrum. **26**, 1034 (1955).

real proportionality of a quartz counter (see Fig. 7c) we show in Fig. 30 the observations made at 810° C with such a counter when a trace of A^{37} was added to the gas. The K-capture peak (which is somewhat complex) has a form practically identical with that for the same radiation (energy ~2.8 kev) observed in a similar tube at room temperature. Very little use of this kind of counter at elevated temperatures has as yet been made but it is clearly an extension which should prove very advantageous both in detection and spectrometry. Thus with difficult low intensity materials like Rb^{87} and K^{40} it should prove its value. Likewise where extremely soft radiation is involved, as in the case of Be^7, decaying largely by Auger emission, the fact that the material can be used in vapour form is of major importance. To a great extent of course this last remark applies equally to the sources Rb^{87} and

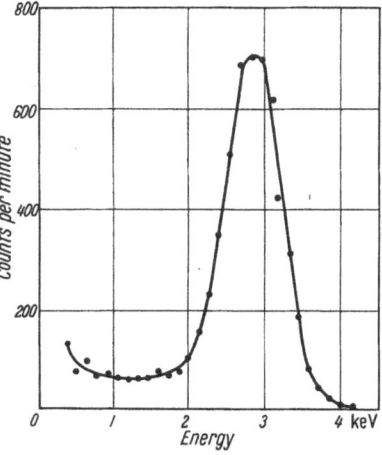

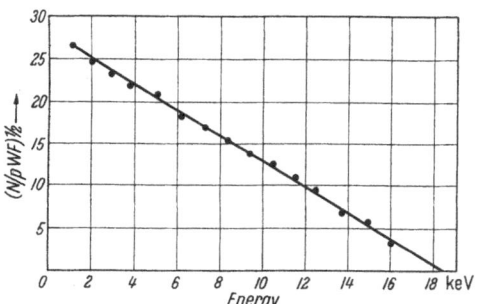

Fig. 29. The spectrum of H^3 showing allowed form to below 1 kev.

Fig. 30. K-peak of A^{37} taken in a quartz counter at 810° C.

K^{40}. In the former there is considerable intensity of beta rays of very low energy, in the latter there is Auger emission. It should be remembered of course that the quartz counter operates from below 200° C to above 800° C. For the full range up to 450° C the somewhat more orthodox tube of Fig. 7b is satisfactory.

32. Spectra of higher energy, and use of high pressure. The application of the method to spectra of higher energy, for sources in gaseous form, is illustrated by reference to the results of MOLJK and CURRAN on C^{14} and S^{35}. Again there is agreement (see Fig. 15) to about 1% between theory and experiment over the energy range from about 3 kev to near the end point (~150 kev). The results on C^{14} are the more interesting in view of the earlier disagreements and the importance of the element on the theoretical side. There is no doubt that up to energies of the order of 200 kev and with the elements which can be made even in small amounts in gaseous form the counter method of examining beta spectra is practicable and accurate. To extend the work appreciably beyond this range the application of suitable magnetic fields is required together with the use of very high pressures. The counter arrangement of Fig. 19d shows how sources have been mounted within rather large cylindrical counters themselves placed in an axial field of several thousand gauss. As an example of this kind of arrangement which was used by COCKROFT and CURRAN we may mention the results on Cs^{137}. Here the spectrum (end-point 610 kev) and the internal conversion electrons (660 kev) were analysed with a source at one end of the tube (Fig. 19d, position C) on the plane perpendicular to the wire. The pressure was seven atmospheres of argon (plus a little methane) and the field was 3000 gauss. In connection with studies at very high pressures the work of FULBRIGHT [6] is particularly notable. Several beta sources of long half-life and relatively high energy (up to about 1.5 Mev) have been studied. Among these radioelements were Be[10], K[40], Y[91], Cs[137], Tl[204] and RaE. Here the sample was mounted in the centre of a very thin aluminium foil (0.2 mg/cm²) held in a metal frame which divided a rectangular cathode cylinder into two equal volumes (1″ × 2.7″ × 6″ long).

The central wire in each of the counting volumes was extremely thin, 1 mil in diameter and they were connected in parallel. This method of employing two proportional counters eliminates to a large degree the distortions of the spectrum that arise in scattering at the source and support. Provided that the energy loss of a beta particle in traversing the supporting foil is not comparable with the energy resolution of the instrument the technique can prove advantageous. As used by FULBRIGHT a second identical unit was mounted alongside the first within a common high pressure container. The second served as a continuous calibrating unit and it was used to observe the internal conversion electrons emitted in the transitions to ground from the metastable state of Ba[137] formed in the decay of Cs[137]. Due to the use of very high pressures (in excess of 2000 lbs per sq.in. on occasions) a high voltage unit giving 20 kv regulated to 0.1% was required. No very special precautions regarding argon purity were necessary but sometimes the calcium purifier failed to clean up the tank supply. To make the gas multiplication less sensitive to voltage changes it was generally advisable to add about 0.5% of methane ot the argon.

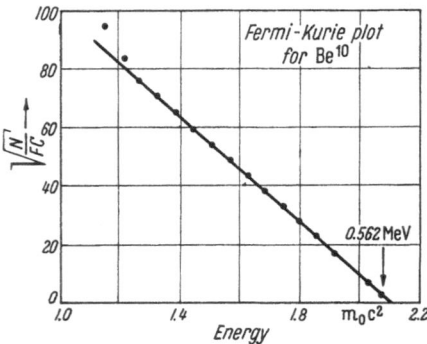

Fig. 31. The plot for Be[10] with the correction $C = C_{2T,2A}$; no.

In practice it was found that a number of factors limited the energy resolution rather severely. Thus the best full width at half-maximum found with Ba[137] was 10%. No doubt the very thin wire was not adequately uniform in cross-section and thicker more uniform wire with higher gas gain would have improved matters. At the low gas gain employed (with the multiplication in the range 2 to 4) the noise contribution was of moment. FULBRIGHT discusses the question of resolution and gain more fully.

An example of the application of the method to the interesting case of Be[10] is shown in Fig. 31. The spectrum was observed with a source too weak for conventional magnetic spectrometry but the agreement of the observations with the theoretical Fermi-Kurie plot for the source, using a $C_{2T,2A}$ correction factor, shows that the spectrum is of the unique D_2 type as predicted by MARSHAK.

II. Study of electron capture, internal conversion, inner bremsstrahlung.

33. Electron capture. Direct methods. The proportional tube is particularly well adapted to the study of the process of electron capture. In the straightforward examination of K-capture isotopes it is important to realise and exploit the features of the tube. Thus WILSON[1] showed for the first time that Ni[59] decayed

---

[1] H. W. WILSON: Phys. Rev. **79**, 1932 (1950); **82**, 548 (1951).

with a long half-life $(8 \times 10^5 \, \mathrm{yr})$ by K-capture while the isotope Ni⁶³ decayed with the emission of soft beta particles[1]. Similar studies of BROWN et al.[2] showed that Ca⁴¹ decayed by K-capture and that the half-life was 1.2×10^5 yr. Such weak K-capture activities are readily studied by the proportional counter. The phenomenon of L capture was first demonstrated with the proportional tube by PONTECORVO et al. and L/K capture ratios can be measured. Moreover the tube offers in principle a very clean method of examining the branching ratio of capture and positron emission as shown by the results of TOWNSEND[3] for Zn⁶⁵ who admitted some vapour of zinc ethyl $Zn(C_2H_5)_2$ containing Zn⁶⁵ to the counter gas. The positrons formed a continuous spectrum on which were superimposed the peaks arising in the K and L capture disintegrations.

Regarding the work of PONTECORVO et al. on L-capture a source of A³⁷ was employed. In a counter filled with argon and methane a peak observed at about 2.8 kev corresponded to the release of the K binding energy of chlorine while a peak near 250 ev was due partly to true L capture and partly to the escape of the K X-rays emitted in K-capture and detection of the associated L radiations. By changing the gas of the counter to xenon the escape of the K radiation was considerably reduced and it was demonstrated that most of the peak at 250 ev corresponded to true L capture. Energy considerations[4] confirmed this conclusion as L capture leads mainly to L_I radiation and K capture and subsequent escape results generally in the expenditure of L_{II} or L_{III} ionization energy in the gas. Their value of the ratio of L to K capture was between 8 and 9%.

34. The study of L and K capture with the multi-wire tube. Very recent experiments by DREVER and MOLJK at Glasgow made use of a new technique based on the form of dating counter described above. This method eliminates much of the uncertainty in certain other procedures.

The K- and L-capture can be detected most easily by the X-rays or Auger electrons emitted in the filling of the vacancies left in the K and L shells. Separate measurements of K and L X-ray or Auger intensities are not themselves sufficient to give a value for the L/K capture ratio since (1) knowledge of the fluorescence yield is required to calculate the total number of K and L transitions. (2) L X-rays or L Auger electrons are emitted with K_α X-rays following a large fraction of the K-capture events. This may make the number of L X-rays observed many times larger than the number of L-captures occurring. This effect can be calculated but the K-fluorescence and the fraction of K X-rays in the K series must be known very accurately.

In the case of Ge⁷¹ LANGEVIN[5] has measured the L and K Auger electrons in a proportional counter filled with propane and a gaseous germanium compound GeH_4. His value of the L/K capture ratio deduced from the measurements was 30% as against a theoretical value of 10%. The experimental value is very critically dependent on the value of the fluorescence yield and this is not known with adequate accuracy so much of the discrepancy could be ascribed to this uncertainty.

In principle the L/K ratio can be measured directly, without making use of the value of the fluorescence yield, if a gas source can be incorporated in the high pressure filling of a large counter. In this way all of the K-capture disintegrations give pulses corresponding to the K-absorption edge whether the decay

[1] H. W. WILSON and S. C. CURRAN: Phil. Mag. **40**, 631 (1949).
[2] F. BROWN, G. C. HANNA and L. YAFFE: Phys. Rev. **84**, 1243 (1951).
[3] J. TOWNSEND: Thesis, Washington Univ., St. Louis 1951.
[4] B. PONTECORVO: Helv. phys. Acta **23**, Suppl. 3, 97 (1950).
[5] M. LANGEVIN: C. R. Acad. Sci., Paris **239**, 1625 (1954).

takes place with X-ray emission, Auger emission, or K_α and L X-rays so that no corrections are required. This method has been used in studying A^{37} but for sources such as Ge^{71} (X-rays of energy 9.2 kev) it is very difficult to operate a sufficiently large counter at the required high pressure. A new form of counter was employed.

The multi-wire counter is shown schematically in cross-section in Fig. 32. Pulses from the central counter pass through an anti-coincidence gate that is closed by pulses from the surrounding counter system which is therefore acting as an anti-coincidence arrangement both for radiations coming from outside and more important from the inner counter into the outer ones. Argon at six atmospheres was used (plus a little methane). The half-thickness of the gas for the X-ray quanta of energy 9.2 kev is 0.8 cm. The gaseous source was $Ge^{71}H_4$.

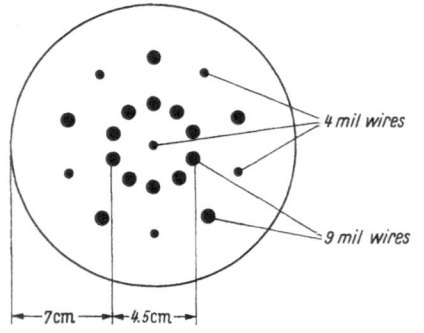

Fig. 32. Multi wire counter assembly for measurement of L/K ratio, used with $Ge^{71}H_4$, operating length of counter between field tubes = 72 cm.

The relative intensities of the K and L peaks for the central counter (with the anti-coincidence circuit in operation) were measured. Neglecting end-effects the possible events when K-capture occurs in the central counter are:

1. K-Auger electron emitted. This event contributes directly to the K-peak.

2. K photon emitted. Two types of event may follow:

(a) If the photon is absorbed in the central counter it contributes to the observed K-peak.

(b) If the photon escapes from the central counter it is absorbed with nearly 100% certainty in the anti-coincidence ring (shortest path across the ring exceeds eight half-lengths). Then any L X-ray or Auger electron emitted simultaneously (by absorption in the central counter) is rejected by the action of the gate circuit so there is no "spurious" L count recorded. This particular K-capture event itself is of course not recorded by virtue of the arrangement but such missing K-capture events in the central counter are almost entirely compensated for by the absorption in the central counter of K photons that are emitted by atoms in the outer ring and escape from the ring. We are making use of the reciprocity of source and detector here. Another way of regarding the system is to consider that for an infinite uniform volume of gas the number of photons absorbed in any given region is equal to the number of photons produced there, and, since the ring is thick as regards absorption, this is true for the central counter. For the compensation to be correct the gate must not be closed by the 1 kev L X-ray pulses occurring in the ring and it is in fact biassed at 3 kev.

Consider events following L-capture in the central counter. Here we have

1. L Auger emission. These events are recorded in the L peak.

2. L photon emission. Most of these are absorbed in the central counter but if any photons escape they are compensated for by those entering from the ring. In these circumstances therefore the experimentally observed L and K capture peaks give exactly the L/K capture ratio when the ring is thick and the counter long. For the counter shown in Fig. 28 the correction factors for the thickness and length were respectively 1 and 5% so uncertainty in the value of the fluorescence yield has a very small effect on the L/K ratio.

35. Fluorescence yield from *K* capture. The same counter system has been used by DREVER to measure the *K* fluorescence yield. The vessel is first used with a methane filling and then with a pressure of four atmospheres of argon (in each case with $Ge^{71}H_4$ present). With the coincidence circuit out of operation the increase in the counting rate of the central counter gives a measure of the fluorescence yield.

The measured value of 13% for the *L/K* is reasonably close to the theoretical one of 10% while the measured fluorescence yield of 51% for Ge is rather close to the presently accepted value of 50%. The success of the proportional counter in this work of DREVER and MOLJK is noteworthy not so much on account of the results (very few radio-active sour-ces can in fact be analysed in the same way) but rather because it illustrates very clearly the power of the tool when its favourable and unfavourable charac-teristics are fully understood and ap-preciated in the planning of the experi-ment.

36. Internal conversion. *L* X-rays were observed by CURRAN *et al.*[1] in the disintegration of RaD. The L_α and L_β peaks corresponded to the radiations of bismuth and were due to the internal con-version of the soft γ-radiation (46 kev) which followed the emission of the soft primary β-particles of RaD (see above).

In the decay of many other isotopes *L* X-rays arising in the internal conversion of γ-radiations have since been detected and in the study of α-sources the soft *L*-radiations have likewise been observ-ed. Thus WEST and DAWSON[2] showed

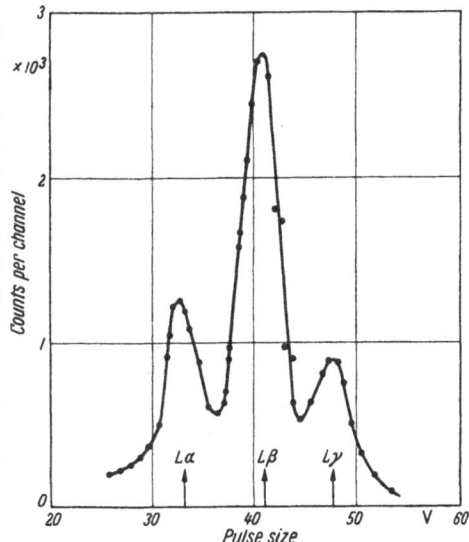

Fig. 33. *L* X-radiations from Pu²³⁹.

that *L*-radiation appeared with an intensity of 4×10^{-2} quanta per α-particle in the decay of Pu²³⁹ and the L_α, L_β and L_γ X-rays were of energies as expected for uranium. The results are shown in Fig. 33 and the soft γ-radiation which gave rise by internal conversion to these *L* peaks was later observed directly. The same phenomenon has appeared with U²³³, Pu²³⁸, Am²⁴¹ and Cm²⁴² and in each case the conclusion is that the α-spectrum exhibits fine structure and the corresponding γ-radiation is internally converted [5].

37. Inner bremsstrahlung. The phenomenon of inner bremsstrahlung has received a considerable amount of study in recent years and the proportional tube lends itself well to the problem. Here we depend to a large extent on the high sensitivity of the spectrometer to the γ-radiation emitted and both theoreti-cally and experimentally this is found to be mainly of relatively low energy. The sources most readily studied are those decaying by pure beta emission, e.g. P³², or by *K*-capture. The source material should preferably be of high specific intensity so that a very thin source giving adequate intensity of γ-radiation can be made. It must of course be mounted on as thin a supporting foil as possible to reduce external conversion of the beta radiation to a minimum. It must be

¹ S. C. CURRAN, A. L. COCKROFT and J. ANGUS: Phil. Mag. **40**, 36 (1949).
² D. WEST and J. K. DAWSON: Proc. Phys. Soc. Lond. A **64**, 586 (1951).

established that any such externally produced radiation is not present to an extent that interferes with the examination of the inner bremsstrahlung. The theoretical treatment of the problem shows that the form of the bremsstrahlung spectrum at low energies is a sensitive function of the parameters and fortunately the proportional tube is best suited to the study of this region.

The interested reader is referred to the original works on this subject[1] [6] but it should be noted that there is on the whole reasonably good agreement between theory and experiment if the experimental difficulties are taken into account.

The phenomenon of inner bremsstrahlung can be explained in terms of the interaction of the field of the nucleus with the beta particle and this leads to the emission of a continuous spectrum. Associated with the continuous spectrum we have radiations characteristic of the daughter atom. The sudden change of nuclear charge accompanying beta decay disturbs the electronic shells and excitation or ionization of the shells may result. In the case of ionization the vacancies in the shell are refilled with the emission of characteristic K and L X-radiation. This inner ionization can be studied with high sensitivity by means of the proportional counter as the radiation is in many cases of low energy. A suitable experimental arrangement for this type of study has been described by RENARD[2] who found for example that the decay of P^{32} was accompanied by the emission of K X-rays of S^{32} of energy 2.3 kev with an intensity of 4.2×10^{-4} photons per beta particle. A number of other sources have been investigated with interesting results. The work can be done effectively with a combination of the technique of scintillation spectrometry and proportional tube counting, the former lending itself to the study of the high energy part of the spectrum of inner bremsstrahlung. This complementary application of the two techniques is illustrated in the work of LANGEVIN-JOLIOT[3].

In connection with K-capture sources it should be remarked that the upper limit of the photon energy spectrum gives a direct measure of the disintegration energy which cannot be found by any other kind of study of the radiation. Usually this energy is rather high to find readily by means of the proportional counter and the scintillation spectrometer is better adapted to the measurement.

III. Measurement of specific ionization and meson studies.

38. The proportional tube has been applied very effectively to study of the specific ionization of charged particles [5]. Even a fast electron at minimum ionization expends sufficient energy in a gas at normal temperature and pressure to give a readily measurable pulse for a track length of a few centimetres. The study of the fluctuations of amplitude of such pulses has been compared with the theory due to LANDAU[4]. A review article by CRANSHAW[5] gives further details of the subject and interesting experimental results have been obtained by ROTH-WELL and by WEST [5]. The specific ionization is important as a means of identifying particles but the large fluctuations in its observed value mean that the identity of a single charged particle cannot be established. Measurements of the specific ionization of μ-mesons by the proportional counter (rectangular in form) were sufficient to establish the logarithmic rise of specific ionization at relativistic energies.

[1] C. S. WU: P. 649 of ref. [6]. An excellent review of the subject of inner bremsstrahlung is given.

[2] G. A. RENARD: J. Phys. Radium **14**, 361 (1953); **16**, 575 (1955).

[3] H. LANGEVIN-JOLIOT: C. R. Acad. Sci., Paris **241**, 286, 1390 (1955).

[4] L. LANDAU: J. Phys. USSR. **8**, 201 (1944).

[5] J. E. CRANSHAW: Progr. Nucl. Phys. **2** (1952).

The applications discussed here may be sufficient to illustrate the versatility of the proportional counter technique and it would be beyond the scope of this article to make a comprehensive list. It is enough to note that fresh applications appear with regularity. We have for example the many forms of counter applied to neutron spectroscopy and the application of the tube to X-ray analysis and to the examination of mesic X-rays are among the many varied uses of the device.

The study of mesic X-rays by WEST and his colleagues is a most interesting example of the adaptability of the proportional tube as a spectrometer. In their early studies WEST, BATCHELOR and BRADLEY[1] applied a special type of tube, filled with xenon to a pressure of seven atmospheres, to the examination of the mesic X-rays produced in carbon by stopping the μ-meson component of cosmic radiation. The case for the use of such detectors as opposed to scintillation devices is made by WEST et al. and the essential features of the counters are described. The use of a special alloy of iron, nickel and cobalt which can be sealed to glass as well as to aluminium (AVES[2]) facilitated the construction of large tubes of aluminium of 1 mm wall thickness. Other points such as uniformity of wire diameter and gas purification were investigated. The experiment was a difficult one due to the low intensity of the cosmic radiation but the method has been applied more recently with much success to the examination of the mesic X-rays produced in light elements by means of beams of μ-mesons created with high energy machines. The energy of the mesic X-radiation even for the light element carbon is 75 kev and the high energies of the radiations make the efficiency of detection low except at high gas pressure and high atomic number, hence leading to the choice of xenon at high pressure. On the other hand the energy resolution of the tube used by WEST was 4.3% half width at half maximum for a quantum energy of 60 kev, a figure which represents a considerable improvement on the performance of a scintillation spectrometer in the same conditions. Increased precision in the energy measurement of the X-radiation corresponds to increased precision in the determination of the mass of the meson, and in the determination of the nuclear coupling constant between nucleons and mesons.

Bibliography.

[1] B. ROSSI and H. STAUB: Ionization Chambers and Counters. New York - Toronto - London: McGraw-Hill 1949.
[2] S. A. KORFF: Electron and Nuclear Counters. New York: Van Nostrand 1946.
[3] D. H. WILKINSON: Ionization Chambers and Counters. Cambridge: Cambridge University Press 1950.
[4] S. C. CURRAN and J. D. CRAGGS: Counting Tubes. London: Butterworths 1950.
[5] D. WEST: Progr. Nucl. Phys. 3 (1953).
[6] Beta and Gamma Ray Spectroscopy, ed. by K. SIEGBAHN. Amsterdam: North Holland Publishing Co. 1955. Section on Proportional Counter Spectroscopy (S. C. CURRAN and H. FULBRIGHT) and various associated subjects.
[7] W. FRANZEN: Theory and Use of Pulse Ionization Chambers. Unpublished Report, Dec. 1951. Princeton University (distributed in mimeographed form).
[8] W. C. ELMORE and M. SANDS: Electronics: Experimental Techniques. National Nuclear Energy Series Div. V, Vol. 1. New York-Toronto-London: Mc.Graw Hill Co. 1949.
[9] A. VAN RENNES: Pulse-Amplitude Analyses in Nuclear Research. Nucleonics 10, 20 (July 1952); 22 (August 1952); 32 (Sept. 1952); 50 (Oct. 1952).
    J. L. W. CHURCHILL and S. C. CURRAN: Advances in Electronics and Electron Physics, Vol. VIII, 1956, contains review article on pulse amplitude analysis pp. 317—362.

---

1 D. WEST, R. BATCHELOR and E. F. BRADLEY: Proc. Phys. Soc. Lond. A 68, 801 (1955).
2 R. AVES: J. Sci. Instrum. 30, 388 (1953).

The Coincidence Method*.

By

S. De Benedetti and R. W. Findley.

With 46 Figures.

I. General and historical introduction.

1. In the study of nuclear radiations and of cosmic rays, it is often of interest to ascertain the time relationship between two or more ionizing events. These events are usually revealed as pulses in radiation detectors and the instrumentation used to determine the time relationship is called a coincidence selector. In the simplest case a coincidence selector gives a signal if, and only if, the detectors respond simultaneously, within the characteristic resolving time of the instrument.

The credit for the introduction and the early development of the coincidence method goes to BOTHE and his collaborators. Their work proved the great power of the method by obtaining the solution of two problems of fundamental importance. In 1925, BOTHE and GEIGER[1] settled the discussions on the nature of the Compton effect by proving the simultaneity of the Compton electron and of the scattered γ-ray; and in 1929, BOTHE and KOLHÖRSTER[2] established the existence of penetrating ionizing particles in the cosmic radiation. In the first of these experiments the detectors were the now almost forgotten point-counters, while in the second the newly discovered Geiger counters were used. The coincidences were detected with a photographic technique: the pulses of the two counters produced the deflection of electrometers and these were photographically recorded, side by side, in a moving film. The simultaneity could be established within 10^{-2} or 10^{-3} sec.

To BOTHE[3] also goes the credit of introducing the first electronic circuit capable of automatically selecting the coincidences and of recording their number. His circuit consisted of a tetrode, whose grids were connected to positive pulses from the counters: only when both grids received a pulse did the plate current flow. This circuit had a resolving time of 1.4×10^{-3} sec. It can be classified as a "series" coincidence circuit, since it operates essentially as two switches in series which have to be simultaneously closed to allow the passage of current.

An important step in the art of coincidence selection is due to ROSSI[4] who, in 1930, introduced a much more flexible circuit, of the "parallel" type. This operates as a number of switches in parallel which have to be all simultaneously open to stop the flow of current. It was realized by connecting in parallel a number of vacuum tubes, with a common plate resistor. Each tube received negative pulses from one of the counters, and the current in the common plate resistor was interrupted only when all tubes were switched off simultaneously.

* Work supported by the United States Atomic Energy Commission.

[1] W. BOTHE and H. GEIGER: Z. Physik **32**, 639 (1925).
[2] W. BOTHE and W. KOLHÖRSTER: Z. Physik **56**, 752 (1929).
[3] W. BOTHE: Z. Physik **59**, 1 (1929).
[4] B. ROSSI: Nature, Lond. **125**, 636 (1930).

Both the series and parallel circuits are still in use and their resolving time can be made as small as 10^{-9} sec. Preference is given to parallel type circuits, which are more easily extended to the selection of coincidences between a large number of counters.

For about 20 years after the introduction of the coincidence method Geiger counters were the best detectors suitable for coincidence work. The resolving time was at first limited by the slow speed of non-self-quenching counters, but soon electronic quenching circuits were devised in order to improve the counter's speed. Noticeable progress was made with the introduction of self-quenching counters, and the coincidence circuits were improved until the limit in the resolution was the one imposed by the speed of response of the counters themselves. This ultimate limit is due to the time of collection of the electrons which have to reach the proximity of the counter's wire before initiating the discharge. Depending on the counter's wire and gas filling the collection time is of the order of 10^{-6} or 10^{-7} sec.

With the introduction of scintillation counting[1,2] the coincidence technique could be much improved and its usefulness was greatly increased. The advantage of the scintillation counters over gas discharge counters is threefold: higher speed, greater efficiency for γ-rays, and energy-proportional response.

The fastest scintillants have a mean life of light emission of a few 10^{-9} sec, and the electronics was soon improved to take advantage of the newly available detection speed. Circuits resolving 10^{-9} sec were developed. In principle it would be possible to obtain higher time resolution by using the very first photon emitted in large scintillations. But, as we will see, there are difficulties of practical nature which prevent the extension of the method to the resolution of times smaller than 10^{-10} sec: among these difficulties are the time of flight of light over lengths comparable to the dimensions of the scintillating sample, and the straggling in the collection time of electrons in the photomultipliers used for the detection of the light pulse.

To enumerate the successful applications of the coincidence technique would be almost equivalent to describing the progress in nuclear physics and cosmic rays in the last 30 years. Nevertheless, it seems worthwhile to mention some of the most remarkable contributions of the coincidence method, chiefly when they are associated with modifications and improvements of the method itself.

One can perhaps divide the applications of the coincidence methods into two classes: those in which the simultaneity of the pulse of the detectors reveals the passage of the same particle through the counters; and those in which different ionizing particles are revealed. To the first class belong the counter "telescopes" which have been used to such great advantage in the study of cosmic rays and of high energy particles emerging from accelerating machines. In the second class one should consider an almost endless variety of arrangements for the study of elementary radiation processes (such as the already mentioned Compton effect), for the observation of secondary particles in the cosmic radiation, for the investigation of radioactive decay schemes, angular correlations, etc.

After the initial work on the Compton effect and on the corpuscular nature of the cosmic radiation the coincidence technique has led to the discovery of the secondary processes of cosmic rays[3]; the coincidences from a telescope were then

[1] H. KALLMANN: Natur und Technik.
[2] I. BROSEX and H. KALLMANN: Z. Naturforsch. 2a, 439 (1947).
[3] B. ROSSI: Phys. Z. **33**, 304 (1932).

used to trigger a cloud chamber[1] and as a result cosmic ray showers were found[2]; an improvement in resolving time permitted the observation of the large atmospheric showers[3]. The main contributions of coincidence techniques in radioactivity have been in the study of decay schemes and in the assignment of quantum numbers of nuclear excited states through angular correlations and polarization measurements. Worthy of mention is the attempt to detect double β-decay[4]. The investigation of electromagnetic interactions has continued after the initial work on the Compton effect. It was proved that the Compton electron is simultaneous with the scattered X-ray within 1.5×10^{-8} sec[5]. The double Compton effect has been detected[6]. The opposite γ-rays of positron annihilation have been studied[7] and their polarization determined[8]. The annihilation in three quanta has been pointed out[9]. The technique has received innumerable applications in connection with accelerators, chiefly when the energy of the radiation is high enough to facilitate the use of counter telescopes. After many scattering experiments performed with coincident counters, the last achievement of the method has been the discovery of the negative proton[10].

It seems appropriate to mention at this point two modifications of the coincidence method which have also had wide applications. The first of these is the anticoincidence technique and consists of the automatic selection of those events which produce the simultaneous discharge of a certain number of counters (in coincidence) provided other counters (in anti-coincidence) are not discharged at the same time. The anti-coincidence technique has been used mainly in connection with counter telescopes. One obvious application is the selection of "single" penetrating cosmic rays, with the exclusion of showers. Another is the selection of particles at the end of their range. For this purpose the coincidence telescope is followed by an absorber after which anti-coincidence counters are located. If these counters efficiently cover the solid angle of the telescope, an anti-coincidence count occurs only when the particle traversing the telescope stops in the absorber. Another application of the anti-coincidence method is the construction of "cosmic ray umbrellas", which reduce the cosmic ray background in a counter used for the observation of a rare event. Anti-coincident cosmic ray protection is often used in connection with radio carbon dating.

Finally, one should describe the method of delayed coincidences. A delayed coincidence instrument responds to events from two counters when these events are separated for a time D, within the error due to the experimental resolving time.

A coincidence circuit is a particular case of a delayed coincidence one, corresponding to $D=0$. Any coincidence selector can be transformed into a selector of delayed coincidences by inserting a circuit element which delays the pulse from one of the counters. One of the most interesting applications of delayed coincidences is the measurement of short mean lives: As an example one can recall the measurement of the μ-meson mean life in cosmic rays, the measurement of the mean life of π and K mesons produced in accelerators, and, in the field

[1] P. M. S. Blackett and G. P. S. Occhialini: Nature, Lond. 130, 363 (1932).
[2] P. M. S. Blackett and G. P. S. Occhialini: Proc. Roy. Soc. Lond. 139, 699 (1933).
[3] P. Auger and R. Maze: C. R. Acad. Sci., Paris 207, 228 (1938).
[4] E. L. Fireman: Phys. Rev. 75, 323 (1949).
[5] W. G. Cross and N. F. Ramsey: Phys. Rev. 80, 929 (1930).
[6] P. E. Cavanagh: Phys. Rev. 87, 1131 (1952).
[7] R. Beringer and C. G. Montgomery: Phys. Rev. 61, 222 (1942).
[8] E. Bleuler and H. L. Bradt: Phys. Rev. 73, 1398 (1948).
[9] S. De Benedetti and R. T. Siegel: Phys. Rev. 85, 371 (1952).
[10] O. Chamberlain, E. Segrè, C. Wiegand and T. Ypsilantis: Phys. Rev. 100, 947 (1955).

of radioactivity, the determination of mean life of α and γ emitters in the micro-second and millimicrosecond region (see Sect. 10). The discovery of the atom of positronium[1,2] is another achievement of the delayed coincidence method.

Delayed coincidences are also used for the selection of particles traversing a telescope according to their time of flight. If the particles have previously been magnetically selected, they have constant momentum and a measurement of time of flight is equivalent to a measurement of mass.

The achievements of the coincidence technique are many and of great importance, but the achievements must not make us forget the limitations. The most fundamental discoveries (such as the Compton effect itself, or the discovery of new particles), are seldom made with counters. Counters in coincidences, even if used in large numbers, do not provide the direct, detailed evidence which is required to prove convincingly an unexpected fact. A single cloud chamber or photo-plate track is self-explanatory while the occurrence of even a large number of coincidences has to be interpreted. It is true that the neutral π meson and the negative proton were discovered with coincidence apparatus: but, possibly, this is because their properties were anticipated by theories, and one knew what to look for.

Counters and coincidence measurements are of great value for quantitative measurements. If one wants to study the frequency of occurrence of a certain effect, the coincidence technique provides an objective, reliable way of selecting and counting the chosen events.

II. Coincidence circuits of moderate resolutions ($\gtrsim 1$ μsec).

2. The plate coupled Rossi stage. In this chapter, we will describe the developments in the art of coincidence selection before the introduction of scintillation counters. A maximum speed of 10^{-7} sec was usable in the coincidence selecting apparatus, and this speed was reached around 1940, thanks to the development in vacuum tube design.

Most of the circuits used are of the parallel type first introduced by Rossi and it seems appropriate to start with a description of this circuit and of some of its modifications.

The circuit of Rossi[3] consists of two triodes whose cathodes are grounded and whose anodes are connected to a positive "battery" voltage V_B through a common plate load resistance R_L (Fig. 1). The grids of the two tubes are also grounded and both tubes conduct in the quiescent condition. Each grid is connected to a counter from which it receives negative pulses. We will suppose that these are large enough to cut off the current in the tube for a time δ.

If the common load resistance R_L is much larger than the internal plate resistance of the tubes R_p, the tubes act as switches which ground the plate, unless both open. When both tubes are cut off the voltage at the plate jumps from 0 to the supply battery value V_B.

A somewhat more precise understanding of the static behavior of the circuit is gained from Fig. 2. In this figure, the plate characteristics of the triodes are shown, together with two load lines: one for load R_L, the other for load $2R_L$. In quiescent conditions the grid voltage is near to zero; the tubes conduct as if each had a load resistance $2R_L$: the quiescent condition is represented by the point A, and V_A is the voltage at the plates. If now one of the tubes receives a

[1] M. Deutsch: Phys. Rev. **82**, 455 (1951).
[2] M. Deutsch: Phys. Rev. **83**, 866 (1951).
[3] B. Rossi: Nature, Lond. **125**, 636 (1930).

pulse and its current is cut off, the second tube carries all the current traversing the plate resistance and operates with load R_L. The operating point moves to S,

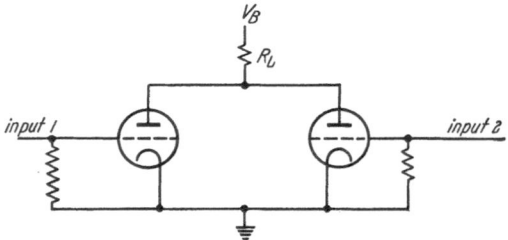

Fig. 1. Schematic diagram of the Rossi circuit.

and the plate voltage is V_s. If instead, both tubes receive a pulse, the operating point moves along the line of load $2R_L$, until the plate voltage reaches the "battery" value V_B.

For small R_L (Fig. 3 a) coincident signals produce plate pulses which are twice as large as those produced by single pulses. A large R_L (Fig. 3 b) is needed for large differentiation between coincidences and single counts. Pentodes are more suitable than triodes for high coincidence differentiation as shown in Fig. 4.

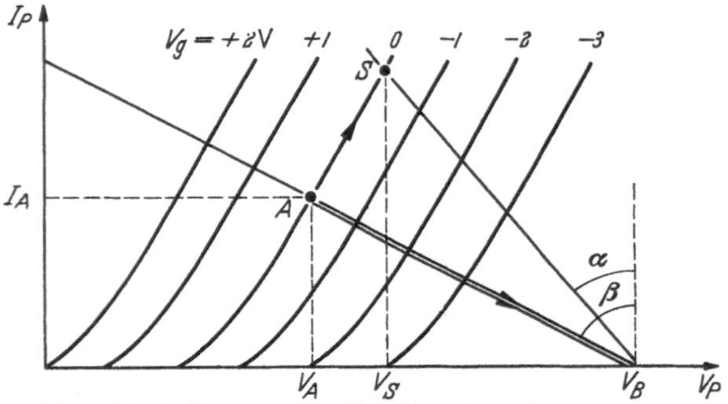

Up to this point, we have considered only equilibrium conditions and our conclusions are valid if the duration of the pulses is long compared to the plate time constant. Wave forms corresponding to this assumption are shown in Fig. 5 a.

Fig. 2. The double arrow $(A \to V_B)$ shows the motion of the operating point when equal negative pulses reach the grid. The single arrow $(A \to S)$ describes the effects of a single pulse. $\alpha = \arctan R_L$, $\beta = \arctan 2R_L$.

In the initial work on coincidence selection it was customary to use large plate resistances ($\approx 10^5$ or 10^6 ohm), in order to obtain a large differentiation between coincident and single pulses. It was soon realized, however, that a large plate resistance corresponds to a long time constant of the plate circuit, and that the transient behavior of the circuit must be considered.

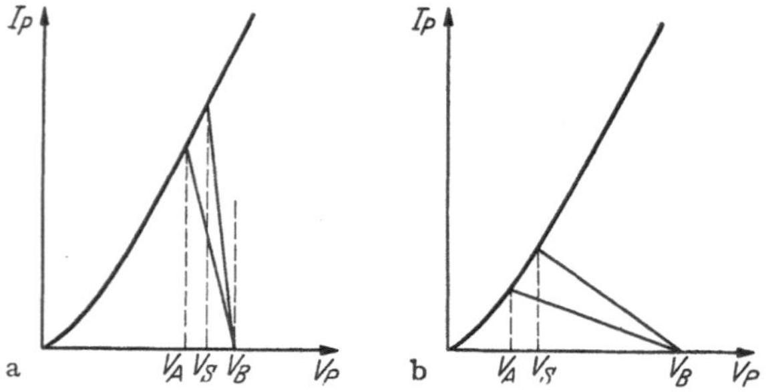

For this purpose, let us call C_p the capacitance of the plate circuit, including both tube plates and associated wiring. Then, if one of the tubes is cut off (single pulse), the plate time constant is

Fig. 3 a and b. Operation of Rossi circuit for (a) small, R_L, and (b) large, R_L.

$$T_s = \frac{R_L R_p C_p}{R_L + R_p}$$

where $R_p = \partial V_p / \partial I_p$ is the internal plate resistance of the tube; while if both tubes are cut off (coincident pulses) the plate time constant is

$$T_c = R_L C_p.$$

Thus the height of a plate signal produced by a single pulse of duration δ_s is

$$H_s = (V_s - V_A)\left(1 - e^{-\frac{\delta_s}{T_s}}\right)$$

and that of coincident grid pulses of duration δ_c is

$$H_c = (V_B - V_A)\left(1 - e^{-\frac{\delta_c}{T_c}}\right).$$

If the grids receive signals of duration δ, but delayed of a time $\varepsilon < \delta$ (so that they are coincident for a time $\delta - \varepsilon$), the plate pulse reaches the height

$$
\left.
\begin{aligned}
H(\delta, \varepsilon) &= (V_s - V_A)\left(1 - e^{-\frac{\varepsilon}{T_s}}\right) + \\
&+ \left[V_B - (V_s - V_A)\left(1 - e^{-\frac{\varepsilon}{T_s}}\right)\right] \times \\
&\times \left(1 - e^{-\frac{\delta - \varepsilon}{T_c}}\right).
\end{aligned}
\right\} \quad (2.1)
$$

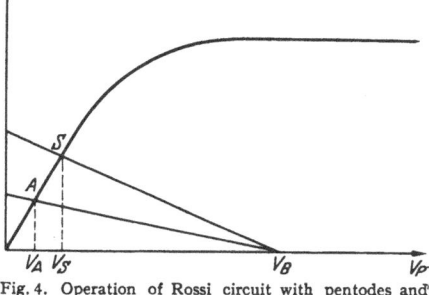

Fig. 4. Operation of Rossi circuit with pentodes and large R_L.

The wave form is shown in Fig. 5b.

Let us now introduce the concept of *resolving time*. If the pulses at the grid have constant duration δ, the resolving time τ is defined as the minimum time interval between the beginnings of pulses at the two inputs which is required for them to be resolved as non-coincident. For short plate time constants $\tau = \delta$. For long plate time constants τ depends on the circuit parameters in a somewhat complicated manner. If the coincidence stage is immediately followed by a fast pulse

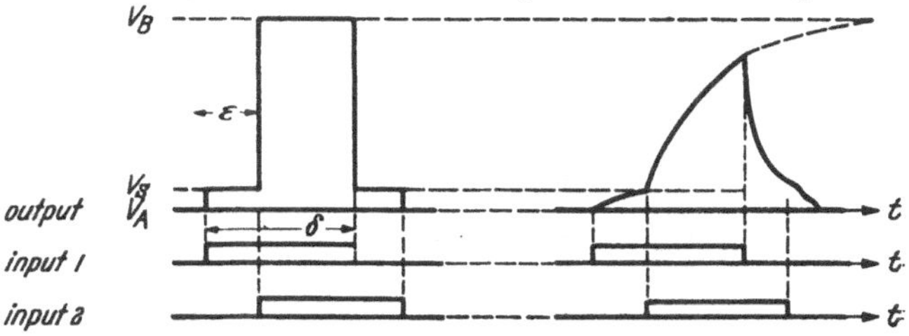

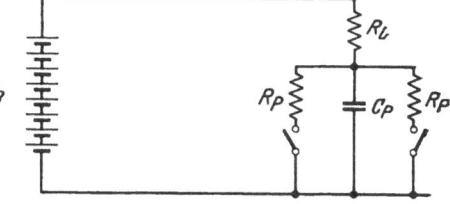

Fig. 5. Output wave forms for partially overlapping signals for (a) short and (b) long plate time constants.

height discriminator which transmits pulses above a certain level H_0, the resolving time is the value of ε which is a solution of the equation

$$H(\varepsilon) = H_0$$

where H is the expression given in (2.1), in which all symbols other than ε are to be considered as parameters.

If the grid pulses are not constant in duration, the resolving time must be averaged over the distribution of pulse durations. Let us suppose, for example,

Fig. 6. Equivalent scheme for a Rossi circuit.

that the pulses in input 1 have duration δ_1, and those in input 2 have duration δ_2. One must then define two resolving times τ_{12} and τ_{21}, according to whether the pulse in 1 precedes or follows the pulse in 2. The overall resolving time of the instrument is then defined as the average

$$\tau = \tfrac{1}{2}(\tau_{12} + \tau_{21}).$$

The case of pulses of various durations in each channel will be discussed in a later section (Sect. 14).

Though the behavior of a coincidence circuit can be predicted only after consulting the characteristic curves of the vacuum tubes, some insight into its properties can be gained by studying the simplified diagram of Fig. 6[1]: This is equivalent to replacing the plate characteristic curve with a straight line passing through the origin with slope $1/R_p$. Introducing the dimensionless parameter $\varrho = R_L/R_p$ and expressing all times τ, δ, T_s, T_c, in units of $R_p C_p$, one has:

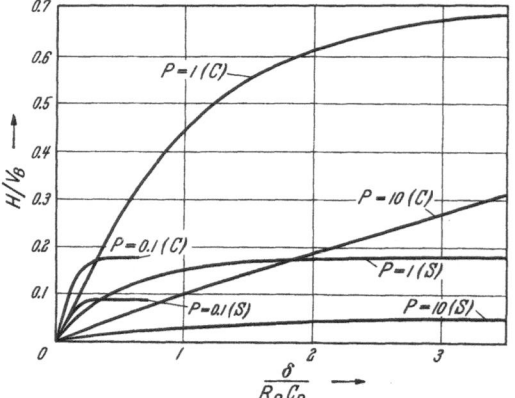

$$\frac{H_s}{V_B} = \frac{\varrho/2}{(\varrho + 1)(\varrho + \frac{1}{2})}\left(1 - e^{-\frac{(\varrho+1)}{\varrho}\delta}\right),$$

$$\frac{H_c}{V_B} = \frac{\varrho}{\varrho + \frac{1}{2}}\left(1 - e^{-\frac{\delta}{\varrho}}\right)$$

where H_s is the pulse height due to a single pulse and H_c the pulse height due to coincident pulses. These functions are plotted in Fig. 7.

Fig. 7. The coincidence (c) and single (s) pulse height (in units of V_B) as a function of the pulse duration δ (in units of $R_p C_p$) for different values of the parameter $\varrho = R_L/R_P$.

From these considerations it appears that, in designing a circuit, one has to compromise between speed and differentiation. When speed is of no interest long pulses can be used, and it might be advisable to choose a pentode and a large plate resistance to obtain a maximum difference between the effects of a single and coincident pulses. But for shorter pulses a smaller plate resistance becomes desirable and a pentode is less advantageous than a triode since it would operate in the horizontal region of its plate characteristic.

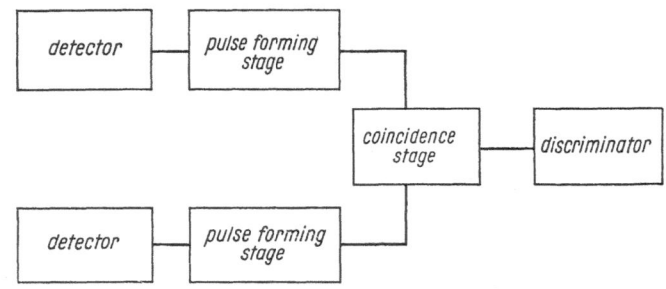

Fig. 8. Block diagram of coincidence circuit with accessory stages.

3. Auxiliary circuits. In the preceding discussion we have supposed that the pulses fed to the coincidence stage are of definite, uniform duration and of sufficient height to cut off the vacuum tubes. We have also assumed that there is a definite pulse height H_0 under which the output pulses are rejected. In order to approximate in practice this state of affairs, a modern coincidence circuit includes a pulse forming stage and a pulse height selector. As a result, the block diagram of the circuit looks as shown in Fig. 8.

The pulse forming stage prepares equalized pulses of sufficient height simultaneous with the event to be timed. Since the collection time of the gas discharge counters produces a "jitter" of 10^{-7} or 10^{-6} sec which impairs the simul-

[1] De Horan, Huijer and Jouker: Physica, Haag **21**, 565 (1955).

taneity between the event and the signal, there is no need of extreme speed in the pulse forming stage: the coincidence stage would lose efficiency if the pulses were shaped with durations shorter than 10^{-6} sec.

The pulse forming stage usually consists of a univibrator (such as the one formed by tubes V_2 and V_3 of Fig. 18) though some experimenters prefer to use a blocking oscillator (Fig. 9). The univibrator has the advantage of being able to produce nearly square pulses whose duration is easily controlled by varying the coupling condenser.

It is advisable to have the coincidence stage followed immediately by a pulse height selector, fast compared to the rise time of the coincident pulses. In the absence of a pulse height selector the resolving time depends on the sensitivity of the subsequent recording stages in a manner which may be difficult to anticipate or to control. If a fast pulse height selector is included in the coincidence stage, the recording stages (scalers, etc.) need only to be fast compared to the average time separation between coincidences.

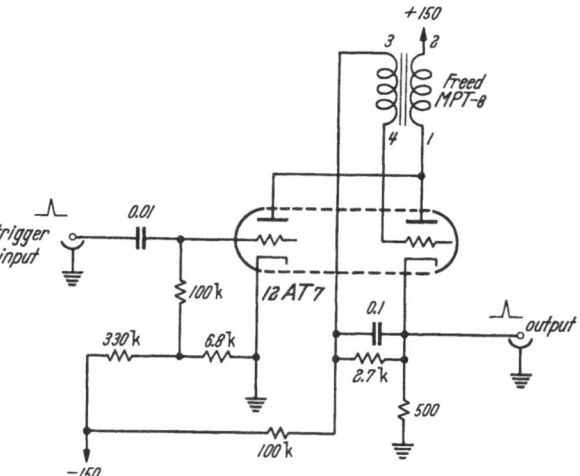

Fig. 9. Blocking oscillator.

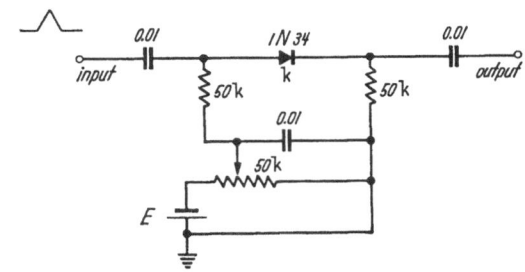

Fig. 10. Biased diode pulse height selector.

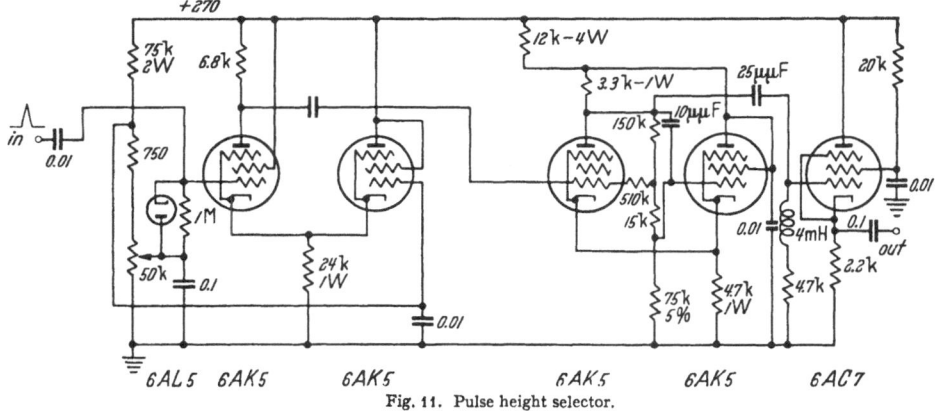

Fig. 11. Pulse height selector.

The simplest pulse height selector can be made with a biased rectifying diode (Fig. 10). A more complex circuit showing a "long tail pair" pulse height selector, followed by a pulse forming stage, is shown in Fig. 11. Fig. 12 shows the Schmitt trigger circuit, which embodies most of the desirable properties of a simple amplitude discriminator. This circuit[1] is a bistable trigger unit, involving two pentodes, whose triggering level can be adjusted over a wide range by means of a potentiometer. All pulses above the required pre-set level, adjusted

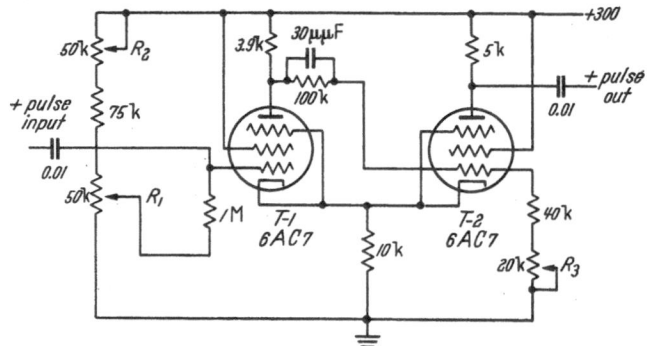

Fig. 12. Schmitt trigger circuit.

by means of a dial calibrated in volts, will produce an output pulse of constant amplitude; by using a regulated power supply stability of the triggering threshold can be held to less than one volt for 100 volts total range.

4. Modifications and applications of the Rossi stage. Several modifications of the original Rossi circuit have been introduced.

In the so called "cathode coupled" Rossi stage the tubes are connected in parallel with a common cathode resistance and a positive grid bias. One obtains in this manner the circuit of Fig. 13a, whose transient behavior can be approximately discussed from the diagram of Fig. 13b.

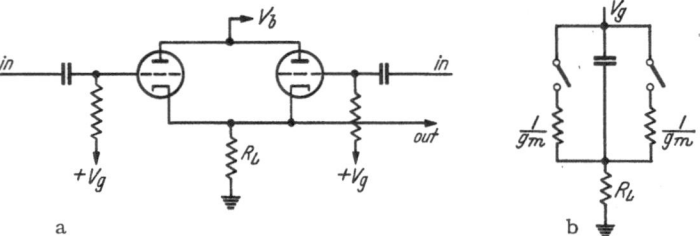

Fig. 13a and b. Cathode loaded coincidence circuit.

The advantage of cathode coupling over the original plate coupling is a gain in speed since the plate resistance R_p is now substituted by the much smaller $1/g_m$. A disadvantage is that the cathode coupled circuit provides no amplification and requires larger imput pulses.

It is also possible to make parallel coincidence selectors by using diodes, and this type of circuit has been widely used after the introduction of crystal diodes. The simplified diagram is shown in Fig. 14. If pulses of a few volts are applied the diodes act as switches and thus perform the same function as the triodes of the original unit. This circuit has the advantage of the very small capacitance of the crystal diodes and the disadvantage of no amplification.

---

[1] W. C. Elmore and M. Sands: Electronics; Experimental Technique, p. 99. New York-Toronto-London: McGraw-Hill Book Co., Inc. 1949.

The output pulse is never larger than the smaller of the two input pulses. If one wants fast operation and good coincidence to single ratio one needs $R_L C$ small and $r \ll R$. Thus, for optimum performance the circuit has a small input impedance, a not too important disadvantage for fast pulse work where large impedances are nowhere tolerated. Since the parallel diode circuit is capable of great speeds we will discuss it in greater detail in connection with its use with scintillation counters (Sect. 9).

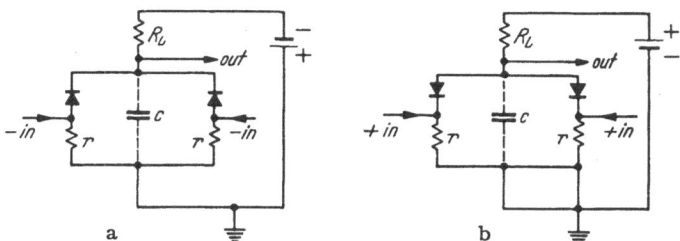

Fig. 14 a and b. Parallel diode coincidence circuit (a) for negative input, (b) for positive input.

All the circuits mentioned above can be easily extended to the selection of multiple coincidences. A discussion of multiple coincidence circuits, would be an obvious generalization of what has been said about double circuits and will not be presented here in detail. Fig. 15 shows two kinds of quadruple circuit operation: Operation (a) (large load resistance) is useful when one wants to accept quadruple coincidences and discard those of multiplicity smaller than 4; a cir-

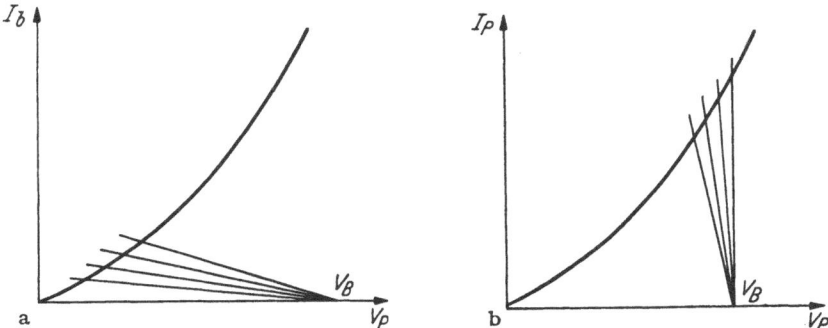

Fig. 15 a and b. Load lines for large and small load resistances in a multiple coincidence circuit.

cuit operating like (b) (small load resistance) can be called an additive mixing stage since it gives an output proportional to the number of counters which have simultaneously discharged. A circuit of this second kind has been used for the determination of particle density in cosmic ray showers. Most often operation (a) is wanted but the requirement of speed may impose the use of a small plate resistance and the less desirable operation (b) is obtained.

5. Anticoincidence and delayed coincidence circuits. Using the circuit elements described in the preceding section it is easy to make anticoincidence or delayed coincidence circuits which are completely satisfactory in the microsecond region.

In order to obtain a reliable anticoincidence circuit one must make sure that the pulses from the anticoincidence counters cancel completely those from the coincidence counters, when produced by simultaneous events. Due to the finite collection time of the counters, and possibly to other reasons, the anticoincidence pulse (counter 2) may occur later than the other (counter 1) and thus be unable to cancel completely the unwanted pulse. In order to prevent this from happening, pulse 1 is first delayed for a time D (see Fig. 16), which must be larger than the counter collection time ($D \approx 0.5\ \mu$sec for large counters); then it is made to produce a negative square pulse of duration δ_1. The anticoincidence counter produces an immediate positive pulse of duration $\delta_2 > \delta_1 + D$. In this way pulse 1

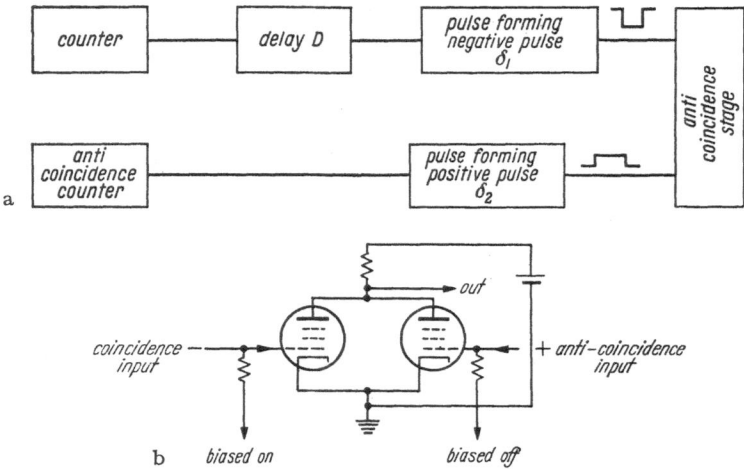

Fig. 16a and b. (a) Block diagram of anticoincidence circuit and (b) schematic diagram of anti-coincidence stage.

is always contained within pulse 2 and can be completely canceled. The delay D can easily be obtained using a delay line of large (≈ 1000 ohm) characteristic impedance.

The anticoincidence stage [1, 2] consists of a Rossi circuit where one of the tubes is biased below cut-off. Its mechanism of operation is clear from Fig. 16b.

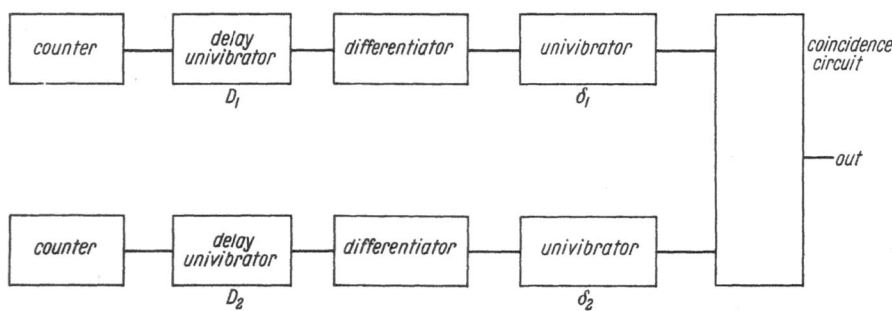

Fig. 17. Delayed coincidence circuit.

Let us now pass to delayed coincidence circuits. Fig. 17 shows the block diagram of a coincidence circuit where delay times and resolving times can be varied by controlling the time constant of univibrators. The delay times are produced by delay univibrators which generate square pulses (for instance positive) of duration D_1 and D_2. These pulses are differentiated in an RC circuit producing sharp signals of opposite sign at the beginning and at the end of the square pulses. The pulses (negative in our example) occurring after time

[1] J. C. Street and E. C. Stevenson: Phys. Rev. **52**, 1003 (1937).
[2] G. Herzog: Rev. Sci. Instrum. **11**, 84 (1940).

D_1 in the first channel and D_2 in the second, start other univibrators whose outputs are fed to the coincidence stage. If these operate with time constants δ_1 and δ_2 a coincidence occurs only if the difference in time Δt between counter discharges satisfies the inequality

$$D_2 - D_1 - \delta_1 \leq \Delta t \leq D_2 - D_1 + \delta_2.$$

A more complete diagram for a three-channel delayed coincidence circuit is shown in Fig. 18. In this circuit the three pulse inputs are fed into cathode

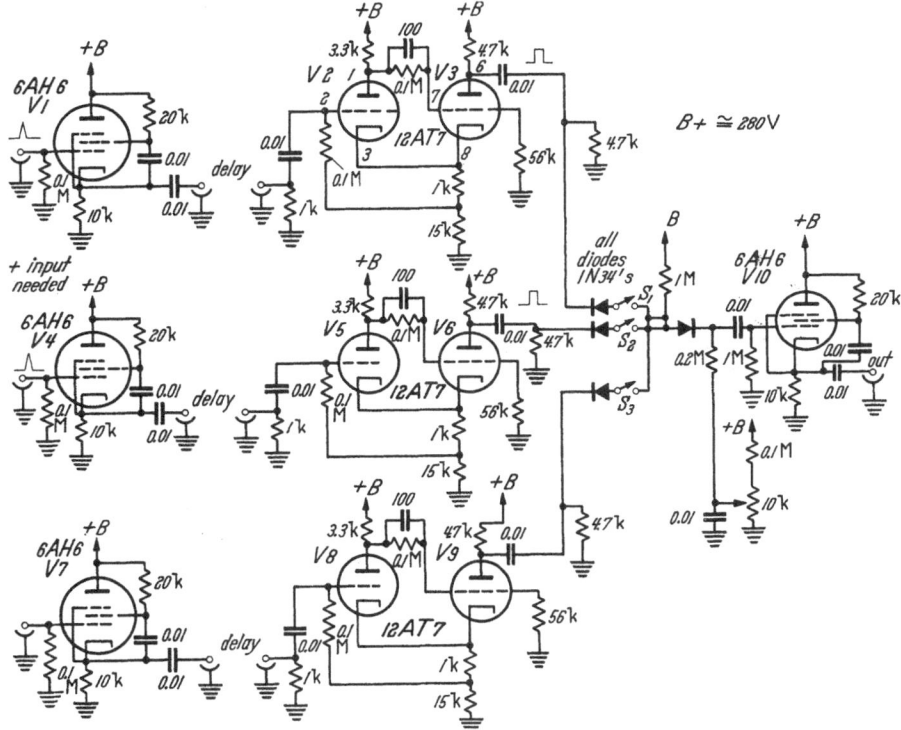

Fig. 18. Three-channel coincidence circuit.

followers which drive delay lines or delay cables of the appropriate lengths. The outputs of these are used to trigger identical univibrators which supply shaped pulses to a diode coincidence circuit. Disabling switches are provided to allow any one of the channels to be disconnected, making the circuit into a two-channel coincidence device. The output pulse from the coincidence diodes is lengthened by a fourth diode, which also acts as a discriminator of controllable bias. The resulting pulse is applied to a cathode follower which furnishes the output. Such a circuit can be made quite reliable, with the advantage that pulses of different amplitudes can be used at the inputs without affecting the action of the coincidence diodes. The channel delays can be quickly and easily changed by inserting lengths of cable into the proper circuit.

The advantage of using delay univibrators, as opposed to delay cables or delay lines is that much longer delays can conveniently be obtained without the attenuation and distortion which results from long delay lines. Furthermore the univibrator is a high impedance input device, and thus is easily driven. On the other hand, the delay cable remains more satisfactory for small time delays.

Multiple delay discriminators capable of measuring the time interval between two events can also be built. The circuit of Fig. 19[1] generates voltage pulses whose amplitudes depend on the time interval between the coincidences of two Geiger counters and the pulses of a third counter.

In the quiescent state V_2 is cut-off while V_1, V_3 and V_4 conduct. The point P is near to ground. When inputs A_1 and A_2 receive simultaneous negative pulses, V_3 and V_4 are cut-off and the voltage at point P starts to rise while the voltage at D is prevented from rising by diode V_5. The voltage at P is approximately

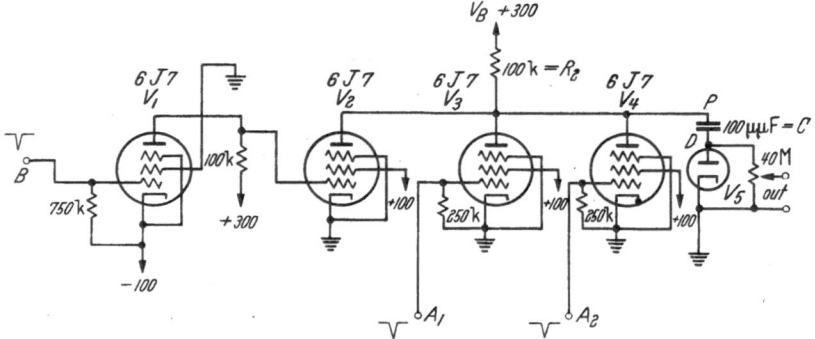

Fig. 19. Multiple delay discriminator.

represented by $V_p = V_B \left(1 - e^{-\frac{t}{R_2 C}}\right)$. If a negative pulse at input B occurs at a later time T_1 the tube V_2 becomes conducting and V_p suddenly decreases to near zero. This potential decrease is fed through the capacitor C to point D, and to the output; the magnitude of the pulse obtained is a measure of the interval between the coincidence of A_1 and A_2 and the negative signal at B.

6. Gating circuits. A coincidence circuit is usually designed to select pulses according to their relationship in time, but does not aim to conserve the other information carried with them, such as pulse height. In some cases, however, it is

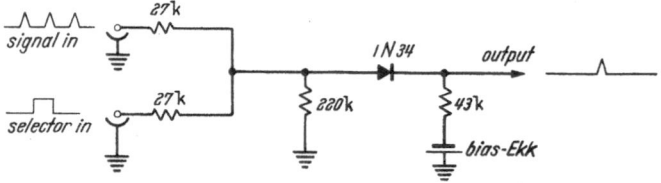

Fig. 20. Basic gating circuit.

of importance to measure the height of a pulse, whenever it is coincident with some other signal: for this purpose, one uses a circuit called a "gate" which blocks the pulses to be studied unless coincident with the auxiliary selector signal. If a coincidence occurs the pulses are transmitted, as far as possible without distortion, to a pulse height analyzer.

A simple example of such a gating circuit is shown in Fig. 20. The selector pulse must be carefully shaped to have the same amplitude as the bias voltage E_{KK}, which must be made as large as the largest signal to be rejected. In this circuit, the selector pulse raises the anode of the diode just to the level of conduction, and the coincident signal pulse, if there be one, is passed through the diode undistorted. Signal pulses not coincident with a selector pulse are rejected.

[1] B. Rossi and N. Nereson: Rev. Sci. Instrum. **17**, 65 (1956).

If the bias is slightly smaller than the selector pulse, a portion of this will be present in the output signal: a coincident signal pulse will appear on top of

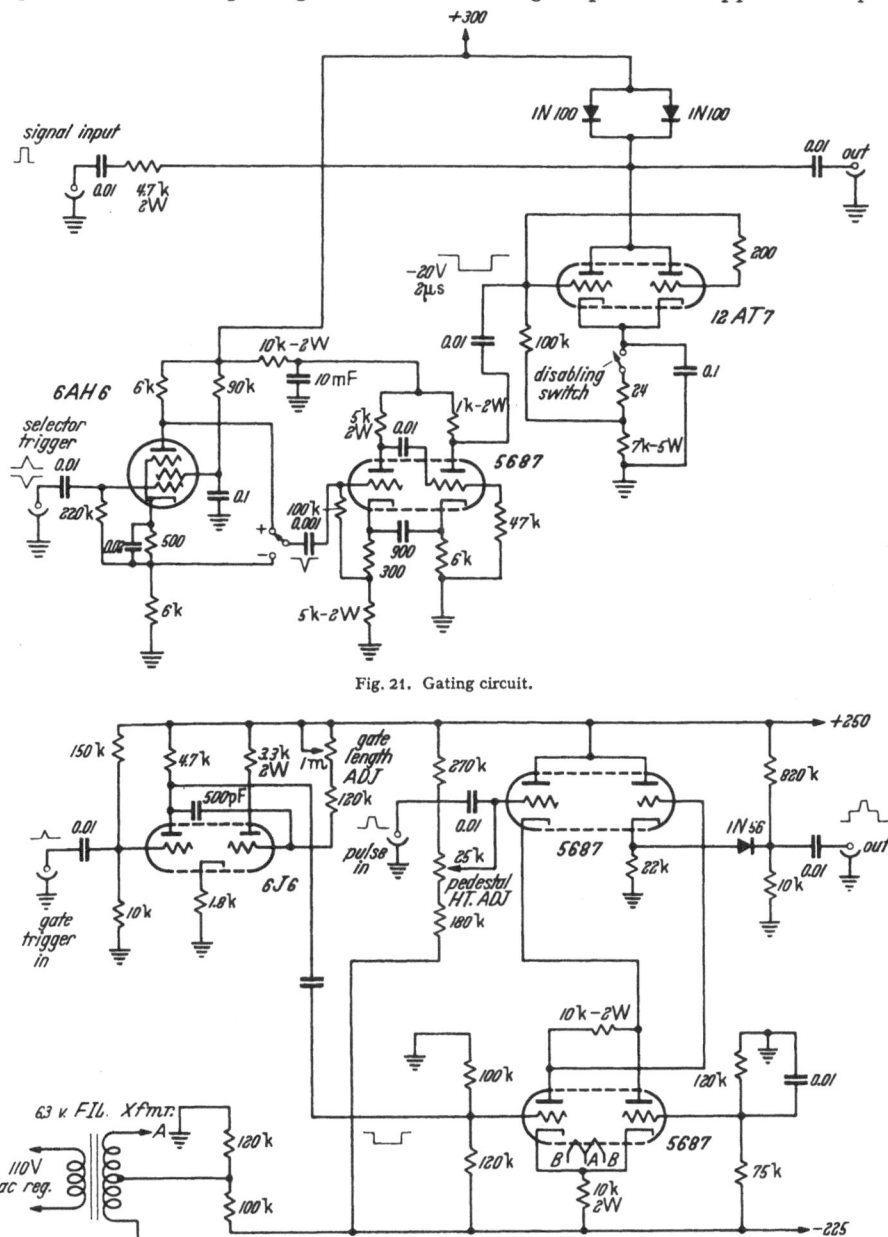

Fig. 21. Gating circuit.

Fig. 22. Linear gating circuit.

the "pedestal" produced by the selector. The existance of such a pedestal is undesirable, particularly if the output is to be used for pulse height selection. For this reason, it is preferable to operate with carefully controlled amplitudes of bias and selector pulse.

A more complex and versatile circuit is shown in Fig. 21. In this circuit, which is designed to be used ahead of a 24 channel pulse height selector, pulses are fed through a 4.7 K resistor to the junction of the two clamping diodes (1 N 100) with the gating tube (12 AT 7). Since the gating tube, in series with the diodes, is normally conducting heavily, the junction point rests at a voltage which is lower, by a volt or two, than the B supply voltage. Any positive input pulse, then is short-circuited to ground by the low impedance of the clamp diodes and the gating tube. However, if the gating tube is turned off by a selector pulse (or by opening the cathode circuit with the disabling switch) the clamping diodes stop conducting and their cathodes rise to the B supply voltage producing a slight pedestal. In these conditions the positive pulses are free to proceed to the output reasonably unchanged in size and shape. The selector pulse is formed by the 5687 univibrator which can be triggered by either a positive or negative pulse. The undesirable pedestal can be kept to a minimum by choosing diodes with very low forward drop.

Fig. 22 shows another gating circuit which has recently been described[1]. In this circuit, which is supposed to be very linear, the selector pulse is made to adjust the grid voltage of a cathode follower through which the signal pulse must travel. A pedestal height of about 3 volts occurs with this circuit.

III. Fast coincidence circuits.

7. Scintillation detectors for fast coincidences. The technique of scintillation counting is at least as old as nuclear physics since the existence of the nucleus itself was established by the scattering of α particles, visually observed as scintillations on a ZnS screen. The use of the technique was limited by the fact that phosphors such as ZnS are not transparent to their own light and can be used only for radiations losing a great amount of energy on a thin surface layer. After the introduction of the gas discharge counters, the method of scintillation counting was virtually abandoned, but the situation has been almost reversed with the discovery of transparent scintillators.

The characteristics of scintillating substances will be given in detail in another part of this volume. The data are usually expressed in terms of ratios with the properties of anthracene, a substance which is, unfortunately, a poor standard of comparison since its scintillating properties vary in a critical manner with the purity of the sample. Taking a somewhat arbitrary average of measurements from different authors one can assume that anthracene emits of the order of 1 photon per 50 ev of energy loss, in the case of weakly ionizing rays; for heavily ionizing radiation, such as α-particles, roughly $1/10$ as much light per ev is obtained. The emission of the photons follows an exponential dependence on time according to a mean life of 3.2×10^{-8} sec.

Other substances are considerably faster than anthracene. For instance a solution of 5 grams of p-terphenyl in a liter of toluene emits 35% of the integrated light output of anthracene with a mean life of 2.2×10^{-9} sec. This corresponds to an initial light output of 3000 photons per mμsec per Mev.

The scintillation light is viewed with a photo-multiplier tube. In particularly favorable conditions it may be possible to collect all the light on the multiplier photo-cathode, and to obtain 1 electron per 10 photons collected. Thus—in the case considered above—the photo-cathode current has an initial value of 300 electrons per mμsec per Mev, or 0.05 μAmp per Mev, and decays with a 2.2×10^{-9} sec time constant.

---

[1] G. S. Stanford and G. F. Pieper: Rev. Sci. Instrum. **26**, 847 (1955).

This current is affected by statistical fluctuations, which are important if the total number of photo-electrons—and thus the energy lost by the radiation—is small. Furthermore, the time spread in the arrival of the photons at the photocathode may be sensibly affected by their different time of flight, if the dimensions of the scintillation and light collecting equipment are of the order of 10 cm or more.

In an average photo-multiplier the current pulse is amplified by a factor $\approx 10^6$. Thus the initial current is ≈ 0.05 A/Mev and the charge deposited by a pulse of 2×10^{-9} sec is $\approx 10^{-10}$ coulomb/Mev. The anode capacitance $\approx 10^{-11}$ farad is charged to ≈ 10 volt/Mev if a large resistance is used. If one wants to maintain a time constant $\approx 10^{-9}$ sec the anode must be grounded with a resistance of ≈ 100 ohms and one expects a pulse ≈ 5 volt/Mev.

These pulse heights are not always achieved, and, if precautions are not taken one obtains about one tenth of this amount. Among the many reasons which may decrease the voltage pulses are the incomplete light collection (particularly important for photo-multipliers having a photocathode recessed deep inside the glass envelope, as the 1 P 21), and the time straggling which decreases the initial current pulse. On the other hand, some experimenters have been able to obtain much larger pulses: for instance BELL et al.[1] succeed in obtaining 3 volt—on a large resistance—per each photo-electron generated at the photo-cathode; with 100% collection efficiency this would correspond to about 3 volt/kev. This remarkable result is obtained by selecting the multipliers capable of standing large overvoltages and having multiplication factors $\approx 10^9$.

The ultimate limit of resolving time obtainable with scintillation counters can be derived from the following considerations due to BELL et al.[1]. The first electrical signal from the event to be timed is the first photo-electron produced at the photo-cathode. Assuming that no significant uncertainties are introduced by the light time of flight nor by the electron multiplication process, there still remains the uncertainty due to the statistical nature of the process of photon and photo-electron emission. The statistical fluctuations have been discussed by POST and SCHIFF[2] who conclude that the first moment about the mean of the time of appearance of the first Q of N photo-electrons is

$$\bar{t} = \frac{Q\tau}{N} \left\{ 1 + \frac{Q+1}{2N} + \cdots \right\}$$

where τ is the mean life of the scintillation. Now, if a 90% efficient coincidence circuit is wanted one must wait until each counter has had a 95% chance of producing at least one photo-electron, and thus a time $3\bar{t}$. Fig. 23 shows the ultimate resolving time obtainable for 90% efficiency as a function of radiation energy (proportional to N) for $Q=1$ (first photo-electron) and for various scintillators.

The circuitry described in Chap. II, is not adequate to take full advantage of the scintillation counter speed. Though the conventional Rossi circuit with sufficiently small resistance will give a ratio 2 between coincident and single pulses of any speed, it can operate reliably under these conditions only if the pulses are large and of equal size and duration. The pulses of the scintillation counters instead, are often small and of variable amplitude.

Triggered circuits generating uniform pulses of speed comparable to the response of best scintillators cannot be built at the present state of the art. Amplifiers can be used in order to increase the pulse height, but only at the price

[1] R. E. BELL, R. L. GRAHAM and H. E. PETCH: Canad. J. Phys. **30**, 35 (1952).
[2] R. F. POST and L. I. SCHIFF: Phys. Rev. **80**, 1113 (1950).

of some loss in speed since the fastest amplifiers available commercially have a rise time of 3×10^{-9} sec. And, at best, amplification may accomplish the purpose of obtaining pulse heights sufficient to cut off the vacuum tubes of a conventional coincidence circuit, but does not equalize the time during which the tubes are cut off.

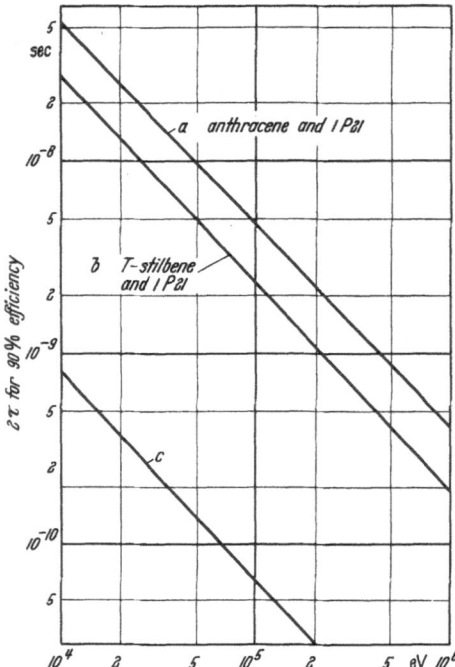

Fig. 23. Minimum resolving time for 90% efficiency of coincidence apparatus. Curves (a) and (b) after Bell *et al.* Curve (c) refers to ideal conditions of light collection and *p*-terphenyl in toluene scintillant.

Only if the pulses at the photomultiplier plate are several volts high, they can be made uniform by equalizing their height with a vacuum tube limiter and their duration with shaping by delay line reflection.

Before selecting the circuit most suited to one's work it may be appropriate to experiment with scintillants and photomultipliers and estimate the pulse height available. Coincidence circuits using vacuum tubes require a pulse height of several volts. If this is not available at the photo-multiplier plate, diode coincidence circuits are advisable since these operate with lower level pulses.

8. Fast coincidence circuits using vacuum tubes. A single modification of the Rossi circuit, capable of operating directly from photo-multipliers with resolving times of 10^{-8} sec has been described by Garwin[1]. This circuit (Fig. 24) requires pulses of about 3 volts, since they must be able to cut off the vacuum tubes; however, the unwanted effects of the variable pulse duration are eliminated through the use of a diode clamp which

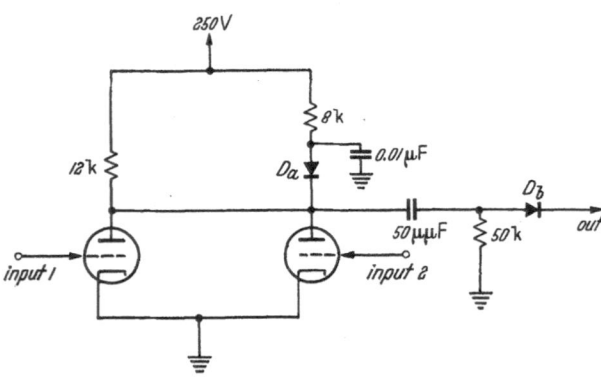

Fig. 24. Garwin coincidence circuit.

prevents the growth of the plate signal produced by long single pulses. The clamping action is exerted by diode D_a, which conducts in quiescent conditions, and thus effectively ties the plates through its small forward resistance (≈ 20 ohm) to a 0.01 μF condenser. In this manner the tubes operate with a 20 ohm plate resistance connected to a voltage slightly higher than the plate quiescent voltage V_a, and the single pulses produce negligible plate voltage variation. But when a coincidence turns off both triodes, the current stops flowing through the diode, and its resistance becomes very high: the entire plate

[1] R. L. Garwin: Rev. Sci. Instrum. **21**, 569 (1950).

voltage and a 12 K load are made available for the coincidence. A second diode, D_b, stretches the plate pulse to make it of a duration comparable to the rise time of the following amplifiers.

In a later paper GARWIN[1] describes a modifications of the same circuit (Fig. 25) in which the diode D_b is biased and operates as a discriminator. The schematic

diagram of a multichannel coincidence—anti-coincidence instrument based on the unit of Fig. 25 is given in the same paper. This is said to work with 1.5 volt pulses and to have a resolving time of 3×10^{-9} sec.

A coincidence circuit of the "series" type (Fig. 26) has been designed by FISCHER and MARSHALL[2] using a 6BN6 gated-beam tube. This tube, originally developed for use as a frequency modulation discriminator, consists of a cathode, a box-like accelerator grid, two control grids (called limiter and quadrature grids) and a plate. The accelerator grid is normally positive and

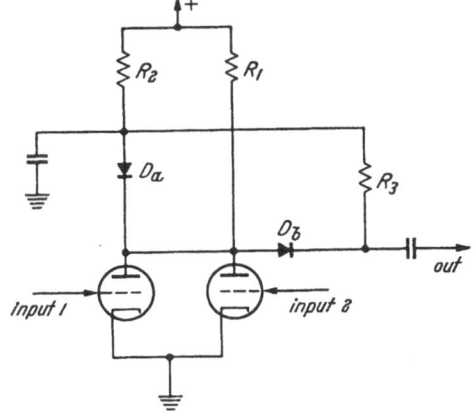

Fig. 25. Garwin circuit with biased diode.

electrons leaving the cathode pass through this and the other grids in a vertical sheet, or beam. Due to the arrangement of the electrodes in the tube, the cathode is shielded from the fields of the control grids whose function is to shift the current from the accelerator to the plate without varying the cathode current. A change

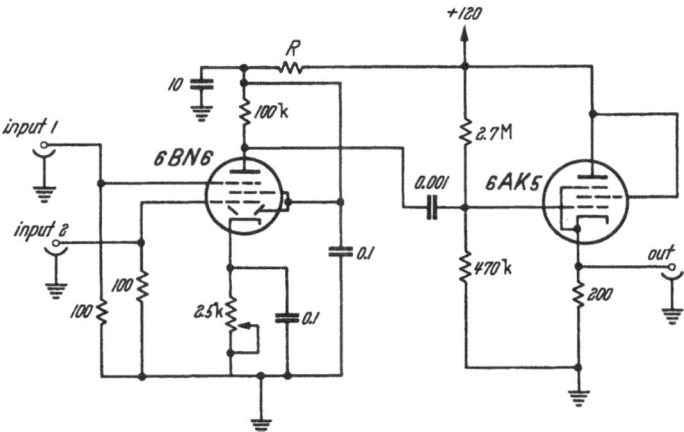

Fig. 26. Gated beam tube coincidence circuit.

of only a few volts in the potential of either control grid will vary the plate current from cut-off to a saturation value depending on available cathode current. The grids act like gates, allowing the flow of plate current only if they are both above cut-off.

In the circuit of FISCHER and MARSHALL the 6BN6 is operated at cut-off by the self-bias produced by the variable cathode resistor, since the cathode

[1] R. L. GARWIN: Rev. Sci. Instrum. **24**, 618 (1953).

[2] J. FISCHER and J. MARSHALL: Rev. Sci. Instrum. **23**, 417 (1952).

current flows to the accelerator. Positive pulses are fed to both grids. When a coincidence occurs, the tube conducts, and the resulting plate pulse is applied to the output through a cathode follower. The resolving time is better than 3×10^{-10} sec for signals from a mercury switch generator.

9. Fast coincidence circuits using crystal diodes as switches. Low level coincidence circuits for scintillation counters using crystal diodes have been first reported by Baldinger et al.[1], and by Morton and Robinson[2]. In 1950, Elmore[3] discussed a particularly simple parallel circuit (Fig. 27) which has been since developed into a practical high speed instrument.

The photo-multipliers anodes A_1 and A_2 are kept slightly positive through the current flowing in the $1\,M$ resistance. Under these conditions the crystal diodes

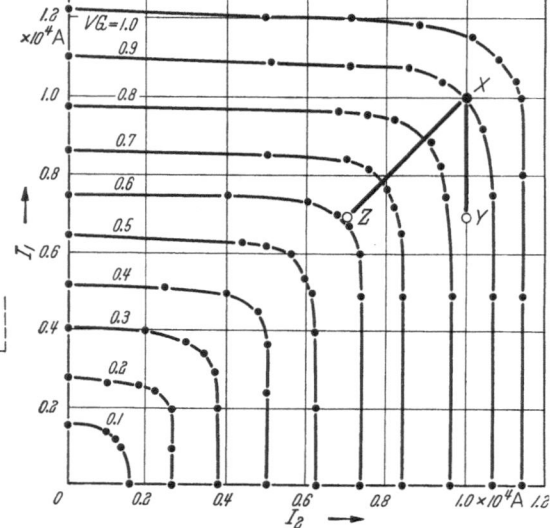

Fig. 27. Low level coincidence circuit.   Fig. 28. Static bias curves for the coincidence circuit.

conduct and the grid of the output amplifier is positive. A small ($\approx 10^{-1}$volt) negative pulse at one of the multiplier plates cuts off the corresponding diode, producing a small change in V_{out}. But, if both plates become negative, both diodes are interrupted and V_{out} reaches the value of the smaller of the two input pulse.

Quantitatively, the static behavior of the circuit can be understood from the curves of Fig. 28. It appears from these that, with a quiescent output voltage of 0.9 volt, a single pulse of 0.03 mA (0.3 volt) or larger changes V_{out} by 0.07 volt while two simultaneous pulses of the same height produce a change of 0.29 volt.

The ability of this circuit to discriminate between coincidence and single low level signals depends on the sharpness of the elbows in the curves of Fig. 28 which is in turn related to the voltage required to cut off the diodes.

Elmore's[3] circuit was designed for use with a NaI scintillator, which has a time constant of light emission of 0.25 μsec, and thus the resistances were chosen to give a time constant of $\approx 10^{-7}$ sec. With slight modifications, however, the circuit can be used to resolve times of the order of 10^{-9} sec.

In the circuit of De Benedetti and Richings[4] (Fig. 29) care is taken to reduce the input time constants, and a third diode is added at the circuit output.

[1] E. Baldinger, P. Huber and K. P. Mayer: Rev. Sci. Instrum. **19**, 473 (1948).
[2] G. A. Morton and K. W. Robinson: Nucleonics **4**, No. 2, 25 (1949).
[3] W. C. Elmore: Rev. Sci. Instrum. **21**, 649 (1950).
[4] S. De Benedetti and H. Richings: Rev. Sci. Instrum. **23**, 27 (1952).

The multiplier's plates are grounded with resistances of ≈ 100 ohm: pulse rise times of $\approx 10^{-9}$ sec are thus obtainable if the multiplier produces a steep current front. In order to reduce the duration of the pulse due to the exponential decay of the scintillation light, the multiplier current is sent to ground by means of a cable whose characteristic impedence equals the plate resistance. In this manner the pulses arrive at the diodes (B_1 and B_2) after the times required for the propagation in cables $A_1 B_1$ and $A_2 B_2$, whose length can be varied if delayed coincidences are wanted. At a later time the reflection from the grounded ends of the cables G_1 and G_2 reach B_1 and B_2; these reflections are of a sign opposite to

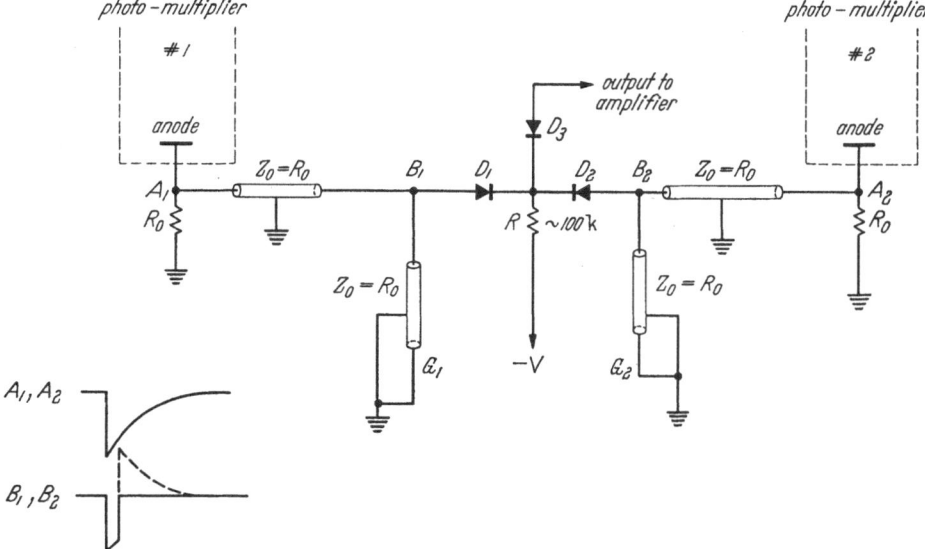

Fig. 29. Fast coincidence circuit of DE BENEDETTI and RICHINGS.

the original pulse which is effectively cancelled as soon as the reflected pulses arrive. If the pulse has a sharp front and a long decay its duration is limited to twice the time of propagation in $B_1 G_1$ and $B_2 G_2$.

The bias current for the diodes is provided through the common resistance R, connected to a negative voltage V. If, to simplify the discussion, the diodes have forward resistance $r \ll R$, and an infinite back resistance, the static voltage at C is $V r/2R$. The parameters are chosen to make this voltage equal to a few tenths of a volt. When a negative pulse (0.1 volt or larger) interrupts the current in one of the diodes, the voltage rises with a time constant rC (C is the stray capacitance at point C) towards a maximum value $V r/R$. If both diodes are cut off by simultaneous pulses the voltage at C starts rising with time constant RC towards the value V: the rise stops when one of the pulses ends, or when C has reached the voltage of the smaller pulses. A single pulse rises as

$$\frac{V r}{2R}\left(1 - e^{-\frac{t}{rC}}\right) \approx \frac{V}{2RC} t + \cdots$$

and a coincidence as

$$V\left(1 - e^{-\frac{t}{RC}}\right) \approx \frac{V}{RC} t + \cdots.$$

Though the initial size of the voltage at C is only a factor 2 greater for coincidences than for single counts, if the pulses last longer than rC ($\approx 10^{-10}$ sec) and are higher than $\frac{Vr}{2R}$ (≈ 0.1 volt), coincident pulses give much larger signals than single ones.

The third diode D_3 is useful because of its non-linearity and of its rectifying action. The non-linearity increases the ratio between coincidences and single pulses. The rectifying action lengthens the pulses to a time comparable to the rise time of the subsequent amplifier (no leak resistance, other than the back resistance of the diode itself needs to be provided) and helps in the discriminating between coincidence and "near misses". In the absence of the lengthening diode two pulses following each other within the amplifier rise time would add, and the selection of fast low level coincidences would be impossible since these are only twice as large as single counts.

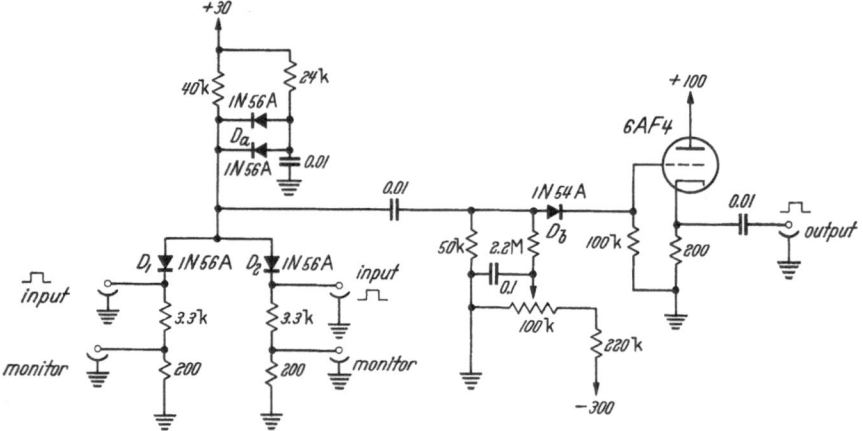

Fig. 30. Crystal diode double-coincidence circuit for positive input pulses.

The circuit has resolving times of the order of 10^{-9} sec and can be operated with pulses of the order of 0.1 volt. It is easily modified to work with positive pulses by inverting the direction of the diodes and the sign of the current through them. The cut-off of the diodes effectively equalizes the single pulse height to about 0.1 volt, but, if large pulses are available, one can insert limiter inverter triodes between the multipliers and the diode.

Further studies and modifications of the parallel diode coincidence circuit have been discussed by Madey[1] who describes among others a circuit for positive pulses of a few mμsec resolving time. Here (Fig. 30) diodes D_1 and D_2 work essentially as in the previously described circuit, while D_a and D_c have a function similar to the diodes in Garwin's circuit. The diode coincidence circuit can be readily adapted to multiple coincidence and anti-coincidences work. The diagram of a multi-channel circuit of this kind which has been used for years by the meson scattering group at Carnegie Institute of Technology is shown in Fig. 31. For this circuit the pulses are pre-amplified, since an input of about 1 volt is needed, and the resolving time is $\approx 10^{-8}$ sec.

10. The circuit of Bell, Graham and Petch. Probably the most successful double coincidence circuit as far as speed and reliability is the one developed at the Chalk River Laboratory by R. E. Bell and his collaborators[2]. This circuit is really an adding circuit (in the sense that the coincident pulses are twice as

[1] R. Madey: Rev. Sci. Instrum. **26**, 971 (1955).

[2] R. E. Bell, R. L. Graham and H. E. Petch: Canad. J. Phys. **30**, 35 (1952).

large as the single ones), followed by a non-linear element to increase the double
to single ratio. It owes its success to a pulse forming circuit which produces
pulses constant in height and duration and simultaneous with the arrival at the
multiplier plate of the first photo-electrons produced at the photo-cathode.

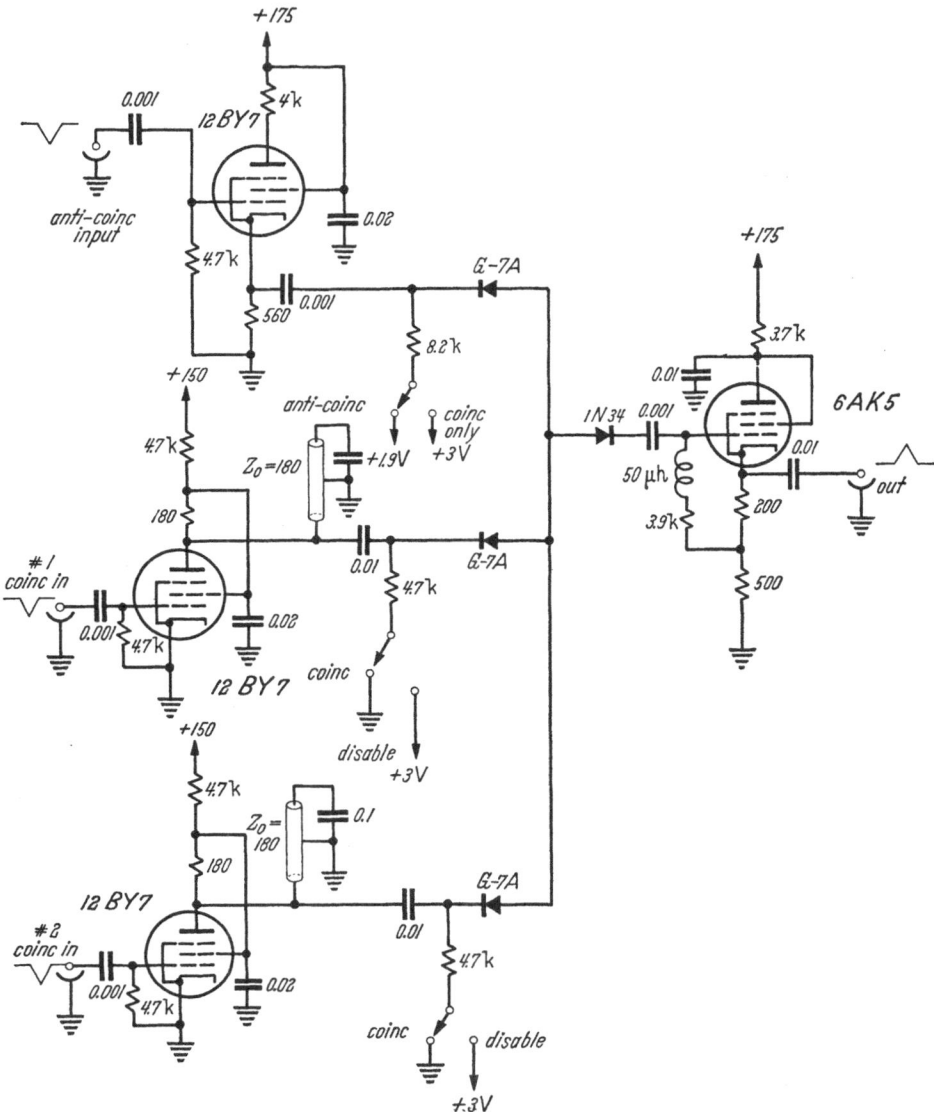

Fig. 31. Multiple coincidence anti-coincidence schematic diagram.

The Chalk River investigators were the first to succeed in operating a 1 P 21
photomultiplier at a voltage of 2400 volt or higher, obtaining gains of at least
2×10^8. Under these conditions each photo-electron ejected from the photocathode
produces at the anode (grounded with a relatively large resistance) a pulse of
3 volts or more, sufficient to cut-off a vacuum tube.

16*

Fig. 32 shows the diagram of the circuit used to select the coincidences between these pulses. The negative photo-multiplier pulses are applied to the grids of tubes V_1 and V_2 which conduct 10 mA under quiescent conditions. The tubes are suddenly cut-off by the pulses due to the first photo-electron and stay cut-off for a long time, depending on the total number of photo-electrons and on the input time constant. The fast rising pulses appearing at the plates of V_1 and V_2 are led into 100 ohm cables whose lengths can be varied if one wants to introduce delays. The end of these cables are joined, and connected to a 50 ohm cable, shorted at the opposite end. Since the impedance at the point where the cables are joined is 25 ohm, a single pulse produces a 0.25 volt signal. This is shaped by its own reflection to a length equal to twice the time length of the

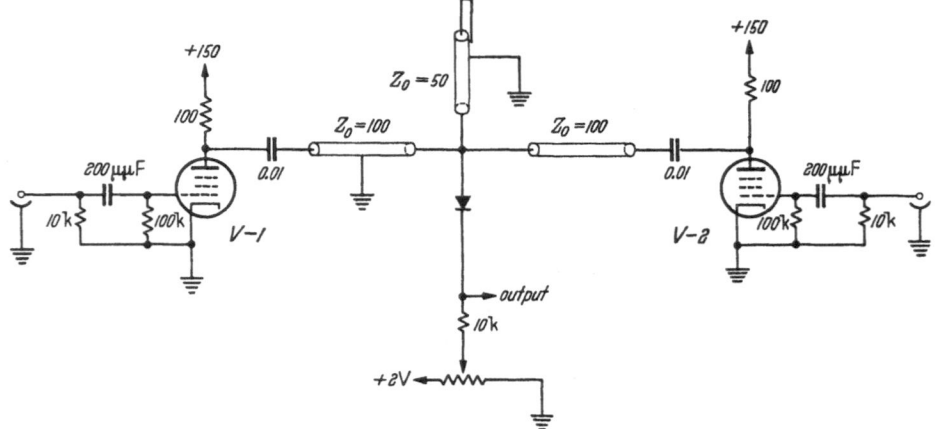

Fig. 32. Coincidence circuit of Bell et al.

50 ohm cable. Further reflections are prevented, since the shaped pulse now sees proper termination. The pulses are thus made uniform in size and duration.

A diode, connected to the junction point, is biased to discriminate against the unwanted 0.25 volt pulse, but to accept the larger signals produced by coincidences. The pulses accepted are stretched and fed to a conventional amplifier.

Some of the tests and work done with this instrument will be reported in Chap. IV. Depending on the length of the shorted cable one obtains resolving times from 1 to 3 mμsec.

11. Miscellaneous fast coincidence circuits. An early use of crystal diodes in millimicroseconds coincidence work is described by Dicke[1] who made use of a tuned radio receiver as detector of the coincident pulses.

The coincidence circuit of Dicke uses the non-linearity of a crystal diode, in conjunction with transmission lines, in an unusual way. As can be seen in Fig. 33 the lines L_1 and L_2—as well as L_3 and L_4—are different in length by $\lambda_0/2$, where λ_0 is the wave length corresponding to the frequence f_0 to which the receiver is tuned. The resistors R_1 and R_2 are chosen so that lines L_3 and L_4 are correctly terminated, and pulses travelling from L_3 through R_1, R_2 to L_4 are severely attenuated.

If a single pulse is fed into either No. 1 or 2 inputs, that pulse, on arriving at A is fed through both R_1 and R_2 to lines L_3 and L_4. At B_1 two pulses arrive displaced by $\dfrac{1}{2f_0}$ second, and the components of frequency f_0 are cancelled, producing no output in the tuned receiver.

[1] R. H. Dicke: Rev. Sci. Instrum. **18**, 907 (1947).

If pulses are fed in coincidence to the two inputs, it can be seen that the first pulse from input No. 2 arrives at B at the same time as the second pulse from input No. 1. Thus a chain of three pulses is produced at B for each coincidence event. The center pulse of these three is larger, and due to the non-linear characteristic of the crystal diode, the frequency components of the center pulse cause an output in the receiver. Consequently, coincidence pulses are the only ones detected except for secondary effects. These effects, such as the internal reflection of the pulses and the difficulty of getting equalized pulses at the inputs, make the operation of the circuit other than ideal for use with scintillation counters.

Another type of coincidence selector is based on a bridge type circuit making use of crystal diodes. Several authors[1-6] have described a variety of ingenious

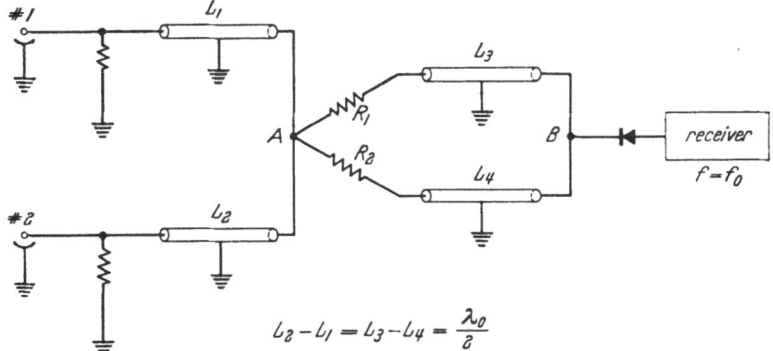

Fig. 33. Coincidence circuit of DICKE.

instruments of this type, but these have not yet received wide application, and the interested reader is referred to the original literature.

12. Auxiliary fast circuitry. The design of fast pulse-forming and discriminating stages presents difficulties which have not been entirely overcome. If one wants to work at the maximum speed obtainable with scintillation counters the only pulse-forming stage is a vacuum tube limiter combined with a delay line shaper, and the only discriminating circuit consists of a biased diode. Though these circuit elements have the advantage of simplicity, they do not include the refinements of the more complex circuits which can be used at slow speeds. As a result, a fast coincidence circuit is a much simpler—though cruder—unit than a slow one, and gains in speed are often accomplished by removing—rather than adding—circuit components.

Fig. 34 shows a fast pulse equalizer consisting of a vacuum tube limiter and of a shorted-cable shaper, of the type mentioned in Sects. 9 and 10. A unit of this kind is particularly effective if the incoming pulses have a sharp rise and a long decay, in which case the output is essentially constant in shape and in amplitude for a large range of inputs. The disadvantage of this method of pulse-forming is that the pulses at the output are smaller than those at the input.

[1] E. BALDINGER, P. HUBER and K. P. MAYER: Rev. Sci. Instrum. **19**, 473 (1948).
[2] G. A. MORTON and K. W. ROBINSON: Nucleonics **4**, No. 2, 25 (1949).
[3] E. F. SHRADER: Rev. Sci. Instrum. **21**, 883 (1950).
[4] A. LUNDBY: Rev. Sci. Instrum. **22**, 324 (1951).
[5] Z. BAY: Rev. Sci. Instrum. **22**, 397 (1951).
[6] K. STRAUCH: Rev. Sci. Instrum. **24**, 283 (1953).

If maximum speed is not desired, one can use triggered circuits, with a gain in pulse height. A circuit of this kind, using a single secondary emission tube, has been described by Moody et al.[1]. The tube (Fig. 35) is normally biased off, and positive pulses produce a blocking oscillator action because of the positive feedback from dynode to grid. The output pulse is about 10 volts in amplitude, with a rise-time of about 2 mμsec. Such a circuit could be made to operate directly from the dynode of a photo-multiplier tube with a considerable gain in pulse height.

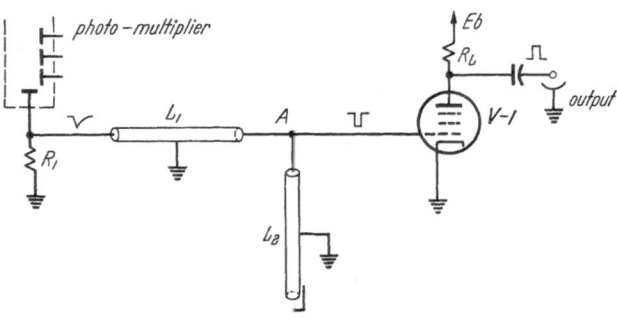

Fig. 34. Basic shaper limiter circuit.

Amplitude discriminators which can measure the fast pulses from scintillation counters form a basic part of the counting equipment required in nuclear research.

One such discriminator described by Moody[1] uses a germanium diode followed by a secondary emission trigger circuit, as shown in Fig. 36. The secondary emission tube is biased in the middle of its normal range, and D.C. negative feedback is provided by resistors R_1, R_2 and R_3 through the cathode follower and diode D_1. Positive feedback is given by dynode to grid coupling, through diode D_3. When positive input pulses are applied to the grid of V_1, and these pulses exceed the bias on D_2, regeneration in V_1 causes it to conduct heavily. After the input pulse disappears, the tube V_1 will regain its original state. The bias on diode D_2 is variable by means of a potentiometer which is used to set the triggering level. The output pulse is 15 volts or more, with a duration of about 0.3 μsec. The discriminator will trigger over a range of from 0.05 to 25 volt input with a stability of better than 0.1 volt for 10 mμsec pulses. The rise-time of the negative output pulse is about 10 mμsec.

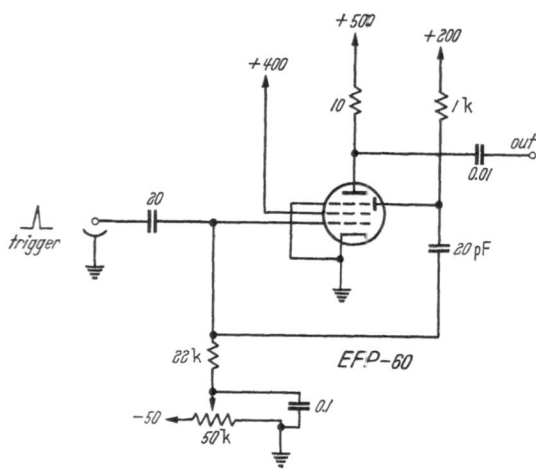

Fig. 35. Triggered blocking oscillator.

A fast discriminator designed for use with a millimicrosecond scaler has been described in a Stanford report by Narud[2]. As can be seen in Fig. 37 the circuit consists of a window amplifier, and a univibrator, both secondary emission tubes. The univibrator, V_2, is held normally cut-off, and triggering is introduced by coupling the dynode of the amplifier to the dynode of the univibrator. The sensitivity of the univibrator is further increased by coupling the amplifier plate to the univibrator cathode. The 1N72 diode serves to offer a high impedence

---

[1] N. F. Moody, G. J. McLusky and M. D. Deighton: Electronic Eng. **24**, 214 (1952).
[2] J. A. Narud: Stanford University HEPL Report No. 34, p. 178, 1955.

to ground for triggering pulses, yet provides a low impedance for cathode current of V_2. The discrimination level is set by the potentiometer which can vary the

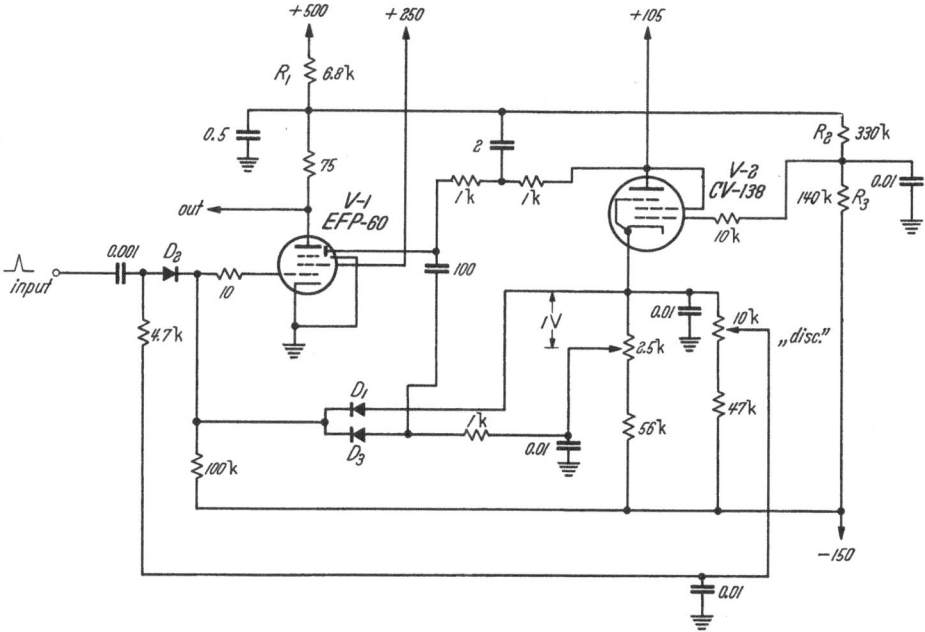

Fig. 36. High speed amplitude discriminator.

bias on the amplifier between −4 and −104 volts. The output pulse is approximately 10 volts in amplitude, negative, and about 25 mμsec in duration, with a

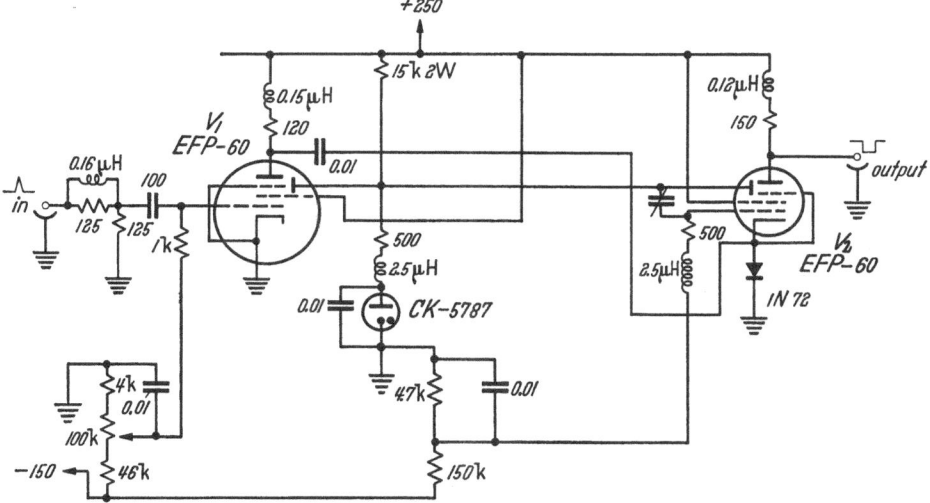

Fig. 37. Fast discriminator circuit.

rise time of 6 mμsec. Minimum triggering level is about 3 volts, and pulses at least 5 mμsec wide should be applied to the input.

13. Chronotrons and fast delay discriminators. The measurement of ultra-short time intervals, using coincidence techniques, has been facilitated by the development of the chronotron. The chronotron is a device for determining the superposition locus of transient pulses traveling in opposite directions on a transmission line, thus measuring the interval between the times of generation of the pulses. This method, introduced by NEWMAN[1], was first used to measure the time lags of spark breakdowns. A similar device (Fig. 38) was used by NEDDERMEYER et al.[2] who applied pulses to a closed loop of coaxial line, the pulse travelling in opposite directions around the loop, and being detected by a series of fixed detectors along the line. A detector at the point of superposition of the pulses will receive a signal proportional to the sum of the heights of the two pulses.

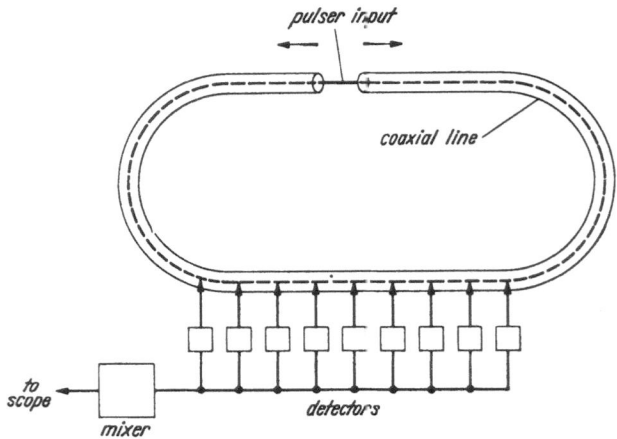

Fig. 38. Chronotron.

Further chronotron work was carried on by KEUFFEL[3],[4] who used this apparatus for measuring the mean lifetimes of μ-mesons. Delays up to 256 mμsec, with a precision of 2 mμsec were measured. In KEUFFEL's circuit, the pulses from the two scintillation counters are first limited in size and duration and are shaped to a height of 10 volts and a width of 25 milli-microsecond. These equalized pulses from the two counters are applied at the opposite ends of a 100 ohm co-axial cable, and the interval between them is determined by finding the position along the cable where they meet. If, for example, the pulses are simultaneous, they meet in the middle. The meeting point is ascertained by means of 10 crystal diodes spaced along the cable at 16 mμsec intervals. The diode nearest the meeting point transmits the largest pulse. The outputs of these diodes (0.5 μsec in duration) are simultaneously displayed on an oscilloscope and photographed. Inspection of the photograph tells between which two detectors the pulses have met, and to which of these two detectors the meeting was closest.

An improved chronotron is described by O'NEILL[5] who has added a circuit for direct reading of the number of pulses coinciding at the different positions on the cable.

Valuable variations of the coincidence circuit include devices for measuring the delay between pulses from two particle counters. Such circuits, called delay demodulators or "time-to-pulse-height-converters", allow easy evaluation of delays between two events. A circuit of this type is described by LEPRI et al.[6] schematically shown in Fig. 39. The principle of operation is shown in Fig. 39a,

[1] M. NEWMAN: Phys. Rev. **52**, 652 (1937).
[2] S. H. NEDDERMEYER, E. J. ALTHAUS, W. ALLISON and E. R. SHATZ: Rev. Sci. Instrum. **18**, 488 (1947).
[3] J. W. KEUFFEL: Rev. Sci. Instrum. **20**, 197 (1949).
[4] J. W. KEUFFEL, F. B. HARRISON, J. N. K. GODFREY and G. I. REYNOLDS: Phys. Rev. **87**, 942 (1952).
[5] G. K. O'NEILL: Rev. Sci. Instrum. **26**, 285 (1955).
[6] F. LEPRI, L. MEZZETTI and G. STOPPINI: Rev. Sci. Instrum. **26**, 936 (1955).

where initially both switches 1 and 2 are closed and the capacitor C is charged to a voltage V_0. If at time t_1 the switch 1 is opened, C begins to discharge, and if the discharge current is held constant, the voltage at A falls linearly with time. Then if at time t_2 the switch 2 is opened, the voltage at A stops decreasing, and remains constant at a value dependent on the time difference between t_1 and t_2. In practice, the switches are replaced by vacuum tubes, as shown in Fig. 39b. The sensitivity obtained with this instrument was about 240 volts per microsecond.

Another circuit for time-to-pulse-height conversion is discussed by WEBER et al.[1]. In this circuit the conversion of time to pulse height is linear down to intervals of 10 mμsec with a precision of about 1 mμsec.

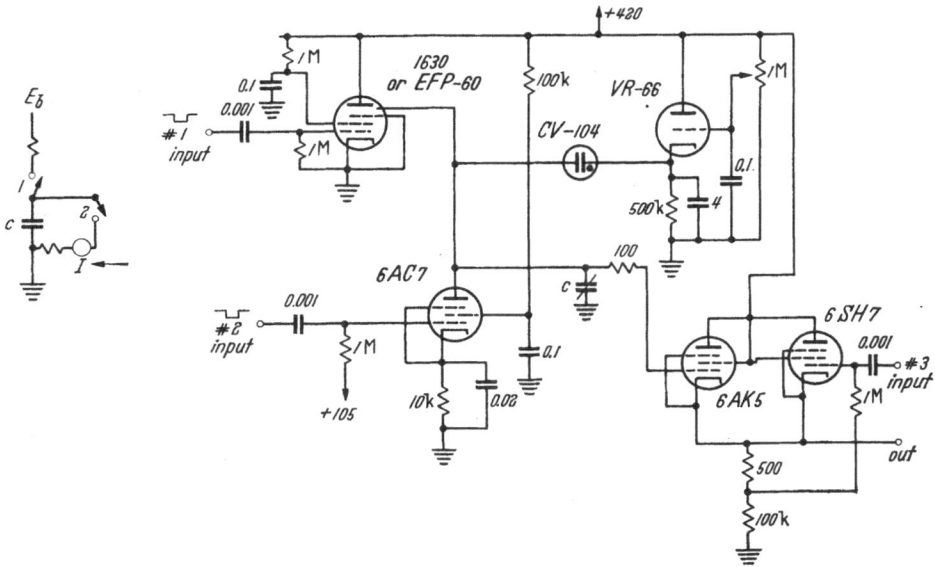

Fig. 39. Delay demodulator.

IV. Coincidence measurements.

14. Measurement of resolving time and resolution curve. The response of a coincidence circuit depends on the pulses used at the input, and it is not always easy to make statements about the circuit performance which are independent on the method of testing. We have seen in Sect. 2 that the resolving time is a simple concept only for input pulses of constant duration δ. In this case it is defined as the minimum time interval between the beginning of pulses at the two inputs which is required for them to be resolved as non-coincident. If the circuit includes a stage forming equalized pulses of duration δ, the resolving time can be measured by observing the output for different delays between the input pulses. In particular, if the pulses fed to the coincidence stage are long compared to the circuit's time constant the resolving time is equal to the pulse duration δ, which can be measured directly on an oscilloscope.

For different pulse durations on the two inputs it is convenient to introduce an average resolving time $\tau = \frac{1}{2}(\tau_{12} + \tau_{21})$, where τ_{12} is the resolving time if the pulse in channel 1 precedes the pulse in channel 2, and τ_{21} is the resolving

---

[1] W. WEBER, C. W. JOHNSTONE and L. CRANBERG: Rev. Sci. Instrum. **27**, 166 (1956).

time if the pulse in channel 2 precedes the pulse in channel 1. This average resolving time, τ, can be measured electronically as described above, or perhaps more conveniently, can be determined from a count of accidental coincidences: if channel 1 and 2 receive N_1 and N_2 independent, random pulses per unit time—as produced by two separate radioactive sources—the accidental coincidence counting rate is

$$R = (\tau_{12} + \tau_{21})N_1 N_2 = 2\tau N_1 N_2.$$

In the case of fast coincidence circuits with scintillation detectors these simple methods fail because of the continuous distribution in pulse height from the detector, and of the difficulty of making reliable high speed pulse-forming circuits. In the absence of an adequate pulse-forming stage the quantities τ_{12}

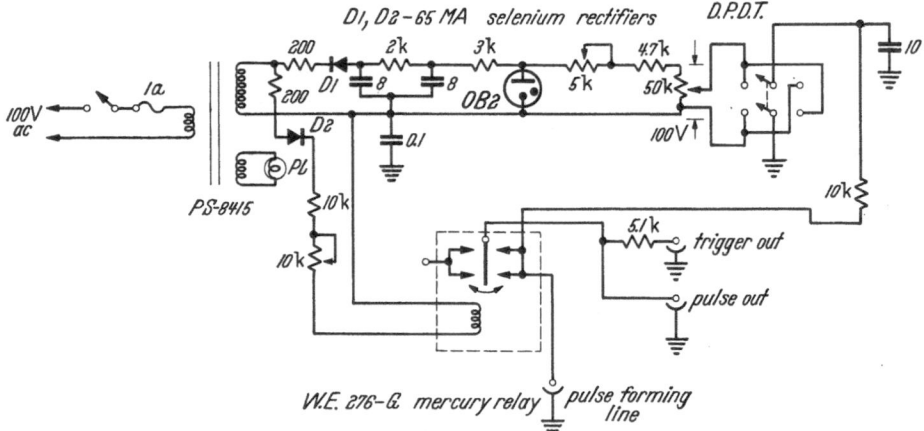

Fig. 40. Mercury relay pulser.

and τ_{21} cannot be defined uniquely and the concept of resolving time has a meaning only as an average over a particular pulse height distribution. For a given pulse height distribution the performance of the coincidence selector can be represented by the so-called *resolution curve* which is obtained by counting the number of coincidences as a function of delays artificially introduced between the simultaneous signals of the two detectors. For constant resolving times the resolution curve reduces to a square of width $\tau_{12} + \tau_{21}$, while, for the maximum speed obtainable for a given detector, the resolution curve often assumes the shape of a Gaussian.

In practice, it is advisable to start the study of the behavior of a fast circuit using the pulses from a fast pulse generator. In this manner one can find how the differentiation between single counts and coincidences varies with pulse height and duration, and one can have some idea concerning resolving time if the pulses are fed to both inputs through different lengths of cable. A practical fast pulse generator convenient for this purpose can be made using a mercury switch. The circuit is shown in Fig. 40.

After using the pulse generator one can perform more complete tests with actual pulses from radiation. The cosmic radiation provides an always available though weak, source of coincidences between radiation detectors, but more conveniently, the circuit can be tested with the coincidence from the annihilation radiations[1]. For this purpose a positron source of a fraction of a millicurie is

---

[1] S. De Benedetti and H. Richings: Rev. Sci. Instrum. **23**, 27 (1952).

located midway between two detectors, which may be kept at a distance of about 50 cm from each other. Since the annihilation γ-rays are emitted in opposite directions they give coincident counts only when the counters and the source are aligned, and contribute only random coincidences if the source is even slightly displaced from the line joining the counters. In this way, one gains rapidly some idea of the performance of the circuit and one can make the first adjustments. Then, by inserting variable cable lengths in the path of the pulses, one can measure the resolution curve.

The ability of the apparatus to resolve small time intervals can be investigated by moving the source of annihilation radiation along the line joining the two counters. In this way the resolution curves are shifted as the time of flight of the radiation to the counters is changed; displacements of a few cm, corresponding to time delays of a few 10^{-10} sec, produce measurable effects in a modern fast circuit (Fig. 41).

A coincidence selector can also be tested with penetrating particles from a high energy accelerator. This kind of test has the advantage that it involves ionizing radiation pulses of nearly uniform height. The pulse height can be controlled to a certain extent by varying the specific ionization of the radiation

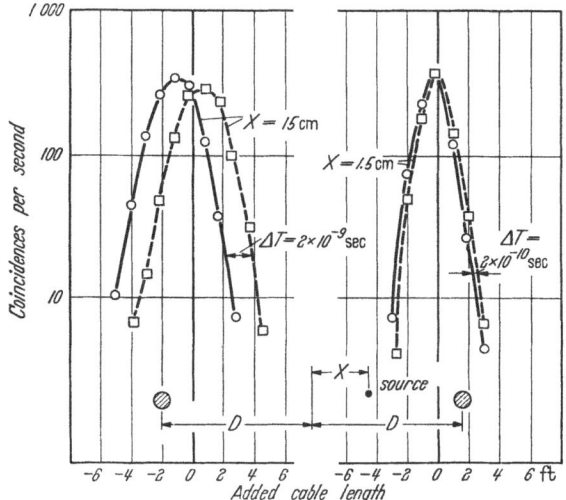

Fig. 41. Shifts caused by time of flight of γ-rays.

or the thickness of the scintillant. However, this is an expensive method of testing and it is used only when needed in connection with some specific experiment.

All these tests give information on the circuit performance under specific testing conditions, but the final tests have to be carried out with the same pulse height distribution used in any particular experiment. For instance, the random coincidence rate occurring in a certain measurement cannot be reliably computed from the single counts, but must be measured directly for each experimental condition, by inserting delay lines as will be described more in detail in Sect. 15γ.

15. Counting errors and corrections. α) *Statistical errors.* All counting measurements are affected by statistical fluctuations whose importance can be evaluated from the random nature of the events selected. If the probability of an event occurring in time dt is $\lambda\,dt$, $\overline{m} = \lambda T$ is the average expected number of events occurring in a time interval T; the probability $P_m(T)$ that m events actually occur in a time T is given by the Poisson distribution

$$P_m(T) = \frac{\overline{m}^m}{m!}\,\mathrm{e}^{-\overline{m}}.$$

The *standard deviation* corresponding to this distribution is

$$\sigma = \sqrt{\overline{m}}.$$

For large m the Poisson formula is equivalent to the a gaussian

$$P_m(T) = \frac{1}{\sqrt{2\pi\overline{m}}}\, e^{-\frac{(\overline{m}-m)^2}{2\overline{m}}}\, .$$

It is customary to express the statistical error of a counting experiment by means of the standard deviation; if C is the measured number of coincidences in a time T the standard deviation is $\approx\sqrt{C}$, and the probability that the true counting rate (the limit of C/T for $T\to\infty$) lies in the interval $(C\pm\sqrt{C})/T$ is 68%. The *probable error* is defined as $0.6745\sqrt{C}$, and the probability of the true rate being in the interval $(C\pm0.6745\sqrt{C})/T$ is 50%. The curve of Fig. 42 expresses the distribution of errors.

In interpreting counting results a good rule of thumb is to consider an effect which is just outside the statistical standard deviation of the background fluctuations as vaguely suggested by the experimental results: if the effect is outside twice the statistical errors one can believe it with reservation; only if the effect is larger than three times the standard deviation, can it be considered as established, since the probability of its being due to background fluctuations is 0.27%.

It may be useful to recall that if $A\pm\Delta A$ and $B\pm\Delta B$ are two independent experimental counting results with their respective errors, the error in the function $f(A\,B)$ is

$$\sqrt{\left(\frac{\partial f}{\partial A}\,\Delta A\right)^2+\left(\frac{\partial f}{\partial B}\,\Delta B\right)^2}\, .$$

If the numbers A and B are not independent the above formula cannot be used. Let us suppose, for example, that one wants to know how many particles of a given beam stop in a certain absorber. The measurement is performed with a telescope formed of 3 counters 1, 2, and 3. One and 2, in coincidence, measure the beam before it reaches the absorber while the coincidences 1-2-3 correspond to particles which have traversed the absorber. Then, if C_{12} and C_{123} are the results of the two successive independent measurements performed for the same length of time T, the number of particles stopped in the absorber is, with its standard deviation $C_{12}-C_{123}\pm\sqrt{C_{12}+C_{123}}$. But if the two measurements are performed simultaneously on the same particles, establishing whether each one of them stops in the absorber, or if the difference is measured automatically by means of an anticoincidence circuit, the same number can be written $C_{12}-C_{123}\pm\sqrt{C_{12}-C_{123}}$.

β) *Counting losses.* In a coincidence apparatus, counting losses may arise from two sources: (1) dead time in the single pulse channels preceding the coincidence stage and (2) dead time in the recording unit following the coincidence stage.

With Geiger counters, which have a dead time of the order of 10^{-4} sec the losses of the first kind may be very important: in a strong radiation field, such as near an accelerator, the Geiger counter may be almost entirely paralized. Scintillation counters, instead, have essentially no dead time, and—apart from dead times arising in the associated equipment—they can be used at all speeds.

The duration of the pulse fed to the coincidence stage, which results in counting losses for single counts is responsible for the random coincidences and as such it may increase—rather than decrease—the counting rate.

Elmore[1] considers two types of counting losses. In the first type, each pulse is followed by a dead time Δ which is not increased if a second—undetected—

[1] W. C. Elmore: Nucleonics 6, No. 1, 26 (1950).

pulse arrives during the dead time of the first. In the second type, the rejected pulse itself is followed by its own dead time. In the first case the fractional counting loss is

$$f = \frac{C_{\text{true}} - C_{\text{meas}}}{C_{\text{true}}} = \frac{C\Delta}{1 + C\Delta}$$

and in the second

$$f = 1 - e^{-C\Delta}$$

where Δ is the dead time following each pulse, and C the counting rate. For $C\Delta \ll 1$ both of these expressions reduced to

$$f = C\Delta$$

and this is the formula of practical interest, since it is advisable to limit the counting rate to the a value such that higher order corrections are not required.

Resolution losses in a scaler are somewhat more complicated; the first stage

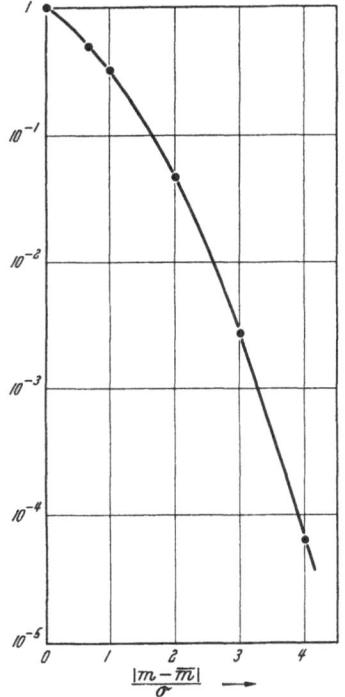

Fig. 42. Distribution of statistical errors. The ordinates express the probability that the error exceeds the value indicated in the abscissa in terms of the standard deviation.

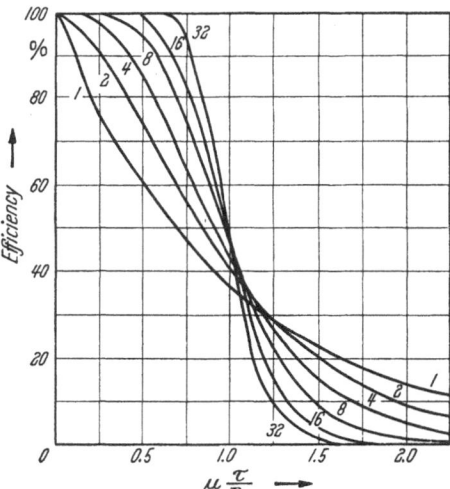

Fig. 43. Efficiency as a function of $\mu \tau/n$, where μ is the statistically averaged input frequency, n the scaling factor and τ the dead time of the register following the scale of n. Each curve is labeled with the appropriate value of n. $\mu =$ statistically averaged input frequency. $\tau =$ resolving time.

of a scaler receives random pulses, but in the subsequent stages the distribution in intervals becomes more and more regular (Fig. 43)[1]. When the number of coincidences to be recorded is large one uses a few stages of fast scaling (10^{-7} sec dead time is realizable and commercially available) followed by slower stages. It is advisable to keep the resolution losses small in all stages but the first.

γ) *Random coincidences.* The random coincidences are another source of error in all coincidence experiments and this number must be subtracted from the counting rates. In the case of constant pulse duration the random rate can be computed from the number of single counts. The double random rate is

$$R = 2\tau N_1 N_2,$$

[1] L. ALAOGLU and N. M. SMITH jr.: Phys. Rev. **53**, 832 (1938).

and the triple coincidences random rate, due to independent events in the three counters is

$$R = 3\tau N_1 N_2 N_3.$$

In most cases, however, the triple random rate given above is negligible compared to that resulting from partially correlated events, consisting of a real coincidence between two of the counters which, by chance, occurs simultaneously with a pulse in the third. When this effect is considered the total random triple rate is

$$R = 3\tau N_1 N_2 N_3 + 2\tau (N_1 C_{23} + N_2 C_{13} + N_3 C_{12})$$

where C_{ij} are the real doubles between counters i and j.

For fast resolving times the number of random coincidences cannot be evaluated accurately from the formulae given above and must be measured for each experimental condition. In the case of double coincidences the measurement involves the use of an artificial delay sufficient to eliminate the simultaneity of the pulses from correlated events. It is advisable to check that the attenuation or distortion introduced by the delay is negligible, by making sure that equal delays do not effect the coincidence rate. For triple coincidences it may be necessary to measure separately each term of the expression above.

The situation becomes more and more complicated with the increase in the multiplicity of the coincidence measured and it is advisable to make the coincidence correction as small as possible, even if this involves the use of extra counters, or long counting times.

δ) The "feasibility" of a coincidence experiment. In order to decide whether a certain effect can be studied with the coincidence method, one usually performs a few elementary computations. The parameters to be chosen are the circuit's speed, the counter's geometry and efficiency, and, when this is under the control of the experimenter, the strength of the source.

Two conditions are to be satisfied to make the experiment possible:

(1) The counting rate must be high enough to obtain sufficiently small statistical error in the time available (the time available is not arbitrarily long, being limited, among others, by the expected survival of the equipment without failure and appreciable drifts);

(2) the total counting rate must be much larger (about a factor 10) than the random rate.

Large counters and strong sources are desirable as far as the first condition is concerned, while the second condition requires small counters and weak sources. Fast circuitry is always preferable if it can be sufficiently stable.

Let us suppose for instance that one wants to measure an angular correlation using two counters each of efficiency ε and covering solid angles ω, and a source of N disintegrations per second. Let τ be the coincidence resolving time, C the coincidence rate and R the random rate. These quantities are connected by the equations

$$N\omega^2 \varepsilon^2 = C,$$

$$2\tau N^2 \omega^2 \varepsilon^2 = R.$$

Supposing that ω, ε and τ are fixed, and that $C_{\min}$ is the minimum counting rate needed to reach the desired statistical accuracy in the available time. Then the conditions $C > C_{\min}$ and $R/C \ll 1$ bracket the source strength between the limits

$$\frac{C_{\min}}{\omega^2 \varepsilon^2} < N \ll \frac{1}{2\tau}.$$

16. Applications of the coincidence method. In this section we want to describe some of the measurements which can be performed by means of the coincidence method. We do not intend to describe the many, sometimes very clever, ways in which coincidence counters can be arranged for particular measurements —such as telescopes, scattering or angular correlation apparatus etc.—but we wish to cover some of the cases in which the coincidence technique is an integral part of a new instrument or a new method of measurement.

α) *Calibration of radioactive sources and measurement of counter efficiency.* The absolute calibration of a radioactive source consists in the determination of the number of disintegrations per second. Such measurement is easy in the case of sources of α-particles, since these travel in straight lines and the geometry of the experiment is immediately determined; but for β-rays the back scattering from the source introduces a large uncertain factor, while for γ-rays the counter efficiency is generally an unknown.

The absolute measurement of a radioactive source can be performed with a coincidence experiment when the decay scheme is simple and known. Let us suppose, for example, that the source emits a simple β-spectrum followed by a γ-ray. It suffices in this case to use two counters (Geiger counters are preferable because of their constant pulse height), one to detect the β-rays, and the other γ-rays. If S_β and S_γ are the single counting rates of the two counters, and C is their coincident counting rate recorded under the same conditions one can write

$$S_\beta = \omega_\beta \, \varepsilon_\beta \, N \,,$$

$$S_\gamma = \omega_\gamma \, \varepsilon_\gamma \, N \,,$$

$$C = \omega_\beta \, \varepsilon_\beta \, \omega_\gamma \, \varepsilon_\gamma \, N,$$

where N is the unknown disintegration rate of the source, ω_β and ω_γ the solid angles covered by the β and γ counters, ε_β and ε_γ their efficiencies. The above equations can be solved to give:

$$N = \frac{S_\beta \, S_\gamma}{C} \,,$$

$$\omega_\beta \, \varepsilon_\beta = C/S_\gamma \,,$$

$$\omega_\gamma \, \varepsilon_\gamma = C/S_\beta .$$

The first of these gives the absolute source strength and the last two yield the efficiency of the counters.

The method, with some complications, can be extended to cases of more complex known decay scheme. When sources of various β and γ emitting substances have been calibrated in this manner, other isotopes can be measured by interpolating the energy dependence of counter efficiency.

β) *Coincidence β-spectrometry.* The coincidence method has been of great help in the study of radioactive decay schemes. For this kind of work the coincidence technique is often combined with the usual techniques of β and γ spectrometry, such as absorption measurements, magnetic spectrometers and scintillation spectrometers.

Let us suppose that two counters are exposed to the radiation from a source emitting β and γ rays with an unknown decay scheme, and that coincidences and single counts are measured as in the preceeding section. In order to obtain information on the decay scheme, the counts are taken as a function of the thickness t of an absorber in front of the β counter. If the ratio $C(t)/S_\beta(t)$ remain constant the γ rays are associated with all the β rays: in the simplest case we are confronted with a simple β spectrum followed by a single γ. But if the ratio $C(t)/S_\beta(t)$ decreases as a function of t, only the softer β's are in coincidence with the γ's and we may be faced with two β spectra of which only one is followed by γ rays.

Instead of using the method of absorption one can measure coincidences within a β spectrometer. For instance, the γ counter may be placed near the source while the β counter detects the magnetically selected β rays. Since a β spectrometer has a small solid angle of acceptance the measurements have to be made with relative strong sources: the high speed and large γ efficiency of the scintillation counters are almost indispensible for this kind of work.

More complete information is obtained by measuring coincidences between two β spectrometers. A typical example of this method is the work on the disintegration scheme of I^{131}. After the β spectrum of this substance had been studied several disintegration schemes were suggested. The problem was settled by the work of P. R. Bell and co-workers[1] who used two scintillation spectrometers in coincidence, followed by the investigations of R. E. Bell and R. L. Graham[2] whose instrument consisted of two coincident β spectrometers of the magnetic lens type. The reader is referred to the original papers for details.

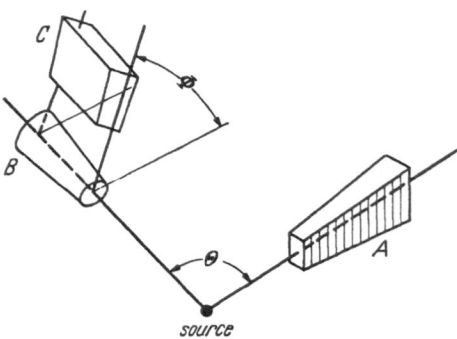

Fig. 44. Schematic diagram of experimental arrangement of polarimeter.

γ) γ-ray polarimeters. Polarimeters for γ-rays of energy around 1 Mev can be built taking advantage of the preferred orientation of Compton scattered γ-rays perpendicular to the electric field of the primary photon. γ-ray polarimetry has been applied with success first to the study of polarization of the annihilation radiation, and then to radiations from radioactive decay[3].

Metzger and Deutsch counted the coincidences between two nuclear γ-rays, while measuring the polarization of one of them. Their instrument is schematically shown in Fig. 44. Two cylindrical scintillation counters, A and B are exposed to the radiation from a source S emitting two γ rays in cascade. A third counter C, can rotate around the axis of B and receives the Compton scattered γ-rays from it. The two counters, B and C, in coincidence, form the polarimeter. From the triple coincidences one can measure the correlation between the polarization of one γ-ray and the angle of emission of the two.

The optimum scattering angle for detecting polarization effects varies from 90° at zero energy to $\sim$87° at 1.5 Mev. The asymmetry ratio (ratio of maximum to minimum counting rate as a function of scattering angle for plane polarized γ-rays) expected under the optimum angle is $\approx$15 at 0.2 Mev, $\approx$5 at 0.5 Mev and $\approx$2 at 1.5 Mev. This ratio is considerably decreased when large solid angles are used, and, in a practical case, varies from 5 at low energy to 2 at 1.5 Mev.

Polarimeters for γ-rays can easily be tested by repeating the experiment on the relative direction of polarization of the annihilation photons: with the effi-

[1] R. R. Bell, J. M. Cassidy, and G. G. Kelley: Phys. Rev. **82**, 103 (1951).
[2] R. E. Bell and R. L. Graham: Phys. Rev. **86**, 212 (1952).
[3] F. Metzger and M. Deutsch: Phys. Rev. **78**, 551 (1950).

cient γ-counters available today, a few minutes of counting are sufficient to verify this effect and to test the polarimeter.

17. Coincidence measurements of time intervals. Before the use of delaying circuit elements became common, the measurement of short time intervals was performed with an "integral method", which consisted in counting the number of coincidences as a function of the resolving time of the selector. With this method some fast α decays in the natural radioactive families were measured. Among these are RaC' (Po^{214}, 2×10^{-4} sec)[1], ThA (Po^{216}, 0.16 sec), ThC' (Po^{212}, 0.3×10^{-7} sec), and Po^{213} (4.2×10^{-6} sec)[2]. The first measurements of the life of the μ meson[3] at rest were also performed with the integral method.

The differential method, making use of a delayed coincidence circuit, was introduced by JACOBSEN and SIGURGEIRSSON[4] in 1943. In 1946 this method was applied to the measurement of short γ lives (short lived isomers)[5,6], and after the advent of scintillation counters, it was extended to shorter and shorter times. In recent years it has been used for the determination of the lives of many of the newly discovered mesons (π mesons[7,8] and K mesons[9,10]) and has opened a new avenue of approach to solid state problems through the study of positron mean lives[11,12].

It is not appropriate here to describe in detail these experiments. The 1952 paper of R. E. BELL et al.[13] still contains the best account of the general features of the experimental methods and of the precautions to be used when one wants to measure time intervals of the same order or smaller than the resolving time of the selector used.

If the mean life to be measured is longer than the resolving time of the instrument, the number of coincidences is an exponential function of delay, which can clearly be separated from the instrumental resolution curve. But, if the mean life is of the same order or smaller than the resolving time, the delays to be measured are difficult to resolve from the instrumental ones.

Since the resolution curves depend on the pulse height spectrum, reliable results can be obtained only with a direct comparison method: all measurements must be comparisons between a coincidence curve obtained with the events under study *(delayed curve)* and a similar curve obtained with simultaneous events *(resolution curve or prompt curve)* having the same pulse height spectrum.

When this rule is followed, any difference between the delayed and the prompt curve outside statistical errors is significant. The shortest life which one can measure depends on the slope (not on the width) of the resolution curve and on the accuracy with which the curves are measured. Though the shortest resolving times obtained are of the order of 10^{-9} sec, under appropriate conditions it is possible to measure lives with an error of the order of 10^{-11} sec.

[1] J. C. JACOBSEN: Nature, Lond. **133**, 565 (1934).
[2] J. V. JELLEY: Canad. J. Res. A **26**, 255 (1948).
[3] F. RASETTI: Phys. Rev. **60**, 1948 (1941).
[4] J. C. JACOBSEN and T. SIGURGEIRSSON: Dan. Mat.-Fys. Medd. **20**, No. 11 (1943).
[5] S. DE BENEDETTI and F. K. McGOWAN: Phys. Rev. **70**, 569 (1946).
[6] S. DE BENEDETTI and F. K. McGOWAN: Phys. Rev. **74**, 728 (1948).
[7] W. L. KRAUSHAAR, J. E. THOMAS and V. P. HENRY: Phys. Rev. **78**, 486 (1950).
[8] M. JAKOBSON, A. SCHULZ and J. STEINBERGER: Phys. Rev. **81**, 894 (1951).
[9] V. FITCH and R. MOTLEY: Phys. Rev. **101**, 496 (1956).
[10] L. ALVAREZ, F. S. CRAWFORD, M. L. GOOD and M. L. STEVENSON: Phys. Rev. **101**, 503 (1956).
[11] S. DE BENEDETTI and H. RICHINGS: Phys. Rev. **85**, 377 (1952).
[12] R. E. BELL and R. L. GRAHAM: Phys. Rev. **90**, 644 (1953).
[13] R. E. BELL, R. L. GRAHAM and H. E. PETCH: Canad. J. Phys. **30**, 35 (1952).

In certain cases the experimental conditions providing the prompt curve are easily realizable. For instance, if a time of flight is measured, it will suffice to put the counters near to each other. But for the study of radioactive decay, the delay time cannot be controlled, and the measurement of the prompt curve requires some special care.

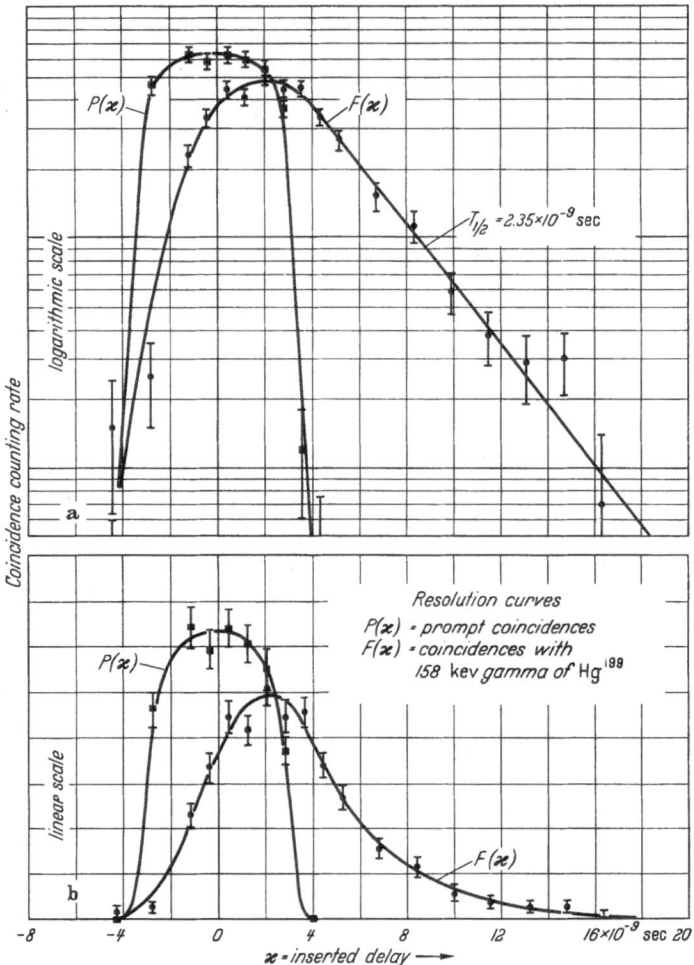

Fig. 45a and b. Delayed coincidence resolution curve $F(x)$ for the 158 kev gamma ray of Hg[199].

R. E. Bell and his collaborators have solved the problem by using the two β-spectrographs already mentioned (Sect. 16β)[1]. The conversion electrons of the γ-rays from the level whose mean life is to be measured are focused on the counters of one of the spectrometers, while the other receives a portion of the β-spectrum. The delayed coincidence curve is obtained in this manner. For the prompt curve a source having a fast line of energy very near to the one investigated is used. For instance, for the measurement of the mean life of the 158 kev excited state in Hg[199], the F-line of ThC (147 kev) was used to obtain the prompt curve. The curves obtained are shown in Fig. 45.

[1] R. E. Bell and R. L. Graham: Phys. Rev. 86, 212 (1952).

It is possible to perform the measurement without the use of a "fast" source, if the role of the two counters is inverted. For the measurement of the mean life of the 412 kev γ-rays of Hg^{198}, only one source (Au^{198}) is placed between the spectrometers A and B. In the first run spectrometer A receives the conversion line to be measured, while B is focussed on the β-spectrum just below the line. In a second run the current in both spectrometers is slightly increased, so that the line is now focussed in B while A received the β-spectrum just above it. The function of the two spectrometers is thus reversed. Since no significant difference was found between the two curves (Fig. 46) the mean life of the 412 kev excited state of Hg^{198} was determined as

$$T = (1.0 \pm 1.7) \times 10^{-11} \text{ sec}^*.$$

The interpretation of the data is made quantitative by the use of simple formulae which connect the delayed curve to the prompt curve and to the mean life to be measured. BAY[1,2] has performed extensive calculations of the moments of the delay curves developing quite general relations. The simplest of these furnishes the law that — in the case of exponentially distributed delays — the "center of gravity" of the delayed curve is displaced by the mean life to be measured, relative to the center of gravity of the prompt curve. Another useful relation derived by NEWTON[3] is that the mean life T is given by

$$T = [F(t_1) - F(t_2)]^{-1} \int_{t_1}^{t_2} [F(t) - P(t)]\, dt$$

where $P(t)$ is the prompt curve, $F(t)$ the delayed curve and t_1 and t_2 are two arbitrarily chosen values of t.

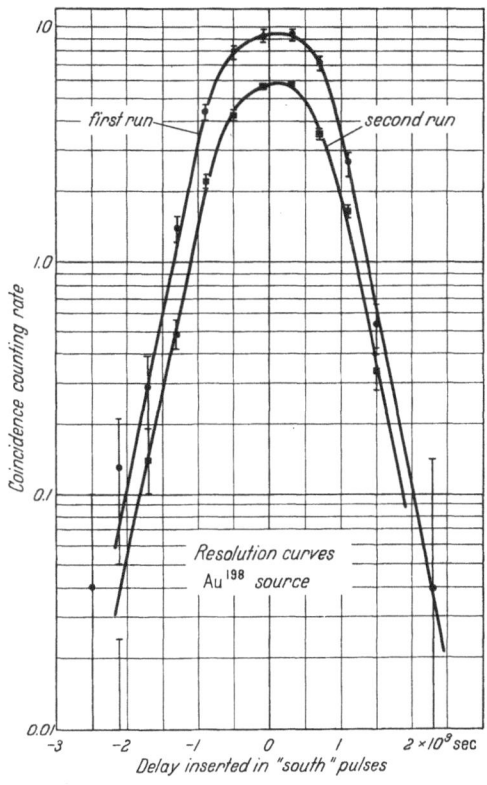

Fig. 46. Self-comparison resolution curves for the 411 kev gamma ray of Hg^{198}.

---

* We are now informed by Dr. A. W. SUNYAR that a more recent measurement yielded the result $(3 \pm 1.5) \times 10^{-11}$ sec. Dr. SUNYAR also uses a comparison method, but the β-spectrometers are replaced by much simpler electronic pulse height selectors. The new result agrees with measurements by the resonance fluorescence method [F. METZGER, Phys. Rev. 98, 200 (1955)].

[1] Z. BAY: Phys. Rev. 77, 419 (1950).
[2] Z. BAY, V. P. HENRI and H. KANNER: Phys. Rev. 100, 1197 (1955).
[3] T. D. NEWTON: Phys. Rev. 78, 490 (1950).

Cloud Chambers.

By

CARL M. YORK.

With 17 Figures.

1. Introduction. For over half a century the Wilson Cloud Chamber has played a unique role in the development of nuclear physics. By making the paths of ionizing particles visible, this device has enabled the nuclear physicist to study the microscopic behavior of the fundamental particles in a direct manner. As a result, many aspects of nuclear phenomena are based on the intuitive application of macroscopic principles of physics rather than upon abstract conceptual ideas.

The long list of contributions to physics by the application of this single type of apparatus is truly impressive. From the first studies of condensation phenomena carried out by C. T. R. WILSON[1] to the more recent discoveries of a number of the new heavy unstable particles[2], the cloud chamber has provided a steady stream of fundamental contributions to the study of physics throughout the years.

In the past decade two other detection devices have become available to the nuclear physicist which have the property of rendering the paths of ionizing particles visible to the observer. These are the photographic emulsions which are sensitive to ionizing particles and more recently the bubble chamber. These techniques have already had a tremendous impact on the study of nuclear phenomena and show promise of continuing to do so. However, it appears at the present time that these techniques essentially complement rather than supplant the cloud chamber as detection devices. For any given experiment depending on the visual observation of the paths of ionizing particles, the problem is to choose one of the three types of apparatus best suited for the work. It is the purpose of this article to describe the operation of the cloud chamber together with some of its advantages and limitations. The techniques dealing with nuclear emulsions and bubble chambers will be treated elsewhere in this volume.

I. The production of tracks in a cloud chamber.

Because the tracks observed in the gas of a cloud chamber are composed of droplets of liquid which have been condensed on ions produced by the traversal of the gas by a charged particle, we shall discuss the theory of the formation of ions first. Then in treating the process of vapor condensation on these ions, it will be found that the vapor must be in a supersaturated state. This in turn leads rather naturally to a discussion of the methods of producing supersaturation, first from the theoretical standpoint and then from the experimental point of view.

---

[1] Cf. C. T. R. WILSON: Phil. Trans. Roy. Soc. Lond. **189**, 265 (1897).
[2] L. LEPRINCE-RINGUET et al.: This Encyclopedia, Vol. XLIII.

a) The ionization process.

2. A theoretical survey. The energy loss of a charged particle passing through matter has been treated classically by BOHR[1]. The quantum mechanical treatment was then given by BETHE and BLOCH[2] and later was modified by FERMI[3]. A detailed treatment of the theory will not be presented here, but an attempt will be made to describe the physical phenomena involved and to present only formulae which are of immediate applicability to the cloud chamber technique[4]. Essentially the calculation considers the electromagnetic interaction of the charged particle with the orbital electrons of the atoms of the material through which it is passing. By imparting energy to these orbital electrons in sufficient amounts to remove them from their bound states the charged particle leaves a trail of ions in its wake and loses energy in the process. The orbital electrons which are ejected in these collisions are divided into two categories, depending upon the energy imparted to them. Those emitted with very low energy are called the *photo-electrons*, while those ejected with sufficient energy to produce ions in subsequent collisions with atoms of the material are called *knock-on-electrons*. The ions formed in a collision with the bombarding particle are called the *primary* ionization and those ions formed by the recoiling knock-on electrons produce the *secondary* ionization. The sum of the primary and secondary ionizations is referred to as the *total ionization* and in comparing the theory to be given below with experiment it will be found necessary to distinguish which of these ionizations has been measured.

Fermi pointed out that very fast particles will have their sate of energy loss modified by the screening effects of the surrounding medium—the so-called "density effect". His original calculation treated the case of a charged particle interacting with isolated atoms having a single resonant frequency. A number of authors have treated the problem in more detail by considering several of the resonant frequencies of the orbital electrons found in the various energy levels of the atoms of the gas. The most recent calculations are due to STERNHEIMER[5] and BUDINI[6] and the work of the former author will be quoted here. The rate of energy loss, (dE/dx), of a charged particle in traversing a thickness, dx, of a material of density, ϱ, is given by different expressions depending upon whether the particle is an electron or a heavier particle. This difference results from a term in the formula which depends on the maximum energy transfer, E_{max}, in a collision of the bombarding particle with an orbital electron. For a particle of mass, μ, and kinetic energy, E, this expression for E_{max} is given by

$$E_{max} = \frac{E^2 - \mu^2 c^4}{\mu c^2 (\mu/2m + m/2\mu + E/\mu c^2)} \tag{2.1}$$

where m is the electron mass and c is the velocity of light. If the particle colliding with the orbital electron is an electron, then μ must be set equal to m, but because of the identity of the particles a factor of $\frac{1}{2}$ must be introduced. Thus

$$E'_{max} = \tfrac{1}{2}(E - mc^2) \quad \text{(for electrons)}. \tag{2.2}$$

---

[1] N. BOHR: Phil. Mag. **30**, 581 (1915).

[2] H. BETHE: Ann. Phys., Lpz. **5**, 325 (1930). — F. BLOCH: Ann. Phys., Lpz. **16**, 285 (1933).

[3] E. FERMI: Phys. Rev. **57**, 485 (1940).

[4] A detailed treatment of ionization is given by B. ROSSI: High Energy Particles, p. 22. New York: Prentice-Hall 1952.

[5] R. M. STERNHEIMER: Phys. Rev. **103**, 511 (1956).

[6] P. BUDINI: Nuovo Cim. **10**, 236 (1953).

Finally, the expression for the energy loss is: for heavy particles

$$-\frac{1}{\varrho}\frac{dE}{dx} = \frac{A}{\beta^2}\left[B + 0.69 + 2\log p/\mu c + \log E_{max} - 2\beta^2 - \delta\right]; \qquad (2.3)$$

and for electrons

$$-\frac{1}{\varrho}\frac{dE}{dx} = \frac{A}{\beta^2}\left[B + 0.43 + 2\log p/mc + \log E'_{max} - \beta^2 - \delta\right]. \qquad (2.4)$$

Here $\beta = v/c$, the usual ratio of particle velocity to that of light *in vacuo*; p is the particle momentum; A and B are constants depending on the material tra-

Table 1. *The constants required to compute the ionization loss of a charged particle traversing a gas*[1].

Gas	A [Mev/gm/cm²]	B	$-C$	a	m	X_1	X_0	ϱ [gm/l] (at N.T.P.)
H_2	0.1524	21.07	9.50	0.505	4.72	3	1.85	0.08988
He	0.0767	19.39	11.18	2.13	3.22	3	2.21	0.17847
N_2	0.0768	17.94	10.68	0.125	3.72	4	1.86	1.25055
O_2	0.0768	17.67	10.80	0.130	3.72	4	1.90	1.42904
Ne	0.0761	17.23	11.72	0.258	3.18	4	2.14	0.90035
A	0.0692	16.09	12.27	0.0255	4.36	5	2.02	1.7837
Kr	0.0661	14.56	13.12	0.0771	3.57	5	2.12	3.708
Xe	0.0632	13.70	13.57	0.150	3.07	5	1.90	5.851
CH_4	0.0958	19.37	9.56	0.0552	4.22	4	1.55	0.7168
$(CH_2)_2$ . . .	0.0876	18.95	9.52	0.0700	3.94	4	1.54	1.2604
$(CH)_2$ . . .	0.0826	18.65	9.95	0.0841	3.91	4	1.61	1.173
CO_2	0.0768	17.82	10.32	0.0865	4.03	4	1.72	1.9769

versed and are tabulated in Table 1; E_{max} is the maximum energy transfer in units of Mev; and δ is the "density effect parameter". δ can be computed from the following convenient relations:

$$\delta = 4.606 X + C + a(X_1 - X)^m \quad \text{for} \quad (X_0 < X < X_1), \qquad (2.5)$$

$$\delta = 4.606 X + C \qquad\qquad \text{for} \quad (X > X_1) \qquad (2.6)$$

where the constants, a, C, and m depend upon the material traversed and are also listed in Table 1. $X = \text{Log } p/\mu c$. X_0 is the value of X which corresponds to that value of the momentum below which $\delta = 0$, while X_1 corresponds to the momentum above which the relation of δ and X can be considered to be linear. The appropriate values of X_0 and X_1 for the various gases are also listed in Table 1. These expressions for δ have no theoretical foundation, but are more useful for computational purposes than the complicated theoretical expressions which they approximate.

Some experiments such as the droplet counting work to be discussed below, do not include the secondary ionization produced by knock-on electrons above a certain maximum energy, E_0, which is much less than the E_{max} or E'_{max} given in Eqs. (2.1), (2.2) above. In this case the energy loss expression becomes:

$$-\frac{1}{\varrho}\frac{dE}{dx} = \frac{A}{\beta^2}\left[B + 0.69 + 2\log p/\mu c + \log E_0 - 2\beta^2 - \delta\right] \qquad (2.7)$$

where E_0 is again expressed in units of Mev. It has been shown experimentally that the energy loss computed in this way is proportional to the number of ions

---

[1] Cf. R. M. SFERNHEIMER: Phys. Rev. **103**, 511 (1956).

per unit length of path, I. Clearly the energy loss depends upon the mean energy required to ionize the substance or the "mean ionization potential". The values of the constants given in Table 1, depend upon the experimentally determined values of the mean ionization potential given by SACHS and RICHARDSON[1]. Because almost all cloud chamber measurements of ionization are limited to exclude clusters of twenty-five ions or more, Eq. (2.7) is plotted in Fig. 1 for $E_0 = 800$ ev, which corresponds to the energy required for the formation of approximately twenty-five ion pairs in helium. The curve in the figure is plotted as the ratio of the energy loss at any given value of $p/\mu c$, to the minimum value of the energy loss or the minimum ionization, I_{min}, rather than in absolute units of Mev/gm/cm². The advantages of this type of relative ionization plot for cloud chamber work will be discussed in connection with ionization determination,

below. Furthermore, it should be emphasized that the energy loss of a charged particle which traverses an absorber depends only upon its velocity and not upon its rest mass.

Now clearly the energy loss of a particle by ionization of the material through which it passes is a statistical process which depends upon the number of collisions it makes in traversing a thickness of the material. The relations given in Eq. (2.3), (2.4), (2.7) express the *average* energy loss, dE, on traversing a distance,

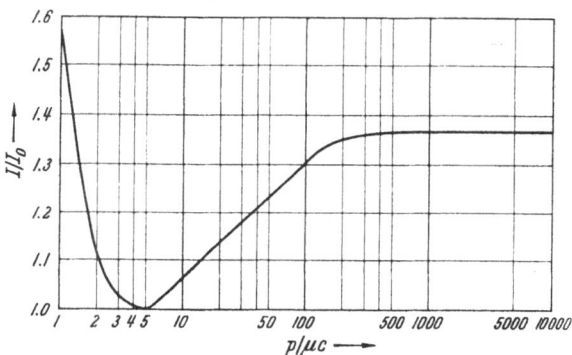

Fig. 1. A plot of the relative ionization in helium as a function of $p/\mu c$. [Eq. (2.7) is plotted using one atmosphere of helium and $E_0 = 800$ ev.]

dx, of the material. One may well enquire as to the amount and distribution of the statistical fluctuations which might be encountered when repeated measurements of the energy loss are carried out. The mathematical solution to the problem is formidable, but has been given by LANDAU[2] and later in greater detail by SYMON[3]. In general the energy loss varies according to a skew or "Landau" distribution, which under certain conditions approximates a symmetric Gaussian distribution. In most work with cloud chambers the energy loss of a particle which traverses the gas in the chamber is very small compared to its kinetic energy. Consequently the theory of SYMON for the case of "thin absorbers" is applicable and in this instance the resulting distribution is of the Landau form. It is important to bear in mind that this implies that the errors in the ionization determinations made with cloud chambers are not symmetrically distributed about the mean, but are biased toward larger values of the energy loss.

b) The condensation of vapor on ions.

3. The classical theoretical treatment. In order that a vapor be condensed into liquid droplets, two basic requirements must be fulfilled. First the vapor must contain *condensation nuclei*, which are small objects suspended in the vapor and around which the vapor molecules can collect to form droplets. Secondly the vapor must be in a supersaturated state so that the droplet can grow. The

---

[1] D. C. SACHS and R. J. RICHARDSON: Phys. Rev. **83**, 834 (1951); l. c. **89**, 1163 (1953).
[2] L. D. LANDAU: J. Phys. USSR. **8**, 201 (1944).
[3] K. R. SYMON: Harvard University, Thesis 1948. This work is discussed in detail by B. ROSSI, High Energy Particles, p. 32. New York: Prentice-Hall 1952.

degree of saturation of a vapor is defined as the ratio of the density of the vapor present to the saturated vapor density at the same temperature. If this ratio is less than unity, the vapor is said to be *undersaturated*; if the ratio is unity the vapor is *saturated*; and if the ratio is greater than unity, it is *supersaturated*. The methods of producing supersaturation in a cloud chamber will be discussed in detail in Sect. d) and e), below. For the moment let us consider some theoretical aspects of the problem of droplet growth and the nature of the condensation nuclei found in the gas of a cloud chamber.

Lord KELVIN showed that as a result of surface tension, the saturated vapor pressure, P_r, over the surface of a drop of radius, r, is greater than that over a flat surface ($r \to \infty$) of the liquid. If P_∞ is the saturated vapor pressure over the flat surface, the condition that the droplet be in equilibrium with the surrounding medium is

$$\log \frac{P_r}{P_\infty} = \frac{M}{RT\varrho}\left(\frac{2\vartheta}{r} + \frac{d\vartheta}{dr}\right) \quad (3.1)$$

where ϑ is the surface tension, ϱ the density and M the gram molecular weight of the liquid, T is the absolute temperature of the system, and R is the gas constant. This relation just gives the condition that the droplet will neither shrink nor grow. In Fig. 2 the variation of P_r/P_∞ as a function of drop radius is plotted

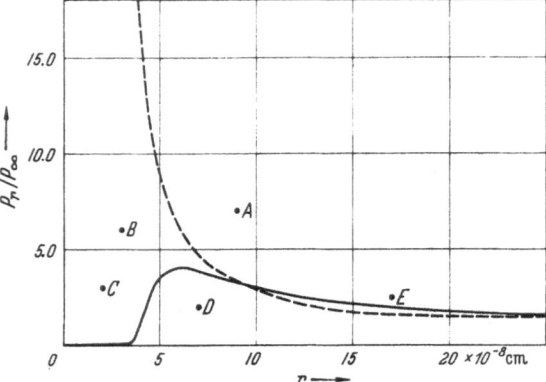

Fig. 2. P_r/P_∞ vs. r (plotted for a water drop at 291° K).

as a broken line for the case of water vapor at 291° K with $d\vartheta/dr = 0$. It is seen that for $r > 10^{-7}$ cm the ratio of P_r/P_∞ is of the order of unity and that for smaller radii the ratio becomes very large. This implies that small droplets evaporate very rapidly unless the pressure of the vapor surrounding them is maintained at a value higher than that indicated by the curve in Fig. 2. Since the density of a vapor is proportional to the vapor pressure, according to BOYLE's law, the ratio of P_r/P_∞ is just equal to the degree of saturation defined above. In this case the ratio is always greater than one so we shall refer to P_r/P_∞ as the supersaturation. Thus the broken curve in Fig. 2 is essentially a plot indicating the degree of supersaturation required to cause uncharged droplets of a given radius to remain in equilibrium with the surrounding vapor.

If a droplet carries an electrostatic charge, Eq. (3.1) is modified by the addition of a term due to the energy of the electrostatic field. According to THOMSON[1] the equation becomes

$$\log \frac{P_r}{P_\infty} = \frac{M}{RT\varrho}\left(\frac{2\vartheta}{r} - \frac{e^2}{8\pi\varepsilon r^4}\right) \quad (3.2)$$

where e is the charge on the drop, ε is the dielectric constant of the surrounding medium, and again $d\vartheta/dr = 0$ for simplicity. The modification in the variation of P_r/P_∞ with r is indicated for a singly charged water droplet in Fig. 2 by the solid curve.

To more readily understand the significance of the two curves shown in Fig. 2, consider droplets which can be represented by points on the diagram according

---

[1] J. J. THOMSON: Applications of Dynamics to Physics and Chemistry, Vol. I, p. 165. London: McMillan & Co. 1888.

to their radius and the degree of supersaturation of the medium surrounding them. A droplet at point A is of such size that vapor will condense upon it whether it is charged or not. Furthermore, if the surrounding vapor is maintained at the indicated degree of supersaturation, the droplet will grow indefinitely and its representative point, A, will move horizontally to the right as its radius, r, increases. A droplet at point B is at a position such that if it is uncharged, it will evaporate. But if it is charged, the supersaturation is sufficient for vapor to condense upon it and it will grow at a constant rate as long as the indicated degree of supersaturation is maintained. Similarly an uncharged drop at point C will evaporate, but a charged drop will grow until its radius is such that it reaches the equilibrium value given by the solid curve. There the droplet size will remain constant and further growth or evaporation is impossible since it is in equilibrium with the surrounding vapor. Point D represents the case in which an uncharged drop will evaporate completely, while a charged drop in the same state will evaporate until it intersects the equilibrium curve and will then remain there.

The foregoing discussion enables one to understand qualitatively the early experimental results of COULIER[1], AITKEN[2], and C. T. R. WILSON[3]. These authors, working with samples of air in sealed containers with sufficient liquid to saturate the volume, found that if an adiabatic expansion of the container's volume were made to supersaturate the gas-vapor mixture, then it was possible to cause the vapor to condense in a cloud of water droplets. If the supersaturation was of the order of 2 or 3 it was noted that the droplets would grow quite large and fall to the bottom of the vessel. Further expansions at the same supersaturation would then fail to produce any further condensation. Since the number of droplets formed at the outset depended upon the amount of dust or smoke in the air, it was concluded that these particles acted as centers about which the vapor molecules tended to cluster when the expansion was made. Because dust particles are of fairly large radius, only a small supersaturation was required for further drop growth, as in the case of point E on Fig. 2 above. When the drops fell to the bottom of the vessel, they attached themselves to the surface and did not diffuse back into the gas. Thus the subsequent expansions failed to produce droplets simply because there were no more dust particles in the gas to act as condensation nuclei.

C. T. R. WILSON found that if a volume of saturated gas, which had had the dust particles swept out of it as just described, was ionized by irradiating it with X-rays or α-particles, it was possible to condense the vapor on these ions if the supersaturation was slightly greater than 4.2. This is just the maximum value of the equilibrium curve for charged drops in Fig. 2. Point B in the figure is representative of the case of droplets formed on ions. It is just this property of being able to condense vapor on ions that has made the cloud chamber such an important tool for the study of the paths of ionizing particles. The minimum value of the supersaturation which will produce condensation on ions is called the "ion limit".

WILSON also found that if the supersaturation was increased beyond this ion limit, then at a certain value a dense cloud was formed. This higher value of the supersaturation is known as the "cloud limit" and sets an upper limit to the degree of supersaturation which can be used for the detection of the trajectories of ionizing particles. The explanation of the onset of this fog lies in the presence of very small condensation nuclei in the gas. These small nuclei can

[1] COULIER: J. Pharm. Chim., Paris 22, 165 (1875).
[2] J. AITKEN: Collected Papers, p. 34. Cambridge 1923.
[3] C. T. R. WILSON: Phil. Trans. Roy. Soc. Lond. 189, 265 (1897).

arise from any of several sources. In expansion chambers, drops which evaporate before falling to the bottom of the chamber will be suspended in the gas and on a subsequent expansion will act as condensation nuclei. As seen in Fig. 2, if the supersaturation is sufficiently large, any small uncharged object will grow into a large drop. Such small neutral objects can consist of statistical aggregates or clumps of molecules of the vapor in the chamber and will always be present[1].

From the preceding discussion it can be seen that in order to observe the trail of ions left in the wake of a charged particle passing through a cloud chamber, it is necessary to adjust the supersaturation in a dust free gas so that condensation will occur only on ions. Although the classical theory presented gives a qualitative understanding of the processes involved, it should be apparent that such a theory is far from being complete. For the very small droplets considered here, the radius is of the same order as the distance of separation of the molecules making up the drop. Furthermore, the number of molecules contained in a droplet of this size is rather small. (e.g. A water drop of radius 6×10^{-8} cm contains about thirty molecules.)[2] The statistical fluctuations on such small numbers will clearly be large and such macroscopic concepts as surface tension are of doubtful use. To set $d\vartheta/dr = 0$, as above, is certain to introduce some error.

The details of the process by which ions attract the surrounding vapor molecules to form a drop are obscure. It has been suggested that the ions attach themselves to the neutral clusters of vapor molecules which are always present and that the resulting aggregate forms an embryonic droplet. In discussing the effects of adding a charge to a droplet of such small size one ought to include the fact that this will lead to an effective surface charge by polarizing the droplet. TOHMFER and VOLMER[3] have studied this effect and find it necessary to include a term in Eq. (3.2) which depends on the dielectric constant of the condensed liquid. However, they point out that the dielectric constant in such a small drop is apt to be quite different from the macroscopic value. Furthermore, such a theory does not explain the experimentally observed fact that positively charged ions are more effective than negative ions as condensation nuclei. Presumably an explanation of this effect is to be found in the suggestion that strongly polar molecules such as water will form an oriented surface layer on the droplet which will depend on the sign of charge; however, no adequate theory of this effect has yet been given. In conclusion we note that although a qualitative description of the condensation phenomena can be presented as above, the detailed theoretical treatment of the problem is far from being complete[4].

c) The production of supersaturation by an adiabatic expansion.

4. The theory. The production of a supersaturated vapor by means of an adiabatic expansion can be described in the following way: Consider a volume, V_1, which contains a noncondensable gas of partial pressure, P_g, and a vapor of partial pressure, P_1, in equilibrium with its liquid phase. If the system is in thermal equilibrium, it will have an absolute temperature, T_1. Let M_1 be the total mass of vapor contained in V_1 at T_1 and let M be its gram molecular weight.

---

[1] For a detailed discussion of these aggregates cf. J. G. WILSON: Cloud Chamber Technique, p. 7. Cambridge 1952.

[2] For a detailed treatment of such small drops cf. FRENKEL: Kinetic Theory of Liquids, Chap. VII. Oxford 1946.

[3] TOHMFER and VOLMER: Ann. Phys., Lpz. **33**, 109 (1938).

[4] It should be noted that C. N. YANG and T. D. LEE, Phys. Rev. **87**, 404 (1952), have made considerable progress in describing the condensation of monatomic gases. Their methods have not yet been extended to the application considered here.

Then if the vapor approximates an ideal gas, one can write

$$P_1 V_1 = \frac{M_1}{M} R\, T_1 \qquad (4.1)$$

where R is the gas constant.

If now the volume of the chamber is expanded from V_1 to V_2 adiabatically, then the temperature T_1 decreases to T_2' as a result of the work performed by the gas during the expansion. This decrease in temperature is given by the well-known[1] relation:

$$\frac{T_1}{T_2'} = \left(\frac{V_2}{V_1}\right)^{\gamma-1}. \qquad (4.2)$$

Here γ is the ratio of specific heats for the mixture of the non-condensable gas and the vapor. RICHARTZ[2] has shown that γ for a mixture can be computed from the relation

$$\frac{1}{\gamma-1} = \sum_i \frac{1}{\gamma_i - 1}\, \frac{P_i}{P_0} \qquad (4.3)$$

where P_i is the partial pressure of the i-th constituent and the sum is to be taken over all constituents of the mixture. $P_0 = \sum_i P_i$, the total pressure of the mixture.

It is worth noting at this point that a variation of the partial pressure of one component of the mixture can be used to vary the γ of the mixture. This effect plays an important role in the operation of high pressure cloud chambers.

After the expansion is completed, but before any condensation occurs, the vapor pressure drops to P_2'. The same mass of vapor M_1 is present in V_2, so that

$$P_2' V_2 = \frac{M_1}{M} R\, T_2'. \qquad (4.4)$$

This state is unstable, since for a temperature of T_2' the mass of vapor M_1 is too great to be contained in the volume V_2. As a result, condensation occurs, and when equilibrium has been established once more, the amount of vapor in V_2 is M_2. The vapor pressure will have dropped from P_2' to P_2 and the temperature will have increased from T_2' to T_2. This temperature increase results from the latent heat of condensation liberated within the volume by the condensation process.

Now the degree of saturation of a vapor has been defined in Sect. 3, above, as the ratio of the density of vapor present to the saturated vapor density at a given temperature. The saturated vapor density can be measured by determining the density of vapor in an enclosed volume containing the liquid and gaseous (vapor) phases in equilibrium with each other. The saturated vapor density at the temperature T_2 is given by $\varrho_2 = M_2/V_2$. The vapor density before condensation is $\varrho_2' = M_1/V_2$ at temperature T_2', and if we assume that $T_2 \approx T_2'$, then the above definition of the supersaturation, S, gives

$$S = \frac{\varrho_2'}{\varrho_2} = \frac{M_1}{M_2}. \qquad (4.5)$$

Substituting from Eqs. (4.1) and (4.4) above, we get

$$S = \frac{P_1 V_1 T_2}{P_2 V_2 T_1}. \qquad (4.6)$$

[1] Cf. P. S. EPSTEIN: Textbook of Thermodynamics, p. 49, Eq. (3.37). New York: John Wiley 1937.
[2] Cf. F. RICHARTZ: Ann. d. Physik 19, 639 (1906).

Using Eq. (4.2) above this becomes

$$S = \frac{P_1}{P_2}\left(\frac{V_1}{V_2}\right)^{\gamma} = \frac{P_1}{P_2}\left(\frac{1}{1+r}\right)^{\gamma} \tag{4.7}$$

where $(1 + r) = \boldsymbol{R}$, the "expansion ratio" and $\boldsymbol{r}$ is the "percentage of expansion".

The saturated vapor pressure is given by an equation of the form

$$\log P = A - \frac{B}{T} \tag{4.8a}$$

or

$$P = A_1 e^{-\frac{B}{T}} \tag{4.8b}$$

to a close approximation[1]. The values of the constants A and B are tabulated for a wide variety of substances[2] and more exact empirical tables of P vs. T for some common liquids and mixtures are available[3].

FLOOD[4] found that the value of the supersaturation at the ion limit or cloud limit depended strongly upon the type of vapor used in the chamber. For example the ion or cloud limit sets in at a lower value of the supersaturation for alcohol than for water and a mixture of the two depresses these limits still farther. A possible explanation of this effect is that either of these limits depends upon the number of droplets formed in a unit volume and for a given supersaturation this number may differ from one liquid to another. The critical supersaturation for a mixture, $S_c = P_r/P_\infty$ given by Eq. (3.1) above has been shown by FLOOD to have the form

$$\log S_c = \mathsf{k}\,\frac{\vartheta}{T}\,\frac{M}{\varrho} \tag{4.9}$$

where k is BOLTZMANN's constant, ϑ is the surface tension of the mixture, and M/ϱ is the molecular volume of the mixture. This relation is in good agreement with the experimental results. BECK[5] has continued these studies and used various mixtures to experimentally determine which would give the best photographs of tracks formed by ionizing particles. Although he recommends a mixture of ethyl alcohol, water, and acetone in the ratio by volume of $2:1:1$, it is a more common practice to use just ethyl alcohol and water in a ratio of $2:1$ by volume. As shown by BECK, the optimum value of this ratio is not critical and gives satisfactory results over a wide range.

A point of immediate interest is the length of time required for a drop to grow to a sufficiently large size so that it may be photographed. The rate of growth of a drop can be computed with the aid of a classical theory. The process to be considered is the following: The vapor surrounding the condensation nucleus diffuses toward it and when the vapor condenses at the surface of the embryonic drop, an amount of heat is liberated equal to the heat of vaporization. This heat in turn tends to increase the temperature of the drop and is removed from the drop by conduction to the surrounding atmosphere. HAZEN[6] has shown that these two processes can be described by means of an approximate equation for the diffusion of vapor toward the drop and another diffusion equation for

---

[1] Cf. P. S. EPSTEIN: Textbook of Thermodynamics, p. 121. New York: John Wiley 1937.

[2] Cf. Handbook of Chemistry and Physics. Cleveland, Ohio, U.S.A.: Chemical Rubber Publishing Company.

[3] Cf. International Critical Tables. New York: McGraw-Hill 1929.

[4] Cf. H. FLOOD: Z. phys. Chem. **170**, 294 (1934).

[5] C. BECK: Rev. Sci. Instrum. **12**, 602 (1941).

[6] W. E. HAZEN: Rev. Sci. Instrum. **13**, 247 (1942).

the conduction of heat away from it. The expressions are:

and

$$\frac{d(r^2)}{dt} = \frac{2D}{\varrho_L} (\varrho_r - \varrho_0) \tag{4.10}$$

$$\frac{d(r^2)}{dt} = \frac{2K}{\varrho_L \lambda} (T_r - T_0) . \tag{4.11}$$

Here ϱ_r and T_r are the vapor density and temperature at the surface of the drop of radius r, while ϱ_0 and T_0 are the values far away from the surface. ϱ_L is the density of the liquid, D is the coefficient of diffusion of the vapor, K is the thermal conductivity, λ the latent heat of condensation, and t the time. If a volume of gas and vapor of density ϱ_1 and temperature T_1 is expanded in the ratio $\boldsymbol{R}$, then with the aid of Eq. (4.8b) the two relations above can be combined to give

$$\frac{d(r^2)}{dt} = \frac{2D\varrho_1}{\varrho_L \boldsymbol{R}} \left[1 - \boldsymbol{R}^\gamma \exp\left\{ -\frac{B}{T_1}\left(1 - \boldsymbol{R}^{1-\gamma} - \frac{\lambda \varrho_L}{2KT_1} \cdot \frac{d(r^2)}{dt} \right)\right\}\right]. \tag{4.12}$$

Measurements of the rate of drop growth have been made by observing the variation in the rate of free fall of droplets in a chamber. With the aid of STO-KES' Law, BRODE and others[1] have estimated that the measured rate of growth is in good agreement with Eq. (4.12) for heavy gases such as nitrogen, but in hydrogen there is a real discrepancy. This latter can presumably be attributed to the fact that in using STOKES' Law, no account is taken of the increased buoyancy of the warm drop which heats the gas immediately surrounding it. For nitrogen the rate of growth $d(r^2)/dt$ is about 5×10^{-6} cm²/sec, so that to produce a droplet of visible size (radius about 10^{-3} cm) requires about one-fifth of a second. In the lighter gases such as helium the time is roughly half of this or one-tenth of a second. This implies that when using an expansion chamber one must delay the time of photography by about one-tenth to one-fifth of a second in order that the drops will be able to grow to visible size. Table 2 gives the

Table 2. *The rates of drop growth in various gases and liquids.*

Author	$\dfrac{d(r^2)}{dt}$ [cm² sec⁻¹ × 10⁶]	Gas	Vapor
HAZEN	5.5	N_2	C_2H_5OH
HAZEN	18	H_2	C_2H_5OH
HAZEN	17	He	C_2H_5OH
HAZEN	10	He	$3:1::C_2H_5OH:H_2O$
HAZEN	4.0	N_2	$3:1::C_2H_5OH:H_2O$
BARRETT and GERMAIN	7.5	Air	H_2O
BARRETT and GERMAIN	4.4	Air	$3:1::C_2H_5OH:H_2O$

experimental results of HAZEN and BARRETT and GERMAIN for the values of $d(r^2)/dt$ in various gases and vapors. The variation of rate of growth with the expansion ratio, $\boldsymbol{R}$, was verified by HAZEN, but the values in the table are only approximate values which hold for final pressures of 1.1 to 1.2 atmospheres when the expansion ratio is adequate to produce condensation on ions.

The rather appreciable time required for drop growth raises the question as to what is the length of time that the supersaturation persists after an adiabatic expansion? It is clear that after the gas in the chamber is chilled by the expansion, it will warm up as a result of heat being conducted to it from the walls of the

---

[1] Cf. R. B. BRODE: Rev. Mod. Phys. **11**, 222 (1939). — W. HAZEN: Rev. Sci. Instrum. **13**, 247 (1942). — O. BARRETT and L. GERMAIN: Rev. Sci. Instrum. **18**, 84 (1947).

chamber. Furthermore, the liberation of the latent heat of condensation through-out the gas due to the formation of drops will also contribute to the warming process. WILLIAMS[1] has calculated the *sensitive time* of a chamber, τ_0, neglecting the effects of condensation. He obtains

$$\tau_0 = 0.77 \frac{\varrho_g C_p}{K(\gamma-1)^2} \left(\frac{V}{A}\right)^2 \left(\frac{\delta r}{r}\right)^2 \tag{4.13}$$

where ϱ_g, C_p, K, and γ are the density, specific heat at constant pressure, thermal conductivity and ratio of specific heats of the gas. V/A is the ratio of volume to surface area of the chamber and $\delta r/r$ is the fractional change in the *percentage expansion ratio*. HAZEN[2] modified this calculation to include the effects of condensation. He has computed a time τ_1, with the aid of the expression for the rate of drop growth, Eq. (4.12), and the density of drops, n.

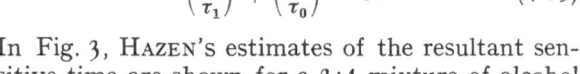

$$\tau_1^{\frac{3}{2}} = \frac{1}{4\pi} \frac{\varrho_g C_p}{K(\gamma-1)} \frac{1}{\frac{2}{3}n(dr^2/dt)^{\frac{1}{2}}} \left(\frac{\delta r}{r}\right). \tag{4.14}$$

The *resultant sensitive time*, $\bar{\tau}$, is finally obtained from the expression

$$\left(\frac{\bar{\tau}}{\tau_1}\right)^{\frac{3}{2}} + \left(\frac{\bar{\tau}}{\tau_0}\right)^{\frac{1}{2}} = 1. \tag{4.15}$$

Fig. 3. The resultant sensitive time as a function of the percentage expansion ratio.

In Fig. 3, HAZEN's estimates of the resultant sensitive time are shown for a 3:1 mixture of alcohol and water in nitrogen. He measured the experimental points in a chamber 30 cm in diameter by moving a collimated β-ray source through the sensitive volume and intermittently illuminating the chamber. By observing the position of the source as it emitted visible tracks, the sensitive time of the chamber was determined. It is clear from Fig. 3 that the resultant sensitive time for a chamber can be several times longer than the time required for drops to grow to visible size.

d) Expansion cloud chambers.

5. Volume defined chambers. One of the most widely used types of expansion cloud chamber follows the original apparatus of C. T. R. WILSON[3]. He employed a vessel with a tightly fitting, movable wall and by regulating the distance of motion of this wall was able to control the supersaturation produced inside the vessel. The motion was controlled pneumatically and the active volume of the vessel was sealed from the outer atmosphere by having both the walls of the vessel and the extended edges of the movable wall remain submerged in a pool of oil or water. This type of seal had the disadvantage that the motion of the piston had to take place in a vertical direction. Various modifications and improvements on this type of chamber were made by BLACKETT[4] who used the apparatus for his classical studies of α-particle scattering.

An important modification of the technique was the use of a flexible rubber sheet clamped to the edges of a solid metal piston and to the walls of the chamber to seal the active volume of the chamber. The rigid piston has the advantage of insuring an uniform expansion of the gas throughout the volume of the chamber.

[1] E. J. WILLIAMS: Proc. Cambridge Phil. Soc. **35**, 512 (1939).
[2] W. E. HAZEN: Rev. Sci. Instrum. **13**, 247 (1942).
[3] C. T. R. WILSON: Proc. Roy. Soc. Lond., Ser. A **189**, 265 (1897).
[4] P. M. S. BLACKETT: Proc. Roy. Soc. Lond. **123**, 619 (1929). — J. Sci. Instrum. **4**, 433 (1927); **6**, 184 (1929).

The only disadvantages are that the inertia of such a solid piston tends to make the expansion time rather long and when used in a magnetic field a heating of the metal due to eddy currents can occur. The effect of a long expansion time upon the accuracy of measurements will be discussed below, but it is sufficient

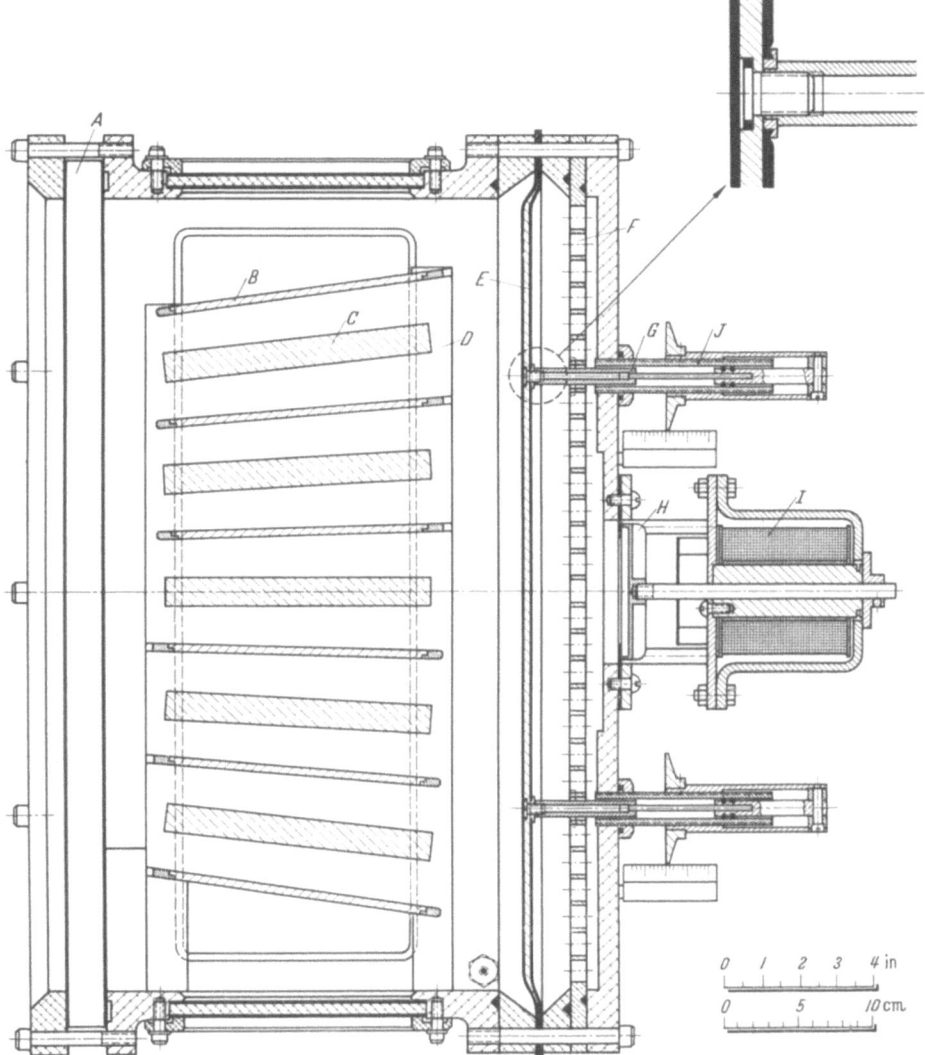

Fig. 4. A multiplate, volume-defined expansion cloud chamber.

to remark that neither of these effects is considered to be a sufficiently serious disadvantage to offset the insurance of a uniform expansion.

An example of a volume controlled expansion chamber is illustrated in **Fig. 4**. Here a cross-sectional view of the chamber used by ALTHAUS and SARD[1] is shown. The metal piston, E, is fastened between two sheets of rubber and in

---

[1] E. J. ALTHAUS and R. D. SARD: Phys. Rev. **91**, 373 (1953).

the expanded position rests against the backing hole plate, *F*. Compressed air is used to force the piston forward until the stops, *G*, on the rods supporting the piston, arrest the motion. The expansion ratio is varied by simply moving the

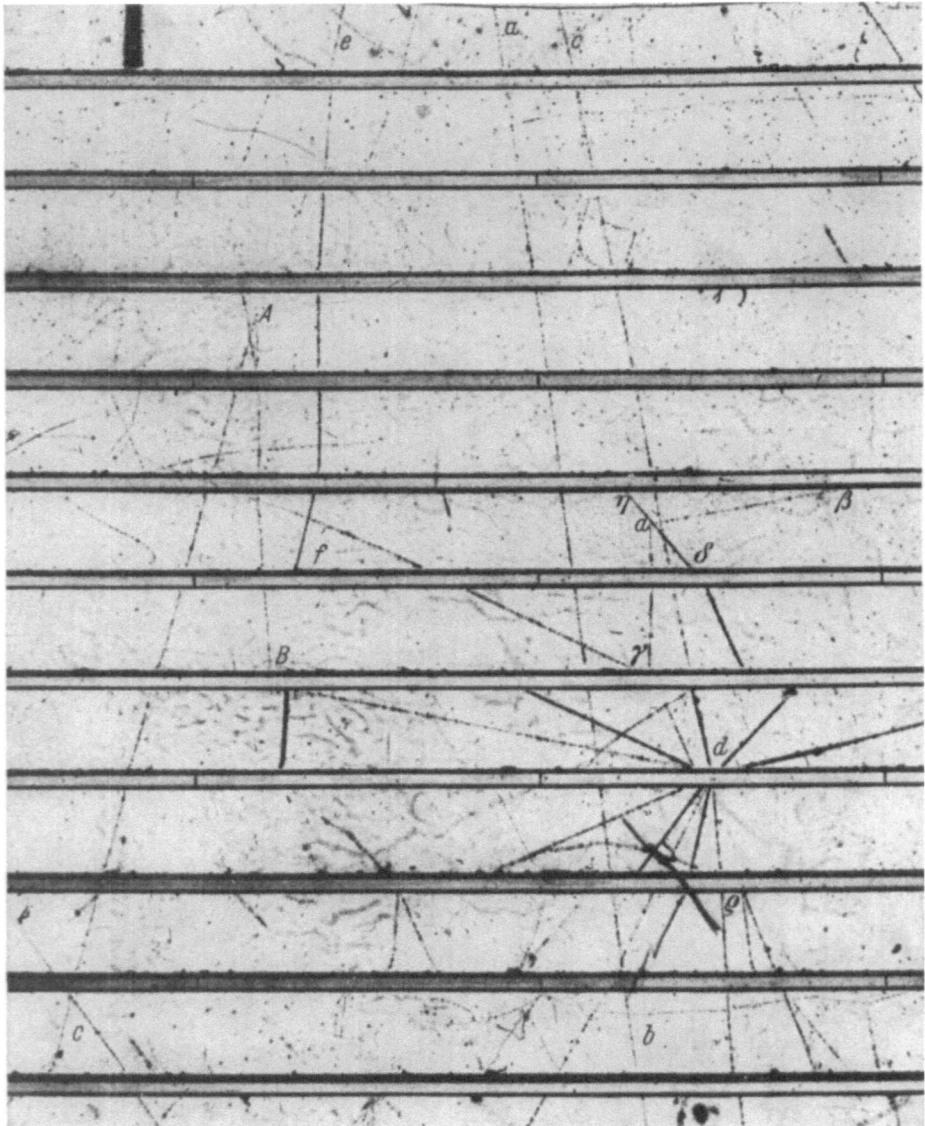

Fig. 5. An example of a photograph taken with the multiplate chamber shown in Fig. 4.

stops back and forth by means of the micrometer screws, *J*. A large expansion valve, *H*, controlled by a solenoid, *I*, is used to exhaust the air from the back volume of the chamber when an adiabatic expansion is required. The active volume of the chamber between the front glass, *A*, and the piston has an array of lead, *B*, and carbon, *C*, plates mounted on a transparent plastic support, *D*.

The plates are mounted at varying angles so that when viewed by the camera only their leading edges are seen. The plates give this type of chamber the name "multi-plate chamber" and its chief application has been in the study of the interaction of charged particles with matter. An example[1] of a photograph taken with this chamber, when only lead plates were mounted in the active volume, is shown in Fig. 5. The particle entering the top of the chamber at point, e, is seen to traverse several plates and come to rest at, f. There is a noticeable increase in its ionization as it is slowed down by penetrating successive plates. The particle entering at c traverses several plates before it undergoes a nuclear interaction at d, giving rise to the emission of a large number of charged particles. At point A, between the third and fourth plates, a neutral particle apparently decays in the gas of the chamber into two oppositely charged fragments whose tracks are seen to extend downward through the chamber to B and C.

6. Pressure defined chambers. Another method of controling the expansion is to divide a chamber into two parts. If one part has its pressure reduced and the intervening wall is suddenly opened, the gas will expand until a pressure equilibrium is established. The supersaturation resulting from such an expansion can then be controlled by the decrease in pressure when the partition is opened. C. T. R. Wilson[2] first developed a chamber which essentially makes use of this principle. The active volume of the chamber is divided into two parts by a screen-wire mesh and hole plate. The back part of the chamber is sealed from the atmosphere by a rubber sheet. This sheet is blown forward against the hole plate by compressed air and thus diminishes the rear portion of the active volume of the chamber. When the compressed air is suddenly released the rubber sheet is forced backward until it comes to rest against a retaining plate. The result is that the gas in the front part of the active volume expands through the screen mesh-hole plate separator and undergoes an adiabatic cooling. It is true that this method of expansion is in the strictest sense not of a pressure defined type, since the change in pressure of the rear portion of the active volume is controlled by the change of volume of this part of the chamber when the diaphram moves. However, it is the most practical way to obtain a pressure defined expansion.

A typical example of this type of chamber, mounted in an iron core magnet, is shown in Figs. 6a and b. This is the apparatus of R. W. Thompson[3] at the University of Indiana. The rubber diaphram is blown forward against the hole plate, T, by compressed air and when the expansion valves, N, are opened, the diaphram is forced backward until it comes to rest against the perforated backing plate, U. The adjustment of the amount of expansion is controlled by changing the position of this backing plate, with the aid of the screws, Z. The perforated plate, V, is faced with black velvet cloth to form a uniform photographic background and separates the front portion of the active volume from the rear. Fig. 7 is an example of a photograph of a penetrating shower taken with this apparatus. The positively charged particle which enters at the upper left is heavily ionizing and decays spontaneously into another charged particle which leaves the chamber at the right[4]. The straight vertical line at the far left is part of a ruled grid of fiducial marks on the front glass. These lines are used

---

[1] Cf. N. F. Harmon: Ph. D. Thesis, Washington University, St. Louis 1955.
[2] C. T. R. Wilson: Proc. Roy. Soc. Lond. **142**, 88 (1933).
[3] R. W. Thompson, J. R. Burwell and R. W. Huggett: Nuovo Cim. **1956**.
[4] The detailed analysis of this photograph has been given by Y. B. Kim, J. R. Burwell, H. O. Cohn, C. J. Karzmark and R. W. Thompson: Phys. Rev. **95**, 661 (1954).

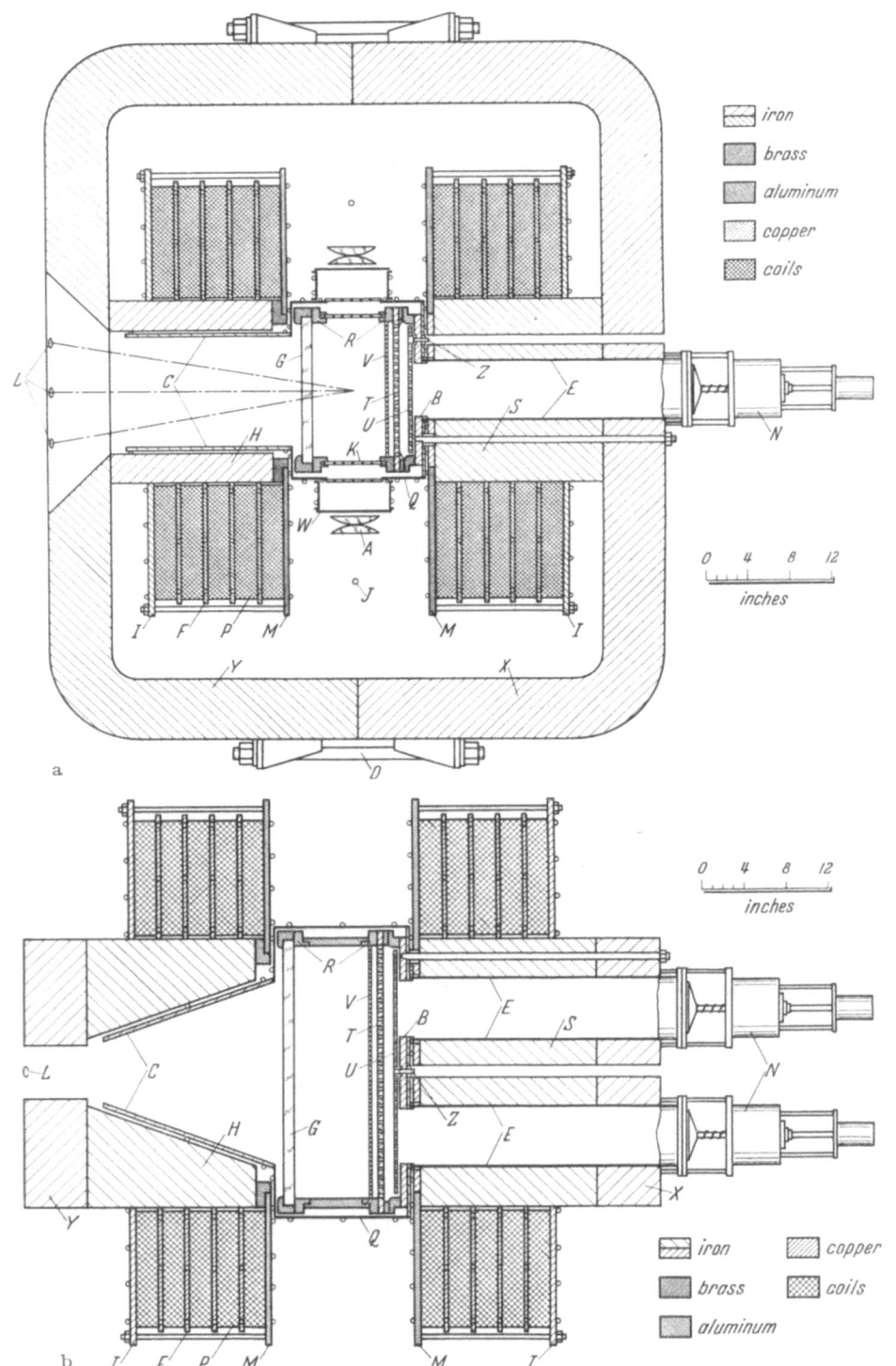

Fig. 6a and b. Top and side views of a pressure-defined expansion chamber in an electromagnet.

to check possible optical distortions as well as for reference in the stereoscopic reprojection of the photographs.

One clear advantage of a rubber diaphram chamber over the metal piston type previously described is that the expansion time is not limited appreciably by the intertial mass of the moving rubber. However, it is difficult to critically damp the motion of such a rubber sheet when it strikes the backing plate and this can cause shock waves and turbulence to be set up in the front portion of the active volume. BRODE and MERKLE[1] have succeeded in damping this motion by placing pads of loosely woven cloth in the expansion valve openings to give a high impedance to the outward flow of air. This has the unfortunate effect that the expansion time is increased, thus losing part of the advantage gained from the use of a sheet of rubber. Another objection which has been raised to this type of expansion mechanism is that when the gas of the chamber is drawn through the various hole plates in the course of an expansion, a high degree of turbulence is set up in the rear portion of the active volume and this in turn can create a large number of condensation nuclei. These are forced into the front section of the active volume upon recompression and produce a dense fog on subsequent expansions. Such objectionable condensation nuclei can be swept out of the chamber after the manner of dust particles, by "slow expansions". That

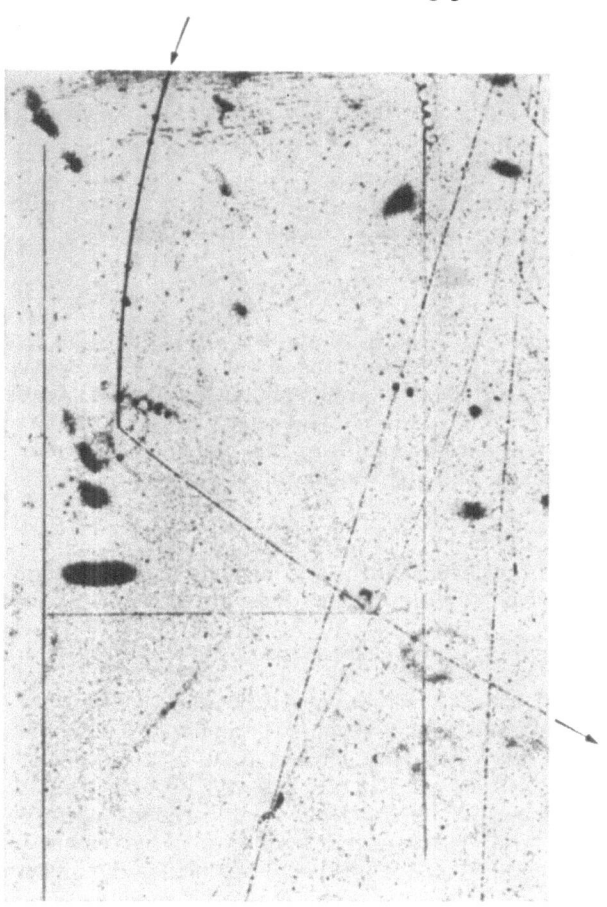

Fig. 7. An example of the decay of an unstable charged particle photographed with the apparatus shown in Fig. 6.

is, by appreciably prolonging the time of expansion, one decreases the amount of supersaturation below the ion limit for a given expansion ratio. These "slow" or subcritically saturated expansions can be used between "fast" expansions to clean the active volume of undesirable condensation nuclei. Many of the solid piston type chambers described above do not require these cleaning expansions, whereas almost all of those employing rubber diaphrams necessitate the use of one or more slow expansions to prepare the chamber for another rapid expansion.

---

[1] R. B. BRODE and T. C. MERKLE: Private Communication.

7. The timing sequence required by expansion chambers. In our previous discussion of the time necessary for droplet growth and the duration of the super-saturation after an adiabatic expansion, the need for proper timing in expansion chamber operation was emphasized. A number of other aspects of the problem must be considered before outlining the general sequence of operations required.

First consider the effect of the time interval between the moment at which an ionizing particle traverses the gas of the chamber and the moment, sometime later, at which the drops formed on the resulting ions have grown to sufficient size to have essentially zero mobility in the gas. In this interval, the ions will diffuse outward from the axis of the original thin cylinder of ions left in the wake of the bombarding particle. The amount of diffusion determines the width of the final column of droplets and will increase with the time interval between formation and immobilization. The diffusion of the ions can be described in terms of the number of ions, $n(r, t)$, to be found in an interval, dr, at a distance, r, from the cylinder axis at the time, t. If none of the ions recombine in the time, t, then[1]

$$n(r, t) = (n_0/4\pi k t) \exp(-r^2/4 k t) \tag{7.1}$$

where n_0 is the total number of ions per unit length of path and k is the diffusion coefficient for the ions in the gas. A photograph of the column of droplets is a projection of the cylinder, onto a plane. To obtain the density of drop images on the film, the above density in space must be integrated over all depths measured at right angles to this plane. If x is now the transverse width of the track image, the distribution of drops in dx at a distance x from the axis is

$$n(x, t) = \left(n_0/\sqrt{4\pi k t}\right) \exp(-x^2/4 k t). \tag{7.2}$$

If the track width is arbitrarily taken as that distance X which contains 90% of the droplet images, then integration of $n(x, t)$ gives the relation that

$$X = 4.68 \sqrt{k t}. \tag{7.3}$$

BLACKETT[2] has measured the track width in air as a function of the time interval between the passage of the ionizing particle and the moment at which the ions are immobilized. Using a value of $k = 0.034$ cm²/sec for the mean diffusion coefficient of the positive and negative ions in air (at N.T.P.), he found that times of the order of 14 milliseconds corresponded to track widths of about 1 mm. This is in reasonably good agreement with the theoretical relation given by Eq. (7.3). It has been shown that the ions are immobilized by drop growth in a time which is very nearly the same as the interval required to complete the mechanical expansion. As a result it has become conventional to estimate the mechanical expansion time of a chamber by the width of the tracks which are formed just at the time of initiating the expansion.

One rather important point has been ignored in the foregoing discussion. As a result of the passage of cosmic ray particles and decay products from radio-active contamination in the walls and gas of the chamber, there will always be a certain number of free ions present in any volume of gas. The recombination of these ions depletes the number being continually formed, but a sufficiently large number is always present to give rise to a general background of droplets when a chamber is expanded. To remove these ions, an electrostatic field of about 10 volt/cm is used to sweep them out of the gas. In chambers with metal

---

[1] Cf. G. JAFFÉ: Ann. Physik **42**, 303 (1913).
[2] P. M. S. BLACKETT: Proc. Roy. Soc. Lond., Ser. A **146**, 281 (1934).

pistons, it is conventional to ground the chamber walls and charge the piston to a rather high voltage. Multiplate chambers alternately charge and ground the metal plates to provide this "clearing" or "sweep" field. Various other schemes such as wire grids, or grids of conducting paint have been used to provide the sweep field. To prevent this electric field from accelerating the diffusion of the ion column formed by a track to be photographed, the sweep field is usually short circuited just as the expansion of the chamber is begun. If the clearing field is not grounded, the positive and negative ion column will be separated, giving a "doubling" of the tracks. HAZEN[1] has used just this effect to advantage in the study of the relative condensation efficiencies of positive and negative ions referred to in Sect. 3.

Prior to the work of BLACKETT and OCCHIALINI[2], cloud chambers were expanded at random and only the tracks formed by the traversal of charged particles during the sensitive time of the chamber could be studied. This sensitive time was seen to be of the order of $1/10$ sec, and for many experimental problems a prohibitive number of photographs was required. The vast improvement in efficiency obtained by BLACKETT and OCCHIALINI using Geiger counters to initiate the expansion has made this or some similar triggering arrangement an almost universally accepted practice. In cosmic ray work and in some experiments with particle accelerators, an array of Geiger or scintillation counters is used in coincidence to select the type of event required for the experiment. The electronic counter circuits can give a triggering pulse within a few microseconds of the traversal of the chamber by the desired ionizing particle. This triggering pulse is used to set off the following sequence of events:

(a) The trigger pulse electronically short circuits the clearing field in a few microseconds.

(b) The expansion (or "pop") valves are opened. An interval of from 5 to 10 milliseconds is required to completely open the valves in current use.

(c) The mechanical expansion of the gas begins as soon as the expansion valves open and, as noted above, should be completed within about 15 to 20 milliseconds.

(d) After a delay of about one hundred milliseconds $(1/10$ sec) the flash lamps are triggered to illuminate the droplets.

(e) After a period of a few seconds the cameras are automatically wound to provide a fresh film, and the number register appearing on each photograph is advanced one digit.

(f) The expansion valves automatically shut after about one second and the chamber begins to recompress. The clearing field voltage is turned on again at this time.

(g) If the chamber requires slow expansions to clean out the re-evaporation and other condensation nuclei formed by the fast expansion, then the first slow expansion is initiated thirty or forty seconds after the trigger pulse. Any subsequent expansions follow at about the same interval and require about fifteen seconds to slowly expand, wait for the droplets formed to fall to the bottom of the chamber, and then recompress.

Usually the chamber is not permitted to repeat this cycle of events for a rather lengthy period (a minute or more) by automatically disconnecting the input trigger pulse during the cycle. The waiting period after the last slow expansion gives the gas in the active volume sufficient time to re-establish thermal

[1] W. E. HAZEN: Phys. Rev. **65**, 259 (1944).

[2] P. M. S. BLACKETT and G. P. S. OCCHIALINI: Proc. Roy. Soc. Lond. **139**, 699 (1933).

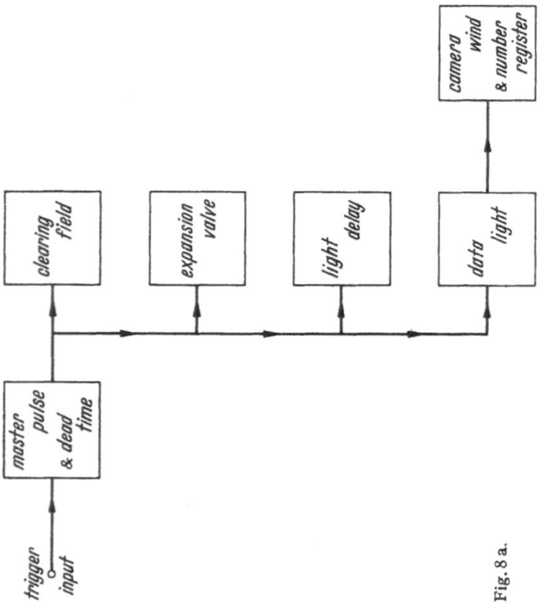

Fig. 8a and b. A typical electronic timing circuit for an expansion cloud chamber. (a) Block diagram and (b) details of the circuits.

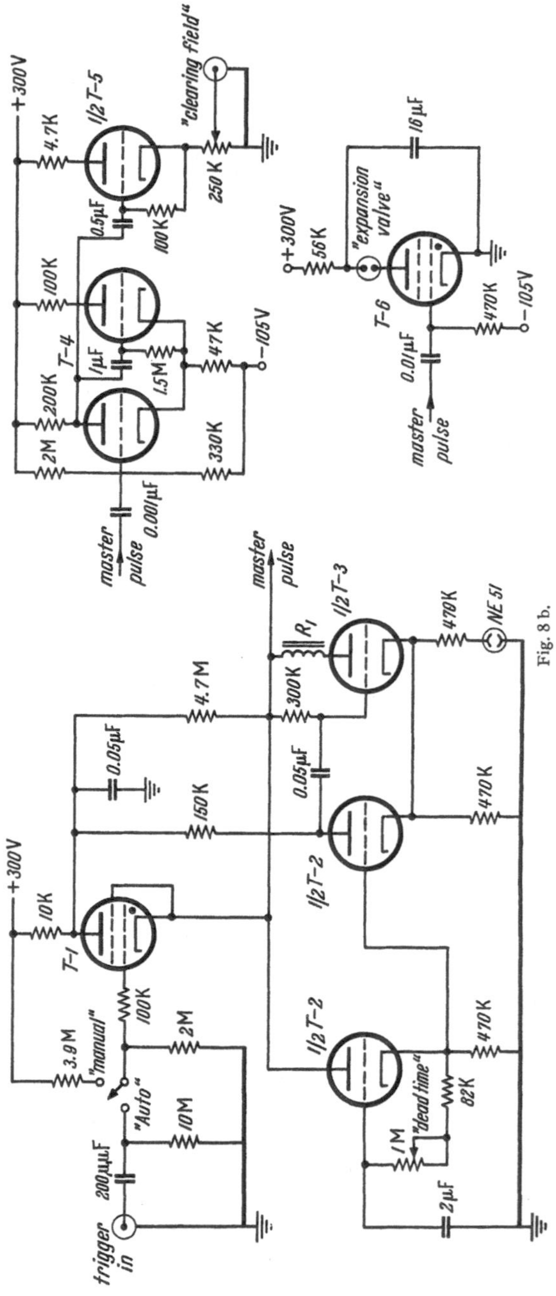

Fig. 8 a.

Fig. 8 b.

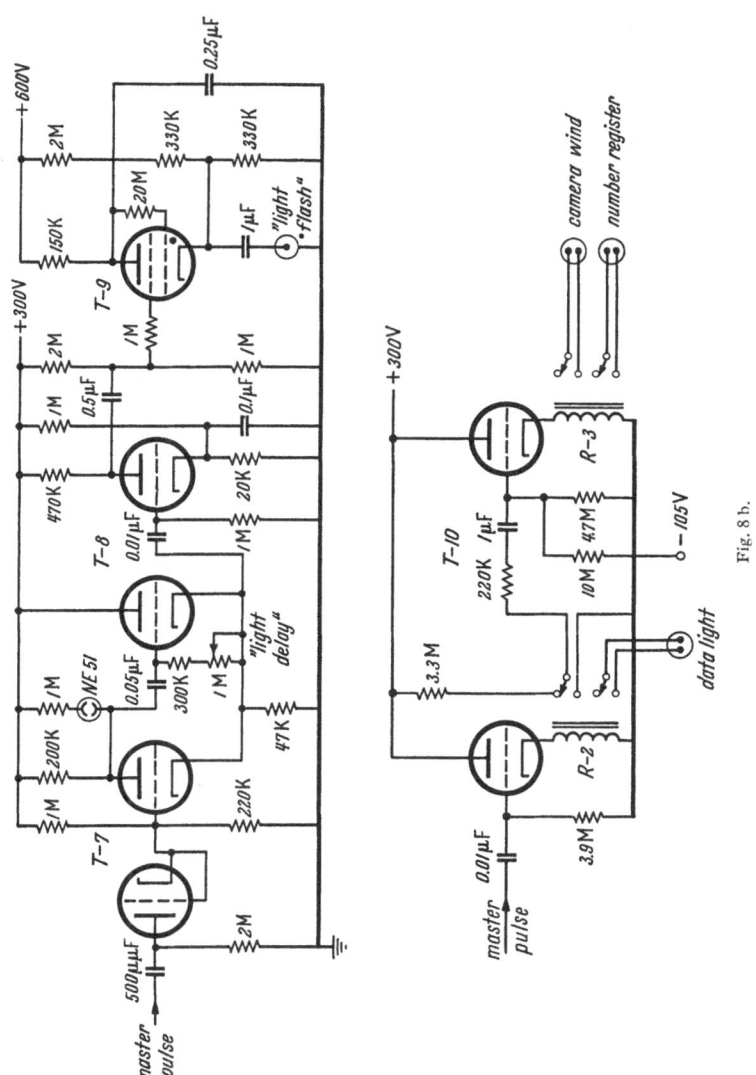

Fig. 8 b.

equilibrium with the surrounding chamber walls. The total period that the trigger pulse is disconnected is called the "dead time" of the chamber and is a measure of the efficiency of the useful running time of any given piece of apparatus.

Fig. 8 illustrates a typical electronic circuit which performs most of the functions outlined above. This particular circuit is a modification of one which was developed by E. W. COWAN[1] at the California Institute of Technology. The block diagram shown in Fig. 8a indicates the functions performed by each part of the circuit. A trigger pulse from a counter coincidence circuit is used to initiate a "master pulse" from a thyratron and simultaneously trip the "dead time" circuit. The master pulse is used to short circuit the clearing field, trip the expansion valves, start the adjustable time delay for flashing the lamps, close first a relay to illuminate a data board and number register viewed by the camera, and finally to close another relay which starts the motor driven camera and advances the number register. The dead times obtained with this circuit are of the order of one to two and one-half minutes and during this period the thyratron giving the master pulse cannot be fired. This circuit is used to control a piston type chamber which does not require slow expansions, so that this function is not included in the timing sequence. When required, the long delays between successive slow expansions can be provided by motor driven cam switches or an electronic circuit which is similar in construction to the "dead time" circuit illustrated here.

8. Thermal effects on expansion chambers. In the preceding paragraph it was noted that after a series of expansions and recompressions of the gas in the active volume of the chamber, a period of time must be allowed for thermal equilibrium to be re-established before continuing the operation of the chamber. One might enquire into the problem of how fluctuations in the temperature of the cloud chamber surroundings affect its operation[2]. If the temperature before expansion, T_1, is essentially the same as that of the air around the chamber, the change in expansion ratio, r, required to offset a change in air temperature, ΔT, can be computed. If the expression in Eq. (4.8b) is substituted into Eq. (4.7), the supersaturation can be written approximately as

$$S \approx \exp\{- B\,r\,(\gamma - 1)/T_1 + 1 + \gamma\,r\}. \tag{8.1}$$

If S is required to be constant for a given change ΔT, then $\Delta S = 0$ and we have the condition

$$\Delta\{- B\,r\,(\gamma - 1)/T_1 + 1 + \gamma\,r\} = 0 \tag{8.1a}$$

or

$$\frac{\Delta r}{r} = \frac{B(\gamma - 1)}{B(\gamma - 1) - \gamma T_1}\,\frac{\Delta T_1}{T_1}. \tag{8.1b}$$

To see what order of magnitude this fractional change of temperature implies, consider an example. For a 3:1 mixture of alcohol and water, $B \approx 5200$. If $T_1 = 293°$ K (20° C), and $\gamma = 1.4$, then $\Delta r/r = 1.25\,\Delta T_1/T_1$. If then $\Delta T = 1°$, the change in $\Delta r/r \sim 0.4\%$, which is a negligible amount.

Although one might correctly conclude that fluctuations in the laboratory air temperature do not greatly affect the expansion ratio required to produce satisfactory tracks, there are other considerations which demand that the temperature of the chamber surroundings be maintained at as constant a value as possible. One of the most striking reasons is that the vapor in the chamber will tend to

---

[1] E. W. COWAN: Private Communication 1953.

[2] E. J. WILLIAMS: Proc. Cambridge Phil. Soc. **35**, 512 (1939).

condense on the inside of the front glass of the chamber as the outside temperature increases. The explanation for this is found in the fact that the glass has a very low heat conductivity compared to the metal chamber walls. The gas in the chamber is warmed by conduction from the walls more rapidly than the front glass, and hence the vapor tends to condense on the cool glass. The resulting film of liquid usually makes photography impossible.

In recent years a great deal of effort has been expended to maintain the temperature of chamber surroundings at extremely constant values. The chamber used by THOMPSON and shown above in Fig. 6 is thermally isolated from the magnet and room air by an elaborate thermal jacket, Q, immediately surrounding the chamber and a shield, C, which extends out into the camera well in the front pole piece of the magnet. Distilled water is circulated from a reservoir through copper tubing attached to the jacket and the water temperature fluctuations are maintained constant to within $\pm 0.02°$ C. Comparable results have been obtained with similar apparatus[1]. The effort to maintain these constant temperatures has grown out of the work of BLACKETT and WILSON[2], who made the first careful study of the distortions of tracks due to motion of the gas inside of the chamber and introduced the use of the thermal jacket to isolate the chamber. The effect of these gaseous distortions and the error they introduce in the various measurements of the trajectories of particles passing through the chamber will be discussed in detail in Sect. 20, below. It is clear that if the gas is undergoing irregular convective motion during the period between particle traversal and the moment of photography, the droplets will be carried along with the gas and will form a distorted track. It has been found that convective currents existing in the gas before the expansion occurs are the primary source of track distortion, and as a result a major effort has been exerted to minimize the temperature gradients causing these currents. A volume of gas at uniform temperature is in a state of neutral stability and very slight temperature variations can initiate convective motion. As a result it has been found useful to impose a small temperature gradient across the vertical dimension of the chamber. By warming the top of the chamber by a slight amount relative to the bottom, a stable equilibrium of the gas is obtained which is less sensitive to slight temperature fluctuations than a chamber at uniform temperature. Furthermore, this type of gradient can be re-established more readily than the uniform temperature after an adiabatic expansion. The vertical gradients required in practice are of the order of a few hundreths of a degree centigrade per centimeter.

9. Variations of the expansion chamber technique. Two novel modifications of the rubber diaphram type chamber were introduced by WILSON and WILSON[3]. The first was a radially expanding type chamber with cylindrical symmetry and the second was to allow this chamber to fall freely. The object was to control or decrease distortion of the tracks formed in the chamber. The radially expanding chamber produced a uniform radial expansion of the gas, while the freely falling chamber ought not to have contained convection currents in the absence of the gravitational field. Neither of these ideas has received widespread useage, although various modifications of one or the other have been employed[4]. The problem of photography when either of these methods was attempted proved very difficult and the uniformity and amount of distortion of the gases has not turned out in practice to be as small as the theory would indicate.

[1] Cf. R. B. LEIGHTON, S. D. WANLASS and C. D. ANDERSON: Phys. Rev. **89**, 148 (1953).
[2] P. M. S. BLACKETT and J. G. WILSON: Proc. Roy. Soc. Lond., Ser. A **160**, 304 (1937).
[3] C. T. R. WILSON and J. G. WILSON: Proc. Roy. Soc. Lond. **148**. 523 (1935).
[4] E. g. C. M. YORK: Phys. Rev. **85**, 998 (1952).

In addition to these ideas, certain experiments have imposed other modifications which are of interest. JOLIOT[1] has operated a chamber which utilizes a gas at less than atmospheric pressure in its active volume. In studies of uranium fission and similar problems in which the range of the observed particles in a gas at atmospheric pressure is so short that the tracks are not measureable, the use of low pressure gases to "magnify" the range has proved invaluable. Such low-pressure chambers can be constructed exactly like one of the previously described chambers, except that the air holding the piston or diaphram in the compressed position is at less than atmospheric pressure and must be exhausted into a large evacuated reservoir instead of the atmosphere. On the other hand, MILLS[2] has used a mechanical expansion mechanism to actuate a small low pressure chamber. When expanded from a total pressure of 45 mm with 40% He and 60% water vapor, he finds that the stopping power of the mixture is reduced to 5.7% of that of air at normal temperature and pressure.

A number of experiments have studied the interaction of particles traversing the chamber with the material in the chamber. The multiplate chamber already described in detail is one solution to the problem of obtaining as many inter-actions per traversal as possible. The multiplate chamber is an extension of the technique introduced by ANDERSON[3] in which he mounted a single metal plate in the center of a chamber in a magnetic field. This gave the desired increase in the number of interactions, together with the ability to determine momenta from curvature measurements in the magnetic field. Although a great deal of work has been performed with arrangements of this type, the immediate dis-advantage of the method stems from the fact that the presence of the plate gives rise to distortions of the gas during an expansion and thus impairs the accuracy of the curvature measurements.

A somewhat different solution to the problem of increasing the amount of material in the chamber was found by MOTT-SMITH[4]. He simply increased the pressure of the gas in the active volume of the chamber, thereby increasing its density and the number of observed interactions. A number of high pressure chambers have been constructed in the years following MOTT-SMITH's work. One of the most elaborate designs is that of JOHNSON, BENEDETTI, and SHUTT[5] which could operate at pressures of the order of 200 to 300 atmospheres. The expansion ratio required to produce tracks in such a high pressure chamber is smaller than that of the conventional type because the ratio of the specific heats, γ, becomes very nearly equal to that of the permanent gas in the chamber as seen from Eq. (4.3). Thus if a monatomic gas such as argon is used at high pressures, the increase in γ is sufficient to decrease the expansion ratio needed to produce a given supersaturation. However, the number of ions per unit path length in the gas increases linearly with pressure, so that very large clearing fields must be used and this can lead to technical difficulties. Furthermore, in the work of JOHNSON et al, serious restrictions were encountered in the long time required for the gas in the active volume to cool to the proper temperature after recompression. When this chamber was filled with argon, a period of up to fifteen minutes was required between expansions in order to establish thermal equilibrium. Presumably these difficulties can in part be overcome by using light

[1] F. JOLIOT: J. Phys. Radium **5**, 216 (1934). — C. R. Acad. Sci., Paris **208**, 647 (1939).
[2] R. G. MILLS: Rev. Sci. Instrum. **24**, 1041 (1953).
[3] C. D. ANDERSON: Phys. Rev. **41**, 405 (1932).
[4] L. MOTT-SMITH: Rev. Sci. Instrum. **5**, 346 (1934).
[5] T. H. JOHNSON, S. D. BENEDETTI and R. P. SHUTT: Rev. Sci. Instrum. **14**, 265 (1943).

gases which have high ionization potentials and high heat conductivities. The construction of a chamber of this type has recently been described by Butler[1].

The low pressure cloud chambers, described above, have been used to detect very short range particles. The short range implied that these chambers had to be operated with random expansions with the inherent loss of efficiency that goes with that procedure. Although a number of workers had mounted a single thin walled ion-chamber or Geiger counter in the sensitive volume of the chamber[2] Hodson, Loria and Ryder[3] extended this idea and used the chamber filling gas itself as the counter. By mounting wire electrodes inside the chamber, they were able to make the counter volume a part of the sensitive region of the chamber and there was no material except the filling gas itself to retard the low energy particles which were sought. The counter, or counters, can be operated as pulse-ion chambers and coincidence techniques plus pulse height discrimination can be used in the selection of the desired events[4].

Another problem which has occupied various workers is how the "dead time" inherent in all expansion chambers can be decreased. This has proved especially important in experiments which utilize artificial accelerators as the source of the particles studied. Gaerttner and Yeater[5] have constructed a small chamber which could be satisfactorily used to photograph tracks as often as every two and one-half seconds. They have introduced the technique of adiabatically "overcompressing" their chamber immediately after photographing the tracks in it. That is, as soon as the lamps are flashed, the piston is mechanically driven to a position which decreases the volume of the chamber to less than the initial volume V_1. This new volume, V_3, can be used to define an "overcompression ratio", V_3/V_1 which is analogous to the expansion ratio. After maintaining the over-compressed volume for a short period, the chamber is allowed to slowly expand to the normal initial volume, V_1. After another brief period of waiting, the chamber is adiabatically expanded once more to be photographed and the cycle repeated. These workers suggest that the overcompression re-evaporates the droplets formed during the fast expansion, thereby decreasing their size sufficiently to enable a strong electric sweeping field to draw the residue to the walls of the chamber. However, this explanation seems incomplete in view of the discussion of re-evaporation nuclei in Sect. 3, above, and the small diffusion constant associated with such large objects as aggregates of thirty vapor molecules. A more complete explanation of the effect of overcompression must include the influence of the slow expansion from the overcompressed volume to the proper initial volume. During this slow expansion drops are observed to grow and fall out of the gas to the bottom of the chamber. Thus essentially a slow expansion is made which has the advantage that the chamber finds itself at the correct pressure, or initial volume, for a subsequent fast expansion. This is to be contrasted with the normal type of slow expansion mentioned earlier, in which the chamber is fully expanded and after a pause while the drops fall out, must be recompressed to the proper volume. After recompressing, another delay must ensue while thermal equilibrium is restored. It seems clear therefore,

[1] C. C. Butler: Report of the Geneva Conference, CERN, June, 1956.

[2] Cf. e. g. H. S. Bridge, W. E. Hazen, B. Rossi and R. W. Williams: Phys. Rev. **74**, 1083 (1948); and R. B. Leighton, C. D. Anderson and A. Seriff: Phys. Rev. **75**, 1432 (1949).

[3] A. L. Hodson, A. Loria and N. V. Ryder: Phil. Mag. **41**, 826 (1950).

[4] H. W. Lewis, W. W. Brown, D. O. Seevers and E. W. Hones: Rev. Sci. Instrum. **22**, 259 (1951).

[5] E. R. Gaerttner and M. L. Yeater: Rev. Sci. Instrum. **20**, 588 (1949).

that the overcompression provides a very efficient type of slow expansion regardless of the detailed mechanism of the clearing process which occurs. Not only does the chamber come to thermal equilibrium rapidly, but also the condensed droplets are re-evaporated before falling to the bottom of the chamber. This last is especially important when large amounts of ionization are detected, because it prevents vapor depletion, or "vapor robbing", in the active volume of the chamber. The high rate of recycling obtained with the small chamber used in the work of GAERTTNER and YEATER cannot be expected to be reproduced if larger chambers are used. Consider Eq. (4.13) above, where the duration of the supersaturation in a chamber was seen to depend upon the square of the ratio of volume to area of the chamber. This time must be short in order to recycle rapidly and will increase as the volume to area ratio increases. This effect was verified by LEIGHTON and YORK[1] who used a chamber with a larger volume to area ratio than the above and were forced to recycle less often as a result.

Another method used to decrease the dead time of cloud chambers has been an attempt to increase the sensitive time of the chamber on each cycle. As just mentioned, the ratio of volume to area can be increased to give a longer sensitive time. On the other hand, this time also increases with the pressure (or density) as also seen in Eq. (4.13). Both of these methods have been successfully used. Various workers have prolonged the sensitive time by making an initial fast expansion in the usual way, and then continuing to expand the volume at a slower rate to maintain the supersaturation[2]. However, none of these methods of prolonging the sensitive time of a chamber have been as effective or as useful as the "continuous" or "diffusion" cloud chambers to be described in the next two sections.

e) Diffusion cloud chambers.

10. The theory of the production of supersaturation by a temperature gradient. The theory of the production of supersaturation by the diffusion of a warm saturated vapor into a cold region has been treated in detail by LANGSDORF[3]. In order that the steady state of diffusion will be stable under the influence of gravity, one considers an enclosed volume which has a vertical temperature gradient maintained across it. For stability in a system composed of a heavy gas and a light vapor, the top must be warm and the bottom cold. The vapor is introduced at the top and, as it diffuses downward, becomes supersaturated due to the decrease in temperature of the surrounding gas. Thus the calculation of the degree of supersaturation requires the solution of a steady-state diffusion problem.

To simplify the calculation one assumes: that the effects of the side walls of the enclosed volume are negligible, so that a one dimensional diffusion equation can be used; that the vapor is a perfect gas in spite of being supersaturated; that the vapor flux and heat flux are in a steady state; and that the effect of absorption and emission of radiant energy by the vapor is negligible. Then the total energy flux, f, across the chamber is the sum of the energy transported by heat conduction and the mass flux of the vapor, c. We can write

$$f = - K \frac{dt}{dx} + c H \tag{10.1}$$

where K is the coefficient of heat conductivity of the gas-vapor mixture, dt/dx is the vertical temperature gradient (x is measured from the bottom upwards)

[1] R. B. LEIGHTON and C. M. YORK: Unpublished, 1953.
[2] E. g. cf. H. MAIER-LEIBNITZ: Z. Physik 112, 569 (1939).
[3] A. LANGSDORF: Rev. Sci. Instrum. 10, 91 (1939). See also R. P. SHUTT: Rev. Sci. Instrum. 22, 730 (1951).

and H is the enthalpy of the vapor. Now $H = C_p t$, the specific heat at constant pressure times the temperature, and $K = K_0 (1 + bt)$ with b as the usual temperature coefficient of heat conductivity. Here $t(x) = T(x) - T(0)$, the difference in absolute temperature between the bottom of the chamber and a point a distance x above it. The boundary conditions on the above differential equation are that $t = 0$, at $x = 0$, and $t = t_1$ at $x = h$ where h is the height of the chamber. Integrating Eq. (10.1) we obtain

$$\frac{x}{h} = \frac{bt + (1 + a\,b\,t_1)\log(1 - t/a\,t_1)}{bt_1 + (1 + a\,b\,t_1)\log(1 - 1/a)} \tag{10.2}$$

where for convenience $a = f/c\,C_p\,t_1$.

To calculate the distribution of the vapor throughout the volume one must assume that the isothermal diffusion equations can be extended to a system which involves a temperature gradient such as this. For one dimensional isothermal diffusion, Kuusinen[1] gives the relations:

$$c = w\,D_1 - k\frac{\partial D_1}{\partial x}, \tag{10.3}$$

$$c' = w\,D_2 - k\frac{\partial D_2}{\partial x}, \tag{10.4}$$

$$w = c\,v_1 + c'\,v_2 \tag{10.5}$$

where $D_1 = M_1 P_1/R\,T$ is the concentration of the vapor; $D_2 = M_2 P_2/R\,T$ is the concentration of the gas; $k = (k_0/P_0)(T/T_0)^{1+\alpha}$ is the diffusion constant; k_0 is the diffusion constant at one atmosphere and $T\,°K$; P_0 is the total pressure of the gas-vapor mixture; and α is the temperature coefficient of diffusivity (in general $0.75 < \alpha < 1$). v_1 and v_2 are the partial volumes per gram of the vapor and gas and $c' = 0$, since there is no net flux of the permanent gas. Thus $w = c\,v_1$, the convective flux associated with diffusion. Now

$$\frac{\partial D_1}{\partial x} = \frac{M_1}{R\,T}\frac{\partial P_1}{\partial x}$$

and

$$\frac{\partial D_2}{\partial x} = \frac{M_2}{R\,T}\frac{\partial P_2}{\partial x}.$$

Then Eq. (10.4) becomes upon substitution

$$\frac{c\,R\,T_0}{M_1 k_0}P_2 = \left(1 + \frac{t}{T_0}\right)^\alpha \frac{d P_2}{d x}. \tag{10.6}$$

If Eq. (10.2) is used to change the variable from x to t, then one can write:

$$\frac{d P_2}{P_2} = \frac{c\,R\,T_0\,K_0 (1 + b\,t)\,dt}{M_1 k_0 (1 + t/T_0)^\alpha (c\,C_p\,t - f)}. \tag{10.7}$$

If $\alpha = 1$, this can be integrated to give:

$$\log \frac{P_2(t)}{P_2(0)} = \frac{T_0}{T_0 + a\,t_1} \cdot \frac{R\,T_0\,K_0}{M_1 k_0 C_P}\left[(1 + b\,a\,t_1)\log(1 - t/a\,t_1) - \right. \\ \left. - (1 - b\,T_0)\log(1 + t/T_0)\right]. \tag{10.8}$$

Now the supersaturation, S, is given by

$$S = \frac{P_1}{(P_1)_S} = \frac{P_0 - P_2}{(P_1)_S} = P_0 - \frac{P_2}{P_2(0)}\frac{P_0 - P_1(0)}{(P_1)_S}.$$

[1] J. Kuusinen: Ann d. Physik **24**, 445 (1935).

If $P_1(0) \ll P_0$, then

$$S = \frac{P_0}{(P_1)_S}\left(1 - \frac{P_2}{P_2(0)}\right) \tag{10.9}$$

where $P_2/P_2(0)$ is given by Eq. (10.8) above.

It is apparent from Eq. (10.9) that the degree of supersaturation is approximately proportional to the total pressure, P_0. Furthermore, the fact the chamber height, h, never appears explicitly in the equations, but only x/h the fractional chamber height, implies that the *form* of S is independent of the height, h. However, if Eq. (10.6) is integrated over x, it is seen that the flux of vapor, c, varies inversely as the height of the chamber. Because the amount of vapor removed by condensation on ions increases directly with h, is is clear that if h is too great, then the vapor flux, c, may be inadequate to maintain the proper degree of supersaturation. Thus for any given rate of ionization in the chamber an optimum chamber height can be determined.

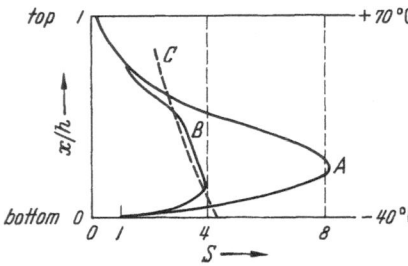

Fig. 9. The supersaturation, S, as a function of the fractional height, x/h, of a diffusion cloud chamber.

A plot of S vs. x/h is given in curve A of Fig. 9. The dashed curve, C, for S_{cr} is the minimum supersaturation required for the condensation of vapor on ions and has been calculated using the relation given by POWELL[1]:

$$\log S_{cr} = C\,\frac{M}{\varrho}\,\frac{\vartheta^{\frac{3}{2}}}{T}. \tag{10.10}$$

Here C is a constant to be determined by the condition $S_{cr} = 2.9$ for $T = 267°$ K, M is the molecular weight of the vapor; ϱ is the density of the vapor; ϑ is its surface tension; and T its absolute temperature. It is clear from the figure that in the region of x/h where S is greater than S_{cr} it will be possible for condensation on ions to occur and in equilibrium operation the actual supersaturation curve is probably of the form B. The shape of the curve for S is very sensitive to the temperature and to the amount of vapor flux. In fact if the temperature is changed by as much as 5 or 10° C, the sensitive region of the chamber can be drastically reduced. It is found that the vapor flux must be of the order of 10^{-6} to 10^{-5} g/cm² per sec for satisfactory operation.

In discussing the theory of diffusion cloud chambers above we have followed the treatment of LANGSDORF and considered only the case of a light vapor diffusing downward through a heavy gas. However, as shown by NIELSEN, et al.[2] this is a needless restriction. These authors point out that the diffusion of a heavy vapor upward through a light gas with the temperature of the top of the chamber less than that of the bottom, leads to a stable operating condition. Another assumption made above is that the side wall effects are negligible. This is not found to be the case in practice, because the vapor supply near the walls is usually inadequate. The result is a decrease in the sensitive region of the chamber near the walls. To reduce this effect, diffusion chambers currently in use are constructed with horizontal dimensions considerably greater than their depth and only the central portion is used for detection.

As noted above, the operation of a diffusion chamber depends on the amount of background ionization in its vicinity. To reduce the number of ions which condense the vapor, electrodes are usually placed inside the chamber to sweep

---

[1] Cf. C. F. POWELL: Proc. Roy. Soc. Lond. **119**, 553 (1928).
[2] C. E. NIELSEN, T. S. NEEDLES and O. H. WEDDLE: Rev. Sci. Instrum. **22**, 673 (1951).

ions out of that portion of the chamber which is undersaturated. These ions can not then diffuse into the region of condensation to disturb the sensitive volume.

Electric fields of up to 130 volt/cm have been used but several authors find them entirely unnecessary when low ionization backgrounds are encountered.

As pointed out by SHUTT[1] the use of mixtures of vapors would seem undesirable in view of the possibility that the components of the mixture could have different rates of diffusion. This might cause a net decrease in the sensitive region of the chamber, and indeed, COWAN[2] reports that mixtures do give results comparable, but slightly inferior, to pure liquids.

11. Diffusion cloud chambers. The diffusion or "continuous" cloud chambers described in the preceding section have several clearcut advantages over the expansion type chamber. Since the chamber is continuously sensitive, there is no need for elaborate timing devices, and because no mechanical expansion is required, the entire construction is greatly simplified. Nevertheless, electric sweep fields must be provided and a time delay between passage of a particle and the moment of photography must be allowed to enable the droplets to grow. Because of these simplifications, most recent experimental work with the high energy accelerators has been performed with the aid of continuous chambers.

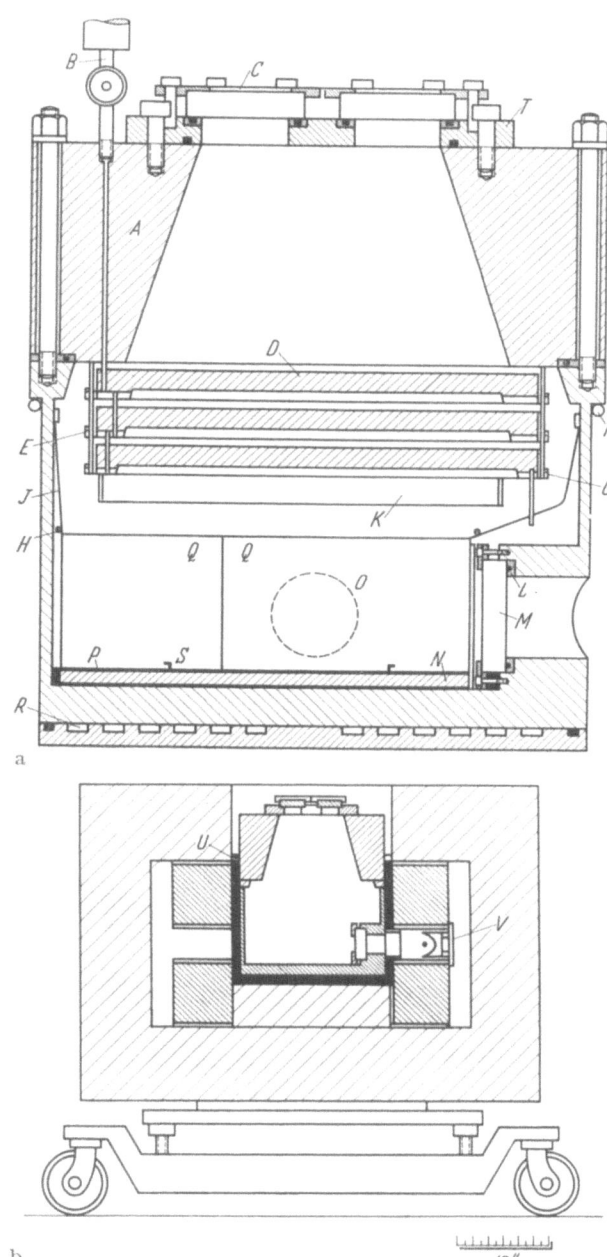

Fig. 10a and b. Detail drawing of a diffusion cloud chamber, (a), and the method of mounting it in an electromagnet, (b).

[1] R. P. SHUTT: Rev. Sci. Instrum. **22**, 730 (1951).
[2] E. W. COWAN: Rev. Sci. Instrum. **21**, 991 (1950).

A representative example of the diffusion chambers now in use is that constructed by SCHLUTER and WRIGHT[1]. This apparatus is operated with 25 to 30 atmospheres of hydrogen and is designed so that the chamber is an integral part of the electro-magnet used with it. Fig. 10a shows a detailed cross-section

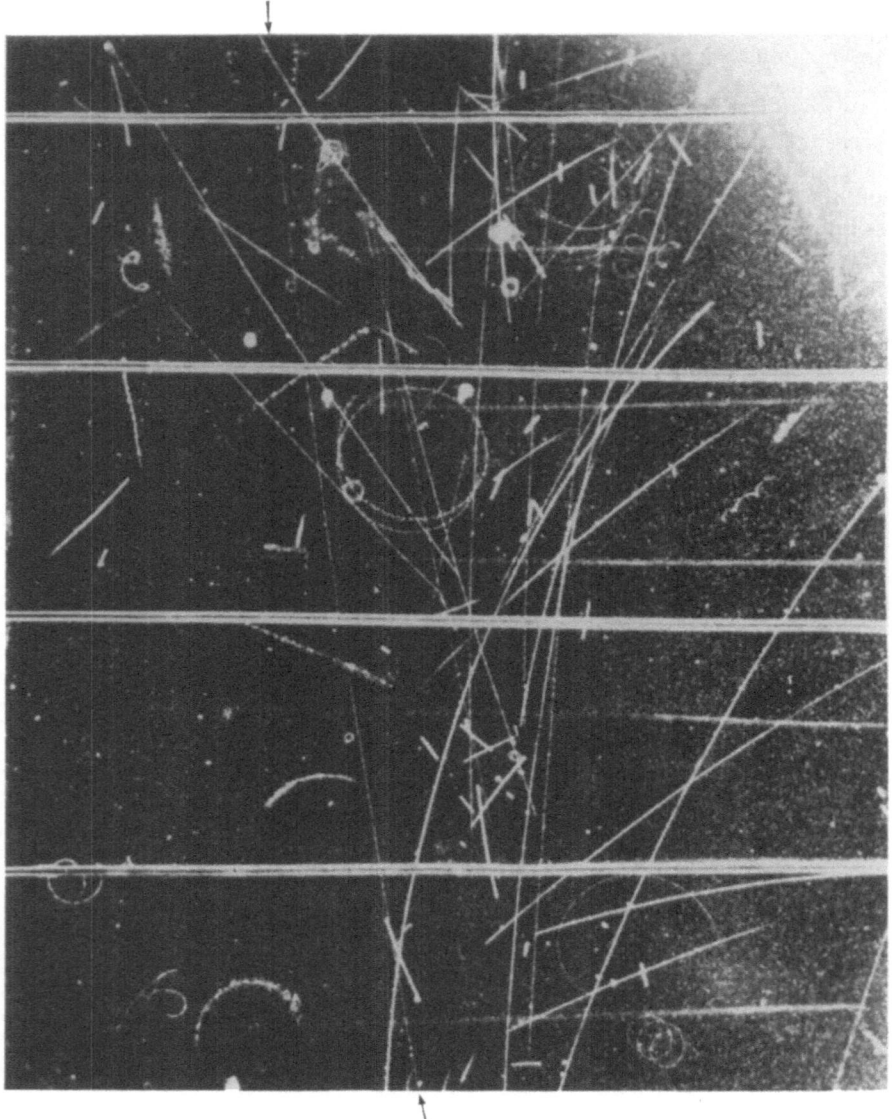

Fig. 11. An example of a photograph taken with the diffusion cloud chamber shown in Fig. 10.

of the chamber itself while Fig. 10b indicates how the chamber unit slips into the electromagnet. In the detailed drawing, methanol is introduced through the valve, B, and fills the trays, D. The trays are warmed by heaters, G, and the warm vapor diffuses downward into the active volume just above the chamber

[1] R. A. SCHLUTER and S. C. WRIGHT: Rev. Sci. Instrum. 26, 1053 (1955).

floor, N, which is refrigerated by passing coolant through the grooves, R, in the base. The active volume is illuminated through the side window, M, and photographed by stereoscopic cameras which are mounted to view the chamber through the windows at C. The chamber floor is maintained at $-80°$ C while the trays are held at $+12°$ C. This gives a sensitive volume of $2\frac{1}{2}$ to 3 inches in depth just above the floor of the chamber. The tolerable ion load in this chamber was found to correspond to about 2 cm of minimum ionizing track in each square centimeter of the photographed area every ten seconds. Higher densities of ionization require longer waiting periods between bursts in order to re-establish the steady state of diffusion. Time delays of approximately 110 milliseconds are used between the passage of ionizing particles and the flash of the lamps.

Fig. 11 shows a typical photograph taken with this chamber when exposed to a pulse of 400 Mev neutrons. The minimum ionization track indicated by the arrows just to the left of the center of the picture illustrates the loss of sensitivity by vapor depletion of such a chamber. Just as the track approaches the second clearing field wire from the bottom of the picture it seems to disappear, then re-appear just past the wire and continue on, uninterrupted across the remainder of the chamber. In the gap in this track is seen a diffuse, and hence old, track crossing the path of the minimum ionizing track. This older track has depleted the vapor density in its immediate vicinity, causing the chamber to be insensitive to the subsequent passage of the minimum ionizing track. Similar cases of "vapor robbing" have been observed in expansion chambers, but are extremely rare.

Although continuous chambers have largely replaced rapid cycling expansion chambers for use with accelerators, they have found only limited application in cosmic ray work. This is largely because of the thin, horizontal detection volume by these chambers. Furthermore, the effects of gaseous distortion on the trajectories of tracks have never been evaluated for most of the currently operated chambers. SARGENT et al.[1] have estimated that "turbulence induced radii of curvature are probably greater than 100 m". Although they neglect to stipulate what length of track was used to arrive at this figure, it does not compare favorably with expansion chamber results, even if the overall diameter of their chamber is used for the track length. This subject will be treated in detail in Sect. C.

II. The photography of cloud chamber tracks.

a) Statement of the problem[2].

There are two essentially different features of a track which give rise to a photographic image. These are individual droplets resulting from the single ions formed by the passage of a charged particle and small clusters of droplets produced by low-energy knock-on electrons along the track. These latter are easily photographed because of their comparatively large size and light scattering efficiency. However, for most practical purposes these clusters are not as useful as the individual droplets. It is found that the geometric trajectories of ionizing particles are most clearly defined and easily measured when the images of the individual drops are resolved on the film. For the determination of ionization by droplet counting, to be discussed below, it is self-evident that the droplets must be resolved. For these two reasons, we shall consider the problem of photographing individual liquid drops.

---

[1] C. P. SARGENT, M. RINEHART, L. M. LEDERMAN and K. C. ROGERS: Phys. Rev. 99, 885 (1955).

[2] The treatment presented here closely follows that given by R. D. SARD: Cosmic Ray Notes, unpublished, Washington University, St. Louis 1955.

The small size of the droplets in a cloud chamber (radius of about 10^{-3} cm at the time of photography) makes it possible to assume point sources of light, and the images produced on the film will be the result of the diffraction effects of the lens aperture, modified by the focusing properties and aberrations of the lens, as well as the properties of the film used to record the image. After a theoretical discussion of the diffraction problem a brief summary of the practical limitations found in applying the theory will be given.

12. The image shape and intensity. Consider a point source on the axis of a lens of focal length f, and diameter D. Let the source be a distance p from the lens. Neglecting lens aberrations, the image produced at a distance q from the lens will be a disk of diameter

$$2.44 \, \lambda \, \frac{q}{D} = 2.44 \, \lambda \, \frac{f}{D} \, (1 + m) \tag{12.1}$$

where λ is the wave length of light emitted by the source, $m = q/p$ is the magnification of the lens, and the lens equation has been used to obtain the expression on the right hand side of the equation. This disk is known as the "Airy" disk and results from the diffraction of the light on passing through the lens aperture[1]. Eighty-four percent of the light passing through the lens aperture can be shown to be concentrated in this disc, at the point of best focus. For planes near this focal plane, the Airy disk is approximately constant in size, but receives a smaller fraction of the light intensity than the disk in the focal plane. For a plane displaced from the focal plane along the axis by an amount such that the optical path difference of an axial ray and the ray from the edge of the lens aperture is $\lambda/4$ (the "Rayleigh limit"), the intensity falls off to 80% of that in the focal plane. For planes displaced such that the path difference is $\lambda/2$ or twice the Rayleigh limit, the intensity in the disk is 45% of that in the focal plane while its diameter is still very nearly that given by Eq. (12.1). For most cloud chamber work it has proved reasonable to assume that this latter criterion of twice the Rayleigh limit is satisfactory for estimating the effective depth of focus of a lens.

For two image points on the axis of a lens a distance δq apart, it can readily be shown that the optical path difference of rays reaching these points from the edge of the lens aperture and from its center is approximately

$$\delta q \, \frac{D^2}{8 q^2} \, . \tag{12.2}$$

If, as suggested above, this path difference can be taken to be twice the Rayleigh limit, then the depth of focus, δq can be written

$$\delta q = 4 \lambda \, (q/D)^2 = 4 \lambda \, (f/D)^2 \, (1 + m)^2 \, . \tag{12.3}$$

Of more immediate interest than δq, is $2 \delta p$, the distance in the object space over which droplets can be resolved. Recalling that $\delta p = - \delta q/m^2$, we can write

$$2 \delta p = 8 \lambda \, (f/D)^2 \left(\frac{1 + m}{m} \right)^2 \, . \tag{12.4}$$

This relation is of immediate interest, because it indicates that to increase the depth over which droplets can be resolved, either the relative aperture of the lens, (D/f), can be decreased, or alternatively the magnification can be decreased. In fact for $m \ll 1$, it is seen that $\delta p \sim 1/m^2$.

However, there are limits to which the lens can be stopped down and the magnification decreased. Let us first consider the light intensity arriving at the

---

[1] Cf. Schuster and Nicholson: Theory of Optics, p. 146. London: E. Arnold 1924.

plane of the film. Let I_0 be the light intensity per unit solid angle emitted by the droplet. Then the total light intensity reaching the lens will be I_0 times the solid angle of the lens subtended at the droplet. As seen above, 84% of this intensity is concentrated in the Airy disk by the lens, and the intensity of light per unit area in the disk will just be

$$\frac{0.84\, I_0\, \frac{\pi D^2}{4p^2}}{\frac{\pi}{4}\,[2.44\,\lambda\,(f/D)\,(1+m)]^2}$$

or

$$\frac{0.14\, \frac{I_0}{\lambda^2} \left(\frac{D}{f}\right)^4}{\left(1+\frac{1}{m}\right)^2 (1+m)^2}. \tag{12.5}$$

Although the fourth power variation of image intensity with lens aperture is never realized in practice due to lens aberrations and losses by reflection, it is clear that if the lens is stopped down to gain depth of field, this is done at the expense of image intensity. Furthermore, for $m \ll 1$, the image intensity is proportional to m². Hence, if the magnification is decreased to gain depth of field, again it is done at the cost of image intensity. Clearly an increase in I_0 will tend to offset this loss and the factors governing the light intensity emitted by a droplet will be discussed in detail in Sect. 15 below.

In addition to the image intensity, there is another limitation to the choice of the magnification and lens aperture. This is simply the requirement that the images of two adjacent drops be readily resolved on the film. If two droplets are separated by a distance d in the chamber, then the centers of their image disks will be separated by a distance $md = \Delta$. For the two images to be resolved clearly, the Airy disks associated with them must be separated by a distance greater than the *radius* of the disk, but of the same order of magnitude. Again neglecting lens aberrations, the minimum image separation will be of the order

$$\Delta_{\min} \approx 1.22\,\lambda\,(f/D)\,(1+m). \tag{12.6}$$

The maximum resolution of the system is the reciprocal of the minimum separation given by Eq. (12.6). It is clear that the maximum resolution diminishes with decreasing relative aperture, (D/f), and decreasing magnification. The ultimate limitation here need not be just the optical resolution computed above, but may be the resolution of the photographic film used to record the images. The resolution of any particular photographic emulsion when developed in a specific way is always given in the technical data supplied by the manufacturer and is a useful guide in estimating the value of $\Delta_{\min}$ which can be achieved in practice. It is sometimes possible to improve on the resolution of the emulsion by special development, but in general no great increase is obtainable.

13. The choice of the aperture and magnification. It is clear from the foregoing discussion that a large number of factors must be considered in solving the problem of photographing cloud chambers and that a compromise adjustment of them must be achieved. Usually several quantities are determined from the outset. The volume of the cloud chamber is usually dictated by the requirements of the experiment to be performed. This determines the depth of field, $2\delta p$, which in turn limits the magnification to be used. To see how this occurs note that

$$2\delta p = \frac{2\delta q}{m^2} = \frac{8}{(1.22)^2}\,\frac{\Delta^2_{\min}}{\lambda\,m^2}.$$

From Eqs. (12.4) and (12.6). This can be written

$$m_{min} = 1.64 \, \Delta_{min} / \sqrt{\lambda \, \delta p}. \tag{13.1}$$

Having thus determined the minimum useful magnification, one can estimate the most convenient film size to be used from the area of the cloud chamber to be photographed. The choice of emulsion to be used depends on the highest possible resolution consistent with a reasonably high light sensitivity or film speed. Having made a choice of emulsion and having used Eq. (13.1) to determine the minimum magnification, it remains to determine the lens aperture to be used from Eq. (12.6).

Unfortunately, the theory described above is not adequate for the practical determination of all of the quantities, encountered in practice, but is useful only as a guide. First it should be emphasized that the treatment of points lying only on the lens axis is a serious limitation to the theory. For adjacent image points lying off the lens axis, the relation in Eq. (12.6) must be modified. If ϑ is the angle between the lens axis and a line drawn from the center of the lens to the images of the points, then Eq. (12.6) must be multiplied by a factor which is approximately $\cos^3\vartheta$ or $\cos\vartheta$ depending on whether the two points are oriented tangentially or radially to the circle drawn in the image plane with center on the lens axis[1]. Thus to maintain an adequate resolution at the edges of the chamber, it is necessary to keep ϑ as small as possible. For a given magnification, this is best accomplished by using a long focal length lens. Another point in favor of the use of long focal length lenses is that the lens aberrations such as astigamatism and curvature of field are not important at small angles to the lens axis. Because the aberrations found in individual lenses vary a great deal and have varying effects upon the lens performance, it is recommended that the lenses to be used be tested by some standard method[2]. It is found that for most well corrected lenses, the above theory predicts their performance within fifteen or twenty percent. It should be borne in mind that the resolution of the lens at the specified aperture should exceed that of the film for the maximum angle of photography. If this condition is not met, the lens resolution rather than that of the film, should be used for Δ_{min} in the determination of m_{min} in Eq. (13.1) above.

14. An application of the theory. Consider the case of a chamber 51 cm × 51 cm × 15 cm. The value of δp is 7.5 cm, and let us assume an average wave length of light to be $\bar{\lambda} = 5 \times 10^{-5}$ cm. The resolution of Linagraph Pan film (Eastman Kodak Co.) is given as 70 lines/mm when developed for 8 min in D-19 developer. Thus $\Delta_{min} = \frac{1}{700}$ cm and Eq. (13.1) gives $m_{min} = 0.118 = 1/8.5$. Now to find the required aperture to be used, this value of m_{min} is to be inserted in Eq. (12.6). If Δ_{min} is again taken as $\frac{1}{700}$ cm, it is to be noted that the optical resolution, as given by Eq. (12.6), is automatically adjusted to match the film resolution. Upon substitution, the relative aperture, (D/f), is found to be 1/21 and the f-number is $f/21$. The image on the film will be a square, 6.0 cm on each side, if the magnification 1/8.5 is used. Since Linagraph Pan film is supplied in rolls 70 mm wide, it is more economical to set $m = 1/7.3$ and use the entire width of the film. Then, the aperture to be used is $f/20.5$. It is clear that the dependence of the aperture on the magnification is very insensitive for small changes in the magnification; however, in view of the strong variation of image intensity with aperture it is advisable to use the maximum possible aperture.

---

[1] Cf. F. E. WASHER: J. Res. Nat. Bur. Stand. **34**, 175 (1945).

[2] Cf. F. E. WASHER and I. C. GARDNER: NBS Circular **1953**, 533.

The maximum angle ϑ for the case of the lens axis passing through the center of the chamber, will be determined by the focal length of the lens used. If the focal length is 12.8 cm and the magnification is 1/7.3, then $p = 106$ cm and the angle measured from the center of the chamber to one of its corners is $\vartheta = 18.8°$. Thus it is necessary to check the resolution of the lens to be used to determine if the resolution is in fact 70 lines per millimeter for aperture, $f/20.5$, at angles up to about 20°.

b) Some other aspects of the problem.

15. The intensity of light scattered from droplets. The intensity of light scattered from a small dielectric sphere has been computed in detail by BLU-MER[1]. The scattered intensity is proportional to the incident intensity of illumination and depends strongly on the angle of scattering. To a lesser extent there is a dependence upon the sphere's radius, the wave length of the incident light and the index of refraction of the sphere. The general form of the angular dependence of the scattered intensity has been measured by WEBB[2] and is shown in Fig. 12. This agrees reasonably well with the predictions of the theory and shows the tremendous increase in intensity obtained when the droplets are viewed near the direction of the incident beam of light. Several authors[3] have made use of this fact to obtain an increase in the scattered light intensity and the only disadvantage encountered is the necessity of keeping the front glass of the chamber exceptionally clean of dust and condensation.

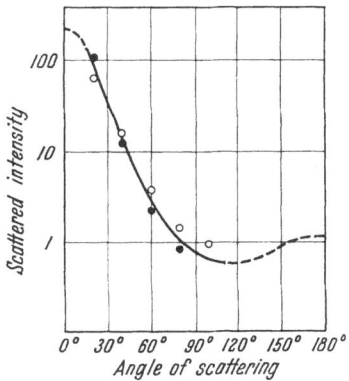

Fig. 12. The intensity of light scattered from droplets of water and an alcohol-water mixture as a function of the angle of scattering.

In many cases the angle of illumination is fixed by some limitation of the rest of the apparatus and an angle of 90° is used most frequently. The only way to increase the light intensity in this case is to increase the intensity emitted by the source. Since current practice is to use a low pressure gaseous discharge tube as a source of light, this implies that the energy dissipated in the flash tube must be increased by operating it at higher voltage or by discharging a larger condenser bank through it. It is standard practice to operate commercially available, xenon filled tubes at from 200 to 300 joules per flash.

16. Some practical requirements. In photographing the liquid droplets in a cloud chamber, it should be stressed that the highest possible degree of contrast between the droplets and the background against which they are seen is desirable. The resolution of the photographic emulsion used in the above discussions, is dependent upon the "contrast ratio" employed. In quoting the resolution of Linagraph Pan film in Sect. 14 above, it should have been noted that the numbers apply to lines which emit thirty times the intensity of the light emitted by the background on which they are drawn. Thus this film has a resolution of 70 lines per mm at a contrast of 30 to 1. For lower contrast the resolution decreases. It is for this reason that the standard resolution test charts are supplied in two different contrasts.

[1] Cf. H. BLUMER: Z. Physik **39**, 195 (1926).
[2] Cf. C. G. WEBB: Phil. Mag. **19**, 927 (1935).
[3] Cf. C. T. R. WILSON and J. G. WILSON: Proc. Roy. Soc. Lond., Ser. A **148**, 523 (1935).

In chambers utilizing small angles of scattered light, the photographic quality suffers from the necessary increase in the intensity of light scattered from the background. For this reason most current work is done with chambers illuminated at angles of about 90° to the direction of photography. This has the advantage that parallel beams of light can be used to illuminate the volume containing the droplets. By carefully adjusting the light source, only a negligible amount of light will be scattered by the rear of the chamber and the front glass. To further reduce the light intensity scattered from the rear of the chamber, it is usually covered either with black velvet cloth to absorb any stray illumination, or with a specularly reflecting coating of lacquer, glass, or plastic.

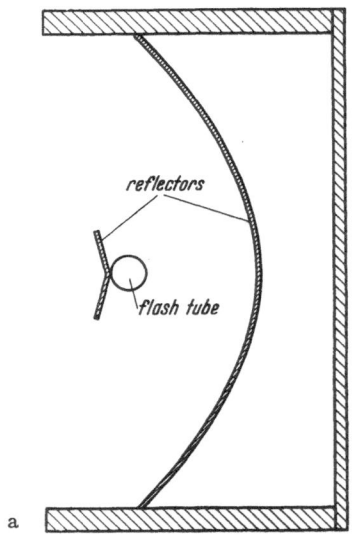 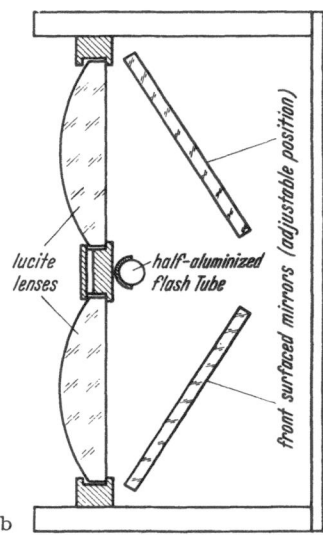

Fig. 13a and b. Examples of light sources used with cloud chambers. (a) A reflector type and (b) a lens type.

The sources used to produce the intense parallel beams of light discussed above differ widely in design, but almost all of them achieve the desired results. Fig. 13 illustrates two of the many types now in use. The first Fig. 13a, shows the cross-sectional plan view of a parabolic reflector type designed by E. W. COWAN[1]. The flash tube is placed slightly out of focus so that the beam from the parabolic reflector converges around the V-shaped reflector in front of the tube. The V-shaped reflector is bent so that the image of the flash tube is concentrated at symmetric points on either side in the edges of the parabolic reflector. This tends to increase the intensity at the edges of the beam where there is a natural decrease in solid angle of the flash tube subtended by a unit area of the parabola. The net effect is to produce an approximately parallel beam of nearly uniform intensity. The only technical difficulty in this design is encountered when one attempts to construct such a parabola with the desired shape and reflectivity. However, this design has proved to be very satisfactory and has been extensively used by the group at the California Institute of Technology.

The design shown in Fig. 13b is a variation of the classical type of light source employing a cylindrical lens. This variation was evolved at the University of Chicago to provide a wide, uniform, well-focussed beam of light. An important

---

[1] E. W. COWAN: Private Communication 1952.

feature was the requirement that the box containing the lenses and lamp be as shallow as possible. Uniformity of illumination is obtained by the symmetry of the lens and mirror system with respect to the flash tube, while the adjustment of the focus of the beam is obtained by translating and rotating the mirrors in their mounting fixtures. The shadow cast by the separation of the lenses is compensated by a slight divergence and overlap of the two beams of light coming from each side. When two of these light sources are placed on opposite sides of the chamber, no detectable decrease in the light intensity in this shadow region is observable.

III. The measurement of cloud chamber tracks.

17. The determination of ionization. The earliest studies of radioactive disintegration with cloud chambers showed that the density of droplets along a track depended markedly on the charge of the particle as well as its velocity. The various theoretical computations of energy loss in the gas could be used to understand the observed variations in a quantitative way. CORSON and BRODE[1] were the first to make an accurate measurement of the ionization of a particle and their experiments established the pattern for a great deal of subsequent work in the field. These authors delayed the expansion of their chamber for several tenths of a second after the traversal of the particle, to allow the ions forming its track to diffuse appreciably before condensation was permitted to occur. The resulting diffuse column enabled the observer to easily count the individual droplets of which it was composed. One of the most recent applications of this technique has been made by GHOSH, JONES, and WILSON[2], who have studied the rise of the ionization curve at relativistic velocities in an oxygen filled cloud chamber. The great disadvantage of this "delayed expansion" technique is that the long period allowed for ion diffusion also permits convection currents to markedly distort the tracks. Thus if it is desirable to measure the momentum of the particle on the same photograph used for droplet counting, these distortions can lead to prohibitively large errors in the momentum determination.

A major development in the technique was introduced by COWAN[3] along lines already suggested by WILSON[4]. By using a light gas (helium) in which the ionic diffusion is large and the number of ions produced per unit length of path is small, COWAN has been able to expand his chamber without a delay and still resolve the individual droplets within the column constituting the track. In fact this technique has been so successful that by adding one part of argon to two parts of helium, the number of droplets per unit length of path has been increased, thus improving the statistical accuracy of the droplet count. An appropriate ionization curve of the form shown in Fig. 1 is then used for the gas mixture. Figs. 14a and b illustrate the type of photograph used by COWAN[5]. Fig. 14b is a greatly enlarged view of the upper portion of Fig. 14a, and shows the individual droplets and clusters which constitute the tracks.

In order to determine the ionization of the charged particles in such a picture the rest mass of at least one of the particles is assumed to be known. A measurement of that particle's momentum determines its relative ionization uniquely

---

[1] D. CORSON and R. B. BRODE: Phys. Rev. **53**, 773 (1938).

[2] S. K. GHOSH, G. JONES and J. G. WILSON: Proc. Phys. Soc. Lond. A **67**, 331 (1954).

[3] E. W. COWAN: Phys. Rev. **94**, 161 (1954).

[4] J. G. WILSON: The Principles of Cloud Chamber Technique, p. 121. Cambridge 1951.

[5] The author is indebted to Professor COWAN for furnishing these unpublished photographs.

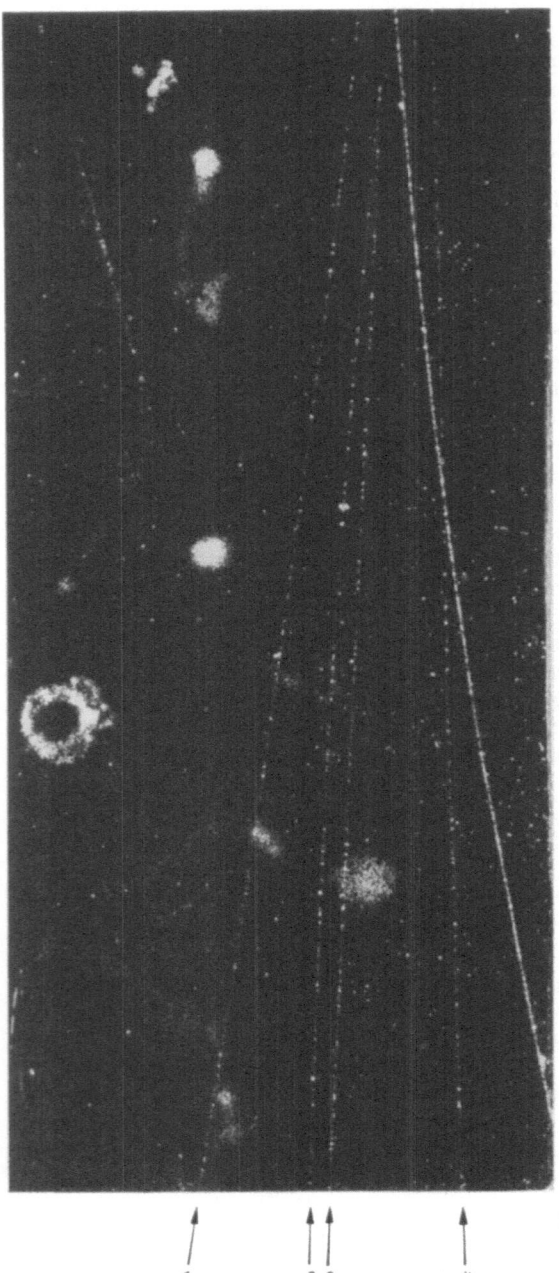

and a count of the number of drops per unit track length can then be used to correlate the ionization and drop-count on a given picture. This calibration procedure is important, because slight variations in temperature, vapor density, and expansion ratio can cause appreciable fluctuations in the condensation efficiency and hence the number of drops per unit path length along the track. Although any particle of known rest mass can be used for this calibration, electron tracks are frequently employed because they are readily identified by their low ionization at low momentum. In Fig. 14a there are no such readily identified electrons, but track No. 3 can be assumed to be a negative π-meson since it is the decay product of a neutral unstable particle. In fact this event can be shown to be the decay of a θ^0-meson by a complete dynamical analysis, and hence this assumption is well justified. If now the ionization of the calibration particle is I_c, and its corresponding number of drops per unit path is N_c, the ionization of any of the remaining unidentified tracks I_u can be determined in terms of its number of ions, N_u, by the simple relation

$$I_u = I_c \frac{N_u}{N_c}. \quad (17.1)$$

It is to be noted that the use of the calibration track determines the proportionality constant between I and N, and eliminates the need for an absolute calibration of the

Fig. 14a. An example of a cloud chamber photograph on which droplet counts and momentum measurements may be performed simultaneously.

energy loss. It is for this reason that the relative ionization plot given in Fig. 1 above, is the most useful form for these measurements. Having determined the relative ionization, I_u, of the unknown particle, a measurement of its momentum does *not* necessarily determine its rest mass uniquely. For a given value of I_u which is less than the high energy "plateau" of the ionization

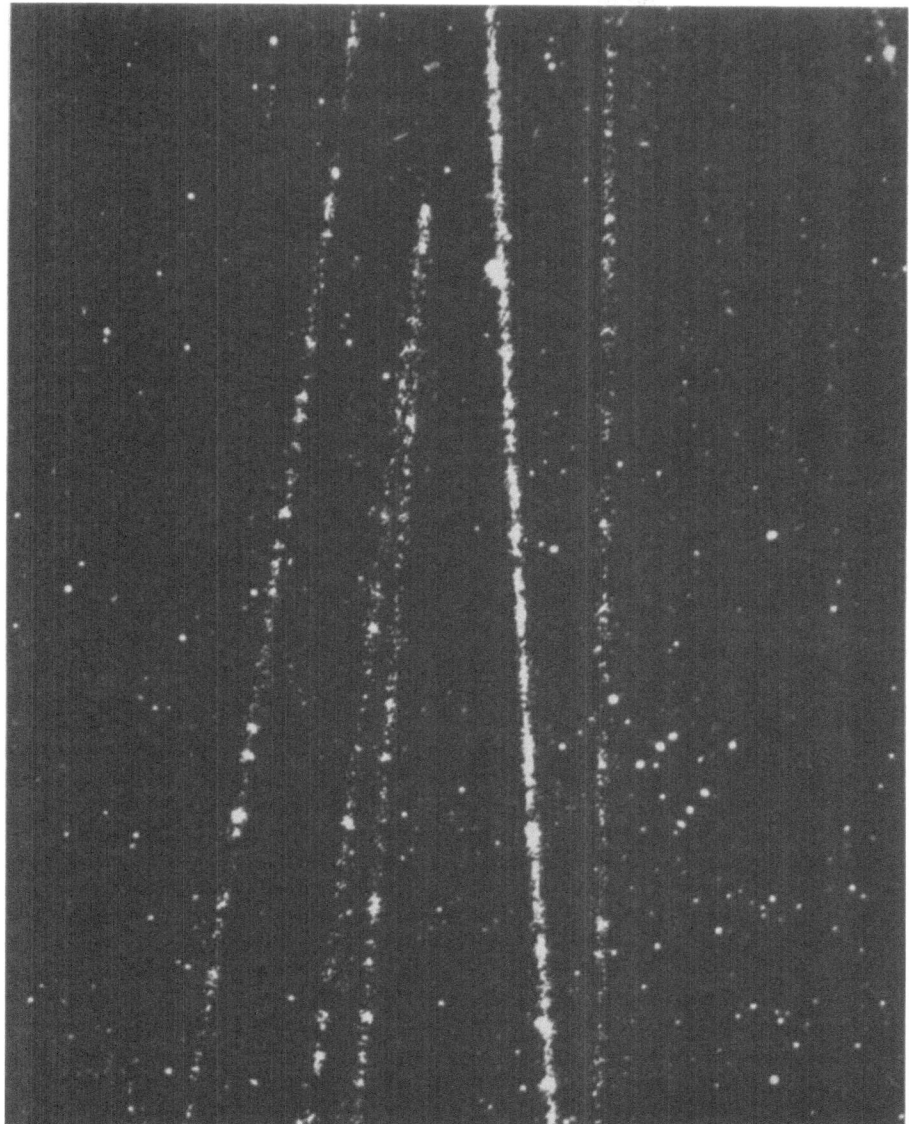

Fig. 14 b. An enlarged view of the upper protion of the photograph showing the individual droplets.

curve in Fig. 1, two values of $p/\mu c$ are possible and in assigning the rest mass, μ, an ambiguity can occur.

The limitation in this method of ionization determination is inherent in the necessity of discarding "blobs" or groups of drops which contain twenty-five or more droplets. Since the number of drops per unit path length is subject to statistical fluctuations, these "blobs" can be created by either just such a fluctuation or by very low energy knock-on electrons which are held close to the parent track by the magnetic field or multiple Coulomb scattering. The actual number of droplets in such clusters becomes difficult to count because they overlap and obscure one another. It is clear then that when the ionization is sufficiently dense to produce an average number of droplets per unit length of track which approaches that of the blobs, the method will break down. This occurs for ionization of the order of 2.5 to 3 times minimum and track No. 5 in Fig. 14a is an example of a heavily ionizing proton track which has too great a droplet density to be counted.

The early work using diffuse columns of droplets was done with a single photograph of the track, and when the drops in the track were counted, a correction for the background drops randomly distributed throughout the chamber had to be subtracted. Using this technique, the correction for overlap of drops has been found to be very appreciable. COWAN has introduced the use of pairs of stereoscopic views of the track and by viewing the image of the drops in a three dimensional space can separate out the background droplets and count only those contained within the track. Furthermore, this approach reduces the overlap correction appreciably in the more heavily ionizing tracks, because a droplet which is obscured in one view can usually be readily seen when viewed from a different angle.

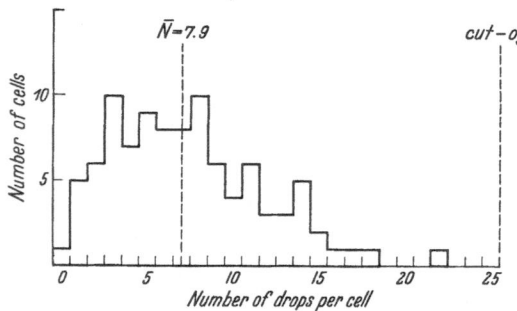

Fig. 15. An histogram of the frequency of occurrence of the number of drops per unit length of track for a particle near minimum ionization.

In applying COWAN's method the two stereoscopic views of the chamber are studied with the aid of low power microscopes at $20 \times$ magnification. A transparent plastic sheet with an appropriate scale is superimposed over the image of the track. The divisions of the scale divide the track into cells which for convenience are of a length just equal to the mean track width. If one plots an histogram of the number of droplets in each cell, a skew distribution of the "Landau type" is observed. An example of such an histogram for an electron track photographed in COWAN's chamber is shown in Fig. 15. In this case no cell contained more than twenty-five droplets, which is the maximum number that can be readily counted. The asymmetric distribution about the most probable value is typical of the Landau type distribution discussed in Sect. 2. The long "tail" of the distribution which gives appreciable numbers of cells with large numbers of droplets, causes the mean value of the number per cell to differ considerably from the most probable value. However, for convenience COWAN uses the mean value, indicated by $\bar{N}$ on the histogram, as the value of N to be substituted into Eq. (17.1) for the ionization determination.

FRETTER, FRIESEN and LAGARRIGUE[1] adopting the general procedure of COWAN, have made a careful study of the errors inherent in the droplet counting method of ionization determination. They have evolved a method of systematically correcting for the overlap of drops and also find that with pure helium in

---

[1] W. B. FRETTER, E. W. FRIESEN and A. LAGARRIGUE, see also W. B. FRETTER and E. W. FRIESEN: Rev. Sci. Instrum. **26**, 703 (1955).

the chamber, it is necessary to correct for the height at which the given portion of track is formed. This latter effect is presumably due to a small but unavoidable vertical vapor gradient across the chamber. The best accuracy in the determination of ionization obtained with a track of 40 cm length in their chamber has been estimated to be about 5%. When it is noted that the momentum measurements made on the same track have as small errors as any previously made with cloud chambers, it is clear that the drop counting technique need not automatically preclude the simultaneous measurement of momentum.

A somewhat different approach to the determination of ionization is the use of a photometer to measure the density of the photographic image on the film[1]. This density must then be related to the ionization of the track being studied. BUTTERWORTH[2] and, independently, BJØRNERUD[3] have found that if the photographic density of an unknown track is compared to that of one whose identity is known, the ionization of the one can be found in terms of the other. Because the photographic image density depends upon the position of the track in the chamber, the film development, the magnification, and a number of other factors, both of these workers have found it necessary to relate the ionization to the image density by means of a calibration curve obtained from measurements on particles of known ionization. BUTTERWORTH did not attempt to measure the density of tracks with ionizations greater than six times the minimum value; however, BJORNERUD extended his measurements to nine times the minimum value and found that above five or six times minimum a marked saturation of the image density began. This is to be expected, since for very dense ion columns the recombination of ions cannot be neglected and a depletion of the vapor in the immediate vicinity of the track will occur. Furthermore the photographic image of the track will saturate on the film giving only an opaque cluster of silver grains. Both of these workers found that it was necessary to measure the density on portions of track which were free of "blobs" just as in the drop counting technique; however, the film does not have to have the individual droplet images resolved upon it in order to apply the photometric method nor does the ionization appear to depend critically on the sharpness of focus of the tracks. The only obvious disadvantages of the method are the need for a separate calibration curve for each different chamber which is used and the care required to select the calibration track in the same part of the chamber as the unknown track. Since the exact size of the rejected blobs is unknown, there is a considerable uncertainty in the value of E'_{max} to be used in computing an ionization curve of the form given by Eq. (2.4). This in turn leads to an uncertainty in the determination of I_c, but the error involved is small due to the logarithmic dependence. The errors in the resulting ionization determinations are of the order of 10 to 15% for tracks of 15 to 20 cm length in the chambers used.

From the foregoing discussion, it is clear that the photometric method extends over a greater range of ionizations than the droplet count method, and can be applied to poorer quality photographs. However, the accuracy thus far attained by photometry is not so great as that of counting the individual droplets.

Very heavily ionizing tracks or tracks in high pressure chambers will not have measurable ionization values by either of the above methods. However, VALLEY and VITALE[4] have suggested a method to use with their high pressure chamber which depends upon the track width. In Eq. (7.2) above we had an expression

[1] N. FEATHER and R. R. NIMMO: Proc. Cambridge Phil. Soc. **24**, 139 (1928).
[2] I. BUTTERWORTH: Phil. Mag. **46**, 884 (1955).
[3] E. K. BJØRNERUD: Rev. Sci. Instrum. **26**, 838 (1955).
[4] G. VALLEY and J. VITALE: Rev. Sci. Instrum. **20**, 411 (1949).

for the distribution of the number of ions in a track as a function of its transverse dimension, x. These authors define the width of a track, w, by the distance separating two points on each edge of the track at which the image density is d_0. They assume that the image density is related to the ionization by the equation

$$d = \log [g\,I\,(x)] \tag{17.2}$$

where g is a constant, depending on the intensity of illumination, the size of the droplets, the angle of scattering of the light, etc. This is the usual definition of image density in terms of the fraction of absorbed light and this latter is simply assumed proportional to the ionization or the density of drops. Now this statement implies that one drop does not obscure another, which is certainly not the case at the center of the track, but is true at the very edge of the track where the density, d_0, is to be measured. If two tracks are formed at the same time, t, but have ionizations I_1 and I_2, their apparent widths are w_1 and w_2 and according to Eqs. (7.2) and (17.2), the image density at the edges of each can be written

$$d_0 = \log \left\{ \frac{g\,I_1}{\sqrt{4\pi k t}} - \frac{w_1^2}{16\,k\,t} \right\} = \log \left\{ \frac{g\,I_2}{\sqrt{4\pi k t}} - \frac{w_2^2}{16\,k\,t} \right\}. \tag{17.3}$$

This can be reduced immediately to the relation

$$\frac{I_1}{I_2} = \exp \left\{ \frac{w_1^2 - w_2^2}{16\,k\,t} \right\}. \tag{17.4}$$

Unfortunately the value of the diffusion constant, k, has not been measured for the various gases at the pressures used in high pressure chambers. However, it is reasonable to assume that the diffusion constant varies linearly with the pressure and this has been used by Burhop[1] in making estimates of the variation of ionization with track width in argon.

Thus far no one has experimentally verified the predicted dependence of ionization on track width in an expansion chamber. However, Schluter[2] using the hydrogen filled diffusion chamber described in Sect. 11 above, has verified that a variation of track width with ionization does occur and can be used to distinguish protons from deuterons with the same momentum. However, he does not attribute this variation in track width to the ionic diffusion, because the mean distances of diffusion are small in a continuous chamber when compared to the measured width of the image. On the other hand, he suggests that the apparent width of the photographic image is the result of the intensity of scattered light which in turn depends upon the number of droplets contained in a segment of track. To check this he photographed a very fine wire mounted in the sensitive region of his chamber and found a variation of image width with the intensity of the illumination used. From these tests he concluded that the scattered light intensity varies approximately as the square root of the ionization. Two methods of measuring the widths of the tracks were used and the results were found to agree quite well. The first method used a template of clear plastic superposed over the image of the track. On this sheet were scribed two lines intersecting at a small angle with a linear scale along one of the lines. When aligned so that the track bisected the angle between the lines, a reading of the scale was taken at the point where the transverse distance separating the two lines was just three times the apparent track width. This criterion of adjusting to three times the track width avoids obscuring the edges of the track, which

[1] E. H. S. Burhop: Nuovo Cim. (2nd Suppl.) **11**, 343 (1954).
[2] R. A. Schluter: Phys. Rev. **96**, 734 (1954).

would occur if the device were adjusted to exactly the track width. The scale was then calibrated in terms of the track width. The other method used to measure the track width was a microphotometer trace of the track image. Since this latter was extremely laborious in comparison with the former, it was used only as a check on the template method. It seems apparent that the determination of ionization from track width will depend not only on the optical effects suggested by SCHLUTER, but in expansion chambers the added influence of the ionic diffusion will affect the result. Hence it seems probable that in order to use this technique reliably, once more comparison tracks will prove necessary, as well as a calibration curve for the chamber in question.

18. The spatial reconstruction of tracks. In many experiments it is essential to know the trajectory of a particle in space, whereas a single photograph only provides a projection of that trajectory onto a plane. A number of schemes for photographing a chamber from several angles simultaneously have been introduced, and each permits the tracks thus recorded to be reconstructed in space. Of all of these methods, the one which is most widely used is the simple stereoscopic pair of photographs. In terms of film economy, accuracy of measurement, ease of film scanning for the desired events, and simplicity of reconstruction in space, this method is by far the most satisfactory. The reconstruction of a track in space from a pair of stereoscopic photographs can be performed in either of two ways and, indeed, some workers find it useful to apply both procedures. The first method entails the measurement of the coordinates of points along the track by placing the film directly on the stage of a comparator microscope. By using reference marks on either the front glass or the photographic background of the chamber, or both, the depth of a point on the track in the chamber can be determined by a simple geometric construction using the two stereoscopic views. Having determined the three rectangular coordinates of a number of points along the track, the reconstruction of the trajectory in the cloud chamber space can then be obtained by applying the appropriate magnification factor.

The second general method used for the spatial reconstruction of tracks employs an optical projection of the track image on the film onto a screen. There are various ways in which this is done, but all use a lens and film positioning system which is as identical as possible to that used in the camera with which the photographs were originally made. In fact, many workers reproject through the *same* camera and film holder used in taking the original picture. This is done to eliminate possible optical distortions originating in the camera lens. The screen onto which the image is projected is opaque or semi-transparent, fixed or adjustable, depending on the method preferred by the particular operator. THOMPSON[1] has described a reprojector which uses a circular ground glass screen with three angular degrees of freedom and a linear motion parallel to the camera axis. By alternately observing the two projected images of a track, the screen can be aligned to the proper angle so that the track lies in the plane of the screen. The position and orientation of the track is then read directly from scales provided for that purpose.

On the other hand, LEIGHTON, WANLASS, and ANDERSON[2] have described a system which uses a fixed screen and treats the two stereoscopic views as projections of the image upon the back piston of their chamber. A sketch of the two track images is made after superposing the images of fiducial marks on the

[1] R. W. THOMPSON, J. R. BURWELL and R. W. HUGGETT: Nuovo Cim. **1956**.
[2] R. B. LEIGHTON, S. D. WANLASS and C. D. ANDERSON: Phys. Rev. **89**, 148 (1953).

piston, and by means of a graphical construction, the position and orientation of a track in the chamber can be obtained. The only apparent advantage of this method over that of the movable screen seems to be the permanent record of each event which the sketches provide. However, for some work the time required for the graphical construction is prohibitive and the analysis is better done with the moving screen type of apparatus. Both reprojection methods yield comparable accuracy of results when carefully applied. It should be noted that the tracks formed in an expansion chamber are distorted uniformly along the axis of the expansion. A suitable correction to the depth measurements must be applied to tracks formed by particles which traversed the chamber before the expansion took place. Furthermore, it should be borne in mind that the optical magnification changes with depth in the chamber and must be taken into account.

THOMPSON has pointed out that the comparator plot of the tracks, while being laborious, gives a set of data from which the trajectories can be computed with the aid of the methods of least squares. Furthermore, this method has the valuable feature of enabling the observer to make a completely objective estimate of the probable errors involved in determining the trajectory. This last is an especially important point when some quantity such as the angle between two intersecting tracks is to be determined together with its error. However, neither the accuracy nor the error determinations of the comparator method have been demonstrated to be appreciably superior to that obtained with the reprojection systems described.

19. The measurement of momentum with the aid of a magnetic field. If a particle with unit charge moves in a plane at right angles to a uniform magnetic field, B, then its path is the arc of a circle of radius, ϱ. If p is the momentum of the particle, then

$$p = 300 \, B \, \varrho \tag{19.1}$$

where p is measured in ev/c, ϱ in cm, and B in gauss. In the preceding discussion of ionization determinations, the need for a measurement of the momentum of the calibration track was stressed. If then the momentum of the unknown track is measured, it can be used together with the ionization to determine the rest mass of the unknown particle. There are also other parameters such as the range in matter which can be used together with the momentum to determine the particle mass. On the other hand, if the rest mass of a particle is known, a measurement of its momentum will determine the velocity of the particle. The velocity can be of fundamental importance in the analysis of physically interesting problems such as dynamic collision processes and decays in flight. This gives some indication of why so much work has been performed with cloud chambers placed in magnetic fields. Several authors[1] have discussed the design of magnets to be used with cloud chambers, but the technical details will not be entered upon here.

The relation given in Eq. (19.1) is not directly applicable to the determination of the momenta of tracks actually photographed in a cloud chamber for several reasons. First, the track does not in general lie in the plane perpendicular to the magnetic field. Because of a component of velocity along the field, the particle describes a helical trajectory. Let ϱ_s be the radius and α the pitch angle of the helix. The momentum is given by

$$p = 300 \, B \, \varrho_s \, (\sin \alpha)^{-1}. \tag{19.2}$$

---

[1] Eg. See J. A. NEWTH: Nuovo Cim. (2nd Suppl.) **11**, 297 (1954).

The angle, α, is sometimes called the "dip" angle when referring to cloud chamber tracks and can be determined by stereoscopic reprojection in space. It must be borne in mind, however, that the true dip angle has been dilated by the expansion of the gas in the chamber, as mentioned above. Now the "apparent" radius of curvature, ϱ_a can be measured on the film and several methods for doing so will be described below. The relation of ϱ_a to the radius, ϱ_s, in the equation is complicated by the fact that the image to be measured is a conical projection of the curve in space onto the plane of the film. Furthermore, the magnification factor, m, varies along a track depending upon the distance of a given point from the camera lens. BARKER[1] has given an approximate treatment of the problem which assumes that the arc length of the track is much shorter than its radius of curvature. In this case the motion can be considered to take place in a plane which is the osculating plane of the mid-point of the arc and describes a circle of radius $\varrho_s \cosec^2 \alpha$. As a coordinate system, he chooses the plane of the front glass of the chamber as the x, y plane with origin at the intersection of this plane with the axis of the camera lens. The z direction is then measured along this axis which is in turn parallel to the direction of the magnetic field, B. The positive z direction is measured toward the rear of the chamber. If the center of the track has coordinates x_0, y_0, z_0 in the chamber, and the well known relation

$$\varrho_a = \frac{L^2}{8s} \tag{19.3}$$

is to be used for determining the apparent radius of curvature from an arc with chord length L and sagitta, s, on the film, then BARKER shows that

$$\frac{1}{\varrho_s} = \left(\frac{8s}{L^2}\right)\left\{\frac{q}{D+z_0}\right\}\left[\frac{1}{\cos^3\beta}\left(1 \pm \frac{y_0 \cot\alpha}{D+z_0}\right)^2\right]. \tag{19.4}$$

Here $(D+z_0)$ is just the object distance and q the image distance, so that the second term is the magnification of the midpoint of the track. D is the distance from the camera lens to the front glass of the chamber. The first term is the circular curvature, while the last expression in brackets is a correction for the conical projection. The angle β is determined from the condition

$$\tan\beta = \frac{x_0 \cot\alpha}{D+z}$$

where z is the depth coordinate of the point on the track nearest to the lens axis.

The magnetic fields used in most of the current cloud chamber work are not strictly axial as assumed in Eq. (19.2), so that the value of B to be inserted in the equation must be determined. Some workers use the average of the field, $\overline{B}$, taken over the total track length, while others use the value of the field at the center of the track, B_c. BARKER quotes some work by M. S. COATES, who has found that unless the field is exceedingly inhomogeneous, $(\overline{B}+B_c)/2$ gives a better estimate of the field than either of the two values taken separately. A more important effect results from a combination of the field and the conical projections. BARKER computes an approximate correction factor of the form $(1 \pm \varepsilon)^{-1}$ to be applied to the field where $\varepsilon = \pm\eta\, x_0/(D+z_0)$. The coefficient η is the ratio of the average transverse component of the field to the average axial component when the field is averaged along the track. Thus the final expression for the momentum is given by

$$p = 300\,\frac{\overline{B}+B_c}{2}\,\frac{\varrho_s}{(1 \pm \varepsilon)\sin\alpha}. \tag{19.5}$$

---

[1] K. H. BARKER: Nuovo Cim. (2nd Suppl.) **11**, 309 (1954).

This correction procedure is essentially equivalent to that used by other workers[1]. The calculations of Leighton and van Lindt, include from the outset the correction for inhomogeneity of the field and are more general than that given here. However, within the approximations used, this somewhat simplified treatment is quite adequate for most practical purposes.

An important systematic correction is encountered in practice which is not included in the above calculation. There it was assumed that the conical projection of the track in space onto the plane of the film was carried out by the camera lens without any distortion being introduced by that lens. Furthermore, the presence of the front glass in the optical path can be shown to give an apparent positive curvature[2], which is usually quite small in practice. Since these corrections depend on the individual lens and front glass used, it is best to determine a calibration curve empirically. Blackett and Brode[3] have photographed a set of fine, straight wires mounted in their chamber at various distances from the lens axis. The apparent curvature as a function of the angle made by the point on the wire closest to the lens axis is plotted and used to determine the proper correction to be applied to all subsequent curvature measurements on tracks. A somewhat different scheme uses a rectangular grid of fiducial marks either on the inside of the front glass of the chamber or on its piston to permanently record any possible distortions on each picture. This grid is also useful in the stereoscopic reprojection discussed earlier. By careful selection of the lens used[4] the amount of optical distortion can be greatly reduced. For example, Thompson[5] using a rectangular reticule on the front glass estimates that the spurious curvatures introduced in his chamber by optical distortions correspond to 0.7 km, or more, in radius.

The measurement of the circular curvature or apparent radius on the film can be performed by any of several methods. However, the technique introduced by Anderson[6] is in many ways the most satisfactory. The photograph of the track is placed upon the stage of a low power comparator microscope which has two transverse movements. The chord length, L, is aligned parallel to one axis of motion. By setting the cross-hair of the microscope on the center of the track at successive intervals along its length and reading off the corresponding transverse displacements of the cross-hair, a series of points can be plotted which represent the particle trajectory. A parabola of the form

$$y = a + b\,x + c\,x^2 \tag{19.6}$$

can be fitted to this array of points by the method of least squares. The apparent radius is then determined from the constant, c, by

$$\varrho_a = \frac{1}{2c}. \tag{19.7}$$

Thompson[7] has given some simplified formulas and tables for the computation of curvatures from such measurements. The use of his method greatly reduces

[1] R. B. Leighton: Unpublished 1951. — V. A. J. van Lindt: Ph. D. Thesis, Calif. Inst. Tech., Pasadena 1954. — R. W. Thompson, J. R. Burwell and R. W. Huggett: Nuovo Cim. 1956. — W. B. Fretter and E. W. Friesen: Rev. Sci. Instrum. 26, 703 (1955).
[2] Cf. N. N. das Gupta and S. K. Ghosh: Rev. Mod. Phys. 18, 225 (1946).
[3] P. M. S. Blackett and R. B. Brode: Proc. Roy. Soc. Lond. 154, 573 (1936).
[4] P. M. S. Blackett: Proc. Roy. Soc. Lond., Ser. A 159, 1 (1937).
[5] R. W. Thompson, J. R. Burwell and R. W. Huggett: Nuovo Cim. 1956.
[6] C. D. Anderson: Phys. Rev. 43, 491 (1933).
[7] R. W. Thompson: Nuovo Cim. 1, 735 (1955).

the labor involved in the application of this technique. Moreover, this method of measurement has proved to be the most accurate and readily permits one to evaluate the error in the curvature. BARKER[1] estimates that if the length of track, L, is divided into a large number of equal intervals the error, $\Delta\sigma$, in determining the curvature, c, is

$$\Delta\sigma \approx 2c \sim 1.76 \, (\delta y/L^2). \tag{19.8}$$

Here the error, $\Delta\sigma$, is expressed in reciprocal meters (m^{-1}); the error, δy, in making an individual setting of the cross-hair on the track center is in microns; and the chord length, L, is in millimeters (mm).

Because the coordinate plotting method just described is rather tedious to apply, a number of workers have introduced devices of one sort or another to speed the measuring process. One of the first such schemes to be introduced was the prism compensator of BLACKETT[2]. The curved track is projected through a lens-prism system onto a screen, and by axially rotating the prism, the image can be distorted into a straight line. The degree of prism rotation can be shown to be related to the curvature of the track and the accuracy of the method depends essentially upon the observer's ability to judge the "straightness" of the image. In fact this is not difficult and the resulting errors in curvature are comparable to those obtained by coordinate plotting.

Another of the many devices designed to simplify the curvature measurement has been recently suggested by LEIGHTON[3] who uses a somewhat different optical system for simplifying the procedure. Again the accuracy of the method compares with that of the coordinate plot and is much simpler to apply. Each such instrument has its merits, but almost no one of these has received widespread usage. However, one other method is considered standard practice in many laboratories. This consists of fitting an arc of known radius ruled on a transparent template to the track by superposing the two. Usually the template has a series of arcs with successively greater radii drawn on it and the best fit to the track is obtained by trial and error. Either the template is superposed directly on the film or the track is reprojected to full scale and the curve fitted to the image. The appropriate magnification factor must then be used when applying corrections of the form given in Eq. (19.4). Again the error of the measurement depends upon the experience of the observer and must be carefully evaluated. On the other hand, this technique is very simple and easy to apply, and is especially recommended for experiments in which a very large number of tracks must be measured.

20. Errors in the determination of momentum. In most cloud chambers currently used for momentum measurements, the dominant error arises from convection currents in the gas. However, it is interesting to enquire about the ultimate limitations to the curvature measurements assuming for the moment that such thermal distortions are negligible. BLACKETT[4] has treated just this problem and has considered the ultimate accuracy to be limited by the diffusion of the ions during the process of track formation and the multiple Coulomb scattering of the charged particles traversing the gas of the chamber.

---

[1] K. H. BARKER: Nuovo Cim. (2nd Suppl.) **11**, 309 (1954).

[2] P. M. S. BLACKETT: Proc. Roy. Soc. Lond. **154**, 564 (1936). For a more complete description see J. G. WILSON: The Principles of Cloud Chamber Technique, p. 101. Cambridge 1951.

[3] R. B. LEIGHTON: Rev. Sci. Instrum. **27**, 79 (1956).

[4] P. M. S. BLACKETT: Nuovo Cim. (2nd Suppl.) **11**, 264 (1954). See also J. G. WILSON: Principles of Cloud Chamber Technique, pp. 94—98. Cambridge 1951.

In treating the error due to ionic diffusion, he considers the distribution of drops about the trajectory of a particle which has traversed the chamber in a straight line. Since the diffusion is a random process, the fluctuations in the distribution of drops will cause an observed deviation from a straight line and can give rise to an apparent curvature. If the resulting curvature is combined with the magnetic field, B, it is shown that the fractional momentum error due to ionic diffusion can be written as

$$\left(\frac{\delta p}{p}\right)_d = 0.13\, L^{-\frac{5}{2}}\, p\, B^{-1}\, (k\,t/n)^{\frac{1}{2}}. \qquad (20.1)$$

Here p is the momentum in ev/c; L is the track length in cm; B is the magnetic field in gauss; k is the ionic diffusion coefficient; t is the expansion time in seconds; and n is the number of droplets formed in each centimeter of the track.

The calculation of the apparent curvature due to multiple Coulomb scattering has been computed by BETHE[1] following the treatment of WILLIAMS[2]. This expression can again be combined with the magnetic field to give the fractional error in the momentum for a singly charged particle.

$$\left(\frac{\delta p}{p}\right)_s = \left(\frac{16\pi}{3}\right)^{\frac{1}{2}} \frac{e\,Z\,N^{\frac{1}{2}}\,A}{L^{\frac{1}{2}}\beta\,B}. \qquad (20.2)$$

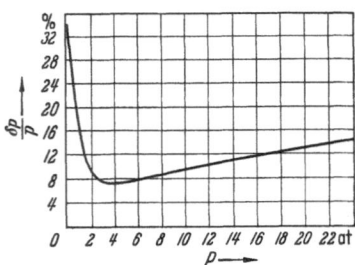

Fig. 16. The variation of the momentum error with operating pressure for an expansion cloud chamber.

Here e is the electronic charge; Z and A are the atomic number and weight of the gas in the chamber; N is the number of nuclei per cm³; and β is the ratio of particle velocity to the velocity of light.

The two expressions in Eqs. (20.1) and (20.2) are statistically independent, so that they can be combined in the usual way by taking the square root of the sum of their squares. The resulting error is then the ultimate limit of accuracy that one can hope to achieve in a momentum measurement. It is given by

$$\left(\frac{\delta p}{p}\right) = \frac{1}{B}\left[\frac{0.0169\,p^2\,k\,t}{L^5\,n} + \frac{16\pi\,(e\,Z\,A)^2\,N}{3\beta^2\,L}\right]^{\frac{1}{2}}. \qquad (20.3)$$

It is to be noted that in the first term, k varies inversely and n directly with the gas pressure, while in the second term N depends directly upon the pressure. Thus, in measuring tracks with some average momentum and length in a chamber, the operating pressure can in principle be adjusted to minimize the error. BLACKETT gives a plot of $(\delta p/p)$ vs. P, the pressure in atmospheres, for an argon filled chamber in a magnetic field of 8000 gauss. One of his curves is reproduced in Fig. 16 for $L = 10$ cm and $p\beta = 10^9$ ev/c. The error is clearly minimized for a chamber pressure of about four atmospheres. As seen from the expression for the error, the curve drops off as P^{-1} for low momenta and after passing the minimum rises as $P^{\frac{1}{2}}$ for higher values. In designing an experiment, the pressure can be chosen to minimize the momentum errors for the appropriate range of track lengths and momenta which are apt to be observed.

As pointed out above, the dominant cause for error in most curvature measurements on cloud chamber tracks is the distortion of the track by motion of the gas. In an expansion chamber the motion of the gas initiated by the expansion itself can be maintained as a simple one-dimensional dilation if care is taken to

[1] H. A. BETHE: Phys. Rev. 70, 821 (1946).
[2] E. J. WILLIAMS: Proc. Roy. Soc. Lond., Ser. A 169, 531 (1939). — Phys. Rev. 58, 292 (1940).

ensure that the piston or diaphram is critically damped as it comes to rest. The expansion, however, produces a temperature gradient between the walls and the adiabatically cooled gas and this gradient gives rise to convective currents. Fortunately, the main effect of these gradients is localized near the chamber walls. As mentioned in the discussion of thermostated cloud chambers, the most serious gas motions are generated by temperature gradients which exist before the expansion is initiated.

To determine the net effect of these convective distortions on the tracks observed in a given chamber, a figure of merit known as the *maximum detectable momentum* has been introduced. This is taken to be the momentum for which the true curvature produced by the magnetic field is just equal to the probable uncertainty of the curvature measurement. To determine the probable uncertainty in the measurement of the curvature, the magnetic field is turned off, or reduced to a low value, and photographs of relativistic particles are taken. In cosmic ray work, μ-mesons with some predetermined minimum range in lead absorber are frequently used, while in work with artificial accelerators other types of particles are employed. After measuring these "no-field" tracks in exactly the same way that all other tracks are measured, an histogram of their apparent curvatures is plotted. Such an histogram is illustrated in Fig. 17 which has been taken from the work of FRETTER and FRIESEN[1].

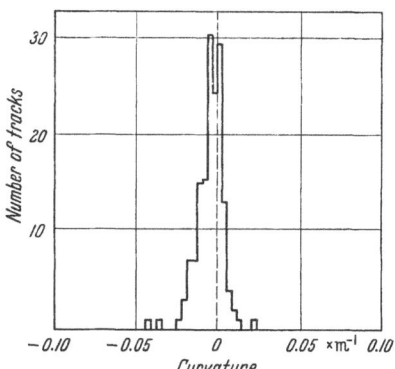

Fig. 17. A frequency distribution of the curvature of no-field tracks used to determine a maximum detectable momentum.

If a Gaussian error curve is fitted to this distribution they find that its peak is at $-1.2 \times 10^3 \, \text{m}^{-1}$ and that its half-width is $\pm 5.6 \times 10^3 \, \text{m}^{-1}$. Since the most probable value is less than the width, no systematic subtraction of this value from the measured values is performed. To obtain the maximum detectable momentum for the chamber, it is noted that the half-width of the no-field distribution corresponds to a radius of 180 m and that the mean value of their magnetic field is 7500 gauss. Then by definition, the maximum detectable momentum is about $4.0 \times 10^{10} \, \text{ev}/c$. This figure is obtained with tracks which average 47 cm in length and since the half-width of the distribution depends upon the track length used, it is conventional to quote both the value of the maximum detectable momentum and the length of track used in its determination. This value of FRETTER and FRIESEN is comparable with the maximum detectable momentum values achieved by earlier workers[2], but is noteworthy in that the droplet counting technique can be applied to the same tracks for ionization determination. The fractional error in any subsequent momentum measurement can be readily expressed in terms of the maximum detectable momentum p_m, by[3]

$$\frac{\delta p}{p} = \frac{p}{p_m}. \qquad (20.4)$$

[1] W. B. FRETTER and E. W. FRIESEN: Rev. Sci. Instrum. **26**, 703 (1955).

[2] P. M. S. BLACKETT and R. B. BRODE: Proc. Roy. Soc. Lond., Ser. A **154**, 573 (1936). — P. M. S. BLACKETT and J. G. WILSON: Proc. Roy. Soc. Lond., Ser. A **160**, 304 (1937).

[3] Cf. For example J. G. WILSON: Principles of Cloud Chamber Technique, p. 73. Cambridge 1951.

Using the maximum detectable momentum of the earlier workers for comparison with the ultimate limitation on the accuracy set by Eq. (20.3), BLACKETT[1] has shown that the two very nearly agree. In fact, using the parameters of his early experiments, Eq. (20.3) predicts that the best accuracy obtainable for a momentum of 10^9 ev/c is about 4.3%, whereas the measured error on a track of this momentum when determined from Eq. (20.4) is about 5%. Similar computations can be made for the results of more recent workers to show that the convective distortions can be very nearly eliminated by proper temperature control of the chamber.

The use of no-field tracks to determine the momentum errors requires some comment. It should be noted that the distribution resulting from the measurement of no-field tracks, if carried out in exactly the same manner as a curvature measurement with the field, automatically has several sources of error folded into it. The error of setting the cross-hair on the midpoint of any section of track is compounded of the judgement of the observer and the possibility of fluctuations in the diffusion of the ions as already described. This latter effect is sometimes referred to as track "noise". In addition, the distribution will include the effects of gaseous distortion, which are of primary interest. On the other hand, the no-field tracks usually have relativistic velocities and the multiple scattering error is negligible. But in subsequent measurements of low momenta, this error must be included and can be estimated from Eq. (20.2). In practice, the no-field tracks selected for use are chosen to lie roughly in the plane of the chamber, and hence of the film, and the various systematic dip corrections are not applied. Therefore, if the maximum detectable momentum is used to estimate the error in the momentum of a steeply inclined track, the result may very well be too small. In principle, this is not a serious limitation but the necessarily tedious measurement of no-field tracks presents an obstacle in most practical situations.

FRETTER and FRIESEN found that the spurious curvature of no-field tracks varied with the track length, L, approximately as $L^{-1.25}$ rather than as L^{-2} as one might expect from Eq. (19.3). If the value of the maximum detectable momentum is appropriately scaled to take account of the track length in determining the error given by Eq. (20.4) and the scattering error is computed from Eq. (20.2), then the two can be combined in the usual way to give a realistic estimate of the error in any given momentum measurement.

21. Range measurements. The distance a charged particle will penetrate into an absorber before being brought to rest is known as its *range* in the material. The range, R, is related to the initial velocity, v, of a particle of mass, m, by a relation of the form

$$R = m\, g\,(v). \tag{21.1}$$

The function g depends upon the constants of the material as well as v and tables of R/m *vs.* v are readily found in the literature[2]. In multiplate cloud chambers the range of a particle is readily measured, if it comes to rest in one of the plates. The particle, $e - f$, in Fig. 5 above has already been pointed out as an example of such an event. The uncertainty in the range of this particle is just equal to the thickness of the plate which arrests its progress. Occasionally, a very slow particle is observed to come to rest in the gas of a cloud chamber, and in high pressure chambers such events are fairly common. In this case the range can be determined much more accurately than in the multiplate chamber. However,

[1] P. M. S. BLACKETT: Nuovo Cim. (2nd Suppl.) **11**, 264 (1954).

[2] See, for example, B. ROSSI: High Energy Particles, p. 35. New York: Prentice-Hall 1952.

the accuracy is limited by the increase of multiple scattering at the end of the track and a number of uncertainties in the mechanism of energy loss at very low velocity. At very low velocity the particle can capture and release orbital electrons making the determination of the effective charge uncertain. These effects together with the fluctuations in the number of collisions per unit path length give rise to the well-known "straggling" of the range.

It is to be noted that the range depends on neither the velocity nor the momentum only, so that a measurement of range and ionization or range and momentum enables one to uniquely determine the rest mass of a particle. A classic example of the range-momentum technique is the first accurate measurement of the rest mass of the μ-meson by FRETTER[1]. Two cloud chambers were used. The first was mounted in a magnetic field to determine the momentum of the particles which were then brought to rest in a multiplate chamber mounted just below the magnet chamber. Similar arrangements have been used more recently to study the new heavy unstable particles[2].

Another variation of the range-momentum method is to observe the change in momentum suffered by a particle in traversing a plate. ANDERSON[3] used this technique to demonstrate the existence of the positron. The method has severe limitations in that only relatively slow particles will lose sufficient momentum in traversing a plate of reasonable thickness to give an accurately determined momentum difference. A plot of the momentum loss in a given plate *vs.* the momentum of the particle can be made and different curves are obtained for particles with different rest masses. This method has been used recently with a slow particle selector[4], but is far too restricted in accuracy to be recommended for general use.

It has proved more reliable in practice to observe the change in ionization as a function of range, rather than to make a single measurement of the ionization to estimate the mass of a particle. POWELL[5] introduced the technique of observing the ionization as a function of the residual range of the particle and used the calculations of WHEELER and LADENBURG[6] to obtain curves of relative ionization *vs.* residual range. A recent application of this technique to the determination of the rest masses of the new unstable particles stopped in a multiplate chamber has been reported by ROSSI[7]. In the earlier work, the ionization determination was estimated visually by simply comparing the density of the various segments of the track in question. However, the later work employs a photometer for the comparison and obtains much more reliable results.

22. Measurements using knock-on electrons.
A knock-on electron was defined above as an orbital electron which has been ejected in a collision between the bombarding particle and an atom of the material with sufficient energy to produce ions by subsequent collision. The very low energy knock-on electrons will produce so few ions that the resulting clusters of ions will be indistinguishable from fluctuations in the primary ionization of the bombarding particle. If E is the minimum kinetic energy which can be imparted to a knock-on electron

---

[1] W. B. FRETTER: Phys. Rev. **70**, 625 (1946).

[2] B. GREGORY, L. LEPRINCE-RINGUET, F. MULLER and C. PEYROU: Nuovo Cim. **11**, 292 (1954).

[3] C. D. ANDERSON: Phys. Rev. **43**, 491 (1933).

[4] C. M. YORK: Phys. Rev. **96**, 1635 (1954).

[5] W. M. POWELL: Phys. Rev. **69**, 385 (1946).

[6] J. A. WHEELER and R. LADENBURG: Phys. Rev. **60**, 754 (1941).

[7] B. ROSSI: Proceedings of the Sixth Annual Rochester Conference 1956 and D. O. CALDWELL and YASH PAL: Rev. Sci. Instrum. **27**, 633 (1956).

so that it can produce a readily detected blob of ionization, then the number of knock-ons per cm of path of the bombarding particle is given very closely by the RUTHERFORD formula[1],

$$N(E) = 2\pi a^2 z^2 \cdot Z N m c^2/\beta^2 E. \tag{22.1}$$

Here a is the classical electron radius $(2.8 \times 10^{-13}$ cm); Z and N are the atomic number and number of atoms per cm³ in the material traversed; z is the charge of the incident particle and β is its velocity in units of the velocity of light. The number of knock-ons per unit path can give an estimate of the velocity of the bombarding particle. That is, from Eq. (22.1) $N = C/\beta^2$ where C is a constant. A number of workers have attempted to use this method for velocity determination, but without success. First the constant C should probably be determined empirically because of the gas-vapor mixture in the chamber, and the difficulty of computing the proper value of E, the minimum detectable knock-on energy. In chambers operated at one or two atmospheres of pressure the number of knock-ons per unit path is so low that the statistical fluctuations in N preclude any reliability in the velocity determination. However, in high pressure chambers[2] there is some hope that the method may prove useful.

A small fraction of the knock-on electrons produced are of sufficiently high energy to produce measurable tracks. If the knock-on total energy, E', momentum, p', mass, m, and angle of ejection, ϑ', are known, then from the dynamics of the collision the velocity of the bombarding particle is given approximately by

$$\beta = \frac{p' c \sec \vartheta'}{E' + m c^2}. \tag{22.2}$$

Here it is assumed that the total energy of the bombarding particle is much greater than the rest energy of the electron. If the angle of scattering, ϑ, of the incident particle can be measured, then in principle its rest mass, μ, could be determined from the approximate relation

$$\mu^2 = \frac{2p'^2}{\vartheta^2 c^2 \beta^2 \gamma^2} \left\{ 1 - \frac{c}{\beta} \frac{p'}{E' + mc^2} \right\}. \tag{22.3}$$

Here $\gamma = (1 - \beta^2)^{-\frac{1}{2}}$ and β is to be determined with the aid of Eq. (22.2). Unfortunately the angle ϑ is of the order of m/μ, so that even for bombarding particles which are μ-mesons, it is so small as not to be measurable. However, the velocity determination from Eq. (22.2) can be combined with a momentum measurement in a magnetic field to determine the rest mass of the incident particle. LEPRINCE-RINGUET and L'HERITIER[3] used this method in analyzing the first case of a particle with mass one thousand times the electron mass ever to be reported. Unfortunately, the occurrence of this type of event is quite rare.

23. Momentum estimates from multiple scattering. Following the introduction of the multiple scattering technique for momentum determination on tracks in photographic emulsions, the group at the Massachusetts Institute of Technology[4] applied the method to tracks observed in their multiplate cloud chamber. By measuring the successive angles of deflection of the track as it emerges from each plate, an estimate of the momentum can be obtained from the average scattering angle. Essentially the method is limited by the small number of angles thus

[1] See B. ROSSI: High Energy Particles, p. 14. New York: Prentice-Hall 1952.
[2] E. H. S. BURHOP: Nuovo Cim. (2nd Suppl.) **11**, 343 (1954).
[3] L. LEPRINCE-RINGUET and M. L'HERITIER: J. Phys. Radium **1**, 66 (1946).
[4] M. ANNIS, H. S. BRIDGE and S. OLBERT: Phys. Rev. **89**, 1216 (1953).

measured, and hence the possibility of large statistical fluctuations in the average angle of scattering. By combining the scattering measurements with the range of stopping particles, these workers have been able to make a rather crude distinction in the rest mass of the particles observed. Several other workers[1] have used this type of momentum determination, but there are a number of theoretical objections to the assumed forms for the scattering distributions used to relate the mean scattering angle to the momentum.

The Coulomb scattering in the chamber gas was seen above to be one of the limiting factors in precise curvature measurements. Several attempts have been made to measure this scattering directly and use it as a measure of the particle momentum[2]. However, no experimental results have been published. BURHOP[3] has suggested using the technique with his high pressure chamber in a magnetic field and gives a detailed estimate of the errors involved. The error in the momentum determination by either multiple scattering or curvature of a μ-meson track 22.5 cm long in a field of 7000 gauss for a chamber operated at 100 atmospheres of argon is of the order of 30% for momenta in the range 70 to 200 Mev/c. This accuracy could be improved by decreasing the operating pressure, increasing the track length, and so on, but indicates the rather severe limitations to the technique.

24. Cascade showers. One of the more important properties of electrons, which has not yet been discussed, is their ability to radiate, or undergo *bremsstrahlung*, in a collision. The photons thus radiated in turn can produce electron-positron pairs, which in turn radiate, and so on, thus producing a "cascade shower" of electrons. The theory of this process is treated in detail elsewhere[4], but it is to be noted that the bremsstrahlung cross section varies inversely as the square of the mass of the particle. As a result, only electrons are light enough to produce cascade showers in appreciable numbers. This simple property has been used to distinguish electrons from heavier particles with considerable success[5].

In multiplate chambers attempts have been made to infer the energy of the electron, or photon, initiating the cascade from the number of secondary tracks observed in the chamber. HAZEN[6] has analyzed photographs taken with the double cloud chamber apparatus installed at the Pic du Midi by the Ecole Polytechnique Group[7]. In the upper chamber the momentum of the electrons is measured with the aid of a magnetic field, while the cascade shower is observed to develop as the electron penetrates 1 cm copper plates mounted in a multiplate chamber below. HAZEN points out that from shower theory one might expect the primary energy, E_0, of an electron, or photon, to be related to the total number of observed shower electrons, N. However, the theory does not take account of Coulomb scattering of the electrons in the solid metal plates, and indeed the number of observed particles is less than that predicted by the theory.

[1] J. B. McDIARMID: Phil. Mag. **45**, 933 (1954).
[2] E. g. R. B. LEIGHTON, J. TEASDALE and C. M. YORK: Unpublished, 1954.
[3] E. H. S. BURHOP: Nuovo Cim. (2nd Suppl.) **11**, 343 (1954).
[4] B. ROSSI: High Energy Particles, Chap. 5. New York: Prentice-Hall 1952.
[5] For example, ROCHESTER and BUTLER argued that the charged particle emitted from the first charged heavy meson decay was *not* an electron because it penetrated 3 cm of lead without producing a cascade shower. Cf. G. D. ROCHESTER and C. C. BUTLER: Nature, Lond. **160**, 855 (1947).
[6] W. E. HAZEN: Phys. Rev. **99**, 911 (1955).
[7] B. GREGORY, A. LAGARRIQUE, L. LEPRINCE-RINGUET, F. MULLER and C. PEYROU: Nuovo Cim. **11**, 292 (1954).

Hence an empirical constant of proportionality is determined for the range of E_0 from 10^8 to 10^9 ev. In this range

$$E_0 = e_0 N \qquad (24.1)$$

with $e_0 = (24 \pm 3)$ Mev. The uncertainty in the determination of the primary energy in this way is expected from the theory to be just the statistical fluctuation, $N^{-\frac{1}{2}}$, in the total number of tracks observed. Experimentally HAZEN finds that this statistical error must be multiplied by a factor of about 1.25 to obtain the observed fluctuations. The total error in the determination of E_0 will be the combination of the statistical error and the error with which the calibration constant, e_0, has been determined.

The application of this technique is limited by the effects of chamber geometry as well as statistics. If part of the shower goes out of the illuminated region of the chamber, it is possible to use that portion which is visible to set a lower limit to the energy, E_0. Such arguments have proved valuable in a number of studies using multiplate chambers[1].

IV. Concluding remarks.

25. In discussing the cloud chamber as a means of detecting ionizing radiation, only the most rudimentary aspects of the ionization and condensation processes have been discussed. The formulae given for computing the ionization loss in gases are in a form which is convenient for computation. The mechanism of condensation of water vapor on ions is not well understood, so that in an effort to maintain a concise treatment the currently accepted ideas on the subject have been sketched, together with the classical description of the process. These processes not only make the detection of ionizing radiation possible, but are the basis of the quantitative measurement of the ionization, and hence the velocity of the particle. Moreover, the time for droplet growth by condensation, is seen to limit the ultimate accuracy with which the trajectory of the particle can be determined. Thus the physical basis of the detection mechanism must be understood, not only to comprehend how a cloud chamber operates, but also to appreciate the inherent limitations of its application to specific experiments.

In describing the several cloud chambers illustrated above, it is hoped that the reader will appreciate that the selection has attempted to be typical rather than comprehensive. Furthermore, the small number of references to experimental work with cloud chambers is designed to act as an introduction to the truly copious literature on the subject. It should be noted that there is probably no such thing as a "standard" cloud chamber apparatus, although several groups use a number of identical chambers, or copies of chambers used in other laboratories. In choosing apparatus to be described, a definite attempt has been made to use contemporary instruments rather than the older designs which have been so thoroughly discussed in the past.

Although no attempt has been made to make a complete list of the contributions to nuclear physics that have been made with cloud chambers, one cannot discuss the various types of measurements that can be performed with this device without continually referring to some of the outstanding physical discoveries of the past half century. In using these discoveries as illustrations of the cloud chamber technique, it is hoped that the emphasis of this article on the cloud chamber as an experimental method has been adequately maintained.

---

[1] E.g. H. BRIDGE, H. COURANT, H. DE STAEBLER and B. ROSSI: Phys. Rev. **99**, 911 (1954).

Acknowledgements.

In conclusion the author would like to express his gratitude to all of those who so generously contributed the illustrations and photographs used in this article. Furthermore he would like to thank the following, who as teachers and co-workers have contributed immeasureably to his knowledge of cloud-chambers over the past ten years: Professors C. D. ANDERSON, K. H. BARKER, P. M. S. BLACKETT, R. B. BRODE, C. C. BUTLER, E. W. COWAN, W. B. FRETTER, W. E. HAZEN, R. B. LEIGHTON, G. D. ROCHESTER and J. G. WILSON.

General references.

[1] GENTNER, W., H. MAIER-LEIBNITZ and W. BOTHE: Nebelkammerbilder. Berlin: Springer 1940. — A complete discussion of the early work done with the cloud chamber technique with many excellent reproductions of photographs and drawings of expansion chambers.

[2] DAS GUPTA, N. N., and S. K. GHOSH: Rev. Mod. Phys. **18**, 225 (1946). — This is a fairly comprehensive review article and it covers the literature very well through 1944.

[3] WILSON, J. G.: The Principles of Cloud Chamber Technique. Cambridge 1951. — A comprehensive monograph on the subject of cloud chambers. It is by far the most complete treatment in print of the expansion type cloud chamber.

[4] ROCHESTER, G. D., and J. G. WILSON: Cloud Chamber Photographs of the Cosmic Radiation. London: Pergamon Press, Ltd., 1952. — An impressive atlas of cloud chamber photographs with excellent discussions of the interpretation and measuring techniques employed in their analysis.

[5] ROSSI, B.: High Energy Particles. New York: Prentice-Hall 1952. — A comprehensive text book which treats the physics of particles and their interactions in a detailed way.

[6] SNOWDEN, M.: Progress in Nuclear Physics, Vol. 3, New York: Academic Press 1953. — An excellent review article on the diffusion cloud chamber.

[7] FRETTER, W. B.: Nuclear Particle Detection (Cloud Chambers and Bubble Chambers). Annual Review of Nuclear Science, Vol. 5, p. 145. Stanford 1955. — A thorough review article with considerable emphasis on diffusion cloud chambers. A comprehensive bibliography of the current literature is given.

The Bubble Chamber.

By

DONALD A. GLASER.

With 10 Figures.

1. Introduction. The bubble chamber is a new type of particle detector that combines many of the advantages of the cloud chamber and the nuclear emulsion for experiments with high energy accelerators. It consists of a closed vessel provided with windows and filled with a liquid at a temperature above the normal boiling point and under sufficient pressure to prevent actual boiling. When the pressure on the liquid is suddenly reduced by some sort of expansion mechanism, the liquid becomes thermodynamically unstable against formation of vapor bubbles. If the degree of instability is sufficient, the formation of bubbles can be nucleated by the passage of ionizing radiation through the chamber. In this way the paths of charged particles are revealed as strings of tiny bubbles which can be photographed to give a permanent record of interesting nuclear events. Bubble chambers now in operation are rectangular or cylindrical in shape, range in sensitive volume from a few cubic centimeters to ten liters or more, and operate with various liquids including hydrogen, helium, xenon, diethyl ether, propane and solutions containing more than one component.

When used with particle accelerators, bubble chambers may be cycled once every few seconds and some special ones have been built to cycle ten times per second or even faster. Tracks can be photographed as soon as a few microseconds after the passage of the particles in most chambers, and after a few milliseconds in the larger hydrogen chambers. Because of this very rapid photography, track distortions due to motion of the liquid are negligible, and the most serious geometric distortions result from optical distortions in the liquid if the temperature is not very uniform across the chamber. Another advantage of the very fast growth of bubbles is that ages of tracks may be estimated on the basis of bubble size to permit one to judge the simultaneity of events not connected by a visible track in a photograph. This estimate can be made with an accuracy of about ten microseconds with hydrocarbon chambers, and a millisecond for hydrogen chambers. Magnetic fields can be used for momentum measurements in bubble chambers containing light liquids, and scattering measurements can be made in heavier liquids.

Measurement of the density of bubbles along tracks can be used to determine particle velocities to about 5 % for a 10 centimeter track and range measurements of high accuracy can be used just as they are in experiments with nuclear emulsions. Interference of background ionizing events can be minimized by adjusting the timing of the expansion and flashlamp cycle so that each photograph records ionizing events occurring only during the beam pulse.

For many kinds of experiments with high-energy particle accelerators, these characteristics of the bubble chamber give it decisive advantages over the cloud chamber and the nuclear emulsion. The combination of large size, high density of the sensitive medium, rapid cycling rate, and ability to accept high beam intensities allows very rapid collection of data on particle interactions and properties. The possibility of precision magnetic momentum measurements, accurate range and velocity measurements, and low geometric distortion permit detailed quantitative analysis of the events observed. By choosing appropriate liquids one may study different types of events; liquid hydrogen for experiments requiring a pure proton target, deuterium for studies involving neutrons, hydrocarbons in cases requiring moderate stopping power and when the target nucleus can be identified as hydrogen or carbon by the reaction kinematics, and xenon for high stopping power and the efficient detection of gamma rays by pair production. Apparently any transparent liquid containing one or more pure components will serve in a bubble chamber, so the choice of the experimenter is limited mainly by factors of convenience such as chemical stability and the pressure and temperature requirements.

2. History of the development of the bubble chamber.

The bubble chamber was invented because a detector of its properties was needed for experiments in high energy particle physics. Underlying the search for a way of providing these desirable properties was the idea that particles passing through condensed matter lose too little energy to be observed directly by macroscopic methods. It was necessary to provide a source of energy that could produce a localized macroscopic disturbance when triggered by the microscopic effects of an ionizing particle. These requirements suggested the use of some type of metastable state which could be upset by the passage of a charged particle, in analogy with the metastabilities exploited in the Geiger counter, the cloud chamber, and the nuclear emulsion. Of the many chemical, electrical, mechanical and thermodynamic metastabilities that are known, one had to choose one that could be established in a solid or liquid of high density, that was readily controllable and led to rapidly reversible changes to permit fast recycling, and that was not upset so quickly or violently as to defy easy photography. On the basis of these requirements the superheated state of a pure liquid was chosen for the initial study.

To aid in the choice of a liquid and the proper experimental conditions for testing this idea, a theory[1] was developed to describe quantitatively the conditions under which ionizing events could nucleate bubbles in a superheated liquid. In close analogy with the theory of cloud chamber operation[2], it was assumed that the concentrations of electrostatic energy in charge clusters in the liquid could be responsible for bubble nucleation in a superheated liquid. Recently collected experimental information[3] on bubble nucleation by ionizing radiations make it seem unlikely that this electrostatic mechanism is correct as proposed originally, though the predictions of the theory agree well with the observed operating conditions of bubble chambers. After the experimental facts are presented, several ideas about the microscopic nucleation mechanism will be discussed and compared.

A prediction of the electrostatic energy calculation was that ionizing radiation should cause boiling in diethyl ether at about 140° C and a pressure of one atmosphere. Since the normal boiling point of diethyl ether is 34.6° C, the required amount of superheat is over 100° C which makes the whole idea of the bubble chamber seem impossible. Fortunately physical chemists have been interested in superheated liquids for a long time, and a search of the literature revealed a

[1] D. A. GLASER: Nuovo Cim. 2, Suppl. No. 2, 361 (1954).
[2] J. G. WILSON: The Principles of Cloud Chamber Technique. London: Cambridge University Press 1953.
[3] D. A. GLASER and L. O. ROELLIG: To be published.

beautifully reported series of careful experiments[1,2] on ultimate attainable superheats in a number of liquids including diethyl ether. Kenrick, Gilbert and Wismer found that below 130° C liquid diethyl ether could be maintained quietly at one atmosphere for several hours, while at 140° C, the liquid erupted at erratic time intervals after being brought rapidly to the high temperature. To demonstrate the "capriciousness" of the phenomenon they quote a typical

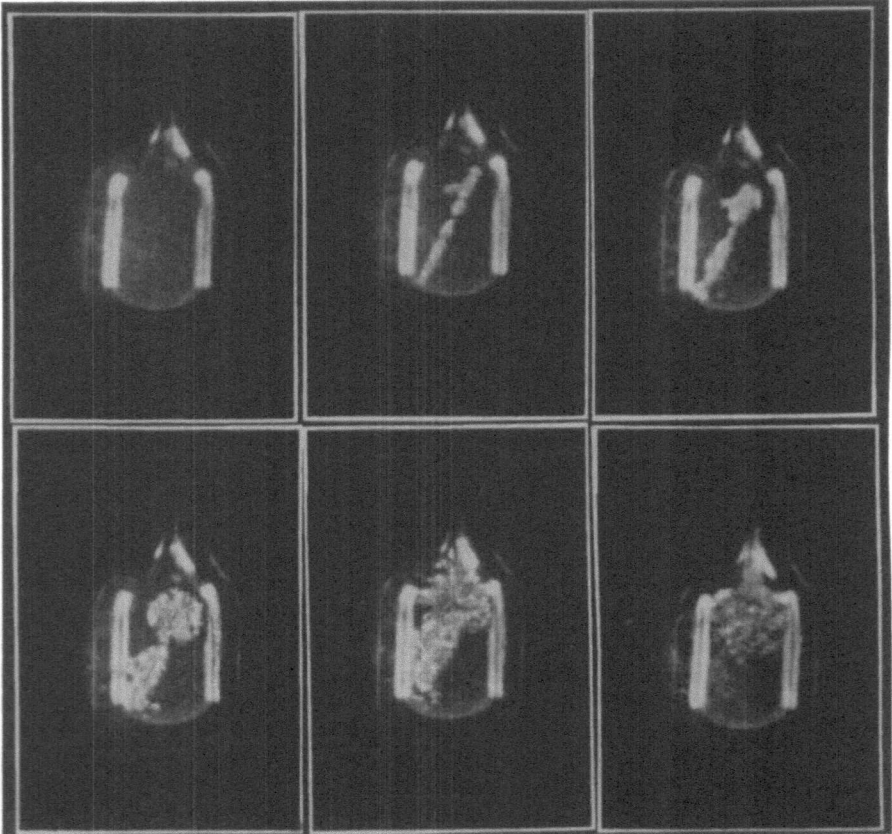

Fig. 1. High speed movies taken at 3000 pictures per second of eruptive boiling initiated in diethyl ether at 135° C by a penetrating particle. The pictures shown were taken at 0, $\frac{1}{3}$, 4, 22, 71 and 250 milliseconds. The first two are consecutive pictures.

series of 30 consecutive "waiting times". When these times were examined, they were found to be consistent with a Poisson distribution corresponding to the random occurrence of a nucleating event which disrupted the small liquid volume with an average waiting time of about 60 seconds. From the reported geometry of the superheating apparatus it was possible to estimate that its "counting rate" for cosmic rays at sea level was also about one count every 60 seconds! On the basis of these very suggestive correlations between predicted and previously observed results, the experiment of Kenrick, Gilbert and Wismer was repeated with controlled sources of ionizing radiation. It was found that

[1] K. L. Wismer: J. Phys. Chem. 26, 30 (1922).
[2] F. B. Kenrick, C. S. Gilbert and K. L. Wismer: J. Phys. Chem. 28, 1927 (1924).

gamma rays could indeed initiate boiling in diethyl ether at about 140° C and one atmosphere pressure[1].

These experiments established the existence of the fundamental phenomenon on which a bubble chamber technology could be based, but it remained to be shown that the explosive eruption of a highly superheated liquid could be photographed at a sufficiently early moment to record accurately the local ionizing events which led to the eruption. Furthermore, it was essential to find out if minimum ionizing particles could be seen in the bubble chamber. A track-forming detector insensitive to minimum ionizing particles would be of little use in studying high energy particle physics.

High speed moving pictures, taken at the rate of 3000 pictures per second, produced the first photographs of tracks in a bubble chamber and gave us some idea of the time scale of the bubble growth[2]. In Fig. 1 is reproduced a sample sequence showing the development of a track in diethyl ether at 135° C. The chamber was a cylindrical pyrex bulb 3 cm long and 1 cm inside diameter, which was kept hot by immersion in a bath of hot mineral oil. A manually controlled piston maintained the pressure necessary to prevent boiling and permitted the rapid pressure drop which allowed the liquid to become rapidly superheated.

To find out if minimum ionizing particles produce tracks, the same pyrex chamber was fitted with an automatic expansion mechanism in which compressed nitrogen actuated a flexible rubber diaphragm to control the pressure on the liquid. This chamber could be cycled continuously at 10 second intervals, and had an average sensitive time of several seconds after each expansion. A flash-lamp triggered by a vertical Geiger counter telescope took a picture whenever a penetrating cosmic ray went through the chamber. Thus it was established that minimum ionizing particles produce usable tracks[3]. It was found that good photographs could be obtained a few microseconds after the particle had passed through the chamber. The great rapidity of the bubble growth is illustrated in Fig. 2 which shows three photographs of cosmic ray tracks taken with various flashlamp delays. The variation of bubble density with temperature for minimum ionizing particle is also illustrated by these pictures.

In order to explore the possibilities of making a non-selective survey of ionizing events in the bubble chamber, methods for photographing all bubble eruptions were sought[4]. One method made use of the mechanical violence of the early stages of bubble growth. A sensitive phonograph pickup with its needle resting against a glass wall of the chamber detected the first vibrations associated with the eruption of the liquid and triggered the flashlamps. Although nuclear interactions were seen in bubble chambers for the first time using this method, only tracks containing large bubbles are obtained because of the low speed of sound in liquids.

Another non-selective triggering method capable in principle of high speed was to illuminate the chamber continuously with a weak steady light source and monitor the light scattered by the liquid with a photomultiplier tube[4]. Changes in the intensity of the scattered light due to bubble formation were used to trigger the flashlamp. This method could probably be refined to yield high quality track photographs if it were required for some special application.

These early experiments demonstrated that all-glass bubble chambers had many desirable characteristics that would make them useful for nuclear physics

---

[1] D. A. GLASER: Phys. Rev. **87**, 665 (1952).
[2] See footnote 1, p. 315.
[3] D. A. GLASER: Phys. Rev. **91**, 762 (1953).
[4] See footnote 1, p. 315.

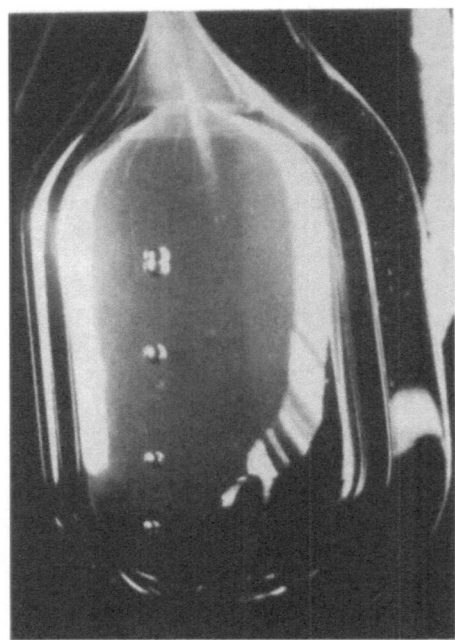

experiments. To be really useful, however, it was necessary to develop a technique for making large chambers containing many liters of liquid rather
than only a few cubic centimeters.
The one-piece glass construction is not
feasible for large chambers because it
is exceedingly difficult to make all-
glass chambers with sufficiently strong
large flat windows. For some experiments in the moderate energy range,
it is possible to use small hydrocarbon
and hydrogen chambers made entirely
of glass, but the size limitation is serious for experiments involving multi-
billion-electron-volt machines. Somewhat larger glass chambers have been
made by supporting the glass walls
hydrostatically with oil at a pressure
chosen to be intermediate between the
high and low pressures encountered during this working cycle[1]. These chambers
were all constructed with very clean,

a

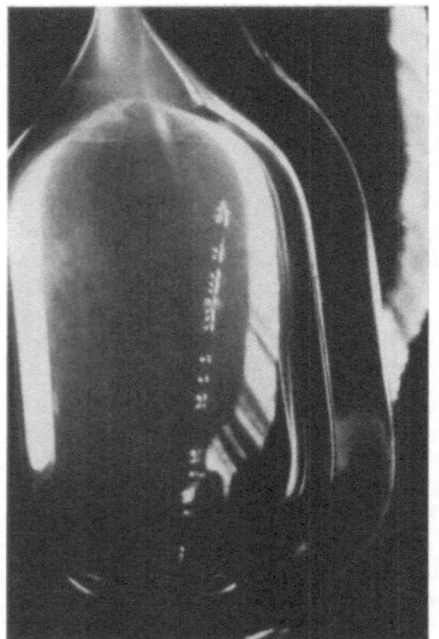

b

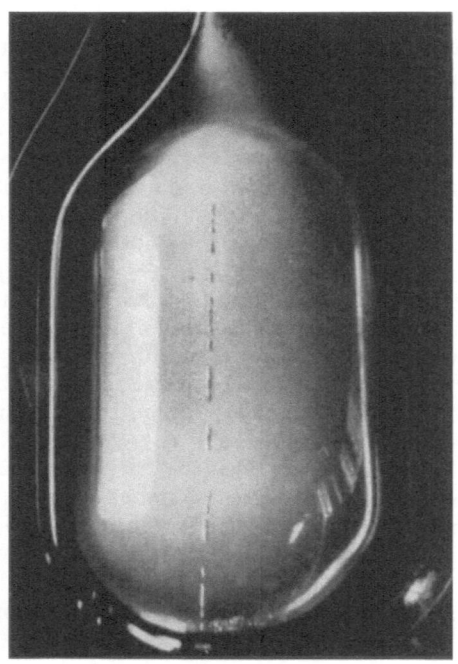

c

Fig. 2a—c. Tracks of penetrating vertical cosmic ray particles in diethyl ether obtained by random expansion and counter-
controlled flash-lamps. (a) 60 microsecond flash delay, 20 microsecond flash duration, 139° C. (b) 10 microsecond flash delay,
20 microsecond flash duration, 140° C. (c) 10 microsecond flash delay, 5 microsecond flash duration, 141° C.

[1] P. Bassi, P. Mittner and I. Scotoni: Nuovo Cim. 2, 1334 (1955).

smooth glass inner surfaces so that nucleation of boiling would begin on ionizing events occurring within the liquid volume rather than at rough places on the boundary surfaces. The basic idea was that it was necessary for the working liquid to remain in a static superheated condition after the expansion until a charged particle penetrated it. It was thought that one could not fabricate large chambers using ordinary metal parts, gaskets and glass windows, because metals and gasket materials nucleate boiling in liquids long before they can be superheated enough to respond to minimum ionizing particles.

A way out of this difficulty depends on the fact that the rate at which vapor is evolved during boiling is severely limited by the rate at which heat can be conducted through the liquid to supply the necessary heat of vaporization. If the volume of the bubble chamber could be increased very quickly by the expansion mechanism, one might therefore be able to maintain a low pressure within the bulk of the liquid even though boiling has begun at some places on the walls of the chamber. Recently developed theories of the growth of bubbles in superheated liquids have emphasized this role of thermal conductivity and heat of vaporization as the major factors limiting the speed of bubble growth[1,2]. These theories have given good quantitative agreement with observations of bubble growth.

J. G. Wood was the first to succeed in operating a bubble chamber fabricated of ordinary materials[3]. Remarkably enough, this first important success was achieved with liquid hydrogen, which involves many difficult experimental problems. It had already been demonstrated by Hildebrand and Nagle[4] that liquid hydrogen could be superheated in clean glass tubes sufficiently to be sensitive to gamma radiation from Co^{60} source. One factor which simplifies the technical problems with hydrogen is that it has a relatively high heat of vaporization and low thermal conductivity, so the evolution of vapor is relatively slow. An exceedingly rapid expansion mechanism is therefore not required. Because no special precautions must be taken to insure cleanliness of the interior of a rapid expansion chamber, such chambers have been called "dirty" chambers, in contrast with "clean" all glass chambers. "Dirty" chambers have now been operated successfully with a variety of hydrocarbons and other liquids.

3. Construction and operation of bubble chambers. Since it is not particularly difficult to construct a workable bubble chamber, and since the technology is still advancing very rapidly, it seems more useful to give general information on the construction and operation of various types of chambers than to describe existing instruments in great detail. We shall discuss the problems involved in constructing and operating chambers filled with organic liquids, hydrogen, helium and xenon as the most important examples.

α) *Organic liquid bubble chambers.* 1. Clean chambers. The clean chambers are made of one piece of round or rectangular pyrex tubing connected to the expansion system by an integral piece of smaller tubing. A seal to the metal parts of the expansion system is made by ◯-ring or other rubber gaskets. Only the liquid in the glass part of the chamber is superheated since the metal parts are operated colder than the liquid in the sensitive volume. Then the low vapor pressure of the liquid in the expansion mechanism does not interfere with the expansion process. Expansion is achieved by a flexible rubber or metal diaphragm or a piston actuated by compressed air. Commercial solenoid valves can be

[1] H. K. Forster and N. Zuber: J. Appl. Phys. **25**, 474 (1954).
[2] M. S. Plesset and S. A. Zwick: J. Appl. Phys. **25**, 493 (1954).
[3] J. G. Wood: Phys. Rev. **94**, 731 (1954).
[4] R. H. Hildebrand and D. E. Nagle: Phys. Rev. **92**, 517 (1953).

used to control the compressed air, although faster magnetic "pop" valves can also be built. In some bubble chambers an hydraulic fluid immiscible with the working fluid has been used to separate the superheated liquid from the expansion system. Glycerine[1], ethylene glycol[2] and other liquids have been used.

Most of these chambers are expanded to about one atmosphere final pressure, but the effects of varying the final pressure have been studied[3,4]. Even negative final pressures have been attained by using a polished glass piston in contact with the hydraulic fluid[5]. Very rapid bubble growth is attained in this way. The largest clean bubble chamber reported thus far has a sensitive volume of 450 cm^3. It has been found that fairly wide variation of the final pressure does not effect the chamber sensitivity very much but does change the bubble growth rate and the length of the sensitive time after each expansion[6,7].

2. "Dirty" chambers. Large "dirty" organic liquid chambers are usually constructed by clamping thick windows of tempered plate glass onto solid metal chamber bodies using O-rings or other commercial gasketing materials. Since boiling is bound to begin at the rough places on the walls and gaskets, it is useful to polish the metal surfaces and use smooth-edged materials to reduce the rate of vapor evolution. The fewer the number of places where boiling begins, the smaller the volume of vapor which is evolved, and the smaller the expansion ratio can be.

In a dirty chamber the fast expansion first relieves the pressure on the liquid and then continues to maintain this low pressure by providing very rapidly the volume required by the newly-forming vapor. For chambers of large volume it is desirable to reduce the hydrodynamic impedance associated with the expansion process in order to achieve very fast expansion. This is usually done by allowing the diaphragm or piston which expands the chamber to serve as one wall of the chamber as in a Wilson cloud chamber, or to communicate with the chamber through a short throat of large open area. Then the expansion mechanism operates at the same temperature as the sensitive liquid, in contrast with the usual practice for clean chambers. To prevent leakage and contamination of the working liquid, large chambers are usually sealed from the expansion mechanism by a flexible diaphragm. Sometimes the motion of the diaphragm is controlled directly by compressed air, sometimes by an O-ring sealed piston which communicates with the diaphragm through an hydraulic fluid, and sometimes by hydraulic fluid controlled by a second diaphragm and compressed air system. A new system in which a piston is controlled by high pressure hydraulic oil is also being tested.

The ultimate desirable characteristics of the expansion mechanism is that it drop the pressure fast, maintain it roughly uniform for at least a few milliseconds, and then sharply compress the vapor in preparation for the next cycle. It is important that the initial pressure drop be fast because tracks formed during this time of varying sensitivity have varying bubble densities and can therefore not be identified readily. Since large accelerators have some uncertainty or "jitter" in the time of beam arrival, it is essential to have a time of uniform

---

[1] D. C. Rahm: Ph. D. Thesis, University of Michigan 1956.

[2] R. J. Plano and I. A. Pless: Phys. Rev. 99, 639 (1955).

[3] P. Bassi, A. Loria, J. A. Meyer, P. Mittner and I. Scotani: Proceedings CERN Symposium on High Energy Accelerators and Pion Physics, Geneva, June 1956, Vol. 2.

[4] G. A. Blinov, Yu. S. Krestinikov and M. F. Lomanov: Proceedings CERN Symposium on High Energy Accelerators and Pion Physics, Geneva, June 1956, Vol. 2.

[5] See footnote 2, p. 320.

[6] See footnote 3, p. 320.

[7] See footnote 4, p. 320.

chamber sensitivity long enough to encompass the useful beam plus its time uncertainty. On the other hand if the sensitive time is made too long, much vapor will be evolved and recompression will be slow. Many of these features of dirty bubble chambers are illustrated in Fig. 3, which shows the operation of a typical dirty diethyl ether chamber.

Rapid recompression is important because it plays a critical role in temperature control and track distortion problems. Because of the rapid photography made possible by the rapid growth of bubbles in organic liquids, there are virtually no track distortions due to actual motion of the bubbles. Distortions which are observed are due rather to optical distortions in the liquid that result from non-uniformity of temperature. Heat is added to the liquid by the irreversible expansion cycle. In a propane chamber of 8 liters volume expanded every 5 seconds this heat can be as much as 200 watts. The time constant for smoothing out the "heat waves" seen visually in such a chamber after recompression may be as long as 15 to 20 seconds since convection and conduction are the principal mechanisms. Although bubble chambers can be recycled once a second or faster, maximum freedom from distortion may require slower rates of cycling. By recompressing the liquid as fast as possible, the heat of vaporization can be put back in almost the same place where it was extracted from the liquid, and also the total work involved in the cycle can be minimized.

Temperature control of large chambers has been achieved by constant temperature air ovens, by liquid baths, by electric heaters in contact with the chamber, and by circulating liquid heat exchangers soldered to the chamber walls. All of these systems can be provided with suitable thermocouple or thermister control to achieve adequate temperature uniformity and stability. Until the problems of temperature uniformity within the liquid are solved, it seems useless to provide highly refined external control systems.

No careful measurement has been made of the volume expansion ratio required for bubble chamber operation. Various chambers have used ratios varying from 1 up to 5%. The minimum useful value depends on the compressibility of the working liquid, which may be considerable when the liquid operates near its critical point. Increasing the ratio beyond the minimum increases the sensitive time of the chamber at the expense of an increase in the work and time required for the recompression.

β) *Hydrogen bubble chambers.* A bubble chamber filled with liquid hydrogen constitutes a pure proton target for the study of production and interactions of elementary particles. Because the interpretation of events in pure hydrogen is simple and unambiguous, much effort has gone into the development of magnetic hydrogen bubble chambers, and highly useful instruments have been constructed in an amazingly short time.

HILDEBRAND and NAGLE[1] first demonstrated that liquid hydrogen could be superheated by the same methods used for other liquids, and that the superheated state was sensitive to ionizing radiation. Their results were obtained with a small clean pyrex chamber of volume 3 cm^3 operating at a temperature at which the vapor pressure of hydrogen is 3 atmospheres. In the same chamber it was also demonstrated that liquid nitrogen is radiation sensitive at higher temperatures and pressures. Bubble tracks in liquid hydrogen were photographed first by WOOD[2] in a "dirty" chamber $2\frac{1}{2}$ inches in diameter using somewhat higher pressures and temperatures than those used in the "clean" hydrogen chamber.

---

[1] See footnote 4, p. 319.
[2] See footnote 3, p. 319.

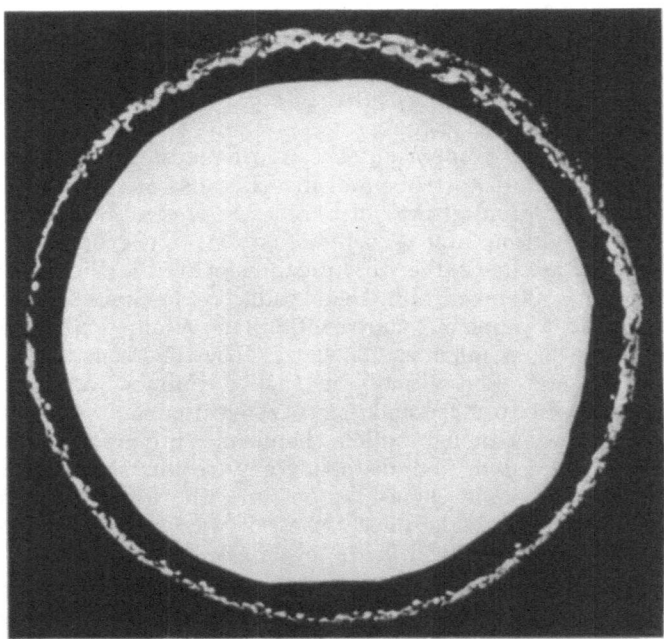

Fig. 3a—h. Operation of a "dirty" diethyl ether chamber 5 cm in diameter and 2.5 cm deep at 154° C. Photographs are taken with a 5 microsecond flash duration at different moments after the beginning of the expansion process. (a) 11 milliseconds, no radiation: Violent boiling occurs at the gaskets but no bubbles have formed in the interior of the liquid or at the glass windows.

Fig. 3b. 12.5 milliseconds, no radiation: Boiling has progressed further and a jet of vapor shoots out of the expansion orifice at the bottom of the chamber.

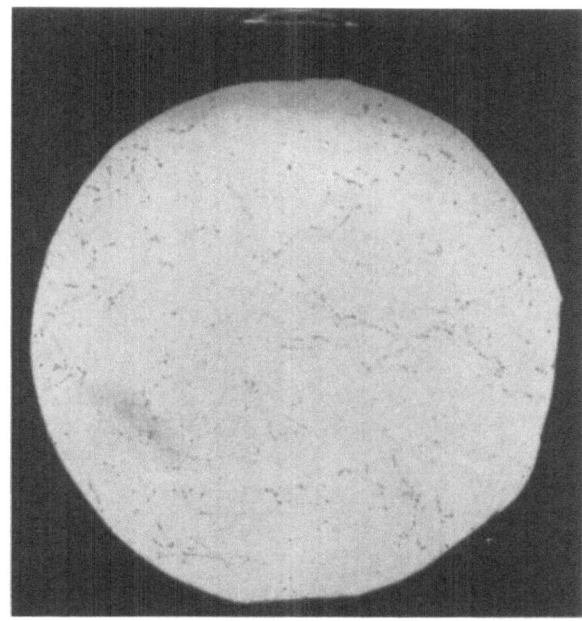

Fig. 3 c. 5 milliseconds, radium source nearby: Tracks of fine bubbles can barely be seen, but the bubble density is subnormal because the expansion is not yet complete.

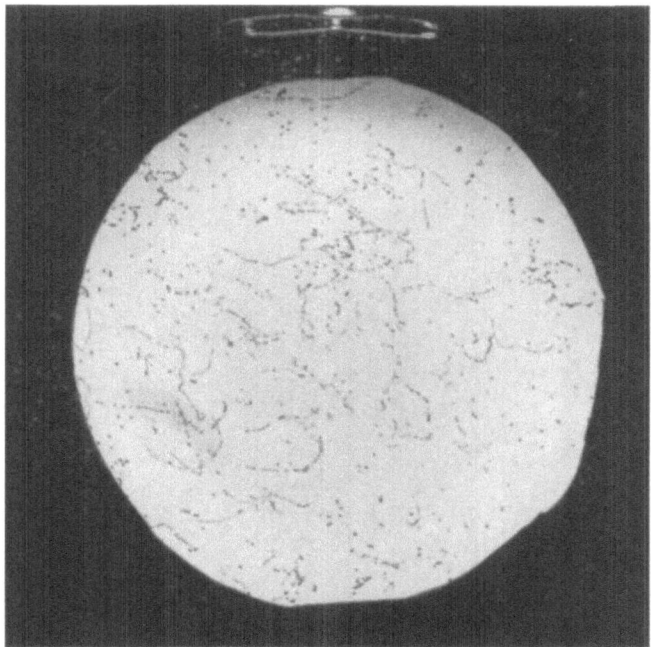

Fig. 3 d. 5.4 milliseconds, radium source nearby: Some fine and some larger tracks are visible, the finer ones showing normal bubble density indicating that the chamber is at full sensitivity.

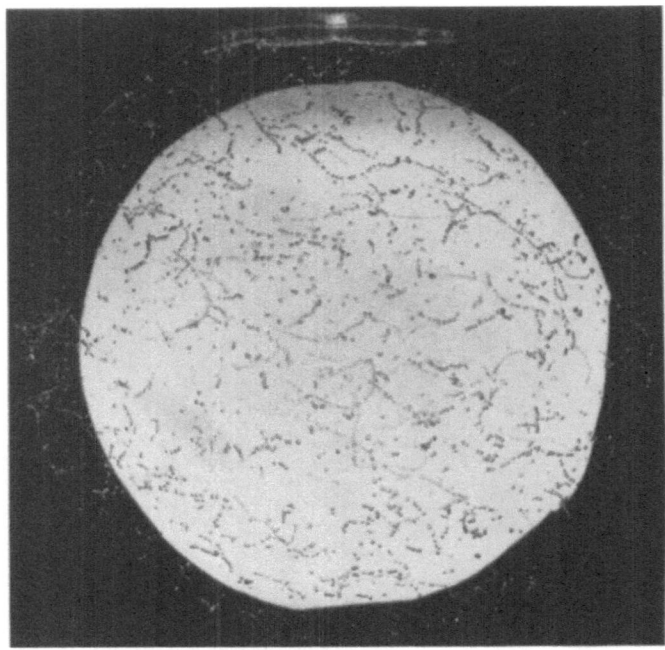

Fig. 3 e. 6.0 milliseconds, radium source nearby: Old tracks have grown considerably, but new fine tracks are still forming since the chamber is still sensitive.

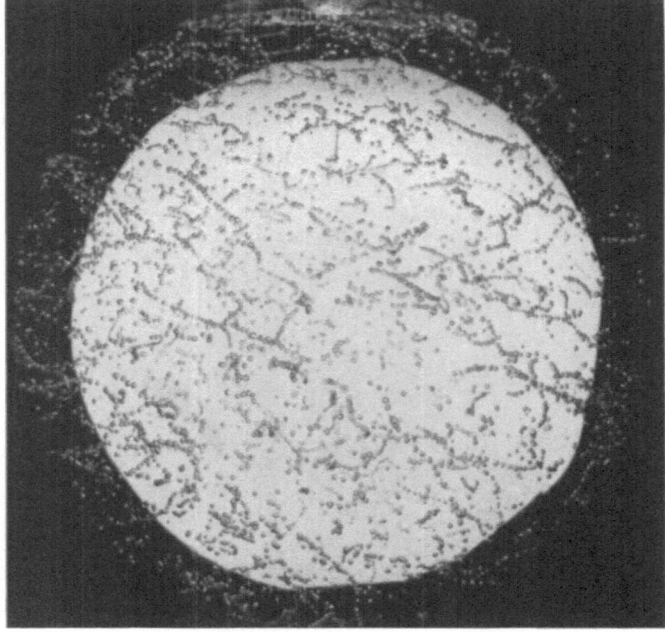

Fig. 3 f. 7.0 milliseconds, radium source nearby: Although the bubbles on the oldest tracks have grown quite large, new tracks are still being formed.

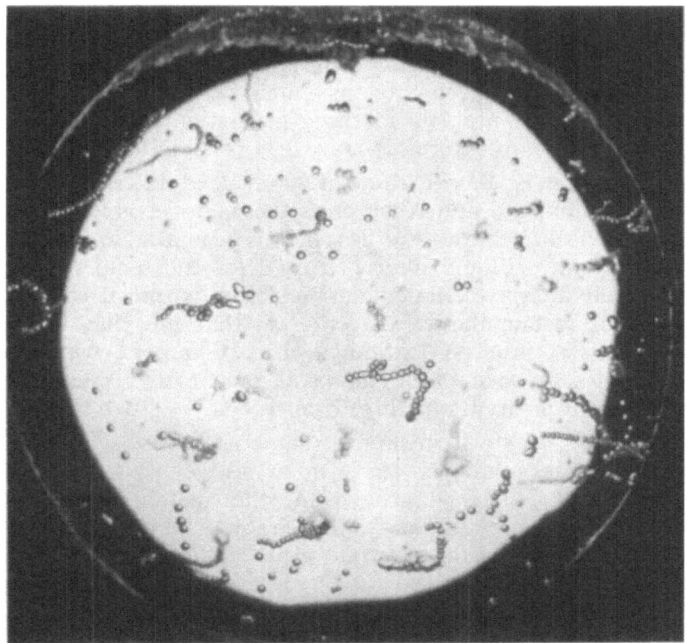

Fig. 3 g. 8.5 milliseconds, weak radium source nearby: New fine tracks are still appearing although the oldest tracks have become too coarse to be useful.

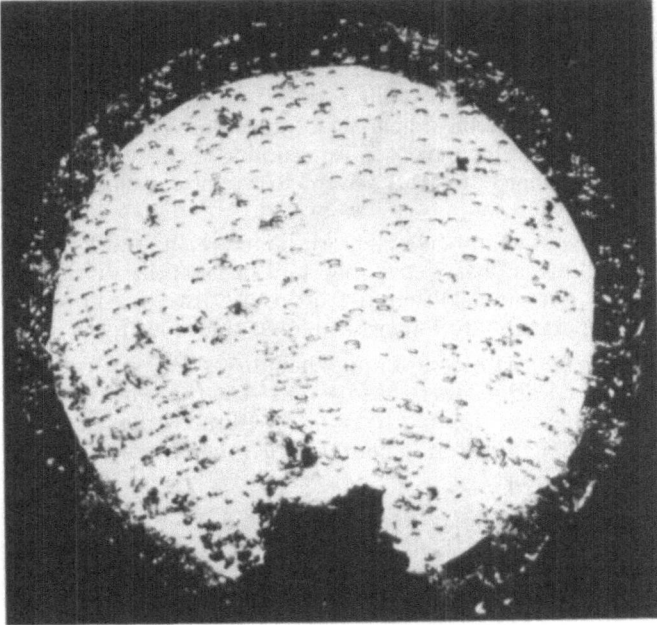

Fig. 3 h. 12.5 milliseconds, weak radium source nearby: Vapor jet from the expansion orifice causes a sudden pressure wave in the chamber. It distorts the tracks and ends the sensitive time abruptly.

Hydrogen bubble chambers operate at a temperature of 27° K and a pressure of about 5 atmospheres. Because of this rather low operating pressure clean all-glass chambers large enough for some low energy meson experiments are feasible[1]. A clean hydrogen chamber is much easier to build and use than a dirty one since a serious problem in the construction of chambers which must operate at very low temperatures is making a reliable leak-proof window seal.

Since wall boiling is almost completely avoided in a clean chamber, the motion required to expand and recompress the chamber is less than for a dirty chamber of the same size. As a result less heat is generated in the cycling process, and the consumption of liquid hydrogen used for the temperature reservoir is lower. Another advantage of the all-glass chamber is that it can be constructed of glass tubing of rectangular cross-section so that four flat viewing windows are available. This permits photography with two cameras whose lines of sight are at right angles to each other, an arrangement that makes possible very accurate localization of events in three dimensions.

Larger chambers up to 10 inches in diameter have been constructed with metal bodies and glass windows, the window seal being made with pure lead O-rings pressed into V-shaped grooves in the metal body[2]. Expansion is achieved in some designs by stainless steel bellows[3] or moving pistons in direct contact with the liquid hydrogen, while other chambers have been controlled by expanding and compressing the gaseous hydrogen which communicates with the working liquid[2]. Many of the technical details of the mechanical design of hydrogen chambers are extremely critical because of the serious safety problem with hydrogen and because of the cryogenics problems which require designs that minimize costly heat leaks into the system.

One method for maintaining the required operating temperature is to immerse the chamber in a bath of boiling liquid hydrogen whose temperature is controlled by controlling the escape pressure of the vapor evolved by the liquid in the boiling bath[3,4]. The whole system operates, in effect, as a liquid hydrogen "pressure cooker". Another system is to connect the chamber thermally with a reservoir of liquid hydrogen boiling at one atmosphere. The thermal connection is some type of variable "heat leak"; a metal bar with heater wire wound on it has been used for this purpose. To minimize heat leaks and liquid hydrogen consumption, the whole apparatus is inclosed in a series of radiation shields, and vacuum jackets.

A liquid hydrogen chamber 4 inches in diameter at the Radiation Laboratory at Berkeley, California, has been operated successfully with liquid deuterium under conditions not much different from those used for liquid hydrogen. The pressure and temperature are somewhat higher than for ordinary liquid hydrogen.

γ) Helium bubble chambers. Liquid helium at a temperature of 4.2° K and a pressure of about 1 atmosphere has been used as the sensitive medium in bubble chambers with volumes up to 110 cm³ [5,6]. Both vapor expansion and direct liquid expansion using a piston in contact with the liquid have been used successfully with these chambers.

---

[1] D. E. Nagle, R. H. Hildebrand and R. J. Plano: Rev. Sci. Instrum. **27**, 203 (1956).

[2] D. Parmentier and A. J. Schwemin: Rev. Sci. Instrum. **26**, 954 (1955).

[3] See footnote 1, p. 326.

[4] See footnote 2, p. 326.

[5] Harth, Fairbank, Blevins and Slaughter: Conference de Physique des Basses Temperatures (Proceedings), Paris 1955.

[6] Fairbank, Earth, Blevins and Slaughter: Bull. Amer. Phys. Soc. **30**, No. 5, 25 (1955).

Many of the constructional and operational details of helium chambers are identical with those of hydrogen chambers, but helium chambers are easier to make and use because of the low operating pressure, the absence of a safety hazard, and the finding that volume expansions of 1% or less are sufficient.

It has been suggested that a particular advantage of helium as a bubble chamber liquid is that the helium nucleus has isotopic spin zero so that simple statements can be made about the interactions of elementary particles in helium which will lead to experimental verification of isotopic spin assignments for these particles[1]. Another particular application of helium chambers is the study of light hyperfragments created in the liquid.

$\delta)$ *Xenon bubble chamber.* In order to be able to detect gamma rays and neutral π mesons in addition to charged particles, a liquid xenon bubble chamber has been developed[2]. Since the radiation length of liquid xenon is 3.1 cm, neutral π mesons and high energy gamma rays can be detected efficiently by production of electron pairs and Compton electrons in the dense liquid.

The first xenon chamber had a diameter of 1 inch and a thickness of $\frac{1}{2}$ inch, was constructed of aluminum and glass like the dirty organic chambers, and was controlled by a rubber diaphragm actuated by compressed air. It operates at about $-19°$ C and a pressure of 26 atmospheres.

In the first tests of this xenon chamber, it was not radiation sensitive, and began to give satisfactory tracks only when the pure xenon has deliberately contaminated with about 2% by weight of ethylene. The ethylene was added because it was known that xenon gas was an efficient scintillation material. Since formation of bubbles in the chamber depends on local concentrations of energy deposited by ionizing events, it was hoped that the ethylene would quench the scintillations by collisions of the second kind, and thus prevent the radiative escape of the ionization energy. It has been found very recently that pure liquid xenon is indeed an excellent scintillator.

In nearly every respect the construction and operation of xenon chambers is the same as that of organic chambers except that refrigeration instead of heating is required. An advantage of xenon compared with other high density or ordinary bubble chamber liquids is that it presents no hazard of explosion, toxicity, or corrosion.

A small clean bubble chamber has also been operated with stannic chloride in order to detect gamma rays efficiently[3]. High operating temperatures and corrosiveness are problems in the use of this liquid.

$\varepsilon)$ *Bubble chambers for studying random events.* Experiments with cosmic rays or with accelerators having essentially continuous beams or long beam pulses cannot be done very efficiently with the bubble chambers described so far because their sensitive times are of the order of milliseconds and their recycling times, seconds. Unfortunately it has been shown that the lifetime of the "latent image" in organic bubble chambers is surely not longer than 1 millisecond, and probably less than 0.1 millisecond[4]. This is too short a time to make counter-controlled expansions feasible for studying random events.

No complete solution of this problem has been found, but some progress has been made. One important advance is the discovery that the sensitive time of

---

[1] E. M. Harth, M. M. Block, W. M. Fairbank, M. J. Buckingham, G. G. Slaughter and M. E. Blevins: Proc. CERN Symposium on High Energy Accelerators and Pion Physics, Geneva, June 1956, Vol. 2.
[2] J. L. Brown, D. A. Glaser and M. L. Perl: Phys. Rev. **102**, 587 (1956).
[3] John Teem: Private communication.
[4] D. A. Glaser and D. C. Rahm: Phys. Rev. **97**, 474 (1955).

a propane chamber may be made as long as 40 milliseconds by operating the chamber at a high initial temperature and pressure and expanding it to some definite final pressure greater than one atmosphere[1, 2]. In this mode of operation the difference between the saturated vapor pressure and the controlled final pressure is less than when the chamber is expanded to one atmosphere. Since the other factors controlling the speed of bubble growth, such as thermal conductivity, surface tension, and heat of vaporization, are not much different for the two modes of operation, the pressure difference is the dominant factor and accounts for the reduced rate of evolution of vapor. Using rapid expansion and recompression valves it may be possible to cycle such a chamber on the order of 10 times per second and attain thereby a duty cycle of 40%. Photographs can then be taken by counter-controlled flashlamps when an interesting event occurs during the 40% of the time that the chamber is sensitive. An alternative way of achieving this rapid recycling is to use a piston driven by a piston rod and rotating crankshaft. The amount of energy involved in this rapid recycling is considerable and severe problems of overheating the sensitive liquid may be encountered.

Another approach to the problem of lengthening the sensitive time per expansion is to use a gas dissolved under high pressure in a liquid as the sensitive medium[3]. When the pressure is rapidly reduced, the solution becomes supersaturated and the gas begins to come out of solution first where ionizing events have occurred in the liquid. The advantage of this two-component system is that the macroscopic growth of the bubble is limited by the diffusion of gas into the bubble. This diffusion of gas slows the bubble growth just as vapor diffusion slows droplet growth in a cloud chamber.

A chamber based on this principle has been operated successfully with a solution of CO_2 in diethyl ether at room temperature and a pressure of about 47 atmosphere[4]. Other gas-liquid systems would probably also work, but the pressures are bound to be high unless low surface tensions can be produced by judicious choice of components, or by operating near the critical point, where the surface tension vanishes. A possible disadvantage of such a chamber is that recompression will be slowed by the diffusion of dissolving gas away from collapsing bubbles into the liquid. Operated as a periodically recycled detector, therefore, there might be no overall gain in the fraction of time the chamber is sensitive.

Several other suggested ways of achieving counter-controlled recording of events may be worth mentioning. If the "latent image" lasts as long as a few microseconds, it might be possible to produce the necessary superheat by dielectric heating of the liquid by a pulse of microwaves. Preliminary tests of this scheme have not produced tracks[5].

4. Microscopic mechanism of bubble nucleation by ionizing events. The operating conditions of the first successful bubble chamber were predicted quite closely by a microscopic model of an ionization-triggered bubble nucleation mechanism. It was supposed that an ionizing particle produces clusters of electric charges of like sign within a liquid, and that the mutual electrostatic repulsion of these charges can fracture a superheated liquid to produce bubble nuclei large enough to grow to visible size using thermal energy from the liquid. Several charges

---

[1] See footnote 3, p. 320.
[2] See footnote 4, p. 320.
[3] G. A. Askarian: J. exp. theor. Phys. USSR. **28**, 636 (1955).
[4] P. E. Argan and A. Gigli: Nuovo Cim., Ser. X **3**, No. 5, 1172 (1956).
[5] L. Bertanza, P. Franzini, G. Martelli and B. Tallini: (CERN).

are required for this model to work, since it can be shown that a single charge on a bubble surface tends to collapse the bubble in a homogeneous liquid dielectric. This is the reverse of the electrostatic effect in cloud chambers in which the growth of a singly-charged droplet is aided by the presence of the charge. An approximate treatment of the effect of electrostatic forces on bubble nucleation was made by assuming n electronic charges to be smeared uniformly over the surface of a potential bubble nucleus. This led to an "electrostatic pressure" effect which predicted that a bubble carrying n like charges would grow if the pressure on the liquid were less than the saturated vapor pressure at the ambient temperature by the amount,

$$P_{\infty}(T) - P > P_n(T) = \frac{3}{2} \left(\frac{4\pi}{n^2 e^2} \right)^{\frac{1}{3}} [\sigma(T)]^{\frac{4}{3}} [\varepsilon(T)]^{\frac{1}{3}} \tag{4.1}$$

in which $P_{\infty}(T)$ is the saturated vapor pressure over a flat liquid-vapor interface, P is the applied pressure, e the electronic charge, and σ and ε the surface tension and dielectric constant of the liquid. Good agreement with experimental observations is obtained with $n = 6$. This formula has had remarkable success in predicting the conditions of operation of many liquids including hydrogen, deuterium, helium, and organic liquids. Attempts have been made to extend the ideas of this theory to the behavior of liquids exposed to radiation in equilibrium states and numerical calculations have been reported for liquids in unstable states[1].

In spite of its practical successes, there are serious objections to this theory. To be effective in rupturing the superheated liquid, the n charges must be deposited inside a sphere of radius

$$r_n = \frac{2\sigma(T)}{P_n(T)} = \frac{2}{3} \left(\frac{2n^2 e^2}{\pi} \right)^{\frac{1}{3}} \sigma(T)^{-\frac{1}{3}} \varepsilon^{-\frac{1}{3}}(T). \tag{4.2}$$

This radius is on the order of 10^{-6} cm under typical operating conditions for organic liquids. One may ask how often a minimum ionizing particle leaves a cluster of six charges *of the same sign* in such a volume. It becomes obvious that it does not happen often enough in primary ionizations to account for the observed number of bubbles, but could conceivably be done by low energy delta rays stopping in the liquid. One can also object that the ions will run away from each other before there is time to make a bubble. Using ordinary mobilities of ions in liquids one finds the escape time to be about 10^{-8} seconds, while a bubble 10^{-6} cm in radius could easily grow in this time. Another possible check on the theory would be furnished if one could verify the temperature dependence on $P_n(T)$ in Eq. (4.1). This has been done in great detail using diethyl ether, benzene, sulphur dioxide and methyl alcohol. Agreement has been found[2] within experimental error with the $\sigma^{\frac{4}{3}}(T) E^{\frac{1}{3}}(T)$ dependence of $P_n(T)$. Finally a contradiction with the theory was found when stopping alpha particles were used as the ionizing agent. To agree with the theory the alpha particle was required to deposit 900 charges *of the same sign* in a region 2×10^{-6} cm in diameter. This is greater than the maximum ionization atteined by a slowing alpha particle, even granting the possibility of the required charge separation! The total energy loss of the alpha particle was sufficient to explain the observed effects, however, so another model for the microscopic mechanism was sought.

A rough estimate shows that at the maximum rate of energy loss of a stopping electron, proton, or alpha particle, sufficient energy is deposited in the

---

[1] L. BERTANZA, G. MARTELLI and A. ZACUTTI: Nuovo Cim. **1**, 324 (1955); **2**, 487 (1955).
[2] See footnote 3, p. 315.

required small volume to raise its temperature at least a few degrees centigrade. With this temperature rise, density fluctuations large enough to produce a viable bubble nucleus are common under the superheat conditions attained in bubble chamber operation[1]. Again one finds the time constant for the decay of this heat pulse by thermal conduction is about 10^{-8} seconds for most liquids. Therefore the thermal and electrostatic theories give about the same lifetime for the latent image.

It remains to describe an efficient process for conversion of the energy loss of a penetrating particle into local heating. Most of the energy lost by an ionizing particle is used in ionizing and exciting the atoms of the liquid. Typical atomic radiation times are of the order of 10^{-8} seconds. On the other hand the average molecular collision time in a liquid is roughly 10^{-12} seconds. Therefore a fairly small cross-section for de-exciting collisions of the second kind is enough to insure that most of the energy is converted into local kinetic energy rather than escaping as radiation. When the xenon bubble chamber refused to work with pure xenon, it was thought that perhaps xenon did not have a large enough cross-section for self de-exciting collisions, and that the excitation energy was escaping as radiation. It was hoped that contamination of the xenon with a substance known to have a large, self-quenching effect might also de-excite the xenon. This was borne out by successful operation of xenon chamber when some ethylene was added. The situation is analogous to the well-known quenching of a gaseous mercury discharge tube by the addition of oxygen. Gaseous and liquid xenon have been found recently to be excellent scintillators. Also the scintillation efficiency of the gas can be destroyed by a small contamination of a gaseous hydrocarbon.

Comparison with the measurements of the temperature dependence of the effect of radiation on liquids can be made by noting that the isothermal reversible production of a bubble nucleus requires an energy

$$W = \frac{16\pi\sigma^3}{3(\Delta P)^2}. \tag{4.3}$$

If we assume that a stopping delta ray provides a "standard heat pulse" in the liquid by converting its energy loss to heat with some efficiency that is roughly independent of the temperature of the liquid, we conclude that

$$\Delta P = P_\infty(T) - P(T) \propto \sigma^{\frac{3}{2}}(T). \tag{4.4}$$

For all the liquids tested this temperature dependence could not be distinguished experimentally from the $\sigma^{\frac{3}{4}}\varepsilon^{\frac{1}{2}}$ result of the electric theory. Both theories give, therefore, about the same temperature dependence, the same lifetime of the latent image, and depend on delta rays for producing bubbles. The electric theory fails quantitatively for heavily-ionizing particles. The thermal theory gives plausible quantitative agreement and can explain the operation of the xenon chamber.

If the thermal theory of bubble chamber operation is taken seriously, it can be used to suggest a way of making a counter-controlled bubble chamber. The general problem is to introduce one slow step in the conversion of ionization and excitation energy into thermal energy in order to gain time enough to superheat the liquid by expansion or heating before the latent image has vanished. Two possibilities are suggested. One is to choose a liquid in which recombination is slow or can be inhibited by the use of an electric field. The other is to use a

---

[1] M. VOLMER: Kinetik der Phasenbildung. Dresden 1939.

liquid that exhibits delayed scintillations like a long-persistence phosphor. Then one can use only a very small concentration of "quenching" liquid so that collisions of the second kind will thermalize the excitation energy very slowly over a long period of time.

5. Photography of bubble chambers. It is possible to photograph a track in most bubble chambers a few microseconds after the passage of the particle because of the rapid growth of the bubbles. Flashlamps capable of short duration flashes ranging from a few microseconds to a few tens of microseconds have been used for bubble chamber photography in order to obtain photographs very soon after the passage of the particle. Early photography reduces the size of the bubbles in the picture and minimizes distortions of the tracks due to motion of the liquid. The necessity for short flash duration limits the total amount of light available for photography and requires the use of moderately efficient illumination arrangements.

To aid in the design of bubble and cloud chamber illumination systems, the angular distribution of light scattered by bubbles and droplets has been computed in the geometrical optical approximation and measured for air bubbles in water and glass spheres in air[1]. The calculation is done by considering a uniform light beam to strike the spherical boundary and computing by SNELL's law the refractions and reflections occurring at the interfaces. The

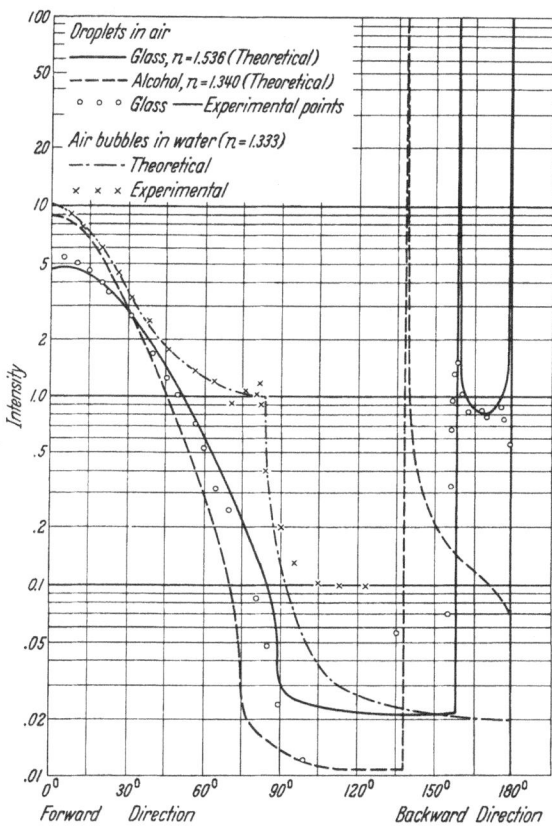

Fig. 4. Scattering of white light by spherical drops and bubbles. Theoretical curves are calculated using geometrical optics and considering several internal reflections and refractions. Measurements of light scattered by air bubbles in water are probably in error at 90° because of the light scattered by the water. Back scattering, which gives rise to rainbow effects with droplets, is absent with bubbles.

method is straightforward and is exactly analogous to the geometrical optical approximation to the theory of the rainbow. Measurements of the angular distribution of the scattered light were made using a photomultiplier tube for a glass sphere in air and for a vertical stream of fine air bubbles generated in water by forcing air through a fine glass capillary nozzle submerged under water. Fig. 4 shows the theoretical and experimental results. Because of the absence of internal critical reflections inside bubbles, the back scattering which gives rise to rainbow effects with droplets is not present for bubbles. Bubbles and droplets are found to exhibit the same generally forward peak, the peak being narrower for

---

[1] D. A. GLASER and R. HARTUNG: Unpublished.

liquids of lower index of refraction. Although bubbles scatter more than droplets at 90°, 90°-illumination, often used for cloud chambers, is probably impractical for bubble chambers if very short flash duration is required. For hydrogen bubble chambers the forward peak is so narrow as to absolutely prohibit 90°-illumination.

Both bright and dark field illumination have been used successfully for various types of bubble chambers. In the bright field arrangement, light from the flashlamp is collected by a lens which illuminates a diffusing screen that is photographed directly through two parallel windows of the chambers. Bubbles appear as dark obstacles on a bright background. Dark field systems use a beam of light that passes through the chamber toward the camera, but is prevented from striking the camera lens by a system of shutters like a "venetian blind" in one system, or by a field lens next to the source which brings the light to a focus away from the camera lens in another system. Both systems take advantage of the large small-angle scattering of light by bubbles. The bright field system has the advantage that the images of large bubbles are symmetric, so measurements can safely be made to the centers of the images. The dark field system is capable of higher contrast.

6. Measurements with bubble chambers. α) *Geometric quantities.* It is possible to determine the position of the center of a bubble image by measurement on the negative with an error corresponding to a distance of 5 microns in the original space. Measurements with hydrocarbon chambers, however, show that track distortions are introduced by the optical inhomogeneity of the liquid that results from temperature non-uniformities produced by the thermal effects of the expansion process. Further errors are introduced by optical and film distortions. It seems possible to come very close to the ultimate limitation mentioned above by installing inside the chamber a set of marks on a precision grid which appears in each picture and furnishes a basis for measuring and eliminating distortions by means of an analytical process of geometric interpolation carried out by an electronic computer. Preliminary work with such a system has been very successful. Most groups working with bubble chambers have chosen to make measurements directly on the negatives and to reduce this data analytically instead of using the analogue reprojection methods of measurement which have been used with cloud chamber pictures. One reason for this is that the higher geometric accuracy attainable by bubble chambers can only be realized by the analytical method, and another reason is that analogue reprojection methods are awkward for liquids with their large indices of refraction. Formulas for reduction of data measured on stereographic pairs of negatives must be fairly elaborate to achieve the attainable accuracy. Such formulas are easily handled by large computers which also calculate true lengths, angles, curvatures, and other geometrical quantities in the original space. The main distortion that cannot be removed easily in the computation is due to optical inhomogeneities of the liquid. These may be reduced by allowing a greater time interval between expansions to insure that thermal equilibrium will be reached. Another possible technique is the use of several cameras to view each event from several different places, making it possible to average out the errors during the analysis calculations.

β) *Coulomb scattering.* Multiple Coulomb scattering interferes with accurate magnetic curvature measurements in hydrogen and helium chambers, provides an important measurement in xenon chambers as in nuclear emulsion, and plays both roles in organic liquid chambers. A convenient way of comparing the effects of Coulomb scattering in various liquids is to compute the average excursion of the midpoint of a track segment from a chord joining its ends. This "sagittal" excursion is given by

$$\delta S_S = \frac{2.14}{p\,\beta\,c}\,\frac{l^{\frac{3}{2}}}{\lambda^{\frac{1}{2}}} \tag{6.1}$$

where l is the length of the segment in centimeters, λ is the radiation length in centimeters, p is the particle momentum in Mev/c and βc is the velocity in cm/sec. In Table 2 are tabulated sample values of the Coulomb scattering for some important liquids.

γ) *Magnetic curvature.* The sagitta of the projected circular arc produced by the motion of a singly-charged particle in a magnetic field of H gauss is

$$\delta S_H = 3.75 \times 10^{-5}\,\frac{l^2\,H}{p\,c} \tag{6.2}$$

where l is the length of the chord in centimeters and $p\,c$ is the particle momentum in Mev.

δ) *Bubble counting for velocity determination.* Bubble counting can be used to determine particle velocities in bubble chambers just as droplet counting is used in cloud chambers and grain counting in nuclear emulsions. The bubble density in propane is found to be not proportional to the total ionization but it obeys the relation

$$b\,\frac{\text{bubbles}}{\text{cm}} = \frac{A}{\beta^2} + B(T) \tag{6.3}$$

where $A = (9.2 \pm 0.2)$ bubbles/cm and B is a function of temperature only[1]. The constant, A, does not vary with temperature in the range 55 to 59.5° C. Because the temperature dependence is contained only in the $B(T)$ term, it was found possible to make the measurements insensitive to temperature variations by using comparison tracks of known velocity. We can form temperature independent differences of bubble densities

$$b_1 - b_2 = A\left(\frac{1}{\beta_1^2} - \frac{1}{\beta_2^2}\right) \tag{6.4}$$

from which β_1 can be found if b_1 and b_2 are measured and β_2 is the known velocity of a comparison track.

For bubble densities up to about 60 bubbles/cm there are essentially no counting errors. A systematic study of errors due to image fusion and actual bubble fusion is now being started. Preliminary microphotometer measurements of the optical density variations along a bubble track make it seem feasible to develop a photoelectric bubble counter. Such a device would save an enormous amount of labor and might also permit a reproducible calibration to be established for tracks too dense for reliable hand counting.

In most cases the principal source of error arises from statistical fluctuations in the bubble production and leads to a fractional error in velocity determination.

$$\frac{\delta\beta}{\beta} = \frac{\beta}{2\sqrt{AL}}\left(1 + \beta^2\,\frac{B}{A}\right)^{\frac{1}{2}} \tag{6.5}$$

in which β is the relativistic particle velocity, L is the track length in centimeters and A and B are the same as in Eq. (6.3). Lowering the operating temperature reduces B and improves the accuracy of the method. At the same time it permits

---

[1] D. A. GLASER, D. C. RAHM and C. DODD: Phys. Rev. **102**, 1653 (1956).

accurate counting of bubbles along tracks of particles with a wide range of velocities. For a 10 cm track segment the error in velocity determination in propane is about 5% for $\beta > 0.5$.

Measurements of other workers[1,2] are in agreement with these results for propane and preliminary measurements in $SnCl_4$[3], and hydrogen give the same general result. Organic liquids do not exhibit a large relativistic rise in total energy loss as measured by scintillation counters, but a relativistic rise in bubble density has been reported for electrons in propane[2]. A relativistic rise has been noticed qualitatively for relativistic electrons in xenon, but no accurate measurements have been made.

Table 1. *Comparison of track-forming detectors. Frequency of rare events in a 50 cm chamber.*

	Type	Density ϱ gm/cm²	Radiation length λ cm	Scattering sagittal δS for 2 Bev proton track 5 cm long microns	Magnetic field required for 10% momentum error in 5 cm relativistic track gauss	Stopping power of a 50 cm chamber gm/cm²	Events/day for σ/nucleon $= 10^{-30}$ cm² $= 1$ microbarn	Events/pulse for σ/nucleon $= 10^{-27}$ cm² $= 1$ millibarn
1	1 atmosphere argon expansion cloud chamber	0.0017	11 600	0.79	2400	0.085	0.015	0.0003
2	20 atmosphere hydrogen diffusion cloud chamber	0.0019	36 300	0.45	1 300	0.095	0.016	0.0003
3	hydrogen bubble chamber	0.05	1 380	2.3	6 900	2.5	0.43	0.008
4	helium bubble chamber	~0.10	963	2.75	8 200	5.0	0.86	0.015
5	propane (C_3H_8) bubble chamber	0.44	108.3	8.2	25 000	22	3.7	0.07
6	$SnCl_4$ bubble chamber	1.5	7.35	31.5	94 000	75	13	0.23
7	xenon bubble chamber	2.3	3.1	48.5	140 000	115	20	0.34
8	nuclear emulsion (AgBr)	4.0	2.8	51.1	150 000	200	34	0.59

In order to estimate the error in a momentum measurement introduced by Coulomb scattering we can consider that a track curvature is determined by a single sagitta measurement δs_H and that an error of the order of δs_S is introduced by scattering. Then the fractional error in momentum can be obtained by combining (6.3) and (6.4) and is approximately

$$\frac{\delta p}{p} = \frac{\delta(\delta S_H)}{\delta S_H} = \frac{\delta S_S}{\delta S_H} = \frac{5.7 \times 10^4}{\beta H \sqrt{l \lambda}}. \tag{6.6}$$

We list in Table 1 the values of the magnetic field strength required to limit this momentum error to 10% for a measurement on a 5 cm segment of the track of a relativistic particle in various detectors.

---

[1] See footnote 3, p. 320.
[2] See footnote 4, p. 320.
[3] See footnote 3, p. 327.

Another comparison that can be made among various track-forming detectors is the probability of seeing a rare event occurring in the sensitive medium. We have calculated the number of times a rare event of cross-section 10^{-30} cm²/nucleon would be seen during an 8 hour running period in each of the detectors listed in Table 1. We have assumed that a beam pulse containing 50 particles passes through the detector every 5 seconds. We have also tabulated the average number of events of cross-section 10^{-27} cm²/nucleon which occur in the sensitive medium during each pulse. This last number is related to the efficiency of scanning for interesting events.

7. Applications of bubble chambers. Many characteristics of bubble chambers make them ideal instruments for experiments with particle accelerators. They are capable of recording large quantities of precision information in a relatively short time. They seem to have no intrinsic limit on the load of beam and background radiation they can record per expansion, since there seems to be no "saturation" of the sensitivity even for very large total ionization in the liquid. The usual goal in planning bubble chamber experiments is therefore to accept as intense a beam as possible consistent with reasonably easy scanning. One must always balance the running time for the experiment against the scanning time required to analyze the data. Reduction of the background is necessary to simplify the scanning. An absolute limit on acceptable beam and background appears when events are obscured and made impossible to measure by the presence of too many irrelevant tracks. The short sensitive time of the chambers and the high degree of reproducibility of the sensitive cycle make it possible to choose operating conditions which minimize background by recording only those ionizing events which are simultaneous with the beam. In a typical experiment with the Cosmotron at the Brookhaven National Laboratory, the expansion of the chamber is begun a few milliseconds before the beam is due to arrive. At full sensitivity a counting gate is opened so that a scintillation counter telescope begins counting beam particles. After the desired number of particles has passed through the chamber the lights are flashed with a few microseconds delay for bubble growth. This arrangement avoids taking useless pictures containing too few beam tracks. Since the Cosmotron beam is usually spread out over several milliseconds, pictures taken this way contain tracks of various ages and hence of various bubble sizes. This has the disadvantage that the tracks do not yield measurements of uniformly high accuracy, but the advantage that associated events in each picture can be recognized by their having tracks of the same bubble size. Some experiments may require very high accuracy, while others profit more from this possibility of measurement of time simultaneity of events. One can choose between these alternatives by using a fast or a slow beam ejection scheme.

In planning a bubble chamber experiment one has at his disposal a wide range of operating liquids. For neutron spectroscopy at energies above a few Mev, liquid hydrogen would be ideal because the range and angle of recoil protons can be measured exactly with the assurance that the recoiling particle is really a free proton. For high energy physics hydrogen offers the same advantages of a pure proton target in addition to permitting accurate magnetic momentum measurements to be made. This is a powerful tool for experiments on the production, scattering and absorption of various types of particles. Similarly a deuterium bubble chamber offers a target nucleus of zero isotopic spin, and the opportunity to observe interactions with neutrons.

Helium bubble chambers may offer the possibility of studying light hyperfragments. There are some theoretical arguments that interactions in helium

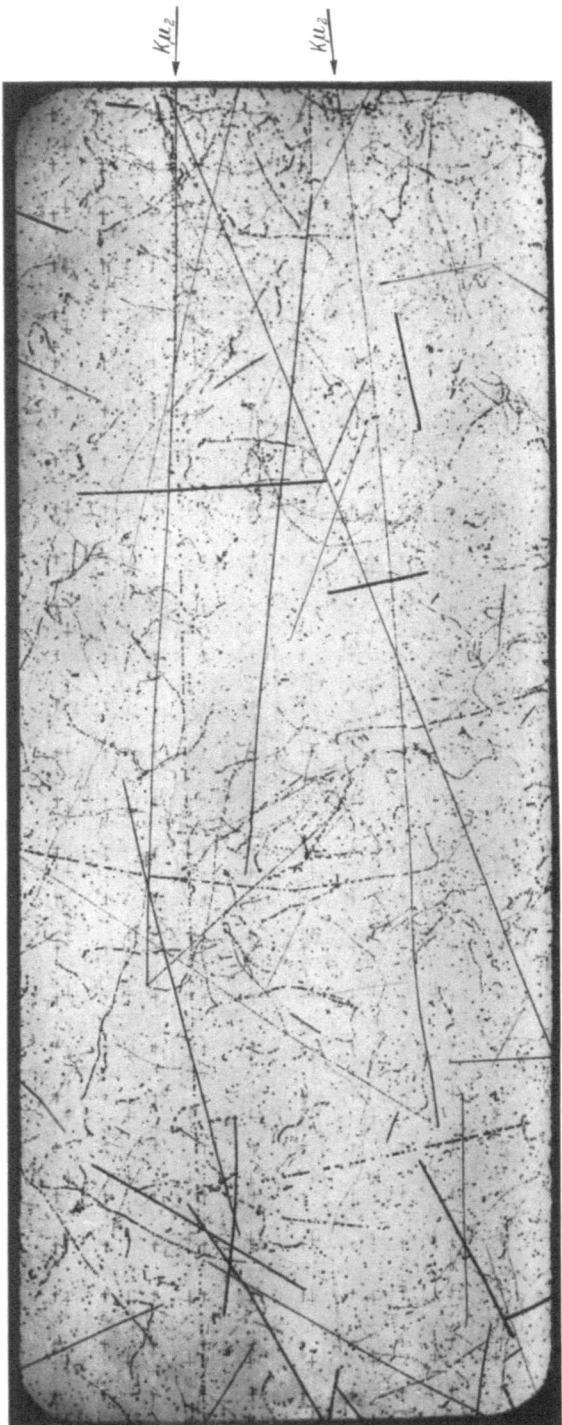

Fig. 5. Two $K\mu_2$ mesons stopping in the University of Michigan 30 cm propane bubble chamber. Coulomb scattering and variation of the bubble density can be compared for the stopping K mesons, the secondary μ mesons and background proton and electron tracks.

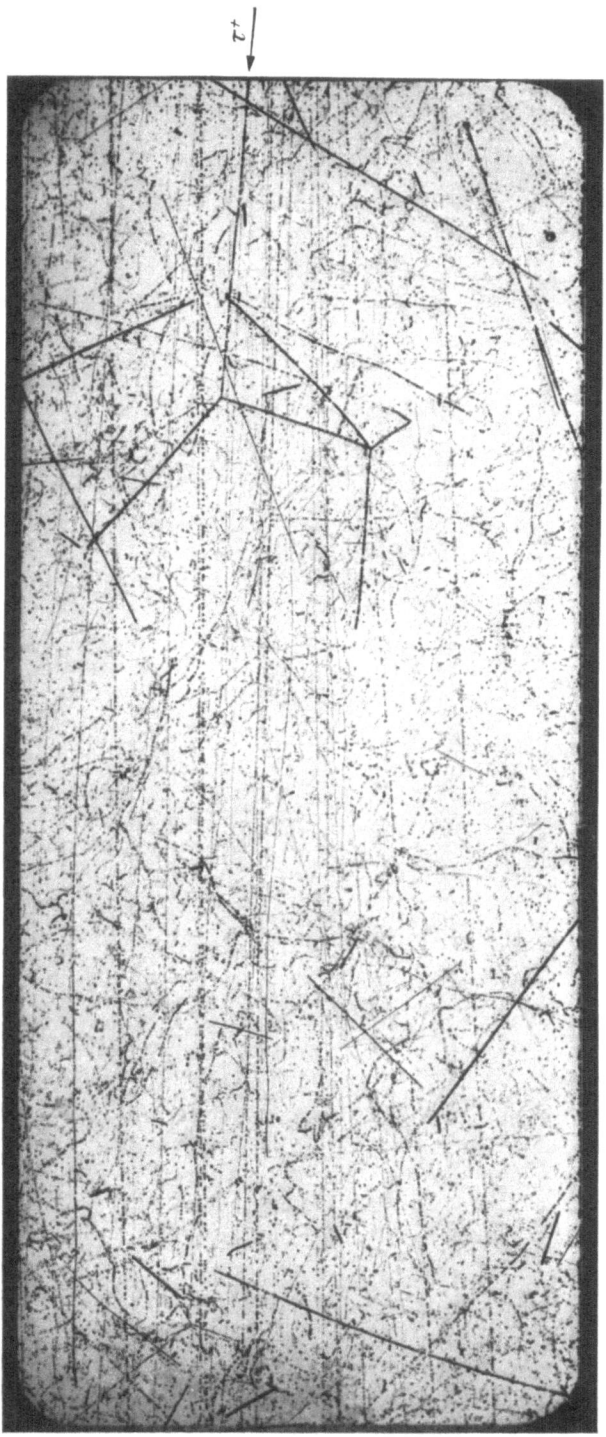

Fig. 6. A τ^+ meson enters the same propane chamber, scatters against a proton whose recoil is seen, and then comes to rest. The two positive π mesons are stopped and easily identified by their characteristic π-μ-e decay. The negative π meson escapes from the chamber.

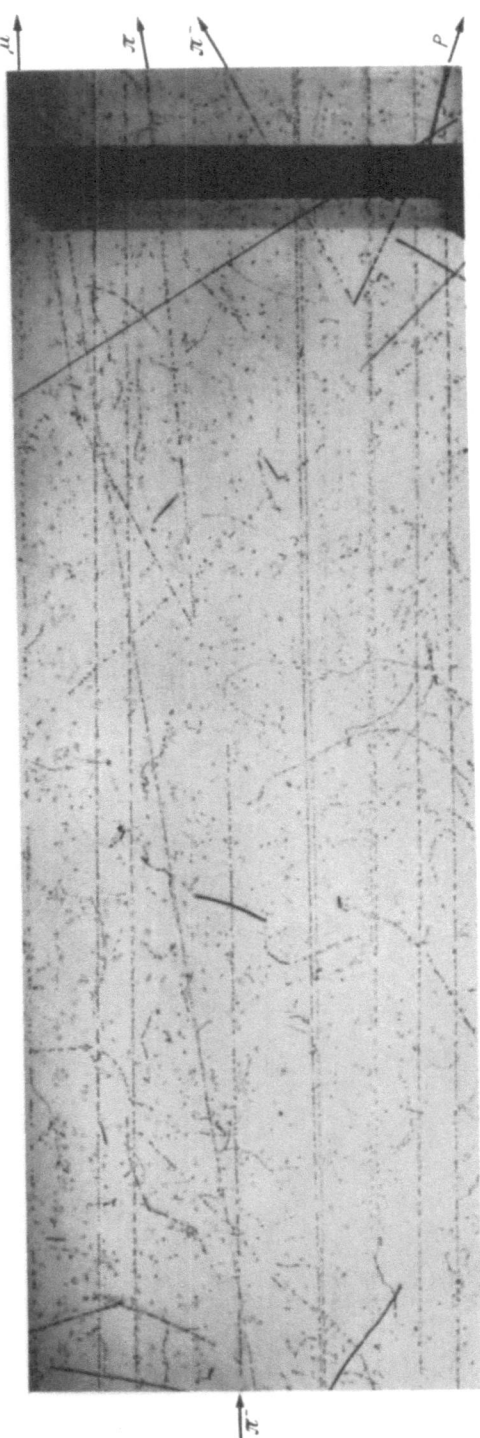

Fig. 7. A 1.1 Bev negative π meson enters the same chamber and strikes a proton producing the reaction.

$$\pi^- + P \rightarrow \Lambda^0 + \theta^0.$$

Followed by $\Lambda^0 \rightarrow P + \pi^-$ and $\theta^0 \rightarrow \pi^+ + \pi^-$. One of the π mesons from the θ^0 decays in flight to produce a μ meson.

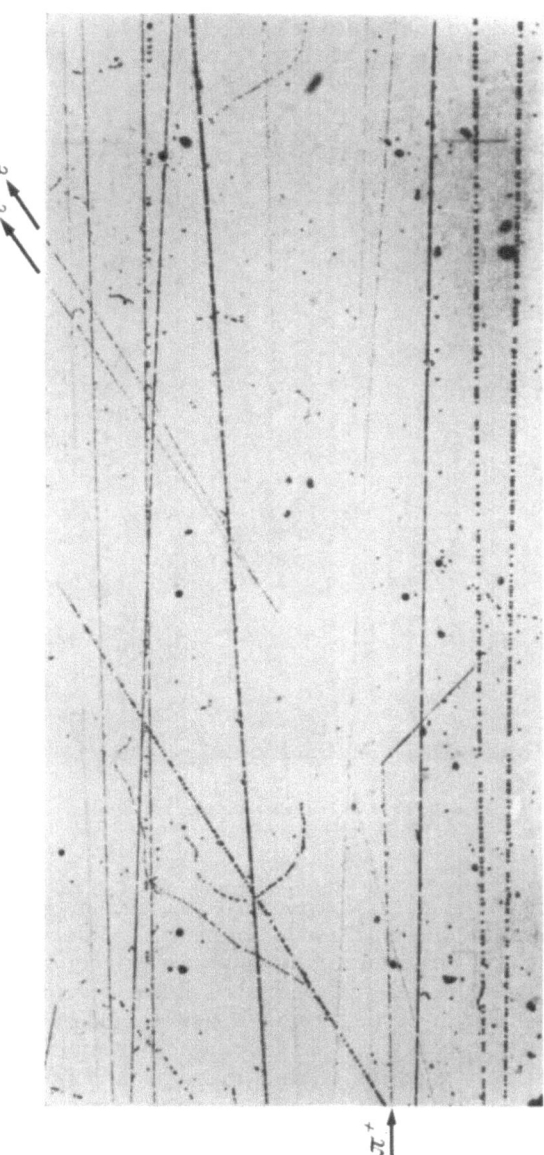

Fig. 8. A photograph of the University of Michigan 15 cm propane bubble chamber selected to illustrate the advantages of a xenon chamber. A positive π meson undergoes a charge exchange scattering against a neutron in a carbon nucleus:

$$\pi^+ + N \rightarrow \pi^0 + P.$$

The recoiling proton is seen and one of the gamma rays from the immediate decay of the neutral π meson materializes to form a high energy pair. In xenon both gamma rays would be seen with a high probability.

will furnish unique information concerning the isotopic spins of the unstable particles.

Organic liquid chambers also present a target rich in protons. In many experiments kinematical considerations are adequate to identify the target nucleus. For these experiments the organic liquid has the advantage of high stopping power which aids in the identification of particles, though momentum measurements are less accurate than in hydrogen. The wide choice of materials for such chambers allows interactions with various light and heavy nuclei to be studied.

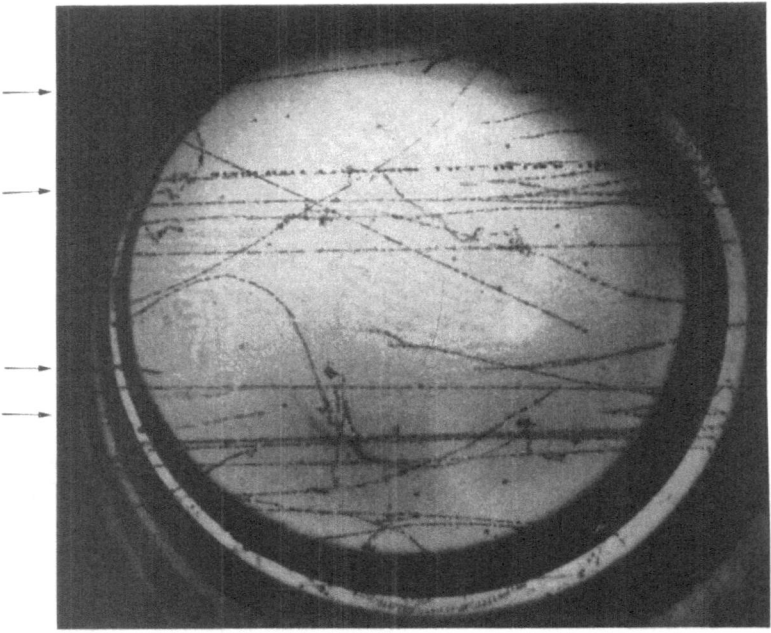

Fig. 9. The University of Michigan 2.5 cm xenon bubble chamber exposed to a high energy π meson beam at the Cosmotron. Several high energy pairs can be seen. Coulomb scattering is large in this chamber.

Finally the heavy liquid chamber, particularly the xenon chamber, offers very high stopping power and efficient detection of neutral π mesons and gamma rays. Magnetic momentum measurements require excessively large fields to achieve precision, but Coulomb scattering will be useful as it is for experiments with nuclear emulsions. The high density of xenon should permit frequent observation of events that are very rare in less dense materials. Since liquid xenon has been found to be a good scintillating material, it may be possible to record both temporal and spatial data about interesting events by monitoring the liquid with a photomultiplier. By dissolving hydrogen-rich compounds in xenon, a supply of free proton targets can be provided if it is desired to observe simple proton interactions.

A number of variations and combinations of simple bubble chambers will probably come into use in the future. There are experiments that could be done by placing metal plates in the liquid and other experiments in which one

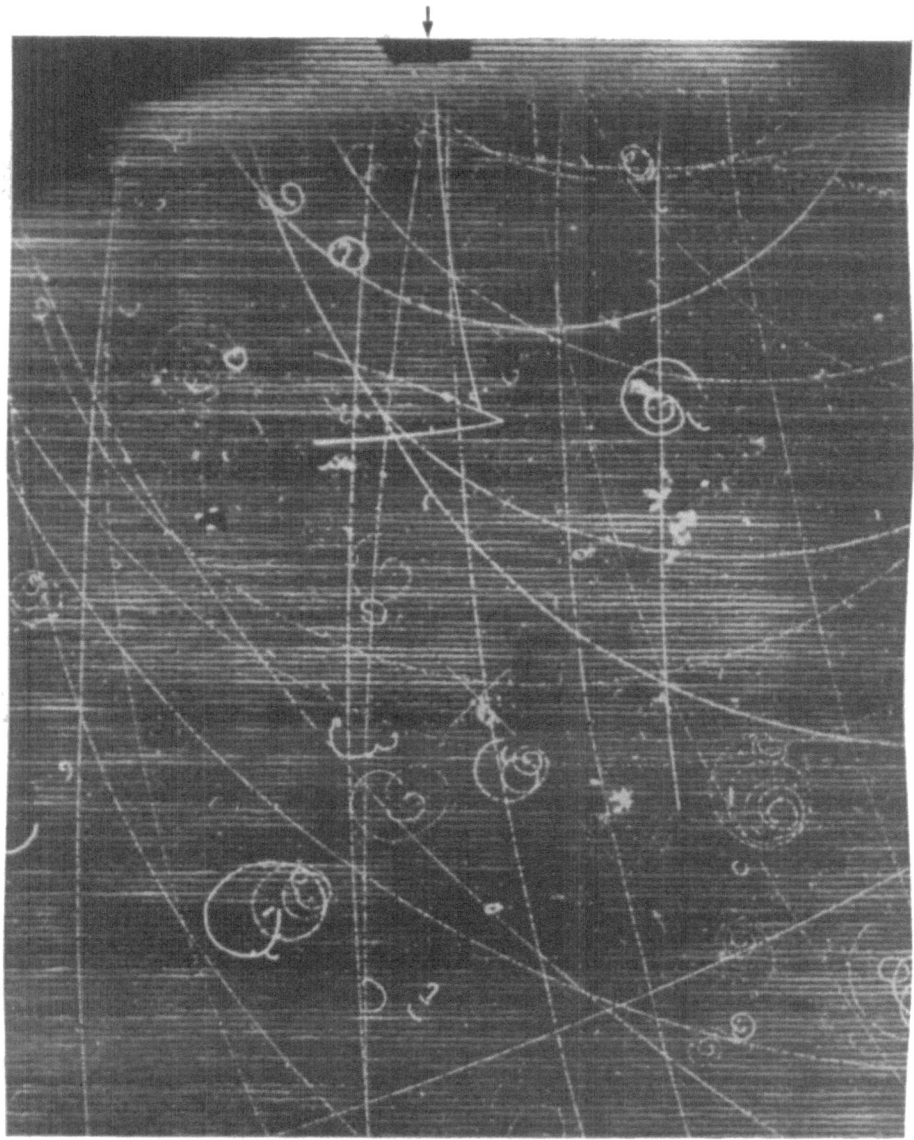

Fig. 10. A beam of negative K mesons enters the 25 cm University of California Radiation Laboratory hydrogen bubble chamber. One K meson produces the interaction:

$$K^- + P \rightarrow \Sigma^+ + \pi^-$$

followed by

$$\Sigma^+ \rightarrow P + \pi^0.$$

Signs of charge and momenta can be observed because of the magnetic field.

type of bubble chamber can be used for recording the production of new particles, and another for observing the characteristics of the particles produced.

Many of the characteristics and possible applications of bubble chambers are illustrated by Figs. 5 through 10.

Nuclear Emulsions.

By

Maurice M. Shapiro.

With 39 Figures.

A. Introduction.

1. Electron-sensitive emulsions. A photographic emulsion which can record the track of any charged particle, regardless of its specific ionization, was achieved by Berriman[1] in 1948. This marked the culmination of decisive advances, made

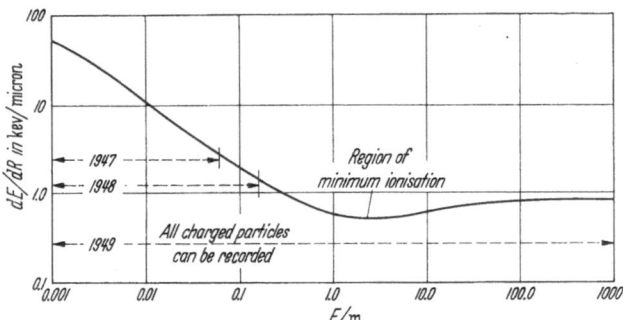

Fig. 1. Rate of energy loss by ionization of a singly charged particle *versus* its kinetic energy in rest-mass units. The dashed lines show the gain in sensitivity of emulsions expressed in terms of the lowest rates of ionization loss which could produce visible tracks. (After Brown *et al.* [*B 16*], and R. H. Herz.)

with emulsions enriched in silver halide, by Demers[2] and by Powell, Occhialini, and others [*P11*][3]. Fig. 1 shows the growth in emulsion sensitivity in the late 1940's, and the corresponding expansion in the spectrum of particle energies which could be investigated with this tool. The photographic technique, nurtured in its infancy by such pioneers as Blau[4], Myssowsky[5], Jdanov[6], and others[7], came of

---

[1] R. W. Berriman: Nature, Lond. **162**, 992 (1948). — The new emulsion, called NT-4, was produced at the Kodak Laboratories, Harrow, England.

[2] P. Demers: Phys. Rev. **70**, 86 (1946). — This work had also been described in reports of the Montreal Laboratory, National Research Council of Canada, P.M. 114-I, July 1945, and P.M. 114-II, September 1945.

[3] The emulsions which these authors investigated and used were produced by C. Waller of Ilford Ltd.

[4] M. Blau: Z. Physik **34**, 285 (1925).

[5] L. Myssowsky and P. Tschishow: Z. Physik **44**, 408 (1927).

[6] A. Jdanov: J. Phys. Radium **6**, 233 (1935).

[7] For a review of early emulsion techniques, their applications, and the literature up to 1940, see reference [*S 5*]. Other, more recent reviews, are cited in the Bibliography at the end of the present article; see, for example [*B 7*], [*G 4*], [*G 5*], [*R 9*], [*T 3*], [*V 7*], [*Y 1*].

age with the dramatic discovery of π-mesons and their decay into muons by LATTES, OCCHIALINI, and POWELL[1]. Today ultrasensitive emulsions provide the nuclear physicist with detectors of proved versatility. In fact, a whole family of emulsions is available for diverse applications to problems in low- as well as high-energy particle physics[2].

This article is concerned with the principles of particle identification, techniques for processing emulsions, and methods of track measurement. We begin, however, with introductory remarks about emulsions, tracks, and some of their uses in studying nuclear phenomena.

2. Distinctive properties of nuclear emulsion. A modern "nuclear" emulsion consists of a high concentration of silver halide crystals embedded in a matrix of gelatine. A charged particle which penetrates this medium activates (i.e., renders developable) many of the AgBr crystals in its path; during subsequent processing, these are converted to Ag grains which form a track. Neutral particles can also be detected, but only by means of their associated charged particles, e.g., through tracks left by their charged secondaries, or through information provided (often from considerations of energy and momentum balance) by trajectories of charged particles participating in the same reaction.

What are some of the special characteristics which make emulsions so useful as detectors? The nuclear emulsion is a dense medium (nearly 4 g/cm^3 under normal conditions), yet, after development, it is transparent. Because of its *high stopping power*, roughly 1800 times that of normal air, energetic particles with ranges of many meters in air can be brought to rest in a few millimeters or centimeters. This characteristic, along with the microscopic width of the tracks, makes it possible to record and observe a great deal of information in a compact volume. The *continuous sensitivity* of the emulsion permits rapid and efficient accumulation of data. In investigations with accelerators, the relatively short duration of exposures conserves the valuable time of nuclear machines. In cosmic ray studies, especially with high altitude balloons, the ability to register a considerable number of relatively rare events in a reasonable time is of great value. Also advantageous in balloon flights is the relatively light weight of the detector. To be sure, some of the larger stacks of emulsion employed lately weigh as much as a counter telescope with its power supply. However, even in high-energy investigations, many problems can still be attacked with relatively light-weight stacks.

The microscopic character of the tracks (for singly-charged particles the track width is less than one micron) permits exceedingly *high angular resolution*. Also, the thinness of an emulsion layer, or even of an emulsion-coated glass plate, makes it possible to employ very refined geometry, e.g., in angular distribution studies. Emulsions owe a good deal of their popularity to their *simplicity* and cheapness. It must be acknowledged, however, that the apparatus required for processing and microscopy, particularly in high-energy experiments, is no longer as simple or as cheap as it formerly was. Moreover, a considerable degree of subtlety now characterizes some of the measurements on tracks and their

[1] C. M. G. LATTES, G. P. S. OCCHIALINI and C. F. POWELL: Nature, Lond. **160**, 453, 486 (1947). — Somewhat earlier, D. H. PERKINS [Nature, Lond. **159**, 126 (1947)] and, independently, POWELL's group, had observed the nuclear capture of negative mesons (not yet distinguished from muons); see Fig. 21.

[2] The literature on emulsion techniques has become so voluminous that only a fraction of it can be cited here. A representative selection is provided in the Bibliography at the end of this article. Extensive lists of papers will be found in some of the references given there. Other references, bearing on specific points, are cited in the footnotes.

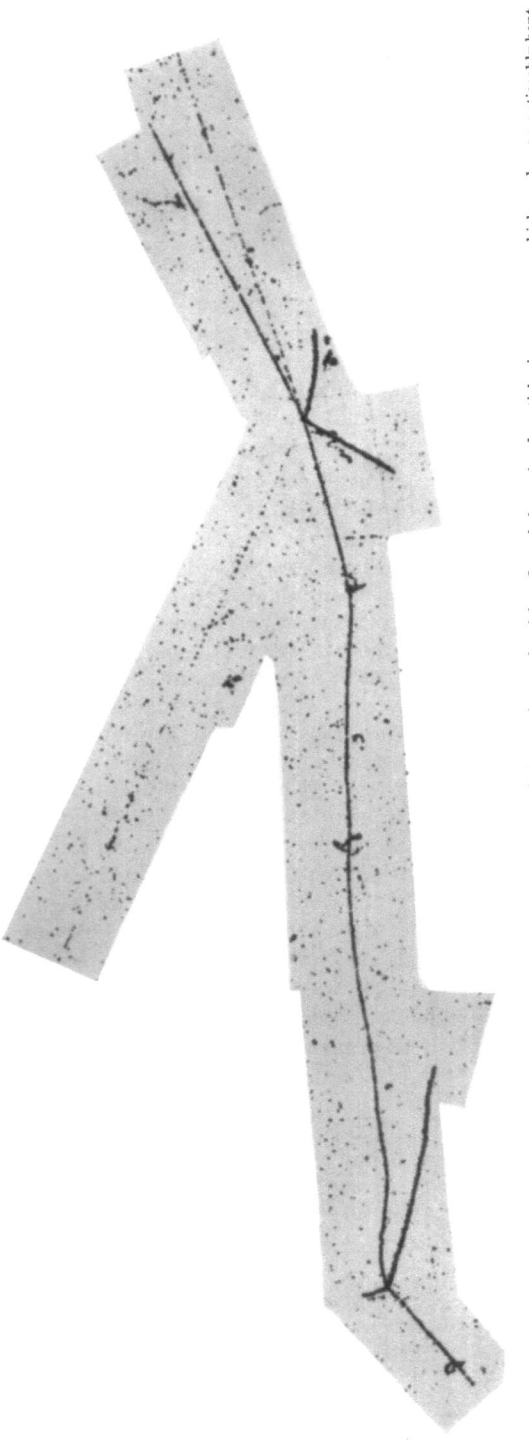

Fig. 2. Nuclear collision of a fast, singly-charged particle (thinnest track) generates the disintegration at the right. One of the emitted particles is a π^- meson, which produces a noticeably bent track, and ends in nuclear capture and a secondary star. (Courtesy Prof. C. F. POWELL.)

interpretation. Nevertheless, the emulsion remains a relatively uncomplicated instrument. Freedom from electronics or "plumbing" in the detector itself, saves time which must otherwise be spent in constructing more elaborate apparatus. In low-energy research, the emulsion is still a *relatively low-cost* detector. In high energy experiments, a single stack may cost hundreds or even thousands of dollars. Yet, in terms of the information that can be concentrated within such a stack, it may still be regarded as inexpensive, as it can provide sufficient data for many man-years of scientific work.

The time required to arrest a particle by ionization in a dense solid, such as emulsion, is short (e.g., $\sim 10^{-10}$ sec for a 40 Mev proton), permitting the observation of very *short-lived phenomena*. The *direct visibility* of particle trajectories is an attractive feature which the nuclear emulsion shares with cloud chambers and bubble chambers. It is illuminating as well as satisfying to "see what happened" in some detail through the configuration of tracks produced by some phenomenon such as collision (Fig. 2) or spontaneous decay (Fig. 3). An idea of the beauty of the method is conveyed by some of the photomicrographs reproduced in this article.

Owing to many of the foregoing properties, nuclear emulsions are widely used for exploratory purposes, e.g. with accelerators, at all energies. They are often exploited in this way in experiments

which are based primarily upon
other types of detectors. Emulsions,
like all detectors, can also be used
in association with other techniques, such as magnetic fields, absorbers, etc.

An inescapable feature of the
nuclear emulsion is that it is a
complex medium, containing a variety of light and heavy elements.
This is a disadvantage in some
types of experiments. The emulsion
has, of course, other limitations
as a detector; these will be evident
from the discussions which follow
(see especially the sections on processing and microscopy).

3. Characteristics of tracks. For
the reader unfamiliar with the
photographic method of detection,
a brief qualitative description of
the varieties of tracks produced by
nuclear particles in emulsion may
be helpful (see Fig. 4). The tracks
are microscopic; only under excep-
tional circumstances are they visible
to the naked eye. They usually
consist of rows of developed silver
grains approximately 0.5 to 0.8 μ
in diameter. Depending upon the
rate of ionization loss of the charged
particle, the sensitivity and other
properties of the emulsion, and the
intensity of development, a track
may be thin, or black, or "gray",
(intermediate in density). A track
may be very straight, indicating
high momentum, or crooked, if it
was produced by a relatively slow
particle[1]. There may be large deflections due to nuclear encounters.
Small-scale deflections are, in general, due to multiple Coulomb scattering. In the electron-sensitive emulsions, there may be few or many
secondary electron tracks, or delta
rays, along the path of the particle.

In length, tracks range all the
way from a short blob of 1 or

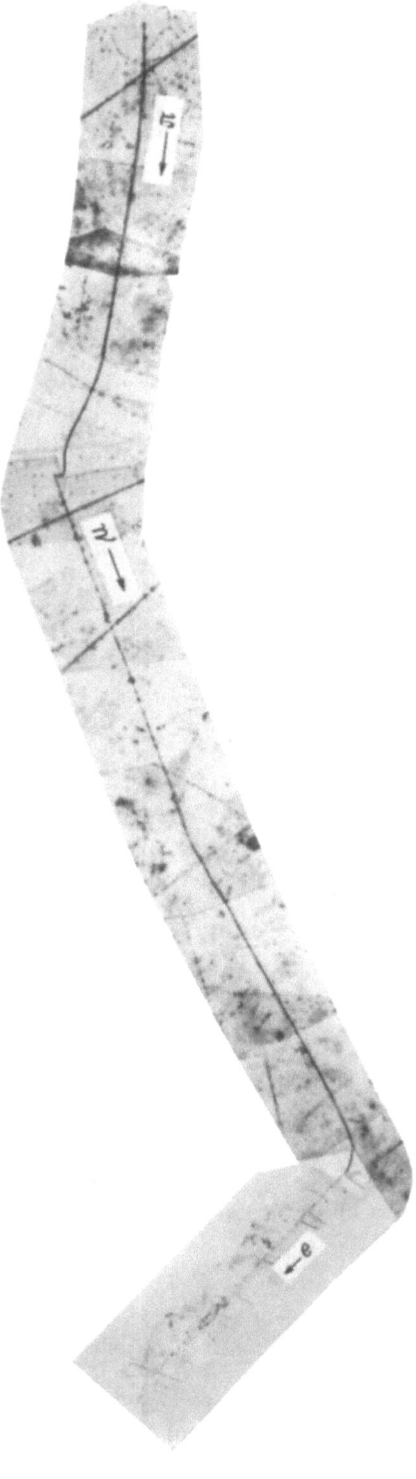

[1] The "crookedness" may also be
the result of spurious distortion effects
in the emulsion.

Fig. 3. Spontaneous decay of a π meson into a muon (and neutrino), followed by decay of the μ meson into an electron (and 2 neutrinos). (U. S. Naval Research Laboratory.)

2 microns, through paths of tens and hundreds of microns, up to ranges of millimeters and centimeters. A trajectory may occur singly, or it may be part of a configuration of tracks. In the latter case, the associated phenomenology may give evidence of the particle's origin, its interaction, its decay in flight, or its decay after coming to rest. Alternatively, the particle may have traversed long distances in the sensitive medium without doing any of the foregoing things, finally passing out

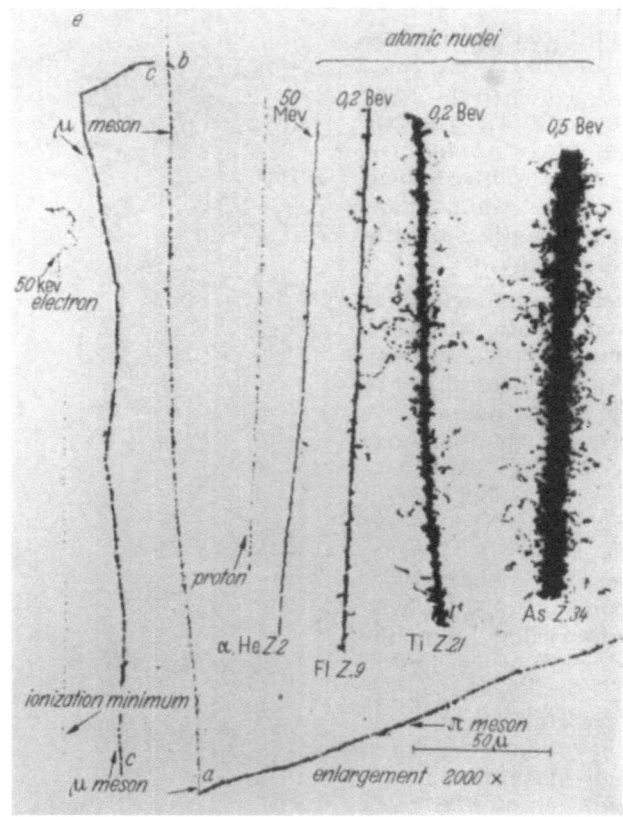

Fig. 4. Tracks of various particles in an electron-sensitive emulsion, showing some of the characteristics described in Sect. 3. The thin track at the lower left of the figure is due to a fast particle with ionization near the minimum. The highly bent track above it is due to a slow electron. Evidence of multiple scattering is visible in all of the tracks shown. The enormous range of ionizations illustrates the great versatility of the electron-sensitive emulsions. For multiply charged particles, the energies cited in the figure are energies per nucleon. Great numbers of delta rays are visible along the tracks at the right, due to highly charged nuclei. (Courtesy Prof. L. LEPRINCE-RINGUET.)

of the emulsion. Tracks may be steep or flat: however thin it may be, the emulsion is a three-dimensional medium, and frequently one has to cope with rather steep trajectories.

4. Some applications to problems in low-energy physics. In "classical" or low-energy nuclear physics, emulsions have been extensively employed for the direct detection of heavy charged particles and electrons, and also of neutrons and gamma rays through their secondary interactions. The angular distribution of charged particles scattered or emitted in nuclear reactions can be determined with high resolution, and their ranges can be measured. Special "scattering cameras" have been devised for use in such experiments[1]. The energy spectra of particles emitted in nuclear disintegrations can be conveniently studied [L2]. In investigations which require accelerators, it is generally possible to accumulate a great deal of information in the emulsion in a relatively short time. This enhances the efficiency of expensive machines by releasing time for additional experiments. Because of their high sensitivity, emulsions have been used in radiochemical studies, (e.g. [P5]), especially for weak activities. Other applications include the determination of half-lives, and of nuclear cross sections.

Slow as well as fast neutrons are detectable by means of their nuclear interactions in emulsion. For slow neutrons, the high cross-section Li^6 (n, α) or B^{10} (n, α) reactions can be utilized by impregnating the emulsion with salts of these elements. The alpha-particle and triton tracks from these reactions are clearly visible, and this makes the method suitable for health monitoring[2]. Fast neutrons are detected either by recoil protons generated in the hydrogen-rich emulsion, or by means of nuclear disintegrations[3]. Because the proton track can be measured from its point of origin, this method of detecting fast neutrons yields better angular resolution than those in which proton radiators of appreciable dimensions are employed. KEEPIN and ROBERTS have studied fast neutrons by loading emulsions with separated Li^6 and measuring the angle between the triton and alpha particle tracks, as well as the ranges[4]. ROSEN [R5] and BARSCHALL et al.[5] have reviewed the measurements of neutron fluxes and neutron energy spectra.

The energies of gamma rays have been measured by loading plates with deuterium, and observing the proton track which results from photodisintegration of the deuteron. This method is useful for photon energies exceeding about 3 Mev, since ~ 2.2 Mev is required to overcome the binding energy of the deuteron, and only half of the remaining energy is available for the proton. Higher energy photons, e.g., X-rays from betatrons, have been studied by means of the electron pairs which they generate in emulsion.

Fission phenomena, such as neutron spectra from fission[6] and the emission of alpha particles[7] have been investigated with emulsions. In some experiments, the emulsion is impregnated with uranium or some other fissionable material. Studies of radioactivity in rocks[8] and in biophysical experiments [Y1], [R9] have been carried out with emulsion techniques[9].

[1] See, for example, J. CHADWICK, A. N. MAY, T. G. PICKAVANCE and C. F. POWELL: Proc. Roy. Soc. Lond., Ser. A 183, 1 (1944). — S. RUBIN: Phys. Rev. 72, 1176 (1947). — L. H. MARTIN, J. C. BOWER, D. N. F. DUNBAR and F. HIRST: Nature, Lond. 164, 310 (1949).

[2] J. ROTBLAT: Nature, Lond. 160, 493 (1947). — M. M. SHAPIRO and J. R. BARNES: Phys. Rev. 73, 1243 (1948). — E. W. TITTERTON: Nature, Lond. 163, 990 (1949).

[3] Health monitoring for fast neutrons utilizes this method. See, e.g. J. S. CHEKA: Nucleonics 12, 40 (June 1954).

[4] G. R. KEEPIN and J. H. ROBERTS: Rev. Sci. Instrum. 21, 163 (1950).

[5] H. H. BARSCHALL, L. ROSEN, R. F. TASCHEK and J. H. WILLIAMS: Rev. Mod. Phys. 24, 1 (1952).

[6] J. E. EVANS: Los Alamos Scientific Laboratory Report LA-1395, July, 1952.

[7] See, e.g., P. DEMERS: Phys. Rev. 70, 974 (1946). — E. O. WOLLAN, C. B. MOAK and R. B. SAWYER: Phys. Rev. 72, 447 (1947).

[8] E.g., by E. E. PICCIOTTO: Bull. Cent. Phys. Nucleaire Bruxelles 1948, Nr. 1 (see also [P6]). — I. H. FORD and E. E. PICCIOTTO: Nuovo Cim. 9, 141 (1952). — F. BEGEMANN, H. v. BUTTLAR, F. G. HOUTERMANS, N. ISAAC and E. PICCIOTTO: Bull. Cent. Phys. Nucleaire Bruxelles, 1952, Nr. 37. — A review of geological and mineralogical applications has been given by YAGODA [Y1].

[9] Further examples of emulsion applications, with references to the literature, will be found in the very useful publications of ROTBLAT [R9], YAGODA [Y1], BEISER [B7], POWELL and OCCHIALINI [P10], HEITLER and KING [H3], MEYER [M8], BLAU [B11], SCHOPPER [S3], and VIGNERON [V3]. A helpful laboratory handbook of nuclear spectroscopy with emulsions has been prepared by ALLRED and ARMSTRONG [A2].

5. Typical applications to high-energy physics; stripped emulsions. Although nuclear emulsions have proved a useful tool in the low-energy nuclear domain, their most distinctive contributions, particularly since the advent of electron-sensitive emulsions, have been to cosmic rays and to high-energy physics in general. Indeed, many of the new unstable particles owe their discovery to the skillful exploitation of these detectors. Among the phenomena which have been investigated in the cosmic radiation are the high-energy disintegrations[1] leading to the emission of nuclear fragments, or the production of mesons, or both [C1], [L4]. The whole field of meson physics has been enriched by the results of many experiments performed with emulsions[2]. The various properties of these particles— their masses, modes of decay, disintegration energies—all these have been explored in cosmic ray studies, and subsequently with the large accelerators[3].

The characteristics of hyperons, neutral as well as charged[4], and the dicovery [D3a] and subsequent investigations[5] of hyperfragments—these are additional examples of phenomena in which emulsions have made important contributions. Recently, with the artificial production of K-mesons and hyperons by the Cosmotron at Brookhaven and the Bevatron at Berkeley, a great deal of effort has gone into studying the interactions and modes of decay of these particles and their secondaries [H13], [B8a], [C6a], [S2].

The heavy primary nuclei of the cosmic radiation were discovered with the help of emulsion [F4], and have been further studied mainly with this tool [H8], [D1], [P4], [W1], [Y2]. Electromagnetic cascades[6] and especially the anomalous photon-induced high-energy showers[7] are striking examples of phenomena in which all the visible tracks are at or very near the minimum of ionization. With the discovery of the antiproton, the evidence for its annihilation (and therefore its antiparticle character) was first provided by studies of the stars produced by antiprotons arrested in emulsion[8].

In the last few years, investigations in high-energy physics have been greatly assisted by the development of the method of stripped emulsion stacks[9]. In studies of heavy mesons and hyperons, secondary effects are often important in

---

[1] M. BLAU and H. WAMBACHER: Sitzgsber. Akad. Wiss. Wien **146**, 623 (1937). — The nuclear interactions of cosmic rays have been reviewed by G. D. ROCHESTER and W. G. V. ROSSER: Rep. Progr. Phys. **14**, 227 (1951).

[2] POWELL [P9] has written a lucid summary of the earlier investigation on π and μ mesons. R. MARSHAK, in his book, Meson Physics (New York: McGraw-Hill Book Co. 1952) provides an admirable account of theory and experiment up to 1952.

[3] LEPRINCE-RINGUET [L4] and DILWORTH, OCCHIALINI and SCARSI [D11] have reviewed the pioneering work on the new unstable particles. A table of L mesons, K mesons, and hyperons with bibliography up to 1955 will be found in [S6]. For proceedings of recent conferences devoted largely to the "elementary particles", see Ref. [B1], [P1], [P8a], [V2], [V2a]. See also Proceedings of the Annual Rochester Conferences on High Energy Physics, distributed by Interscience Publishers, Inc., New York.

[4] An interesting example is the precise determination of the mass and decay energy of the Λ° hyperon [F4a]. See also Footnote 3.

[5] E.g., W. F. FRY, J. SCHNEPS and M. S. SWAMI: Phys. Rev. **101**, 1526 (1956).

[6] A. G. CARLSON, J. E. HOOPER and D. T. KING: Phil. Mag. **41**, 701 (1950). — M. KOSHIBA and M. F. KAPLON: Phys. Rev. **97**, 193 (1955).

[7] M. SCHEIN, D. HASKIN and R. G. GLASSER: Phys. Rev. **95**, 855 (1954).

[8] O. CHAMBERLAIN W. W. CHUPP, G. GOLDHABER, E. SEGRÈ, C. WIEGAND, E. AMALDI, G. BARONI, C. CASTAGNOLI, C. FRANZINETTI and A. MANFREDINI: Nuovo Cim. **3**, 447 (1956). — Phys. Rev. **101**, 909 (1956).

[9] P. DEMERS [D6]. — B. STILLER, M. M. SHAPIRO and F. W. O'DELL: Bull. Amer. Phys. Soc. **1951** [Phys. Rev. A **85**, 712 (1952)] and Ref. [S11]. — C. F. POWELL: Phil. Mag. **44**, 219 (1953).

identifying the parent particle. The incentive for applying thick emulsions in high energy experiments is, of course, to increase the observable path length and thereby to enhance the probability of observing related events, e.g., the production as well as the nuclear capture or spontaneous decay of a meson. By employing *stacks* of "pellicles", (i.e., stripped emulsions) the thickness can be increased at will simply by adding layers. Difficulties in processing and microscopy are not excessive if the individual layers are of moderate thickness. Accordingly, the technique of "emulsion blocks" has played a decisive part in determining the masses, and elucidating the modes of decay of the new mesons and hyperons [*P11a*]. The value of large emulsion stacks is well exemplified by the "G-Stack Collaboration Experiment", which contributed heavily to our knowledge of the *K* mesons [1].

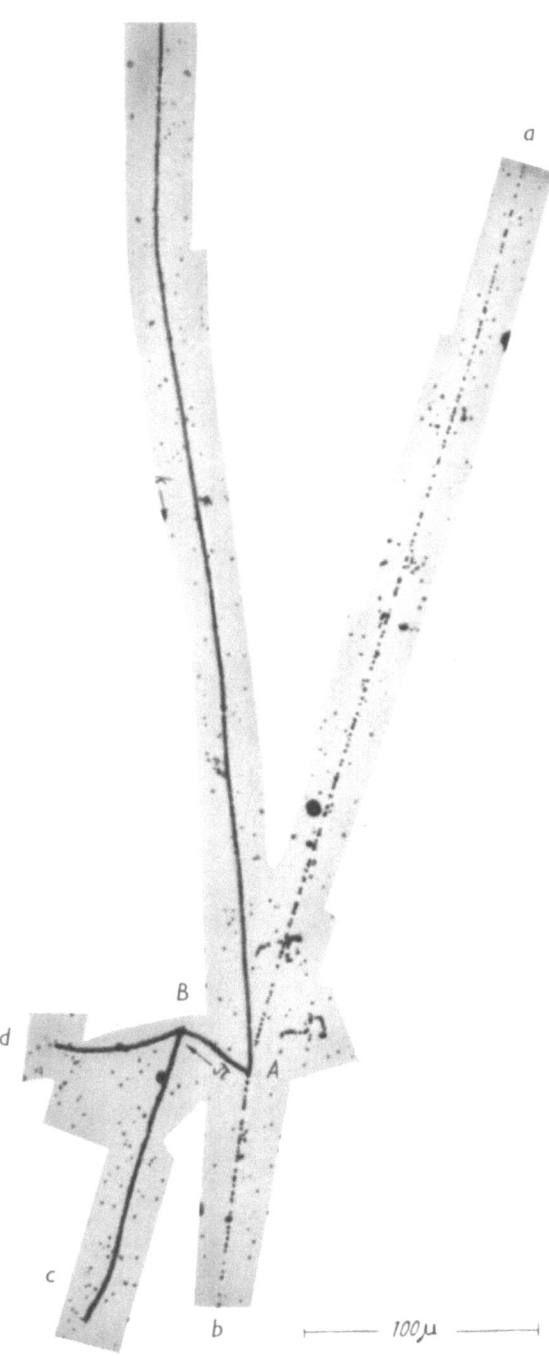

Fig. 5. Decay of a τ meson (marked *k*) of mass ∼966 m_e, into three π mesons (*a*, *b* and track *A B*). One of the pions is captured by a nucleus at *B*, producing a disintegration with the emission of slow particles *c* and *d* [*B16*]. (Courtesy Prof. C. F. POWELL.)

---

[1] This experiment is also a classic example of international collaboration in science: eight laboratories in five countries participated. J. H. DAVIES, D. EVANS, P. E. FRANÇOIS, M. W. FRIEDLANDER, R. HILLIER, P. IREDALE, D. KEEFE, M. G. K. MENON, D. H. PERKINS and C. F. POWELL; J. BØGGILD, N. BRENE, P. H. FOWLER, J. HOOPER, W. C. G. ORTEL and M. SCHARFF; L. CRANE, R. H. W. JOHNSTON, and C. O'CEALLAIGH; F. ANDERSON, G. LAWLOR, and T. E. NEVIN; G. ALVIAL, A. BONETTI, M. DI CORATO, C. DILWORTH, R. LEVI SETTI, A. MILONE, G. OCCHIALINI, L. SCARSI, and G. TOMASINI; M. CECCARELLI, M. GRILLI, M. MERLIN, G. SALANDIN and B. SECHI: Nuovo Cim. 2, 1063 (1955).

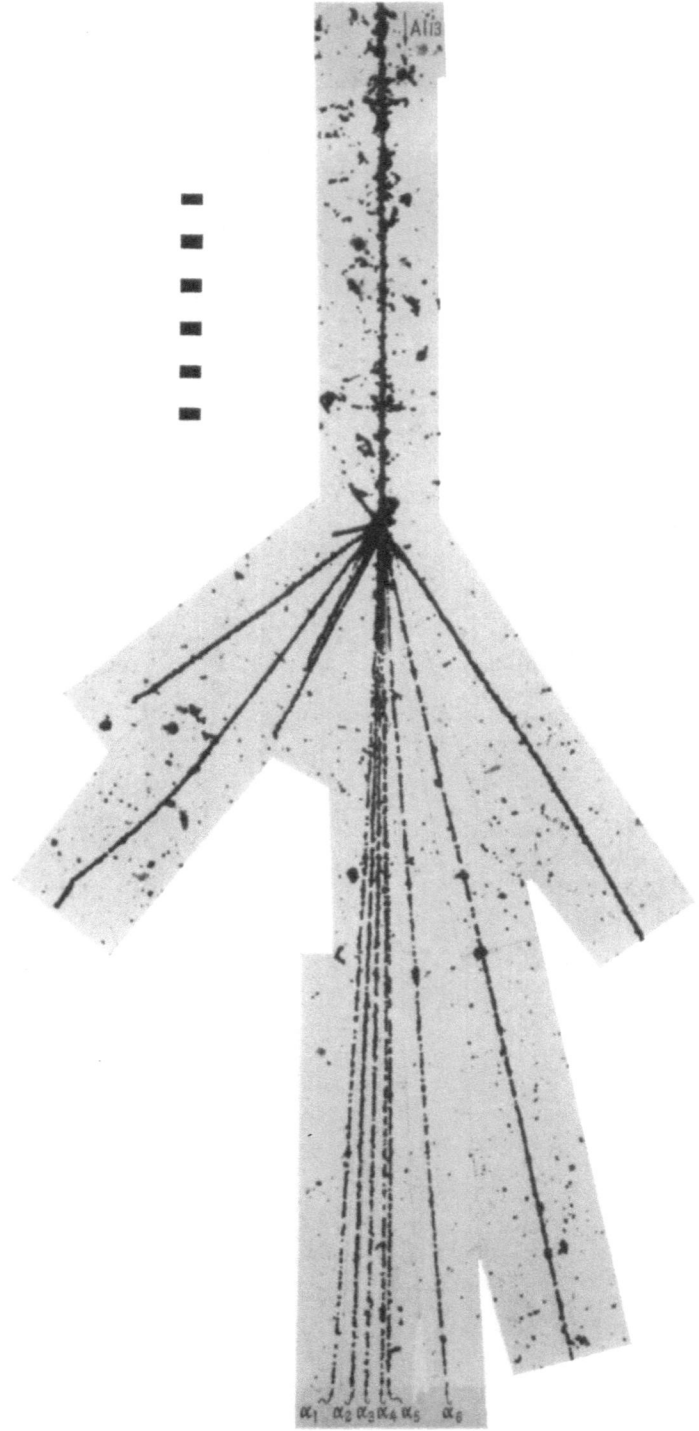

Fig. 6. A fast cosmic-ray Al nucleus is fragmented in collision with an emulsion nucleus. Six highly energetic α particles ($\alpha_1 - \alpha_6$) are emitted, as well as a number of low-energy disintegration products. Scale: 1 div = 10 μ. (Courtesy Prof. C. F. POWELL.)

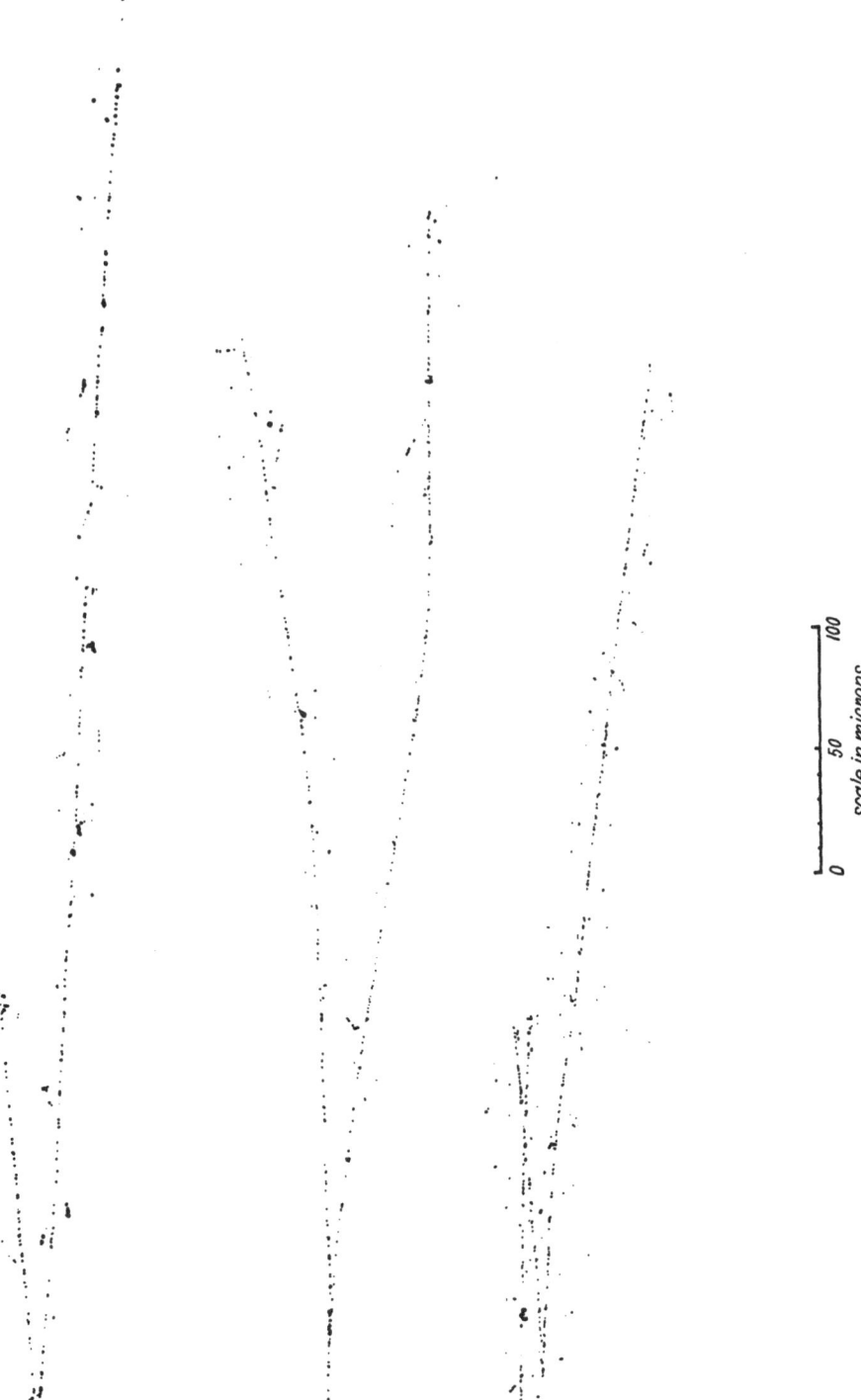

scale in microns

Fig. 7. Electron pairs generated in nuclear emulsion by X-rays. (Courtesy Dr. R. H. HERZ.)

B. Principles of particle identification and velocity measurement.

6. Introduction. The identification of a particle responsible for a given track is often accomplished by inspecting and measuring a configuration of associated tracks, e.g., a decay process (Fig. 5), a nuclear interaction (Fig. 6), or the production of an electron pair (Fig. 7). In other instances it is necessary or desirable to deduce the mass and charge of a particle from measurements on its *own* track. These properties, together with the velocity (hence also momentum and energy), are ordinarily determined by exploiting two basic phenomena—*ionization* and multiple Coulomb *scattering*. In addition to these inevitable effects of the particle's passage through matter, one may also choose to bend its path *magnetically*, in order to ascertain the sign of its charge and—with very strong fields—to measure its momentum. We shall discuss in turn the manifestations of these several processes, including the residual *range* of particles brought to rest by ionization. In this chapter we shall dwell on principles of particle identification, postponing to a later section the detailed description of techniques of measurement.

I. Ionization and range.

7. Types of ionization effects in emulsion. Charged particles produce developable tracks in traversing a photographic emulsion by the same basic process— ionization—which signals their passage through most other nuclear detectors. Historically the earliest index of ionization in emulsions—and still one of the most widely used—is the grain (or blob) density. At low ionizations, up to twice minimum, when only a fraction of the halide crystals is activated, most of the silver grains in the developed track are resolvable, and the number of grains per unit path can be counted fast. To be sure, even in very thin tracks some grains merge into clumps, so that two can be mistaken for one. Therefore, "blob" counting, in which no effort is made to resolve grains which have coalesced, gives greater reproducibility. At ionizations exceeding $2I_{min}$, and up to 5 or $6I_{min}$, measurement of blob length coupled with the convention that this length is proportional to the number of grains, gives a rather objective result. It is, however, less efficient than other methods, originally devised to cope with still higher ionizations, when tracks become ever blacker, with fewer and smaller gaps. These gaps then provide a measure of the ionization loss based either upon their mean size or, on what may be called the "linear opacity", i.e. the fraction of a track segment which is black. The latter differs from areal opacity, usually measured photoelectrically, which depends on the track's projected width, or profile, as well. Finally, the more energetic electrons, or delta-rays, which leave secondary tracks of their own along the path of the particle, supply yet another measure of the rate of ionization loss—one which is especially useful for multiply charged particles.

8. Some results of ionization theory. Although the theory of energy loss by ionization[1] is treated elsewhere in this Encyclopedia, it will be useful to summarize some important results of this theory. The average energy loss I_a per unit path by ionization may be written[2]

$$-\frac{dE}{dx} \equiv I_a = \frac{2\pi n_e e^4 z^2}{m c^2 \beta^2}\left[\ln\frac{(2 m c^2 \beta^2 \gamma^2)\,T}{w^2} - 2\beta^2\right] \tag{8.1}$$

---

[1] For references to the theory of ionization loss up to 1948, see N. Bohr: Kgl. danske Vid. Selsk., mat.-fys. Medd. **18**, No. 8 (1948), and [*B 14a*]. Among later reviews, see, e.g. [*P 12*], [*U 1*].

[2] We employ the expression given by B. Rossi: High Energy Particles, p. 24. New York: Prentice-Hall, Inc. 1952. References to the basic works by H. A. Bethe, C. Møller, F. Bloch and others will be found in Rossi. See also J. Ashkin and H. A. Bethe: Passage of Radiations through Matter. In: Experimental Nuclear Physics, Vol. I, Part II (edit. by E. Segrè). New York: John Wiley & Sons. 1953. The "density effect", not included in Eq. (8.1), is treated below.

where n_e = number of electrons per unit volume,

β = velocity of particle in units of the velocity c of light,

$\gamma = (1 - \beta^2)^{-\frac{1}{2}}$,

ze = particle's charge; m = rest mass of the electron,

w = average ionization potential of atoms in the medium,

T = maximum energy transfer to an electron; for a heavy particle of rest mass M, it is given by[1]

$$T = \frac{(\gamma^2 - 1)\, M\, c^2}{\dfrac{M}{2m} + \dfrac{m}{2M} + \gamma}.$$ (8.2)

When $M \gg m$, and $\beta\gamma \ll M/m$, the maximum energy transfer may be approximated by

$$T \approx 2\, m c^2\, \beta^2\, \gamma^2.$$ (8.2a)

With this approximation, Eq. (8.1) becomes[2]

$$I_a = \frac{4\pi\, n_e\, e^4\, z^2}{m\, c^2\, \beta^2} \left[\ln \frac{2\, m\, c^2\, \beta^2\, \gamma^2}{w} - \beta^2\right].$$ (8.3)

Eq. (8.1) applies to heavy particles. For electrons, the corresponding equation is[3]

$$\left.\begin{array}{l} \underset{\text{electrons}}{I_a} = \dfrac{2\pi\, n_e\, e^4\, z^2}{m\, c^2\, \beta^2} \times \\[2mm] \times \left[\ln \dfrac{(m\, c^2\, \beta^2\, \gamma^2)\, E}{2\, w^2} + \dfrac{9}{8} - \beta^2\right]. \end{array}\right\}$$ (8.4)

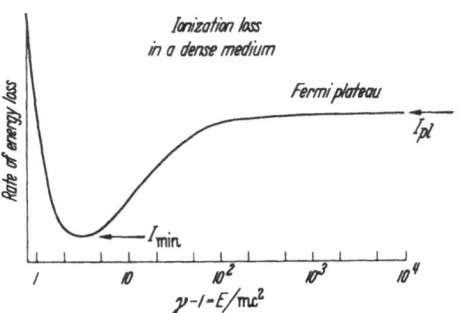

Fig. 8. Qualitative illustration of the relativistic rise in the rate of energy loss by ionization, showing the effect of polarization in a condensed substance.

The maximum energy transfer to an electron is E, the kinetic energy of the incident electron; however, the incident and struck particles are indistinguishable, so T must be replaced by $E/2$.

Fig. 8 shows qualitatively, for heavy particles, the dependence of I_a on $\gamma - 1 = E/M c^2$, the kinetic energy measured in rest-mass units. The latter is, of course, a function of the velocity alone. The minimum at $E/M c^2 \approx 3$ is followed by a nearly logarithmic rise at higher relativistic velocities. This rise is due to two effects: an augmentation in the transverse radius of action of the particle due to the Lorentz transformation of its Coulomb field, and an increase in the maximum energy which can be transferred to electrons in the medium.

Actually, this increase in I_a is limited, in condensed substances, by the polarization of the medium, as predicted by SWANN and shown quantitatively by FERMI[4]. The displacement of nearby electrons by the passing particle shields the remote electrons from its Coulomb field, so that a diminution in the logarithmic rise in ionization loss results. In fact, at sufficiently high velocities (corresponding to $\gamma > 100$) the ionization loss approaches a plateau value. The polarization effect depends primarily on the electron density in a medium. It has, however, come to be known simply as the "density effect", because it is important

---

[1] H. J. BHABHA: Proc. Roy. Soc. Lond. **164**, 257 (1937).

[2] The statement is frequently made that, for a given medium, I_a depends only on β and z, and is independent of the particle mass M. This is true in approximation (8.2a) which is adequate over a very wide range of energies. It is sometimes overlooked, however, that at sufficiently high energies I_a does depend on the mass because it depends on T [cf. Eqs. (8.1) and (8.2)].

[3] Cf. Eq. (42) in Ref. [S8].

[4] E. FERMI: Phys. Rev. **56**, 1242 (1939); **57**, 485 (1940).

mainly in condensed materials. FERMI treated the electrons in the medium as classical oscillators set in motion by the moving particle, and he assigned a single frequency to all the electrons. His theory was extended by WICK[1], HALPERN and HALL [H 1a], BOHR [B 14a], SCHÖNBERG[2], and STERNHEIMER [S 7a], who showed that it is necessary to construct a multifrequency theory.

To take account of the density effect, it is necessary to subtract a correction, Δ, from the brackets of Eqs. (8.1), (8.3), and (8.4). Values of Δ for emulsion have been computed by STERNHEIMER[3] as a function of the ratio $\beta\gamma$ of momentum to mass. For convenience in computation, we also employ the following notation: $a = \frac{2\pi e^4}{m c^2}$, and $B = \ln \frac{2 m c^2}{w^2}$. Then the density-corrected energy-loss I_d per unit path can be written[4]

$$I_d = \frac{a\, n_e\, z^2}{\beta^2} \left[B + 2\ln \beta\gamma + \ln T - 2\beta^2 - \Delta \right] \tag{8.1a}$$

and, with approximation (8.2a),

$$I_d = \frac{2a\, n_e\, z^2}{\beta^2} \left[B + 2\ln \beta\gamma + \ln w - \beta^2 - \frac{\Delta}{2} \right]. \tag{8.3a}$$

9. **Restricted rate of ionization loss.** In comparing ionization theory with experiment, it is often necessary to distinguish between the *average* space rate of ionization loss and the *most probable* rate. Owing to the statistical nature of the collision process, fluctuations in the rate of energy loss give rise to a skewed distribution of ionization losses around the most probable one: there is a high-energy tail, due to occassionally large transfers of energy to electrons[5,6]. In emulsion, yet another kind of ionization loss should be calculated for comparison with grain densities—the "*restricted* rate" of loss, I_r, arising only from energy transfers to electrons up to a few kev which are absorbed in AgBr crystals along the track. In other words, I_r is the loss due only to distant collisions, and it is given by[7]

$$I_r = \frac{2\pi\, n_e\, e^4\, z^2}{m c^2 \beta^2} \left[\ln \frac{(2 m c^2 \beta^2 \gamma^2)\, T_0}{w^2} - \beta^2 - \Delta \right]. \tag{9.1}$$

T_0 is the upper limit of delta-ray energy corresponding to the maximum energy deposited in a single grain.

The difference between the restricted rate of loss and the average (total) rate is shown in Fig. 9. The solid curves show the theoretical *average* rate of ionization loss I_d for protons (curve I) and electrons (curve II) in nuclear emulsion as a function of the kinetic energy in rest-mass units. The calculation of curve III, the restricted rate of energy loss for all singly-charged particles, was done for AgBr alone rather than for the composite emulsion, since the Ag grain density

[1] G. C. WICK: Ric. sci. **11**, 274 (1940). — Nuovo Cim. (9) **1**, 302 (1943).

[2] M. SCHÖNBERG: Nuovo Cim. **9**, 210, 372 (1952).

[3] In an early paper [S 7a] STERNHEIMER used the ionization potentials of BAKKER and SEGRÈ. Subsequently [S 7b], he computed new values of Δ based on CALDWELL's ionization potentials. See also [S 8] for slight corrections to his energy loss expressions in [S 7a].

[4] To convert the units of I_d (or I_a) from ergs/cm to Mev/cm, multiply by 6.242×10^5 Mev/erg. When it is desired to express the energy loss in Mev/g cm⁻², then I_d or I_a should also be divided by the density ϱ of the medium.

[5] L. LANDAU: J. Phys. USSR. **8**, 201 (1944).

[6] K. R. SYMON: Thesis, Harvard University 1948. — Some of SYMON's results are given by B. ROSSI in "High Energy Particles", Prentice-Hall, Inc., New York 1952.

[7] H. A. BETHE: Z. Physik **76**, 293 (1932).

is determined, to a good approximation, by the energy deposited in the halide crystals. The value $T_0 = 5$ kev was used here[1]. Curve III includes only that part of the energy loss which was expended in transforming AgBr grains; hence this type of curve is to be compared with experimental grain or blob densities. On the other hand, for computing the energy lost by a particle in a given path length in emulsion, hence for calculating a range-energy relation, the type of loss in curve I (or II) is relevant.

The higher average rate of energy loss of protons than electrons at a given value of $\gamma - 1$, is due to the higher maximum possible energy transfer from the fast particle to the electrons in the medium. This distinction does not enter in the case of Curve III, since the same upper limit of energy transfer applies to various particles regardless of mass. Thus, a single curve adequately describes the restricted rate of loss of all singly-charged particles as a function of γ, or of any parameter which depends on the velocity alone.

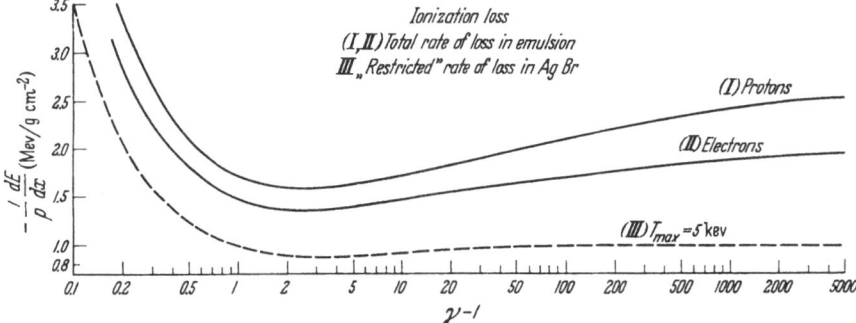

Fig. 9. Rate of ionization loss of charged particles in emulsion is plotted against the kinetic energy in rest-mass units (the velocity function E/Mc^2). When the "restricted" rate of loss is computed, the curves for protons and electrons both collapse to Curve III.

10. Relativistic increase of grain density in emulsion. Failure to distinguish between the restricted rate and the average rate of ionization loss was partly responsible for the early experimental difficulties in establishing the existence of a rise in grain density with increasing relativistic energies in nuclear emulsion. When electron-sensitive emulsions became available, evidence for the rise was sought in a number of laboratories by measuring the grain densities in tracks due to fast particles[2]. This was apparently done under the implicit assumption that the magnitude of the increase above the minimum of ionization should be 30 or 40%. No effect was observed exceeding the experimental error, which was believed to be $\sim 10\%$ or less. Hence, it was supposed for a time that there is no relativistic rise in emulsion. MESSEL and RITSON[3] then pointed out that in theoretical calculation of the ionization loss, it is inappropriate to use the maximum energy transfer T [e.g., in Eq. (8.1a)], since the more energetic secondary electrons by no means spend all of their energy in AgBr grains along the track.

In 1950 PICKUP and VOYVODIC reported indications of a relativistic rise of $\sim 10\%$[4]. Later VOYVODIC and others at Bristol extended this work, and found

---

[1] There is some uncertainty as to the best value of T_0, and also of w, the mean ionization potential to use in the calculations. The computed values of I_r are, however, insensitive to the choice of these parameters, since they enter only in the logarithm.

[2] References are given in [P 12], see especially pp. 74—78.

[3] H. MESSEL and D. M. RITSON: Phil. Mag. 41, 1129 (1950).

[4] E. PICKUP and L. VOYVODIC: Phys. Rev. 80, 89 (1950).

a rise of $\sim 8\%$[1]. Investigation of electron tracks in emulsions established the existence of an ionization plateau at high energies[2]. However, these measurements were not extended down to the minimum of ionization, and therefore could not yield the ratio I_{pl}/I_{min} (ionization at plateau)/(ionization at minimum).

Using cosmic-ray tracks, SHAPIRO and STILLER[3] measured the magnitude of the rise by comparing the Ag grain densities, n, of two groups of particle tracks, the thin shower tracks emerging from nuclear explosions and the "primary" tracks of high-energy stars with associated shower multiplicities $N_s \gtrsim 5$. The former group originates primarily from pions and, from their known energy spectrum, their grain density distribution was expected to have a peak at n_{min}. Similarly, the latter group is attributable mainly to protons so energetic that their histogram should display a peak at or near the plateau value n_{pl}. The

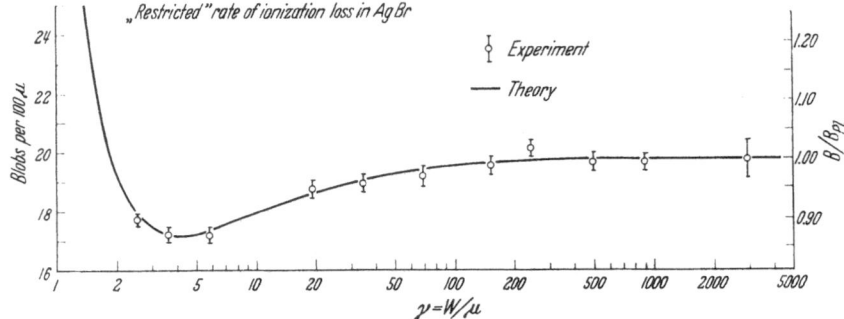

Fig. 10. Relativistic rise of the rate of energy loss by ionization in emulsion. Blob-density is plotted against the tota l energy of singly charged particles in rest-mass units ($\mu = M c^2$). The scale at the right shows the normalization of the blob count to the Fermi plateau. Experimental data [$S\,10$] are compared with theory as explained in the text.

theoretically expected peaks were observed, and an experimental value of the ratio $n_{pl}/n_{min} = 1.12 \, {}^{+\,0.04}_{-\,0.03}$ for G.5 emulsion was deduced therefrom. This could be compared with their calculated ratio (for the *restricted* rates of loss) I_{pl}/I_{min} in AgBr, of 1.14.

Later these authors replaced the high-energy protons by electrons with $\gamma > 100$, and again obtained well separated peaks from which the magnitude of the relativistic rise was deduced to be 12% [$S\,10$]. Repetition of the measurements on a plate from a different batch, exposure and processing gave an electron peak 13% above that for the star-shower tracks[4].

Apart from the question of the ratio of blob densities, B_{pl}/B_{min}, it was important to delineate the shape of the curve B vs. γ by measuring B in the intermediate region between minimum and plateau on identifiable tracks whose velocity was independently known, say from their multiple Coulomb scattering. DANIEL et al.[5] drew a "best-fit" curve through their normalized data, and those

---

[1] The results were published in a paper by R. R. DANIEL, J. H. DAVIES, J. H. MULVEY and D. H. PERKINS: Phil. Mag. **43**, 753 (1952).

[2] D. R. CORSON and M. R. KECK: Phys. Rev. **79**, 209 (1950). — I. B. McDIARMID: Phys. Rev. **84**, 851 (1951). — A. H. MORRISH: Phil. Mag. **43**, 533 (1952). — MORRISH subsequently measured tracks of pions, as well, at two energies, $\gamma = 1.85$ and 2.60 [Phys. Rev. **91**, 423 (1953)].

[3] M. SHAPIRO and B. STILLER: Phys. Rev. **87**, 682 (1952).

[4] Grain counts were replaced by "blob counts" to increase the objectivity. Experienced observers agree within statistical error in their blob densities (B) though not necessarily in their grain densities (n).

[5] R. R. DANIEL, J. H. DAVIES, J. H. MULVEY and D. H. PERKINS: Phil. Mag. **43**, 753 (1952).

of Voyvodic and Danysz *et al.* From this curve they concluded that the saturation of n (or B) already sets in at $\gamma = 10$ or 20, much faster than theory predicts. Stiller and Shapiro [S 10] obtained the results in Fig. 10 from measurements on cosmic-ray electrons, pions, and protons. Each point represents, on the average, a count of over 7000 blobs. Their data are altogether consistent with the slow rate of rise calculated according to the Halpern-Hall-Sternheimer theory, and suggest that the polarization plateau is reached at values of $\gamma > 100$, as predicted by this theory.

Their best estimate of $B_{\rm pl}/B_{\rm min}$ based on this work is 1.143 ± 0.03. For comparison with theory, they computed I, according to Eq. (9.1), for two values of T_0, 2 kev and 5 kev. The calculated ratio $I_{\rm plat}/I_{\rm min}$ for $T_0 = 2$ kev (the value used in Fig. 10), is 1.152; for $T_0 = 5$ kev, the ratio is 1.137. For w, the mean ionization potential of AgBr, constants based on experiments by Bakker and Segrè[1] were used. Sternheimer's expressions for the density-effect function δ for AgBr, were employed. (This function is called Δ in our notation.)

Fleming and Lord[2] used accelerator-generated pions with $\gamma \leq 2.61$, which enabled them to compare grain densities with ionization loss at energies below the minimum of the theoretical curve. They also measured tracks of electrons from muon decay, plotted their average blob-densities at a single energy $\gamma = 67.6$, and normalized their pion data so that this electron point fits a theoretical curve. Combining their results with those of other workers[3,4] they concluded that the data are in reasonable agreement with theory. Michaelis and Violet[4] measured grain densities of 293 Mev electrons from a synchrotron and 2.8 Mev electrons from a linear accelerator. At these two energies they obtained grain densities in the ratio 1.087 ± 0.010.

It would be desirable to measure the rate of ionization loss in emulsion over a wide range of γ (say, for $1.1 < \gamma < 300$) for a *single* species of particle accelerated to known energies. At this writing, proton beams are available only at energies $\gamma < 8$, and π-mesons at $\gamma < 35$. Electrons have, of course, been accelerated up to energies of hundreds of rest masses. However, at *low* electron energies, there are formidable obstacles to precise determination of grain densities. In the region of the theoretical minimum, electrons have energies < 3 Mev and mean scattering angles $> 7°/(100\,\mu)^{\frac{1}{2}}$. Owing to their strong scattering, such electrons (a) seldom stay in the emulsion over a sufficient path length; (b) when they do, the inclusion of spurious background grains in a highly bent track is more probable than in a straight track; (c) even the measurement of true length of track is subject to appreciable error; (d) there is danger of shifting unwittingly to a neighboring electron track which crosses the one in question; (e) if one selects relatively straight and flat segments of such electron tracks for counting, there is a tendency to choose the more easily visible segments, i.e., those of higher grain densities than average. This type of selection could yield an excessively high n for low-energy electrons, and thus a low value of $n_{\rm plat}/n_{\rm min}$[5].

The low value of the relativistic rise ($\sim 8\%$) reported in some of the early work, and the rapid rise of the grain density in apparent contradiction to theory, can be accounted for by the method of grain-density normalization which was employed. In order to correct for variations in n between various plates, between different areas of the same plate, and as a function of depth in the emulsion,

[1] C. J. Bakker and E. Segrè: Phys. Rev. **81**, 489 (1951).

[2] J. R. Fleming and J. J. Lord: Phys. Rev. **92**, 511 (1953).

[3] A. H. Morrish: Phys. Rev. **91**, 423 (1953).

[4] R. P. Michaelis and G. B. Violet: Phys. Rev. **90**, 723 (1953).

[5] The possibility of this last source of systematic error was called to the author's attention by Dr. W. H. Barkas.

the observed n of a track was divided by the density n_0 of a nearby "reference" track, and the ratio n^* was plotted against the product momentum $\times$ velocity (deduced from multiple scattering). The reference track selected was that of the "primary" which generated the same star, or that of a neighboring star. Whenever possible, events were chosen with primary particles for which $\gamma > 10$. The method thus implicitly *assumed* that the plateau of grain density already sets in at $\sim\gamma = 10$, and the resulting plot of n^* vs. pv necessarily reflects this assumption.

Since all forms of the polarization theory agree that the ionization loss has a plateau value at sufficiently high energies, a fair method of normalization for comparison with ionization theory dictates a choice of reference plateau particles in the far relativistic region, say $\gamma > 100$. If, instead, "reference" particles of lower energy are employed, the values n^* so obtained can easily be too high, leading to apparent saturation at values of γ much lower than theory predicts. Moreover, $n_{\min}/n_0$ will also be too high, so that $n_{\mathrm{plat}}/n_{\min}$ will be underestimated.

In cosmic ray exposures, the most readily available particles having plateau ionization are electrons $(\pm)$ with $E > 50$ Mev. These can be recognized phenomenologically when they occur as the result of pair formation within the emulsion, and a rapid, crude measurement of their multiple Coulomb scattering suffices to establish whether they have enough energy. Alternatively, there are singly-charged primaries—usually protons or pions—of narrow-angle "jets" from whose angular distribution one can deduce that the primary energy $\gamma > 100$. Except for their scarcity, these primaries are also suitable reference particles.

BROWN[1] has considered the effect of fluctuations in the energy loss. According to his analysis, the intensity of development of the emulsion could affect the magnitude of the relativistic rise in n (or B), and the shape of the curve n vs. γ. However, at low ionizations ($< 2 I_{\min}$, where grain counting is useful) and for usual intensities of development, emulsion measurements indicate[2] a proportionality of n to I. Also, *moderate* changes in the level of B_{pl} will affect the ratio $B_{\mathrm{pl}}/B_{\min}$ inappreciably. Hence it seems reasonable to compare grain density measurements with theoretical I_r values, at least for thin tracks. It would be remarkable if the agreement so obtained were purely fortuitous.

A part of the energy loss of a charged relativistic particle consists of Cerenkov radiation, as noted by FERMI[3] and further elucidated by A. BOHR [B14a]. The possibility that some of this radiation might escape from the AgBr crystals along the path of a particel in a nuclear emulsion and thus detract from the observable relativistic rise in grain density has been considered by MESSEL and RITSON[4], SCHÖNBERG[5], HUYBRECHTS[6], and BUDINI[7]. STERNHEIMER[8] concludes that the energy loss in the form of Cerenkov radiation contributes only about 2% of the relativistic rise in the total ionization loss. Hence the Cerenkov radiation may be neglected in comparing the theory with measurements on grain-density in emulsion.

11. Range versus energy. When a known particle has been arrested by ionization loss in emulsion, its residual range R is the most precise index of its

[1] L. M. BROWN: Phys. Rev. **90**, 95 (1953).

[2] See, e.g., Ref. [F2].

[3] E. FERMI: Phys. Rev. **57**, 485 (1940).

[4] H. MESSEL and D. M. RITSON: Phil. Mag. **41**, 1129 (1950).

[5] M. SCHÖNBERG: Nuovo Cim. **9**, 210, 372 (1952).

[6] M. HUYBRECHTS and M. SCHÖNBERG: Nuovo Cim. **9**, 764 (1952).

[7] P. BUDINI: Phys. Rev. **89**, 1147 (1953), and [B17].

[8] R. M. STERNHEIMER: Phys. Rev. **89**, 1148 (1953); a more detailed analysis is given in [S8].

energy E. Or, if the particle's identity is unknown, its range combined with other properties measured as a function of range (e.g., ionization loss, multiple scattering), provides a knowledge of its charge and mass. The precautions to be observed in reliable range determinations are discussed later. We shall here assume that an adequate range measurement has been made, and that the density of the emulsion under the conditions of the exposure is known. Then, in deducing the energy corresponding to a given range, one can, in principle, rely on existing $R-E$ measurements, or on a theoretical $R-E$ relation, or on a suitable combination of the two. As we shall see, the last method is frequently employed, especially at higher energies, where few experimental $R-E$ data are available.

Let the rate of energy loss of Eq. (8.3 a) be written in the form[1,2]

$$I_d = \frac{n_e z^2}{\beta^2} F(w, \beta) = \left| \frac{dE}{dx} \right|. \tag{11.1}$$

From this the range is obtained by integrating $\dfrac{dE}{|dE/dx|}$ along the residual path of the particle:

$$R = \int_E^0 \frac{dE}{I_d} = \frac{M}{n_e z^2} \int_\beta^0 F_1(w, \beta)\, d\beta. \tag{11.2}$$

For an emulsion of fixed composition—and this implies a definite moisture content—n_e is fixed and w depends only on β, so that we may write

$$R = \frac{M}{z^2} F_2(\beta). \tag{11.3}$$

Thus, for a class of particles with the same charge (e.g., for singly charged particles), R/M is a function of the velocity alone. For other values of z, the quantity Rz^2/M is a function of the velocity alone[3].

Eq. (11.3) may be rewritten

$$\frac{R}{M} = \frac{1}{z^2} F_3\left(\frac{E}{M}\right), \tag{11.3a}$$

a relation which is exploited in constructing $R-E$ tables for the various kinds of particles from measurements on any single type (e.g., protons).

12. *R-E* and density. The chemical composition of an emulsion affects its stopping power and hence the range of particles brought to rest in it. As seen from Eq. (8.1), the property of the medium which primarily determines the rate of energy loss is the electron density n_e, while the mean ionization potential w, occurring in the logarithm, is of secondary consequence. Hence in correcting for *minor* changes in emulsion composition, it suffices to take account of the new electron density, neglecting the small change in w.

Suppose that, through change in composition, the density ϱ of an emulsion is altered by $\Delta\varrho$ and its electron density n_e by Δn_e. The change in the number of electrons per gm is then $\Delta\left(\dfrac{n_e}{\varrho}\right)$. If the ratio of concentrations of heavy to

---

[1] At non-relativistic energies, where the "density effect" is negligible, it suffices to use I_a of Eq. (8.3).

[2] At extremely low velocities, the function F should, rigorously, include a charge-dependent term to take account of the capture and loss of electrons which occur near the very end of the range. For singly-charged particles this effect is negligible, and for helium nuclei its contribution to the range is of the order of one micron. For other multiply-charged ions (e.g., fragments or hyperfragments emitted from stars), the effect is appreciable. We shall, however, neglect it in the present discussion, and take account of it later.

[3] See, however, the preceding footnote.

light elements is unchanged, then, despite variation in chemical constitution,

$$\frac{\Delta n_e}{n_e} \approx \frac{\Delta \varrho}{\varrho} \quad \text{and} \quad \Delta\left(\frac{n_e}{\varrho}\right) \approx 0. \tag{12.1}$$

If, however, the density is changed by variation of moisture content[1], then Eqs. (12.1) no longer hold, since ordinary hydrogen (H^1) is decidedly richer in electrons per gram than other nuclides.

The manufacturer of nuclear emulsion may try to keep the composition constant from batch to batch, but he has no control over the moisture content at the time of exposure. In fact, because gelatine and glycerine are hygroscopic, the most variable constituent of emulsion is apt to be water. This will affect the exact value of the range. The variation in density of emulsion under change in relative humidity is discussed in Sect. 21. From Table 7 it is evident that the density departs by roughly $\pm 5\%$ from that at "normal relative humidities" ($\sim$50 to 60%) for very dry and moist emulsion, respectively.

It is instructive to consider the effect of change in moisture content on the electron density, and hence on the range. If R is the range in cm, let $\Lambda = \varrho R$ be the range in g/cm². In Sect. 21 it is shown that, for G.5 emulsion at relative humidities between zero and $\sim$84% the electron density n_e is given by Eq. (21.5). Accordingly,

$$\frac{\Delta n_e}{n_e} = 0.90 \frac{\Delta \varrho}{\varrho}. \tag{12.2}$$

Since R is inversely proportional to n_e, we have

$$\frac{\Delta R}{R} = -0.90 \frac{\Delta \varrho}{\varrho}. \tag{12.3}$$

On the other hand, Λ varies as ϱ/n_e; hence,

$$\frac{\Delta \Lambda}{\Lambda} = \frac{\Delta \varrho}{\varrho} - \frac{\Delta n_e}{n_e} = 0.10 \frac{\Delta \varrho}{\varrho}. \tag{12.4}$$

Thus Λ is comparatively insensitive to changes in relative humidity. For example, a rather thorough desiccation of normal emulsion which increases its density by $\sim$5%, augments Λ by only 0.5%. More usual values of $\Delta\varrho/\varrho$ are $\lesssim 2\%$, and the resulting changes in Λ are $\lesssim 0.2\%$, which is less than the uncertainty in available R-E calibrations.

In practice, what is directly measured is the length R, and accurate conversion to Λ depends on precise knowledge of the density. Irrespective of whether R or Λ is used, the best way to ensure valid comparisons of range between various experiments is to *measure* the density of the emulsion under the conditions of exposure in every case. This is necessary even if a particle's range is being measured only to learn its exact energy. It is doubly important in making a series of R-E measurements for the purpose of establishing an R-E relation. The latter will be most useful if it is valid at some known, "standard" density. Ranges measured at other densities are then readily reduced to the "standard" ranges in an R-E table or graph.

Some recent R-E calibrations have included density determinations[2]. Most earlier ones have not done so, and this is perhaps their principal limitation. To

---

[1] The absolute moisture content of emulsion is apparently not well known. For fairly dry emulsion (3.94 g/cm³), values of $\sim$0.06 to 0.08 g/cm³ can be gleaned from the literature [W4], [G1]. What has been measured more precisely is the *change* in weight and volume with varying environmental humidity from those of emulsion at some "normal" R. H. ($\sim$50 to 60%); see Table 7, and [O4].

[2] See, for example, [B4], [H2], and [C1a].

be sure, many of the authors have described the conditions of humidity under which their emulsion was exposed. This information, however, is not without ambiguity, since long periods of time are required for a thick emulsion to reach equilibrium with air at some new relative humidity. The fact that emulsion is subject to hysteresis in its absorption and loss of moisture, compounds the ambiguity. Some experimenters have assumed that their plates were thoroughly dry because they had been kept in vacuum for some hours before exposure. Actually, it has been shown[1] that it takes many weeks for a thick emulsion to lose all of its moisture.

13. *R-E* relation at proton energies < 40 Mev. Before extensive range-energy measurements were available in emulsion, it was customary to deduce ranges in emulsion from those in air by computing the stopping power of emulsion relative to that of air as a function of the particle velocity. This method was adopted, for example, by CÜER [*C 7*], WEBB [*W 3*], and WILKINS [*W 4*].

A good deal of experimental *R-E* work has been done, mainly on protons and α-particles, at energies < 40 Mev. In modern Ilford emulsions (mostly C. 2), some of the earlier measurements were made by LATTES, FOWLER and CÜER [*L 2*], BRADNER, SMITH, BARKAS and BISHOP [*B 15*][2], ROTBLAT [*R 10*], CATALA and GIBSON [*C 3*], CÜER and JUNG [*C 8*], and PANOFSKY and FILLMORE[3]. The energies have usually been determined from the Q-values of well-known nuclear reactions, and sometimes by bending the ions in a magnetic field. Data at energies < 1 Mev have been obtained by FARAGGI [*F 1*] and by NERESON and REINES [*N 1*].

In applications of emulsion to problems in low-energy physics (e.g., with Van de Graaff accelerators), the plates are commonly exposed singly (i.e., unstacked) in vacuum. Partly for this reason, it became customary to give *R-E* values for "dry" emulsion. Recently, GIBSON, PROWSE and ROTBLAT [*G 1*] have made new observations and combined these with their earlier data for dry C.2 emulsion at proton energies from 2 to 21 Mev. By the term "dry" they mean that the emulsion was kept in vacuum long enough for its moisture content to approach a constant value. They estimate that their emulsion density was "about 3.94 g/cm³". Table 1 reproduces the results of GIBSON *et al.* Considerable

Table 1. *Range energy relation for protons in "Dry" Ilford C. 2 emulsion*[4].

Range (microns)	Energy (Mev)	Range (microns)	Energy (Mev)	Range (microns)	Energy (Mev)	Range (microns)	Energy (Mev)	Range (microns)	Energy (Mev)
40	2.01	440	8.80	840	12.83	1240	16.12	1640	18.98
80	3.13	480	9.25	880	13.18	1280	16.42	1680	19.25
120	4.01	520	9.69	920	13.53	1320	16.72	1720	19.52
160	4.79	560	10.12	960	13.87	1360	17.01	1760	19.79
200	5.48	600	10.54	1000	14.21	1400	17.30	1800	20.05
240	6.12	640	10.94	1040	14.54	1440	17.59	1840	20.31
280	6.72	680	11.34	1080	14.86	1480	17.88	1880	20.57
320	7.28	720	11.72	1120	15.18	1520	18.16	1920	20.82
360	7.82	760	12.10	1160	15.50	1560	18.44	1960	21.07
400	8.32	800	12.47	1200	15.81	1600	18.71	2000	21.32

[1] See [*O 4*]. It is worth noting that the phrase "absolutely dry" has dubious significance when applied to emulsion, for when its temperature is raised, moisture continues to be driven off even when the emulsion is baked at about 110° C. Its properties have then changed so markedly that it is no longer emulsion in the ordinary sense.

[2] BRADNER *et al.* were the first to extend *R-E* measurements well beyond proton energies of 13 Mev.

[3] W. K. H. PANOFSKY and F. L. FILLMORE: Phys. Rev. **79**, 57 (1950).

[4] After GIBSON, PROWSE and ROTBLAT [*G 1*].

weight should be attached to these extensive measurements, carried out under rather uniform conditions with high internal precision. However, their data would be even more valuable if the authors had been able to measure their emulsion density under the conditions of exposure. Instead, they apparently deduced it from information supplied by the manufacturer. Therefore, an attempt at exact comparison with results of others is hampered by slight ambiguity[1].

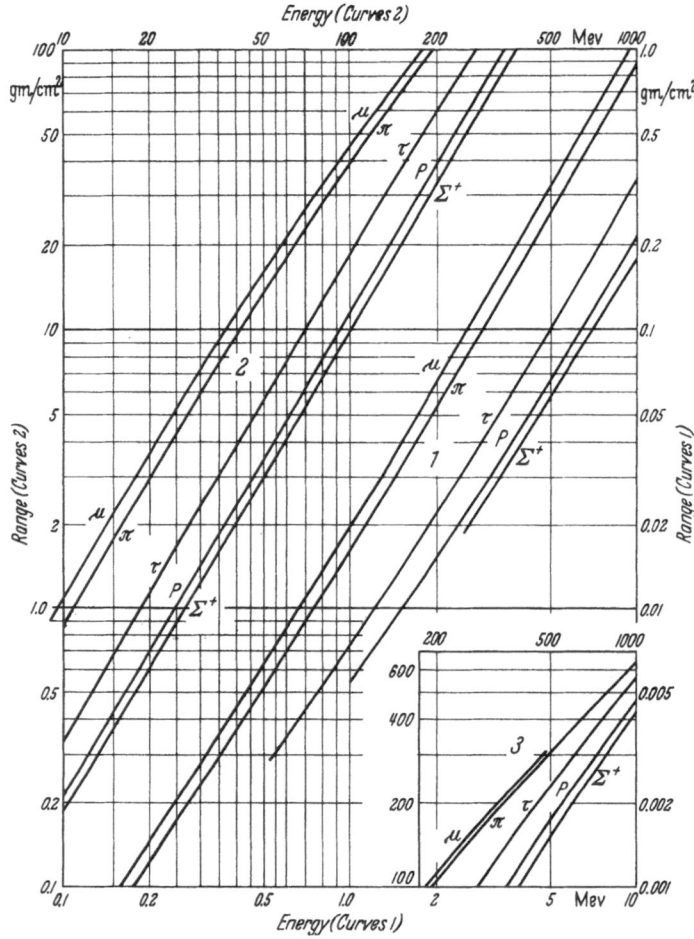

Fig. 11. The range-energy data of Table 3 for G.5 emulsion at a density of 3.815 g/cm³ are plotted, with the ranges expressed in g/cm³ (instead of microns or cm) for muons, pions, τ-mesons, and Σ⁺ hyperons, as well as protons. The ranges were computed from proton ranges in Table 2. It should be noted that the sets of curves 1, 2 and 3 have different scales.

More recently Barkas[2] constructed a semi-theoretical R-E curve for proton energies up to 40 Mev, and adjusted it to fit the measurements of Barkas, Smith and Heckman[3] in G.5 emulsion of known density. The measured ranges were reduced to those at a standard density of 3.815 g/cm³, and the results are

---

[1] See below.

[2] Private communication.

[3] W. H. Barkas, F. M. Smith and H. H. Heckman: University of California Radiation Laboratory Report 3513, August, 1956. Only 3 of 8 experimental points are given in this report.

given in Part I of Table 2 (see columns 2 and 7). Table 3 gives the range versus energy for various singly-charged particles: τ mesons, π mesons, μ mesons, and Σ^+ hyperons, as well as protons. The values were computed, using Eq. (11.3a), from BARKAS' relation for protons. The same data, with ranges Λ in g/cm², are plotted in Fig. 11.

The ranges of GIBSON *et al.* are consistently about 2.5% lower than those of BARKAS between 3 and 21 Mev. However, from the difference in nominal densities we should expect a difference of $\sim$2.9% [cf. Eq. (12.3)]. Hence the two sets of ranges differ by only $\sim$0.4% — a discrepancy which would vanish if the density of GIBSON's emulsion were actually 3.92 rather than the estimated value 3.94 g/cm³. Altogether, the agreement is remarkably good[1].

Among *R-E* calibrations for Eastman emulsions are those of RICHARDS, JOHNSON, AJZENBERG and LAUBENSTEIN [R2], STEIGERT, TOOPS and SAMPSON[2], and GAILAR, SEIDLITZ, BLEULER and TENDAM[3].

When the ratio of gelatine to AgBr is significantly altered, as in Ilford's "diluted" emulsions, a separate *R-E* calibraton is needed. LEES, MORRISON, and ROSSER [L3] have provided this for energies up to 9 Mev at two dilutions, "G.5×2" and "G.5×4".

VIGNERON [V4] has given an excellent theoretical treatment of the *R-E* relation in emulsion. By analysis of the experimental data then available for C.2 emulsion, he was able to derive a semi-empirical *R-E* curve extending from very low energies up to $\sim$220 Mev. The data he used came from many sources, and had been obtained under a variety of conditions. Moreover, *R-E* calculations at low energies suffer from some uncertainty because the effect on the stopping power of the binding of electrons in the medium is difficult to compute. Indeed, a definitive theory of the penetration of charged particles of low energy remains to be developed. Thus, it was an achievement for VIGNERON to obtain a good fit to the bulk of the experimental data. BARKAS finds that in order to obtain agreement with his own ranges at the "standard density" 3.815, VIGNERON's ranges require increments varying from 1 to 3% in the energy interval from 3 to 40 Mev.

14. Range-energy relations at higher energies. What was probably the first good extrapolation of the *R-E* relation for protons to energies above 40 Mev, was provided by VIGNERON [V4]. His curves, which have been computed for deuterons, tritons, and alpha particles, as well as protons[4], are reproduced in Fig. 12.

At the higher energies (above $\sim$30 Mev) VIGNERON's choice of parameters amounts to a selection of electron density $n_e = 1.045 \times 10^{24}$ electrons/cm³. In the light of present knowledge of G.5 and C.2 composition, this corresponds to a mass density $\varrho = 3.815$ g/cm³.

---

[1] As shown in Sect. 21, the fact that the emulsion was of type G.5 in one experiment, and C.2 in the other should make an imperceptible difference ($< 0.1\%$), provided that ranges are compared at the same density. More generally, it may be remarked that few of the figures given in the literature for the differences in stopping power (usually reported as $< 2\%$) between various types of emulsions, can be used with confidence, since there is generally no assurance that each of the emulsions was first brought into equilibrium with the same well controlled environmental humidity. Nor, usually, have the authors specified the exact densities of the emulsions being compared.

[2] F. E. STEIGERT, E. C. TOOPS and M. B. SAMPSON: Phys. Rev. **83**, 474 (1951).

[3] O. GAILAR, L. SEIDLITZ, E. BLEULER and D. J. TENDAM: Rev. Sci. Instrum. **24**, 126 (1953).

[4] The reader should note, however, that the ranges may be in error by two or three percent at energies up to $\sim$15 Mev, if they are assumed to apply to emulsion of the same electron density (1.045×10^{24} per cm³) as the higher energy data. VIGNERON himself called attention to the possibility of errors of this magnitude in view of uncertainty in the moisture content.

FAY, GOTTSTEIN, and HAIN [$F\,1a$] have published a range-energy table based on selected data at energies up to 40 Mev, and have fitted those data to a power law

$$\frac{E}{\text{Mev}} = (0.281 \pm 0.005)\left(\frac{R}{\mu}\right)^{(0.568 \pm 0.003)} \tag{14.1}$$

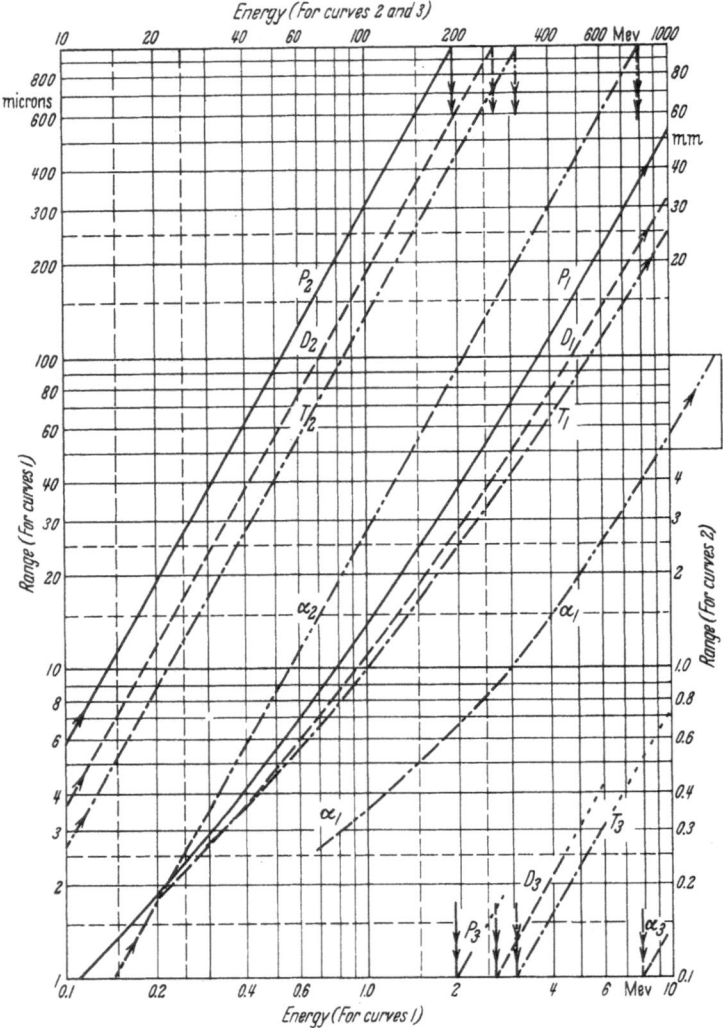

Fig. 12. The range-energy curves of VIGNERON [$V\,4$] for protons, deuterons, tritons, and α-particles in Ilford C.2 emulsion. At the higher particle velocities (for proton energies > 40 Mev) the data apply to an emulsion of electron density 1.045×10^{24} electrons/cm³. At low energies, up to 15 or 20 Mev, the ranges may be in error by 2 or 3% if they are assumed to refer to emulsion of the same electron density. Curves 2 and 3 are probably good to 1 or 2%.

(where μ = micron) which they extrapolated up to a proton energy of ~ 390 Mev. The constants in this power law agreed independently with those computed by MENON and O'CEALLAIGH[1] for VIGNERON's range-energy calculations.

FAY *et al.* realized, however, the appreciable inaccuracies introduced at the higher energies, particularly above 200 Mev, in using the same value of the

[1] M. G. K. MENON and C. O'CEALLAIGH: Phil. Mag. **44**, 1291 (1953).

exponent, when the actual exponent approaches unity at proton energies of several Bev.

Empirical formulas of the form $E = A R^\eta$ have been widely used to fit experimental R-E data for emulsions, and they played a particularly useful role in the low-energy region, where calculation of the rate of energy loss is subject to uncertainty. Also, because of the slow variation with energy of the "range-energy index" η (cf. column 10 of Table 2), a single formula may give a fairly good fit over an energy interval as wide as $\sim$100 Mev. Nevertheless, η does vary appreciably with energy; hence, a better correlation of R-E measurements, especially when these encompass a wide range of energies, is obtained by fitting the data to an R-E relation derived from the theory of energy loss, as has been done by VIGNERON [V 4], BARKAS and YOUNG [B 5], and BARONI, CASTAGNOLI, CORTINI, FRANZINETTI, and MANFREDINI [B 6].

BARKAS[1] extended the R-E calculations up to proton energies of $\sim$30 Bev for C.2 emulsion at a density of 3.815 g/cm^3. BARONI et al. calculated an R-E curve for C.2 emulsion up to proton energies of 5 Bev, using a density of 3.92 g/cm^3. They found agreement (within 2%) with VIGNERON's ranges between 100 and 200 Mev, and also, up to considerably higher energies, with those of DANIEL, GEORGE, and PETERS[2]. The latter group showed that, at low energies, ranges R_e in emulsion are simply related to those in Al and Pb (R_{Al} and R_{Pb}), at the same energy, by the simple expression $R_e^2 = C R_{Al} R_{Pb}$. Assuming that the coefficient C is independent of energy, and using the value[3] of $C = 5.075$, they constructed emulsion curves giving proton ranges up to 100 cm.

Notwithstanding the painstaking calculations of R-E relations at high particle velocities, the need has become increasingly apparent for R-E measurements at energies well above 40 Mev. This requirement has been stimulated by rapid developments in the physics of heavy mesons and hyperons, and by the growing use of large emulsion stacks designed to stop the long-range secondaries of these unstable particles. The first experimental R-E determinations in emulsion above 100 Mev were made by HEINZ [H 2], and by CARVALHO and FRIEDMAN [C 1a]; in each experiment the mean range of a single group of protons was measured in Ilford emulsions as well as in copper. The former work yielded a range of (92.68 ± 0.25) g/cm^2 for 342.5 Mev protons in C.2 emulsion at an average density of (3.81 ± 0.01) g/cm^3. The latter determination was for G.5 emulsion of density (3.85 ± 0.01) g/cm^3; and for (208 ± 4) Mev protons, it gave a range of (41.3 ± 0.2) g/cm^2, a value exceeding that expected from HEINZ's measurement by 1.2% [C 1a]. In each case, the proton energy was deduced from the observed range in copper.

Recently BARKAS and his collaborators [B 4] have begun an experimental program of range measurements in emulsion of known density at high energies. They used mainly π mesons with equivalent proton energies up to 700 Mev. The momenta of these mesons were measured magnetically with high precision. BARKAS applied the new experimental data in constructing a revised R-E table[4]

---

[1] He subsequently revised the R-E table given in Ref. [B 5] in the light of new high-energy measurements, as described below.

[2] R. R. DANIEL, E. C. GEORGE, and B. PETERS: Proc. Ind. Acad. Sci. 41, 45 (1955).

[3] This value of C applies when R_e is expressed in mm, and the ranges in Al and Pb expressed in g/cm^2. The value is based mainly on the emulsion results of GIBSON et al. [G 1] and HEINZ [H 2], together with the R-E curves for Al and Pb of E. P. GROSS; see D. J. X. MONTGOMERY, Cosmic Ray Physics. Princeton University Press. 1949. Appendix E.

[4] W. H. BARKAS: University of California Radiation Laboratory, UCRL 3384, April, 1956. Dr. BARKAS very kindly supplied the author with the original IBM data sheets which included rates of energy-loss and the velocity functions β, γ, $\beta\gamma$, and $\beta^2\gamma$.

Table 2. *Quantities useful in nuclear emulsion measurements*[1].

Normalized range, energy loss and other properties of heavy particles tabulated as functions of the velocity (for columns 6 and 7, $\varrho = 3.815$ g/cm³; $n_e = 1.045 \times 10^{24}$ electrons/cm³).

1	2	3	4	5	6	7	8	9	10
Velocity	Proton kinetic energy	Total energy	Momentum and magnetic curvature	Multiple Coulomb scattering	Rate of energy loss	Residual range	Rate of energy loss	Residual range	Range-energy index
$\beta = \dfrac{v}{c}$	$\varepsilon = \dfrac{EM_p}{M}$ Mev	$\gamma = \dfrac{W}{Mc^2}$	$\beta\gamma = \dfrac{p}{Mc}$	$\beta^2\gamma = \dfrac{pv}{Mc^2}$	$\iota = \dfrac{I}{z^2}$ Mev/cm	$\lambda = \dfrac{z^2 R' M_p}{M}$	$\iota' = \dfrac{\iota}{\varrho}$ Mev/g cm⁻²	$\lambda' = \lambda\varrho$ g cm⁻²	$\eta = \dfrac{\lambda\iota}{\varepsilon}$

Part I: Proton energies $<$ 40 Mev.

0.046 1324 2	1.000	1.001 066	0.046 1815 9	0.002 130 468	498	14.2 μ	130	0.005 42	0.707
0.065 1904 9	2.000	1.002 132	0.065 3294 6	0.004 258 859	317	40.3	83.1	0.015 4	0.639
0.079 7771 9	3.000	1.003 198	0.080 0322 8	0.006 384 750	243	76.7	63.7	0.029 3	0.622
0.092 0456 4	4.000	1.004 263	0.092 4380 6	0.008 508 520	201	122.2	52.6	0.046 6	0.613
0.102 8290	5.000	1.005 329	0.103 3770	0.010 630 15	172	176.4	45.1	0.067 3	0.607
0.112 5535	6.000	1.006 395	0.113 2733	0.012 749 31	150	238.9	39.4	0.091 1	0.599
0.121 4755	7.000	1.007 461	0.122 3818	0.014 866 39	136	308.2	35.6	0.118	0.598
0.129 7602	8.000	1.008 527	0.130 8666	0.016 981 27	124	385.4	32.4	0.147	0.596
0.137 5224	9.000	1.009 592	0.138 8415	0.019 093 82	113	470.1	29.5	0.179	0.588
0.144 8468	10.00	1.010 658	0.146 3906	0.021 204 22	105	561.8	27.4	0.214	0.588
0.161 6249	12.50	1.013 323	0.163 7782	0.026 470 63	88.2	827.0	23.1	0.315	0.588
0.176 7037	15.00	1.015 988	0.179 5288	0.031 723 40	77.1	1127	20.2	0.430	0.579
0.190 4886	17.50	1.018 652	0.194 0416	0.036 962 71	68.6	1470	18.0	0.561	0.576
0.203 2435	20.00	1.021 317	0.207 5759	0.042 188 45	61.9	1851	16.2	0.706	0.573
0.226 3515	25.00	1.026 646	0.232 3828	0.052 600 20	52.0	2735	13.6	1.043	0.569
0.246 9990	30.00	1.031 975	0.254 8968	0.062 959 25	45.2	3767	11.8	1.437	0.567
0.265 7668	35.00	1.037 304	0.275 6810	0.073 266 87	40.1	4938	10.5	1.884	0.566

Part II: Proton energies $\geq$ 40 Mev.

0.283 0339	40.00	1.042 633	0.295 1006	0.083 523 49	36.81	0.6249 cm	9.648	2.384	0.5750
0.291 1908	42.50	1.045 298	0.304 3812	0.088 633 01	35.13	0.6944	9.209	2.649	0.5740
0.299 0657	45.00	1.047 963	0.313 4097	0.093 730 09	33.63	0.7672	8.814	2.927	0.5733
0.306 6806	47.50	1.050 627	0.322 2070	0.098 814 63	32.27	0.8431	8.457	3.216	0.5727
0.314 0556	50.00	1.053 292	0.330 7921	0.010 388 71	31.03	0.9221	8.133	3.518	0.5722
0.325 3978	54.00	1.057 555	0.344 1261	0.111 9779	29.26	1.055	7.670	4.025	0.5717
0.336 2279	58.00	1.061 818	0.357 0130	0.120 0317	27.72	1.195	7.266	4.561	0.5714
0.346 5963	62.00	1.066 082	0.369 5000	0.128 0673	26.36	1.343	6.910	5.126	0.5712
0.356 5451	66.00	1.070 345	0.381 6263	0.136 0670	25.15	1.499	6.593	5.718	0.5712
0.366 1101	70.00	1.074 608	0.393 4250	0.144 0369	24.07	1.661	6.309	6.339	0.5713
0.379 8053	76.00	1.081 003	0.410 5709	0.155 9370	22.64	1.919	5.935	7.320	0.5716
0.392 7949	82.00	1.087 398	0.427 1245	0.167 7723	21.41	2.191	5.611	8.360	0.5721
0.405 1518	88.00	1.093 792	0.443 1524	0.179 5440	20.33	2.479	5.328	9.458	0.5727
0.416 9373	94.00	1.100 188	0.458 7096	0.191 2531	19.38	2.782	5.079	10.61	0.5734
0.428 2033	100.0	1.106 584	0.473 8427	0.202 9011	18.53	3.098	4.857	11.82	0.5742
0.462 5305	120.0	1.127 900	0.521 6882	0.241 2967	16.29	4.253	4.269	16.23	0.5772
0.492 7732	140.0	1.149 217	0.566 3032	0.279 0590	14.64	5.551	3.839	21.18	0.5806
0.519 7617	160.0	1.170 534	0.608 3984	0.316 2222	13.39	6.981	3.510	26.63	0.5843
0.544 4826	180.0	1.191 850	0.648 4650	0.352 8185	12.40	8.535	3.251	32.56	0.5882
0.566 169	200.0	1.213 167	0.686 8580	0.388 8780	11.61	10.20	3.042	38.92	0.5921
0.586 3535	220.0	1.234 484	0.723 8437	0.424 4283	10.95	11.98	2.869	45.70	0.5961
0.604 8954	240.0	1.255 800	0.759 6278	0.459 4953	10.39	13.85	2.725	52.86	0.6001
0.622 0045	260.0	1.277 117	0.794 3726	0.494 1034	9.925	15.82	2.601	60.37	0.6041
0.637 8523	280.0	1.298 434	0.828 2090	0.528 2750	9.521	17.88	2.496	68.23	0.6081
0.652 5810	300.0	1.319 750	0.861 2440	0.562 0314	9.170	20.02	2.404	76.40	0.6121
0.666 3099	320.0	1.341 067	0.893 5662	0.595 3920	8.862	22.24	2.323	84.86	0.6160
0.679 1412	340.0	1.362 384	0.925 2509	0.628 3761	8.591	24.54	2.252	93.61	0.6200
0.691 1621	360.0	1.383 700	0.956 3612	0.661 0006	8.350	26.90	2.189	102.6	0.6239
0.702 4479	380.0	1.405 017	0.986 9514	0.693 2819	8.135	29.33	2.132	111.9	0.6278
0.713 0648	400.0	1.426 334	1.017 068	0.725 2357	7.942	31.81	2.082	121.4	0.6317
0.723 0703	420.0	1.447 651	1.046 753	0.756 8760	7.768	34.36	2.036	131.1	0.6355
0.732 5155	440.0	1.468 967	1.076 041	0.788 2168	7.610	36.96	1.995	141.0	0.6393
0.741 4455	460.0	1.490 284	1.104 964	0.819 2708	7.466	39.62	1.957	151.1	0.6430
0.749 9007	480.0	1.511 601	1.133 550	0.850 0503	7.335	42.32	1.923	161.5	0.6467
0.757 9171	500.0	1.532 917	1.161 824	0.880 5664	7.215	45.07	1.891	171.9	0.6504
0.772 7597	540.0	1.575 551	1.217 522	0.940 8519	7.004	50.70	1.836	193.4	0.6576
0.786 1964	580.0	1.618 184	1.272 211	1.000 207	6.824	56.49	1.789	215.5	0.6646
0.798 4107	620.0	1.660 817	1.326 014	1.058 704	6.670	62.42	1.748	238.1	0.6715
0.809 5552	660.0	1.703 451	1.379 038	1.116 407	6.537	68.48	1.713	261.2	0.6782
0.819 7580	700.0	1.746 084	1.431 367	1.173 374	6.421	74.65	1.683	284.8	0.6848

[1] See explanatory notes at the end of this table.

Table 2. (Continued.)

1	2	3	4	5	6	7	8	9	10
Velocity	Proton kinetic energy	Total energy	Momentum and magnetic curvature	Multiple Coulomb scattering	Rate of energy loss	Residual range	Rate of energy loss	Residual range	Range-energy index
$\beta = \dfrac{v}{c}$	$\varepsilon = \dfrac{E M_p}{M}$ Mev	$\gamma = \dfrac{W}{Mc^2}$	$\beta\gamma = \dfrac{p}{Mc}$	$\beta^2\gamma = \dfrac{pv}{Mc^2}$	$\iota = \dfrac{I}{z^2}$ Mev/cm	$\lambda = \dfrac{z^2 R' M_p}{M}$ cm	$\iota' = \dfrac{\iota}{\varrho}$ Mev/g cm^{-2}	$\lambda' = \lambda\varrho$ g cm^{-2}	$\eta = \dfrac{\lambda\iota}{\varepsilon}$
0.829 127 4	740.0	1.788 718	1.483 075	1.229 658	6.320	80.93	1.657	308.8	0.6912
0.837 755 9	780.0	1.831 351	1.534 225	1.285 306	6.231	87.31	1.633	333.1	0.6974
0.845 722 8	820.0	1.873 984	1.584 871	1.340 362	6.152	93.77	1.613	357.7	0.7035
0.853 096 9	860.0	1.916 618	1.635 061	1.394 865	6.083	100.3	1.594	382.7	0.7095
0.859 937 7	900.0	1.959 251	1.684 834	1.448 852	6.021	106.9	1.578	407.9	0.7153
0.866 296 9	940.0	2.001 884	1.734 226	1.502 355	5.966	113.6	1.564	433.3	0.7210
0.872 220 5	980.0	2.044 518	1.783 270	1.555 405	5.917	120.3	1.551	459.0	0.7265
0.877 748 3	1 020.0	2.087 151	1.831 993	1.608 029	5.874	127.1	1.540	484.9	0.7320
0.882 916 0	1 060.0	2.129 785	1.880 421	1.660 254	5.835	133.9	1.529	511.0	0.7373
0.887 754 9	1 100.0	2.172 418	1.928 575	1.712 102	5.800	140.8	1.520	537.2	0.7424
0.893 383 5	1 150.0	2.225 710	1.988 412	1.776 415	5.761	149.5	1.510	570.2	0.7487
0.898 590 2	1 200.0	2.279 001	2.047 889	1.840 213	5.727	158.2	1.501	603.4	0.7548
0.903 417 2	1 250.0	2.332 293	2.107 034	1.903 530	5.695	166.9	1.493	636.8	0.7606
0.907 901 1	1 300.0	2.385 585	2.165 875	1.966 400	5.668	175.7	1.486	670.4	0.7662
0.912 074 3	1 350.0	2.438 877	2.224 437	2.028 852	5.644	184.6	1.479	704.1	0.7716
0.915 965 5	1 400.0	2.492 168	2.282 740	2.090 911	5.622	193.4	1.474	738.0	0.7769
0.922 999 9	1 500.0	2.598 752	2.398 648	2.213 952	5.587	211.3	1.464	806.1	0.7870
0.929 175 1	1 600.0	2.705 335	2.513 730	2.335 696	5.560	229.2	1.457	874.5	0.7966
0.934 627 2	1 700.0	2.811 918	2.628 096	2.456 290	5.539	247.3	1.452	943.3	0.8056
0.939 466 2	1 800.0	2.918 502	2.741 834	2.575 861	5.523	265.3	1.448	1 012	0.8142
0.947 647 5	2 000.0	3.131 669	2.967 718	2.812 351	5.505	301.6	1.443	1 151	0.8301
0.954 263 0	2 200.0	3.344 836	3.191 853	3.045 867	5.497	338.0	1.441	1 289	0.8446
0.959 691 2	2 400.0	3.558 003	3.414 584	3.276 946	5.499	374.4	1.441	1 428	0.8577
0.964 201 7	2 600.0	3.771 170	3.636 168	3.506 000	5.505	410.7	1.443	1 567	0.8697
0.967 991 5	2 800.0	3.984 337	3.856 804	3.733 354	5.516	447.0	1.446	1 705	0.8807
0.971 207 1	3 000.0	4.197 504	4.076 645	3.959 267	5.530	483.2	1.450	1 843	0.8907
0.974 400 1	3 500.0	4.730 421	4.623 514	4.519 023	5.572	573.3	1.461	2 187	0.9127
0.981 785 4	4 000.0	5.263 338	5.167 468	5.073 345	5.620	662.7	1.473	2 528	0.9310
0.985 005 1	4 500.0	5.796 255	5.709 341	5.623 730	5.669	751.2	1.486	2 866	0.9464
0.987 439 4	5 000.0	6.329 173	6.249 674	6.171 174	5.718	839.1	1.499	3 201	0.9596
0.990 258 7	5 800.0	7.181 840	7.111 879	7.042 060	5.793	978.0	1.519	3 731	0.9769
0.992 224 2	6 600.0	8.034 508	7.972 033	7.910 044	5.864	1115	1.537	4 255	0.9909
0.993 649 3	7 400.0	8.887 175	8.830 735	8.774 654	5.930	1251	1.554	4 772	1.002
0.994 715 4	8 200.0	9.739 843	9.688 372	9.637 172	5.991	1385	1.570	5 284	1.012
0.995 533 7	9 000.0	10.592 51	10.545 20	10.498 10	6.048	1518	1.585	5 791	1.020
0.996 314 5	10 000	11.658 35	11.615 38	11.572 57	6.113	1682	1.602	6 419	1.029
0.997 367 2	12 000	13.790 01	13.753 71	13.717 50	6.229	2007	1.633	7 655	1.041
0.998 025 7	14 000	15.921 68	15.890 25	15.858 88	6.328	2325	1.659	8 870	1.051
0.998 464 7	16 000	18.053 35	18.025 64	17.997 96	6.414	2639	1.681	10 070	1.058
0.998 772 0	18 000	20.185 02	20.160 24	20.135 48	6.490	2949	1.701	11 250	1.063
0.998 995 5	20 000	22.316 69	22.294 27	22.271 88	6.558	3255	1.719	12 420	1.068
0.999 163 1	22 000	24.448 36	24.427 90	24.407 46	6.620	3559	1.735	13 580	1.071
0.999 292 0	24 000	26.580 03	26.561 21	26.542 41	6.676	3860	1.750	14 720	1.074
0.999 393 3	26 000	28.711 70	28.694 28	28.676 87	6.727	4158	1.763	15 860	1.076
0.999 474 3	28 000	30.843 37	30.827 15	30.810 94	6.775	4454	1.776	16 990	1.078

This table, based on data kindly supplied by Dr. W. H. BARKAS, applies to heavy particles (i.e., those massive compared to electrons): The quantities in columns 1, 3, 4 and 5 are clearly functions of the velocity alone. The numbers tabulated in the remaining columns can be applied directly to protons. However, they are really more general. Thus, consider another heavy particle with mass M and charge ze having the same velocity as that of a proton. Then, as shown in Sect. 11, the energy E, range R', and rate of energy loss I of the particle can be reduced to those of the proton by the simple transformations

$$\varepsilon = \frac{E M_p}{M}, \qquad \lambda = \frac{z^2 M_p R'}{M}, \qquad \text{and} \qquad \iota = \frac{I}{z^2},$$

respectively. Thus, ε, λ, and ι (and therefore η), may be considered functions of the velocity alone[1]. Following BARKAS, we have used Greek symbols to denote this velocity dependence.

---

[1] This statement is subject to slight qualification by the weak dependence of ι and λ on the rest mass at very high velocities (discussed in footnote 2, p. 353). The dependence on the charge at very low velocities (cf. Sect. 15) has been taken into account.

Notes to Table 2 (Continued).

It may be noted that the values in column 2 are given to fewer significant figures than warranted by our knowledge of the proton mass. This is done because the value of M_p used in BARKAS' *IBM* calculations was the 1953 value, 1.67243×10^{-24} g, rather than the more recent value 1.67239×10^{-24} g, given by E. R. COHEN, J. W. M. DuMOND, T. W. LAYTON, and J. S. ROLLETT, Rev. Mod. Phys. **27**, 363 (1955). The tabulated ε values form a set consistent with the corresponding numbers in columns 1, 3, 4 and 5. In columns 6 to 10, only as many significant figures have been retained as seem justified by the present state of experiment and theory.

Further remarks on Table 2 appear in the text.

Table 3. *Emulsion range versus energy for protons, τ mesons, π mesons, μ mesons, and Σ^+ hyperons, at density 3.815 g/cm³.*

$r = R/M$ in units of microns/electron mass. Ranges $R < 1$ cm are expressed in microns, longer ranges in cm. All energies are given in Mev.

r	Protons		τ mesons		π mesons		μ mesons		Σ^+ hyperons	
(μ/m_e)	E_p	R_p	E_τ	R_τ	E_π	R_π	E_μ	R_μ	E_{Σ^+}	R_{Σ^+}
0.00773	1.000	14.2 μ	0.5262	7.47 μ	0.1487	2.11 μ	0.1125	1.60 μ	1.268	18.0 μ
0.0219	2.000	40.3	1.052	21.2	0.2974	5.99	0.2250	4.53	2.535	51.1
0.0417	3.000	76.7	1.579	40.4	0.4460	11.4	0.3376	8.63	3.803	97.2
0.06655	4.000	122.2	2.105	64.30	0.5947	18.17	0.4501	13.75	5.070	154.9
0.09607	5.000	176.4	2.631	92.83	0.7434	26.23	0.5626	19.85	6.338	223.6
0.1301	6.000	238.9	3.157	125.7	0.8921	35.52	0.6751	26.88	7.606	302.8
0.1679	7.000	308.2	3.684	162.2	1.041	45.82	0.7876	34.68	8.873	390.7
0.2099	8.000	385.4	4.210	202.8	1.189	57.30	0.9002	43.36	10.14	488.5
0.2560	9.000	470.1	4.736	247.4	1.338	69.89	1.013	52.90	11.41	595.9
0.3060	10.00	561.8	5.262	295.6	1.487	83.53	1.125	63.21	12.68	712.1
0.4504	12.50	827.0	6.578	435.2	1.859	123.0	1.406	93.05	15.85	1048
0.6138	15.00	1127	7.893	593.0	2.230	167.6	1.688	126.8	19.01	1429
0.8006	17.50	1470	9.209	773.5	2.602	218.6	1.969	165.4	22.18	1863
1.008	20.00	1851	10.52	974.0	2.974	275.2	2.250	208.3	25.35	2346
1.490	25.00	2735	13.16	1439	3.717	406.6	2.813	307.7	31.69	3467
2.052	30.00	3767	15.79	1982	4.460	560.1	3.376	423.9	38.03	4775
2.689	35.00	4938	18.42	2598	5.204	734.2	3.938	555.6	44.37	6259
3.403	40.00	6249	21.05	3288	5.947	929.1	4.501	703.1	50.70	7921
4.178	45.00	7672	23.68	4037	6.691	1141	5.063	863.2	57.04	9725
5.022	50.00	9221	26.31	4852	7.434	1371	5.626	1038	63.38	1.169 cm
6.909	60.00	1.269 cm	31.57	6676	8.921	1886	6.751	1427	76.06	1.608
9.049	70.00	1.662	36.84	8743	10.41	2470	7.876	1870	88.73	2.106
11.43	80.00	2.099	42.10	1.104 cm	11.89	3120	9.002	2362	101.4	2.660
14.04	90.00	2.578	47.36	1.357	13.38	3834	10.13	2901	114.1	3.268
16.87	100.0	3.098	52.62	1.630	14.87	4607	11.25	3486	126.8	3.928
24.86	125.0	4.565	65.78	2.402	18.59	6787	14.06	5136	158.5	5.787
34.05	150.0	6.251	78.93	3.289	22.30	9294	16.87	7034	190.1	7.924
44.33	175.0	8.139	92.09	4.283	26.02	1.210 cm	19.69	9158	221.8	10.32
55.57	200.0	10.20	105.2	5.369	29.74	1.517	22.50	1.148 cm	253.5	12.93
80.77	250.0	14.83	131.6	7.804	37.17	2.205	28.13	1.669	316.9	18.80
109.1	300.0	20.03	157.9	10.54	44.60	2.977	33.76	2.253	380.3	25.38
173.3	400.0	31.82	210.5	16.74	59.47	4.730	45.01	3.580	507.0	40.33
245.5	500.0	45.07	263.1	23.72	74.34	6.701	56.26	5.071	633.8	57.13
323.7	600.0	59.43	315.7	31.28	89.21	8.837	67.51	6.687	760.6	75.34
406.6	700.0	74.65	368.4	39.28	104.1	11.10	78.76	8.400	887.3	94.63
493.0	800.0	90.53	421.0	47.64	118.9	13.46	90.02	10.19	1014	114.8
582.3	900.0	106.9	473.6	56.26	133.8	15.90	101.3	12.03	1141	135.5
673.7	1000	123.7	526.2	65.10	148.7	18.39	112.5	13.92	1268	156.8
766.9	1100	140.8	578.8	74.10	163.5	20.94	123.8	15.84	1394	178.5
861.5	1200	158.2	631.5	83.23	178.4	23.52	135.0	17.80	1521	200.5
957.1	1300	175.7	684.1	92.47	193.3	26.13	146.3	19.77	1648	222.8
1054	1400	193.4	736.7	101.8	208.2	28.76	157.5	21.77	1775	245.2
1151	1500	211.3	789.3	111.2	223.0	31.41	168.8	23.77	1901	267.8
1248	1600	229.2	842.0	120.6	237.9	34.08	180.0	25.79	2028	290.6
1347	1700	247.3	894.6	130.1	252.8	36.76	191.3	27.82	2155	313.4
1445	1800	265.3	947.2	139.6	267.6	39.45	202.5	29.86	2282	336.3
1544	1900	283.5	999.8	149.2	282.5	42.15	213.8	31.89	2408	359.3
1643	2000	301.6	1052	158.7	297.4	44.84	225.0	33.94	2535	382.3
1742	2100	319.8	1105	168.3	312.2	47.55	236.3	35.98	2662	405.4
1841	2200	338.0	1158	177.9	327.1	50.25	247.5	38.03	2789	428.4
1940	2300	356.2	1210	187.4	342.0	52.96	258.8	40.08	2915	451.5
2039	2400	374.4	1263	197.0	356.8	55.66	270.0	42.12	3042	474.5
2138	2500	392.5	1316	206.6	371.7	58.36	281.3	44.17	3169	497.6
2237	2600	410.7	1368	216.1	386.6	61.06	292.5	46.21	3296	520.6

Table 3 is based on recent *R-E* calculations of W. H. BARKAS for protons (cf. columns 2 and 7 of Table 2). The following mass values were used: π, 273.0 m_e; μ, 206.6 m_e [S6]; τ, 966.2 m_e; Σ^+ 2327.4 m_e. As in Table 2, each row pertains to a single velocity. Velocity functions other than E or R can be found, for a given row, by looking them up opposite the corresponding E_p or R_p in Table 2. The first column gives the "equivalent electron range", (in microns/electron mass),

$$r = \frac{R_p}{m_p} = 5.44624 \times 10^{-4} R_p$$

where m_p is the proton rest mass expressed in electron masses. The tabulation of r is included for convenient reference to results in [C2].

---

for proton energies ≥ 40 Mev. Using an energy-loss law of type (8.3) for proton energies up to ~800 Mev in G.5 emulsion at a density 3.815 g/cm³, the value of the mean ionization potential w was adjusted to fit the new range observations. This procedure yielded the following formula for the rate of ionization loss of heavy, singly charged particles.

$$\iota = \frac{0.5325}{\beta^2} \left[\ln 3159 \beta^2 \gamma^2 - \beta^2\right] \text{Mev/cm}. \tag{14.2}$$

The constants correspond to an electron density of 1.045×10^{24} electrons/cm³, and a value for w of 323 ev[1,2]. Barkas extended the calculations above ~800 Mev up to 30000 Mev taking into account the density effect, i.e. using a formula of type (8.3a) with constants proposed by STERNHEIMER [S7].

Part II of Table 2 gives in columns 6 and 7, respectively, BARKAS' results on the rate of energy loss ι in Mev/cm, and the range λ in cm expressed as (velocity) functions of the proton energy[3]. Columns 8 and 9 were derived from 6 and 7, respectively, using the density 3.815 g/cm³ for conversion. Column 10 tabulates the "range-energy index" $\eta = \lambda \iota / \varepsilon$, which was computed from columns 2, 6, and 7. η is the value of the exponent in the power-law approximation to the R-E relation in the vicinity of a given energy[4]. Knowledge of η facilitates range interpolation by use of the relation

$$\frac{\lambda^{\eta_1}}{\varepsilon} = \frac{\lambda_1^{\eta_1}}{\varepsilon_1} \tag{14.3}$$

where ε is an energy in the vicinity of a tabulated energy ε_1. While it is not easy to assess the standard errors in the range values given in Part II of Table 2, the following estimates are considered conservative: at $\varepsilon < 700$ Mev, the values are probably good to ~1% or better; at higher energies, within ~2%.

Columns 3 to 5 of Table 2 give values of the following useful velocity functions: γ, the total energy in rest-mass units; $\beta\gamma$, the momentum in units of Mc; and $\beta^2\gamma$, which expresses the product momentum × velocity in rest-mass units, and occurs in calculation of multiple Coulomb scattering.

Table 3 gives the range vs. energy in G.5 emulsion of density 3.815 g/cm³, for various mesons and Σ^+ hyperons. The table was constructed from the R-E data for protons of Table 2, using the following values of the mass ratios: $\tau/p = 0.5262; \pi/p = 0.1487; \mu/p = 0.1125; \Sigma^+/p = 1.2676$. (Further information is given in the Notes to Table 3). In Fig. 11, the same data are plotted, with the ranges expressed in g/cm², instead or microns or cm, for energies up to ~1000 Mev.

---

[1] This differs from the value 332 eV previously used [B5] for the mean ionization potential.

[2] The following adjustment was also required: because the energy loss formula is unsatisfactory below 40 Mev, BARKAS added an arbitrary integration constant to the ranges at higher energies so as to make them fit the empirical data, including the measured range of the muon in π-μ decay (at an equivalent proton energy of 36.55 Mev).

[3] See notes to Table 2.

[4] If we write $\varepsilon = k \lambda^\eta$, then $\iota \equiv \left|\frac{d\varepsilon}{d\lambda}\right| = \eta k \lambda^{\eta-1}$ and $\lambda \iota = \eta \varepsilon$.

More detailed calculations of K- and L-shell corrections by Dr. BARKAS dated March 26, 1957, lead, for the most part, to slightly longer ranges than those in Tables 2 and 3. The main increases occur for ε near 1 Mev (by $\sim$1.4%), and near the interval 50 to 400 Mev (by an average of $\sim$0.7%). At other energies, the change is less than 0.5%.

15. R-E relations for slow electrons and multiply charged particles. Measurements on tracks of slow electrons are difficult because the tracks are so highly bent (see Fig. 13). When the early electron-sensitive emulsion NT2A, precursor of type NT4, was developed by BERRIMAN at the Kodak Research Laboratories in England, the ranges of electrons up to about 80 kev were measured in it by Herz. He found that 80 kev was approximately the limiting electron energy to which these plates were sensitive. ROSS and ZAJAC [R8] carried out similar measurements, and later, when NT4 plates became available, they extended the R-E relation for slow elec-

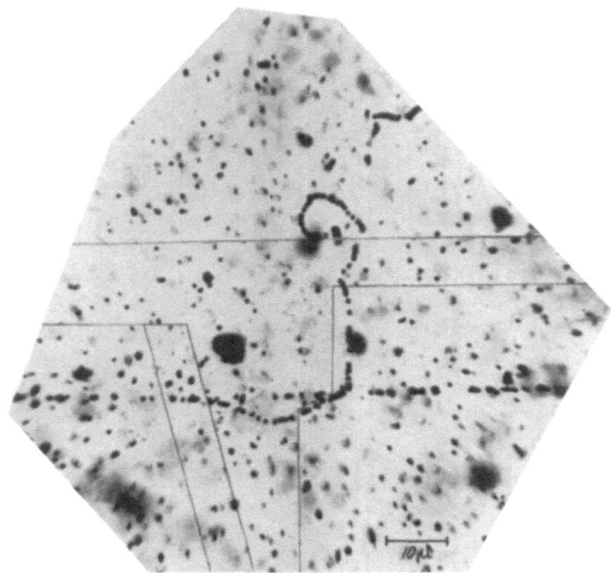

Fig. 13. Collision of a fast, singly charged particle with an electron in Kodak NT4 emulsion. The high energy (horizontal) track shows an ionization loss very close to the minimum. The secondary electron has an energy of about 130 kev. (Courtesy R. H. HERZ[1].)

Energy (kev)	Range (μ)
147	95.4 ± 1.2
200	$141 \ \pm 6$
250	$201 \ \pm 8$

trons up to 250 kev[2]. BLUM [B13a] measured the ranges of electrons at energies below 25 kev, in G.5 emulsion. His results, together with those of ZAJAC and ROSS, up to energies of about 100 kev are shown in Fig. 14. At energies above 100 kev, the following ranges were reported by ZAJAC and ROSS (see Table above).

For particles of charge ze, it was shown in Sect. 11 [cf. Eq. (11.3)] that the quantity Rz^2/M is a function of the velocity alone. If this were exactly true, it would be unnecessary in column 7 of Table 2 to replace the range R by a modified range R'. Actually, the phenomenon of electron pickup by positive ions of low velocity increases the effective charge and extends the range, thereby making Eqs. (11.3) and (11.3a) inexact. Range measurements on nuclei of known energy with $z \geq 2$ have been carried out by FARAGGI[3], MILLAR and CAMERON[4], MILLER[5], and BARKAS[6]. All of this work was done in Ilford emulsion; FARAGGI and BARKAS

[1] A. C. COATES and R. H. HERZ: Phil. Mag. **40**, 1088 (1949).

[2] B. ZAJAC and M. A. S. ROSS: Nature, Lond. **164**, 311 (1949).

[3] H. FARAGGI: C. R. Acad. Sci. Paris **229**, 1223 (1949).

[4] C. H. MILLAR and A. G. W. CAMERON: Phys. Rev. **78**, 78 (1950).

[5] J. F. MILLER: University of California Radiation Laboratory Report UCRL 1902, July, 1952.

[6] W. H. BARKAS: Phys. Rev. **89**, 1019 (1953), see also [B5].

used C.2; MILLAR and CAMERON, and MILLER used E.1; the latter also used D.1 emulsion for a study of fission fragments.

Using data available up to 1950, WILKINS [W4] constructed a graph showing the variation of the quantity $R z^2/M$ with the velocity function E/M for ions with charges 1—9. BARKAS[1], upon analyzing the additional data available up to 1953, concluded that the range of a multiply charged positive ion can be given by an expression of the form

$$R = \frac{M \lambda}{M_p z^2} + 0.12 \frac{M z}{M_p} \text{ microns}. \tag{15.1}$$

Here λ is the proton range at the same velocity, and the first term is the range of the ion which is calculated from the proton range by neglecting the effect of

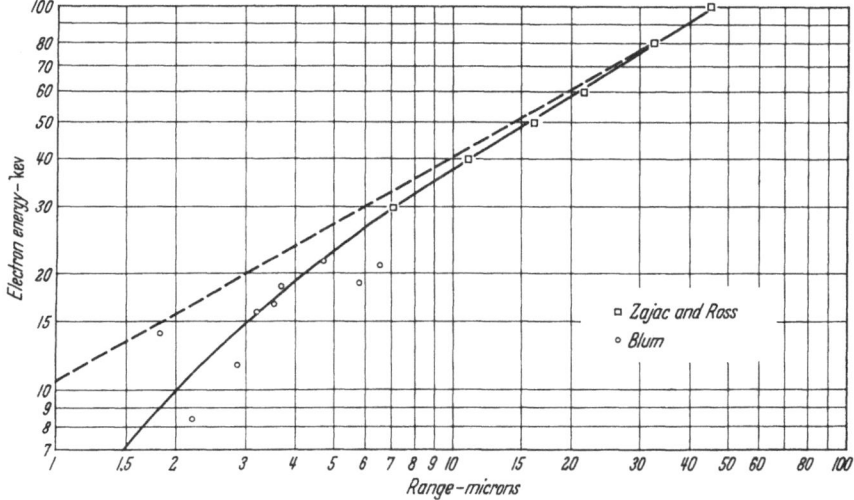

Fig. 14. R-E relation for low-energy electrons in emulsions sensitive to particles at minimum ionization: Ilford G.5 plates, Blum; Kodak NT4 plates, ZAJAC and Ross. Solid curve is that proposed by BLUM [B13a]; dashed curve is based on R-E data for protons [B15]. (After VOYVODIC [V7]).

electron capture. The second term represents the extension of the range due to the loss of effective charge. If we now define R' as the difference $R - 0.12 \frac{M z}{M_p}$ microns, then it will be seen that λ, the velocity-dependent proton range tabulated in column 7 of Table 2 is given by $\frac{z^2}{M} R' M_p$. Although, for a boron nucleus, the range extension is only about 6 μ, the extension for a stripped iron nucleus would exceed 160 μ, if Eq. (15.1) could be extrapolated from carbon up to iron.

II. Multiple Coulomb scattering.

16. Theories of scattering. While penetrating matter, a charged particle suffers a series of elastic collisions with the Coulomb fields of nuclei adjacent to its path[2]. The succession of individual scatterings, each of which ordinarily deflects the particle very slightly, leads to an observable deviation in its path. Along

---

[1] See footnote 6, p. 370.

[2] The theory of multiple Coulomb scattering shows that for a fast particle ($\beta \approx 1$) the frequency of such collisions in nuclear emulsion is ∼3000 collisions per millimeter. This implies an average of ∼1 collision per AgBr grain. For slower particles, the frequency is higher (cf. [M11], [S4], or [G6], Fig. 2).

an emulsion track of reasonable length, many such deviations can be measured, thanks largely to the high concentration of Ag and Br atoms. The statistics so obtained provide valuable data for particle identification, as well as for estimation of energy and related properties.

The differential probability for an individual scattering process has been calculated by Williams[1], who took account of the modifications required in the Rutherford scattering law by two effects: the screening of the Coulomb field of the nucleus by the orbital electrons, and the finite size of the nucleus. Goudsmit and Saunderson[2] treated the same problem using a different, and somewhat more exact mathematical formulation. Molière [M 11], [B 8] gave a rigorous quantum mechanical treatment of the effect of electron screening, and obtained a result valid for all momenta.

Given a cross-section for a single small-scattering process, the next problem is how a number of such events combine to produce the resultant change in direction of the particle. More precisely, it is desired to calculate the frequency distribution of the angular deviations for a given path length, and thence the mean absolute deflection (or root mean square deflection) per unit path. This has been done by the foregoing authors, and by Snyder and Scott[3]. It was found that the small-angle scattering events generally combine to give a small deflection, and that the probability distribution of these *"multiple-scattering"* deflections approximates a Gaussian function. On the other hand, a large bend in a track is typically due to a single large-angle scattering, with only a trivial contribution from small-angle processes. The appreciable probability of the larger deflections results in a significant *"single-scattering* tail" to the frequency distribution, and has led to cut-off procedures in the statistical treatment of scattering measurements.

For application to the measurement of tracks in emulsion, the essential result of multiple scattering theory may be expressed as follows [M 11], [G 6]:

$$\vartheta = 2 e^2 \left[\sum_i N_i Z_i^2 \right]^{\frac{1}{2}} f(v, t) \frac{t^{\frac{1}{2}} z}{p v}, \tag{16.1}$$

$$= C \frac{t^{\frac{1}{2}} z}{p v} \tag{16.1a}$$

where ϑ is the mean absolute projected angular deflection for a penetration distance t of a particle with charge ze, velocity v, and momentum p;

N_i is the number of atoms with atomic number Z_i, per cm³ of emulsion;

$f(v, t)$ is a slowly varying logarithmic function[4, 5].

Thus, insofar as the particle's properties are concerned, ϑ varies directly as the charge, and inversely as the product momentum × velocity. The coefficient C

[1] E. J. Williams: Proc. Roy. Soc. Lond., Ser. A **169**, 531 (1939).

[2] S. Goudsmit and J. L. Saunderson: Phys. Rev. **57**, 24 (1940); **58**, 36 (1940).

[3] H. S. Snyder and W. T. Scott: Phys. Rev. **76**, 220 (1949).

[4] The function $f(v, t)$ can be written in the form $(\ln \vartheta_{max} - \ln \vartheta_{min})^{\frac{1}{2}}$, where ϑ_{max}, the maximum single-scattering angle, is fixed by the finite size of the nucleus, and ϑ_{min} is an effective minimum angle which depends on the screening of the Coulomb field by the orbital electrons. Since this shielding effect, in particular, has been calculated differently in the various formulations of scattering theory, various expressions for $f(v, t)$ appear in the literature. That of Molière (cf. [G 6], Eq. (8)) has been used by many emulsion workers.

[5] $f(v, t)$ has a slight dependence on z which may be neglected here. It also depends, in general, on the particle's spin [see, for example, N. F. Mott, Proc. Roy. Soc. Lond., Ser. A **124**, 425 (1929)]; however, since we are dealing with small deflections, this effect may also be ignored.

depends primarily on the properties of the medium and, insensitively, on the particle velocity and the distance of penetration. The important effect of this distance on ϑ is given by the factor $t^{\frac{1}{2}}$.

17. Methods of scattering measurement: constant cell and constant sagitta. Suppose that a high-energy charged particle has traversed a path in emulsion such that its energy has remained nearly constant throughout. The corresponding section of track can be divided into segments, or "cells", of equal length t, and the change in direction of the tangent to the track from cell to cell can be determined. Alternatively, the angles between successive chords provide a measure of the multiple scattering[1]. What is ordinarily measured is not the space angle, but its projection on the plane of the emulsion surface. On the average, the former is $\sqrt{2}$ times as large as the latter.

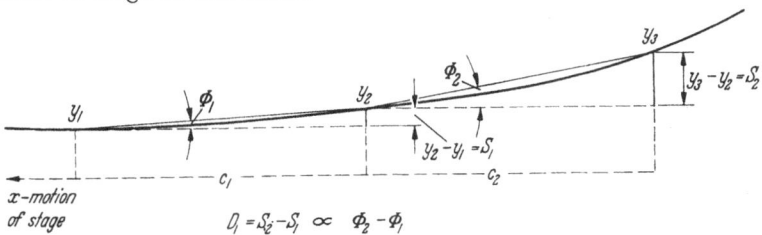

Fig. 15.   Schematic illustration of FOWLER's coordinate method of measuring multiple Coulomb scattering. The dark curve represents a segment of track containing two cells. D_1 is the second difference $(y_3 - y_2) - (y_2 - y_1)$. It is proportional to the angle between the projected chords to the track in cells c_1 and c_2, since the angle is small. In the constant-cell method, successive cells such as c_1, c_2 are of equal length. In the constant-sagitta method, the cells are unequal but so related that, on the average, D is the same for successive pairs of cells.

Two basic methods have been developed for precise determination of mean scattering angles in emulsion. Among the "constant-cell" methods, the angular technique and the coordinate technique have both been employed. For particles arrested by ionization in emulsion, the "constant-sagitta" method has come into widespread use.

In principle, the angular method $[G\,3]$ involves a direct determination, e.g., with a microscope goniometer, of the projected angles between tangents to the track in successive cells. Actually, the track is not a smooth curve, but a set of Ag grains and blobs whose projected centroids are distributed more or less at random about the true trajectory. Hence, in practice, the "tangent" is approximated by a straight-line fit to the grains in a half-cell, and the angle used is the one between alternate half-cells. It has been shown[2] that this "smoothing" gives an angle 0.96 times that between tangents. The angular method is capable of high precision $[G\,3]$, $[L\,5]$, but it is rather laborious.

A major advance was the introduction by FOWLER $[F\,2]$ of the coordinate method, sometimes called the sagitta method. This procedure allows quick determination of the angles between successive chords. The emulsion is oriented on the stage so that the track to be measured is aligned approximately parallel to one of the stage motions, say, the x-direction. Then, with a screw micrometer, the stage is recurrently displaced by a suitable distance (cell length) so that a new position of the track (cell boundary) is centered in the field of view. Two such cells are shown schematically in Fig. 15, in which the track is represented by an idealized smooth curve[3]. At each cell boundary, the y-coordinate of the track

---

[1] S. LATTIMORE: Nature, Lond. **161**, 518 (1948).

[2] Y. GOLDSCHMIDT-CLERMONT, D. T. KING, H. MUIRHEAD and D. M. RITSON: Proc. Phys. Soc. Lond. **61**, 183 (1948).

[3] The vertical scale in Fig. 15 is greatly expanded with respect to the horizontal scale in order to make angles φ_1, φ_2 readily visible.

is read; in practice, the numbers y_i are scale readings in an eyepiece reticule. From these readings, first differences $S_i = y_i - y_{i-1}$ can be computed; each is proportional to the angle φ_i between a chord and the x-direction, and the value of S_i depends, of course, on the arbitrary alignment of the track. The *second differences*, $D_i = S_i - S_{i-1}$, on the other hand, are proportional to the angular deviations $\varphi_i - \varphi_{i-1}$ between adjacent cellular chords, the angles being sufficiently small to make this a very good approximation.

Let α be the arithmetic mean of a set of projected angles, without regard to sign, between the adjacent chords in a section of track. The mean angle ϑ between *tangents* [cf. Eq. (16.1)] is thus replaced by α. It can be shown[1] that $\alpha = \sqrt{2/3}\,\vartheta$. Thus, if we define $K' \equiv \sqrt{2/3}\,C$, Eq. (16.1a) may be replaced by

$$\alpha = \frac{K' z t^{\frac{1}{2}}}{p v} . \tag{17.1}$$

If α is expressed in degrees, then the mean absolute second difference $\overline{D}$ is given (to sufficient approximation) by

$$|\overline{D}| = \frac{\pi}{180}\,\alpha t = \frac{\pi K' z t^{\frac{3}{2}}}{180\, p v} . \tag{17.2}$$

Since the cell length t is a disposable parameter, direct comparison of mean deflections[2] for different tracks requires that α be normalized to some standard cell length. In applying the constant-cell method, the length $100\,\mu$ has been generally adopted as standard. Let s represent the cell length expressed in units of $100\,\mu$. Also, let α_s denote the mean deflection for a cell of length s units. Then we may write

$$\alpha_s = \frac{K z s^{\frac{1}{2}}}{p v} \tag{17.3}$$

and

$$\alpha_1 = \frac{K z}{p v}\,(100\,\mu)^{\frac{1}{2}} = \frac{\alpha_s}{(s/100\,\mu)^{\frac{1}{2}}} . \tag{17.4}$$

If the angles are expressed in degrees and $p v$ in Mev, then the scattering coefficient K is expressed[3] in $\dfrac{\text{Mev deg}}{(100\,\mu)^{\frac{1}{2}}}$.

It is sometimes convenient to write $p v = (M c^2)\beta^2\gamma$, so that

$$\alpha_1 = \frac{K z\,(100\,\mu)^{\frac{1}{2}}}{(M c^2)\,(\beta^2\gamma)} . \tag{17.5}$$

At non-relativistic velocities, the product $p v$ is, of course, just twice the kinetic energy, whereas at extremely high velocities ($\beta \approx 1$), $p v$ is equal to the energy of the particle. At either extreme, if one knows K and z, a measure of α_s gives directly[4] an evaluation of the particle's energy. If the particle's identity

---

[1] See, however, [M 11], [B 7a], and [S 4] for more precise values.

[2] Henceforth, in our discussion of multiple scattering, the word "deflection" or "angle" will refer to the absolute value of the projections of these quantities on the emulsion surface, unless the contrary is specified.

[3] The somewhat anomalous use of the nameless unit $100\,\mu$ can lead to confusion. Because this usage is widespread, the following should be noted: although $K = K'$ [the scattering coefficient of Eq. (17.1)], the latter is expressed in units of $\dfrac{\text{Mev deg}}{\mu^{\frac{1}{2}}}$, so its numerical value in these units is $\dfrac{1}{10}$ that of K. Thus, e.g., when $K = 25\,\dfrac{\text{Mev deg}}{(100\,\mu)^{\frac{1}{2}}}$, $K' = 2.5\,\dfrac{\text{Mev deg}}{\mu^{\frac{1}{2}}}$.

[4] Except for corrections due to "noise" and distortion.

is known, this is true for any velocity. If, however, M is not known, then, for intermediate velocities, α_s alone cannot give the particle energy, but must be combined with measurement of some other track parameter, e.g., ionization or range.

Although K has often been referred to as the "scattering constant", it actually includes the function $f(v, t)$ [cf. Eq. (16.1)] and thus changes slowly with β and s. For $\beta \approx 1$ and $s = 1$, MOLIÈRE's theory leads to a value $K = 25.8$ deg Mev $(100 \,\mu)^{-\frac{1}{2}}$ [$S4$]. However, this *scattering coefficient* K, which is based on an average of *all* deflections including the large ones due to single scattering, is not the coefficient which is of interest in most experiments. In order to take account of the single-scattering tail in the frequency distribution, it is customary to truncate the

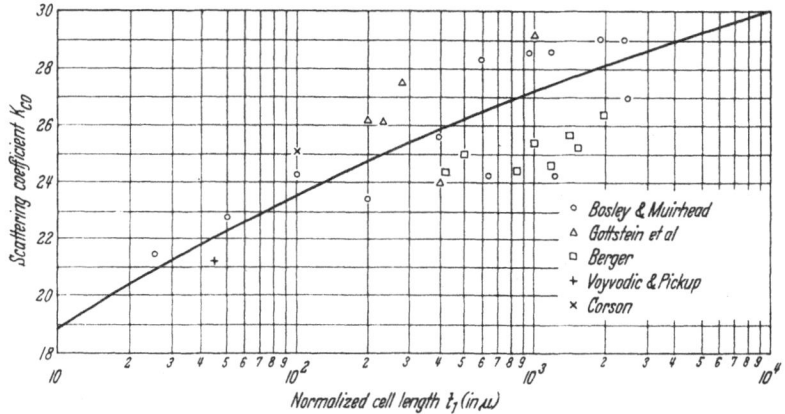

Fig. 16. The scattering coefficient K_{co} (with cut-off at 4 times the mean), expressed in deg Mev $(100\,\mu)^{-\frac{1}{2}}$, is plotted against the normalized cell length t_1 (cf. Eq. (17.7)), after VOYVODIC [$V7$]. Various experimental values of K_{co} are also shown.

experimentally observed distribution at a value 4 times the mean. Therefore, a different scattering coefficient, K_{co}, is appropriate in relating pv to α by this procedure. The theoretical value of K_{co} corresponding to the value of K above, is 23.2 deg Mev $(100 \,\mu)^{-\frac{1}{2}}$.

In terms of $|\overline{D}_s|_{co}$, the mean absolute second difference with cut-off at four times the mean, the following relation, which parallels Eq. (17.4), is applicable:

$$(\alpha_1)_{co} = \frac{K_{co}\, z\, (100\mu)^{\frac{1}{2}}}{p\, v} = \frac{(\alpha_s)_{co}}{(s/100\,\mu)^{\frac{1}{2}}} = \frac{180\,|\overline{D}_s|_{co}\,(100\,\mu)^{\frac{1}{2}}}{\pi\, s^{\frac{3}{2}}}. \qquad (17.6)$$

The coefficient K_{co}, like the corresponding factor without cut-off, is a function—albeit a slowly varying function—of the cell length and velocity. To show the variation of K_{co} as a function of the first of these variables alone, it is useful to define an "equivalent cell length" (p. 274 of [$V7$]) applicable to particles with $\beta = 1$, and related to the true cell length at other velocities by the expression

$$t_1 = \frac{1.30\, t}{\beta^2 + 0.30}. \qquad (17.7)$$

In Fig. 16 the coefficient K_{co} is plotted as a function of the normalized cell length t_1. The experimental values of K_{co} obtained by various observers[1] using particles

[1] W. BOSLEY and H. MUIRHEAD: Phil. Mag. **43**, 63 (1952). — K. GOTTSTEIN *et al.* [$G6$]. — BERGER [$B7a$]. — L. VOYVODIC and E. PICKUP [$V8$]. — D. R. CORSON: Phys. Rev. **80**, 303 (1950); **84**, 605 (1951).

of known energy are shown in this figure. Although there is considerable spread in the data, the ensemble of results tends to confirm the Williams-Molière theory.

As we have seen, the constant-cell scattering method is applicable when the particle is sufficiently fast so that the relative decrement in its energy from beginning to end of the available track is not too large. For slow particles, and especially those arrested in the emulsion by ionization loss, this criterion is not satisfied, for the magnitude of scattering changes rapidly with range. The simplest technique available for coping with this problem is the "constant sagitta" method, devised for identification and mass measurement of slow particles [B9], [D9], [H10][1, 2]. In this method the length of scattering cell t is varied as a function of the range R according to a predetermined scheme $t(R_i)$, so as to keep the mean absolute second difference $|\bar{D}|$ constant along the track.

To determine a "cell scheme" which conforms to this requirement, we first make several simplifying assumptions: (a) the range-energy relation is describable over the entire residual range by a single power law; (b) the particle is non-relativistic so that the classical approximation $pv = 2E$ suffices; and (c) the scattering coefficient K' [cf. Eq. (17.1)] remains constant. Although these assumptions are oversimplifications, it is convenient to construct cell schemes based upon them, and later to apply the required corrections (as given, e.g., by DILWORTH et al. [D9]) to the observed mean second difference.

If we now replace pv by $2E$, Eq. (17.2) can be rewritten:

$$t = \left(\frac{360}{\pi}\frac{E}{K'z}|\bar{D}|\right)^{\frac{2}{3}}. \tag{17.8}$$

To deduce $t\,(R, M, z)$, we employ the power law approximation to the R-E relation. For protons, $\varepsilon = A\,\lambda^\eta$ (ε = energy, λ = range), and, for other particles[3],

$$E = A\left(\frac{M}{M_p}\right)^{1-\eta}z^{2\eta}R^\eta. \tag{17.9}$$

Combining the last two equations,

$$\begin{aligned}t &= \left[\frac{360}{\pi}\frac{A|\bar{D}|}{K'}\left(\frac{M}{M_p}\right)^{1-\eta}z^{2\eta-1}\right]^{\frac{2}{3}}R^{\frac{2\eta}{3}}\\ &= QR^{\frac{2\eta}{3}}\end{aligned} \right\} \tag{17.10}$$

where Q is the coefficient preceding the range factor in Eq. (17.10). Since K' is assumed constant, the condition that $|\bar{D}|$ be fixed along the track of a given type of particle is that the product $tR^{-\frac{2\eta}{3}}$ be constant. If the same scheme $t(R)$ is now applied to the track of another particle, with different M and/or z, then $|\bar{D}|$ will still be constant along the track; it will, however, have a new value.

Although in a given cell scheme Q must be fixed, various schemes can be constructed, each with a single value of Q. Thus one can, in effect, choose a value of $|\bar{D}| = D_a$ of magnitude convenient for a known type of particle, and then, using the

---

[1] Reported at the Bagnères Conference, 1953, by C. DILWORTH, S. GOLDSACK, L. HIRSCHBERG; S. BISWAS, E. C. GEORGE, D. LAL, YASH-PAL, and B. PETERS [B1], p. 136.

[2] Before the development of this method, MENON and ROCHAT [M5] had devised a procedure for evaluating the scattering coefficient of slow particles. As a means of measuring mass, their method is rather more complicated than the constant sagitta technique.

[3] Since $E = \left(\frac{M}{M_p}\right)\varepsilon$, and $\lambda = \frac{M_p}{M}z^2R'$ (cf. Notes for Table 2), Eq. (17.9) follows. The difference between R' and the actual range R is small for particles of low z (cf. Sect. 15). The effect of replacing R' by R in Eq. (17.9) is negligible compared to the errors in measuring D.

latter as a standard, employ the same cell scheme for identification or mass determination of some other particle. It has been shown by BISWAS *et al.* [*B9*] that the error introduced by using the same cell scheme for two particles which differ in mass by a factor of 6 is less than 3 %, an error which ordinarily would contribute negligibly to the overall uncertainty in the measurement. Therefore, reasonable results may be expected even when measurements are made according to a scheme with arbitrary Q.

Suppose that a particular cell scheme of this type has been applied to scattering measurements on the tracks of two stopping particles (or groups of particles) with masses M_1, M_2 and charges z_1, z_2, respectively, and that mean values $|\overline{D}_1|$ and $|\overline{D}_2|$ were obtained. Then, inspection of Eq. (17.10) shows that

$$\frac{|\overline{D}_1|}{|\overline{D}_2|} = \left(\frac{M_2}{M_1}\right)^{1-\eta} \left(\frac{z_2}{z_1}\right)^{2\eta-1}. \tag{17.11}$$

Thus, if the ratio of charges is known, the ratio of masses can be deduced from the scattering measurements. A common situation in high energy physics is that both particles are singly charged. Then experimental determination of the two mean second differences yields the mass ratio. Thus, one might obtain the ratio of a hyperon mass to the proton mass, or of a heavy meson to the π meson.

As can be seen in column 10 of Table 2, the range-energy index varies with residual range. A value of η which gives a good fit to the R-E relation over a wide range of energies [e.g., that in Eq. (14.1)] may not be the best choice for construction of a constant-sagitta cell scheme, since even in a long track, most of the cells occur at the shorter ranges, corresponding to, say $\beta < 0.3$. Hence a reasonable choice of η appears to be ~ 0.58 (cf. [*D9*]). (The corresponding empirical value of the coefficient A is 0.25.) With this choice of η, the basic condition for the constant sagitta cells becomes

$$t \propto R^{0.387} \tag{17.10a}$$

and Eq. (17.11) becomes

$$\frac{|\overline{D}_1|}{|\overline{D}_2|} = \left(\frac{M_2}{M_1}\right)^{0.42} \left(\frac{z_2}{z_1}\right)^{0.16}. \tag{17.11a}$$

Thus, $M \propto |\overline{D}|^{-2.38}$, and the sensitivity of the mtehod for mass determination is accordingly limited. However, when scattering measurements are required for tracks of slow particles, the constant sagitta technique is the most convenient: $|\overline{D}|$ is easily calculated, a single cut-off can be applied to all the data, and the "noise" can be readily computed.

Various useful cell schemes, tabulating lengths t_i of successive cells at residual ranges R_i have been given by several authors [*B9*], [*D9*], [*F1a*], [*G1a*].

18. Mass determination from combinations of various parameters. Although an "elementary particle" has many properties, including spin, magnetic moment, statistics, etc., two characteristics—the rest mass and charge, including the sign of the charge—ordinarily suffice to identify it uniquely[1]. The nature of the track produced by the particle is determined, for most practical purposes, by its mass, velocity, and charge. In principle, therefore, three independent quantities, such as ionization, range, and scattering, must be measured in order to identify the particle. Actually, the charge is often known (e.g., $z = 1$ when the ionization is less than 4 times the minimum), so that it frequently suffices to measure two parameters.

---

[1] This, at least, was true until the discovery of K mesons, which pose a special problem: the particles undergo various modes of decay but their masses appear indistinguishable. What is more, many of them have the same lifetimes. It is not certain at this writing which of the K-meson disintegrations are simply alternative modes of decay of the same particle.

For charged particles which come to rest by ionization loss in nuclear emulsion, two types of methods of mass estimation have been employed. One is based on the variation of multiple Coulomb scattering with residual range (S-R), the other on ionization versus range (I-R).

We have seen in the last section how scattering is combined with range in the constant-sagitta technique. S-R methods have the advantage that they do not depend on the average intensity or the gradient of development. How-

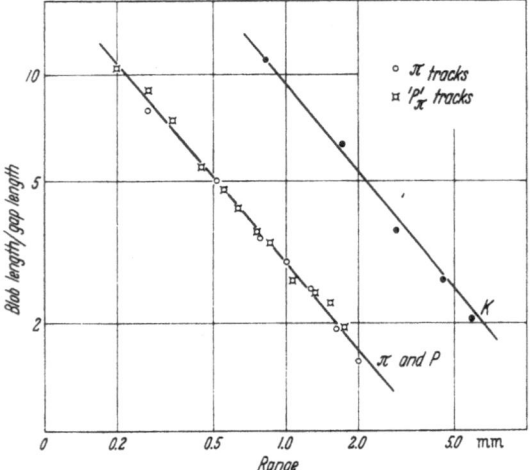

ever, their inherent precision is low. Under favorable conditions, methods which combine *ionization* with range or with scattering are capable of higher precision in mass determination [1].

As outlined in Sect. 7 and further described in Sects. 37—41, there are various measurable track characteristics $f(I)$ which depend on the rate of ionization loss I (e.g., grain density or mean gap length). From Eq. (11.1), we can write $I = z^2 f_1(\beta)$, and from Eq. (11.3), $R z^2/M = F_2(\beta)$. Thus, for a given charge, the velocity can be deduced from an ionization measurement alone (except at extremely relativistic energies, where the Fermi plateau precludes this). Moreover,

$$\frac{R}{M} = f(I), \qquad (18.1)$$

Fig. 17. Application of an ionization-range method to mass estimation of K mesons. Proton data, labeled P_π, were combined with pion data by reducing the proton ranges by the known π/p mass ratio. The ionization parameter employed was the ratio of blob-length to gap-length per unit path. At any given value of this parameter, the experimental ratio of K to π masses is given by the ratio of the respective residual ranges [2].

where $f(I)$ denotes some measurable track parameter which depends on I. To determine the rest-mass M_1 of a particle which ends its path in emulsion, $f(I)$ is measured as a function of its residual range R_1. This is also done for a particle of known mass M_2 as a function of its range R_2. Then, for any particular value of $f(I)$,

$$\frac{M_1}{M_2} = \left[\frac{R_1}{R_2}\right]_I. \qquad (18.2)$$

This method is widely used, and yields good results in the absence of systematic errors due to differences in the intensity of development. Fig. 17 illustrates an application of the mass-range ratio method to measurement of the mass of a group of K mesons by comparison with those of protons and pions. The particular parameter employed in these measurements, the ratio of blob length to gap length per unit path, is especially useful when only short tracks are available.

Other applications of I-R methods have involved grain density vs. range [3], mean gap length vs. range [4], gap counts vs. range [$K2$], and photometric density vs. range [5]. Fig. 18 shows a comparison of the K meson and μ meson masses by the first of these methods.

[1] P. H. FOWLER and D. H. PERKINS: [$M10$], p. 340.

[2] F. W. O'DELL and M. M. SHAPIRO: U. S. Naval Research Laboratory Quarterly on Nuclear Science and Technology, April 1, 1955. — Phys. Rev. **99**, 641 (A) (1955).

[3] E.g., [$H7$], or C. DAHANAYAKE, P. E. FRANCOIS, Y. FUJIMOTO, P. IREDALE, C. J. WADDINGTON and M. YASIN: Nuovo Cim. **1**, 888 (1955).

[4] M. G. K. MENON and C. O'CEALLAIGH: Proc. Roy. Soc. Lond., Ser. A **211**, 292 (1954).

[5] See Appendix for references.

It frequently happens, especially in high-energy investigations, that a particle does not end its path in the emulsion by ionization loss, so that its residual range is not directly measurable. If its energy is not changing too rapidly, then the usual method of estimating the particle's mass is by measuring its mean ionization loss and scattering $(I\text{-}S)$. One can then use Eq. (17.5), which we rewrite:

$$M c^2 = \frac{K z (100\,\mu)^{\frac{1}{2}}}{\alpha_1 (\beta^2 \gamma)}. \qquad (18.3)$$

The rest mass $M c^2$ is in Mev provided that α_1 is expressed in degrees and K in $\frac{\text{Mev deg}}{(100\,\mu)^{\frac{1}{2}}}$. The velocity function $\beta^2 \gamma$ (numerically equal to the number of rest masses in the product $p v$) is evaluated from the rate of ionization loss for a par-

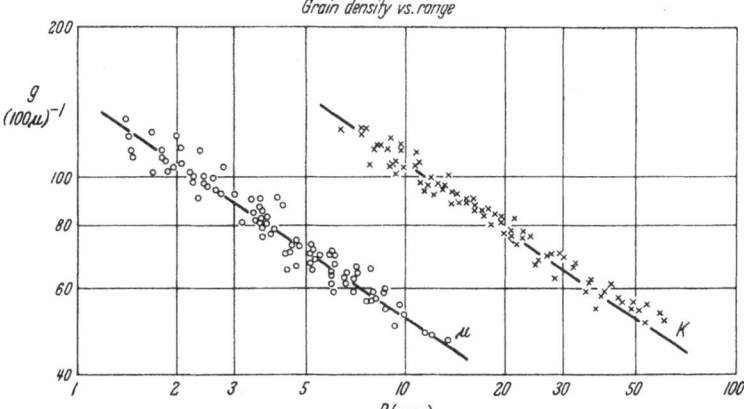

Fig. 18. Grain density *versus* residual range of K meson and μ meson tracks observed in G.5 emulsions exposed to the cosmic radiation. The convention was employed that the number of grains in a blob is proportional to its length. For a given grain density, the estimated ratio of the K and μ masses is equal to the ratio of the ranges[1].

ticle of known charge. This measurement, together with that of the mean scattering angle α_1, gives the mass. In order to determine $\beta^2 \gamma$ from the measurement of some ionization parameter, a calibration curve[2] is required which gives the variation of that parameter $f(I)$ with velocity (or with some function of the velocity, such as $\gamma - 1$) for a known type of particle. For best results the calibration should be carried out in the same batch of emulsion.

Many investigations have been carried out with the $I\text{-}S$ method. Among these are inquiries into the nature of the "shower" particles generated by high energy cosmic-ray collisions[3] and, in particular, the identification and mass estimation of K mesons and hyperons associated with these events[4].

[1] N. Seeman, M. M. Shapiro and B. Stiller: Proc. Fifth Ann. Rochester Conf. on High Energy Physics, Jan. 1955, New York: Interscience Publishers, p. 113ff.

[2] Examples of excellent calibration curves for grain- and blob-density are those of G. Alexander and R. H. W. Johnston: Nuovo Cim. 2, 363 (1957).

[3] Fowler [F 2] applied this method systematically, as did others [C 1] of the Bristol school, e.g., R. R. Daniel, J. H. Davies, J. H. Mulvey, and D. H. Perkins: Phil. Mag. 43, 753 (1952). See also the work of Danysz and Yekutieli reported in the same paper, and of Voyvodic ([V 7], Fig. 10).

[4] A number of experiments in which the primary unstable particles or their secondaries, or both, were identified by $I\text{-}S$ methods are described in the Proceedings of the Padua Conference of 1954 [P 1]; see especially the communications of Baldo et al., Bøggild et al., Fowler and Perkins, Bonetti et al, Daniel et al., Rosendorf et al., and Hoang et al. For further examples, see D. T. King, N. Seeman and M. M. Shapiro: Phys. Rev. 92, 838 (1953); and M. F. Kaplon, J. Klarmann and G. Yekutieli: Phys. Rev. 99, 1528 (1955). The magnitude of the relativistic rise in grain density was important in the work of Kaplon et al. This was also determined by the $I\text{-}S$ method; cf. [S 10].

For a particle which does not come to rest in the emulsion, another method of mass estimation is available, provided that there is an appreciable change in its rate of ionization loss along the track. ROSENDORF and YEKUTIELI ([P1], p. 416) have given an analytical procedure for utilizing the change in grain density when the latter is at least ∼1.3 times the value at plateau.

III. Magnetic methods.

19. Magnetic deflection. The attempt to measure particle momenta by magnetically bending their tracks in emulsion[1,2] encounters several difficulties: in order to get a useful measurement of the radius of curvature, a strong magnetic field, or long tracks, or both are required. In addition, other causes of deflection, such as multiple Coulomb scattering, may mask the magnetic deviation. For these reasons, it was arranged in the early magnetic experiments that the bending should take place principally in a layer of air sandwiched between two layers of emulsion, hence the name "sandwich method". This technique was applied by BARBOUR [B2], FRANZINETTI [F3a], and GOLDSCHMIDT-CLERMONT and MERLIN[3]. Using a magnetic field of 34000 gauss, DILWORTH et al. [D10] investigated the possibilities of bending the tracks *inside* an emulsion. With this field strength, and reasonably long tracks, they were able to determine the sign of the charge on the particles. Momentum measurements require more intense fields.

It is instructive to compare the magnitude of the systematic deflection due to a magnetic field with that of the random deflection due to multiple scattering. For a field H normal to the plane of the emulsion surface, the magnetic deflection in a projected track of length t is given by

$$\vartheta_m = C_m \frac{H z t}{p} \tag{19.1}$$

where C_m is a constant[4]. Dividing this by the mean scattering angle given in Eq. (16.1a)[5], we find the ratio of magnetic to scattering deflections:

$$\frac{\vartheta_m}{\vartheta} = \frac{C_m}{C} H v t^{\frac{1}{2}}. \tag{19.2}$$

It will be recalled that C changes very slowly with v and t. Thus, the relative importance of magnetic bending grows with increase in field intensity, particle velocity, and length of track.

MOYAL[6] has discussed the motion of multiply scattered particles in a uniform magnetic field, and deduced maximum likelihood estimates of the magnetic and scattering deviations. He found that for every value of H, there is a critical value of the momentum below which the scattering determination of p is more accurate than the magnetic one. DILWORTH et al. [D10] have published curves which show the length of track required to make ϑ_m as large as ϑ for a relativistic particle. At 30000 gauss, this length is about 5 mm. With fields more intense

---

[1] C. F. POWELL and S. ROSENBLUM: Nature, Lond. **161**, 473 (1948).

[2] W. J. BATES and G. P. S. OCCHIALINI: Nature, Lond. **161**, 473 (1948).

[3] Y. GOLDSCHMIDT-CLERMONT and M. MERLIN: Nuovo Cim. **7**, 220 (1950).

[4] With H in gauss, t in microns, p in Mev/c, and ϑ_m in degrees, the numerical value of C_m is $3 \times 10^{-8} (180/\pi)$.

[5] From the definitions given in Sects. 16 and 17, the factor C in Eq. (19.2) is related to the scattering coefficient K by $C = \sqrt{3/2}\,K$, where K is of the order of $25 \dfrac{\text{Mev deg}}{(100\,\mu)^{\frac{1}{2}}}$.

[6] J. E. MOYAL: Phil. Mag. **41**, 1058 (1950).

by an order of magnitude, it should be possible to increase the magnetic bending relative to scattering deflection by the same factor, so that momentum estimates on 1 cm tracks become feasible. FURTH [F6] has shown how to combine the magnetic and multiple-scattering information contained in a track so as to obtain optimum determination of momentum, charge, or mass. His analysis demonstrates that for highly relativistic particles, in a field of 100000 gauss, the effective track length is enhanced by a factor 2.3 if one includes the scattering information. For the same particles in a considerably stronger field (say $H \gtrsim 300000$ gauss), the omission of scattering data will lower the precision only slightly.

Before the advent of billion-volt accelerators, the application of pulsed magnetic fields to the low-intensity and randomly directed particles of the cosmic radiation would have been of dubious value, since high intensity magnets must be operated with a small duty factor. However, the availability of intense, pulsed beams of high energy particles has stimulated interest in the use of very strong, transient magnetic fields ($\sim$hundreds of kilogauss) for bending particle tracks inside emulsion. In addition, the feasibility of the magnetic method is improved by the development of stripped-emulsion stacks, in which longer tracks are more easily registered.

In preliminary experiments FURTH and WANIEK[1] produced a magnetic field of 160000 gauss pulsed in such a way that it could be synchronized with an accelerator beam. A bank of capacitors discharged 7800 joules into a pair of Helmholz coils, in a time of 1 to 10 milliseconds. They subsequently attained fields up to 650000 gauss, and concluded from their experiments that "almost arbitrarily high fields can be produced in sufficiently hard and well insulated coils, if a correspondingly large and low-inductance source of energy is available".

The importance of the magnetic method in high energy physics is likely to grow, since determinations of multiple scattering suffer from large errors due to spurious scattering at energies of several Bev/nucleon.

C. Techniques.

I. Characteristics of emulsions.

20. Composition and density[2]. Modern nuclear emulsion is about equally divided in volume between silver halide and lower-Z materials (gelatine, glycerine, water). By weight, the AgBr accounts for approximately 80%[3]. The precise ratio of heavy to light elements varies with the moisture content. Since the development of the earliest Ilford concentrated emulsions, the manufacturer has tried to keep the ratio of heavy to light elements as nearly constant as possible[2]. In this way it was hoped that the range-energy relations and multiple Coulomb scattering would be much the same in the different types of emulsion, and reasonably constant with time. Although there has been no intentional change in the composition, there have been progressive changes in the figures

---

[1] H. P. FURTH and R. W. WANIEK: Nuovo Cim. **3**, 1350 (1955).

[2] Since the most recent data on emulsion composition have not yet been published at the time of this writing, some of the information in this section is based on private communications from Mr. C. WALLER of the Ilford Laboratory to the writer and others. Further information was obtained from Ref. [I1] and data on Eastman emulsions were supplied by Dr. JOHN SPENCE.

[3] These figures do not apply to emulsions diluted with additional gelatine, or to those impregnated with various elements. For normal Kodak NT4 emulsion, R. W. BERRIMAN ([M10], p. 272) gives the ratio of silver halide to gelatine as 5.6 by weight, and 1.08 by volume.

on composition provided by the manufacturer, as improved knowledge of the contents became available. Because these changes have led to some confusion in the literature, it should be emphasized that the early data, usually supplied for C.2 emulsion, were based only on knowledge of the manufacturing process. Analytical results were then unavailable.

A complete chemical analysis of nuclear emulsion has not yet been carried out, but the halide concentration of G.5 emulsion has recently been determined with considerable precision; also the gel phase, separated by centrifuging, has been analyzed. Table 4 gives the mean composition of 40 batches of Ilford G.5 emulsion in equilibrium with air at 58% relative humidity and normal room temperature[1]. A measure of the variations from batch to batch is provided by the "limits" quoted in the table; these represent two standard deviations from the mean, and are not experimental errors in measurement.

Table 4. *Composition and related properties of G. 5 emulsion at relative humidity 58%.*

Constituent	Atomic weight	Concentration[2] g/cm^3	N Atoms/cm^3 $\times 10^{-22}$	$n_e = NZ$ Electrons/cm^3 $\times 10^{-22}$	σ_1 barns	$N\sigma_1$ cm^2 $\times 100$	σ_2 barns	$N\sigma_2$ cm^2 $\times 100$
$_{47}$Ag	107.880	1.817 ± 0.029	1.015	47.69	1.395	1.416	1.025	1.040
$_{35}$Br	79.916	1.338 ± 0.020	1.009	35.31	1.144	1.154	0.841	0.848
$_{53}$I	126.92	0.0120 ± 0.0002	0.00570	0.302	1.558	0.0089	1.144	0.0065
$_6$C	12.010	0.277 ± 0.006	1.389	8.34	0.323	0.449	0.237	0.330
$_1$H	1.008	0.0534 ± 0.0012	3.192	3.19	0.0619	0.198	0.0455	0.145
$_8$O	16.000	0.249 ± 0.005	0.937	7.50	0.391	0.367	0.287	0.269
$_7$N	14.008	0.074 ± 0.002	0.318	2.23	0.358	0.114	0.263	0.084
$_{16}$S	32.066	0.0072 ± 0.0002	0.0136	0.217	0.623	0.0084	0.457	0.0062
Mean density Σ (incl. H)		3.828 ± 0.035	7.879	104.79		3.715		2.729
Σ (excl. H)		3.774	4.687	101.60		3.517		2.584

In addition, Table 4 provides other useful information derived from the composition. For example, the electron density is important in calculating the rate of energy loss by ionization, hence the range-energy relation (see Sect. 12). In the last four columns, geometric cross sections σ_1 and σ_2, and macroscopic cross sections $N\sigma_1$ and $N\sigma_2$ are given for each emulsion constituent, and for the following two values of the nuclear radii: $r_1 = 1.4 \times 10^{-13} A^{\frac{1}{3}}$ and $r_2 = 1.2 \times 10^{-13} A^{\frac{1}{3}}$, where A is the mass number of the nucleus. The last two columns are included in view of recent evidence for smaller nuclear radii. The total macroscopic cross section $\Sigma N\sigma_1$, including hydrogen, corresponds to a mean free path of 26.9 cm, or 103.0 g/cm^2 in emulsion. The mean atomic number, $\overline{Z} = \dfrac{\sum\limits_i N_i Z_i}{\sum\limits_i N_i}$, is 13.30, close to that of aluminium; the mean excluding hydrogen is 21.67. It should be remembered that these numbers, and most of the columns in Table 4, refer to G.5 emulsion in equilibrium with a particular R.H., 58%.

Emulsions in the series C.2, E.1 and D.1 differ only slightly in average chemical composition from G.5, i.e. in the ratio of bromine to iodine. At 58% R.H.,

---

[1] Letter from C. Waller to E. E. Gross, dated 24 January 1956. The same composition is given in [11]. In the latter reference, the overall density appears as 3.8378, through a typographical error. The number intended was 3.8278.

[2] The figures in column 3 were provided by C. Waller in January 1956, and are based upon 40 batches of G. 5 emulsion analyzed in 1955. The "limits" are based upon the variations from batch to batch, and represent two standard deviations.

the mean concentrations of these elements are: Br, $1.324 \, g/cm^3$; I, $0.052 \, g/cm^3$. The manufacturer knows of no reason to expect other differences in composition between the C.2 series and the G.5 type.

Table 5 summarizes the relative proportions of heavy and light elements in G.5 emulsion at 58% R.H. by weight and by macroscopic geometric cross section. Figures are given including and excluding hydrogen, respectively.

Table 5. *Relative weights and geometric cross sections of heavy and light elements in G.5 emulsion at 58% R.H.*

	Including hydrogen		Excluding hydrogen	
	Relative weight %	Relative[1] $\sum_i N_i \sigma_i$ %	Relative weight %	Relative $\sum_i N_i \sigma_i$ %
Ag, Br, I	82.7	69.4	83.9	73.3
C, N, O, S	15.9	25.0	16.1	26.6
H	1.4	5.3		

As an indication of the degree of agreement between the more recent tables on G.5 composition provided by Ilford, it may be useful to compare the following figures, given for G.5 at 58% relative humidity, in August, 1955[2] with the information provided in Table 4 above, which is dated January, 1956.

Concentrations in g/cm^3.

Ag	1.831 ± 0.030	H	0.0533 ± 0.0009
Br	1.349 ± 0.020	O	0.249 ± 0.004
I	0.012	N	0.073 ± 0.001
C	0.276 ± 0.005	S	0.007

Density 3.850 ± 0.041

The same communication[3] gives an idea of the maximum variation observed from batch to batch at various relative humidities:

Relative humidity	Density (g/cm³)		
	Lowest batch	Mean	Highest batch
Dry (H_2SO_4)	4.033	4.062	4.113
32%	3.922	3.962	4.002
58%	3.811	3.851	3.892
84%	3.592	3.630	3.671

An appreciable fraction of the apparent spread is believed to be due to experimental error in the measurement of the density. Note that the mean densities at relative humidities other than 58% also differ somewhat from those reported in 1956; cf. Table 4.

The composition changes, of course, with variation in the moisture content which may be induced by prolonged exposure to air at a different relative humidity. This variation affects *all* the concentrations given in Table 4. The mean density of the emulsion, as well as the partial densities of its constituents at a

---

[1] Macroscopic geometric cross section.
[2] Letter from C. WALLER to A. H. ROSENFELD.
[3] See previous footnote.

humidity other than 58%, can be computed if one knows the incremental water content.

Table 6 provides data on the constitution by weight, and the atomic composition of Eastman Kodak Emulsions. The numbers are based on several batches of Nuclear Track Emulsions which were equilibrated at 50% relative humidity before analysis. The "NTB series" includes NTB, NTB2, and NTB3. The figures in this table "should be regarded only as representative for the emulsions from which they were determined". According to the manufacturer, "they are not suitable for quantitative use. Where accurate figures are required, measurements should be made on the specific emulsions being used" [E 1]. It should also be remarked that in improving the adhesion of their emulsions to glass, the Eastman laboratories have modified their composition over that of previous emulsions [E 1][4].

21. Relative humidity, density, hydrogen concentration, and electron density. In its absorption and desorption of moisture, emulsion shows hysteresis. Hence, in order conveniently to reproduce measurements of density as a function of R.H., it is advisable to start from some standard condition of "normal" R.H. (not far from 50%). Table 7 shows the loss or gain in weight and volume of 1 cm³ of G.5 emulsion at 58% R.H. when brought to equilibrium with air at other humidities [I 1]. Equilibrium is considered to be reached when the emulsion weight remains constant with time. For 200 μ emulsions between one and two weeks' time was allowed. To attain the condition R.H. = 0, the emulsion was dried over concentrated sulphuric acid. The densities and hydrogen concentrations were calculated from columns 2 and 3. It should be noted that water apparently does not occupy its normal volume in emulsion, but rather 1 gram of water occupies 0.84 cm³ in emulsion at least over the range of R.H. between 15 and 84%, according to WALLER. OLIVER [O 4] confirms this non-additivity of volumes at R.H. < 50% [5].

The variation of density with relative humidity in Ilford nuclear emulsion, according to several observers, is shown in Fig. 19. The G.5 data labeled "Waller" are those in Table 4. OLIVER's measurements [O 4] were made on C.2 emulsion. The 1949 Ilford brochure does not specify the type of emulsion, but the data in that pamphlet were mostly for C.2. The point labeled "Wilkins" was not given explicitly by that author [W 4] but may be deduced from his Figs. 1 and 2.

Table 6. *Composition of Eastman Kodak nuclear track emulsion* [E 1][1].

	Types NTA and NTB [2]		Type NTC	
	Percentage by weight	Atomic ratio [3]	Percentage by weight	Atomic ratio [3]
Ag	42.3	1.0	31.7	1.0
Br	30.4	0.97	22.8	0.97
I	1.4	0.03	1.0	0.03
C	9.2	1.95	17.0	4.8
H	1.6	4.3	2.9	9.7
O	12.3	1.95	19.8	4.2
N	2.8	0.5	4.8	1.2

---

[1] The figures shown here should be regarded only as representative for the emulsions from which they were determined. They are not suitable for quantitative use. Where accurate figures are required, measurements should be made on the specific emulsions being used [E 1].

[2] Group NTB includes NTB, NTB2, and NTB3.

[3] Relative to the number of Ag atoms.

[4] The information was kindly communicated by Dr. JOHN SPENCE of the Eastman Kodak Laboratory. The abbreviated name "Eastman" is often used to distinguish the emulsions made in the Rochester laboratories from those of Kodak Ltd. in Harrow, England.

[5] As this is being written, W. BARKAS is investigating the matter further. His tentative results indicate that the average change in volume per unit change in weight is 0.94, and he considers it possible that the asymptotic value of this ratio (if one waits long enough for equilibrium to be reached) may be unity (private communication).

Agreement between the various observers is excellent over the range $50\% < \text{R.H.} < 75\%$, but not so good at R.H. $< 50\%$. The apparent differences, if real, would be puzzling[1].

From the data in Table 7, one can deduce the following expression[2,3] for the hydrogen concentration ϱ_H in g/cm³ as a function of the density ϱ in g/cm³:

$$\varrho_H = 0.1684 - 0.03003 \, \varrho. \tag{21.1}$$

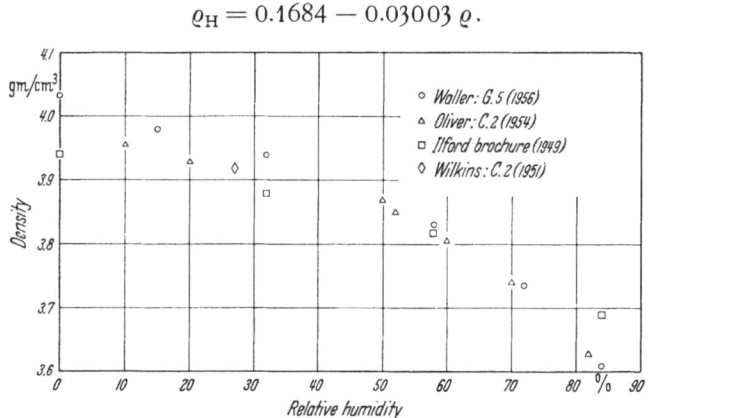

Fig. 19. Density of Ilford G.5 and C.2 emulsions *versus* relative humidity.

In an analogous manner it may be shown that ϱ_0, the partial density of oxygen in the emulsion is related to ϱ by

$$\left. \begin{aligned} \varrho_0 = 1.402 - \\ -0.3013 \, \varrho \text{ g/cm}^3. \end{aligned} \right\} \tag{21.2}$$

To deduce the electron density n_e as a function of ϱ, when the change in density is brought about only by change in moisture content, we first note that $n_e = (n_e)_H + (n_e)_O + (n_e)_r$. The three quantities on

Table 7. *Effects of change in equilibrium humidity on 1 cm³ of G.5 emulsion at 58% R.H.*

R.H. %	Gain in weight mg	Gain in volume cm³ × 10³	Density g/cm³	Hydrogen concentration mg/cm³
0	−85	−72	4.033	47.4
15	−71	−60	3.997	48.4
32	−47	−40	3.939	50.2
58	0	0	3.828	53.4
72	+43	+36	3.736	56.2
84	+108	+91	3.608	59.9

[1] The main difference between types C.2 and G.5 is in grain size, the former having smaller AgBr crystals. However, it is doubtful that the grain size affects the equilibrium water content of emulsion. In fact, C. WALLER finds that the gel phase without any halide at all, behaves similarly to the complete emulsion.

[2] This formula differs from that of BARKAS and YOUNG [B 5] who assumed additivity of volumes, and used older data on the composition of "absolutely dry" C.2 emulsion. The H concentration given in those data is inconsistent with that in the more recent data of Table 4.

[3] Derivation of Eq. (21.1): Starting with 1 cm³ of emulsion for which the density ϱ_1 and hydrogen concentration ϱ_H^1 have been determined, let m_w gm of water be added. Then the new density is

$$\varrho = \frac{\varrho_1 + m_w}{0.843 \, m_w + 1} \frac{\text{g}}{\text{cm}^3},$$

and the new partial density of H is

$$\varrho_H = \frac{\varrho_H^1 + 0.1119 \, m_w}{0.843 \, m_w + 1} \frac{\text{g}}{\text{cm}^3}$$

where 0.1119 is the fraction by weight of hydrogen in water. Elimination of m_w, and substitution of the values of ϱ^1 and ϱ_H^1 at 58% R.H. leads to Eq. (21.1).

the right denote respectively the electron densities in emulsion of hydrogen, of oxygen, and of all the remaining elements combined. Also,

$$(n_e)_H = 0.598 \, \varrho_H \times 10^{24}; \quad (n_e)_O = 0.301 \, \varrho_O \times 10^{24}, \atop (n_e)_r = 0.2670 \, \varrho_r \times 10^{24}, \right\} \quad (21.3)$$

where the coefficients were computed from columns 3 and 5 of Table 4. By definition,

$$\varrho_r = \varrho - (\varrho_H + \varrho_O). \quad (21.4)$$

Combining Eqs. (21.1) to (21.4), the electron density is given by

$$n_e = (0.1037 + 0.2468 \, \varrho) \times 10^{24} \, \frac{\text{electrons}}{\text{cm}^3}. \quad (21.5)$$

The dependence of n_e on ϱ is insensitive to small differences in emulsion composition such as those between batches, or even between different types of emulsion. Thus, Eq. (21.5) is useful, to well within the uncertainties of emulsion composition, for the C.2 series (C.2, E.1, D.1), as well as G.5.

At R.H. 58%, Ilford [I1] reported in March, 1956 the following differences between C.2 and G.5: Type C.2 has 0.014 g/cm³ less bromine and 0.040 g/cm³ more iodine than G.5 (an overall difference in density of $\sim(0.7 \pm 0.4)$ %[1]; earlier estimates of this difference had been as low as ~0.2%). Applying these differences to the composition given in Table 4, one obtains for C.2 emulsion at R.H. 58%, $n_e = 1.0543 \times 10^{24}$ electrons/cm³, and $\varrho = 3.854$ g/cm³. These values satisfy Eq. (21.5)—which was derived for G.5 composition—to better than 0.1%. This close agreement in n_e vs. ϱ has an important consequence for the relative rates of ionization loss and hence for relative ranges, of charged particles in the two types of emulsion. As shown in Sect. 12, these parameters are considerably less sensitive to the mean ionization potential w of the emulsion than to n_e. Moreover, in view of the very slight difference in composition, w differs negligibly between C.2 and G.5. Thus, at a given density, the rates of ionization loss expressed in Mev/g cm⁻² should be indistinguishable in the two series of emulsions, and the same applies to ranges expressed in g cm⁻², within presently attainable accuracies in the determination of these quantities. (This does not, of course, apply to emulsions diluted in gelatine.)

22. Mean size of halide crystals[2] and of silver grains. It is necessary to distinguish between the undeveloped AgBr crystals in the sensitive emulsion, and the considerably larger developed silver grains in the processed emulsion. Electron photomicrographs of the AgBr crystals in the G.5 emulsion show that these are approximately spherical, and that their size distribution is sharply peaked about the mean[3]. Most measurements of the mean crystal diameter in G.5 have yielded values[4] fairly close to 0.3 μ, although a value of 0.17 μ has also been reported[5].

---

[1] A better knowledge of the difference awaits further work on the chemical analysis of C.2-type emulsion.

[2] The term "crystal" is often used loosely in the literature for the AgBr grain in the unprocessed emulsion. It is also employed here for the sake of brevity, and to help distinguish the undeveloped halide grain from the developed silver grain. The term should be interpreted in the context of this article as a small granule consisting of an aggregate of minute AgBr crystals, rather than as a single crystal.

[3] E. PICKUP [P7], found that 78% of the crystals have diameters lying within ~20% of the mean size.

[4] Mean diameters of 0.27 μ, 0.26₅ μ, and 0.31 μ, respectively, have been reported by E. C. DODD and C. WALLER [M10], p. 266, PICKUP [P7], and C. GEGAUFF and J. BOISSIER [J. Phys. Radium **17**, 162 (1956)].

[5] G. BARONI and C. CASTAGNOLI: Nuovo Cim. **7**, 364 (1950).

Indirect evidence for a diameter of $\sim 0.3\,\mu$ in G.5 emulsion comes from estimates (e.g. [F 3]) of $0.2\,\mu$ for the mean path length of a particle in a single crystal— a distance which should be two-thirds of the diameter for spherical crystals.

For Kodak NT4 emulsion, which is similar in sensitivity to G.5, a mean diameter of $\sim 0.4\,\mu$ has been reported ([M 10], p. 272). In addition, measurements of crystal diameters for other types of emulsions[1] have yielded the following mean values: Ilford C.2, $0.16\,\mu$; D.1, $0.12\,\mu$; DEMERS' fine-grained emulsion, $0.08\,\mu$.

During chemical processing, an activated crystal grows into a silver grain of roughly twice the original diameter by "physical development", i.e., by accretion of additional silver from the solution. If the development is very strong, the growth may be by as much as a factor of ~ 3. Thus, in G.5, typical Ag grain diameters are $\sim 0.6\,\mu$, but mean sizes of ~ 0.5 to $\sim 0.8\,\mu$ are not uncommon. In a uniformly developed emulsion, the distribution in final grain size, like that in crystal diameter, is a rather sharp distribution.

A useful relation between mean crystal diameter d, fractional volume of silver halide C, and the highest attainable density of halide grains n_{max} in a given emulsion, has been derived by JDANOV[2]:

$$n_{max} = \frac{150\,C}{d}\ \text{grains}/100\,\mu,\tag{22.1}$$

where d is in microns. With C close to 0.5 for most modern nuclear emulsions, the theoretical value of n_{max} for the electron-sensitive emulsions lies between 200 and 300 grains per $100\,\mu$; for Ilford C.2, ~ 400; and for DEMERS' fine-grained emulsion, ~ 1000.

23. Fading of the latent image. The latent image[3] produced by a charged particle in a nuclear emulsion tends to fade with the passage of time[4, 5, 6] leaving a diminished density in the developed track. This effect is more pronounced for emulsions such as the Ilford C.2 than for electron sensitive ones. The enhanced image stability of the NT4, G.5, and NTB3 emulsions is correlated with their lower threshold of developability [W 3] and with their larger halide grain size ([M 10], p. 290). On the average, a particle traverses a longer path length in a larger crystal and deposits more energy in it.

Since these sensitive emulsions have become available, and balloon flights have tended to replace cosmic-ray exposures of long duration, the problem of latent image regression has become less acute. Nevertheless, it can still be a source of error in ionization measurements[7], particularly when there is an undue delay before processing. Thin tracks tend to fade more rapidly than dense ones; thus, even within a track along which the ionization changes, the effect is more pronounced at the "fast end". The rate of fading increases exponentially with temperature [A 1]; this has been interpreted[8] by considering the effect of temperature on the velocity of a chemical reaction between a gas and solid.

[1] W. KNOWLES and P. DEMERS: Phys. Rev. **72**, 535 (1947) (A).

[2] A. JDANOV: J. Phys. et Radium **6**, 233 (1935).

[3] A discussion of theories of latent image formation is beyond the scope of this review. The reader is referred to the fundamental paper of R. W. GURNEY and N. F. MOTT: Proc. Roy. Soc. Lond., Ser. A **164**, 151 (1938). — W. F. BERG: Rep. Progr. Phys. **11**, 248 (1948) and the more recent theories of J. W. MITCHELL and others [M 10].

[4] M. BLAU: Akad. Wiss. Wien **140**, 623 (1931).

[5] J. LAUDA: Akad. Wiss. Wien **145**, 707 (1937).

[6] The effect is analogous to the regression of the latent image produced by light. Cf. L. WINAND and L. FALLA: Bull. Soc. Sci. Liège **18**, 184 (1949).

[7] K. KRISTIANSSON: Nature, Lond. **173**, 78 (1954).

[8] A. BEISER: Phys. Rev. **81**, 153 (1951).

The presence of an oxidizing agent tends to desensitize an activated crystal[1]. A series of studies[2] has shown that fading is accelerated in an oxygen atmosphere, and retarded in vacuum or in nitrogen. Moreover, the rate of fading rises exponentially with moisture content, so that, owing to the hygroscopic quality of emulsion, the effect depends strongly on the relative humidity. It is also found that fading is more rapid near the surface than in the interior of the emulsion. These observations are well accounted for by the fading mechanism of ALBOUY and FARAGGI ([A 1] and [M 10], p. 290). They suggest that in the presence of moisture, atmospheric oxygen converts silver atoms in the sensitivity centers to silver ions:

$$4\,Ag + O_2 + 2\,H_2O \rightarrow 4\,Ag^+ + 4\,OH^-.$$

According to this hypothesis, we should expect that the p_H of the emulsion affects its rate of fading. Tracks in an alkaline emulsion, with its excess of OH^- ions, should fade more slowly than one with $p_H < 7$. This is indeed true; in fact, before a good explanation for fading was available, it had already been found that borax-impregnated emulsions of the C.2 type have a more stable image than unloaded plates[3,4].

24. Influence of temperature on sensitivity; effects of pressure and contact. Emulsions exposed at temperatures below 0° to 20° C exhibit reduced grain densities in tracks of densely-ionizing particles. This applies to emulsions of moderate sensitivity[5,6] such as Ilford C.2 or Kodak NT2a, as well as to Ilford G.5 [L10]. COSYNS et al.[7] suggested exploiting this effect to reduce the accumulation of background by storing emulsions at low temperatures until they are ready for exposure. WANIEK reports a deviation from the monotonic decrease of grain density with progressive cooling of emulsion below room temperature, i.e., at −40° C he finds a local maximum in sensitivity[8]. To explore the possibility of using G.5 emulsions at low temperatures in very intense magnetic fields, WANIEK exposed pellicles 200 μ thick in a cryostat at temperatures as low as 4° K, to 96 Mev protons. At liquid nitrogen temperatures he observed only a 25% reduction in sensitivity below that at room temperature.

For weakly ionizing particles, the temperature effect is greatly diminished. Thus, in Kodak NT4 emulsions, COATES and HERZ[9] found a negligible loss in grain density of tracks near the minimum of ionization between room temperature and −80° C. LORD [L10] reported a similar result down to −50° C. More recently, RITSON and SCHLUTER[10] have demonstrated the feasibility of recording

[1] W. F. BERG: Ann. Rep. Progr. Chem. **39**, 49 (1942).

[2] J. LAPALME and P. DEMERS: Phys. Rev. **72**, 536 (1947). — H. YAGODA and N. KAPLAN: Phys. Rev. **71**, 910 (1947). — G. ALBOUY and H. FARAGGI [A 1]. — K. B. MATHER: Phys. Rev. **76**, 486 (1949). — W. HÄLG and L. JENNY, Helv. phys. Acta **24**, 508 (1951).

[3] For further discussion of the theory of fading, see P. DEMERS, J. LAPALME and J. THOUVENIN: Canad. J. Phys. **31**, 295 (1953).

[4] An extensive report on fading in G.5 emulsion appeared after the present article was written: cf. G. LEIDE: Ark. Fysik **11**, 329 (1957).

[5] C. DILWORTH: Cosmic Radiation. London: Butterworth Publ. 1949.

[6] E. M. DOLLMANN: Rev. Sci. Instrum. **21**, 118 (1950).

[7] M. COSYNS, C. DILWORTH and G. OCCHIALINI: Note Nr. 6, Centre de Phys. Nucléaire, Bruxelles, January 1949.

[8] R. W. WANIEK: Bull. Amer. Phys. Soc. **1**, 219 (1956). — There is also an indication of a maximum near −40° C in data plotted by COSYNS et al. (cf. preceding footnote).

[9] C. A. COATES and R. H. HERZ: Phil. Mag. **40**, 1088 (1949).

[10] D. M. RITSON and R. A. SCHLUTER: Ann. Prog. Rpt. of MIT Laboratory for Nuclear Science (June, 1955, to May, 1956). These authors report that liquid hydrogen wets the emulsion surface, but does not chemically affect it. They encountered no difficulty in adhesion of the pellicles to the mounting glass.

tracks near the ionization plateau in G.5 emulsion in contact with liquid hydrogen. One-half hour after immersion in the liquid hydrogen, pellicles were exposed to 300 Mev x-rays. The emulsions were kept at 14° K for one hour, and then warmed up to room temperature in two hours. Electron and positron tracks in the developed plates showed a grain density of 20 grains per 100 microns, as compared with 28 grains per 100 microns in a set of control emulsions exposed to the beam at room temperature.

The uniform application of *pressure* does not have a pronounced effect on modern nuclear emulsions [G 4]. In fact, pressures up to several hundred atmospheres can be tolerated. However, the handling often involved in preparing stacks of stripped emulsion, e.g., in printing surface grids for alignment, is apt to leave scratches and other surface darkening which later interfere with visibility. In some experiments it is desirable to expose emulsions under vacuum. As the emulsion dries out, however, it tends to peel off the glass backing, frequently pulling laminae of glass with it. When practicable, the use of emulsions without backing solves this problem. Alternatively, plates can be "plasticized" by immersion in a glycerine solution.

Emulsion enters into chemical reaction with various metals upon contact, and severe fogging results. Prolonged contact with aluminum or iron can spoil an emulsion completely by chemical decomposition. However, at temperatures as low as $-85°$ C, emulsion may be in contact with Al without much blackening[1].

II. Types of emulsion.

25. Selection of emulsion for nuclear research. Nuclear emulsions are commercially available in a wide range of sensitivities. Table 8 lists the principal types produced by the Eastman Kodak Co. (Rochester, N. Y.), Ilford Ltd., and Kodak Ltd. (Harrow, England). Estimates are given of the maximum velocity ($\beta_{max} = v_{max}/c$) of singly-charged particles which can be recorded in each and the highest proton energy ε_{max}. The corresponding limiting energies for other singly-charged particles are proportional to their respective masses. Omitted from the table for lack of quantitative information on β_{max} are the two least sensitive emulsions, designed specifically for recording the tracks of fission fragments in the presence of a background of less highly ionizing radiation. These are Eastman's NTC and Ilford's D.1 plates. Also omitted are Ilford's "diluted emulsions", described below.

Table 8. *Sensitivities of commercially available emulsions*[2].

	β_{max}	ε_{max} (Mev)
Eastman[3]		
NTB3 . .	1	∞
NTB2 . .	0.7	375
NTB . .	0.3	50
NTA . .	0.2	20
Ilford[4]		
G.5 . . .	1	∞
G. Special	0.4	80
G.0 . . .	0.1	5
C.2 . . .	0.3	50
E.1 . . .	0.12	7
Kodak[5]		
NT4 . . .	1	∞
NT2a . .	0.6	200
NT1a . .	0.2	20

The figures quoted in Table 8 provide only a rough indication of the limiting ionization rates which can produce readily discernible tracks. Not only emulsion sensitivity, but also intensity of development and density of background can, of

---

[1] M. DEBEAUVAIS-WACK: Nuovo Cim. **10**, 1590 (1953).

[2] β_{max} = estimated maximum velocity of singly charged particles which can be recorded; ε_{max} = corresponding proton energy.

[3] Eastman Kodak Laboratories, Rochester, N. Y. [E 1].

[4] Ilford Laboratories, Ilford, London [I 1].

[5] Laboratories of Kodak, Ltd., Harrow, England.

course, influence the visibility of tracks, and especially of the thinner ones. The factors affecting track recognition have been investigated by Coates ([M 10], p. 320), Berriman ([M 10], p. 272), and Beiser[1].

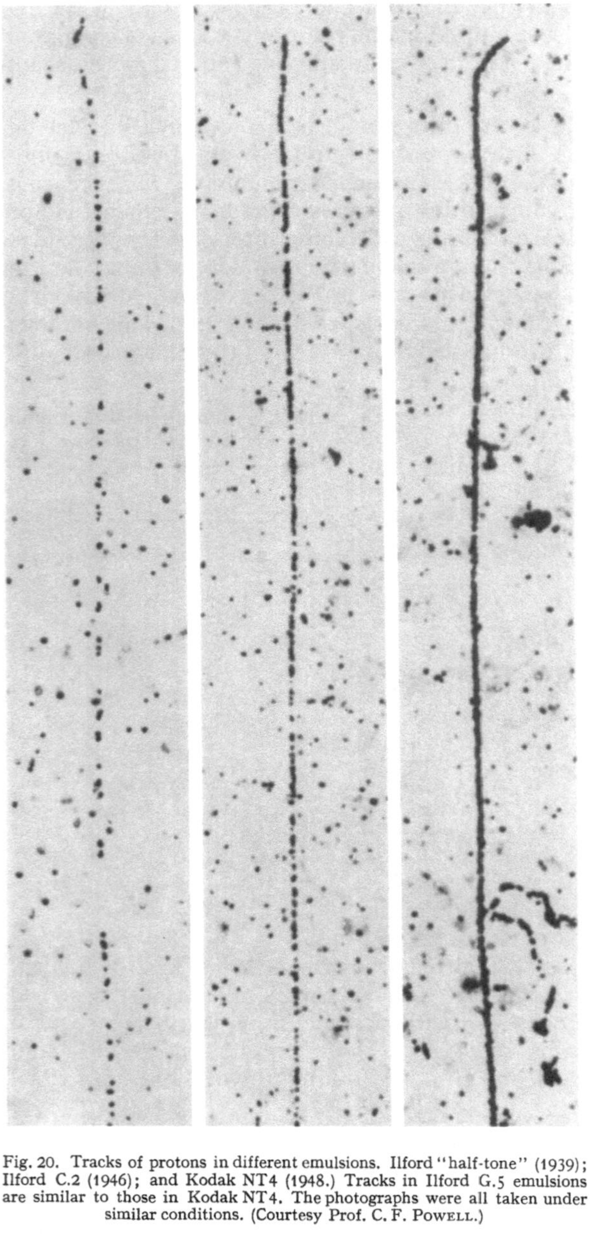

The choice of emulsion for a given application is governed by such considerations as the nature and energy of the particles to be detected, the nature of the background radiation, and how important it is to discriminate between two or more kinds of particles. The more sensitive the emulsion, the more likely it is to accumulate a background of tracks and random grains between the times of manufacture and use. In particular, the electron-sensitive plates, which record virtually all of the ionizing cosmic-ray particles incident upon them, may already have considerable background at the time of use. Therefore, there are good reasons for choosing an emulsion with sensitivity not greatly in excess of that required to detect the particles of interest. Thus, if the most energetic particles to be recorded are 25 Mev protons or 50 Mev deuterons, then it suffices to use C.2 or NTB emulsions; to employ instead one of the ultra-sensitive emulsions is to invite needless background difficulties. As another example, suppose it be desired to record 4 Mev protons in the presence of gamma radiation. E.1, or NTA, or NT1a plates might then be chosen in preference to those of higher sensitivity, since the latter would acquire a higher background due to the electron secondaries generated in the emulsion or glass by the gamma rays. Similarly, although fission fragments

Fig. 20. Tracks of protons in different emulsions. Ilford "half-tone" (1939); Ilford C.2 (1946); and Kodak NT4 (1948.) Tracks in Ilford G.5 emulsions are similar to those in Kodak NT4. The photographs were all taken under similar conditions. (Courtesy Prof. C. F. Powell.)

[1] A. Beiser: Rev. Sci. Instrum. 23, 500 (1952).

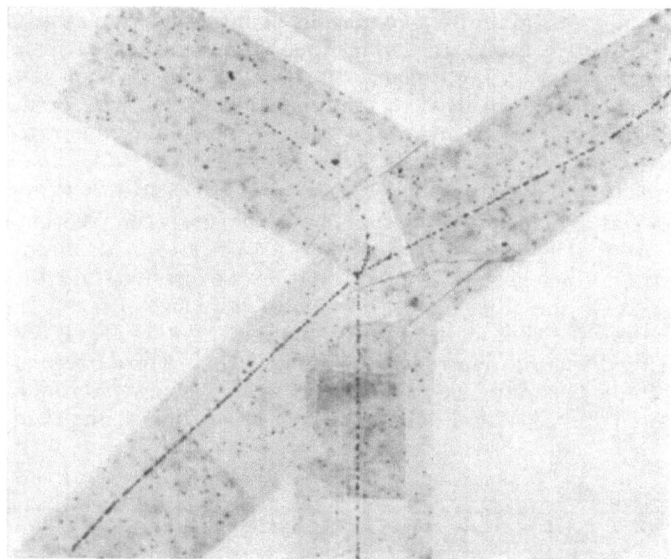

Fig. 21. Photomosaic of a meson-induced star in an Ilford B.1 emulsion. (The latter was similar in sensitivity to the C.2 type.) This picture was probably the first to be published of a star generated by the nuclear capture of a light meson. The tracks may be compared with those of Fig. 22. (After Perkins [1].)

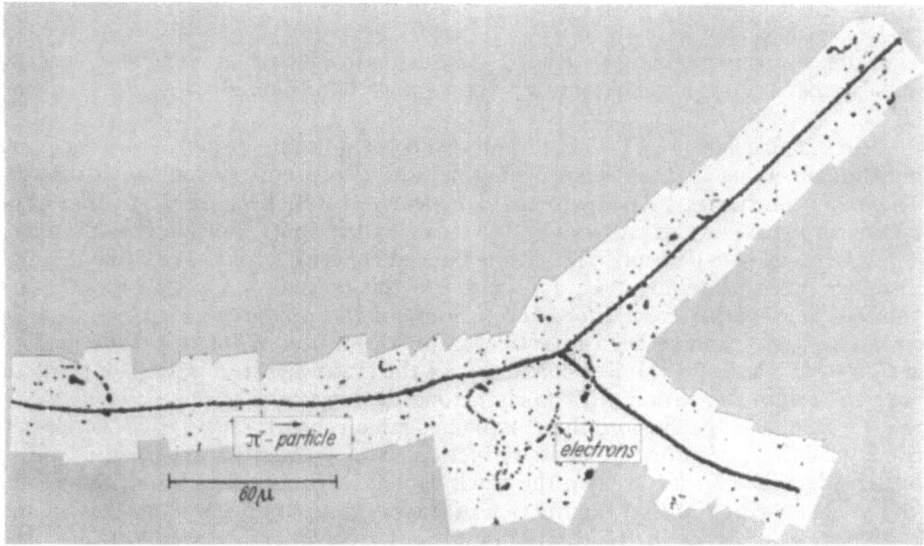

Fig. 22. Star produced by the nuclear capture of a π^- meson in a G.5 emulsion. The tracks of the slow particles emitted from the disintegration, as seen in this electron-sensitive emulsion, may be contrasted with those of a similar event in Fig. 21. (Courtesy Prof. C. F. Powell.)

with their high rate of ionization loss can produce tracks in any nuclear emulsion, it is often advantageous to select the least sensitive ones for this purpose, e.g., G.0, D.1, or NTC.

Until experience has been acquired with the various types of emulsions, the following rules of thumb may be helpful: for typical low-energy applications

[1] D. H. Perkins: Nature, Lond. 159, 126 (1947).

in the nuclear physics laboratory, emulsions of intermediate sensitivity, such as the C.2 or NTB types are worth trying first. For recording electron tracks and for most investigations in high energy physics, as well as exploratory research in general, an emulsion sensitive to minimum ionization is recommended. The different appearance of tracks in various types of emulsion is illustrated in Figs. 20 to 22. In particular, Figs. 21 and 22 show how a "sigma star", i.e., one induced by meson capture, looks in G.5 emulsion, and in a less sensitive type (Ilford B.1).

The G.Special and the G.0 emulsions have the same composition as G.5, and the same mean undeveloped grain diameter, 0.3 μ. They can, however, be more easily processed in thick layers without formation of excessive stain, than emulsions of the C.2 type. On the other hand, the smaller grain size of the C.2, E.1 and D.1 emulsions makes possible better discrimination. C.2 or NTB emulsions are widely used for detecting neutrons by means of their proton recoils. E.1, NTA and NT1a emulsions have been used for investigating photonuclear effects. For densely ionizing particles, such as the heavy primary nuclei of the cosmic

Table 9. *Some properties of Ilford "diluted emulsions"* [1].

. Designation [2]	$\dfrac{\text{Halide vol.}}{\text{Total vol.}}$	Shrinkage factor	Density g/cm^3	n_{min} $(100\,\mu)^{-1}$
"Normal" (G.5)	0.49	2.30	3.9	36
2 × normal. . .	0.35	1.67	3.2	33
4 × normal. . .	0.23	1.34	2.5	21
8 × normal. . .	0.13	1.17	2.0	10

radiation, or the fragments with $z > 2$ which are ejected in some high energy nuclear disintegrations, unsaturated tracks may be obtained with G.0 emulsion [H5].

The ultrasensitive NT4, G.5, and NTB3 emulsions, which have been of inestimable value in high energy physics, are also useful in radioautography [3]. They are contributing uniquely to biological research because of their ability to record the tracks of β particles and to localize their point of origin in a specimen.

A series of G-type emulsions with enhanced concentrations of gelatine relative to silver halide (roughly 2×, 4×, and 8× the normal G.5 ratio of ~1:1 by volume) is available from Ilford [I1]. Some of the properties of these *"diluted emulsions"* which have been investigated by DODD and WALLER([M10], p. 266) are listed in Table 9. The last column gives the grain densities obtained in tracks near minimum ionization with fairly strong development of the emulsions. In the "2×normal" emulsion there is some enhancement rather than a loss in track visibility, despite the slight reduction in n_{min}, because the random grain background falls off faster than the track density with increasing dilution. DODD and WALLER also find that the fraction of traversed AgBr grains which is rendered developable is higher in the diluted emulsions than in the normal G.5, i.e., the apparent sensitivity of individual halide grains is higher.

Diluted emulsions are used mainly for experiments in which an enhanced concentration of the lighter elements (e.g., hydrogen) is desired. However, even when the increase in concentration is appreciable, as in the 8×normal

---

[1] After DODD and WALLER ([M10], p. 266).

[2] The numbers 2, 4, 8 are the approximate values of the ratio of gelatine volume to halide volume in the respective emulsions, relative to the ratio in "normal G.5".

[3] W. P. NORRIS and L. A. WOODRUFF: The Fundamentals of Radioautography. Annual Rev. Nucl. Sci. **5**, 297 (1955).

emulsion, it is sometimes insufficient to compensate for the reduced scanning efficiency [R7]. Another consequence of low AgBr content is a reduction in multiple Coulomb scattering (the coefficient K is smaller). Thus, diluted emulsions may prove advantageous for magnetic bending of particle trajectories, especially at lower energies when scattering tends to mask magnetic deflections.

Sensitive emulsion is sometimes required for low-intensity applications in which the usual levels of accumulated background, due mainly to cosmic rays, cannot be tolerated. For such experiments Ilford emulsions are available in the form of gel which may be melted (at $\sim 50°$ C) and poured by the individual user. GEORGE and EVANS[1] have prepared layers of G.5 from gel in underground tunnels in order to study the stars produced by the penetrating μ mesons of the cosmic radiation. YAGODA has poured "castings" of emulsion, up to 3 mm thick[2], and used them for cosmic-ray exposures at high altidudes [Y2].

In a quite different field, that of electron track radiography, liquid G.5 emulsion has been used to estimate the content of radioactive nuclides in biological specimens, and to localize them with precision[3,4]. KING et al. employed this technique to investigate the uptake of P^{32} by protozoa; they were able to count beta-particle tracks from individual micro-organisms containing only a few hundred atoms of the radioisotope. The absence

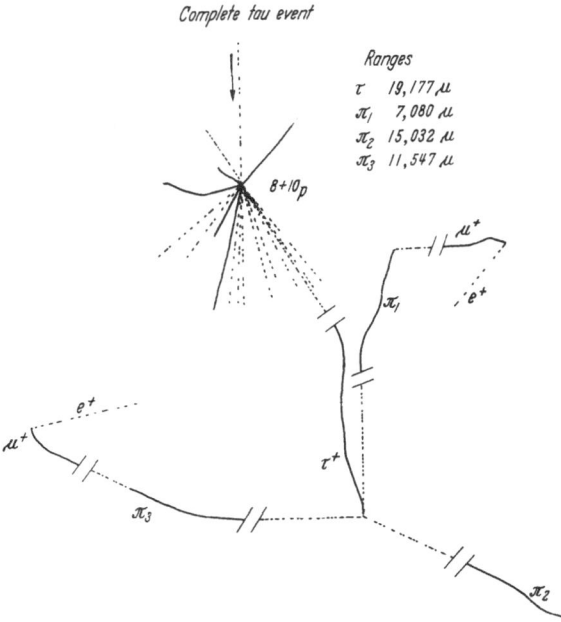

Fig. 23. Projection drawing of a "complete τ event" observed at the Naval Research Laboratory[5] in a stack of stripped emulsions exposed at high altitude. The τ^+ meson emerging from a high energy distintegration, decays into three pions. Of these, the two positive ones decay into muons, and the latter into electrons. The negative pion is captured without producing a visible star. Pellicle stacks make it possible to trace such connected phenomena as these.

of extraneous electron tracks made this possible. Similarly, the freedom from background which can be attained in freshly poured emulsions is an invaluable aid in the detection of x-rays by means of the electron pairs which they produce. Instructions on the use of gel are provided by the manufacturer. Further advice on "home pouring" and drying of gel has been given by ROSENFELD et al. [R7].

Emulsions are obtainable commercially in "pellicle" form, i.e., stripped from their glass base. The role of stripped-emulsion stacks in high-energy physics has been discussed in Sect. 5. Blocks consisting of scores of layers, each several hundred microns thick, are now extensively used both in cosmic-ray research and in experiments with the larger accelerators. Whenever it is desirable to

[1] E. P. GEORGE and J. EVANS: Proc. Phys. Soc. Lond. A **63**, 1248 (1950); **64**, 193 (1951).
[2] H. YAGODA: Phys. Rev. **79**, 207 (1950); **80**, 753 (1950).
[3] D. T. KING, J. E. HARRIS and S. TKACZYK: Nature, Lond. **167**, 273 (1951).
[4] M. BLUNDELL and J. ROTBLAT: Nature, Lond. **167**, 645 (1951).
[5] N. SEEMAN, M. M. SHAPIRO and B. STILLER: Unpublished.

increase the amount of contiguous sensitive volume, uninterrupted by inter-vening glass, pellicle stacks are likely to be helpful. Among their advantages is the visibility of a series of genetically related processes. This is illustrated by the projection drawing of a "complete tau event" shown in Fig. 23.

Non-commercial emulsions. Very few successful attempts have been made by individual experimenters, outside of industrial photographic laboratories, to produce emulsions suitable for nuclear research. DEMERS was the first to prepare an emulsion highly concentrated in silver halide[1], which he employed for cosmic-ray research [D6]. HÄLG and JENNY produced an emulsion of mo-derate sensitivity [H1]. Later JENNY succeeded in making an electron-sensitive emulsion by adding a derivative of 3.5 dimethyl-pyrazole, after extensive ex-perimentation with various chemical sensitizers [J1]. DEMERS has evolved a fine-grained emulsion sensitive to particles at minimum ionization, and described his method of preparation in some detail [D7]. The very small crystal size, $\sim 0.08\,\mu$, which leads to a developed grain $\sim 0.15\,\mu$ in diameter, permits better discrimination between different types of particles at moderate and high rates of ionization loss than that attainable with commercial electron-sensitive plates. Somewhat higher precision in multiple-scattering measurements on "stiff" tracks is also possible with a fine-grained emulsion since one type of noise is reduced: the projected centroids of the tiny grains have a smaller mean deviation from the path of the particle.

Some of the difficulties associated with emulsion techniques, e.g., the strains which produce distortion, are due to certain properties of gelatine. If a superior substitute could be found to serve as a matrix for the silver bromide, these troubles might be eliminated. Attempts in this direction have recently been made by DEMERS and SCHWERIN[2] using polyvinyl alcohol as a substitute for gelatine. (Their main purpose appears to have been to alter the composition of the light elements in the emulsion, in particular to eliminate the nitrogen.) Although their work thus far has not achieved a practical emulsion, they have made a useful start in this direction.

III. Treatment before exposure.

Although emulsions, and particularly the more sensitive ones, are generally used as soon as possible after manufacture, they must frequently be stored first. Storage underground reduces the cosmic ray intensity, but does not protect against natural gamma radiation from surrounding rocks, which is even more serious. At sea level, the total background is about 0.1 roentgen per year, and a two-month's accumulation of slow-electron tracks, at the rate of $4000/cm^3$ per day, produces an extremely heavy background in an emulsion of the G.5 type [R7]. Lead shielding alone is usually inadequate, since lead tends to harbor traces of radioactivity. Three or four inches of iron surrounded by a similar thickness of Pb provide a fairly effective shield. A temperature near 10° C, and a relative humidity near 50% are suitable for storage.

26. Impregnation and sandwiching. For the detection of neutrons, the study of fission processes, or the investigation of other reactions, emulsions can be "loaded" with various elements by immersion in suitable solutions. Or, the substance of interest may be introduced into the emulsion at the time of manu-

---

[1] P. DEMERS: Phys. Rev. **70**, 86 (1946).

[2] P. DEMERS and A. K. SCHWERIN: Comm. à l'Ass. Canad. Fr. Avanc. Sci., 3 Nov. 1956. Ann. Assoc. Canad. Fr. Avanc. Sci. **23** (1957).

facture; this has been done with salts of Li, B and Bi[1,2]. Emulsions have been impregnated with deuterated salts[3] for photodisintegration experiments and with uranium salts[4] for studies of fission. A much higher concentration of deuterium can be incorporated into emulsion by soaking in heavy water[5]. For loading by immersion, buffered solutions (p_H close to 7) are used to minimize fading and other difficulties. Some desensitization is apt to result from bathing the emulsion, but most of the sensitivity can be restored by thorough drying.

In other experiments, metal foils[6] or layers of gelatine[7] (without halide) have been sandwiched between sensitive layers of emulsion. Target elements have also been embedded in emulsion in the form of wires[8] [*M 7*], of solid grains suspended in layers of gelatine ([*M 10*], p. 265), and inside capillary tubes[9] for the study of interactions in various materials [*M 7*].

27. Eradication. Before exposure it is possible to remove most of the accumulated background of developable tracks and fog by using one of a number of oxidizing agents. Thin emulsions ($< 100\ \mu$) may be immersed in a weak solution of chromic acid or chromic trioxide and then allowed to dry[10]. In another method of "defogging", the latent background grains are first developed and then suppressed by oxidation[11] with a solution of $KMnO_4$. Hydrogen peroxide has also been employed for oxidizing the latent image[12]. In eradicating background by any of these methods, one runs the risk of reducing the sensitivity of the emulsion to an extent which depends upon the concentration of the oxidizing agent and the duration of immersion in solution or exposure to vapor. In general, the more sensitive the type of emulsion employed, and the greater its thickness, the more difficult it is to eliminate background without destroying some sensitivity. In fact, some of the agents mentioned above have been used deliberately to produce emulsions which are relatively insensitive. For thick emulsions, the same problems which complicate the task of development also lead to difficulties in eradication. Hence defogging techniques have been applied mainly to emulsions $\lesssim 200\ \mu$ thick.

Adequate eradication without serious loss of sensitivity is attainable by exposure to warm water vapor. This has been done for C.2 plates up to $200\ \mu$

---

[1] The advantages of using Li- or B-loaded plates for slow-neutron monitoring were demonstrated by M. M. Shapiro and J. R. Barnes: Phys. Rev. **73**, 1243 (1948).

[2] Yagoda [*Y1*] has used Li and B together in $Li_2B_4O_7$ for neutron detection.

[3] W. M. Gibson, L. L. Green and D. L. Livesey: Nature, Lond. **160**, 534 (1947).

[4] K. Lark-Horovitz: Phys. Rev. **59**, 941 (1941). — N. A. Perfilov: C. R. Acad. Sci. URSS. **47**, 623 (1945). — P. Cuer, M. Morand and E. Cotton: Cahiers de Phys. **22**, 72 (1944). — L. B. Borst and J. J. Floyd: Phys. Rev. **70**, 107 (1946). — E. Broda: J. Sci. Instrum. **24**, 136 (1947). — L. L. Green and D. L. Livesey: Nature, Lond. **158**, 272 (1946). — P. Demers: Canad. J. Res. **25**A, 223 (1947). — S. T. Tsien, Z. W. Ho, L. Vigneron and R. Chastel: Nature, Lond. **159**, 773 (1947).

[5] G. Goldhaber: Phys. Rev. **74**, 1725 (1948).

[6] I. Barbour and L. Greene: Phys. Rev. **79**, 406 (1950). — P. E. Hodgson: Phil. Mag. **42**, 82 (1951).

[7] J. B. Harding: Nature, Lond. **163**, 440 (1949). — P. E. Hodgson: Phil. Mag. **42**, 955 (1951).

[8] M. Danysz and G. Yekutieli: Phil. Mag. **42**, 1185 (1951).

[9] Cylindrical emulsions surrounding such tubes have been described by A. Bonetti and G. P. S. Occhialini: Nuovo Cim. **8**, 725 (1951).

[10] N. A. Perfilov: C. R. Acad. Sci. URSS. **42**, 258 (1944); **43**, 14 (1944). — Powell et al. [*P11*]. — P. Demers: Canad. J. Res. **25**, 223 (1947).

[11] L. N. Liebermann and H. H. Barschall: Rev. Sci. Instrum. **14**, 89 (1943).

[12] H. Yagoda and N. Kaplan: Phys. Rev. **73**, 634 (1948).

thick[1], at higher temperatures for electron-sensitive emulsion of similar thickness[2], and even for 400 μ layers of G.5[3]. The improvement in recognizability

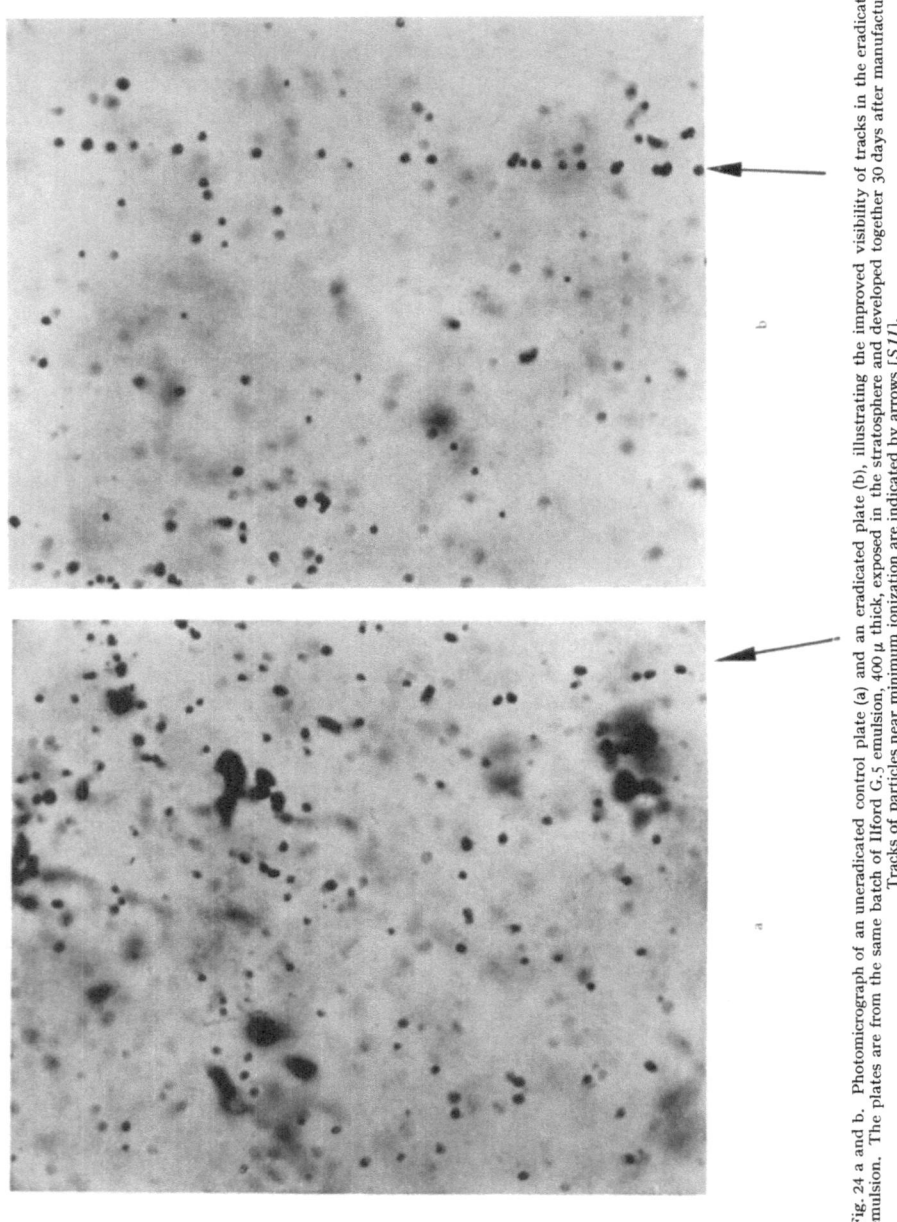

Fig. 24 a and b. Photomicrograph of an uneradicated control plate (a) and an eradicated plate (b), illustrating the improved visibility of tracks in the eradicated emulsion. The plates are from the same batch of Ilford G.5 emulsion, 400 μ thick, exposed in the stratosphere and developed together 30 days after manufacture. Tracks of particles near minimum ionization are indicated by arrows [S11].

[1] M. Wiener and H. Yagoda: Rev. Sci. Instrum. 21, 39 (1950).
[2] G. Albouy and H. Faraggi: C. R. Acad. Sci. Paris 230, 1351 (1950). — Rev. Sci. Instrum. 22, 532 (1951).
[3] B. Stiller, M. M. Shapiro and F. W. O'Dell: Phys. Rev. A 85, 712 (1952). See also [S11].

of thin tracks is illustrated in Fig. 24, and the restoration of sensitivity near minimum ionization is shown in Fig. 25. A useful byproduct of the treatment

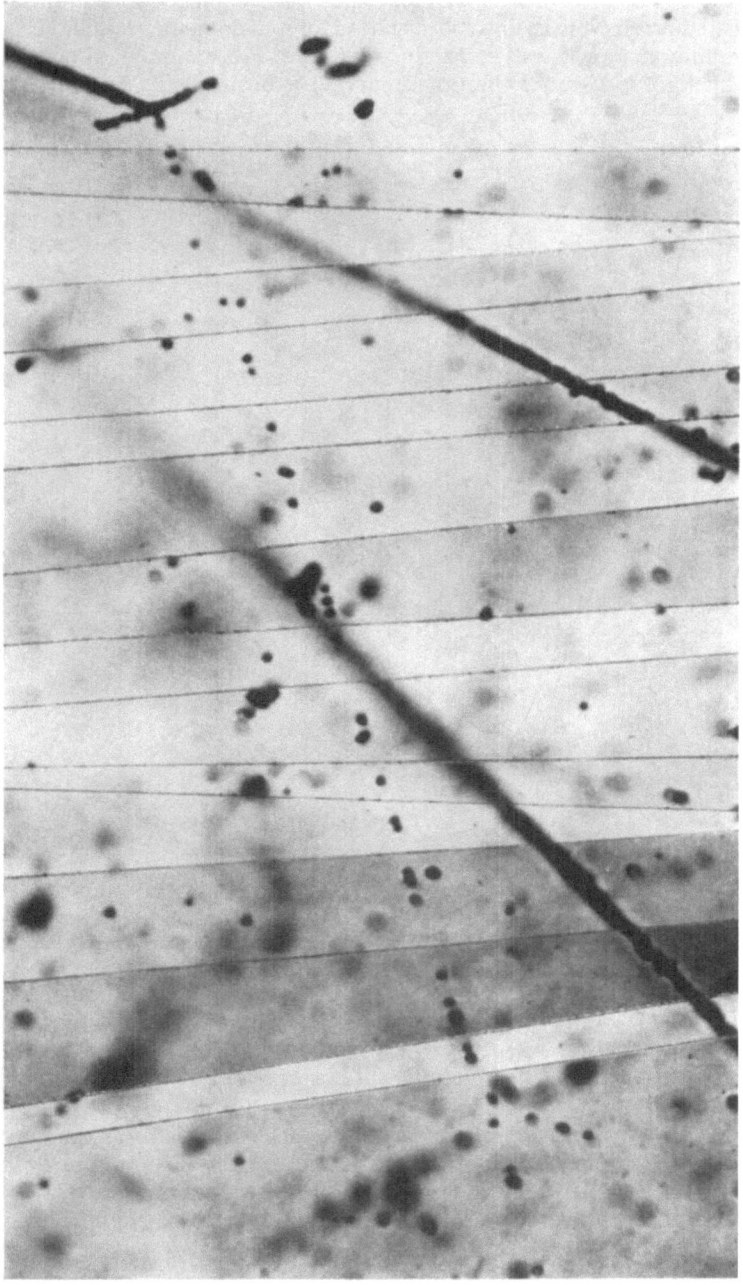

Fig. 25. Photomicrograph of a "hammer track" produced by the disintegration of Be^8 into 2 α-particles, following the beta decay of the parent Li^8. It was observed near the bottom of a 400 μ Ilford G.5 emulsion which had been exposed in the stratosphere after eradication. Visibility of the near-minimum electron track shows the restoration of sensitivity following eradication [S 11].

with water vapor is the reduction of track distortion [S 11]. Fig. 26 compares the magnitude of the distortion vector, measured by the Cosyns-Vanderhaeghe method [C 6], in an eradicated plate and a control plate from the same batch.

28. Preparation of pellicle stacks for exposure. When stripped emulsion stacks are used to permit continuous observation of phenomena from layer to layer, certain preliminary measures can reduce tedium. For example, in order to follow no more background tracks than necessary, it is desirable to rearrange the stack

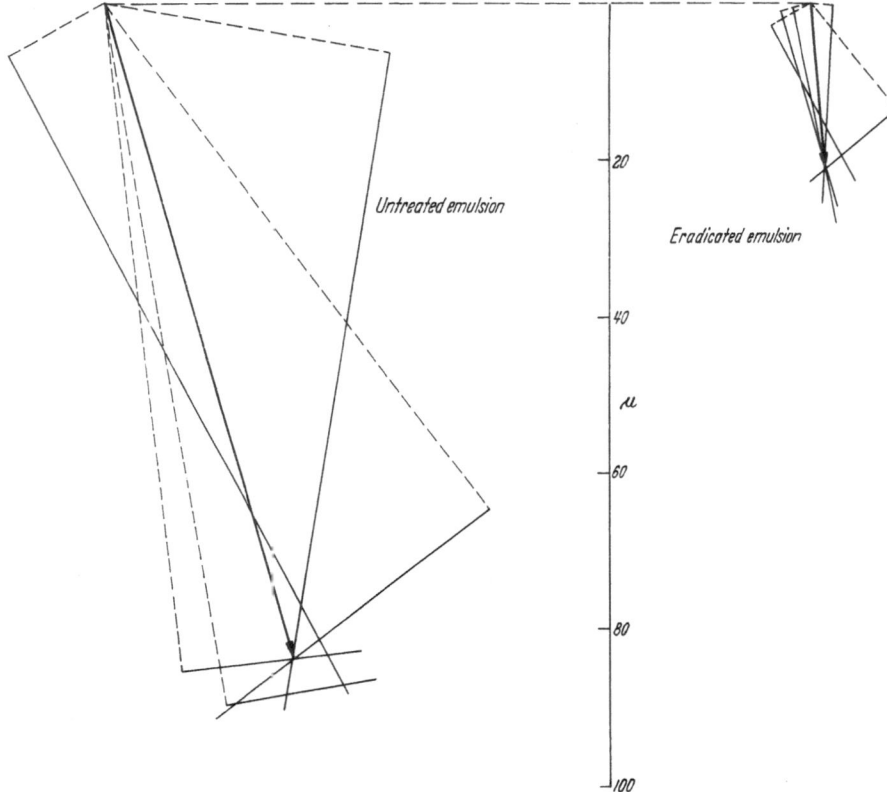

Fig. 26. Comparison of the distortion vector in an uneradicated control plate and an eradicated plate from the same batch of emulsion [S 11]. Distortion was measured by the method of COSYNS and VANDERHAEGHE [C 6].

so that pellicle surfaces adjacent during shipment are separated during exposure. Reshuffling ensures that tracks of charged cosmic-ray particles which penetrated several layers during air transit, or at any time since the stack was packed, will not be followable and hence not be confusable with tracks registered in the intentional exposure.

Preliminary steps in alignment. Because much of the effort in microscopy is devoted to "tracing tracks through", schemes which facilitate this work can save much time. Among the procedures used for alignment[1-3] we shall first describe some which are carried out before exposure or development:

[1] R. W. BIRGE, L. T. KERTH, C. RICHMAN, D. H. STORK and S. L. WHETSTONE: Techniques for Handling and Processing Emulsion Stacks, University of California Radiation Laboratory Report UCRL 2690. 1954.

[2] G. GOLDHABER, S. J. GOLDSACK and J. E. LANNUTTI: Method for Alignment of Stripped Nuclear Emulsions, University of California Radiation Laboratory Report UCRL 2928. 1955.

[3] E. SILVERSTEIN and W. SLATER: J. Sci. Instrum. **33**, 381 (1956).

(a) *Machining* the edges of the stack; this can be done by clamping the stack, e.g. between two slabs of bakelite or lucite, and using a milling machine. Having uniform-sized pellicles with straight edges and right-angled corners helps in the post-processing task of cutting the glass plates (to which pellicles are usually cemented before development), or of gluing the glass base to a plastic frame.

(b) *X-rays.* The edges of the stack are exposed through a Pb slit to a thin pencil of x-rays in several positions. The matching x-ray marks serve as fiducial lines.

(c) *Holes punched* in corresponding positions near the edges may be used instead of x-rays. These holes ($\frac{1}{8}$ to $\frac{1}{4}$ inch in diameter) serve as guides in stack assembly and in printing coordinate grids.

(d) *Photographic grids.* An array of thin lines or small dots on a master negative, with a matrix of numbers or letters, is contact-printed on one surface of each pellicle in a reproducible manner, so that corresponding coordinates will match on successive layers. A grid spacing of 500 μ or 1000 μ between dots or lines has been employed.

(e) *A grid of nylon threads* is soaked in a solution of an alpha-emitting nuclide (e.g., Po), and dried. Then lines, which are due to the composite effect of many α-particle tracks, are printed by contact [D3].

(f) In emulsions exposed to the cosmic radiation at very high altitudes, the dense tracks of *heavy primary nuclei* provide fiducial marks for alignment. Complete reliance on these heavy tracks obviates the special preparations required in methods (a) to (e). However, it does not facilitate the job of tracing tracks as well as some of the other schemes.

One or more of the foregoing procedures will serve as a basis for the final steps in pellicle alignment which must be taken after processing[1].

IV. Exposure, and preparation for development.

The special conditions under which emulsions can be exposed are as diverse as the types of experiments performed with them. (See Sects. 4 and 5.) Environmental conditions at the time of exposure, and the state of the emulsion itself— e.g., the density, temperature, relative humidity, presence of background radiation—should be controlled if possible, or should at least be well known. When precise measurements are to be made on tracks, it is useful to determine the density and thickness of representative samples of the emulsion, both before and after exposure.

29. Mounting pellicles before development. Prior to 1951, pellicle stacks were rarely used, owing mainly to the difficulties in controlling the dimensions of the stripped emulsions during processing. Lateral swelling by 20% or more of the original size, followed by variable shrinkage, discouraged their use. In that year a successful method was evolved[2] for mounting thick pellicles onto glass after exposure, but prior to processing, so that the stripped emulsions can be developed like ordinary plates. Procedures were devised which yield emulsions free from lateral swelling and relatively free from blistering. Methods for processing pellicles have also been developed by the Bristol group[3], the Bombay group [L1], and in other laboratories.

---

[1] See the last paragraph of Sect. 32.

[2] Reported at APS Meeting, Chicago, 1951, by B. STILLER, M. M. SHAPIRO, and F. W. O'DELL: Phys. Rev. A **85**, 712 (1952). See also Phys. Rev. A **91**, 496 (1953), and Ref. [S11].

[3] C. F. POWELL: Phil. Mag. **44**, 219 (1953).

From experiments on "free processing", LAL et al. ([L1], Appendix) concluded that the method appears unpromising for emulsion blocks. Subsequently OREAR [O5] described techniques for developing very thick pellicles ($\sim$1200 μ), unmounted on glass[1]. He reported that the dried emulsions are virtually free of distortion. Offsetting this advantage is the fact that when a set of pellicles is processed free, the individual emulsions are not all restored to the same size, or even shape, after drying [R7]. Thus, alignment is possible only over a limited part of each pellicle. This difficulty can, however, be mitigated by printing a reference grid on each emulsion before processing. The benefits to be derived from "free processing" are sufficiently important so that further efforts to improve this technique may be expected. These advantages are (a) a sharp reduction in the time required for processing, because solutions can diffuse in and out through both surfaces; (b) a probable improvement in the uniformity of development with depth; and (c) decisive reduction in track distortion.

Currently, the general practice is to attach pellicles to glass *before* processing. A mounting procedure which has been applied to emulsions up to 600 μ thick with good results is the following[2]: Two solutions, consisting of (A) wetting agent in water, and (B) *gelatine* and wetting agent in water are boiled, then cooled to room temperature[3]. Glass plates, specially treated[4] for good adhesion to emulsion, are immersed in solution A. The glossy surface[5] of each pellicle is moistened with the gelatine solution B (but not immersed in it), the emulsion is then positioned on a wet glass plate, and attached to it with a rubber hand roller. After the mounted pellicles have been left to dry for about 15 hours, they are ready for development.

The foregoing sketchy outline of the method omits details of the mounting technique which can spell the difference between success and failure: The cooled solution A is poured into two containers; one of these, A-1, is used for cleaning the glass plates, both sides of which are swabbed with cotton moistened in the solutions. Scrupulous cleanliness of the glass is required to prevent blistering or "bubbling" (see also Ref. [D3]). A glass plate is then immersed in the second container, A-2. In mounting a pellicle it is desirable to keep its air surface free of the gelatine solution. This can be done as follows: the pellicle is placed on a clean flat surface with the glossy side down. With a camel's-hair brush dipped in solution A, a picture-frame border is painted around the periphery of the pellicle surface. A thin ($\sim$0.001 inch) sheet of polyethylene, slightly larger than the pellicle, is placed over the surface. It adheres at the border, and keeps the surface dry during the subsequent stages. The pellicle is turned over and its uncovered, glossy surface is moistened generously with solution B, using a wide camel's-hair brush. Promptly, a clean glass plate is removed from container A-2, allowed to drain for a moment, and lowered onto the pellicle. The glass first makes contact along the length of one edge of the pellicle, with the plate held at a small angle to the emulsion plane. Then the plate is lowered in such a way that the line of contact progress along the pellicle. The glass is gently pressed to the emulsion for a few seconds, and then the plate is turned over so that the polyethylene sheet faces up. A rubber print roller (preferably

---

[1] The method was apparently applied to pellicles of modest lateral dimensions (a few inches).

[2] This method, in use at the Naval Research Laboratory, is a modified version of that described in [S11].

[3] A suitable wetting agent, "Aerosol OT", is a dioctyl ester of sodium sulfosuccinic acid. It is available commercially as a 25% aqueous solution which we shall simply call Aerosol. The solutions used in mounting are prepared as follows: Solution A: 4 cm^3 of Aerosol in one liter of distilled water. Solution B: 20 g of gelatine is added to one liter of distilled water at room temperature, and dissolved by heating. To this, Aerosol is added, and the solution filtered.

[4] Available from Ilford, Ltd., Ilford, London.

[5] The emulsion surface to which the glass plate should be attached is the one which was in contact with the polished glass on which the liquid emulsion had been poured in manufacture. This surface can be identified as the glossier of the two. It adheres better to the glass; furthermore, it is less likely to acquire a film of metallic silver which tends to deposit on emulsion surfaces during processing.

a heavy one) is run over the sheet with gentle, uniform pressure, starting at the center and rolling toward each edge. The plastic cover provides protection against abrasion. (An alternative method of rolling employs a clothes wringer, or mangle, on which the spring tension is adjustable.) Next, the polyethylene is promptly covered with a piece of plate glass, and on top of this is put a well-distributed weight ($\sim$0.2 lb/sq. in.), having the same outside dimensions as the plate glass. After ten minutes, the weight and covering plate glass are removed, and the plastic sheet stripped off the emulsion. The mounted pellicle is allowed to dry in a horizontal position for $\sim$15 hours[1] before development, which is then carried out as for ordinary glass-backed nuclear emulsions.

V. Processing.

30. Difficulties and special requirements. The considerable thickness and high silver-halide concentration of nuclear emulsions introduces problems in their chemical processing which are a good deal more troublesome than those encountered in ordinary photographic development. The various chemical solutions must penetrate great thicknesses of emulsion, and then diffuse out again. Not only is this process time-consuming, but it introduces the danger of non-uniformity in depth. During the fixing stage, the principal constituent of the emulsion, AgBr ($\sim$80% by weight), is dissolved away. This leads at first to a very spongy emulsion which, during subsequent drying, collapses to 40% or less of the original, unprocessed thickness. The total shrinkage from the swollen state during processing to the final dry state involves an even greater change. During fixing, washing, and drying, it is very easy for distortions to arise.

For thicker emulsions, total processing times of the order of one or two weeks are not unusual. Certain properties of the developed emulsion which affect its degree of usefulness for quantitative microscopy depend strongly on the methods of processing employed. Among these properties are the degree of distortion, uniformity, visibility, discrimination, and durability. Good processing results in: (a) a minimum of track distortion which interferes with measurements of multiple Coulomb scattering; (b) reasonably uniform development in depth, so as to obviate difficult corrections in ionization measurements along a track; (c) adequate grain density so that the tenuous tracks of weakly ionizing particles can be discerned; (d) low fog background of random silver grains; (e) transparency, i.e., freedom from collodial silver or other causes of stain; (f) sufficient discrimination between different rates of ionization for the problem at hand; and (g) stability against spontaneous stripping of the processed gelatine layer from its glass base. Additional desiderata, which apply to stacks of pellicles, are (h) constancy of lateral dimensions; and (i) freedom from blistering.

31. Development. There are probably as many different recipes in use for developing nuclear emulsions as there are laboratories which use them. The possible combinations of ingredients, concentrations, temperatures, and times are legion, and a number of them have given satisfactory results. No experimenter is likely to be satisfied until he has compared a fair variety of methods for himself. Nevertheless, some guiding principles and a few tested formulas may prove helpful.

---

[1] For a drying time of 100 hours, E. J. BURGE, J. H. DAVIES, I. J. VAN HEERDEN, and D. J. PROWSE [Nuovo Cim. **5**, 1005 (1957); Fig. 1] find a bubble density of $\sim$3 per 100 sq. in. This is the same as the average frequency encountered in 6 stacks (with a total area of 11300 sq. in.) at the Naval Research Laboratory using the mounting method described here, in which a much shorter drying time ($\sim$15 hours)has been employed. The incidence of blistering was not further reduced in the NRL method by extending the drying time. The long drying time which BURGE et al. found necessary may be due to the immersion of the pellicles in their mounting procedure. If so, several days of drying time could be saved and the likelihood of fading reduced, by avoiding immersion.

Whatever method is adopted, the equipment for carrying it out must be capable of sufficiently close control so that a high degree of reproducibility can be achieved in the conditions of development. This applies particularly to the most sensitive emulsions, in which the action of particles near minimum ionization is probably close to the sensitivity threshold of the photographic process. Hence, in following a given routine of development, some of the conditions are rather critical, contrary to the latitude in ordinary dark-room practice. For example, the temperature should be kept within narrow limits ($\pm 0.5°$ C). The requirement of reproducibility, and the arduous demands of day-and-night sequences of processing, speak in favor of semi-automatic equipment.

Choice of developer and processing method is governed by the type and thickness of emulsion and the purpose of the investigation at hand. For *"thin"* *emulsions* (say 25 to 100 microns) of type C.2 or NTB, relatively uncomplicated methods suffice. The most widely used developer is Kodak D19b (or Ilford ID-19) which contains both elon and hydroquinone. In typical formulas, the stock solution[1] of D-19b is diluted with 3 parts of water, and the plates are developed at 18 to 20° C for about 20 to 40 minutes. Concentration of developer is less critical than time of development, and the latter in turn is less critical than temperature. In fact, the concentration may be varied by a factor of 2 to 4, in order to obtain the desired results. If, for example, the thin tracks of lightly ionizing particles are to be accentuated, then "strong" development (high concentration, relatively long time, or high temperature) is required. The advantage so gained is to some extent offset by heightened fog background. If, on the other hand, it is important to discriminate between particles of different— and moderately high—ionizing power, then weaker development is employed. This may entail some sacrifice in visibility of thin tracks.

After development, the plates are rinsed in water for a minute, and then transferred to a 0.5% acetic acid stop bath for 15 to 20 minutes. Toward the end of this interval it is wise to examine the plates (using safelight) to see whether they look black. If they do, the cause may be a thin surface deposit of silver which is removable by wiping gently with moist cotton. When this surface film appears, failure to remove it results in opaque-looking plates. If the deposit is not wiped off while the plates are wet, it may still be subsequently removed by swabbing the dry emulsion with a piece of chamois leather moistened with acid or xylol.

The plates are then fixed in a 35% thiosulphate solution for about 50% longer than the time required for clearing. Finally, they are "plasticized" (see below), washed until a test for hypo is negative, then allowed to dry. If edge distortion proves troublesome, it can be reduced by using "guard rings" or alcohol drying, as described below.

For *development of thicker emulsions,* traditional methods are inadequate. Slow diffusion of heavy organic molecules of developer through thick layers of gelatine and silver halide can lead to non-uniform development. Furthermore, low distortion and high ratio of track density to background grains are more difficult to attain in thick layers. The main question is how to achieve adequate

---

[1] A liter of Kodak D-19b stock solution is prepared as follows:

Distilled water (at about 50° C)	500 cm³
Elon (metol)	2.2 g
Sodium sulfite, anhydrous	72 g
Hydroquinone	8.8 g
Sodium carbonate, anhydrous	48 g
Potassium bromide	4 g
Cold water up to	1000 cm³

(Elon: monomethyl p-aminophenol sulphate).

penetration of the solution without overdeveloping the surface layers. This problem can be solved in principle by separating the penetration of the developer from the process of development *per se*—first allowing the developer to diffuse into the emulsion under conditions which discourage active development, then permitting development to proceed. Two methods of accomplishing this— variation of the p_H of the developer, and change in its temperature—have been proposed. The first, due to BLAU and DE FELICE [*B 12*], works satisfactorily for emulsions $\lesssim 200\,\mu$ thick. It exploits the fact that the rate of development can be slowed down by lowering the p_H[1]. Two solutions are used in turn. The first lacks the usual alkaline constituent, hence the developer permeates the emulsion but is ineffectual as a reducing agent. The plates are then immersed in the second solution, which consists of ordinary developer enriched in alkali. In this bath active development occurs.

The second method, due to DILWORTH, OCCHIALINI, and PAYNE[2], is the one which has been generally adopted, especially for emulsions thicker than $\sim 200\,\mu$. It depends on the fact that reaction rates are reduced considerably more than diffusion rates when the temperature is lowered. Thus the emulsion is first immersed in cold ($\sim 5°$ C) developer long enough to assure penetration, then the solution is warmed up and kept at, say, $20°$ C during the actual development. In this way, reasonable uniformity in depth can be achieved without introducing excessive fogging.

Plainly, this "temperature-cycle" method requires that little or no development occur during the "cold stage". Hence, the more quickly penetration is accomplished during this stage, the better. In addition [*D 2*], [*D 12*], the developer should have a p_H fairly close to that of the emulsion itself (i.e., rather low alkalinity), and stability at temperatures up to $\sim 30°$ C. Moreover, it should not produce excessive background fog or stain in the emulsion. While no single developing agent excels in all these respects, amidol (p-diaminophenol) meets the requirements sufficiently well [*D 12*] so that it has come into widespread use[3]. DAINTON *et al.* [*D 2*] have compared the characteristics of amidol with those of Azol (para-aminophenol) and an elon-hydroquinone developer (D 19b). They found that amidol penetrates a good deal more rapidly than the others. This reduces the amount of development which can take place during the cold soaking stage. Besides, amidol can act in an acid environment, and this is of value in avoiding the excessive swelling and concomitant deformation induced by an alkaline developer in the warm stage [*D 12*][4]. Since the time of penetration increases with thickness of emulsion, it is clear that the duration of the cold stage must likewise increase. On the other hand, this may require a change in the composition of the developer solution, in order to prevent active development of the surface layers during the longer penetration time. In fact, the Brussels group [*D 12*] found it helpful, for thicknesses exceeding $300\,\mu$, to add a restrainer such as KBr, which extends the time during which the developer is inactive. For emulsions thicker than $400\,\mu$, the addition of boric acid lowers the p_H of the solution and gives fair uniformity of development with depth. An alternative acidifier, employed by the Bristol group [*D 2*] is sodium bisulphite. For emulsions

---

[1] For investigations of the effects of p_H upon development, see F. ALBUOY and H. FARAGGI: J. de Phys. **10**, 105 (1949). G. W. STEVENS ([*M 10*], p. 310) and Nature, Lond. **162**, 526 (1948).

[2] C. C. DILWORTH, G. P. S. OCCHIALINI and R. M. PAYNE: Nature, Lond. **162**, 102 (1948).

[3] The advantages of amidol have also been confirmed by A. J. HERZ [J. Sci. Instrum. **29**, 60 (1952)], and by STILLER *et al.* [*S 11*].

[4] A modest amount of swelling by soaking in distilled water prior to the cold-developer bath is, however, allowed in order to facilitate the permeation of the developer (see below).

up to 400 μ in thickness, a p_H of ~7.2 to 7.4 gives good results. For thicker emulsions, developers with p_H of ~6.4 to 6.7 have been found suitable.

Presoaking bath. The speed of penetration of cold (5° C) developer into thick emulsion ($\geq$ 400 μ) is approximately doubled if the permeation stage is preceded by soaking in distilled water cooled to ~5° C. Therefore, this procedure has become fairly standard. Although the practice in many laboratories has been to start the presoaking at room temperature and cool the water down to 5° C, the Chicago group [R7] precools the emulsion in a refrigerator, and then immerses it directly into cold solution. This prevents excessive swelling and reticulation. The time allowed for presoaking should be about the same as that for penetration (see below).

Permeation stage. In principle, the temperature-cycle method permits only negligible development during this cold stage. This has worked out well for emulsion thicknesses up to ~400 μ, for which penetration times are relatively short. However, for greater thicknesses, uniformity of development from top to bottom has been increasingly difficult to achieve. This is apparently due to an increase in sensitivity of G.5 emulsions during the past few years, which results in some "cold" development during the period required for penetration. Accordingly, it is desirable to keep the temperature at this stage even lower than 5° C in an effort to inhibit development.

The time required for the cold developer to permeate the emulsion uniformly is longer than that required for the first arrival of developer at the lower surface [D2]. A useful rule of thumb at 5° C seems to be to allow 20 to 25 minutes for each 100 μ of depth [B13d].

Developer formulae. The following amidol solutions are modified versions of Brussels formulae[1], which have given satisfactory results in several laboratories:

For emulsions up to 400 μ in thickness:

Stock solution: 1 liter of distilled water
8 cm³ of a 10% solution of KBr.

Just before using add: 12 g sodium sulphite (anhydrous),
2.8 g amidol.

For emulsions $\geq$ 600 μ thick [R7]:

Same as above, except for addition of 25 g boric acid (anhydrous) to the stock solution.

The stock solution should be stored at 5° C; it will keep for months. The sodium sulfite serves as a preservative, and should be added first, while stirring the cold stock solution to prevent formation of a hard, insoluble mass.

Active development. In the temperature-cycle method, actual development takes place by warming up the cold developer with which the emulsion has been impregnated. If the plate is simply left in the solution during this stage, fresh developer may enter and produce stronger development near the surface. To prevent this, the "hot-plate" or "warm-dry" method is commonly employed: the plate is transferred from the cold-developer solution to a metallic surface with which it makes good thermal contact. (Excess droplets of developer are gently blotted from the surface.) Alternatively, the cold developer is removed from the tray or tank, and the latter warmed to the desired temperature. In

---

[1] See [D12] and [B13d]. The principal change is the reduction in amidol concentration from 4.5 to 2.8 g per liter, and corresponding diminutions in other ingredients. The lower concentration of amidol gives a considerable reduction in random grain background [D12], [R7] as compared with slight reduction in grain densities. The recipes given here should be considered as representative samples, rather than definitive formulae.

either case, the warm container is covered so that only a very shallow air space is permitted above the surface of the emulsion. It is desirable that the cover be kept at the same temperature as the "hot-plate" itself, as is done in the "dry rack" of Fig. 28. In this way, a uniform temperature is maintained throughout the emulsion.

Warm temperatures ranging all the way from 15 to 32° C have been reported in the literature for amidol developers. The time required for good development varies sharply with the temperature, from ∼ two hours down to ∼20 minutes; it also depends, of course, on the degree of discrimination desired. Temperatures of 23 to 26° C have been commonly employed for warm development, with times between 20 minutes and one hour. At these temperatures, the latter period usually gives "full development" (approximately 30 to 35 grains per 100 μ for tracks near minimum ionization in G.5 emulsion), whereas 20 minutes leads, in most batches of emulsion, to "underdevelopment", i.e., near-minimum grain densities of the order of 20 grains per 100 μ. Visibility, even of the more tenuous tracks, is generally satisfactory under these conditions[1], but the degree of distortion often leaves something to be desired.

There is some evidence that strains are sometimes "frozen" into emulsion during manufacture, e.g., in the dyring stage[2]. When this happens, the strains are, of course, still present at the time of exposure. During processing, these strains may be relieved, with resulting deformation of tracks; this is more likely to occur, the higher the temperature. Supporting this view is the observation that emulsions which are eradicated by exposure to warm water vapor prior to irradiation have much less distortion than uneradicated plates from the same batch ([S 11]; see also Fig. 26). It would appear that strains present in the untreated emulsion are relieved under the conditions of temperature and humidity employed in eradication. In any event, it is generally agreed that low-temperature development tends to reduce distortion[3]. In fact, HOPPER et al.[4] found that distortion can be diminished to a negligible amount by developing emulsions at 15° C (and drying with an alcohol-water solution of increasing concentration — a procedure to be described below). The time they allowed for development at this temperature was 120 minutes. In emulsions developed at 27° C, they observed distortion ∼4 times as large as that at 15° C, and the distortion at 20° C was almost as great as that at 27°.

Fox and WANIEK[5] have used a very low temperature of development, 4° C, with D 19 developer diluted 1:3 parts of water. The time required for developing a 400 μ emulsion or a 1000 μ pellicle is ∼4.5 hours. Other workers who have tried developing (with amidol) at temperatures below 10° C find[6,7] that while distortion is low in the processed emulsions, the fog background is excessive.

---

[1] Coates, using a metol developer, observed the effect of stain on the visibility of near-minimum tracks. He reports that there is a critical time of development (varying with the batch of emulsion) which should not be exceeded if the appearance of stain is to be avoided ([M 10], p. 320ff.). Coates also found that visibility is a function of the ratio of the square of grain density in a track to the areal density of background grains.

[2] BONETTI et al. [B 13c] called attention to this source of distortion, and compared it to the strains in unannealed glass.

[3] See discussion, Bagneres Conference ([B 1], p. 66). (In this discussion, Prof. C. F. POWELL pointed out that probably the stresses in *stripped* emulsion are large and would lead to distortion.)

[4] V. D. HOPPER, Y. K. LIM and M. C. WALTERS: Austral. J. Phys. **7**, 288 (1954).

[5] R. FOX and R. W. WANIEK: Nucleonics **13**, No. 7, 52 (1955).

[6] B. STILLER and F. W. O'DELL: Private communication; see also HOPPER et al., reference cited above.

[7] A. J. HERZ and M. EDGAR: Proc. Phys. Soc. Lond. A **66**, 115 (1953).

Yagoda[1] has employed an "isothermal" method of processing which altogether eliminates the warm stage, for emulsion layers up to 2 mm in thickness. He reports some reduction in distortion, and fairly uniform depth development for pellicles processed "free" (except for a layer near the surface). Less satisfactory results were obtained with thick emulsions coated on glass.

Stop-bath. In order to stop development at the end of the "warm stage" in the temperature cycle method, the temperature of the emulsion is first lowered, not too rapidly, to about 10° C. Then, a stop-bath, consisting of a 0.2 to 0.5% solution of acetic acid, is introduced at the same temperature, and the solution is cooled down to ~5° C at which temperature it remains for the rest of the stop-bath period. The duration of this period is about the same as that of the cold-soaking developer bath, i.e., the permeation time. After the stop-bath, some workers wash the plates in running water before fixing [*D 12*]. On the other hand, there are those who omit the stop-bath altogether.

32. Fixing, washing, and drying. The process of dissolving the large mass of unactivated silver bromide alters profoundly the mechanical properties of the emulsion. Until this stage, and despite immersion in a succession of solutions, the emulsion has remained fairly rigid, particulary if unduly high temperatures have been avoided during the warm stage. During fixing, the removal of about 80% by weight and about 50% by volume of the emulsion converts it to a jelly-like substance which must be treated with the greatest care during the remainder of the processing in order to minimize the danger of distortion.

Thus far, no satisfactory substitute has been found for sodium thiosulphate ("hypo") as the fixing agent. Since fixing is one of the most time-consuming stages in processing, it is highly desirable to employ a concentration of hypo which clears the emulsion in the shortest possible time. Various figures are reported in the literature, but it appears that a concentration of 300 to 350 g of hypo in a *total volume* of 1 liter of final *solution* gives the best results[2]. Other ingredients are sometimes added to the fixing bath in order to increase emulsion transparency, reduce the clearing time, or suppress swelling. To achieve greater clarity, 20 g or more of sodium bisulphite per liter is added by some workers. The lower p_H of the resulting solution also reduces swelling. Bonetti *et al.* [*B 13c*] suspect that bisulphite may damage the developed silver grains during the long fixing times, and therefore prefer to control swelling by the use of sodium sulphate (approximately 7 to 10%). However, a modest quantity of sodium bisulphite (~7 g per liter) gives clarity without introducing other difficulties [*S 11*]. The addition of a small amount (2 to 5 g) of ammonium chloride hastens fixation. Under some conditions, the developed silver grains can be attacked by the corrosive action of hypo, especially in the upper layer of the emulsion. An effective antidote to corrosion is the addition of several (2 to 8) grams of silver per liter of hypo solution[3].

[1] H. Yagoda: Rev. Sci. Instrum. **26**, 263 (1955).

[2] Y. Prakash: Indian J. Phys. **29**, 569 (1955). See also R. W. Birge *et al.*, UCRL 2690, 1954, who report the recommendations of G. Goldhaber. At first sight, these figures may seem to conflict with the "40% hypo solution" recommended elsewhere in the literature. Actually, the question of optimum concentration has been confused by the different ways in which the concentration is specified, i.e., whether the amount of hypo added to a liter of *water* is given, or the amount of hypo per liter of final *solution*. Where a 40% hypo solution has been mentioned, this has generally referred to the addition of 400 g of hypo to 1 liter of *water*. Since the latter corresponds approximately to 330 g of hypo per liter of *final solution*, there appears to be general consensus regarding the optimum concentration [*B 13c*], [*D 2*], [*S 11*].

[3] Reported by· G. P. S. Occhialini, for the Brussels group ([*B 1*], p. 61 ff.). A method of intensifying a corroded or underdeveloped image is discussed in the same report.

To prevent stagnation of the fixing solution, it is circulated slowly between the fixing vessel and a reservoir tank, maintaining gentle laminar flow across the plates, and avoiding vigorous agitation. Replenishment of the hypo is also accomplished gradually, to avoid abrupt changes in concentration, by replacing only part of the fixing solution in the storage tank at several intervals. For the same reason, washing is preceded by gradual dilution of the fixing solution, i.e., after the plates have cleared, cold filtered tap water is dripped slowly into the fixing tank until the fixing solution has been slowly replaced by water. If this procedure is omitted, and the transition from hypo solution to wash water is made suddenly, then strains due to osmotic pressure can easily produce distortion. In the fixing bath itself, the emulsions need be left only long enough to clear. The customary procedure of allowing an additional time of about half that required for clearing, in order to assure complete removal of the silver bromide, is obviated by the dilution stage, during which hypo is still present. The latter stage lasts for approximately 50 to 70% of the time required for complete clearing in the concentrated fixing solution. Typical clearing times for 400 μ G.5 emulsions are of the order of 24 to 36 hours: for 600 μ emulsions, of the order of 2 or 3 days.

For emulsions thicker than 400 μ, mounted on glass, it is not uncommon to find blisters or bubbles forming during the fixing stage. When these are detected in their incipient phase, BONETTI et al. [B 13 c] have found that by lowering the temperature and adding some sodium sulphate to the fixing bath, the tendency to further bubbling may be controlled. If blisters nevertheless arise, they can be repaired, at least to some extent, by the following procedure: a fine hypodermic needle is gently inserted into the blister, and the liquid inside it drawn out. It must be acknowledged that this measure of last resort is only a palliative and not a real cure.

In the last stages of dilution, the addition of a minute quantity of acetic acid further helps to prevent swelling, and often improves the clarity of the emulsion. Throughout the periods of fixing and dilution, the solutions are kept at 5° C. A certain amount of washing (i.e., elimination of hypo) has, of course, taken place during the dilution period. In the final washing stage, cold water (at ∼5° C) is slowly dripped in until a permanganate test shows no trace of hypo. Maintenance of low temperature throughout the washing stage helps prevent distortion, and lessens the likelihood of blister formation. The duration of the washing period is of the same order as the clearing time.

The tendency of thick emulsions to strip from their glass base once they are stored after drying, or in use in the microscopy laboratory, is a familiar trouble to emulsion workers. Of the various counter-measures which have been adopted, two are especially effective: impregnation with a plasticizer, and storage as well as use under controlled conditions of humidity. Before drying the emulsions, therefore, they are immersed in a plasticizing solution of 1 to 5% glycerine in water. Some workers have added still more glycerine (up to ∼10%) in order to reduce the amount of subsequent shrinkage of the emulsion when it dries. A soaking time of about 1 hour suffices for the more dilute glycerine solutions.

Drying. During this stage the emulsion shrinks from a swollen state down to a thickness which is smaller than that of the original emulsion by a factor of 2.5 or more. The drying tends to occur most rapidly near the edges of the plate, and this results in maximum deformation of tracks near the periphery. When plates are dried in air, therefore, even with the use of devices such as "guard rings" of wet strips of emulsion, it is exceedingly difficult to secure uniform shrinkage throughout the emulsion. A method which leads to decisive reduction in the distortion introduced during drying, is to employ alcohol-water solutions of increasing concentration [B 13c]. It is desirable that a few percent of glycerine be added to these solutions as well. The plates may be kept for about 2 hours in each of a series of solutions at ~5° C, in which the alcohol concentration is increased by about 15% [1]. HOPPER et al. [2] have shown that plates dried in this way have a distortion which is only 1/5 as large as plates dried in air. They report, in fact, that when alcohol drying, as well as a warm-stage temperature of ~15° C is employed for development, the error in scattering measurements due to distortion is much smaller than that due to "noise".

In the course of dyring, the emulsion shrinks by a large factor, since the mass of unused silver halide had been dissolved away in the fixing bath. For accurate measurement of lengths and angles it is necessary to know the "shrinkage factor", i.e., the ratio of thicknesses before and after processing. ROTBLAT and TAI [3] have investigated the corrections needed for shrinkage, and VIGNERON [4] has employed the tracks of α-particles from ThC' to measure this quantity. The effect of glycerine on the degree of contraction has been treated by MIGNONE [M 9] and others [5]. A thorough discussion of the subject of shrinkage has been given by BEISER [B 7].

Once the emulsions are dry, their surfaces may be swabbed with xylene or alcohol to remove silver deposits. They are next covered with a protective coating which will help keep them from drying out unduly, and peeling off the glass if they should be accidentally subjected to a sharp decline in relative humidity. A strippable plastic coating, which can be replaced if damaged, may be used [6]. Alternatively, the plates may be dipped in a solution of thinned Duco cement [7]. Together with the precautions cited above, i.e., immersion in a plasticizing solution before drying, and maintenance of humidity control in the microscopy laboratory, this will safeguard the emulsions from being wrecked by stripping off the glass [8].

For pellicle stacks, several methods preliminary to alignment were described above, such as matching the edges of the stack, printing a photographic grid, etc. If the pellicles were mounted on glass plates, they can be aligned in one of the following ways, once they are dry: (a) Careful cutting of the glass margins with a diamond scribe, using a suitable jig. This procedure cannot be relied upon to match the plates better than to within ~0.2 to 0.5 mm. (b) Metal tabs glued to the corners of the plate. These provide reference edges for matched positioning on the microscope stage [9]. (c) Plastic rectangular frames [L 1] whose outside dimensions are slightly larger than the plates, while their inside dimensions are slightly smaller. The first emulsion plate is glued to one of these frames. The plate carrying the next emulsion is glued to another frame with a slowly drying cement. Its position can be adjusted for about 30 minutes under the microscope so as to assure good alignment. If distortion is low, an accuracy of about 10 microns can be achieved from emulsion to emulsion.

[1] Some workers prefer a logarithmic increase of concentration with time which considerably extends the total duration of drying.

[2] V. D. HOPPER, Y. K. LIM and M. C. WALTERS: Austral. J. Phys. **7**, 288 (1954).

[3] J. ROTBLAT and C. TAI: Nature, Lond. **164**, 835 (1949).

[4] L. VIGNERON: J. Phys. Radium **10**, 305 (1949).

[5] M. K. JURIC and Z. A. SMOKOVIC: Rev. Sci. Instrum. **23**, 564 (1952).

[6] One such plastic is Eronel "spray-peel No. 725 xi clear". Eronel Services, Inc., Milford, Conn., U.S.A.

[7] E. I. duPont de Nemours and Co., Inc., Wilmington, Del., U.S.A.

[8] If the glass backing of a valuable emulsion should be broken, it may be desired to remove the emulsion from the glass and remount it. A method of doing this has been described by C. DAHANAYAKE: Nuovo Cim. **1**, 1251 (1955).

[9] R. T. BIRGE et al., UCRL 2690. 1954.

33. General precautions in processing; apparatus for processing. Throughout the processing it is highly desirable to keep the plane of the emulsions horizontal. Particularly during fixation and subsequent stages, when the emulsion is soft and spongy, the plates should not stand vertically, if serious distortion is to be avoided. An example of the "choppy" type of distortion which can result from this practice is shown in Fig. 27. The tracks have discontinuities as though the emulsion had been locally torn. A type of distortion which is less obvious but more widely prevalent is the C-shaped (or S-shaped) distortion which can arise in many ways even when the plates are kept horizontal during processing. As discussed above, these distortions are most likely to arise as a result of excessive temperature during the warm stage, or of uneven drying. It is prudent to avoid shocks of any type, e.g., sudden changes in p_H or temperature, and mechanical rocking. Temperature shocks, in particular, may produce reticulation. In the temperature cycle method, fairly rapid changes in temperature occur during the warm stage. However, provided the temperature is not too high, the emulsion is still fairly stiff, and not yet jelly-like. Hence, these changes are not necessarily a serious source of distortion. However, it seems prudent to use as

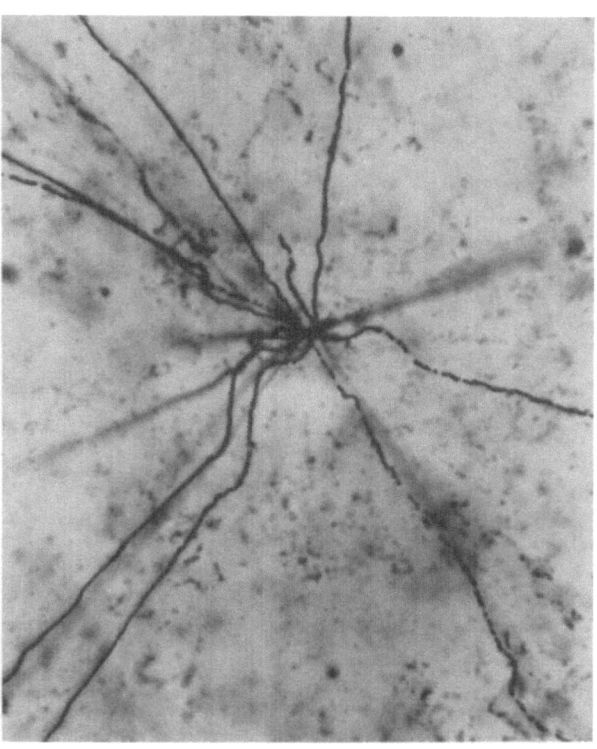

Fig. 27. "Choppy" distortion resulting from mechanical disturbance of an emulsion during the fixing stage. The discontinuities in the tracks are as severe as though the emulsion had suffered a local tearing. The same effect can be produced by fixing thick emulsions with their planes vertical.

low a "warm temperature" as is consistent with other requirements in order to minimize troubles due to temperature variation.

The purest available chemicals should be used as ingredients in the various solutions, and some of the latter should be prepared shortly before use; this applies especially to developers. Despite the large quantities of hypo which are required to fix stacks of thick emulsions, the temptation to use commercial grade hypo sometimes leads to stained or discolored plates. Chemically pure thiosulphate should be employed. Gentle circulation, rather than strong agitation should be applied in bringing fresh solution to the emulsions. The growth of bacteria in the gelatine, especially when it is wet, can lead to serious damage. A small quantity of disinfectant[1] may be added to the solutions, and especially to the more lengthy

---

[1] BONETTI et al. [B 13 d] recommend ∼0.1% solution of "Santobrite" as the disinfectant. YAGODA uses ∼0.2 g of phenol per liter in the last stages of washing (private communication).

baths if bacteria prove troublesome. The alcohol used in drying is itself, of course, a good disinfectant.

In the design of *processing equipment* certain requirements should be kept in mind: it should be possible to admit solutions into the processing vessel, and to let them flow out without removing the emulsion plates; likewise, it should be possible to change the temperature without removing the plates. Circulation of solutions should be achieved by gentle laminar flow, rather than by strong agitation. Good temperature control, at least to $\pm 0.5°$ C, is necessary. Surfaces

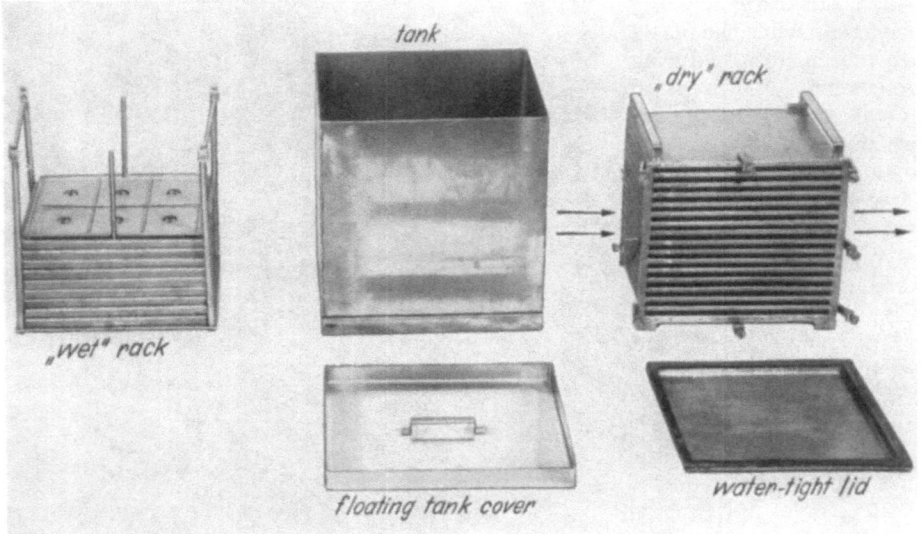

Fig. 28. Apparatus for processing a number of nuclear emulsions in a compact volume. The emulsion plates lie on shelves in the "wet" rack which is immersed in the tank during the various stages of processing, except that of active development. In the latter stage, the "dry" rack is immersed in a constant temperature bath. Plates are transferred to the dry rack by sliding the shelves removed from the wet rack into narrow watertight compartments accessible from the front of the dry rack when the cover is removed. Warm water flows above and below each compartment through shallow channels between the compartments. The direction of flow (indicated by arrows) is perpendicular to the direction in which the shelves are inserted.

with which the processing solutions come in contact must be resistant to chemical corrosion, e.g., glass or good quality stainless steel. The apparatus should permit a high degree of reproducibility from one round of processing to the next. Clearly this desideratum can best be realized by building in a considerable amount of automatic control over the processing routine.

In planning a facility for developing thick nuclear emulsions, one of the necessary choices is that between "two-dimensional" and "three-dimensional" processing vessels, i.e., whether the emulsion plates are to lie in a single layer or in some stacked arrangement on shelves. The former style has the advantage that there is less need for concern over vertical gradients of concentration in the design of the system for circulating the solutions. Also, the method lends itself to the application of commercially available trays, if considerations of expense preclude the fabrication of more elaborate processing equipment. The limitations of the "two-dimensional" method are the following: in developing large batches of emulsion, which is a typical requirement associated with the pellicle stacks used for high-energy physics, the total area of emulsion surface may run to several square meters. This necessitates extensive darkroom areas, and correspondingly extensive temperature baths. The large trays, or the considerable number of them, or both, tend to produce a rather unwieldy situation in the darkroom. Also, a larger volume of processing solutions must be prepared. It is possible instead to employ multiple-layered racks [S 11] such as those shown in Fig. 28. This can be done even for the "warm-dry" stage of development by means of a dry rack such as that shown in the same figure, in which warm water circulates in channels between the shelves without coming in contact with the plates. The compactness of the "three-dimensional" type of processing equipment makes only modest demands on darkroom space, and conduces to great convenience in handling the emulsions.

For fixing, it is highly desirable to circulate the hypo solution slowly between the fixing tank and a reservoir, or storage vessel. Replenishment is carried out in the latter tank, thereby minimizing concentration shocks, and the solution is circulated by means of a self-priming stainless steel pump.

Techniques for processing unmounted pellicles, free from glass, are described by Rosenfeld et al. [R7], and by Fox and Waniek[1].

VI. Microscopy.

34. Nuclear track microscopes. In recent years several manufacturers have developed microscopes suitable for scanning and measurements required in the application of nuclear emulsions. Several of these instruments are illustrated in Fig. 29. Part (d) of that figure shows a microscope developed by G. T. Zorn et al.[2]; one of its distinctive features is the very large stage, required to accommodate the large sheets of emulsion often used in high-energy physics. Among the limitations of standard microscopes which makes many of them unsuitable for use with emulsions, are the excessive "stage noise", the very short working distance of their objectives, the small size of the mechanical stage, and the lack of suitable screw micrometers for range and scattering measurements.

The Cosyns stage. The translatory motion of a typical microscope stage, though sufficiently straight for most purposes, departs from rectilinearity by amounts (~ 0.2 to 0.5μ) which limit the momenta measurable by multiple scattering. Even when the ways in which the stage slides have been machined with extreme precision so as to avoid curvature, the tiniest dust particles settling upon the lubricated surfaces can produce transverse displacements which result in spurious angular readings. To overcome this difficulty, M. Cosyns [C 5] has designed a radically new type of stage in which linear motion is achieved by flexure rather than by rolling or sliding. Fig. 30 shows the basic elements of his design, which has been incorporated in microscopes built for nuclear research by Fratelli Koristka of Milan. The stage is reported to be capable of reproducible rectilinear motion to within 0.05 micron. This is comparable to the precision with which the cross-hairs of an ocular micrometer can be centered on a Ag grain.

The stage of the Koristka R4 model nuclear track microscope can accomodate plates up to a square foot in area. It is equipped with hydraulic transmission for the Z-drive, high precision micrometric and gonimetric eyepieces, long-working-distance objectives, a magnetomechanical rotator for the plates, and cell divisors for stepwise motion during scattering measurements.

35. Scanning equipment. In scanning, a binocular microscope with inclined eye pieces is recommended for avoidance of eye strain. High quality lenses are

[1] R. Fox and R. W. Waniek: Nucleonics **13**, No. 7, 52 (1955).

[2] G. T. Zorn: Rev. Sci. Instrum. **27**, 628 (1956). — Cf. also paper by I. Kalberg, D. M. Haskin, M. Schein and G. T. Zorn: Rev. Sci. Instrum. (to be published).

important for quantitative work, and helpful even in qualitative inspection. It is advantageous to use a microscope in which the fine-focus knob is as close as possible to the table level. This knob is constantly manipulated during microscopy, so that fatigue is reduced when the observer's forearm can rest on the table. "Hyperplane" or "periplanatic" oculars give a flatter image than ordinary ones. For some purposes, e.g., in scanning for easily observable events, wide-angle

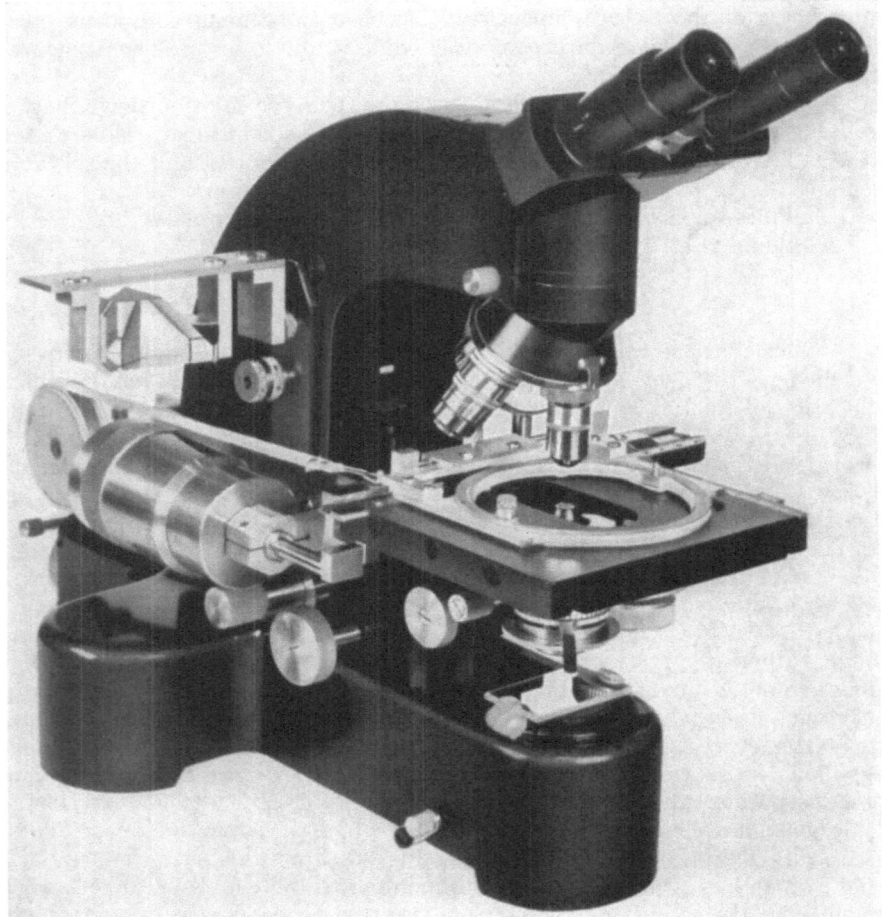

Fig. 29a — d. Several microscopes used for scanning and measurement. (a) Leitz Ortholux. A precision screw micro meter has been attached by the user in order to facilitate multiple scattering measurement.

eyepieces are useful. In this type the field is enlarged at the expense of some fuzziness in the periphery.

Some research microscopes are equipped with a built-in system of illumination; this is less likely to get out of adjustment than one which is not integral with the microscope frame. Bright-field illumination is adequate for nearly all applications, though a dark-field condenser is sometimes used to advantage. A blue filter reduces the size of diffraction haloes and improves resolution; a green filter further contributes to viewing comfort. Correct adjustment of the whole optical system is at least as important as good posture in minimizing fatigue.

Most microscopes are equipped with several objectives ranging in focal length from about 2 to 16 mm and in magnification from approximately 10 to 100. Apochromatic or fluorite lenses are desirable; these contribute to precision as well as comfort in observation. The higher magnification objectives are apt to be of the oil immersion type with high numerical aperture. At least one such lens is essential for examination of details, and for measurements. The limited depth

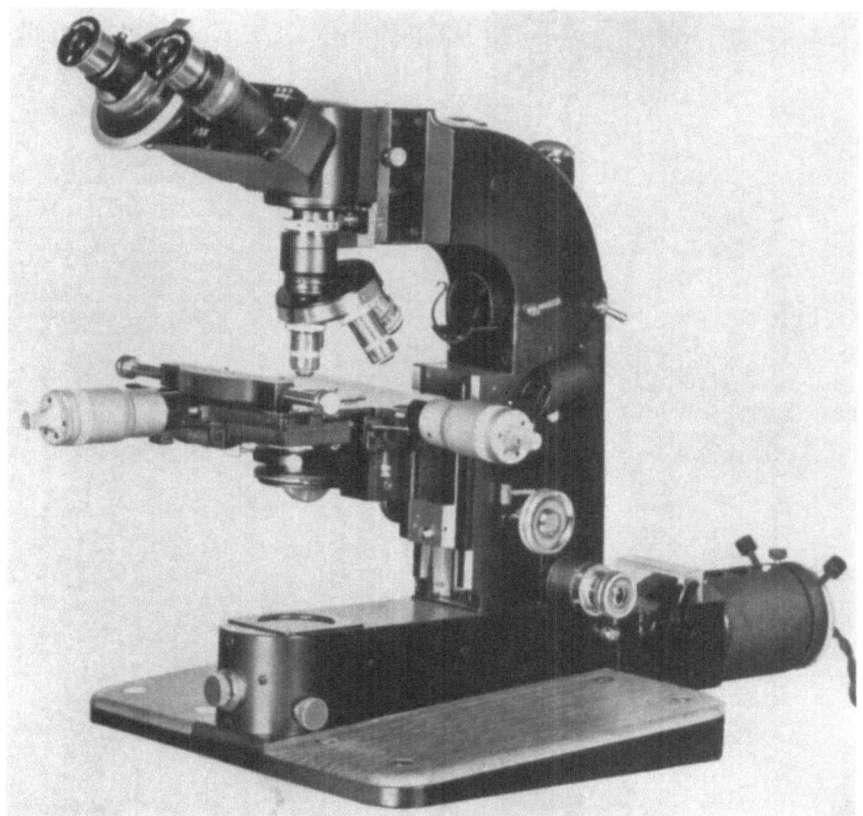

Fig. 29 b. Nuclear research microsope built by COOKE, TROUGHTON and SIMMS.

of focus in such an objective is very useful; fewer background silver grains appear to confuse the field of view; it is easier to ascertain whether a grain belongs to a given track; and depth measurements are more precise. Good immersion oils do not damage nuclear emulsion.

The medium- and high-magnification objectives commonly have a very limited working distance between the front surface of the lens and the object. When thick emulsions are employed, this can be a severe limitation. However, oil immersion objectives with long working distances (1.5 to 3.0 mm) suitable for use with very thick emulsions have recently become available from LEITZ, KORISTKA, and COOKE, TROUGHTON and SIMMS. Their magnifications vary from $\sim$20 $\times$ to $\sim$55 $\times$, and their numerical apertures, from 0.65 to 1.05. At a magnification of 100$\times$, KORISTKA has designed an objective with w.d. 0.53 mm, and N.A. 1.32.

c

Fig. 29c and d. (c) BAUSCH and LOMB microscope with special stage for scattering measurement, and large micrometer drum to facilitate precise setting of cell lengths, and reading of ranges. (d) Microscope for accommodating emulsion plates up to 16″ × 10″ in size. *A*. Precision dove-tail slides with a 21 × 15 cm movement. *B*. Ribbed cast iron base. *C*. Rotatable super stage. *D*. Magnetic clutch arrangement for plate alignment. *E*. Cast iron column for optics support. *F*. Leitz binocular microscope body and vertical focus control. *G*. Light source. *H*. Condenser assembly. *J*. Flexible cable for vertical focusing. *K*. Metric dial indicator for measurements in vertical direction. *L*. Metric dial indicator for measurements along *x*-coordinate direction.

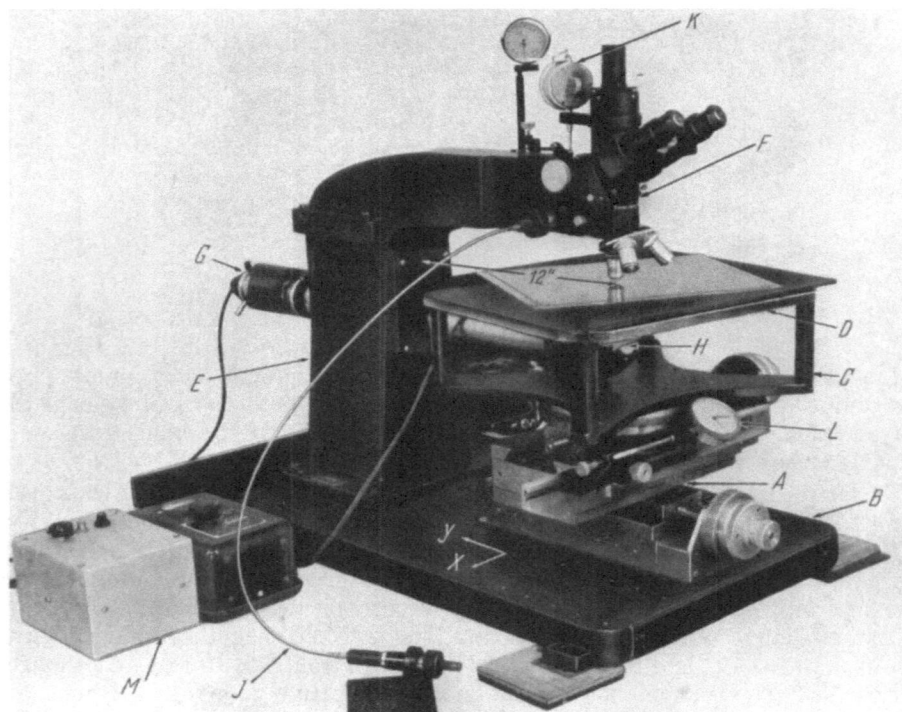

d

A method of achieving even longer working distances up to 1 cm, is the use of a *reflecting microscope* invented by BURCH[1]. Both its large working distance

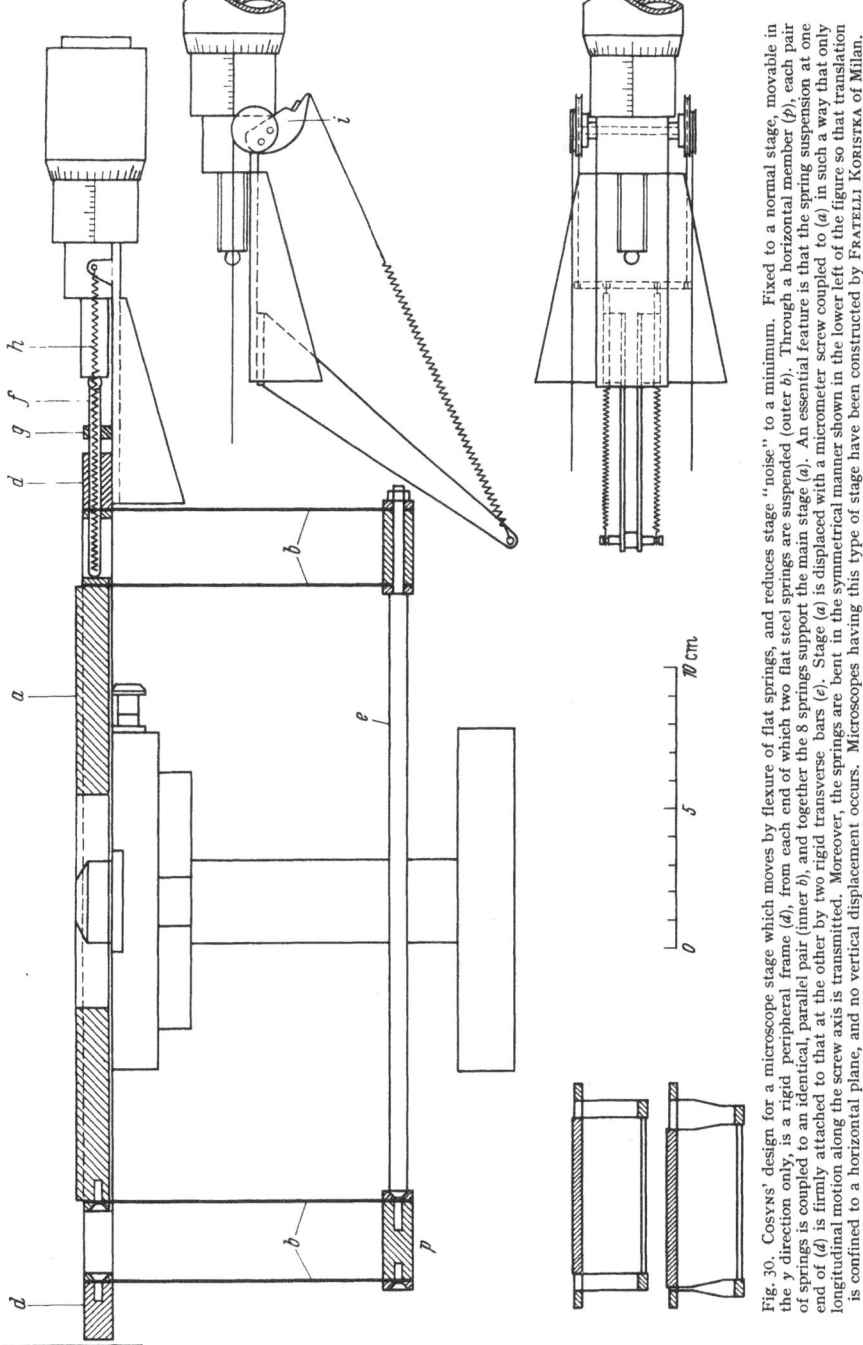

Fig. 30. COSYNS' design for a microscope stage which moves by flexure of flat springs, and reduces stage "noise" to a minimum. Fixed to a normal stage, movable in the y direction only, is a rigid peripheral frame (d), from each end of which two flat steel springs are suspended (outer b). Through a horizontal member (p), each pair of springs is coupled to an identical, parallel pair (inner b), and together the 8 springs support the main stage (a). An essential feature is that the spring suspension at one end of (d) is firmly attached to that at the other by two rigid transverse bars (e). Stage (a) is displaced with a micrometer screw coupled to (a) in such a way that only longitudinal motion along the screw axis is transmitted. Moreover, the springs are bent in the symmetrical manner shown in the lower left of the figure so that translation is confined to a horizontal plane, and no vertical displacement occurs. Microscopes having this type of stage have been constructed by FRATELLI KORISTKA of Milan.

[1] C. R. BURCH: Proc. Phys. Soc. Lond. **59**, 41 (1947). — Nature, Lond. **152**, 748 (1943). — R. L. DREW: Conf. on "Examination of Metals by Optical Methods", Brit. Iron and Steel Research Asn., May, 1949.

and its freedom from chromatic abberation are due to the replacement of lenses with reflecting surfaces. Fig. 31a illustrates the optical system, which consists of an aspherized concave mirror and a spherical convex mirror. The former has a radius of curvature of ≈ 5 cm, and the latter ≈ 1 cm. In air the working distance to the object plane is ≈ 14 mm, and the numerical aperture (which determines the resolution) is 0.65. An oil immersion arrangement is possible which increases the N.A. to 0.98, but decreases the working distance to ≈ 5 mm (cf. Fig. 31b). With this instrument it is possible to examine very thick emulsions, or to inspect an emulsion through the glass backing, or to view simultaneously the components of an emulsion "sandwich". Since the Burch microscope is completely achromatic over the spectral range from ultra-violet to infra-red, an image once in focus remains so for every wave length. Though of great potential value, this instrument has hitherto been little used for nuclear emulsions. Not only has it been generally unavailable, but also sharp images are difficult to see through great thicknesses of emulsion with any microscope, since the light is scattered on its way through the intervening layers.

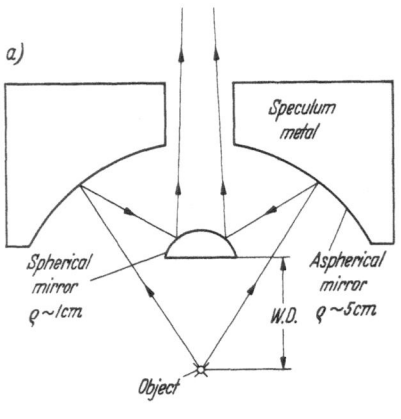

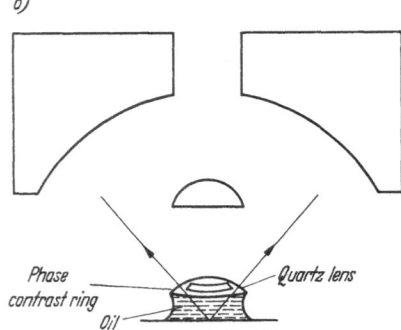

Fig. 31 a and b. (a) Optical system of the Burch reflecting microscope. (b) Adaptation to oil-immersion viewing.

The Cartesian coordinates on most *mechanical stages* can be read to 0.1 mm. While this suffices for location of tracks, it is valuable in many types of measurements (in particular, for multiple scattering) to have a much more precise micrometer screw which can set the stage to within 1 micron. Some research microscopes are also provided with a rotating stage, but unfortunately the Cartesian stage is usually mounted on top of the circular one, rather than *vice versa*. Thus, although the projection of a track on the emulsion plane can be orientated as desired, it cannot then be displaced along its own direction simply by motion of the "x" or the "y" screw alone. This becomes possible only when the circular stage is superimposed upon, and moves with, a stage capable of rectilinear motion.

For measurement of angles projected on the plane of the emulsion, an eyepiece goniometer (protractor) is useful, but not essential. Consider the two principal types of angle measurements: (a) projected angular deviations along a track for the determination of a particle's momentum from its multiple scattering; (b) angle between two tracks. For type (a), a high precision goniometer, preferably one provided with auxiliary low-power microscope for reading the angle, is required, if the angular (tangent) method of scattering measurement, rather than the sagitta method, is employed. For (b), the angle of interest is usually not the projected angle, but the true space angle between two tracks. This can be readily obtained from the components of a segment of each track, including, of course, the component normal to the emulsion plane, corrected for shrinkage. Since the latter measurement cannot ordinarily be avoided even when a goniometer is employed, this accessory is not of great help in measuring angles of type (b), except when the microscope is equipped with a tilting stage.

Crude measurements of depth can be made with almost any microscope, since a scale reading nominally in microns is commonly engraved on the fine adjustment focussing knob. In practice, the vertical motion of the optical system (or stage) is seldom found to be linear in the scale readings. Therefore, for accurate depth measurements, a special device, such as a dial gauge of the type used by machinists for high precision work, can be attached to the frame of the microscope. The dial indicator has the additional advantage that it can be read conveniently. Apart from its use in measuring depth, a dial gauge can alternatively be mounted so that it is actuated by horizontal displacement of the stage, and thus utilized to measure the projected lengths of tracks parallel to one of the stage axes. A good micrometer screw permanently attached to the stage is superior for this purpose, but when it is unavailable, a dial gauge serves reasonably well. The gauge is particularly helpful when, in experiments with controlled beams, the exposure can be arranged so that all or many of the tracks of interest are nearly parallel. This method of measuring length can be employed even when the tracks are not parallel to an edge of the plate, provided that the microscope has a circular stage mounted on a rectangular one, as described above. The plate can then be rotated, and the track projection aligned so that its length (e.g., residual range) can be measured with the dial gauge.

A wide assortment of eyepiece reticules and scales is available commercially. A square reticule is helpful in systematic scanning, and can be used for rapid estimates of "x" and "y" components; a scale divided into 50 to 100 divisions is needed for various measurements on tracks. For calibration of eyepiece scales or other reticules, a "stage micrometer", usually divided into 0.01 mm divisions, can be used. A more finely divided spectroscopic grating is useful at the higher magnifications.

Photomicrography and microprojection of tracks. The basic equipment required is a photographic bellows, ground glass screen, and plate or film holder, i.e., a camera with lens removed. The camera is rigidly mounted over the monocular tube of the microscope so that an image will remain in focus when the ground glass screen is replaced by sensitive film. Since most tracks dip appreciably with respect to the emulsion plane, the process of getting a high quality photomicrograph is a time-consuming task, even with special equipment. The use of oil immersion objectives with high numerical aperture yields a sharper image, and the limited depth of field reduces the number of background grains in focus, thereby enhancing the visibility of the track. Under these conditions, however, only a short segment of track is in focus at a time, and many separate exposures, leading finally to a photomosaic, are required. Commercial apparatus for photomicrography must ordinarily be modified for convenience in making photomosaics.

Microprojection has been used as an aid in measurement and for the training of microscopists. Some of the earliest measurements of multiple Coulomb scattering were made by projecting the track image on graph paper, sketching the track, and measuring angular deviations from the drawing. This method has been generally supplanted by the improved techniques now available. Projection is particularly useful, of course, when it is desired to demonstrate an event or a method of measurement to several observes at once.

36. Auxiliary instruments for track microscopy. The considerable time and effort expended in the microscopic examination of emulsions has led to the invention of instruments for easing and speeding the work. Most of these devices have been designed either to facilitate *detection*, or to enhance the speed, convenience, and precision of *measurement*, once an event of interest has been located. Several instruments of each type will be described.

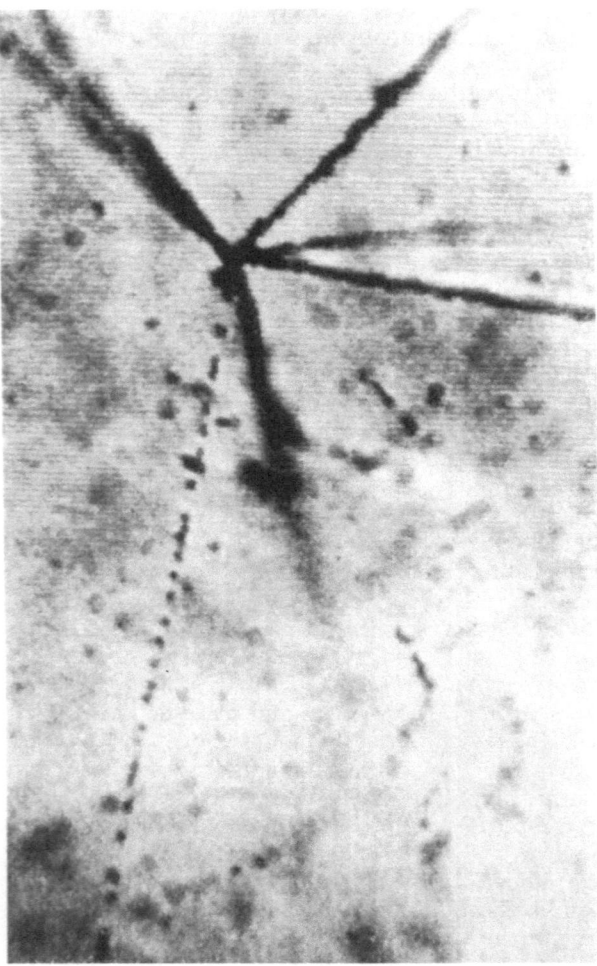

Fig. 32. Appearance of cosmic-ray star on a television screen. (Courtesy E. PICKUP.)

A method of illumination has been devised by LAND et al.[1] which depends upon the slow rotation of a planar darkfield beam to illuminate successively the tracks at various azimuths. As a given track is momentarily illuminated by light perpendicular to it, it scatters the light more strongly than the background grains, and thus calls attention to itself by "blinking", i.e., alternatively gleaming and disappearing. The system operates at low magnifications (50 to 100×, overall), the field of view is ∼3 mm in diameter, and the depth of field is ∼50 to 100 microns. Tracks with reasonably high grain density, and particularly stars which include several such tracks can be found rapidly with this device. Thin tracks are easily missed, however.

An electronic scanner for counting fairly dense tracks of particles was built by D. T. WILLIAMS[2]. An apparatus for scanning by projection, with the stage motor-driven both in the emulsion plane and perpendicular to it, was developed by MASKET and WILLIAMS[3] to lighten the tedium of systematic searching. AMALDI and his collaborators[4] devised an electronic scanner which lends itself especially to probing emulsions exposed

[1] E. H. LAND, G. BIRD and W. A. SHURCLIFF: J. Opt. Soc. Amer. **40**, 61 (1950).

[2] D. T. WILLIAMS: Batelle Memorial Institute Report T-33 (1950).

[3] A. V. MASKET and L. B. WILLIAMS: Rev. Sci. Instrum. **22**, 113 (1951). — Another type of scanning device has been described by P. V. C. HOUGH and R. O. WINDER: Bull. Amer. Phys. Soc. [II] **1**, 291 (1956).

[4] E. AMALDI, C. CASTAGNOLI and C. FRANZINETTI: Nuovo Cim. **4**, 1165 (1956).

to beams of particles with essentially parallel paths. Independently, GOLDSACK and VAN DER RAAY[1] constructed a photoelectric scanning instrument capable of measuring semiautomatically the angular distribution of tracks to about $\pm\frac{1}{4}$ degree. Tracks near minimum ionization can be detected with this apparatus. Television techniques have been applied to nuclear microscopy by ROBERTS and YOUNG[2] and by PICKUP[3]. Fig. 32 shows the appearance of a televised cosmic-ray star at a single focal depth.

An ingenious and rather complex apparatus suitable for a variety of semi-automatic measurements on tracks was invented by BLAU, RUDIN, and LINDEN-BAUM [B 13]. It is able simultaneously to count grains, make scattering measurements, and determine ranges, as well as to record these quantities. Automation in nuclear track microscopy is demonstrating its value in facilitating the search for specific types of phenomena which are not too complicated, and in speeding up measurement. Thus far, however, semi-automatic methods of scanning have not been widely adopted. No matter how ingeniously contrived, these instruments do not yet offer serious competition to the human eye in the recognition of complex and interesting phenomena in emulsions.

Considerable attention has been devoted to various improvements in the design of mechanical stages. Among these are a microscope superstage with a high precision screw[4], a device for rotating emulsion plates about the optical axis of the objective[5], and a tilting stage [B 3]. Further examples of auxiliary equipment are reflection goniometers which permit angular measurements of high precision[6,7], and an optical system for attaining long working distances in microscopy[8].

37. Measurement of ionization: various parameters. Among the measurable characteristics of tracks which depend at least partly on the rate of ionization loss, are the following:

n = grain density, the number of Ag grains in a length t of track;

B = blob density, the number of Ag clusters in t;

h = the number of gaps[9] in t; $(h = B)$;

G = length of a gap;

L = total gap length, i.e., sum of gap lengths in t;

L_B = total blob length $= t - L$;

h_ε = number of gaps in t which exceed ε in length;

L_ε = sum of gap lengths in t which exceed ε;

$\overline{G}$ = mean gap length (O'Ceallaigh parameter);

$\overline{G}_\varepsilon$ = mean value of gap lengths exceeding ε;

g = $1/\overline{G}$ = coefficient of the exponent of the gap length distribution (Fowler-Perkins parameter);

[1] S. J. GOLDSACK and H. VAN DER RAAY: J. Sci. Instrum. **33**, 135 (1956).

[2] F. ROBERTS and J. Z. YOUNG: Nature, Lond. **167**, 231 (1951); **169**, 963 (1952).

[3] E. PICKUP, Television Viewing and Automatic Scanning for Photographic Emulsions. Unpublished report, National Research Council, Canada; and private communciation.

[4] T. M. PUTNAM and J. F. MILLER: Rev. Sci. Instrum. **23**, 760 (1952).

[5] A. K. DYSON, F. C. GILBERT, C. O. HERRALA, C. E. VIOLET and R. WHITE: UCRL-4672, March, 1956.

[6] M. G. E. COSYNS: Bull. centre phys. nucleaire univ. libre Bruxelles, No. 30. 1951.

[7] B. RANKIN: Rev. Sci. Instrum. **25**, 496 (1954).

[8] J. DYSON: Proc. Phys. Soc. Lond. B **62**, 565 (1949).

[9] The term "gap density" is avoided in this article, as it has been used, rather confusingly, to describe two different parameters, h and L/t.

n_0 = grain density at plateau ionization. Similarly, the subscript zero will denote plateau values of the other variables;

$n^* = n/n_0$. Similarly, the asterisk will be used for other variables to denote the ratio of a track parameter to its value at plateau ionization.

L/t = "lacunarity", the fractional gap length.

Thus far we have dwelt mainly on the grain density n as a measure of the rate of ionization loss of a particle. For the usual intensities of development, the value of n_0, the grain density at the Fermi plateau in G.5 emulsion lies between $\sim$20 and 35 per 100 μ. On the other hand, there is evidence [H4], [F3] that the number of AgBr crystals per 100 μ of unprocessed emulsion is of the order of 300. Since the developed Ag grains are commonly $\sim$0.5 to 0.8 μ in diameter, saturation in n can occur at values as low as $\sim$4n_0. Subjectivity in grain counting becomes, of course, very serious as saturation is approached, and, even for thin tracks, the reproducibility is limited. This difficulty can be circumvented by employing the convention that the grain count for a blob is proportional to its length. The blobs may be measured with a filar micrometer, or, instead of literally measuring each blob, the observer can view the track against an eyepiece scale, and convert immediately from estimated blob length to grain count[1]. In this way, one can attain an adequate degree of reproducibility. However, the ease and speed of simple grain counting is lost. For gray or dense tracks, ionization parameters other than n can yield the required information more efficiently.

Blob counting[2] provides an adequate solution[3] for densities up to $\sim$2n_0. At higher ionizations, B becomes insensitive; it reaches a broad maximum, then declines as blobs continue to coalesce into larger blobs. However, as we shall see below, blob counting can be combined with gap counting to give a method which is applicable over a wide range of ionizations. HODGSON[4] suggested exploiting the *gaps* in dense tracks. Measuring the lacunarity L/t, he showed that groups of particles of different mass are statistically resolvable even at very short ranges.

38. Method of mean gap length. O'Ceallaigh found that the frequency distribution of gap lengths follows an exponential law[5]

$$p(G)\, dG = \frac{1}{\overline{G}}\, \epsilon^{-G/\overline{G}}\, dG, \tag{38.1}$$

where $\overline{G}$ is the mean gap length. In practice it is convenient to measure the integral size distribution which is, of course, also exponential (see Fig. 33). $\overline{G}$ suffers from the limitation that the number of smallest, barely resolvable gaps counted depends on the optical conditions. Hence it is preferable[6] to

---

[1] For examples of the use of n versus residual range in mass determination, see C. F. PoWELL ([V2], p. 182); and M. M. SHAPIRO, N. SEEMAN and B. STILLER, Proc. Fifth Rochester Conference, January, 1955, Interscience Publishers, Inc., New York, pp. 114—115; see Fig. 18 in this article.

[2] L. VOYVODIC, Report on Bristol Conference on Heavy Mesons, 1951. — A. H. MORRISH: Phil. Mag. **43**, 533 (1952). — The effect of processing on estimation of ionization by blob counting is discussed by C. O'CEALLAIGH ([P1], p. 412).

[3] This is also true, as we shall see, for exceedingly dense tracks, provided the charge on the particle is not too high. Here, however, it is more convenient to count the gaps.

[4] P. E. HODGSON [H9]; see also [R1]. — G. BELLIBONI and M. MERLIN: Nuovo Cim. **8**, 349 (1951).

[5] Ref. [B1], p. 73.

[6] M. G. K. MENON and C. O'CEALLAIGH: Proc. Roy. Soc. Lond. **221**, 292 (1954). — R. M. TENNENT: Ph. D. thesis, London University, 1953.

measure $\overline{G}_\varepsilon$, the mean length of gaps exceeding a suitable lower limit ε. As a consequence of Eq. (38.1), $\overline{G}$ may then be deduced simply from

$$\overline{G} = \overline{G}_\varepsilon - \varepsilon. \tag{38.2}$$

Now $-\overline{G}$ is the reciprocal slope of the semi-logarithmic curve $\ln h_1 = \ln h - \dfrac{\varepsilon_1}{\overline{G}}$,

where h_1 is the number of gaps per unit length exceeding ε_1. Then $\overline{G}$ can be determined from two counts h_1 and h_2 of gaps exceeding ε_1 and ε_2 respectively:

$$\overline{G} = \frac{\varepsilon_2 - \varepsilon_1}{\ln (h_1/h_2)}. \tag{38.3}$$

Assuming that the gap length distribution remains exponential down to the very smallest visible gaps, it should be possible to replace h_1 with $h = B$, the mean number of gaps per unit length, i.e., the blob density. This variant of the mean gap method was suggested and explored by FOWLER and PERKINS [F 3], who adopted as an ionization parameter the quantity $g \equiv 1/\overline{G}$, the coefficient of the exponent of the gap length distribution. In order to minimize the subjective (and optical) variability involved in counting the barely resolvable gaps or blobs, they normalized the observed g values with respect to g_0, the value for plateau tracks. The ratio $g^* = g/g_0$ was found to be highly reproducible. Moreover, g^* appeared to be independent of development, at least for ionizations corresponding to proton ranges > 1 cm. For denser tracks, this was no longer true. Finally, for singly charged particles, g^* varied approximately as the mean total rate of energy loss of the particle[2].

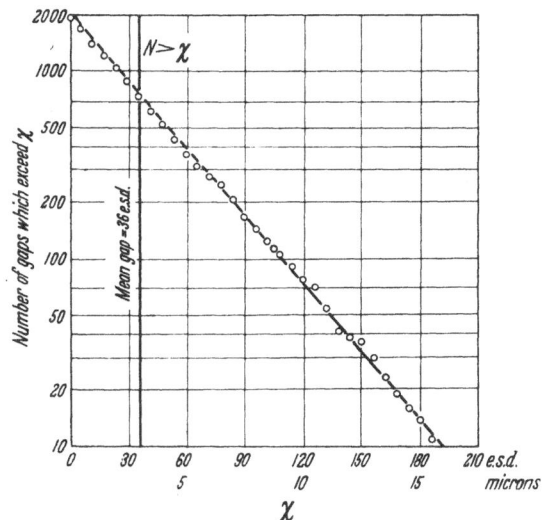

Fig. 33. Integral distribution of gap lengths in the track of a pion. The number of gaps which exceed χ is plotted against χ. (e.s.d. = eyepiece scale division.) (After MENON and O'CEALLAIGH[1].)

If ε_2, the larger of the two lower limits, is chosen to be ~ 2.5 times the mean gap length, high accuracy in g can be achieved relatively fast. The efficiency of the method is due first, to the speed of blob counting; secondly, to the relatively small number of long gaps, which may be rapidly counted without too frequent pause for decision on borderline gap lengths. A useful rule of thumb appears to be that the number of large gaps to be counted is $\sim \frac{1}{4}$ the number of blobs. Fig. 34 shows results obtained by FOWLER and PERKINS with protons. The normalized coefficient g^* is plotted against residual range[3].

---

[1] M. G. K. MENON and C. O'CEALLAIGH: Proc. Roy. Soc. Lond., Ser. A **221**, 292 (1954).

[2] Although the smallest gaps deviate from an exponential distribution, this deviation does not vitiate the empirical utility of the method. It merely results in a measured g which differs slightly from $1/\overline{G}$.

[3] The utility of counting longer gaps alone (h_ε) vs. range, for particle discrimination, has been discussed by F. T. GARDNER and R. D. HILL, Nuovo Cim. **2**, 820 (1955). Generally, h_ε by itself is not a precise index of velocity, since it depends on the final grain diameter, and is sensitive to gradient in development ([B 1], p. 77).

39. Lacunarity. For ionizations exceeding $\sim 6\, I_{\text{plat}}$ (corresponding to $\beta < 0.3$ and proton ranges $\lesssim 9\,$mm), the mean gap method has very limited precision

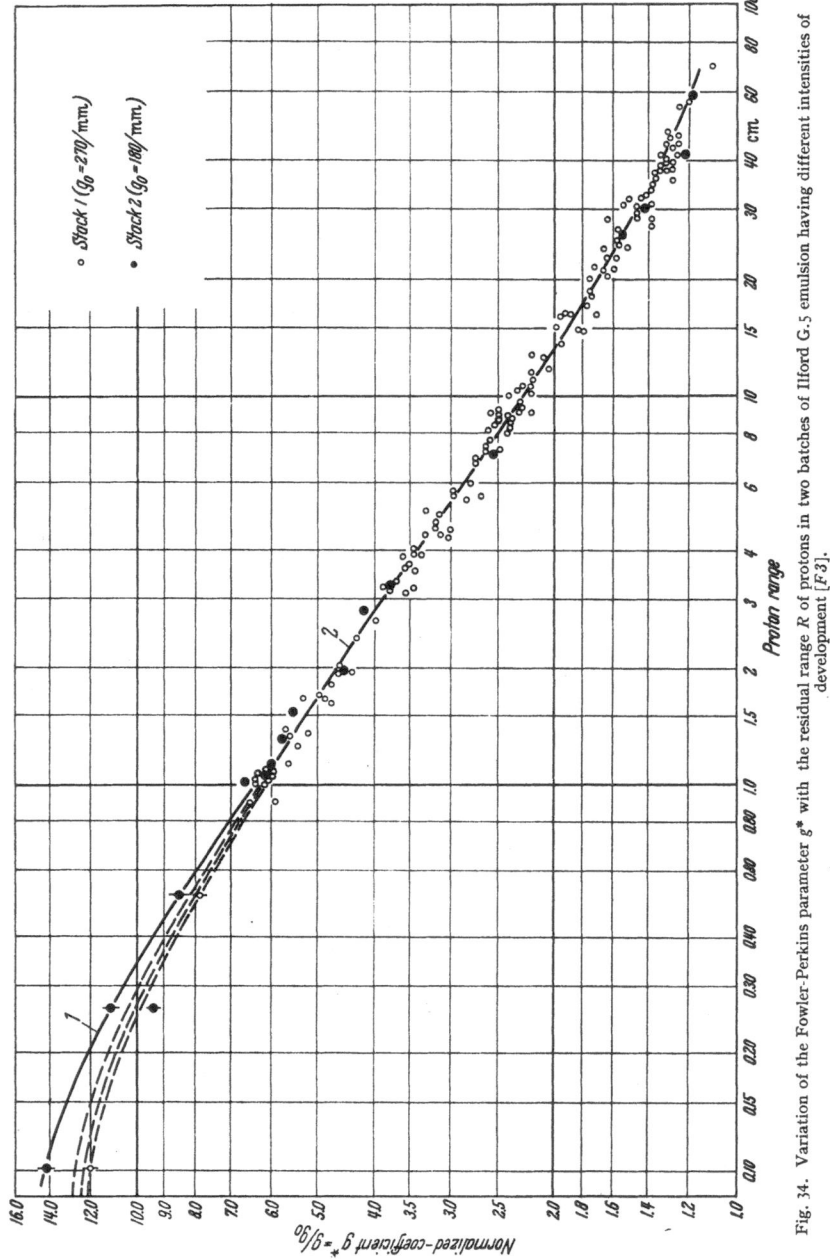

Fig. 34. Variation of the Fowler-Perkins parameter g^* with the residual range R of protons in two batches of Ilford G.5 emulsion having different intensities of development [F3].

when applied to single tracks of slow particles. To be sure, the number of "long" gaps per unit length remains intrinsically a sensitive parameter, provided that enough long gaps are available. This is true in a statistical study when the gap data may be drawn from many tracks in each range interval. It also applies to light, relativistic, multiply charged ions like Li or Be, where adequate track length of essentially uniform density is available. However, for the rapidly changing ionization of a singly charged particle near the end of its range, the available number of gaps within a narrow interval of ionization is too meager.

An alternative parameter is the lacunarity L/t introduced by HODGSON. O'DELL and SHAPIRO[1] have explored a variant of this method for mass determination of short-range particles (1 mm $\lesssim R_p \lesssim 1$ cm). Plotting L_B/L, the ratio of blob length to gap length, versus range, for protons, pions, and K mesons, they found that the relationship conforms to an approximate power law down to small ranges $(R_p \approx 1$ mm$)$[2]. Lacunarity can be measured fairly fast, since no record is required of individual gap lengths, and one need not count gaps nor pause to judge when a minimum gap size has been exceeded. Moreover, the method lends itself readily to semi-mechanization, using a simple motor-drive technique like that of D. RITSON [R3].

40. Comparison of ionization parameters. CASTAGNOLI, CORTINI, and MANFREDINI [C2] have compared critically the various parameters associated with ionization. They plotted each parameter $f(r)$ against the residual range $r = R/M$ measured in the units microns/electron mass[3]. In judging which are the best variables to use in the various intervals of range, they applied the following criteria: (a) ratio of slope of the $f(r)$ curve to the relative fluctuation of the variable in question for a given track length; (b) reproducibility from observer to observer; (c) independence of intensity of development. Their recommendations may be summarized as follows[4]: For $0 < r < 5$ μ/m_e, use $\bar{B}$ or $\bar{L}$, for $5 < r < 10$, use $\bar{L}$ or $\bar{G}$; for $10 < r < 300$, use $\bar{G}$.

41. Techniques for measuring mean gap length, lacunarity, etc. For carrying out these measurements, it is desirable to have a microscope equipped with a rotating stage so that the tracks may be aligned along the x or y direction of motion, as well as an easily controllable micrometric movement for the stage in at least one of these directions. In addition, a filar micrometer eyepiece is desirable. As an illustration, consider the measurement of total blob length L_B. Suppose that the fine movement of the stage is along the x direction (see Fig. 35). Then the track is first aligned parallel to the x movement. The hairline in the eyepiece, whose initial position is at the left in the field of view, is moved to the right across a gap. The track is then displaced to the left until the length of the next blob has crossed the hairline. This procedure is repeated until the hairline has traversed the field of view. The total gap length is read from the eyepiece micrometer, and the blob length from the stage micrometer. This type of procedure has been used by DELLA CORTE [D4] to measure mean gap length as well. In this case, one must count the gaps as well as accumulate their length on the micrometer drum.

---

[1] F. W. O'DELL and M. M. SHAPIRO: Phys. Rev. A **99**, 641 (1955); U. S. Naval Research Laboratory Quarterly on Nuclear Science and Technology, April 1, 1955, p. 13. (Progress Report for the Period January — March 1955.)

[2] The degree of precision attainable is indicated by the following mass-ratio determinations: from 10π tracks and 6 proton tracks, a p/π ratio of 6.96 ± 0.45 was obtained, compared with the best value, 6.73; from 10 tracks of K-mesons and 7 tracks of protons, a p/K ratio of 1.92 ± 0.15, compared with the accepted value 1.90 for p/τ. The measurements were made at residual ranges $R < 2$ mm for pions, < 6 mm for K-mesons, and < 12 mm for protons. See Fig. 17.

[3] See Fig. 1 of Ref. [C2]. In Table 3 of the present article, column 1 gives the range r in these units for convenient comparison with the work of CASTAGNOLI et al.

[4] Note that the range intervals for a particle of mass M electron masses, are given by $R = rM$ microns.

Instead of measuring $\overline{G}$ directly, one may use the integral method which, according to O'CEALLAIGH [O 2] is about 5 times faster. The observer counts the number of gaps whose length exceeds some minimum value ε_1, then re-

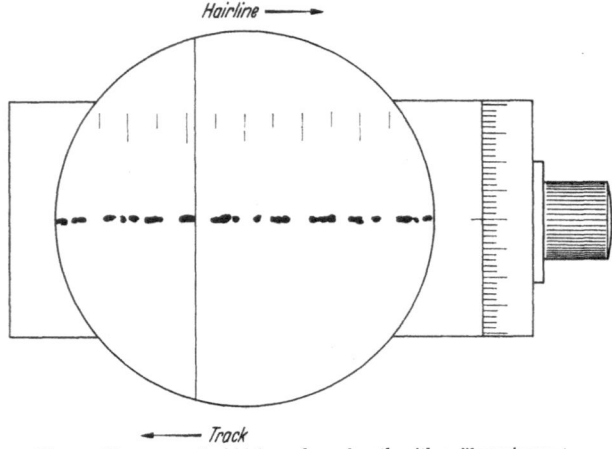

peats the procedure for at least one other mini- mum value ε_2. $\overline{G}$ is then given by Eq. (38.3), or, if high precision is re- quired, and more than two points have been plotted in the integral gap-size distribution, the best slope $1/\overline{G}$ may be obtained by least squares. In carrying out the measurements, O'CEAL- LAIGH employed a special reticule mounted in the image plane of the filar micrometer eyepiece. This consists of a "spider web"

Fig. 35. Measurement of blob- and gap-length with a filar micrometer eyepiece.

having hairlines spaced at convenient intervals, corresponding to various cut- off values of gap size. The web arrangement is shown in Fig. 36.

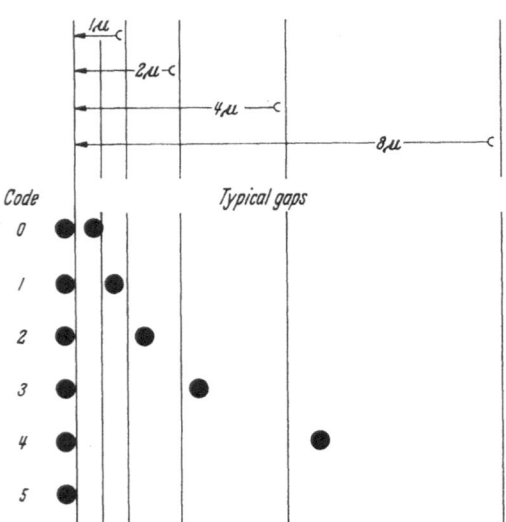

Fig. 36. O'CEALLAIGH's method of determining mean gap length from an integral distribution of gap-sizes.

In order to speed up the fairly laborious measurements of ioniza- tion, a number of electro-mechan- ical schemes have been devised. In RITSON's method [R 3] a motor drive is attached to the micro- scope stage in such a way that the track is driven past a hairline at a slow, uniform rate. The observer is provided with two counters, one of which runs con- tinuously, marking the passage of time, hence the passage of a known length of track. The other counter runs only when a button is depressed, and this is done by the observer from the moment that the hairline touches one end of the gap until it has traversed the gap.

Various elaborations of this procedure have been developed, and some of these are capable of extracting several types of data on ionization simultaneously. An example is the apparatus of BARONI and CASTAGNOLI ([P 1], p. 364), with which it is possible to measure concurrently the parameters h, h_ε, and L, as well as the number of delta rays in a given length of track. Another semiautomatic system, devised by HOOPER and SCHARFF [H 11], enables one to record the lengths of individual gaps without writing these down. Fig. 37 shows their arrangement, in which an insulated wheel with 24 evenly spaced brass contacts is mounted on the drum of an eyepiece micrometer, and connected through a microswitch A and a power supply to a suitable counter. When a gap is being traversed, the observer depresses switch A, and the number of contacts passed as the hairline moves from one end of the gap to the other is printed by the counter on a strip of paper. This record enables a detailed analysis of the observed gap-length distribution to be made later, should it be desirable. Similar devices are in use at the University of Rochester and the Naval Research Laboratory.

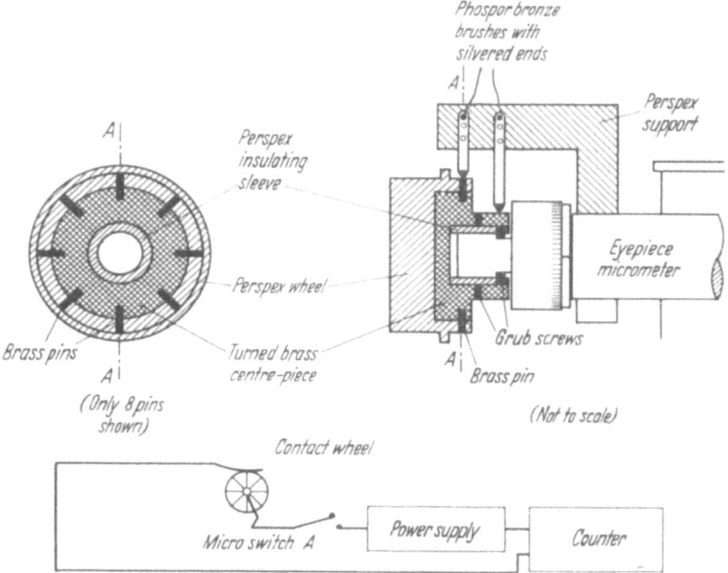

Fig. 37. Apparatus of HOOPER and SCHARFF for semi-automatic measurement of mean gap length in nuclear emulsion. (Only 8 of 24 brass contact pins are shown in the sketch.) The length of gap traversed by a hairline in the eyepiece is inferred from the number of contacts indicated by the counter.

42. Theories of track formation: variable-spacing and constant-spacing models.

A number of investigators have explored the process of track formation with a view to accounting quantitatively for the measurable characteristics[1] of tracks. A good theory of this process would make it possible to assess the relative merits of the different indices of ionization listed in Sect. 7 for measuring a particle's rate of energy loss. Such a theory would predict the values of these ionization parameters as a function of (a) the characteristics of the undeveloped emulsion (e.g., the distribution in space and in size of the AgBr crystals); (b) the average probability for the charged particle to activate a crystal—a probability which depends, of course, on emulsion sensitivity as well as on the particle's rate of energy loss; and (c) the degree of development. Various models of track formation

---

[1] In speaking of track characteristics in this section, we shall not be concerned with the deflections in a track which result from scattering, but only with the degree of blackening — i.e., the distribution of developed silver grains along the track. The effects of Coulomb scattering are treated elsewhere in this article. Also, it is beyond the scope of this monograph to discuss theories of formation of the latent image. For this subject, the reader is referred to the proceedings of two conferences on photographic sensitivity [M 10] and [M 12]. Additional references will be found in those volumes.

have been proposed by DEMERS[1], BLAU[2], and HAPP, HULL, and MORRISH[3]; their ideas have influenced the more recent theories which we shall discuss here[4].

Consider first the state of affairs in the unprocessed nuclear research emulsion. Since the silver halide occupies very nearly one-half of the emulsion volume, the crystals must form a rather closely packed configuration like that in the upper part of Fig. 38. During development, the activated crystals grow considerably in size (cf. Sect. 22); the remaining crystals are dissolved by the fixing process. However, an unactivated crystal adjacent to one which has been rendered developable, is partly invaded or completely enveloped by the process of physical development which enlarges the activated neighbor.

Neither the actual spatial distribution of crystals in the unprocessed emulsion, nor the detailed process of crystal growth and blob formation during development, is known experimentally. Hence, in trying to account for the observed properties of the track formed by a given particle, one must begin by assuming some type of crystal distribution. Should one postulate a completely random distribution, or a fairly regular one? Although each of these assumptions is an idealization, it is instructive to calculate the consequences of each extreme type of model, and to compare them with experiment.

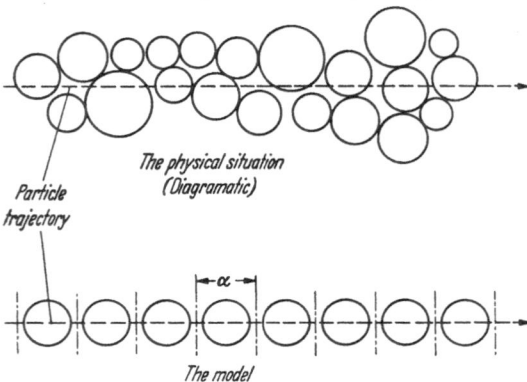

Fig. 38. Array of idealized halide "crystals" according to the constant-spacing model (after HERZ and DAVIS [H4]).

A "variable spacing model" was proposed by O'CEALLAIGH[5], and further explored by FOWLER and PERKINS [F3]. A "constant spacing model" was adopted by DELLA CORTE[6], and investigated more rigorously by HERZ and DAVIS [H4]. Employing a "one-dimensional approximation", BLATT [B10] subjected a variety of models to mathematical analysis, deriving expressions for the values of several ionization parameters and their standard deviations.

Variable spacing model. From the observed exponential distribution in gap length G [see Eq. (38.1)], O'CEALLAIGH[7] was led to postulate that the gaps between the original crystals are also distributed exponentially. Experimentally it is known that the distribution in G is *not* exponential, but it becomes asymptotically so for large G. Thus, for practical purposes it is exponential for thin tracks, and even for dense ones when h_ε is counted rather than h (this cuts out gaps $< \varepsilon$). For thin tracks, O'CEALLAIGH verified experimentally the relation $\overline{G} = \overline{G}_\varepsilon - \varepsilon$

[1] P. DEMERS: Canad. J. Res. A **25**, 223 (1947).

[2] M. BLAU: Phys. Rev. **75**, 279 (1949).

[3] W. W. HAPP, T. HULL and A. H. MORRISH: Canad. J. Phys. **30**, 669 (1952).

[4] *Note added in proof.* See also the Proceedings, First International Conference on Corpuscular Photography, held at the University of Strasbourg, July 1—6, 1957.

[5] [B1], p. 73; [O2]; [P1], p. 412; see also, M. G. K. MENON and C. O'CEALLAIGH, Proc. Roy. Soc. Lond., Ser. A **221**, 292 (1954); and R. H. W. JOHNSTON and C. O'CEALLAIGH, Phil. Mag. **45**, 424 (1954).

[6] M. DELLA CORTE, M. RAMAT and L. RONCHI jr.: Nuovo Cim. **10**, 509 and 958 (1953), also [D4].

[7] [B1], p. 73. FOWLER and PERKINS [F3] found that the distribution is exponential up to very high rates of energy loss, i.e., for *dense* as well as thin tracks.

[Eq. (38.2)], which follows from his assumptions. He concluded, furthermore, that the value of $\overline{G}_\varepsilon$ is *independent of the degree of physical development*. Hence, no correction for variations with depth (or from plate to plate) should be required. To the extent that this is true, the parameter $\overline{G}_\varepsilon$ has a very great advantage over other indices of ionization, for it mitigates the common difficulty of non-uniform development. Actually it has not been proved that this is true for very dense tracks[1].

Constant spacing model. Since the variations in crystal size are small, and the AgBr crystals are rather closely packed, it is worth while to explore a model in which fluctuations are neglected and an orderly array of crystals is assumed. The lower part of Fig. 38 illustrates the model adopted by HERZ and DAVIS [*H 4*]. The particle which makes a track is assumed to traverse a succession of equally spaced spherical crystals of uniform size; the spacing α thus constitutes a "cell" containing a single crystal.

We shall now compute the blob density $B = h$ according to this model. Let p be the probability that a crystal be rendered developable, which depends upon the rate of energy loss of the passing particle. If the crystal is activated, it grows during development to a diameter[2] $\gamma\alpha$. Let Γ be the integral part of γ (e.g., for $2.0 \leq \gamma < 3.0$, $\Gamma = 2$). Then the smallest possible gap is $\alpha - (\gamma - \Gamma)\alpha$. This is called a "first-order gap". The next larger gap is $2\alpha - (\gamma - \Gamma)\alpha$ in length, and the i-th order gap is

$$G_i = i\alpha - (\gamma - \Gamma)\alpha. \tag{42.1}$$

Note that Γ is the number of unactivated crystals in the first-order gap, $\Gamma + 1$ is the number in the second-order gap, and $\Gamma + i - 1$ the number in the i-th-order gap. Therefore, the probability for an i-th order gap to start with any specified cell is the probability of finding $(\Gamma + i - 1)$ unactivated crystals bounded by two developable ones, i.e., $p^2(1-p)^{+i-1}$. In a track length t there are t/α cells, and thus the expected number of i-th order gaps is

$$h_i = \frac{t}{\alpha}\, p^2 (1-p)^{\Gamma+i-1}. \tag{42.2}$$

The *total* number of gaps, $h(= B)$, is derived by summing over all values of i:

$$\left. \begin{aligned} h &= \frac{t}{\alpha} p^2 \sum_{i=1}^{\infty} (1-p)^{\Gamma+i-1} \\ &= \frac{t}{\alpha} p (1-p)^{\Gamma}. \end{aligned} \right\} \tag{42.3}$$

Similarly, the *total gap length* in t is

$$\left. \begin{aligned} L &= \sum_{i=1}^{\infty} G_i h_i = t p^2 (1-p)^{\Gamma-1} \sum_{i=1}^{\infty} (1-p)^i (i + \Gamma - \gamma) \\ &= t(1-p)^{\Gamma} [1 - p(\gamma - \Gamma)]. \end{aligned} \right\} \tag{42.4}$$

The *mean gap length* $\overline{G}$ is simply $\overline{L}/h$. Thus,

$$\overline{G} = \alpha \left[\frac{1}{p} - (\gamma - \Gamma) \right]. \tag{42.5}$$

---

[1] The results of FOWLER and PERKINS [*F 3*] shown in Fig. 34 are perhaps an indication to the contrary. It is not clear, however, whether the dependence of g^* on development at high ionizations necessarily implies that this is true of $\overline{G}_\varepsilon$. The parameter g depends on the frequency of the very smallest gaps, as well as on that of the large ones.

[2] Note that the size of the original crystal does not appear explicitly in this formulation. Instead the final Ag grain diameter is expressed in units of the cell spacing α.

If we confine our attention to "long gaps", i.e., those of order $i \gtrless s$, then their lengths exceed ε, where

$$\varepsilon = \alpha\,(s - \gamma + \Gamma). \tag{42.6}$$

Their number $h_{i \gtrless s}$ in length t, their total length $L_{i \gtrless s}$, and their mean gap length $\bar{G}_{i \gtrless s}$, respectively, are then derived like their counterparts in Eqs. (42.3) to (42.5), and given by Eqs. (42.7) to (42.9):

$$\left.\begin{aligned} h_{i \gtrless s} &= \frac{t\,p^2}{\alpha}\,(1 - p)^{\Gamma-1} \sum_{i=s}^{\infty} (1 - p)^i \\ &= \frac{t\,p}{\alpha}\,(1 - p)^{\Gamma-1+s}, \end{aligned}\right\} \tag{42.7}$$

$$\left.\begin{aligned} L_{i \gtrless s} &= \sum_{i=s}^{\infty} G_i\,h_i \\ &= t\,p^2\,(1 - p)^{\Gamma-1}\left[\sum_{i=s}^{\infty} i\,(1 - p)^i + (\Gamma - \gamma)\sum_{i=s}^{\infty} (1 - p)^i\right] \\ &= t\,(1 - p)^{\Gamma-1+s}\,[1 + p\,(s - 1 + \Gamma - \gamma)], \end{aligned}\right\} \tag{42.8}$$

$$\bar{G}_{i \gtrless s} = \frac{L_{i \gtrless s}}{h_{i \gtrless s}} = \alpha\left[\frac{1}{p} + (s - 1) + (\Gamma - \gamma)\right], \tag{42.9}$$

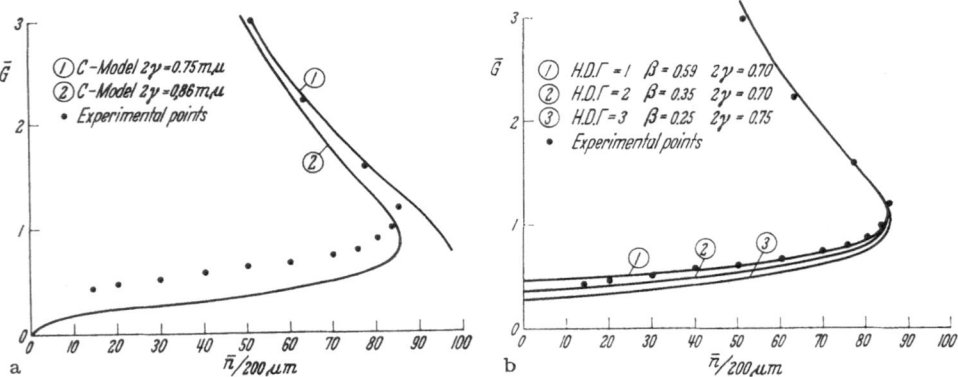

Fig. 39a and b. Comparison of two models of track formation with experiment. (a) Variation of mean gap length with blob density according to the variable-spacing model (solid curves). (b) The same measurements are compared with predictions of the constant-spacing model. (After CASTAGNOLI et al. [C2].)

and by comparison with Eq. (42.5) it is seen that

$$\bar{G}_{i \gtrless s} = \bar{G} + \alpha\,(s - 1). \tag{42.9a}$$

To deduce an expression for the grain density n, which should be applicable even for dense tracks, we first obtain the total blob length L_B in t, and then divide it by the grain diameter $\gamma\alpha$:

$$L_B = t - L = t\,[1 - (1 - p)^{\Gamma}\{1 - p\,(\gamma - \Gamma)\}]. \tag{42.10}$$

Finally,

$$n = \frac{L_B}{\gamma\,\alpha} = \frac{t}{\gamma\,\alpha}\,[1 - (1 - p)^{\Gamma}\{1 - p\,(\gamma - \Gamma)\}]. \tag{42.11}$$

Comparison of the models with experiment. From measurements on tracks having a wide range of mean gap lengths, HERZ and DAVIS obtained excellent agreement with their model in blob density B as well as total gap length L. Independent measurements by CASTAGNOLI et al. [C 2] also confirmed the predictions of the constant-spacing model, while finding agreement with the variable-spacing model only for tenuous tracks (large values of $\bar{G}$). Fig. 39 shows how the two models fare in comparison with measurements of mean gap length and blob density.

43. Other techniques and problems. In order to keep the length of this article within bounds, several methods of track measurement and certain sources of error have been mentioned only briefly. These include track photometry, the "profile method", delta rays and "thin-down", a new multiple scattering parameter, semiautomatic devices for scattering measurement, emulsion distortions, types of spurious scattering and noise, and range straggling. Literature on these topics is cited in the Appendix which follows the Bibliography.

Acknowledgments.

The author is grateful to Dr. W. H. BARKAS for communicating his range-energy tables before publication, to Mr. C. WALLER for data on Ilford emulsions, and to his colleagues at home and abroad who have generously supplied photomicrographs. He takes pleasure in acknowledging the help of Dr. R. G. GLASSER and Messrs. B. STILLER, F. W. O'DELL, G. FAZIO and E. WITTERHOLT, and in thanking AGNES G. LEAMY and DOROTHEA G. CHERVENAK for their competent typing of the manuscript.

Bibliography[1].

[A 1]   ALBOUY, G., et H. FARAGGI: Regression of the latent image due to charged particles. J. Phys. Radium **10**, 105 (1949).

[A 2]   ALLRED, J. C., and A. H. ARMSTRONG: Laboratory handbook of nuclear spectroscopy. Los Alamos Laboratory Report LA-1510, Rev., 1953.

[A 3]   AMALDI, E., C. CASTAGNOLI, G. CORTINI e C. FRANZINETTI: Life-time measurements of unstable charged particles of cosmic radiation using emulsions. Nuovo Cim. **12**, 668 (1954).

[A 4]   AMALDI, E., C. CASTAGNOLI e C. FRANZINETTI: An electronic scanner for nuclear emulsions. Nuovo Cim. **4**, 1165 (1956).

[B 1]   Bagnères Congress Proceedings, I.U.P.A.P. Conference on Cosmic Radiation 1953. (Unpublished.) See also report by M. M. SHAPIRO: Science (Lancaster, Pa.) **118**, 701 (1953).

[B 2]   BARBOUR, I.: Magnetic deflection of cosmic-ray mesons using nuclear plates. Phys. Rev. **78**, 518 (1950).

[B 3]   BARBOUR, I.: A pantograph and tilting stage for use with nuclear plates. Rev. Sci. Instrum. **20**, 530 (1949).

[B 4]   BARKAS, W. H., P. H. BARRETT, P. CÜER, H. H. HECKMAN, F. M. SMITH and H. K. TICHO: High velocity particle ranges in emulsion. Phys. Rev. **102**, 583 (1956).

[B 4a] BARKAS, W. H., F. M. SMITH and W. BIRNBAUM: Range straggling in nuclear track emulsion. Phys. Rev. **98**, 605 (1955).

[B 5]   BARKAS, W. H., and D. M. YOUNG: Emulsion tables. I. Heavy-particle functions. UCRL Report 2579, Rev. 1954.

[B 6]   BARONI, G., C. CASTAGNOLI, G. CORTINI, C. FRANZINETTI and A. MANFREDINI: On the range-energy relation for protons in nuclear emulsions. C.E.R.N. publication BS-9, Geneva 1954.

[B 7]* BEISER, A.: Nuclear emulsion technique. Rev. Mod. Phys. **24**, 273 (1952).

[B 7a] BERGER, M. J.: Multiple scattering of fast protons in photographic emulsions. Phys. Rev. **88**, 59 (1952).

---

[1] This bibliography is necessarily confined to a small fraction of the literature on nuclear emulsions and their uses. Other papers are cited in the footnotes, and many more references will be found in bibliographies provided in the papers listed here, particularly in those marked with an asterisk (*).

[B8]   Bethe, H. A.: Molière's theory of multiple scattering. Phys. Rev. **89**, 1256 (1953).

[B8a]   Birge, R. W., D. H. Perkins, J. R. Peterson, D. H. Stork and M. N. Whitehead: Decay characteristics and masses of positive K mesons produced by the
Bevatron. Nuovo Cim. **4**, 834 (1956).

[B9]   Biswas, S., E. C. George and B. Peters: An improved method for determining
the mass of particles from scattering *versus* range. Proc. Indian Acad. Sci. A **38**,
418 (1953).

[B9a]   Biswas, S., B. Peters and Rama: Scattering measurements in nuclear emulsion
and their application to measuring the charge of primary cosmic ray nuclei. Proc.
Indian Acad. Sci. A **41**, 154 (1955).

[B10]   Blatt, J. M.: Theory of tracks in nuclear research emulsion. Austral. J. Phys.
8, 248 (1955).

[B11]   Blau, M.: Possibilities and limitations of the photographic method in nuclear
physics and cosmic radiation. Acta phys. Austriaca **3**, 384 (1950).

[B12]   Blau, M., and J. A. de Felice: Development of thick emulsions by a two-bath
method. Phys. Rev. **74**, 1198 (1948).

[B13]   Blau, M., R. Rudin and S. Lindenbaum: A semi-automatic device for analyzing
events in nuclear emulsions. Rev. Sci. Instrum. **21**, 978 (1950).

[B13a] Blum, J. M.: Photoelectrons in nuclear track plates. J. Phys. Radium **12**, 860 (1951).

[B13b] Bøggild, J. K., and M. Scharff: Stage noise and spurious scattering. Reversing
sagitta method. See [P1], p. 374.

[B13c] Bonetti, A., C. C. Dilworth et G. P. S. Occhialini: Processing nuclear emulsions.
II. After-development techniques. Bull. Cent. Phys. Nucléaire Univ. Libre Bruxelles
1951, No. 13 (b).

[B13d] Bonetti, A., C. C. Dilworth and L. Scarsi: Brochure on nuclear emulsions,
"Pocket Book" series. London: George Newnes, Ltd. (to be published).

[B14]   Bonetti, A., e G. P. S. Occhialini: Cylindrical emulsions. Nuovo Cim. **8**, 725 (1951).

[B14a] Bohr, A.: Atomic interaction in penetration phenomena. Kgl. danske Vidensk.
Selsk., mat.-fys. Medd. **24**, No. 19 (1948).

[B15]   Bradner, H., F. M. Smith, W. H. Barkas and A. S. Bishop: Range-energy relation
for protons in nuclear emulsion. Phys. Rev. **77**, 462 (1950).

[B16]   Brown, R., U. Camerini, P. H. Fowler, H. Muirhead, C. F. Powell and D. M.
Ritson: Observations with electron-sensitive plates exposed to cosmic radiation.
I. Decay of μ-mesons. II. Further evidence for the existence of unstable charged
particles of mass $\sim 1000\, m_\varrho$. Nature, Lond. **163**, 47, 82 (1949).

[B17]   Budini, P.: On the energy lost by a relativistic ionizing particle in a material medium
and on the Cerenkov radiation. Nuovo Cim. **10**, 236 (1953).

[C1]   Camerini, U., W. O. Lock and D. H. Perkins: The analysis of energetic nuclear
encounters occurring in photographic emulsions. Progress in Cosmic Ray Physics,
Vol. 1, Chap. 1. Amsterdam: North Holland Publ. Co. 1952.

[C1a]   Carvalho, H. G. de, and J. I. Friedman: Range of 208 Mev protons in G.5 emulsion. Rev. Sci. Instrum. **26**, 261 (1955).

[C2]   Castagnoli, C., G. Cortini e A. Manfredini: On the measurement of ionization
in nuclear plates. Nuovo Cim. **2**, 301 (1955).

[C3]   Catala, J., and W. M. Gibson: Range-energy relation for protons and alpha-
particles in photographic emulsions. Nature, Lond. **167**, 551 (1951).

[C4]   Ceccarelli, M., and G. T. Zorn: On photometry in thick nuclear emulsions. Phil.
Mag. **43**, 356 (1952).

[C5]   Cosyns, M. G. E.: Microscopes for the measurement of scattering. Bull. Cent.
Phys. Nucléaire Bruxelles **1951**, No. 30.

[C6]   Cosyns, M. G. E., et G. Vanderhaeghe: Distortion of nuclear emulsions. Bull.
Cent. Phys. Nucléaire Bruxelles **1951**, No. 15.

[C6a]   Crussard, J., V. Fouché, J. Hennessy, G. Kayas, L. Leprince-Ringuet, D. Morellet e F. Renard: K-mesons in emulsions exposed to a 6.2 GeV proton beam.
Nuovo Cim. **3**, 731 (1956).

[C7]   Cüer, P.: Stopping power of photographic emulsions. C. R. Acad. Sci., Paris **223**,
1121 (1946).

[C8]   Cüer, P., et J. J. Jung: Experimental R-E curve for $0-5$ Mev protons in emulsion.
J. Phys. Radium **12**, 52 S (1951).

[D1]   Dainton, A. D., P. H. Fowler and D. W. Kent: A new method of determining
the charge and energy of heavy nuclei in the cosmic radiation. Phil. Mag. **42**, 317
(1951).

[D2]   Dainton, A. D., A. R. Gattiker and W. O. Lock: The processing of thick photographic emulsions. Phil. Mag. **42**, 396 (1951).

[D 3] DANIEL, R. R., G. FRIEDMANN, D. LAL, Y. PAL and B. PETERS: On the construction of large nuclear emulsion block detectors. Proc. Indian Acad. Sci. A 40, 151 (1954).

[D 3 a] DANYSZ, M., and J. PNIEWSKI: Delayed disintegration of a heavy nuclear fragment. Phil. Mag. 44, 348 (1953).

[D 4] DELLA CORTE, M.: The grain density and the process of track formation in nuclear emulsions. Part III. Nuovo Cim. 12, 28 (1954).

[D 5]* DEMERS, P.: Ionographie—Les Emulsions Nucléaires, Principes et applications (to be published).

[D 6] DEMERS, P.: Cosmic-ray investigations with special photographic emulsions. Canad. J. Res. 28, 628 (1950).

[D 7] DEMERS, P.: Cosmic ray phenomena at minimum ionization in a new nuclear emulsion having a fine grain, made in the laboratory. Canad. J. Phys. 32, 538 (1954).

[D 8] D'ESPAGNAT, B.: Theoretical introduction to the methods of measuring multiple scattering in photographic emulsions. J. Phys. Radium 13, 74 (1952).

[D 9] DILWORTH, C., S. J. GOLDSACK e L. HIRSCHBERG: Determination of the mass of slow particles by the constant sagitta method. Nuovo Cim. 11, 113 (1954).

[D 10] DILWORTH, C., S. J. GOLDSACK, Y. GOLDSCHMIDT-CLERMONT and F. LEVY: The magnetic deflection of fast charged particles in the photographic emulsion. Phil. Mag. 41, 1032 (1950).

[D 11]* DILWORTH, C., G. P. S. OCCHIALINI and L. SCARSI: Heavy mesons. Annual Rev. Nucl. Sci. 4, 271 (1954).

[D 12] DILWORTH, C., G. P. S. OCCHIALINI and L. VERMAESEN: On processing nuclear emulsions. See Ref. [M 10], p. 297.

[E 1] Eastman Kodak Co., Research Laboratories: Kodak nuclear track plates and Kodak nuclear track pellicles. Leaflet issued August 1956.

[E 2] EKSPONG, A. G.: On multiple scattering measurements in nuclear research emulsions. Ark. Fysik 9, 49 (1954).

[F 1] FARAGGI, H.: Experimental determination of R-E relations and the stopping power of nuclear emulsion for charged particles of low energy. C. R. Acad. Sci. Paris 230, 1398 (1950).

[F 1 a] FAY, H., K. GOTTSTEIN e K. HAIN: Numerical tables of relations frequently used in nuclear emulsion work. Suppl. Nuovo Cim. 11, 234 (1954).

[F 2] FOWLER, P. H.: Nuclear transmutations produced by cosmic-ray particles of great energy. III. (Includes description of the author's coordinate method of multiple scattering measurement.) Phil. Mag. 41, 169 (1950).

[F 3] FOWLER, P. H., and D. H. PERKINS: Measurement of ionization in nuclear emulsions. Phil. Mag. 46, 587 (1955).

[F 3 a] FRANZINETTI, C.: On the masses of charged particles of the cosmic radiation. Phil. Mag. 41, 86 (1950).

[F 4] FREIER, P., E. J. LOFGREN, E. P. NEY, F. OPPENHEIMER, H. L. BRADT and B. PETERS: Evidence for heavy nuclei in the primary cosmic radiation. Phys. Rev. 74, 213 (1948).

[F 4 a] FRIEDLANDER, M. W., D. KEEFE, M. G. K. MENON and M. MERLIN: On the mass of the Λ° particle. Phil. Mag. 45, 535 (1954).

[F 5] FRY, W. F., and G. R. WHITE: Range distribution of μ-mesons from π-meson decay in photographic emulsion. Phys. Rev. 90, 207 (1953).

[F 6] FURTH, H. P.: Magnetic analysis of scattered particles. Rev. Sci. Instrum. 26, 1097 (1955).

[G 1] GIBSON, W. M., D. J. PROWSE and J. ROTBLAT: Range-energy relation in nuclear track emulsions for protons of energy up to 21 Mev. Nature, Lond. 173, 1180 (1954).

[G 1 a] GLASSER, R. G.: Relation between multiple Coulomb scattering and residual range in nuclear emulsion. Phys. Rev. 98, 174 (1955).

[G 2] GOLDSACK, S. J., and H. B. VAN DER RAAY: An automatic scanner for nuclear emulsions. J. Sci. Instrum. 33, 135 (1956).

[G 3] GOLDSCHMIDT-CLERMONT, Y.: On the measurement of scattering in the photographic plate. Nuovo Cim. 7, 331 (1950).

[G 4]* GOLDSCHMIDT-CLERMONT, Y.: Photographic emulsions. Annual Rev. Nucl. Sci. 3, 141 (1953).

[G 5]* GOTTSTEIN, K.: Die Durchführung und Auswertung von Messungen in kernphotographischen Emulsionen. Vorträge über Kosmische Strahlung; W. HEISENBERG, ed., p. 494 ff. Berlin: Springer 1953.

[G 6] GOTTSTEIN, K., M. G. K. MENON, J. H. MULVEY, C. O'CEALLAIGH and O. ROCHAT: Observations on the multiple scattering of ionizing particles in photographic emulsions I. The value of the scattering constant. Phil. Mag. 42, 708 (1951).

[H1]   HÄLG, W., and L. JENNY: Herstellung und Eigenschaften einer photographischen Emulsion zum Nachweis geladener Teilchen. Helv. phys. Acta **21**, 131 (1948).

[H1a]   HALPERN, O., and H. HALL: The ionization loss of energy of fast charged particles in gases and condensed bodies. Phys. Rev. **73**, 477 (1948).

[H2]   HEINZ, O.: Range and specific ionization of high-energy protons in nuclear emulsions. Phys. Rev. **94**, 1728 (1954).

[H3]   HEITLER, H. K., and D. T. KING: The study of atomic processes with the photographic plate. Experientia (Basel) **6**, 281 (1950).

[H4]   HERZ, A. J., and G. DAVIS: The characteristics of tracks in nuclear research emulsions. Austral. J. Phys. **8**, 129 (1955).

[H5]   HERZ, A. J., and C. WALLER: A new nuclear research emulsion. Phil. Mag. **43**, 592 (1952).

[H6]   HERZ, R. H.: Electron tracks in photographic emulsions. Nature, Lond. **161**, 928 (1948).

[H7]   HOANG, T. F.: Method of mass estimation by grain density-residual range. Suppl. Nuovo Cim. **1**, 186 (1955).

[H8]   HOANG, T. F.: Sur les noyaux lourds primaires du rayonnement cosmique. Thesis, Université de Paris 1950.

[H9]   HODGSON, P. E.: Gap measurement as a method of analyzing cosmic ray stars in emulsions. Phil. Mag. **41**, 725 (1950).

[H10]   HOLTEBEKK, T., N. ISACHSEN and S. O. SÖRENSEN: The determination of the mass of energetic helium isotopes emitted in nuclear explosions. Phil. Mag. **44**, 1037 (1953).

[H11]   HOOPER, J. E., and M. SCHARFF: Apparatus for measuring mean gap length in nuclear emulsions. Document BS-12, C.E.R.N., Geneva 1954.

[H12]   HOPPER, V. D., Y. K. LIM and M. C. WALTERS: The measurement and reduction of distortion in thick emulsions. Austral. J. Phys. **7**, 288 (1954).

[H13]   HORNBOSTEL, J., and E. O. SALANT: Interactions of negative K mesons. Phys. Rev. **102**, 502 (1956).

[I1]   Ilford photographic emulsions for nuclear research (brochure of Ilford, Ltd.), March 1956.

[J1]   JENNY, L.: Preparation and sensitization of electron-sensitive plates. See Ref. [M10], p. 259.

[K1]   KAPLON, M. F., B. PETERS, H. L. REYNOLDS and D. M. RITSON: The energy spectrum of primary cosmic radiation. Phys. Rev. **85**, 295 (1952).

[K2]   KAYAS, G.: Method of estimation of masses by gap counting. Suppl. Nuovo Cim. **1**, 200 (1955).

[K3]   KRISTIANSSON, K.: The accuracy of photoelectric mass determinations. Suppl. Nuovo Cim. **12**, 394 (1954).

[L1]   LAL, D., Y. PAL and B. PETERS: The preparation of large nuclear emulsion detectors and their application to the study of K-mesons and hyperons. Proc. Indian Acad. Sci. A **38**, 277 (1953).

[L2]   LATTES, C. M. G., P. H. FOWLER and P. CÜER: Range-energy relation for protons and α-particles in Ilford emulsions. Nature, Lond. **159**, 301 (1947).

[L3]   LEES, C. F., G. C. MORRISON and W. G. V. ROSSER: The range-energy relation for protons and alpha-particles in diluted Ilford G.5 emulsions. Proc. Phys. Soc. Lond. A **66**, 13 (1953).

[L4]*   LEPRINCE-RINGUET, LOUIS: Mesons and heavy unstable particles in cosmic rays. Annual Rev. Nucl. Sci. **3**, 39 (1953).

[L4a]   LEPRINCE-RINGUET, L.: Cosmic-Ray Laboratory of École Polytechnique: Étude de mesons lourds à l'arrêt dans les emulsions photographiques. Suppl. Nuovo Cim. **1**, No. 3, 169 (1955).

[L5]   LEVI SETTI, R.: Contribution to the methods of measurement of scattering in the photographic plate. Nuovo Cim. **8**, 96 (1951).

[L6]   LINDHARD, J., and M. SCHARFF: Energy loss in matter by fast particles of low charge. Kgl. danske Vidensk. Selsk., mat.-fys. Medd. **27**, No. 15, 30 (1953).

[L7]   LIPKIN, H. J., S. ROSENDORFF e G. YEKUTIELI: A new multiple scattering parameter. Nuovo Cim. **2**, 1015 (1955).

[L8]   LOHRMANN, E., e M. TEUCHER: Spurious scattering in nuclear emulsions. Nuovo Cim. **3**, 59 (1956).

[L9]   LONGCHAMP, J. P.: A contribution to the systematic study of photographic emulsions used in nuclear physics. Ann. Phys., Paris **10**, 201 (1955).

[L10]   LORD, J. J.: Altitude and latitude variation of nuclear disintegrations produced in the stratosphere by cosmic rays. Phys. Rev. **81**, 901 (1951).

[M1]   MABBOUX, C.: A method of measuring the multiple scattering of high-energy particles in a photographic emulsion. C. R. Acad. Sci. Paris **232**, 1091 (1951).

[*M 2*] MAJOR, J. V.: A quick method of distortion measurement in nuclear emulsions. Brit. J. Appl. Phys. **3**, 309 (1952).

[*M 3*] MARGUIN, G.: Treatment of nuclear emulsions from 100 to 1000 μ thick. J. Phys. Radium **14**, 43 (1953).

[*M 4*] MEES, C. E.: The theory of the photographic process. New York: MacMillan 1945.

[*M 5*] MENON, M. G. K., and O. ROCHAT: Observations on the multiple scattering of ionizing particles in photographic emulsions. V. Scattering measurements on tracks of slow protons. Phil. Mag. **42**, 1232 (1951).

[*M 6*] MERLIN, M.: Magnetic deflection in nuclear emulsions. Suppl. Nuovo Cim. **11**, 218 (1954).

[*M 7*] MEULEMANS, G., G. P. S. OCCHIALINI and A. M. VINCENT: The wire method of loading nuclear emulsions. Nuovo Cim. **8**, 341 (1951).

[*M 8*]* MEYER, P.: The photographic emulsion in nuclear physics. Naturwiss. **35**, 369 (1948).

[*M 9*] MIGNONE, G.: Elimination of contraction in nuclear emulsions. Nuovo Cim. **8**, 896 (1951).

[*M 10*]* MITCHELL, J. W. (Editor): Fundamental mechanisms of photographic sensitivity. Proceedings of a symposium held at the University of Bristol, March 1950, London: Butterworths Scientific Publications. 1951. In a session devoted to nuclear emulsions (pp. 259—345), the following topics were discussed: Preparation of electron sensitive plates, "sandwich" emulsions, diluted emulsion, track recognition, latent image production and fading, processing, p_H buffers, grain density and ionization loss, sensitivity, shrinkage, range-energy relation, measurements and particle identification. The contributors were: G. ALBOUY and H. FARAGGI, R. W. BERRIMAN, M. BOGAARDT and L. VIGNERON, A. C. COATES, P. CÜER, C. C. DILWORTH, G. P. S. OCCHIALINI and L. VERMAESEN, E. C. DODD and C. WALLER, P. H. FOWLER and D. H. PERKINS, L. JENNY, M. MORAND and L. VAN ROSSUM, F. A. ROADS, J. ROTBLAT and C. T. TAI, G. W. STEVENS.

[*M 11*] MOLIÈRE, G.: Theory of scattering of fast charged particles. Z. Naturforsch. **2a**, 133 (1947); **3a**, 78 (1948); **10a**, 177 (1955).

[*M 12*]* MORAND, M., et A. VASSY (ed.): Colloque sur la sensibilité des cristaux et des émulsions photographiques, Paris, September 1951. Editions de la Revue d'Optiques, Paris 1953.

[*M 13*] MORELLET, D.: Method of estimation of mass by photometric density-residual range (photometer with two slits). Suppl. Nuovo Cim. **1**, 209 (1955).

[*N 1*] NERESON, N., and F. REINES: Nuclear emulsions and the measurement of low energy neutron spectra. Rev. Sci. Instrum. **21**, 534 (1950).

[*O 1*] OCCHIALINI, G. P. S.: On the identification of high energy particles in electron-sensitive plates. Suppl. Nuovo Cim. **6**, 413 (1949).

[*O 2*] O'CEALLAIGH, C.: Measurement of ionization in photographic emulsions by the technique of mean gap length. Document BS-11 C.E.R.N., Geneva 1954.

[*O 3*] O'CEALLAIGH, C., and O. ROCHAT: Observations on the multiple scattering of ionizing particles in photographic emulsions. III. Statistics of the sampling distributions of second differences and the technique of overlapping cells. Phil. Mag. **42**, 1050 (1951).

[*O 4*] OLIVER, A. J.: Measurements of the effects of moisture in nuclear track emulsion. Rev. Sci. Instrum. **25**, 326 (1953).

[*O 5*] OREAR, JAY: The preparation of emulsion chambers suitable for quick tracing of all tracks. Rev. Sci. Instrum. **25**, 1023 (1954).

[*O 6*] ORKIN-LECOURTOIS, A.: Method of estimation of mass by multiple scattering-residual range. Suppl. Nuovo Cim. **1**, 222 (1955).

[*P 1*]* Padua, Proceedings of International Congress on Unstable Particles and High Energy Cosmic Ray Events. Suppl. Nuovo Cim., Ser. IX, **12**, No. 2, 163 (1954).

[*P 2*] PERKINS, D. H.: Emission of heavy fragments in nuclear explosions. Proc. Roy. Soc. Lond., Ser. A **203**, 339 (1950).

[*P 3*] PERFILOV, N. A.: Destruction of the latent image of α-particle tracks in photographic plates. C. R. U.S.S.R. **42**, 258 (1944).

[*P 4*] PETERS, B.: The nature of primary cosmic radiation. Progress in Cosmic Ray Physics. Amsterdam: North Holland Publishing Co. 1952.

[*P 5*] PICCIOTTO, E.: Photographic sandwich emulsion technique in nuclear chemistry. C. R. Acad. Sci. Paris **228**, 2020 (1949).

[*P 6*] PICCIOTTO, E.: Utilisation des émulsions liquides dans l'étude de la radioactivité des roches. Bull. Soc. belge Geol. Paleontal. et Hydrol. **165**, 257 (1956).

[*P 7*] PICKUP, E.: Grain size in Ilford G.5 emulsions. Canad. J. Phys. **31**, 898 (1953).

[*P 8*] PICKUP, E., and L. VOYVODIC: Relativistic increase in ionization of charged particles in photographic emulsion. Phys. Rev. **80**, 89 (1950).

[P8a]* Pisa, Proceedings of International Conference on Elementary Particles, June, 1955. Suppl. Nuovo Cim., Ser. X 9, No. 2, 135—1078 (1956). (Includes noteworthy applications of emulsions to problems in high-energy physics.)

[P9]* POWELL, C. F.: Mesons. Rep. Progr. Phys. 13, 350 (1950).

[P10] POWELL, C. F., and G. P. S. OCCHIALINI: Nuclear physics in photographs. Oxford: Clarendon Press 1947.

[P11] POWELL, C. F., G. P. S. OCCHIALINI, D. L. LIVESEY and L. V. CHILTON: A new photographic emulsion for the detection of fast charged particles. J. Sci. Instrum. 23, 102 (1946).

[P11a] POWELL, C. F.: Recent advances in our knowledge of heavy mesons and hyperons. Pisa Conference Proceedings ([P8a], p. 337).

[P12] PRICE, B. T.: Ionization by relativistic particles. Rep. Progr. Phys. 18, 52 (1955).

[R1] RENARDIER, M., H. MOUCHARAFIEH and M. MORAND: The identification of tracks ending in electron-sensitive emulsions. C. R. Acad. Sci. Paris 231, 848 (1950).

[R2] RICHARDS, H. T., V. R. JOHNSON, F. AJZENBERG and M. J. W. LAUBENSTEIN: Proton range-energy relation for Eastman NTA emulsions. Phys. Rev. 83, 994 (1951).

[R3] RITSON, D. M.: Some K-particle mass measurements. Phys. Rev. 91, 1572 (1953).

[R4] ROEDERER, B.: On the mass estimation of charged particles in nuclear photographic emulsions by means of the variable-cell method. Nuovo Cim. 2, 135 (1955).

[R5] ROSEN, L.: Nuclear emulsion techniques for the measurement of neutron energy spectra. Nucleonics 11, July issue p. 32, August issue, p. 38 (1953).

[R6] ROSENDORF, S., and G. YEKUTIELI: A new method for mass measurement of fast, charged particles. Suppl. Nuovo Cim. 12, 416 (1954).

[R7] ROSENFELD, A. H., M. BACKUS, J. FRIEDMAN, W. F. FRY, D. HASKIN, J. LACH, R. LUX, M. ORANS, J. OREAR, R. SILVERSTEIN, W. SLATER, F. SOLMITZ, R. SWANSON and H. TAFT: How to devil up emulsion. University of Chicago 1955.

[R8] ROSS, M. A. S., and B. ZAJAC: Range-energy and other relations for electrons in Kodak nuclear plates. Nature, Lond. 162, 923 (1948).

[R9]* ROTBLAT, J.: Progress in nuclear physics, Vol. I. (FRISCH, O. R., Ed.): Photographic emulsion technique, p. 37ff. London: Butterworth-Springer Ltd. 1950.

[R10] ROTBLAT, J.: Range-energy relation for protons and alpha-particles in photographic emulsions. Nature, Lond. 167, 550 (1951).

[S1] SCHEIN, M., D. M. HASKIN e R. G. GLASSER: Nucleon-nucleus interactions at very high energy. Suppl. Nuovo Cim. 12, 355 (1954).

[S2] SCHEIN, M., D. M. HASKIN e R. G. GLASSER: Heavy unstable particles produced by pions in the Berkeley Bevatron. Nuovo Cim. 3, 131 (1956).

[S3] SCHOPPER, E.: The photographic plate as a measuring instrument in nuclear physics. Phys. Bl. 1950, No. 3, 113.

[S4] SCOTT, W. T.: Mean-value calculations for projected multiple scattering. Phys. Rev. 85, 245 (1952).

[S5]* SHAPIRO, M. M.: Tracks of nuclear particles in photographic emulsions. Rev. Mod. Phys. 13, 58 (1941).

[S6]* SHAPIRO, M. M.: Mesons and Hyperons, Sec. 8L of American Institute of Physics Handbook. New York: McGraw Hill Book Co., Inc. 1957. Also in Amer. J. Phys. 24, 196 (1956).

[S6a] SÖRENSEN, S. O. C.: The determination of the charge of heavy particles emitted during the explosive disintegration of nuclei. Phil. Mag. 40, 947 (1949).

[S7] STERNHEIMER, R. M.: The density effect for the ionization loss in various materials. Phys. Rev. (a) 88, 851 (1952); (b) 103, 511 (1956).

[S8] STERNHEIMER, R. M.: The energy loss of a fast charged particle by Čerenkov radiation. Phys. Rev. 91, 256 (1953).

[S9] STEVENS, G. W.: Temperature coefficient of swelling and development in thick emulsions for nuclear research. Photogr. J., B 90, 129 (1950).

[S10] STILLER, B., and M. M. SHAPIRO: Ionization loss at relativistic velocities in nuclear emulsion. Phys. Rev. 92, 735 (1953).

[S11] STILLER, B., M. M. SHAPIRO and F. W. O'DELL: Techniques for processing thick nuclear emulsions. Rev. Sci. Instrum. 25, 340 (1954).

[T1] TAKIBAEV, ZH. S.: Introduction of fine particles of a substance into thick nuclear emulsions. Zh. éksper. teor. Fiz. 24, 229 (1953).

[T2] TENNENT, R. M.: Two methods of measuring the degree of straightness of a microscope stage. J. Sci. Instrum. 30, 89 (1953).

[T3]* TEUCHER, M.: Technique and application of nuclear emulsions. Ergebn. exakt. Naturwiss. 28, 407 (1955).

[T4] TIDMAN, D. A., E. P. GEORGE and A. J. HERZ: The production of delta-rays in nuclear-research emulsions. Proc. Phys. Soc. Lond. A 66, 1019 (1953).

[U 1]   UEHLING, EDWIN A.: Penetration of heavy charged particles in matter. Annual Rev. Nucl. Sci. **4**, 315 (1954).

[V 1]   ROSSUM, L. VAN: Determination of the masses of unstable heavy particles by photometry of their tracks. C. R. Acad. Sci. Paris **240**, 747 (1955).

[V 2]*   Varenna, Lectures at the International School of Physics, 1953. Suppl. Nuovo Cim., Ser. IX **11**, No. 2, 141 (1954).

[V 2 a] Varenna, Lectures on Elementary Particles, 1954. Suppl. Nuovo Cim., Ser. X **2**, No. 1 (1955).

[V 3]*   VIGNERON, L.: The use of photographic emulsions for the detection and study of nuclear phenomena. J. Phys. Radium **14**, 121 (1953).

[V 4]   VIGNERON, L.: General calculation of the range-energy relation for particles in any emulsion. Numerical application to Ilford C.2 emulsion. J. Phys. Radium **14**, 145 (1953).

[V 5]   VIOLET, CHARLES E.: Multiple scattering of 2.39 Mev electrons in nuclear emulsion. UCRL Report 4704. 1956.

[V 6]   FRIESEN, S. V., and L. STIGMARK: Photo-electric mass determinations in nuclear emulsions. III. Description of the new experimental arrangements. Ark. Fysik **8**, 121 (1954).

[V 7]*   VOYVODIC, L.: Particle identification with photographic emulsions, and related problems. Progress in cosmic ray physics, Vol. 2, p. 217ff. Amsterdam: North Holland Publishing Co. 1954.

[V 8]   VOYVODIC, L., and E. PICKUP: Multiple scattering of fast particles in photographic emulsions. Phys. Rev. **85**, 91 (1952).

[W 1]   WADDINGTON, C. J.: The alpha-particle component of the cosmic radiation. Phil. Mag. **45**, 1312 (1954).

[W 1 a] WALDESKOG, B.: Photoelectric determination of the charge of heavy primaries in the cosmic radiation. Ark. Fysik **7**, 475 (1953).

[W 2]*   WALLER, C.: Nuclear emulsion technique. A review of progress, 1948—1953. J. Photogr. Sci. **1**, 41 (1953).

[W 3]   WEBB, J. H.: Photographic plates for use in nuclear physics. Phys. Rev. **74**, 511 (1948).

[W 4]   WILKINS, J. J.: Range-energy relations for Ilford nuclear emulsions. Atomic Energy Establishment (Harwell) Report, G/R 664. 1951.

[W 5]   WINAND, L., et C. BEETS: Eradication of proton tracks in G.5 nuclear emulsions. Bull. Soc. Roy. Sci. Liège **21**, 115 (1952).

[Y 1]*   YAGODA, H.: Radioactive measurements with nuclear emulsions. New York: John Wiley & Sons, Inc. 1951.

[Y 2]   YAGODA, H.: Observations of stars and heavy primaries in emulsions flown in a rocket. Canad. J. Phys. **34**, 122 (1956).

[Y 3]   YEKUTIELI, G.: Estimation of $p\beta$ by multiple scattering. Publication BS-9 of C.E.R.N. Geneva 1954.

[Z 1]   ZORN, G. T.: A microscope for large nuclear plates. Brookhaven National Laboratory, February 1956.

Appendix: Supplemental bibliography.

The following references, arranged topically, relate to serveral emulsion techniques and problems which were treated sketchily or omitted for the sake of brevity.

Track photometry. Summary, ROSSUM, L. VAN [V 2], p. 212; see also [V 1]. — DEMERS, P., and R. MATHIEU: Phys. Rev. **75**, 1327 (1949) (A); Canad. J. Phys. **31**, 97 (1952). — FRIESEN, S. VON, and K. KRISTIANSSON: Nature, Lond. **166**, 686 (1950); Ark. Fysik **4**, 505 1952). — CECCARELLI, M., and G. T. ZORN: Phil. Mag. **43**, 356 (1952). — KAYAS, G., et D. MORELLET: C. R. Acad. Sci., Paris **234**, 1359 (1952). — DELLA CORTE, M., e M. RAMAT: Nuovo Cim. **9**, 605 (1952). — FRIESEN, S. VON, and L. STIGMARK [V 6]. — DELLA CORTE, M.: Nuovo Cim. (10) **4**, 1565 (1956). — FRIESEN, S. VON: Ark. Fysik **8**, 305 (1954). — KRISTIANSSON, K. [K 3]; see also Arkiv Fysik **10**, 447 (1956). — KAYAS, G., et D. MORELLET: J. Phys. Radium **14**, 353 (1953). — MABBOUX, C., and D. MORELLET [P 1], p. 405ff.

"Profile method", in which the projected thickness of a track is measured along closely-spaced range intervals: BONETTI, A., C. DILWORTH, M. L. LADU e G. P. S. OCCHIALINI: Atti Acad. Naz. Lincei **17**, 311 (1954). — ALVIAL, G., A. BONETTI, C. DILWORTH, M. LADU, J. MORGAN e G. P. S. OCCHIALINI: Nuovo Cim., Suppl. **4**, 244 (1956).

Delta rays and "thin-down". BRADT, H. L., and B. PETERS: Phys. Rev. **74**, 1828 (1948). — DAINTON, A. D. *et al.* [D 1]. — DAINTON, A. D., P. FOWLER and D. W. KENT: Phil. Mag. **43**, 729 (1952). — VOYVODIC, L.: Canad. J. Res. A **28**, 315 (1950); see also [V 7], p. 258ff. — HOANG, T. F., et D. MORELLET: C. R. Acad. Sci., Paris **231**, 695 (1950). — HOANG, T. F.:

Nature, Lond. **167**, 644 (1951); J. Phys. Radium **12**, 739 (1951); see also [H 8]. — SÖRENSEN, S. O. C. [S 6a]. — J. CRUSSARD: Thesis Université de Paris 1952. — CÜER, P., et J. LONCHAMP: C. R. Acad. Sci., Paris **236**, 70 (1953). — DEMERS, P., et Z. L. WASIUTYNSKA: C. R. Acad. Sci., Paris **235**, 474 (1952) and Canad. J. Phys. **31**, 480 (1953). — J. HÉBERT: Thesis submitted to University of Montreal 1955. — O. B. YOUNG and W. C. BALLOWE: Amer. J. Phys. **24**, 157 (1956). — TIDMAN, D. A. *et al.* [T 4].

A new multiple scattering parameter. LIPKIN, H. J., S. ROSENDORFF e G. YEKUTIELI: Nuovo Cim. (10) **2**, 1015 (1955). — ROSENDORFF, S., and Y. EISENBERG: To be published. — ROSENDORFF, S.: Thesis, Hebrew University, Jerusalem 1957.

Semiautomatic devices for multiple scattering measurements. STILLER, B., e F. I. LOUCKES jr.: Nuovo Cim. (10) **4**, 642 (1956). — See also [B 13].

Emulsion distortions, types of spurious scattering, and noise. "Large-scale distortion": COSYNS and VANDERHAEGHE [C 3]. — MAJOR [M 2]. — CAULTON, M.: Rev. Sci. Instrum. **24**, 569 (1953). — APOSTOLAKIS, A. J., and J. V. MAJOR: Brit. J. Appl. Phys. **8**, 9 (1957). — Spurious scattering and noise: GOLDSCHMIDT-CLERMONT [G 3]. — LEVI SETTI [L 5]. — FOWLER [F 2]. — BONETTI *et al.* [B 13c]. — D'ESPAGNAT [D 8]. — TENNENT [T 2]. — GOTTSTEIN, K.: Nuovo Cim. **12**, 619 (1954); see also [G 5], p. 505 ff. — BØGGILD, J. K., e M. SCHARFF: Nuovo Cim., Suppl. **12**, 374 (1954). — DiCORATO, M., D. HIRSHBERG e B. LOCATELLI: Nuovo Cim., Suppl. **12**, 381 (1954). — STODIEK, W.: Nuovo Cim. (10) **2**, 467 (1955). — ROEDERER [R 4]. — HUYBRECHTS, M.: Thesis, Université Libre de Bruxelles 1955; Nuovo Cim. Suppl. (10) **4**, 903 (1956). — APOSTOLAKIS, A. J., J. V. MAJOR: Nuovo Cim. (10) **5**, 337 (1957). — "Small-scale distortion": BISWAS *et al.* [B 9a]. — FAY, H.: Z. Naturforsch. **10a**, 572 (1955). — LOHRMANN and TEUCHER [L 8]. — BRISBOUT, F. A., C. DAHANAYAKE, A. ENGLER, P. H. FOWLER e P. B. JONES: Nuovo Cim. **3**, 1400 (1956). — Relative scattering measurements: SEEMAN, N., e R. G. GLASSER: Nuovo Cim. (10) **4**, 703 (1956); see Appendix, p. 711.

Range straggling in emulsion. BARKAS, W. H., F. M. SMITH and W. BIRNBAUM: Phys. Rev. **98**, 605 (1955).

Detection of Neutrons*.

By

H. H. BARSCHALL.

With 5 Figures.

Since neutrons do not interact appreciably with electrons, they are always detected through effects caused by their collision with nuclei. Almost every type of nuclear interaction of neutrons results in a directly observable effect which has been used for the detection of neutrons. In the first part of the present article these various types of neutron-induced effects will be discussed; the use of the interactions in various detecting devices will be treated in part B, and applications to specific problems in neutron physics will be the subject of part C.

A. Interactions used for detection.

Most neutron interaction cross sections vary rapidly with energy, and a knowledge of the interaction cross section as a function of energy is important in the design of detectors. A summary of neutron cross sections is contained in a report prepared under the auspices of the U.S. Atomic Energy Commission's neutron cross section advisory group [11]. Some of the cross sections of particular interest to neutron detection are reproduced in the present article.

Since neutrons are detected by means of secondary effects, most detectors of neutrons have a low detecting efficiency. Consequently, the usefulness of a detector is greatly influenced by its efficiency. The other principal considerations in the choice of a detector are its response as a function of neutron energy and its sensitivity to other radiations, primarily γ-rays.

1. **Elastic scattering.** Except for slow neutrons the most widely used interaction for the detection of neutrons is elastic scattering. This interaction is usually observed through the ionization produced by the recoiling nucleus. If a neutron of rest mass μ and kinetic energy T_n collides elastically with a nucleus of rest mass M, it transfers to the bombarded nucleus a kinetic energy

$$T_R = \frac{2\mu(\gamma+1)\,M\,T_n\cos^2\varphi}{M^2 + 2M\mu\gamma + \mu^2(\gamma^2\sin^2\varphi + \cos^2\varphi)} \tag{1.1}$$

where γ represents the total energy (kinetic plus rest energy) of the incident neutron in units of μc^2 and φ is the angle that the recoiling nucleus makes with the incident neutron. In the non-relativistic limit ($\gamma = 1$) this expression approaches the more familiar form

$$T_R = \frac{4\mu M T_n}{(M+\mu)^2}\cos^2\varphi. \tag{1.2}$$

The angle of recoil φ in the laboratory system of reference is related to the center-of-mass scattering angle ϑ by

$$\cot\frac{\vartheta}{2} = \frac{(\mu\gamma+M)\tan\varphi}{\sqrt{\mu^2 + 2\mu M\gamma + M^2}} \tag{1.3}$$

---

* This article was written in 1956.

or in the low-energy limit $\vartheta/2 = \pi/2 - \varphi$. The kinematics of elastic collisions is treated in detail in a number of references, for example in [19a] and in a report by BLUMBERG and SCHLESINGER[1].

Since φ may vary from 0° (maximum energy transfer) to 90° (grazing collision, no energy transfer), the recoiling nucleus may have energies between

$$T_R^{\max} = \frac{2(\gamma + 1) M \mu T_n}{M^2 + 2 M \mu \gamma + \mu^2}$$

and 0. Neutrons of a given energy produce, therefore, a distribution of energies of recoil which depends on the angular distribution of the elastic scattering. This energy distribution measured in the laboratory system is the same as the angular distribution with respect to cos ϑ of the scattered particles in the center-of-mass system[2]. The simplest case which will occur at low enough energies is that the scattering is isotropic in the center-of-mass system. Then all energies of recoil between 0 and $T_R^{\max}$ are equally probable. The relation between recoil energy and angular distributions may also be used to determine angular distributions of scattered neutrons from observations of energy distributions of recoiling nuclei.

If one wishes to deduce information about the energy of the incident neutrons from the energy of recoil, it is easiest to observe those recoils which make a fixed angle with the incident neutrons. It is possible, however, to obtain the energy distribution of the incident neutrons $N(T_n)$ from the energy distribution of the nuclei $R(T_R)$ recoiling in all directions provided the differential scattering cross section σ_s is known as a function of neutron energy and scattering angle. If the scattering is isotropic in the center-of-mass system, the neutron energy distribution may be obtained from the recoil distribution by simple differentiation.

$$N(T_n) = - \frac{T_R}{n \sigma_s} \frac{dR}{dT_R}, \tag{1.4}$$

where n is the number of scattering nuclei per cm².

Observations of recoils enable one not only to deduce the distribution in energy of neutrons, but also offers the most general accurate method for absolute determinations of fast neutron flux. The method is based on the fact that total neutron cross sections may be measured in a transmission experiment without the necessity of performing an absolute flux determination. If elastic scattering is the only interaction which takes place, a measurement of the total number of recoils from a known number of nuclei allows a determination of the incident neutron flux. Frequently only the number of recoils emitted into a known solid angle is measured. Such a measurement together with a knowledge of the angular distribution of the scattered neutrons will also give a determination of the neutron flux.

The choice of nuclide to be used as recoil detector depends on many considerations. Since the recoil energy decreases with increasing atomic weight, the lightest nuclides are favored, as their higher energy makes them more readily observable. If it is desired to construct a detector without need of making an absolute flux determination, usually high detecting efficiency is of greatest interest. At the lowest neutron energies (below 0.5 Mev) this consideration strongly favors protons because of the high neutron-proton scattering cross section (see Fig. 1). In addition, at low neutron energies the distinction between ionizing events produced by recoils and those produced by electrons from γ-rays offers a problem

[1] L. BLUMBERG and S. I. SCHLESINGER: U.S. Atomic Energy Commission Report LADC 2121, 1955.
[2] H. H. BARSCHALL and J. L. POWELL: Phys. Rev. 96, 713 (1954).

which is most easily overcome if the recoils have the highest possible energy.
At neutron energies around 1 Mev helium has advantages because of the broad
resonance in the neutron-helium scattering at that energy (Fig. 1). Added advantages of helium are the shorter range of the recoil α-particles and the greater ease
of operation of high-pressure counters filled with an inert gas rather than with
hydrogen. At energies above the resonance in helium, deuterium has some desirable characteristics. While the deuterium cross section is in this energy range
about the same as the hydrogen cross section (Fig. 1), the angular distribution
is such as to produce a peak in the recoil energy distribution at the high energy
end of the distribution (Fig. 2). Because of this fact and because of the shorter
range of recoiling deuterons compared to protons, it is sometimes possible to

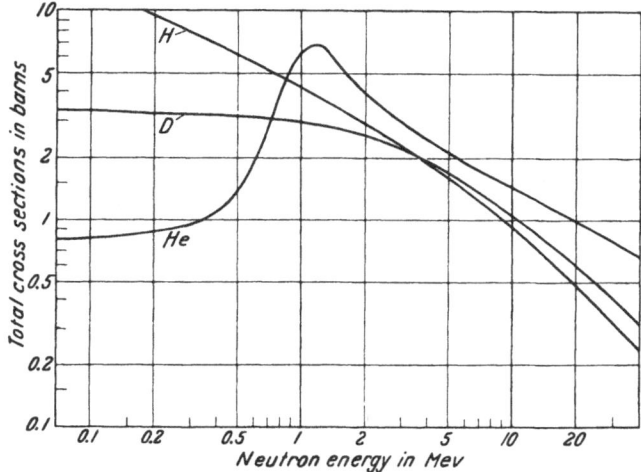

Fig. 1. The total cross sections of hydrogen, deuterium, and helium as a function of neutron energy [11].

realize better discrimination against γ-rays for a given detecting efficiency for
a deuterium-filled counter than for a hydrogen-filled counter. Owing to the
anisotropic scattering of neutrons by deuterons, deuterium-filled counters also
can be more readily given directional properties than hydrogen-filled counters.
Fig. 2 shows some typical angular distributions for the scattering of neutrons
by deuterium and helium[1-4]. The center-of-mass angular distributions plotted
against the cosine of the scattering angle are also the recoil energy distributions
which one would observe.

   In recent years organic scintillators have become the most widely used recoil
detectors. Such detectors are limited to the observation of recoiling protons.
They usually have very high detecting efficiency but do not always enable one
to distinguish clearly between events caused by neutrons and by γ-rays.

   Nuclides heavier than helium are rarely used as recoil detectors. The main
reason is that the ratio of the ionization produced by the recoils to that produced
by electrons becomes rapidly smaller as the mass of the recoil nuclei increases.
This is due not only to the smaller energy transfer to the recoils but also to increased scattering of the electrons by the heavier nuclei which results in longer
effective path lengths.

[1] R. K. ADAIR, A. OKAZAKI and M. WALT: Phys. Rev. 89, 1165 (1953).
[2] J. C. ALLRED, A. H. ARMSTRONG and L. ROSEN: Phys. Rev. 91, 90 (1953).
[3] R. K. ADAIR: Phys. Rev. 86, 155 (1952).
[4] J. D. SEAGRAVE: Phys. Rev. 92, 1222 (1953).

When the recoil particle method is used for absolute flux measurements, protons are almost always chosen for several reasons. The cross section for neu-

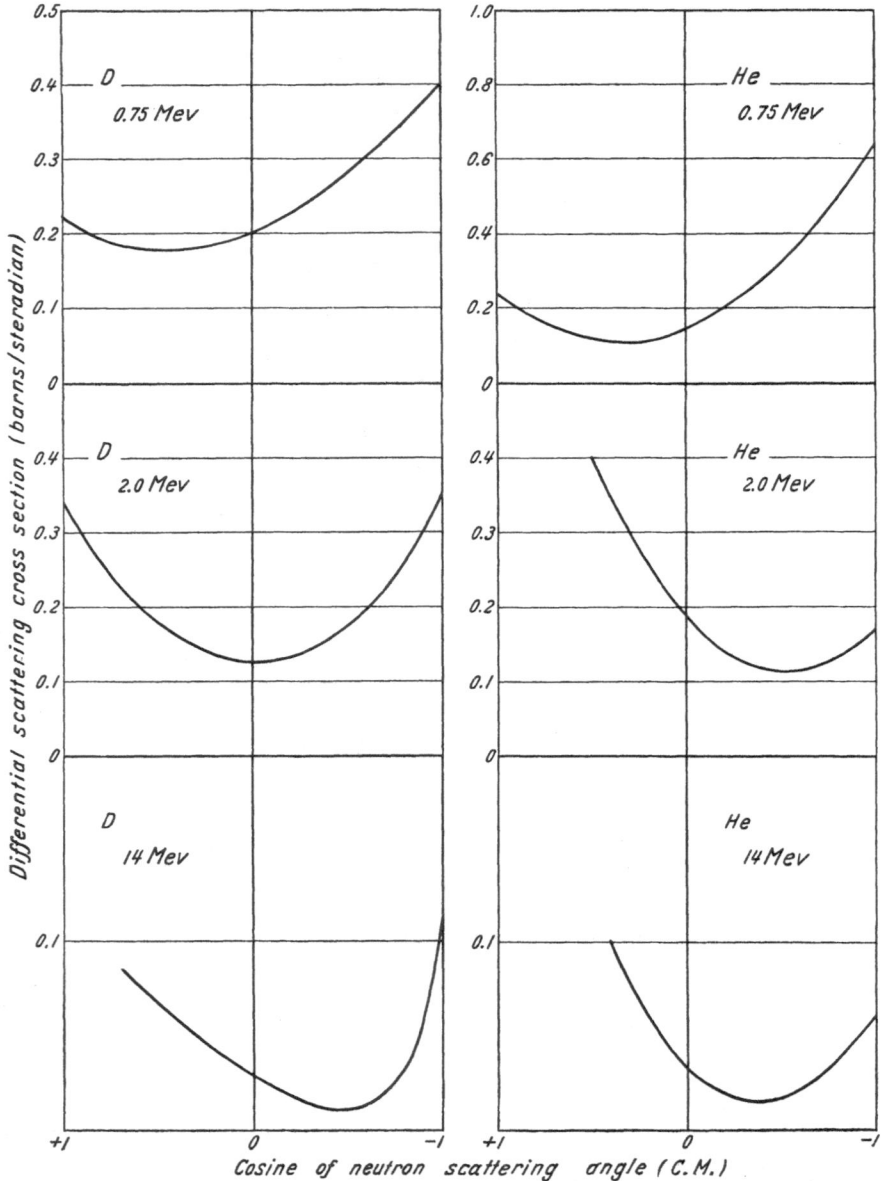

Fig. 2. The angular distributions of neutrons scattered elastically by deuterium and helium at neutron energies of 0.75, 2.0, and 14 Mev [11]. The cross sections are plotted against the cosine of the scattering angle in the c.m. system. The distributions are also the expected energy distributions of the recoils in the laboratory system.

tron-proton scattering has been measured more accurately than any other neutron scattering cross section, and so has the angular distribution of the scattering. At energies at which there are no precise measurements it is possible to compute the cross section and the angular distribution, since the variation with energy of the phase shifts for neutron-proton scattering is well known. Furthermore, the cross section varies smoothly and slowly with energy so that uncertainties in neutron energy do not introduce large errors into the flux measurement. For hydrogen, elastic scattering is the only possible neutron interaction except for radiative capture which has a negligible cross section for fast neutrons. This is important, since the recoil particle method of flux measurement requires the use of a nuclide with which neutrons interact only by elastic scattering.

The detecting substance may be in either solid or gaseous form. The solid, in form of a foil, has the advantage of defining the origin of the recoils although the gas may also be placed in a well-defined volume by confining it by means of a thin foil. At low neutron energies (below 1 Mev) the range even of recoil protons is, however, sufficiently short that the energy loss in a solid radiator is great enough to make the sensitivity of the detector very low. Similarly the energy loss in a foil separating hydrogen gas from the ionization detector does not allow enough ionization to occur beyond the foil to make the detection convenient. Consequently observation of all the recoils occurring in a gas is usually employed at low neutron energies while radiators are usually used at higher energies.

For the purpose of an absolute flux determination it is necessary to know the number of scattering nuclei accurately. The properties of the light gases are well enough known to permit a determination of the number of nuclei from the gas pressure. For the case of hydrogen or deuterium compounds it is necessary to know the composition of the radiator accurately. Polyethylene or glycerol tristearate are widely used for hydrogenous radiators.

If one observes recoils emitted in all directions, it is always necessary to bias the ionization detector, i.e., to reject all events which produce less ionization than a fixed bias B. This necessity arises from the fact that any detector has some background noise and that electrons produced by γ-rays will produce some ionization. The energy response of such a biased recoil detector may be calculated from the scattering cross section as a function of energy and angle. A rough idea of the energy response of a biased proton recoil detector in which no collimation is used may be obtained by making the following simplifying assumptions: neutron-proton scattering is isotropic in the center-of-mass system and the scattering cross section varies as $T_n^{-\frac{1}{2}}$ which is a fairly good approximation for neutron energies from a few hundred kev to several Mev. With these assumptions the sensitivity of the detector is proportional to

$$T_n^{-\frac{1}{2}}(1 - B/T_n). \tag{1.5}$$

The sensitivity is zero below $T_n = B$, reaches a maximum for $T_n = 3B$ and differs from the maximum value by less than 25% in the energy range from $1.6B$ to $9.5B$[1].

At higher neutron energies the neutron-proton cross section decreases more rapidly with increasing neutron energy and consequently the sensitivity of a biased proton recoil counter, after it has gone through its maximum, will fall off more rapidly with neutron energy.

It is frequently desirable to record the entire pulse height distribution rather than only the number of pulses above a fixed bias. From the pulse height distribution, information about the energy distribution of the incident neutrons may be deduced, and the neutron flux can be ascertained more accurately than from the number of pulses above a fixed bias.

---

[1] H. H. BARSCHALL and H. A. BETHE: Rev. Sci. Instrum. **18**, 147 (1947).

For a radiator which is thicker than the range of the most energetic proton recoils the sensitivity may also be calculated readily, if one makes the additional assumption that the range of the proton recoils varies as $T_R^{\frac{3}{2}}$. The sensitivity is then given by

$$T_n^{-2}(T_n^{\frac{3}{2}} - B^{\frac{3}{2}})^2. \tag{1.6}$$

This function is zero and has zero slope at $T_n = B$, increases at first quadratically with T_n, and, at energies large compared to B, linearly with T_n. Because of the rapid increase of the sensitivity with neutron energy such a detector is useful for the investigation of the most energetic neutrons in a given distribution[1].

2. Exothermic reactions. For the detection of slow neutrons exothermic reactions in which protons or α-particles are produced have been widely used.

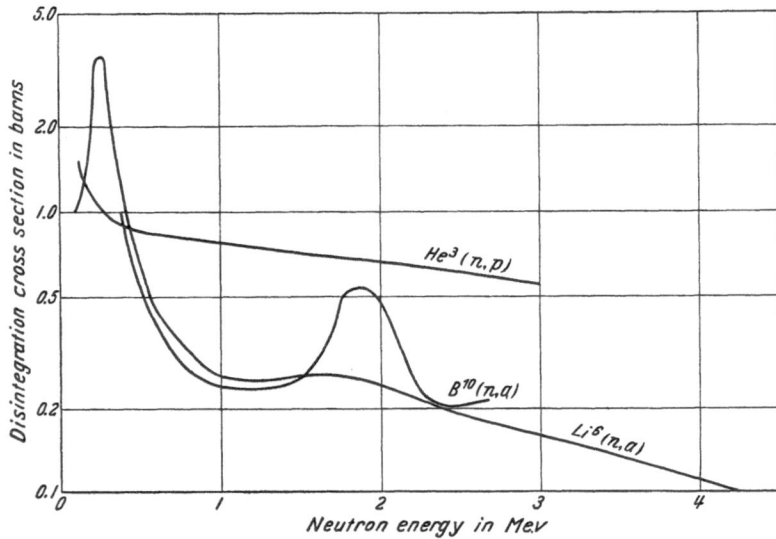

Fig. 3. The cross sections for the disintegration of He³, Li⁶, and B¹⁰ as a function of neutron energy [11].

Such reactions are also applied to the detection of fast neutrons. Exothermic reactions of this type that have large cross sections at low energies are:

$$\text{He}^3 + n \rightarrow \text{T} + p + 0.764 \text{ Mev}; \quad \sigma = 5400 \text{ b}. \tag{2.1}$$

$$\text{Li}^6 + n \rightarrow \text{T} + \alpha + 4.780 \text{ Mev}; \quad \sigma = 945 \text{ b}. \tag{2.2}$$

$$\text{B}^{10} + n \rightarrow \text{Li}^7 + \alpha + 2.792 \text{ Mev}; \quad \sigma = 4010 \text{ b}. \tag{2.3}$$

Behind each reaction the reaction cross section for a neutron velocity of 2200 m/sec is given [11]. Another reaction of this type which has been proposed for the detection of neutrons but has a much smaller cross section is

$$\text{N}^{14} + n \rightarrow \text{C}^{14} + p + 0.627 \text{ Mev}; \quad \sigma = 1.75 \text{ b}. \tag{2.4}$$

The disintegration cross sections of He³, Li⁶, and B¹⁰ are shown as a function of neutron energy in Fig. 3 [2,3].

Interest in the He³ (n, p) reaction has increased recently as larger quantities of He³ have become available. He³ is produced in the decay of tritium, and this

---

[1] See footnote 1, p. 441.

[2] Batchelor, Aves and Skyrme: Rev. Sci. Instrum. **26**, 1037 (1955).

[3] Petree, Johnson and Miller: Phys. Rev. **83**, 1148 (1951).

has been the usual source of the material. If the He^3 is introduced into a counter, even small impurities of tritium may make the counter inoperative. He^3-filled counters have been operated successfully by purifying the He^3 over hot calcium and thereby reducing the tritium concentration to about 1 part in 10^8 [1].

Desirable properties of the $He^3 (n, p)$ T reaction are the smooth variation of the cross section with energy, the absence of excited states of T resulting in a fixed kinetic energy of the reaction products for a given neutron energy, the absence of other neutron-induced reactions producing charged particles, at least at neutron energies which have so far been studied. The fact that He^3 is a gas which is suitable for use in counters allows the construction of detectors of high efficiency.

The $Li^6 (n, \alpha)$ reaction, on the other hand, lacks some of these desirable features. The cross section exhibits a high peak at a neutron energy of 0.25 Mev and perhaps another anomaly at a neutron energy of 1.5 Mev [2]. Competing reactions which produce charged particles occur at higher energies. The reaction $Li^6 (n, p) He^6$ has a Q value of -2.77 Mev and $Li^6 (n, d) He^5$ -2.43 Mev. At a neutron energy of 14.2 Mev the cross sections for reactions induced in Li^6 are (26 ± 4) mb for α-particle emission, (6 ± 2) mb for proton emission, and (77 ± 9) mb for deuteron emission [3]. Since no gaseous compound of Li suitable for counter filling is available, it is not conveniently possible to prepare counters of high detecting efficiency. An advantage of the $Li^6 (n, \alpha)$ reaction is its high reaction energy. This helps to distinguish reaction products from recoils and γ-ray-induced events. For the observation of disintegrations Li^6 has been imbedded in photographic emulsions for the purpose of studying neutron spectra.

The $B^{10} (n, \alpha)$ reaction has been used most widely for the detection of slow neutrons. The high cross section of the reaction at thermal energies, the availability of B^{10} either in ordinary boron of which it constitutes almost 20% or isotopically separated, and the existence of both solid and gaseous compounds are responsible for the wide use of B^{10}. The gaseous compound which is suitable for counter filling is BF_3. While BF_3 is commercially available, the purity of commercial grade BF_3 is frequently not high enough for satisfactory counter operation. Various methods for the preparation of pure BF_3 have been described in the literature. The most widely adopted method involves the conversion of tank BF_3 into calcium fluoborate [9b], and subsequent regeneration of BF_3 by thermal decomposition of this compound. As the complex, highly enriched in the B^{10} isotope, is available from the U.S. Atomic Energy Commission, it has become customary to prepare counter fillings from this complex.

The preparation of thin films of elemental boron is discussed in Ref. [9a]. A method which has been successful is the thermal decomposition of diborane gas on a metal surface which is heated to a temperature of about 900° C. Only metals which do not react with boron and into which the boron does not diffuse should be used as backing of the film. Tantalum appears to be the most satisfactory choice.

The (n, α) cross section of B^{10} is inversely proportional to the neutron velocity at the lowest energies, but deviates from this relation at a few hundred kev. A marked peak in the cross section occurs at 1.9 Mev. At thermal energies only 6.5% of the disintegrations produce Li^7 in its ground state, most of the disintegrations leading to the 0.48-Mev excited state of Li^7. The branching ratio varies

---

[1] See footnote 1, p. 442.

[2] F. L. RIBE: Phys. Rev. **103**, 741 (1956).

[3] G. M. FRYE jr.: Phys. Rev. **93**, 1086 (1954).

with energy and appears to reach a peak around 2 Mev. At this energy Li[7] is produced about twice as often in its ground state as in the excited state[1,2].

At higher neutron energies the energy distribution of the reaction products is further complicated by the occurrence of other neutron-induced reactions: $B^{10}(n, p)$, $B^{10}(n, d)$, and $B^{10}(n, t)$. Recent observations[3] indicate that at neutron energies around 5 Mev a large number of $B^{10}(n, t)$ or possibly $B^{10}(n, p)$ reactions occur. If the boron is in the form of BF_3, the $F(n, \alpha)$ reaction which has a reaction energy of -1.49 Mev may also interfere with the observation of the boron disintegrations.

The γ-rays emitted in the decay of the excited state of Li[7] formed in boron disintegrations have also been employed for the detection of slow neutrons[4] particularly for time-of-flight measurements. Detection of the γ-rays has the advantage that a small sample of solid boron may be used to give high detecting efficiency. The small size of the sample defines the position of the detector better than would be possible with a BF_3 counter of comparable efficiency. An additional advantage is that the rise time of the pulses from a γ-ray scintillation detector is by a factor of 10 shorter than what can be achieved with a BF_3 counter so that flight times may be better defined.

Whenever any of the described reactions is used for the detection of fast neutrons difficulties are experienced because of the effect of thermal and epithermal neutrons. The cross section at the very low neutron energies is so high that even a small number of slow neutrons may produce many more disintegrations than the fast neutrons. Such slow neutrons are usually present in the neighborhood of a fast neutron source even if great precautions are taken to keep moderating materials away from the source. Some of these slow neutrons are probably produced by slowing down in air. The counting rates caused by slow neutrons may be substantially reduced by covering the detector with cadmium sheet in such a manner that no neutrons can get into the counter without passing through cadmium. Even a cadmium covered detector may still show an appreciable counting rate from epithermal neutrons. The magnitude of this counting rate depends on the distance of the detector from the source and on the position of the source with respect to neighboring objects. If the detector is used only to monitor fast neutrons of a fixed energy, the counting rate produced by fast neutrons may be increased with respect to the background by placing a layer of moderating material, usually paraffin, between the cadmium cover and the detector. This moderator will reduce the energy of the fast neutrons to a range in which the interaction cross section is high. Another way in which the detector can be made insensitive to slow neutrons is to take advantage of the lower energy of the disintegration products preduced by slow neutrons. In principle, it should be possible to bias out the lower energy events. In practice, this has been really successful only for the $He^3(n, p)$ reaction.

The dependence of the energy of the disintegration products on the energy of the incident neutrons can also be used to measure the energy distribution of the neutrons. Only if both disintegration products are observed simultaneously will the energy of the products determine the neutron energy uniquely. In addition to disintegrations, the incident neutrons will also produce recoils by elastic scattering. This imposes restrictions upon neutron energy spectra which can be analyzed from observations of disintegration energies, *i.e.*, the energy distribution

[1] PETREE, JOHNSON and MILLER: Phys. Rev. **83**, 1148 (1951).
[2] BICHSEL, HALG, HUBER and STEBLER: Phys. Rev. **81**, 456 (1951).
[3] JAMES, KUBELKA, HEIBERG and WARREN: Canad. J. Phys. **33**, 219 (1955).
[4] DUCKWORTH, MERRISON and WHITTAKER: Nature, Lond. **165**, 69 (1950).

of the recoils should not overlap the energy distribution of the disintegrations. In principle it would be possible to unfold even the overlapping distributions provided all the cross sections involved are well known, but it is not clear that such a procedure would have any advantage over the analysis of a recoil energy distribution in a material in which no charged particle reactions are induced by neutrons.

Although the $N^{14}(n, p)$ reaction was first proposed for the application in neutron spectroscopy[1], this reaction did not prove suitable because of the rapid variation of the cross section with energy which is caused by energy levels in N^{15}. In addition, the $N^{14}(n, \alpha)$ reaction complicates the energy distribution of the disintegration products. The $B^{10}(n, \alpha) Li^7$ reaction suffers from the difficulty of the production of Li^7 nuclei in the excited state. Some success in measuring neutron energy distributions has been achieved with Li loaded photographic emulsions. But the best results have been obtained with a He^3-filled proportional counter[2].

If for a disintegration induced by fast neutrons only one of the products is observed, its kinetic energy will depend on its direction of emission with reference to the incident neutron. The distribution in energy of one of the products for a fixed neutron energy is related to the angular distribution of the products in a manner analogous to that discussed in Sect. 1. The distribution in energy of one of the product particles of a reaction measured in the laboratory system is the same as the angular distribution in the zero-momentum system expressed in terms of $\cos \vartheta$ [3].

3. Fission. Because of the large amount of energy released in the fission process, observation of fission offers a convenient method of detecting neutrons. Fission detectors are almost completely insensitive to γ-rays and no other neutron-induced events which might take place in the detector are likely to be confused with fission processes. Three nuclides which are available in sufficient quantities undergo fission for slow neutrons: U^{233}, U^{235}, and Pu^{239}. For fast neutrons a great number of nuclides undergo fission, but only some of them have been used as fission detectors. Table summarizes some of the properties of these nuclides [*11*].

Table 1. *Properties of fission detectors.*

Nuclide	Thermal cross section	Threshold (Mev)	Cross section at 3 Mev (barns)	Half-life (years)
Th^{232}	< 0.2 mb	1.3	0.14	1.4×10^{10}
Pa^{231}	10 mb	0.5	1.1	3.4×10^4
U^{233}	530 b	—	1.9	1.6×10^5
U^{234}	< 0.6 b	0.4	1.5	2.5×10^5
U^{235}	580 b	—	1.3	7.1×10^8
U^{236}	—	0.8	0.85	2.4×10^7
U^{238}	< 0.5 mb	1.2	0.55	4.5×10^9
Np^{237}	19 mb	0.4	1.5	2.2×10^6
Pu^{239}	750 b	—	2.0	2.4×10^4

In addition, the fission of Bi has been used for the detection of neutrons of energy above 50 Mev[4].

The nuclides which are fissionable for slow neutrons have fission cross sections which do not change rapidly with neutron energy in the energy range from 0.1 to 3 Mev. Therefore, the cross section at 3 Mev given in column 4 of Table 1 is characteristic of a large energy range. The other fissionable nuclides have a characteristic threshold at which the cross section rises rapidly, although usually

[1] B. T. FELD: Phys. Rev. **70**, 429 (1946).

[2] BATCHELOR, AVES and SKYRME: Rev. Sci. Instrum. **26**, 1037 (1955).

[3] H. H. BARSCHALL and J. L. POWELL: Phys. Rev. **96**, 713 (1954).

[4] C. WIEGAND: Rev. Sci. Instrum. **19**, 790 (1948).

somewhat irregularly. At an energy a few hundred kev above the threshold the cross section tends to flatten out again. The cross section at 3 Mev given in column 4 of Table 1 is therefore also characteristic of a wide energy range. Since the rise at threshold extends over several hundred kev, the energy which is taken as the threshold is somewhat arbitrary. The thresholds listed in column 3 of Table 1 are in the neighborhood of the energy at which the cross section has a value of one-tenth that in the flat region above threshold. Even those nuclides which exhibit a fission threshold may have a small fission cross section for slow neutrons. This cross section is so small, however, as not to interfere with the usefulness of the nuclide as a threshold detector.

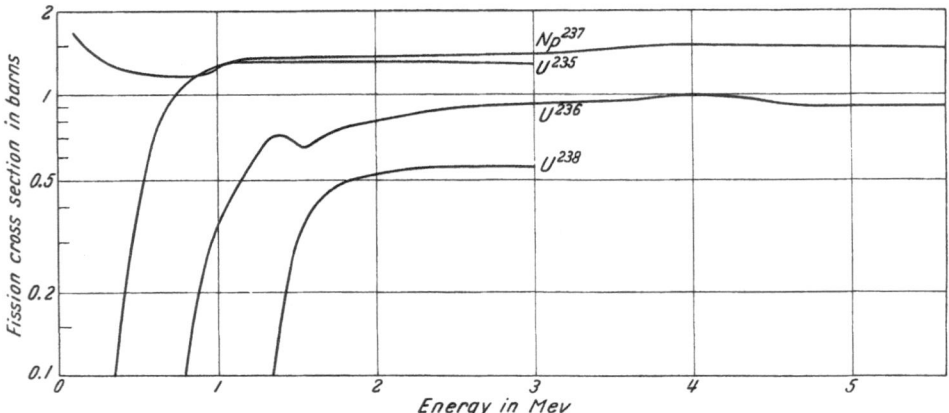

Fig. 4. The fission cross sections of U^{235}, U^{236}, U^{238} and Np^{237} as a function of neutron energy [11].

In the last column the half-lives of the fissionable nuclides are given. The half-life is of importance, because long-lived substances are much preferable as detectors. All the nuclides listed emit α-particles. If the rate of α-particle emission is very high, it may not be possible to resolve the α-particle events and pile-up may result. For short-lived nuclides pile-up of α-particles may be indistinguishable from fission events. It is usually inconvenient to use substances with a half-life of less than 10^6 years, *i.e.*, which emit more than 3×10^4 α-particles per milligram per second into a solid angle of 2π steradians. Consequently U^{235} is most suitable as a slow neutron detector, and U^{236}, U^{238}, and Np^{237} as fast neutron detectors. Fig. 4 shows the fast neutron fission cross sections of these nuclides [11]. The relatively small fission cross section of Th^{232} has weighed against the use of this nuclide in spite of its long half-life.

It is possible to discriminate against α-particles by taking advantage of the rapid fall-off of the ionization density along a fission fragment track, while α-particles ionize most heavily at the end of their range. In a very shallow ionization chamber fission fragments will, therefore, produce relatively very much more ionization than α-particles. This effect may be enhanced by covering the foil of fissionable material with a collimator to exclude α-particles which travel at a small angle with respect to the plane of the foil.

In the discussion of neutron detection by exothermic (n, p) and (n, α) reactions it was mentioned that the effect of thermal and epithermal neutrons produces a background in counting of fast neutrons. The same applies to detectors in which the active material is fissionable for slow neutrons. For this reason the threshold fission detectors are preferable for counting neutrons of energies above 0.5 Mev.

Although a gaseous compound of uranium is available (UF_6), its properties are undesirable for counter fillings. Electron collection (see Sect. 6) is not possible in this gas, and it is highly corrosive so that it attacks most types of insulators. In addition, it is toxic and rather unstable. Fission detectors, therefore, always use foils of the metal or oxide. Because of the short range of fission fragments, such foils must be very thin in order that most of the fissions be observable. Self-absorption in the foil will be small, if the thickness is of the order of 50 $\mu g/cm^2$. For use of fission detectors as monitors, it may not be important to avoid self-absorption in the foil, and thicker layers have the advantage of higher detecting efficiency. Since the range of the fission fragments is about half that of the α-particles emitted by most fissionable materials, the layer should, however, not be much more than 1 mg/cm^2.

Methods of preparation of thin foils are described in Ref. [9a]. Layers of the oxides of U, Np, and Pu are usually prepared by electrodeposition. Another method for preparing foils is the Zapon spreading technique in which the nitrate of the substance to be deposited is mixed with a dilute solution of Zapon lacquer in an organic solvent. This solution is painted onto a metal surface and then ignited to remove organic substances and convert the nitrate to oxide. Evaporation of the metal is another way in which thin layers of uranium have been prepared. Electrodeposition can be carried out quantitatively to determine the amount of fissionable material on the foil. The weight determination may be checked by weighing the foil before and after deposition. In addition, the number of active nuclei may be determined by α-particle counting although this method requires an accurate knowledge of the isotopic composition of the foil or a careful pulse height analysis.

Usually only one of the two fission fragments is observed, the other fragment going into the thick backing of the foil. For slow neutrons the two fragments travel in opposite directions so that for a thin flat foil all fissions should be observable. If fission is induced by fast neutrons, the two fragments will no longer move in exactly opposite directions because of conservation of momentum. The number of fissions observed will then depend upon the orientation of the foil with respect to the incident neutrons.

While in most cases the occurrence of fission is detected by the ionization produced by the moving fragments, it is occasionally desirable to detect the radioactivity of the fragments instead. To accomplish this a catcher foil, such as cellophane or aluminum, is placed close to a foil of fissionable material, and the activity collected on the catcher foil is counted after the neutron bombardment. Care must be taken that the catcher foil is not contaminated by the natural radioactivity of the fissionable material or the radioactivity caused by radiative capture. This method is not useful for absolute flux determination, but allows neutron counting by fission detection in places where ionization chambers cannot be readily employed because of spatial limitations or because high neutron or γ-ray flux interferes with the operation of ionization chambers.

4. Radiative capture. The compound nucleus formed in the interaction of a neutron with a nucleus may decay into its ground state by the emission of one or more γ-rays. This process may serve for the detection of neutrons in two ways. The capture γ-rays may be detected or, if the product nucleus is radioactive, the induced activity may be observed.

Radiative capture is particularly important for slow neutrons, since for slow neutrons, besides reemission of the incident neutron, γ-ray emission is the only energetically possible mode of decay of the compound nucleus except for the few cases mentioned previously. At low energies, the probability for reemission of a neutron is proportional to the neutron velocity, while the probability of γ-ray emission is independent of the neutron energy. As a consequence, it is to be expected that the cross section for radiative capture will, on the average, vary as $1/v$ at the lowest neutron energies, where the neutron width is usually small compared to the radiation width, and at somewhat higher energies as $1/T_n$ as neutron and radiation widths reach comparable magnitudes. Actually, however, the effect of resonances predominates at the lowest energies and results in very large fluctuations in the capture cross section with energy. The general trend of a rapid decrease of the capture cross section with energy is nevertheless sufficient to make radiative capture an important interaction for the detection of slow neutrons only.

Table 2. *Nuclides used for detection of neutrons by induced activity.*

Capturing nuclide	Half-life of product	Radiation emitted (energies in Mev)	Thermal activation cross section (barns)	First resonance (ev)
Mn[55]	2.6 hr	β^- up to 2.8 γ 0.82, 1.77, 2.06	13.2	340
In[115]	54 min	β^- up to 1.0 γ 2.09, 1.49, 1.27, 1.09, 0.41, 0.14	145	1.46
I[127]	25 min	β^- endpoint 2.1 γ 0.45 6% *E C*	5.5	21
Au[197]	2.7 day	β^- up to 1 γ 0.41	99	4.9

Other nuclides which have occasionally been used are Rh[103] (4.5 min and 44 sec) and Dy[164] (1.3 min and 139 min).

Detection of the γ-rays from radiative capture has been applied in some cases, the most important of which has been the use of a cadmium compound in large liquid scintillators. This will be discussed more in Sect. 8. Another application of the detection of γ-rays from radiative capture has been described by ALBERT and GAERTTNER[1].

Induced radioactivity has been used extensively for detecting slow neutrons. The most important practical requirements for the choice of a radioactive detector are that the half-life is in a convenient range (several minutes to several days) and that the capture cross section is large. The nuclides which have been used most frequently are shown in Table 2 together with their half-lives, types of radiation emitted, and thermal activation cross section. In addition the position of the lowest resonance is given.

In all cases periods other than those listed may be excited by fast neutrons. In[116] has an additional 13-sec period which is activated by slow neutrons. If one allows, however, a few minutes to elapse between irradiation and counting, the 13-sec period decays to a negligible intensity. The less abundant isotope In[113] (abundance 4.2%) produces activities of 72 sec (thermal activation cross section 2 barns) and 49 days (56 barns). The longer of these activities does not interfere with measurements of the 54-min period provided the neutron bombardment is restricted to a few hours, while the shorter period has a small cross section.

---

[1] R. D. ALBERT and E. R. GAERTTNER: Rev. Sci. Instrum. **26**, 572 (1955).

The procedure for carrying out an activation is to bombard a sample for a time t_b in a constant neutron flux. The sample is then transferred to the counter, and after a time t_w a count is started, and the number of counts, C, that occur during a time t_c is determined. Since these times may vary for different irradiations, it is customary to calculate from the data the saturated activity A_s which is the counting rate the same sample would have given if it had been bombarded for an infinite time in the same neutron flux,

$$A_s = C \, e^{\lambda t_w} (1 - e^{-\lambda t_b})^{-1} (1 - e^{-\lambda t_c})^{-1}, \tag{4.1}$$

where λ is the disintegration constant.

If a thin sample which contains n nuclei is placed into a neutron flux of Nv neutrons per cm² per sec, the number of neutrons captured per sec will be $n \, Nv \, \sigma_c$ where σ_c is the capture cross section for these neutrons. For substances for which the capture cross section varies as $1/v$ the number of neutrons captured per sec will be proportional to the neutron density N at the sample.

Activation measurements are frequently carried out with the sample surrounded by a Cd absorber which prevents thermal neutrons from reaching the sample so that primarily resonance neutrons produce activation. Such a detector when placed into a moderating medium measures the density of neutrons which pass per second through the energy region of the resonance. It is possible to determine the activity produced by the thermal neutrons alone by subtracting the activity of a cadmium covered sample from that of a bare sample. Before this subtraction is carried out, a correction has to be applied for the absorption of the resonance neutrons in cadmium.

If a weak activity is produced in a large sample, the activity may be concentrated to facilitate counting. Since the product of radiative capture is an isotope of the element used for detection, the concentration cannot be carried out by ordinary chemical means, but the Szilard-Chalmers method[1] must be used. This method is based on the fact that one or more γ-rays are emitted by the compound nucleus in its decay to the radioactive product. When the γ-rays are emitted, the product nucleus recoils with sufficient energy that it may break a chemical bond. By using a suitable chemical compound it is then possible to separate the radioactive nuclei from the stable ones. This method has been applied both to concentrating Mn and I activities. Mn is used in the form of calcium or potassium permanganate dissolved in water. Calcium is preferred because less activity is produced in calcium than in potassium. The radioactive Mn^{56} is precipitated as MnO_2 [9b]. Similarly iodine may be used in the form of ethyl iodide as was done by Szilard and Chalmers in their original experiment. In this case the I^{128} is extracted with an aqueous reducing solution and precipitated as silver iodide.

5. Endoergic reactions. For the detection of fast neutrons it is convenient to use reactions which cannot be induced by slow neutrons, such as some of the fission reactions which have been mentioned. There are many other neutron-induced reactions which have thresholds at almost any desired neutron energy. All reactions in which more than one neutron is emitted and most reactions in which charged particles are emitted are threshold reactions; the most important exceptions have been discussed in Sects. 2 and 3. A detector with a threshold close to the energy of the primary neutrons will be most effective in reducing the background caused by neutrons which are degraded in neighboring material.

---

[1] L. Szilard and T. A. Chalmers: Nature, Lond. **134**, 462 (1934).

The occurrence of a reaction may be observed either by means of the ionization produced by a charged particle emitted in the reaction or by means of the induced activity if the final nucleus is radioactive. Since the threshold of the reaction is usually chosen to be close to the energy of the neutrons that are to be detected, the charged particles which are emitted will have a low energy and will therefore be difficult to detect, particularly because of the presence of recoil nuclei. For this reason the measurement of induced activity is normally employed. The same considerations as were discussed in the case of detection of radiative capture by radioactivity apply to the choice of suitable threshold detectors. In

Table 3. *Neutron-induced threshold reactions.*

Reaction	Half-life of radioactive product	Radiation emitted	Effective threshold (Mev)	Reference to cross section measurement
$C^{12}(n, 2n) C^{11}$	20.4 min	β^+ 1 Mev	22	a)
$O^{16}(n, p) N^{16}$	7.3 sec	β^-, γ	12	b)
$Al^{27}(n, p) Mg^{27}$	10 min	β^-, γ	3	[11]
$P^{31}(n, p) Si^{31}$	2.7 hr	β^- 1.5 Mev, γ	2	c)
$S^{32}(n, p) P^{32}$	14 days	β^- 1.7 Mev	2	c) d)
$Cl^{35}(n, \alpha) P^{32}$	14 days	β^- 1.7 Mev	3	e)
$Ni^{58}(n, 2n) Ni^{57}$	36 hr	β^+, γ	13	f)
$Cu^{63}(n, 2n) Cu^{62}$	10 min	β^+	12	f)
$Mo^{92}(n, 2n) Mo^{91}$	15.5 min	β^+ 3.4 Mev	14	a)
$I^{127}(n, 2n) I^{126}$	13 days	β^-, γ	11	g)
$Tl^{203}(n, 2n) Tl^{202}$	300 hr	EC, γ	12	f)

a) BROLLEY, FOWLER and SCHLACKS: Phys. Rev. **88**, 618 (1952).
b) H. C. MARTIN: Phys. Rev. **93**, 498 (1954).
c) R. RICAMO: Nuovo Cim. **8**, 383 (1951).
d) T. HÜRLIMANN and P. HUBER: Helv. phys. Acta **28**, 33 (1955).
e) ADLER, HUBER and HÄLG: Helv. phys. Acta **26**, 349 (1953).
f) H. C. MARTIN and B. C. DIVEN: Phys. Rev. **86**, 565 (1952).
g) H. C. MARTIN and R. F. TASCHEK: Phys. Rev. **89**, 1302 (1953).

Table 3 some of the threshold detectors which have been used are listed, together with the half-life of the radioactive product, the type of radiation emitted by the product, and the effective threshold for the reaction as measured in the references given in the last column. For all the detectors listed there are available some cross section measurements, although in some instances the measurements cover only a small energy range. Since half-lives of the order of 10 min are most convenient, the threshold detectors which have been most useful are: $Al(n, p)$, $Cu(n, 2n)$, and $C(n, 2n)$.

As in the case of radiative capture, activities other than the desired one are likely to be excited, and the decay of the activity should be followed through several half-lives to ascertain that no other activities are present. For $(n, 2n)$ reactions there is always the possibility that the same activity might be produced by a (γ, n) reaction. Since the cross section for (γ, n) reactions is usually much smaller than for $(n, 2n)$ reactions, and since in the neighborhood of most fast neutron sources the number of high-energy γ-rays is smaller than that of fast neutrons, many (γ, n) reactions are likely to occur only if neutrons are produced by a high-energy electron accelerator. More troublesome is the case in which one desires to observe the reaction $Z^A (n, 2n) Z^{A-1}$ and the isotope Z^{A-2} is also stable. In this event the same activity may be produced by the reaction

$Z^{A-2}(n, \gamma) Z^{A-1}$. Because of the presence of slow neutrons in the neighborhood of most neutron sources, the presence of the $Z^{A-2}(n, \gamma) Z^{A-1}$ reaction should be checked by noting whether surrounding the detector with cadmium reduces the intensity of the activity. A measurement of the cadmium difference is advisable in any event in order to determine whether any activities caused by radiative capture are present.

For the evaluation of measurements it is important to know the energy dependence of the response of the detector. The cross section near the threshold for the reactions in which charged particles are emitted depends on the orbital angular momentum of the emitted particles and the probability for penetration of the Coulomb barrier. In Table 12 of Ref. [19b] there are given for some of the reactions listed in Table 3 the Gamow factors calculated under the assumption that the particles are emitted with zero orbital angular momentum. According to these calculations the energy at which the probability of escape of the charged particle is 0.1 may occur at much higher energy than the threshold calculated from the masses. For example for the $P^{31}(n, \alpha)$ reaction which has a threshold of 0.9 Mev, the "effective threshold" just defined occurs according to this calculation at 6.6 Mev. In light elements the actual variation of the cross section with energy has little similarity to that calculated from the barrier penetration, but is determined largely by the effect of resonances. For instance, the $N^{14}(n, \alpha)$ reaction has a threshold of 0.3 Mev calculated from the masses. At 1 Mev it is barely possible to observe the reaction, the cross section is 30 mb at 1.4 Mev, 10 mb at 1.5 Mev, 150 mb at 1.8 Mev, and 30 mb at 2 Mev. Similar resonance effects are observed in other reactions of light nuclei. A detector which exhibits such resonances may yield experimental results which might be difficult to interpret if the neutrons which are detected do not have a well-defined energy.

In $(n, 2n)$ reactions the excitation energy of the compound nucleus is sufficiently high that resonance effects should not be important and statistical theory should be applicable. At an energy at which the $(n, 2n)$ reaction becomes energetically possible the cross section for inelastic scattering $(n, n\gamma)$ should begin to drop off in such a way that the sum of $\sigma(n, n\gamma)$ and $\sigma(n, 2n)$ remains constant, until the $(n, 2n)$ reaction replaces inelastic scattering. In turn, at a higher energy at which the $(n, 3n)$ reaction becomes energetically possible, the $(n, 3n)$ reaction will replace the $(n, 2n)$ reaction. Not many measurements on $(n, 2n)$ cross sections have been carried out over a wide range of energies, but the results appear to be in reasonable agreement with calculations based on statistical theory.

Another type of endoergic reaction which could be applied to the detection of fast neutrons is inelastic scattering. This process is most easily observed if the nucleus is formed in an isomeric state. Examples of such isomers are In^{115} which has a half-life of 4.5 hours and Au^{197} with a half-life of 7.5 sec. The threshold for excitation of these states is about 0.5 Mev[1,2]. The cross section for excitation of In^{115m} reaches a peak of 0.4 b at 3 Mev, while for Au^{197m} a maximum of 1.3 b is reached at about 2.5 Mev. In view of the large cross sections these isomers should be useful threshold detectors, but they have so far not been used for the purpose.

B. Detecting devices.

6. Ionization chambers. Ionization chambers are used in two distinct ways. Either the average ionization current produced by many events is measured (integrating chamber) or individual ionization events are amplified and their

---

[1] A. A. Ebel and C. Goodman: Phys. Rev. **93**, 197 (1954).
[2] Martin, Diven and Taschek: Phys. Rev. **93**, 199 (1954).

number and magnitude are recorded (pulse operation). For neutron detection the former method has only limited application because it does not permit readily discrimination against γ-rays and X-rays.

Attempts have been made to reduce the effect of γ-rays by observing simultaneously the ionization current in a second identical chamber which has the same sensitivity for γ-rays as the first but a different response to neutrons. For example, one chamber may be filled with hydrogen, the other with deuterium [17].

Integrating ionization chambers have been applied to the measurement of fast neutron flux by observation of recoiling protons. For this application it is possible to eliminate the effect of protons which lose part of their energy outside the sensitive volume of the chamber by coating the walls of the chamber with a material of the same chemical composition as the filling gas. The density of ionization in such a "homogeneous" ionization chamber will be essentially the same as that in an infinite volume of the gas. Background caused by ionization events other than proton recoils may be balanced out by employing two homogeneous ionization chambers. One of the chambers may be filled, for example, with C_2H_4 and have walls of polythene, while the other may be filled with C_2D_4 and have walls of heavy paraffin wax.

Apart from the application just discussed ionization chambers have been operated for neutron detection in such a way as to detect individual ionization events originating either in the filling gas of the chamber or caused by particles ejected from a radiator. Ionizing particles detected in an ionization chamber may be either recoiling nuclei or the charged products of nuclear reactions. Since high detecting efficiency is desirable, it is preferable to have the entire gas filling of the chamber act as source of ionizing particles rather than a radiator whose effective thickness is limited by the range of the charged particles. However, many elements do not have gaseous compounds at normal temperatures, and even if there are gaseous compounds, their electrical properties may render them undesirable for use in an ionization chamber. Other reasons for the choice of a solid radiator may be that one may wish to observe only particles moving in a given direction with respect to the incident neutrons or that the range of the particles is very long. Since it is usually necessary that the range of the particles is short compared to the dimensions of the chamber, very long range particles produced in the gas filling may require the use of impractically high gas pressures.

It is normally desirable that all the ionization produced by a charged particle in the sensitive volume of an ionization chamber be collected on the electrodes. This condition is called saturation; its failure is usually caused by recombination of the positive and negative charges before they reach the collecting electrodes. Since neutrons are detected in ionization chambers through the ionization produced by heavy particles, the type of recombination which is important is the so-called columnar recombination in which charges of opposite sign in the same track neutralize each other. As would be expected, columnar recombination is much more probable if the track is parallel to the electric field in the chamber than if the track is at right angles to the field. Therefore, if the chamber is operated below saturation the charge collected depends on the direction of the track in the chamber, and it becomes difficult to interpret the observed ionization current or pulse height distribution. In addition, below saturation, the electric charge collected depends on the applied collecting voltage so that this voltage must be kept very constant.

In discussing the behavior of ionization chambers it is important to distinguish whether the carriers of negative charge in the chamber are electrons or negative ions. This depends upon the gas in the chamber. It affects both the recombination and the time of collection of the negative charges. Electrons move sufficiently fast that columnar recombination of electrons with positive ions is under normal conditions believed to be an unimportant effect[1]. On the other hand, if negative ions are collected, columnar recombination becomes rapidly more serious as the gas pressure increases so that at pressures not far above atmospheric electric fields of several kilovolts/cm are needed to produce saturation. There is a practical limitation of about 10 kV to the voltage which can be applied to a chamber without introducing some electrical noise caused by breakdown of the insulators or corona. Chambers have been constructed in which high electric fields are applied to extended volumes without raising the voltage too high by introducing a number of parallel electrodes which are alternately connected to the high voltage supply and the detecting circuit. These electrodes are made transparent to the charged particles producing the ionization by the use of grids of fine wire or foils.

In order to obtain saturation it is advantageous to choose a gas filling in which electrons are the carriers of negative charge. To accomplish this a gas must be used which does not form negative ions. Among such gases are the inert gases, hydrogen, nitrogen, and carbon dioxide. On the other hand, many gases and vapors such as oxygen, water vapor, carbon tetrachloride, and halogens form negative ions very readily. Even small impurities of these electronegative gases will attach the electrons.

Techniques for the purification of gases for counter fillings have been described extensively [8]. Unless the chamber has been thoroughly baked out to remove absorbed gases, it may be necessary to purify the filling gas continuously in order to maintain the concentration of electronegative gases at a sufficiently low level. This may be accomplished in some cases by circulating the gas by convection through a chamber containing a getter. For inert gases and nitrogen, calcium turnings at a temperature around 300° C are an effective getter; for hydrogen, magnesium may be used.

Frequently chambers used for neutron detection are filled with hydrogen or deuterium at high pressure. For such fillings it is particularly important to exclude electronegative impurities. One part in 10^7 of oxygen in a filling of 10 atmospheres of hydrogen prevents electron collection[2]. Very pure hydrogen may be prepared by admitting the hydrogen through a palladium leak, by thermal decomposition of uranium hydride, by distillation from liquid hydrogen, or by a combination of these methods. Of these methods the thermal decomposition of uranium hydride appears to be the simplest and most effective[3]. Care should be taken, however, to prevent the very finely powdered pyrophoric uranium from producing radioactive contamination in the counter.

In a pulse ionization chamber a voltage will be induced on the detecting electrode when the ions in the track begin to separate. The induced voltage will increase until all the charges have reached the electrodes of the chamber and will be equal to the charge of one sign of the primary ions divided by the capacity of the chamber provided no ions are lost by recombination or diffusion. The collection time of the ions depends on the gas in the chamber, its pressure, the collecting voltage, and the dimensions of the chamber. For heavy ions it will be of the order of some milliseconds for most chambers used in neutron studies. In addition, the collection time will depend somewhat on the position of the ionizing track with respect to the electrodes. If the amplified voltage is to be proportional to the charge produced in the chamber and independent of the

---

[1] See, however, E. H. BELLAMY and W. R. HOGG: Phil. Mag. 1, 722 (1956).
[2] WILSON, BEGHIAN, COLLIE, HALBAN and BISHOP: Rev. Sci. Instrum. 21, 699 (1950).
[3] W. N. ENGLISH: Rev. Sci. Instrum. 22, 598 (1951).

collection time, all time constants of the amplifier must be large compared to the largest collection time or, in most practical cases, the time constants must be of the order of 100 milliseconds. Until about 1942 linear amplifiers used for neutron detection had indeed such long time constants, and their operation was one of the main problems in neutron detection. Not only are such amplifiers very likely to pick up and amplify the frequency of the a.c. line voltage, but they also amplify acoustical vibrations occurring in the ionization chamber which acts as a condenser microphone. The ionization events must be separated by times large compared to the time constant of the amplifier so that two events are not amplified at the same time. Since γ-rays are present wherever there are neutrons, pile-up of γ-ray-induced ionization events is likely to interfere with neutron detection when a slow amplifier is used.

Because of these difficulties slow amplifiers are at present rarely used for neutron detection. Instead of observing the slow motion of positive and negative ions, one observes the motion of electrons ("electron collection"). There remain, however, cases in which the gas filling of a chamber has to be of such a nature as to preclude the motion of free electrons, and in such cases slow amplifiers have to be used. Examples are measurements of neutron-induced disintegrations when intensity considerations make it impractical to deposit the element to be investigated in the form of a foil and all gaseous compounds of the element form negative ions.

Two simultaneous developments brought about the introduction of electron collection in pulse ion chambers for neutron detection: development of fast amplifiers and the realization that it was possible to maintain fillings of ionization chambers free enough of electronegative gases to avoid electron attachment. For electrons the collection times are of the order of microseconds rather than milliseconds as they are for heavy ions. Amplifiers with microsecond time constants are almost free of the difficulties caused by microphonics and line frequency pick-up, and pile-up becomes less probable so that much faster counting rates may be maintained and a much higher γ-ray intensity can be tolerated.

The drift velocity of electrons in a given electric field and at a fixed pressure depends markedly on the nature of the gas. In pure inert gases the drift velocity is particularly slow because the electrons will reach a high agitation energy even in small electric fields. This is due to the fact that the first excitation level of inert gas atoms lies at a high energy, and the electrons will not be able to transfer small amounts of energy to the gas atoms. Since the drift velocity of the electrons is inversely proportional to the square root of their agitation energy, a reduction of the agitation energy will result in an increase in the drift velocity. To accomplish this a small amount of a polyatomic gas which does not form negative ions such as N_2, CO_2, or CH_4 may be added to the inert gas. Electrons can excite low lying levels in these molecules, and the drift velocity is thereby markedly increased. Because of the large pulses produced by neutrons in nitrogen and hydrogen, CO_2 is preferred as an admixture to inert gases for neutron detection. On the other hand, it has been shown[1] that if 2 to 4% of nitrogen are added to argon as much as 0.5% of oxygen may be present in the mixture without producing appreciable electron attachment. In the nitrogen-argon mixture the agitation energy of the electrons is below 1 ev where the electron capture cross section of oxygen is very small.

During the time of collection of electrons the positive ions remain almost stationary. This has the effect that the voltage pulse induced on the collecting electrode is proportional to the work done by the electric field in moving the

---

[1] U. FACCHINI and A. MALVICINI: Nucleonics, April **1955**, 36.

electrons from their point of formation to the collecting electrode. As a consequence, the voltage pulse which is amplified depends not only on the number of electrons collected but also on the distance through which they move. The dependence of the voltage pulse on the position of the ion track is the principal disadvantage of electron collection, but several methods for minimizing this difficulty are effective.

While the induced voltage depends on the distance through which the electrons move before they are collected, the induced current does not depend on this distance but only on the drift velocity of electrons. The initial induced current at a time before any electrons have been collected is, therefore, proportional to the charge and independent of its point of origin. This initial current may be measured by differentiating the voltage pulse by means of a delay line or a short RC time constant in the amplifier[1]. The differentiation reduces the signal to noise ratio to such an extent, however, that this method has rarely been applied.

O. R. FRISCH suggested a method for overcoming the difficulties with electron collection that is particularly suitable when the ionizing particles originate in a foil. A grid is introduced between the collecting electrode and the region in which the ionization occurs. This grid which is kept at an intermediate potential serves as an electrostatic shield and prevents the motion of the electrons from inducing a voltage on the collecting electrode until they have passed through the grid. Only the motion of the electrons between the grid and the collecting electrode produces a voltage pulse. The height of the pulse will then be independent of the position of the track provided no part of the track is in the region between the grid and the collector. If the tracks originate in the filling gas of the chamber, this method can in general not be applied. It is possible even in this case, however, to make the pulse height distribution more similar to the distribution of the number of ions by keeping the volume between grid and collector small compared to the total volume of the chamber.

Another method which reduces the distortion of the pulse height distribution involves non-uniform electric fields in the chamber. If, for example, a thin wire serves as the collecting electrode in a cylindrical chamber, most of the work on the electrons is done in the neighborhood of the wire. Under these conditions the height of the induced voltage pulse does not depend strongly on the position of the track in the chamber for most of the chamber volume. A spherical geometry would be even better, but a chamber with spherical symmetry and a small collecting electrode is difficult to realize.

When the charged particles originate in a thin foil, their number is sometimes not much larger than the number of pulses that are observed when the foil is removed. One source of such background pulses is due to recoils of light nuclei present on the walls of the chamber, particularly hydrogen in the form of water vapor. It has been found that this background may be reduced materially by lining the walls of the chamber with gold or platinum sheet 0.2 mm thick. The metal sheets are first cleaned by boiling in nitric acid and then heated to red heat[2]. Heavy elements for the wall lining also help to reduce the number of disintegrations induced by fast neutrons.

Another source of background is the occurrence of disintegrations in the filling gas. Upon bombardment with sufficiently energetic neutrons the nuclei of any filling gas will emit charged particles. At the lower neutron energies the disintegration cross sections of krypton and xenon are, however, negligibly small.

[1] R. SHERR and R. E. PETERSON: Rev. Sci. Instrum. **18**, 567 (1947).
[2] J. H. COON and R. A. NOBLES: Rev. Sci. Instrum. **18**, 44 (1947).

If care is taken to reduce these backgrounds, chambers with a hydrogenous radiator have very good directional properties for fast neutrons. The number of pulses observed when the proton recoils from the radiator enter the sensitive volume of the chamber may be 100 times greater than when the chamber is inverted[1].

Whenever the ionizing particles originate in the gas filling, some of the tracks starting in the sensitive volume will leave the sensitive volume, others starting outside the sensitive volume will enter the sensitive volume, and some particles will hit the electrodes of the chamber. Even when the particles are produced in a foil, some tracks may not lie entirely within the gas volume; for example, the wall of a cylindrical counter may intercept tracks from a radiator placed on the wall of the cylinder. All these effects in which part of the energy loss of a particle occurs outside the sensitive gas volume are called wall effects. While the design of a chamber should be such as to minimize wall effects, it is frequently necessary to calculate their magnitude. In Fig. 5 range-energy curves for protons, deuterons, and α-particles in air at atmospheric pressure are shown as an aid in estimating the importance of wall effects. The curves are taken from Ref. [18b].

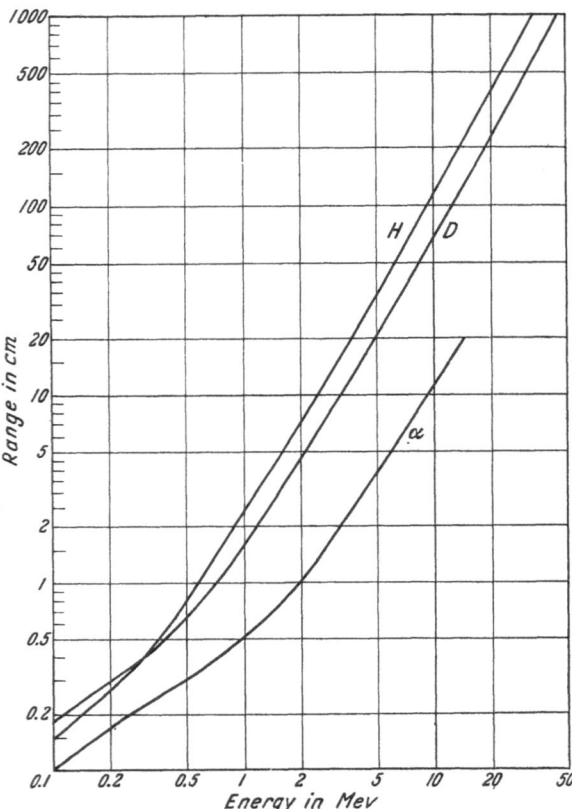

Fig. 5. Range-energy relations for protons, deuterons, and α-particles in air at atmospheric pressure and 15° C [18b].

Calculations of wall effects have been carried out for a number of specific geometries and neutron interactions:

—Cylindrical parallel plate chamber, neutrons incident parallel to axis producing proton recoils in a gas [17].

—Cylindrical chamber with axial electrode, neutrons incident parallel to axis producing either proton recoils in a gas or charged particles of fixed range[2] [17].

—Cylindrical chamber, fissionable material on outer electrode [17]. Particles of fixed range intercepted by plane, cylindrical, or spherical boundary [18].

In all these calculations it is assumed that the sensitive volume has a well-defined boundary. Special precautions have to be taken to realize this assumption in an ionization chamber. In a parallel plate chamber a guard-ring may be used to surround the collecting electrode. This guard-ring is insulated from the collector but kept at closely the same potential. For cylindrical chambers

[1] J. H. COON and R. A. NOBLES: Rev. Sci. Instrum. **18**, 44 (1947).
[2] SKYRME, TUNNICLIFFE and WARD: Rev. Sci. Instrum. **23**, 204 (1952).

in which the cylindrical electrode is used as collector a similar arrangement is possible. It is, however, customary to employ the center wire as collector. For this construction the ends of the center wire may be surrounded by a conducting sleeve, which is insulated from the wire and approximately at the same potential as the wire. Actually such a sleeve will distort the electric field unless it is kept at a potential appropriate to its position in the electric field[1].

There are limitations in energy or range of particles that may be conveniently detected in an ionization chamber. The amplifier noise will set a lower limit to the energy of particles which may be detected. While it is possible to reduce amplifier noise by a variety of techniques[2,3], ionization chambers are difficult to use when the energy of the charged particles is below about 0.1 Mev. Even at somewhat higher energies a proportional counter is preferable. The other limitation applies to particles of long range. In order to stop them in the gas of the ionization chamber high pressures are required, at which electron attachment and saturation difficulties arise. For the detection of particles which have a range longer than about 0.5 meters in air at atmospheric pressure scintillation detectors or counter telescopes are much more convenient than ionization chambers. Ionization chambers are, however, useful for the detection of fissions because of the high energy and short range of the fission fragments. When fissions are counted, it is frequently unnecessary to know the energy distribution of the fragments so that distortions of the distribution because of electron collection effects may not interfere with the counting of the fission events.

Ionization chambers in which fissions are counted have low detecting efficiencies because the fissionable material is normally not available in a gaseous compound and the fission fragments have such a short range that fissionable foils have to be very thin. In order to increase the amount of fissionable material which can be placed into a chamber of given volume, multi-plate chambers have been constructed. Alternate electrodes are connected so that they form two sets, one serving as high voltage electrode, the other as collecting electrode [17].

Even larger amounts of fissionable material may be used in the so-called Spiral Chambers developed by W. C. BRIGHT [17]. The fissionable material is deposited on both sides of each of two long strips of aluminum foil. The two foils are wound into two concentric spirals and form the two electrodes of an ionization chamber. Two or three sewing threads serve as spacers and insulators between the two foils. The spiral is inserted into a small container which is filled to about 10 atmospheres with argon. With a spacing of 0.5 mm between the foils useful foil deposits of 360 cm² area have been fitted into a cylinder 2.5 cm in diameter and 2.5 cm high.

7. Proportional counters. Proportional counters are employed widely in neutron detection for observations of ionizing particles of energies from a few kev to several Mev. The usual design of such proportional counters is the same as that of cylindrical ionization chambers, i.e., one electrode consists of a thin straight wire, the other of a concentric cylinder. In proportional counters the central wire is kept at a positive potential with respect to the outer electrode.

Electrons approaching the central wire reach velocities sufficient to produce additional ionization in the immediate neighborhood of the wire. As the secondary positive ions move outward, they induce a voltage pulse which is then amplified.

[1] A. L. COCKROFT and S. C. CURRAN: Rev. Sci. Instrum. **22**, 37 (1951). — PETREE, JOHNSON and MILLER: Phys. Rev. **83**, 1148 (1951).

[2] E. BALDINGER and W. HAEBERLI: Ergebn. exakt. Naturw. **27**, 248 (1953).

[3] W. C. ELMORE and M. SANDS: Electronics. New York: McGraw-Hill Book Company 1949.

If the counter is operating properly, the induced voltage pulse should be proportional to the number of primary ions. Normally gas amplifications of the order of 20 are employed. The gas amplification is the average number of secondary ions formed per primary ion. For the detection of particles of very low energy gas amplifications as high as 10^4 have been used. The advantage of gas amplification over electronic amplification is that the limitations imposed by electronic amplifier noise are eliminated. While the process of gas amplification is statistical and therefore introduces a spread in pulse heights, this spread is appreciable only for processes of very low energy and will not result in any confusion between ionization pulses and noise.

The other advantage of a proportional counter over an ionization chamber in which electrons are collected is that the pulse height is independent of the radial position of the particle track as long as there is no electron attachment and no lack of saturation, since all the positive ions which produce the voltage pulse start from the immediate neighborhood of the center wire. In order that the pulse height be also independent of the axial position of the track, it is necessary for the gas amplification to be the same all along the wire. For this reason the wire should be uniform in diameter. Variations in gas amplification are likely to occur at the ends of the wire. In some designs the wire is supported by metal rods of sufficiently large diameter that no multiplication takes place along the rods. Such support rods distort the electric field near the end of the wire resulting in reduced gas amplification. In addition, the electric field is not radial in this region so that the sensitive volume is not a right circular cylinder. A better design is, therefore, to surround the center wire by a conducting sleeve which is kept at a potential appropriate to its position in the electric field as was mentioned in the description of ionization chambers.

While the height of the induced pulse may be made independent of the position of the particle track, the position of the track will determine the time lag between the passage of the ionizing particle through the counter and the start of the voltage pulse. This lag is given by the time it takes an electron to travel from its point of formation to the wire.

The gas multiplication of the counter increases as the applied voltage is increased, or the gas pressure is reduced, or a smaller central wire is used. After the onset of gas multiplication, it increases approximately exponentially with increasing voltage over a wide region of voltages. The rate of increase with voltage depends on the gas filling. For inert gases, hydrogen, and nitrogen of high purity and at low pressures the rate of increase of the multiplication with voltage is so rapid that the counters become difficult to operate stably. In order to stabilize such counters, it is convenient to add a few percent of a polyatomic gas, such as CO_2 or CH_4. In Ref. [17] the variation of gas multiplication with applied voltage is given for several gases and gas mixtures frequently used in neutron detectors.

While the voltage necessary to produce a given gas multiplication decreases as the diameter of the center wire is decreased, too small counter wires introduce difficulties. Even small irregularities in very thin wires, specks of dust or lint will produce local fluctuations of gas multiplication which in extreme cases produce a Geiger counter behavior in some parts of the counter. Another disadvantage of small wires is that the electric field is very high near the wire and very low near the cylinder. As a consequence the field in much of the gas volume is small enough to give the electrons a low velocity and consequently the collection time becomes long. Best results are usually obtained with wires of diameters between 0.05 and 0.1 mm.

The principal applications of proportional counters to neutron detection are: observation of recoils in hydrogen, deuterium, and helium; observation of ionizing particles originating in a foil; counters filled with boron trifluoride. For the observation of proton recoils the counter may be filled either with hydrogen gas or a gaseous hydrocarbon. Hydrogen has the advantage that no recoils other than protons will occur and that the ratio of pulse heights caused by protons to that caused by electrons is most favorable. On the other hand, hydrocarbons may be a more satisfactory counter gas when no great effort to purify the gases is made. Furthermore for hydrocarbons much lower gas pressures are needed for a given hydrogen concentration and for a given stopping power. The disadvantages of hydrocarbons are carbon recoils and relatively larger pulses caused by γ-rays. Actually pulses caused by carbon recoils appear to be appreciably smaller than what would be expected on the basis of the respective energies of proton and carbon recoils. According to [1] the energy expended per ion pair for carbon ions in methane is 40% greater than the energy expended per ion pair for protons in methane. In addition to methane, propane and butane are suitable gases for proportional counters. All these hydrocarbons are commercially available in high enough purity so that they frequently need no further purification before they are admitted to the counter.

Proportional counters for the detection of proton recoils in hydrogen and methane have been constructed and investigated most carefully at the British Atomic Energy Research Establishment [1][1]. In these studies observed pulse height distributions of proton recoils from monoergic neutrons were compared with distributions calculated taking into account wall effects. Very good agreement between the calculated and observed distributions was obtained for neutrons of energies between 0.05 and 2.2 Mev. Small deviations at the lowest energies may be due to uncertainties in the range energy relation for protons below 0.1 Mev.

In pure hydrogen it is possible to obtain gas multiplication even at very high pressures. WILSON et al.[2] describe a proportional counter, with a 25 μ central wire and an outer cylinder of 1 cm radius, which was filled with hydrogen to a pressure of 20 atmospheres. When 6 kV were applied to the counter, a gas gain of about 40 was achieved. To accomplish this, special care had to be taken in the purification of the hydrogen, and the wire had to be flashed at high temperature. With this counter it was possible to count low energy protons in the presence of an intense background of secondary electrons produced by γ-rays by using a large amplifier band width to amplify the initial short rise of the pulses (0.05 μsec). This could not have been done with an ionization chamber because the noise of such an amplifier is too great.

If a proportional counter is to be used for the observation of charged particles originating in a foil, relatively simple designs of the counter are satisfactory, since the tracks will occur only in a limited volume, so that field distortions at the ends do not matter. A suitable counter design is described by COON and NOBLES[3]. Even in this case the sensitive volume should be well defined to limit the size of pulses caused by γ-rays. Precautions for reducing background pulses caused by ionizing particles other than those which one desires to observe have been discussed in the section on ionization chambers.

Boron trifluoride has been used widely for filling proportional counters, but many difficulties in its use have been reported. While the gas in very pure form

[1] SKYRME, TUNNICLIFFE and WARD: Rev. Sci. Instrum. **23**, 204 (1952).
[2] WILSON, BEGHIAN, COLLIE, HALBAN and BISHOP: Rev. Sci. Instrum. **21**, 699 (1950).
[3] J. H. COON and R. A. NOBLES: Rev. Sci. Instrum. **18**, 44 (1947).

is a good counter gas, its strong chemical activity appears to introduce complications. Reactions with even small traces of water vapor may produce hydrogen fluoride which in turn may react with glass in the counter to produce silicontetrafluoride and more water. Counters which when freshly filled show good pulse height distributions, sometimes deteriorate later on. It has been found[1] that this deterioration may be prevented by filling the counter with pure BF_3 and then letting it "soak" for a long time (up to three months). Upon refilling such counters are very stable. A possible explanation of this behavior is that a layer is formed on the walls which prevents further reactions. The same authors[1] describe methods for preparing BF_3 for counter fillings, designs of BF_3 counters, calculations of counting efficiencies, and methods of testing.

There is evidence that the gas multiplication in BF_3 proportional counters shows a strong temperature dependence[2]. It has been reported[3] that BF_3 proportional counters deteriorate when subjected to high counting rates. According to this reference, the pulse height distribution changes in an irreversible manner after 10^8 counts. A possible explanation may be the formation of electronegative fluorine gas from the dissociation of BF_3 or the presence of small amounts of water vapor in the gas.

The difficulties in operating BF_3 counters increase as the gas pressure is increased. Satisfactory operation of BF_3 counters has been reported at pressures of 100 cm Hg[4]. It is possible to operate BF_3 counters at even higher pressures to obtain higher counting efficiencies[5], but such counters usually do not give good pulse height distributions.

8. Scintillators. For many applications in neutron physics detectors are needed in which the range of charged particles from neutron interactions is small compared to the sensitive volume of the detector. Ionization chambers and proportional counters in which this condition is satisfied require frequently such high gas pressures that their operation is difficult. The scintillation detector solves this problem. In addition, many scintillators produce much faster pulses than ionization chambers and proportional counters, a fact which is of particular value in coincidence counting.

As the maximum possible range of charged particles in the detector is increased, the distinction between the heavy charged particles which one usually tries to observe in neutron detection and electrons from γ-rays becomes increasingly difficult. For scintillators this difficulty is accentuated by the fact that, for a given energy, electrons produce much more light than heavier charged particles[6]. Low energy α-particles may produce ten times less light than electrons of the same energy. At high ionization densities the response of many scintillators is more nearly proportional to the range of the particles than to their energy. The sensitivity to γ-rays severely limits the applicability of scintillators to neutron detection.

As in the case of gas-filled counters, neutrons may be detected with scintillators either by interactions taking place inside the scintillator or outside the scintillator. In the latter application a charged particle originating in a foil enters a scintillator in which neutrons do not produce observable light pulses directly.

[1] I. L. FOWLER and P. R. TUNNICLIFFE: Rev. Sci. Instrum. **21**, 734 (1950).
[2] LOCKWOOD, WOODS and BENNETT: Rev. Sci. Instrum. **25**, 446 (1955).
[3] SOBERMAN, KORFF, FRIEDLAND and KATZENSTEIN: Rev. Sci. Instrum. **24**, 1059 (1953).
[4] TONGIORGI, HAYAKAWA and WIDGOFF: Rev. Sci. Instrum. **22**, 899 (1951).
[5] PETERSON, BARSCHALL and BOCKELMAN: Phys. Rev. **79**, 593 (1950).
[6] TAYLOR, JENTSCHKE, REMLEY, EBY and KRUGER: Phys. Rev. **84**, 1034 (1951).

Interactions which have been used inside scintillating materials for neutron detection are: scattering by protons and disintegrations of Li^6, B^{10}, and S. In these interactions the charged recoil or disintegration particles are observed. In addition, it is possible to activate a constituent of the scintillator and observe the activity after the neutron irradiation is terminated. Examples of nuclides activated in this manner are C and I.

For neutron detection the organic scintillators originally developed by KALL-MANN[1] are particularly useful. They are employed in solid or liquid form. In addition, inorganic scintillators both in solid or gaseous form are applied to problems of neutron detection.

For the observation of particles originating in a foil inorganic scintillators are preferable, since they do not contain hydrogen and therefore do not produce proton recoil pulses. Among the solid inorganic scintillators Ag-activated ZnS or Tl-activated NaI or CsI are normally used. NaI has a faster decay time (0.2 μsec) than CsI, on the other hand the deliquescence of NaI makes it inconvenient to handle. NaI and CsI have the advantage over ZnS that large transparent crystals may be grown.

When a faster response is required, the recently developed detectors in which scintillations are produced in an inert gas[2] are advantageous, since they have response times of less than 10^{-8} sec. Most of the light from these gaseous scintillators is of such high frequency that it does not penetrate through the glass of the photomultiplier tube. To avoid this difficulty an organic wavelength shifter may be inserted to lower the frequency of the light or a small amount of nitrogen may be added to the inert gas or a quartz window photomultiplier tube may be used.

More frequently, however, neutrons are detected through interactions in the scintillator itself, primarily by observation of proton recoils in organic scintillators. Of these anthracene has the highest efficiency for conversion of absorbed energy into photons, and it has a decay time of about 3×10^{-8} sec. Stilbene has about half the conversion efficiency of anthracene, but it has an about four times shorter decay time. Plastic scintillators which have conversion efficiencies and decay times similar to stilbene are most convenient, since they may most easily be shaped to any desired configuration.

For organic scintillators of large size, liquids have advantages over solids. While a great variety of liquid scintillators have been tried, a solution of terphenyl in toluene has been most widely adopted. Since this solution absorbs its own fluorescent radiation, the addition of a wavelength shifter (alpha naphthyl phenyl oxazole) improves the transparency of very large volumes[3]. Liquid scintillators have about the same light conversion efficiencies as stilbene, but the response times are somewhat shorter.

The principal problem in applying organic scintillators to neutron detection is their sensitivity to γ-rays. A number of different methods have been proposed to reduce the pulse height of γ-ray induced interactions compared to that caused by proton recoils. HARDING[4] pointed out that an increase in the relative sensitivity to neutrons could be obtained by using layers of scintillator of thickness of the order of the maximum proton range interspersed with layers of a hydrogenous material of thickness of the maximum electron range. A proton would then lose

[1] H. KALLMANN: Natur u. Technik, Juli **1947**.
[2] C. EGGLER and C. M. HUDDLESTON: Phys. Rev. **95**, 600 (1954). — G. P. BOICOURT and J. E. BROLLEY jr.: Rev. Sci. Instrum. **25**, 1218 (1954).
[3] REINES, COWAN, HARRISON and CARTER: Rev. Sci. Instrum. **25**, 1061 (1954).
[4] G. N. HARDING: Nature, Lond. **167**, 437 (1951).

all its energy in the phosphor, while an electron could never lose more than a small fraction of its energy. Harding constructed a scintillator consisting of a powder of activated KI. This powder was suspended in α-bromonaphthalene which has the same refractive index as KI at the wavelength of the fluorescent light. With this scintillator it was possible to detect 2.5-Mev neutrons with much higher efficiency than 2.6-Mev γ-rays.

Hornyak [1] constructed a γ-ray insensitive neutron detector following the same idea by molding a dispersion of ZnS grains into Lucite. This detector is practically insensitive to γ-rays, but its detecting efficiency is not much higher than what can be achieved by gas-filled counters and the pulse heights do not vary as rapidly with neutron energy as would be expected for proton recoils. The latter property is a disadvantage if discrimination between different neutron groups is needed. There is evidence [2] that this detector records predominantly disintegrations in sulphur rather than proton recoils. Numerous similar detectors have been proposed in which ZnS or KI are suspended in paraffin, polystyrene, or other organic materials [3].

Another design based on the same principle is to embed small pieces of a plastic phosphor in Lucite or quartz [4]. Here the relative size of the phosphor and non-scintillating material may be chosen to give the best neutron detecting efficiency attainable in the presence of a given γ-ray energy distribution. In testing such scintillators for sensitivity to γ-rays it is necessary to use γ-rays of the same maximum energy as are likely to be present when the neutrons are detected. In particular, electron pairs from high energy γ-rays will give larger pulses than what is observed if a low energy γ-ray source is used in testing.

Inhomogeneous detectors designed to reduce γ-ray sensitivity will give pulse height distributions quite different from those expected in a large homogeneous scintillator, and it is in general not possible to predict what the pulse height distribution for neutrons of a given energy will be.

For the detection of neutrons in the absence of γ-rays large organic scintillators may be used, and calculated pulse height distributions may be compared with observed distributions. This has been done by Cross [5] for 2.5-Mev neutrons from the D-D reaction and 14-Mev neutrons from the D-T reaction. Both anthracene and stilbene scintillators were employed in these experiments. In order to calculate the expected pulse height distribution, it is necessary to know the dependence of the light output on proton energy. The choice of size of detector is influenced by the desirability that the range of the most energetic recoil protons be small compared to the dimensions of the detector. On the other hand, if the detector is too large, there is an appreciable probability that an incident neutron will produce a second recoil proton after it has been scattered once. The latter effect is more difficult to calculate than the effect of protons leaving the scintillator.

Cross has calculated in detail the wall or boundary effects for neutrons incident normal to one of the faces of a rectangular crystal whose dimensions are larger than the range of the most energetic recoiling protons. Besides proton recoils other interactions of neutrons will produce pulses in an organic scintillator. Recoils from elastic scattering by carbon will have such a short range that they

[1] W. F. Hornyak: Rev. Sci. Instrum. **23**, 264 (1952).

[2] G. R. Keepin: Rev. Sci. Instrum. **25**, 30 (1954).

[3] H. B. Frey: Rev. Sci. Instrum. **21**, 886 (1950). — W. S. Emmerich: Rev. Sci. Instrum. **25**, 69 (1954). — Seagondollar, Esch and Cartwright: Rev. Sci. Instrum. **25**, 689 (1954). — Bernhard Brown: Rev. Sci. Instrum. **26**, 970 (1955).

[4] McCrary, Taylor and Bonner: Phys. Rev. **94**, 808 (1954).

[5] W. G. Cross: Phys. Rev. **83**, 873A (1951).

will normally not be detected. Inelastic scattering of high energy neutrons by carbon may produce 4.4-Mev γ-rays, but these will have a small probability of interacting within the detector as long as the detector is small. In addition, high energy neutrons may induce the reactions $C^{12}(n, n)\, 3\alpha$ and $C^{12}(n, \alpha)\, Be^9$. The largest pulses from these reactions for 14-Mev neutrons should be less than 15% of the largest proton pulses. When all these effects were taken into account, the agreement between observed and calculated pulse height distributions was good down to pulse heights of about one-quarter of the maximum proton recoil pulse height.

The γ-ray sensitivity of organic scintillators may be reduced by a method similar to that described in Sect. 6 for homogeneous ionization chambers[1]. Two scintillators are prepared which contain the same total number of electrons, but which differ by a large factor in their hydrogen content. After suitable adjustment of the amplification of the output pulses from the two scintillators, the difference in the pulse height distributions from the two scintillators will be produced by proton recoils from fast neutrons.

Slow neutrons may be detected with scintillators by the Li (n, α) or the B (n, α) reactions although other reactions such as fission have occasionally been used. Li-containing scintillating crystals may be prepared as first suggested by HOFSTADTER[2] in the form of LiI which may be activated either with Tl or Eu. LiI is hygroscopic and the crystals are therefore usually kept hermetically sealed in transparent containers. In good crystals the pulses from Li disintegrations by thermal neutrons are uniform in size. The pulse heights show a large spread, however, when the disintegrations are induced by monoergic fast neutrons. This is believed to be caused by the fact that the conversion of energy into light is not the same for the tritons and α-particles produced in the disintegrations. Since for fast neutrons different fractions of the total energy are carried by the two product particles in different events, a spread in pulse heights may result. Although it was hoped at one time that LiI scintillators would make good fast neutron spectrometers, this application does not now appear feasible.

A variety of ways has been proposed for incorporating boron into scintillating materials. A mixture of one part of B_2O_3 and five parts of ZnS is a satisfactory scintillation detector for slow neutrons. A layer of 70 mg/cm^2 of this mixture is reported to have a 30% efficiency for thermal neutrons and is insensitive to γ-rays[3]. Similar detectors have been prepared by incorporating ZnS into fused B_2O_3[4]. Most convenient and commercially available are boron containing plastic scintillators. Boron compounds have also been incorporated into liquid scintillators, such as boron oxide or methyl borate in terphenyl toluene[5]. The properties of various ZnS phosphor mixtures for neutron scintillation counting have been compared by KOONTZ et al.[6].

For these detectors, as for all neutron scintillation detectors, it is important to prevent their responding to γ-rays. The cross sections for slow neutrons of boron and lithium are sufficiently high that enough neutron detecting efficiency may be obtained with thin layers of phosphors in which electrons do not lose enough energy to give large pulses. If detectors of this type are to be designed for fast neutrons, larger sizes have to be used for good detecting efficiencies. It

[1] I. B. BERLMAN and L. D. MARINELLI: Rev. Sci. Instrum. **27**, 858 (1956).

[2] HOFSTADTER, McINTYRE, RODERICK and WEST: Phys. Rev. **82**, 749 (1951).

[3] D. E. ALBURGER: Rev. Sci. Instrum. **23**, 769 (1952).

[4] GUNST, CONNOR and BAYARD: Rev. Sci. Instrum. **26**, 894 (1955).

[5] J. E. DRAPER: Rev. Sci. Instrum. **22**, 543 (1951). — G. E. THOMAS jr. and C. O. MUEHLHAUSE: Nucleonics **11**, 44 (January 1953).

[6] KOONTZ, KEEPIN and ASHLEY: Rev. Sci. Instrum. **26**, 352 (1955).

then becomes necessary to apply coincidence techniques to discriminate against
γ-rays.

Some of the constituents of phosphors become radioactive under neutron
bombardment, and this β-activity may be observed subsequent to the neutron
bombardment. The 20-min activity of C^{11} has been observed in anthracene
crystals when they were bombarded with very energetic neutrons[1]. In this way
an organic scintillator forms a highly efficient threshold detector. Similarly the
13-day activity of I^{126} may be observed in alkali iodide scintillators so that these
are efficient threshold detectors at somewhat lower energy[2]. In addition to the
13-day I^{126} activity, the 25-min activity of I^{128} is formed by radiative capture in
alkali iodide crystals. Since this activity tends to build up whenever such crystals
are used for neutron detection, it frequently forms an undesirable background.

9. Coincidence techniques. Throughout the discussion of neutron detectors
it has been emphasized that events other than those which one wishes to detect
tend to cause an undesirable background. If events in two or more detectors are
observed simultaneously, such backgrounds may be reduced or eliminated. Two
different approaches are possible. In the first method the neutron produces a
charged particle which is detected in two or more counters. Since the charged
particles have a high speed, they produce effectively simultaneous counts in the
several detectors and coincidences are observed. The other approach is to observe
two interactions produced by the same neutron. In this case there may be a
measurable time delay between the two interactions, particularly if the neutron
is slowed down appreciably in the first interaction. The desired interactions may
then be selected on the basis of the expected delay, i.e., delayed coincidences
may be observed. For the effectiveness of any of these coincidence techniques it
is essential that the detector have a fast response.

Neutron detectors in which the same charged particle is observed more than
once were introduced by AMALDI[3] and are frequently called counter telescopes.
This name implies that charged particles traveling in a prescribed direction are
selected. In most of these devices a proton recoil originates in a foil of hydrogenous material or a cell filled with hydrogen gas. The proton then produces,
in the simplest case, simultaneous counts in two detectors. Each of these detectors
by itself will give a high counting rate in the neutron flux because of interactions
taking place in the gas with which it is filled or in the material out of which it
is made. It is necessary to calculate how many accidental coincidences are to be
expected on the basis of the resolving time of the coincidence circuit and of the
counting rates of the individual detectors. If this number is an appreciable
fraction of the number of protons which one wishes to observe, it may be necessary
to use more than two counters. In many cases threefold and sometimes fourfold
coincidences are recorded[3]. The discrimination against background may be
improved if the energy of the protons is known so that those protons can be selected
which lose a given amount of energy in each counter.

Counter telescopes have very low detecting efficiency, because it is necessary
that the radiator have a thickness small compared to the range of the charged
particles to be detected. The lower the energy of the particles the more serious
is this limitation. In order to increase the detecting efficiency one may employ
the gas in a counter as the radiator in a coincidence arrangement. Counter
telescopes are used primarily as fast neutron spectrometers and flux meters.
These applications will be discussed in Sects. 14 and 17.

[1] J. SHARPE and H. C. STAFFORD: Proc. Phys. Soc. Lond. A **64**, 211 (1951).
[2] H. C. MARTIN and R. F. TASCHEK: Phys. Rev. **89**, 1302 (1953).
[3] AMALDI, BOCCIARELLI, FERRETTI and TRABACCHI: Naturwiss. **30**, 582 (1942).

Devices in which two interactions induced by the same neutron are observed have likewise usually very low sensitivity and serve primarily as neutron spectrometers. Large liquid scintillators enable one, however, to observe two successive interactions of the same neutron with very high efficiency[1]. For this application cadmium in the form of cadmium propionate is added to a large volume of liquid scintillator (see Sect. 8). A fast neutron entering the solution produces one or more proton recoils effectively instantaneously. The neutron is slowed down in the solution until, some milliseconds later, it is captured by Cd at which time a γ-ray pulse will occur. Neutron counts are identified by observing delayed coincidences.

This device has several applications. It may be used for the measurement of the strength of weak neutron sources. In addition, it can be employed to determine the number of neutrons given off simultaneously such as in fission. For this purpose advantage is taken of the fact that the γ-rays from the capture of different neutrons will occur at different times. It may also be used in conjunction with a pulsed neutron source to distinguish between elastic scattering (one neutron, no γ-ray), inelastic scattering (one neutron, γ-radiation), and radiative capture (no neutron, γ-radiation). In this way the cross sections for these processes, particularly for radiative capture, may be measured.

10. Cloud chambers. Although in much of the early work in neutron physics the neutrons were detected in cloud chambers, this method of detection is now applied only infrequently. The reason is that cloud chambers have a much larger physical size than other detectors and, for most applications, have to be constructed of heavy materials which scatter and absorb the incident neutrons. In addition the evaluation of cloud chamber tracks is very tedious. The difficulties are particularly great when long range particles have to be observed such as proton recoils from energetic neutrons. For such problems cloud chambers have been replaced almost entirely by photographic emulsions.

There are, however, a few applications in neutron detection where a cloud chamber is the most satisfactory detector. An example of this was the measurement of the low energy end of the energy spectrum of neutrons produced in fission. Below 500 kev neutron energy, tracks of proton recoils in photographic emulsions cannot be measured accurately, and the detection with counters of a small number of low energy neutrons in the presence of many faster neutrons would be a very difficult task. In this experiment[2] the cloud chamber had a diameter of 30 cm and was about 20 cm deep. It was designed to have as small an amount of scattering material in it as possible. To this end, the chamber walls were made of 0.5 cm thick glass. In order to get long proton tracks from neutrons of low energy, the chamber was filled with hydrogen gas at $\frac{1}{3}$ atmosphere and water vapor. With this mixture 50-kev neutrons produced proton tracks of maximum length of 7 mm which could be measured reliably. The neutron spectrum could be determined in the energy interval from 50 to 700 kev.

Another recent example in which a cloud chamber was employed for detecting neutrons was a study of the elastic and inelastic scattering of 14-Mev neutrons by oxygen[3]. In this experiment a chamber 24 cm in diameter and with a 1-cm thick glass wall was filled with $\frac{1}{5}$ atmosphere of oxygen and water vapor. A study of the oxygen recoil tracks permitted a measurement of the angular distribution of the elastically scattered neutrons and of the cross sections for elastic

[1] REINES, COWAN, HARRISON and CARTER: Rev. Sci. Instrum. **25**, 1061 (1954). — DIVEN, MARTIN, TASCHEK and TERRELL: Phys. Rev. **101**, 1012 (1956).

[2] BONNER, FERRELL and RINEHART: Phys. Rev. **87**, 1032 (1952).

[3] J. P. CONNER: Phys. Rev. **89**, 712 (1953).

and inelastic scattering of the neutrons. Because of the short range of the oxygen recoils it would have been difficult to analyze the recoils by other methods.

While cloud chambers are restricted to the study of those neutron interactions in which short range charged particles are produced, it appears likely now that very long range charged particles could be most conveniently observed in a similar way in a bubble chamber[1]. Bubble chambers can have quite small sizes, and the walls may be thinner than those of cloud chambers. Their detecting efficiency is very high. In particular, liquid hydrogen bubble chambers are likely to be useful for measuring the flux and the energy spectrum of very energetic neutrons.

11. Photographic emulsions. Observation of the tracks produced by charged particles in photographic emulsions has been widely applied for a long time to the detection of neutrons. There are essentially three ways in which photographic plates serve for neutron detection: (a) Observation of proton recoils in the emulsion, (b) Observation of products of reactions taking place inside the emulsions, most frequently in nuclides with which the emulsion has been impregnated, (c) Observation of charged particles originating in a radiator and entering the emulsion from the outside.

Photographic emulsions have some advantages over other methods of neutron detection. Their small physical size allows them to be placed close to the neutron source without impairment of definition of the detector position. Because of their small mass and the absence of associated equipment the photographic plates will not alter the neutron flux appreciably by their presence. They are continuously sensitive and all types of events are registered simultaneously. Distinction between neutron and γ-ray induced events is normally possible in a photographic emulsion without any difficulty. Finally photographic emulsions require a minimum of equipment.

When photographic plates are used as detectors, exposures to the neutron source require in most cases relatively little time and a large amount of information is contained in the exposed plate. This advantage is offset by the tedious work involved in the measurement of the tracks in the plate. In deciding whether or not to carry out measurements with the photographic emulsion technique, one must weigh the possible saving in exposure time against the plate reading time. Photographic emulsions are usually advantageous when the interactions to be investigated are manifold or complicated or when a weak radioactive source is being studied.

One of the important applications of the photographic plate method to neutron detection is the measurement of spectra and flux of fast neutrons. Many of the techniques for this application were developed by Rosen and are summarized in his review articles [16]. The details of the technique depend on the nature of the neutron source. The simplest case is that of an intense point source of energetic neutrons. Here method (c) may be applied: a radiator containing either hydrogen or deuterium and a photographic plate are placed into a vacuum chamber. Proton or deuteron recoils produced in the radiator are observed in the emulsion. The angle between the recoils and the photographic plate, and the thickness of the emulsion are chosen so that the recoils stop in the emulsion. With this arrangement there will be recorded in the emulsion not only the recoils originating in the radiator but also proton recoils produced in the emulsion. To prevent the latter recoils from obscuring the desired tracks it may be necessary to shield the plate from the neutron source. In analyzing the plate one looks

---

[1] D. A. Glaser: Nuovo Cim. Suppl. **11**, 361 (1954).

for tracks which start from the surface of the emulsion and which are in a direction consistent with having originated in the radiator. The distribution in range and hence in energy of the recoils may then be determined. After the distribution in energy of the proton recoils is established, this distribution may be transformed into a neutron energy distribution by taking into account the differential cross section for scattering of neutrons by protons or deuterons. From the geometry and the number of nuclei in the radiator the absolute neutron flux may be calculated.

For point sources of lower intensity and also for lower neutron energies method (a) is more efficient, *i.e.*, the emulsion serves both as radiator and detector. Although this method has been applied to the study of neutron spectra for a long time, its application to absolute flux measurements has been developed only recently[1]. The spectrum is determined by measuring the ranges of the tracks which originate in a definite volume of the emulsion and which proceed at an angle with respect to the incident neutrons of less than a prescribed value (perhaps 15°). For an absolute flux measurement it is necessary to know in addition the hydrogen concentration in the emulsion, the emulsion volume which is analyzed, and the solid angle of acceptance of proton recoils. A correction has to be applied for tracks leaving the emulsion. The probability for a track to leave the emulsion will depend on the energy of the recoiling protons. In order to reduce this probability the plate is exposed in such an orientation that the neutron source is approximately in the plane of the plate, but not exactly, so that the neutrons are attenuated as little as possible in the plate.

Occasionally one wishes to measure a neutron spectrum or a flux from an extended source, for example inside a reactor or moderator. Here it is not possible to deduce the neutron energy from the proton recoil energy since one does not know the direction of the incident neutrons. If one analyzes all the tracks in a given volume of the emulsion, one may deduce from this energy distribution the distribution in energy of the neutrons by differentiation according to Eq. (1.4) [2]. This method tends, however, to give only rough results because of the inaccuracy inherent in the differentiation method.

Constituents of the emulsion other than hydrogen will react with neutrons also. In particular, neutrons of energy above 10 Mev produce the break-up of C^{12} into three α-particles, in most cases through an excited state of Be^8. Furthermore, disintegrations may be observed by loading the emulsion with a nuclide not usually present in the emulsion.

For measurements of neutron energies a photographic emulsion may be loaded with Li^6 [3]. From a measurement of the energy and direction of the disintegration products (triton and α-particle) it should be possible to determine both the energy and direction of the neutron producing the disintegration. In order to deduce the neutron spectrum it is necessary to know also the cross section for the Li^6 (n, α) reaction and the angular distribution of the products. If the neutron spectrum extends to high energies, the competing Li^6 (n, p) and Li^6 (n, d) reactions may introduce complications. In the experiments which have been carried out, emulsions were loaded with 0.2 mg/cm^2 of Li^6. Special techniques have to be applied to make it possible to distinguish clearly between proton recoils, α-particles, and tritons. This is accomplished by underdeveloping the plate, use of special developers, and delayed developing which produces differential fading of the image. Even when these techniques are applied, the

[1] E. R. GRAVES and L. ROSEN: Phys. Rev. **89**, 343 (1953).

[2] N. NERESON and F. REINES: Rev. Sci. Instrum. **21**, 534 (1950).

[3] G. R. KEEPIN jr., and J. H. ROBERTS: Rev. Sci. Instrum. **21**, 163 (1950).

point at which the disintegration takes place is frequently difficult to identify, particularly when the triton and α-particle travel in approximately opposite directions. For point neutron sources the direction of incidence of the neutrons is known so that it is possible to check whether the measured triton and α-particle energies and directions are consistent with the known direction of incidence of the neutrons. Attempts have been made to apply the detection by Li-loaded emulsions to extended sources[1]. This has so far not been entirely successful because of the difficulty just mentioned.

Nuclear emulsions have also been impregnated with B^{10}. The $B^{10}(n, α) Li^7$ reactions may be studied, but because of the fact that many of the disintegrations produce Li^7 in an excited state it is not possible to deduce uniquely the energy of the incident neutron from such a measurement. At high neutron energies one may observe the reaction $B^{10}(n, 2α) T$, but this reaction has such a small cross section that the number of these events will be very small compared to the number of proton recoils occurring in the emulsion.

Nitrogen which is contained in nuclear emulsions likewise undergoes disintegrations when bombarded by neutrons. The cross section for nitrogen reactions is highest at thermal energies. This fact is utilized in radiation monitoring. A photographic emulsion is partly enclosed in cadmium which absorbs thermal neutrons. If proton tracks corresponding to the 0.6 Mev Q-value of the N (n, p) reaction appear in the uncovered but not in the covered part of the emulsion, they can be identified as being due to thermal neutrons and the thermal neutron dose may be determined. These tracks may, however, be difficult to recognize in the presence of many proton recoils from fast neutrons.

12. Foils. Induced radioactivity from radiative capture of neutrons or other neutron reactions is normally measured with the aid of foils. After irradiation the foil is wound around a Geiger counter and the β-activity is counted. For the counting of very weak activities the cosmic ray background of the Geiger counter may be reduced by surrounding the counter by a ring of Geiger counters which are connected in anti-coincidence with the counter which detects the activity[2].

For neutron cross-section measurements it is frequently necessary to carry out absolute β-counting. The techniques for determining the absolute β-activity are treated in another article. Usually the measurement is carried out by comparison with a Ra-DEF source of known activity. Necessary precautions include the use of as large an acceptance angle of the counter as possible to reduce the effect of electron scattering in the sample, and of thin samples and thin counter windows to reduce the effect of absorption. In some cases, such as Au^{198}, absolute counting may be most readily performed by applying a coincidence method, since Au^{198} emits in 99% of the disintegrations a β-ray followed by a 0.41-Mev γ-ray. The activity is counted simultaneously with a β-ray counter and a γ-ray counter. In addition, the number of coincidences between the two counters is recorded. From these three rates the absolute β-activity may be calculated [13][3].

In many experiments in neutron physics it is sufficient to compare the activities of foils irradiated under different conditions. Even in this case precautions have to be taken to make the results meaningful and reproducible.

When slow neutrons are detected, the foils are frequently not thin for the neutrons, nor are they thin for the emitted electrons. It is therefore advisable

---

[1] Roberts, Solano, Wood and Billington: Rev. Sci. Instrum. **24**, 920 (1953).

[2] R. Tangen: Det Kgl. norske Vid. Selsk. Skr. **1946**.

[3] J. V. Dunworth: Rev. Sci. Instrum. **11**, 167 (1940).

either to count both sides of a foil or to count always the same side. Normally in an experiment several foils of the same material are counted and these foils are intended to give the same activity in the same flux. For this to be satisfied the foils must not only have the same mass but also the same area and should be of uniform thickness. In addition the counter should have uniform wall thickness, and the foil has to be placed into a reproducible position with respect to the counter [9b].

The activity of the foil may be counted with a scintillation counter rather than a Geiger counter. Since Geiger counters respond to every electron that enters the sensitive volume, little is gained by the use of a scintillation counter for β-counting, while the scintillator has the disadvantage of giving a continuous distribution of pulse heights resulting in variations of counting rate for small changes of amplifier gain or bias setting. On the other hand, scintillators have a much higher efficiency for detecting γ-rays than Geiger counters. If the thickness of the foil is not limited by self-absorption of neutrons, it may be advantageous to use thicker foils and achieve higher counting rates by counting γ-rays with a scintillator. This is particularly applicable to the detection of fast neutrons when the activated nuclide decays with γ-ray emission.

When foils are used in a medium in which slow neutrons are diffusing, the presence of the foil may depress the neutron intensity over what it would be in the absence of the detector. Corrections for this effect have been discussed by several authors[1] [9b].

C. Applications to various types of measurements.

In this part of the article the application of neutron detectors to specific problems in neutron physics will be discussed. While in the preceding part detectors were discussed according to their principle of operation, in the present section they will be ordered according to applications such as measurements of spectrum, flux, or source strength.

13. Measurement of relative intensity. The simplest requirement which may be imposed upon a neutron detector is that it record a number proportional to the number of neutrons impinging upon it provided the neutron spectrum remains constant. Such detectors are frequently called monitors. Many neutron experiments may be performed with monitors, such as measurements of total cross sections in transmission experiments, measurements of differential scattering cross sections, and measurements of inelastic collision cross sections by sphere transmission experiments [2].

Desirable characteristics of monitors are high detecting efficiency for neutrons and low detecting efficiency for γ-rays. Additional properties which may have to be considered are the response to neutrons of energies other than those which one wishes to observe and the response of the detector to neutrons incident in different directions. For example in a measurement of an inelastic collision cross section by sphere transmission the detector should have a response which falls off rapidly as the neutron energy decreases, and it should have the same response to neutrons incident in all directions. In contrast, in a measurement of a differential elastic scattering cross section it may be desirable that the detector respond to neutrons incident in one direction with much higher efficiency than to neutrons incident in other directions.

[1] C. W. TITTLE: Phys. Rev. **80**, 756 (1950). — Nucleonics **8**, 5 (June 1951); **9**, 60 (July 1951). — E. D. KLEMA and R. H. RITCHIE: Phys. Rev. **87**, 167 (1952). — M. W. THOMPSON: J. of Nucl. Energy **2**, 286 (1956).

All detectors which have been mentioned previously may be used as monitors. Which detector is most suitable depends entirely on the experiment to be performed and the energy of the neutrons to be measured. Because of the tedious process of reading tracks, cloud chambers and photographic emulsions are hardly ever employed as monitors. Monitors which are used most commonly are listed in Table 4 together with the energy range in which they are most useful. In addition, qualitative information is given in the table regarding the sensitivity

Table 4. *Properties of neutron monitors.*

Detector	Useful neutron energy range	Neutron detecting efficiency	γ-ray response	Response to thermal and epi-thermal neutrons	Directional properties
He³- and BF₃-filled counters. Boron containing scintillators	Below 0.2 Mev	High	Minimal	Yes	Essentially none
Radiative capture foil activation	Below 0.1 Mev	Moderate	Essentially none	Yes	None
γ-rays from boron disintegrations	Below 0.1 Mev	Very high	Very high	Yes	None
U²³⁵ fission	Below 0.1 Mev	Moderate	Essentially none	Yes	None
Gas-filled recoil counter	0.1 to 5 Mev	High	Low if used at appropriate pressure	No	None or directional depending on design
Recoil scintillator	Above 1 Mev	Very high	High	No	Essentially none unless specially shaped
Recoil detector using hydrogenous radiator	Above 1 Mev	Very low for thin radiator, moderate for thick radiator	Minimal	No	Strong discrimination between forward and backward hemisphere
Threshold fission	0.3 Mev above threshold and higher	Low	Essentially none	No	None
Foil activation by endoergic reactions	Above 1 Mev	Moderate	Slight	No	None

of the various types of detectors to slow neutrons, fast neutrons, and γ-rays. In the last column of Table 4 some indication is given as to how these detectors respond to neutrons incident in different directions.

Most detectors detect neutrons with essentially the same efficiency irrespective of their direction of incidence except recoil detectors. Since all the recoils are emitted into the forward hemisphere, recoil detectors using radiators give a very different response to fast neutrons traveling in opposite directions. For some experiments it is important to have an arrangement which gives a high sensitivity for neutrons incident within a small angle and a very low sensitivity for neutrons incident in other directions. To accomplish this neutron collimators have been developed. Their design depends on the energy of the neutrons which one wishes to collimate. For experiments on neutrons of low energies the collimator is usually a mixture or solution of a boron compound and water or paraffin. The neutrons are slowed down in the hydrogenous substance and absorbed in boron[1]. For higher energies the degradation of the neutrons occurs more rapidly by inelastic collisions in heavier materials than by elastic collisions with protons. A collimator which has been used for 14-Mev neutrons consists of 40 cm of iron backed by 15 cm of paraffin [16]. For applications in which the detector is only a monitor alterations of the neutron spectrum by the collimator do not matter.

14. Measurement of fast neutron spectra. The determination of the spectrum of fast neutrons is a difficult problem on which much work has been done but which so far has no entirely satisfactory solution. In an unpublished report SWARTZ[2] has summarized the situation as of 1954.

There are essentially three methods available for measuring fast neutron energies: measurements of the energy of the products of a nuclear reaction, usually $He^3(n, p)$ or $Li^6(n, \alpha)$, measurements of the energy of recoils produced in elastic collisions usually with hydrogen, and measurements of the time of flight of neutrons.

The most useful reaction for measuring neutron energies is $He^3(n, p)$. The energy release in this reaction is high enough that there is little danger of confusion of disintegrations with electrons produced by γ-rays and is sufficiently low that good energy resolution may be attained for neutrons of energies of a few hundred kev. A disadvantage of He^3 is that recoils from elastic scattering are so energetic that an ambiguity arises when the spectrum under investigation contains neutrons of energies above 1 Mev. Furthermore, the long range of the protons produced in the disintegrations necessitates high stopping powers of the filling gas in order to reduce wall effects. For a cylindrical counter 5 cm in diameter a gas pressure of 8 atmospheres of krypton is needed in addition to the He^3 to keep the wall effect at 10%. A He^3-filled proportional counter was found to give a spread in pulse heights of disintegrations of about 7% when it was irradiated with 1-Mev neutrons[3].

By far the largest amount of work has been done on the proton recoil method for measuring spectra and practically every recoil detector which has been discussed has been applied to the problem.

The oldest method for measuring proton recoil energies is to analyze tracks in a cloud chamber or photographic plate. Because of the tedious work involved in such observations many attempts have been made to apply counting techniques to the energy determinations. The counter telescope previously mentioned is the device most frequently employed for this purpose. The efficiency of a counter telescope is proportional to the radiator thickness and to the square of the sine of the acceptance angle of recoil protons. As the efficiency is increased the energy resolution decreases, and a compromise between these two factors has to be made.

The energy of the protons may be determined by adding the energy losses in the counters forming the telescope. Rather than measuring the energy loss in each counter, it is preferable to arrange for the proton to lose most of its energy in one of the counters and employ the other counters only to reduce the background. In one such device the protons (or other charged particles produced in a neutron interaction) first pass through two low pressure proportional counters and then enter a NaI scintillator which serves to measure their energy[4]. In another similar device the protons from a thin hydrogenous radiator are detected

[1] C. T. HIBDON, A. LANGSDORF jr., and R. E. HOLLAND: Phys. Rev. **85**, 595 (1952).
[2] C. D. SWARTZ: U.S. Atomic Energy report NYO-3863 (1954).
[3] BATCHELOR, AVES and SKYRME: Rev. Sci. Instrum. **26**, 1037 (1955).
[4] F. L. RIBE and J. D. SEAGRAVE: Phys. Rev. **94**, 934 (1954). — C. H. JOHNSON and C. C. TRAIL: Rev. Sci. Instrum. **27**, 468 (1956).

after collimation in a proportional counter near the radiator. Their energy is then measured in an ionization chamber with a Frisch grid[1]. In order to avoid measuring recoils which leave this ionization chamber, another ionization chamber is placed behind the first and connected in anti-coincidence with the proportional counter and the main chamber.

Instead of measuring the energy of the charged particles in a counter telescope, it is also possible to use such a device for measuring the range of the charged particles by means of absorbers[2]. The proton recoils travel through two or three proportional counters filled with gas at low pressure and are stopped in absorbers. An additional proportional counter is connected in anticoincidence with those ahead of the absorber. By varying the thickness of the absorber, the range of the protons may be determined.

The range of the recoil protons may be measured electronically[3]. This is accomplished by letting the proton first travel through two ionization chambers connected in coincidence. The proton then enters a third chamber in which it is stopped by the gas. This third chamber has a collecting electrode shielded by a grid from the region in which the recoil proton is stopped and located beyond this region. From the difference in time between the coincidence in the first two chambers and the signal observed in the third chamber it is possible to deduce the range of the proton in the third chamber.

In order to increase the detecting efficiency, especially at low neutron energies, one may use hydrogen gas in one of the counters also as the radiator. In one such device coincidences between two hydrogen filled proportional counters are measured[4]. Absorbers between the two counters serve to determine the range of recoil protons originating in one of the counters and traveling in the direction of the incident neutrons. A modification of this detector which was designed for measurements of neutrons in the energy range from 50 kev to 1 Mev consists of two proportional counters with parallel axes filled with methane to a pressure of a few cm Hg[5]. A brass block separating the two counters has 400 holes which restrict recoil protons to an angle of 28° with respect to the direction of the incident neutrons. Coincident pulses in the two counters are added. From the sum of the pulse heights the neutron energies can be deduced. The energy resolution of this device was about 10%.

A somewhat different approach to the problem was taken by GILES[6]. Proton recoils are observed in a cylindrical proportional counter 60 cm long and 2.5 cm in diameter the axis of which is oriented in the direction of the incident neutrons. The cathode of the counter is formed by a screen of 12 parallel wires. Concentric with this counter is an aluminum tube, 7.5 cm in diameter. Between the wire screen and this tube are six counting wires which are connected together and form the anode of an anticoincidence counter. Only those proton recoils are counted which stay entirely in the central counter. This arrangement forms a radiator thick compared to the range of the recoil protons and effectively collimates the recoils. When this detector was filled with 1 atmosphere of methane, 2.5-Mev neutrons were counted with an efficiency of 4×10^{-4} and the energy resolution was about 5%. A disadvantage of the detector is that protons which

---

[1] N. NERESON and S. E. DARDEN: Phys. Rev. **89**, 775 (1953).

[2] AMALDI, BOCCIARELLI, FERRETTI and TRABACCHI: Naturwiss. **30**, 582 (1942). — D. L. HILL: Phys. Rev. **87**, 1034 (1952). — BROLLEY, COON and FOWLER: Phys. Rev. **82**, 190 (1951). — R. G. COCHRAN and K. M. HENRY: Rev. Sci. Instrum. **26**, 757 (1955).

[3] J. R. HOLT and A. E. LITHERLAND: Rev. Sci. Instrum. **25**, 298 (1954).

[4] D. C. WORTH: Phys. Rev. **75**, 903 (1949).

[5] G. J. PERLOW: Rev. Sci. Instrum. **27**, 460 (1956).

[6] R. GILES: Rev. Sci. Instrum. **24**, 986 (1953).

have recoiled at a large angle with respect to the neutrons and do not have a long enough range to leave the central counter, will produce a background of small pulses.

Instead of letting the recoils originate in a thick gas volume it is also possible to use an organic scintillator to serve simultaneously as radiator and as the first counter. A detector of this type has been suggested by MOZLEY and SHOEMAKER[1]. Two anthracene crystals were mounted in such a manner that the first crystal was the source of protons while the second defined the angle of proton recoil. The pulses from both scintillators were added electronically to give the energy of the proton recoils. To reduce background counts from electrons a proportional counter was placed between the two crystals. The proportional counter was biased to respond only to protons. Only those events were recorded in which there was a triple coincidence between all three counters.

A modification of this detector has been proposed[2] in which only one of the crystals acts as radiator while NaI is chosen for the other, and in which the pulses from the two crystals are applied separately to the horizontal and vertical deflecting plates of an oscilloscope. This device gave appreciably better neutron energy resolution than could be obtained with the previous one, because it enables one to take into account that the light pulse from a scintillator is not proportional to the energy of the protons.

Because of the small probability of detecting a neutron in a counter, methods based on observing the same neutron twice have been made possible only by the development of high efficiency scintillators. In the first device of this type[3] recoil protons were selected which were produced by head-on collisions of fast neutrons. To accomplish this, coincidences were measured between protons recoiling in an organic scintillator and pulses in a second crystal arranged at 90° to the primary neutron beam. Only delayed coincidences were measured which were caused by neutrons traveling from the first to the second crystal with energies less than 30 kev. For primary neutron energies of several Mev the recoil proton carries more than 99% of the primary neutron energy and a measurement of the proton energy therefore allows a direct determination of the neutron energy. For the detection of the slow scattered neutron a NaI crystal was surrounded by 0.5 cm of silver. Capture of the neutrons in either iodine or silver results in γ-rays which could be detected in the NaI crystal.

Modifications of this detecting device have been proposed in which the scattering angle of the neutrons in the first crystal was about 45° rather than 90°[4]. In these devices the second neutron detector serves to select those events in the first detector in which the neutrons have been scattered through a given angle. Since the scattered neutrons still have a high energy, they may be detected by proton recoils occurring in an organic scintillator. To increase the detecting efficiency a number of such detectors may be placed on a circle around the first detector. Coincidences between the two detectors may also be caused by γ-rays scattered in one scintillator and detected by the other. The γ-ray background can be reduced by requiring that the pulse from the second detector be delayed by the time it takes the scattered neutrons to travel from one detector to the other.

Rather than observing the pulse height distribution of the proton recoils in the first detector one may determine the time of flight of the neutrons from the

[1] R. F. MOZLEY and F. C. SHOEMAKER: Rev. Sci. Instrum. **23**, 569 (1952).

[2] CALVERT, JAFFE and MASLIN: Proc. Phys. Soc. Lond. A **68**, 1017 (1955).

[3] BEGHIAN, ALLEN, CALVERT and HALBAN: Phys. Rev. **86**, 1044 (1952).

[4] CHAGNON, MADANSKY and OWEN: Rev. Sci. Instrum. **24**, 656 (1953). — J. E. DRAPER: Rev. Sci. Instrum. **25**, 558 (1954).

first to the second detector in order to measure the energies of the primary neutrons[1].

In some recoil spectrometers, for example when photographic plates are used with an external radiator, it is necessary to shield the detector from the neutron source. For this purpose collimators are frequently employed. When one wishes to deduce the primary neutron spectrum from the observations, he has to know whether the spectrum has been altered by the collimator. Some studies of this effect show that a properly designed collimator does not affect the fast neutron spectrum appreciably. Such investigations have been performed for 2.5-Mev neutrons[2] and 14-Mev neutrons [16].

Measurements of neutron energies by their time of flight have been carried out for slow neutrons for many years, but the application of the time-of-flight method to fast neutrons has become possible only more recently. For neutrons of energies around 1 Mev and flight paths of the order of meters, flight times of the order of 10^{-8} sec have to be measured. Only scintillation detectors have a fast enough response for such measurements and development of such fast electronic circuits has offered technical difficulties which have only recently been overcome.

The usual procedure is to produce bursts of neutrons by sweeping a beam of accelerated particles across a target or by modulating the beam in some other manner. Alternatively it is possible to detect associated particles produced in the same reaction as the neutrons to define the time at which the neutron leaves the target. α-particles from the $D+T$ reaction[3] have served for this purpose and He^3-particles from the $D+D$ reaction may be used.

Several electronic circuits have been developed for measuring the short time interval between the production and detection of the fast neutrons. The two pulses may be sent in opposite directions along a delay line and the point at which they meet may be determined[4], delayed coincidences may be observed, or the time interval may be converted into a slow pulse which has a height proportional to the interval[5]. All these circuits are complicated and require much equipment. On the other hand the time-of-flight method appears to be the best method for measuring spectra from inelastic scattering of neutrons[6]. It is obviously not applicable to the measurement of spectra from radioactive neutron sources.

In Table 5 the properties of the most promising fast neutron spectrometers are summarized. The table is intended to give only qualitative information, since the actual performance of a spectrometer will vary greatly depending on the details of its design for a specific problem. The efficiency of neutron detection is the number of recorded events that can be evaluated per incident neutron. In general it is possible to increase the efficiency by decreasing the energy resolution (or increasing the energy spread). The efficiency of detection is, in most cases, a rapid function of the neutron energy, and the indicated efficiency is, therefore, to be interpreted only as an order of magnitude estimate at a typical energy for which the spectrometer is useful. Sensitivity to γ-rays likewise depends frequently on the details of design.

In Table 5 spectrometers in which all recoils are registered and the neutron spectrum is deduced by differentiation of a pulse height distribution are not

[1] G. C. Neilson and D. B. James: Rev. Sci. Instrum. **26**, 1018 (1955).

[2] Segel, Swartz and Owen: Rev. Sci. Instrum. **25**, 140 (1954).

[3] G. K. O'Neill: Phys. Rev. **95**, 1235 (1954).

[4] Neddermeyer, Althaus, Allison and Schatz: Rev. Sci. Instrum. **18**, 488 (1947). — G. K. O'Neill: Rev. Sci. Instrum. **26**, 285 (1955).

[5] Weber, Johnstone and Cranberg: Rev. Sci. Instrum. **27**, 166 (1956).

[6] L. Cranberg and J. S. Levin: Phys. Rev. **103**, 343 (1956).

Table 5. *Properties of fast neutron spectrometers.*

Method	Most useful neutron energy range (in Mev)	Efficiency of detection	Energy spread	γ-ray sensitivity	Remarks	Reference
He^3 (n, p) counter	0.1 to 1	10^{-4}	0.05 Mev	Moderate	No neutrons above 1 Mev must be present	Ref. 3, p. 471
Hydrogen-containing cloud chamber	0.05 to 1	2×10^{-3}	0.05 Mev	Insensitive	Evaluation tedious	Ref. 2, p. 465
Coincident proportional counters	0.05 to 1	10^{-5}	10%	Fairly insensitive		Ref. 5, p. 472
Collimated proton recoils in gas	0.5 to 5	4×10^{-4}	5%	Moderate	Only part of spectrum can be evaluated with one filling	Ref. 6, p. 472
Proton recoils in photographic emulsion	0.5 to 20	2×10^{-3}	0.1 Mev	Insensitive	Evaluation tedious	[16]
Thin polyethylene radiator with proton collimator	Above 1	10^{-5} at 1 Mev 10^{-4} at 10 Mev	10%	Insensitive	Several radiators and fillings are needed to cover broad spectrum	Ref. 4, p. 471
Detection of neutrons scattered by hydrogen	Above 1	10^{-4}	10%	May be minimized by requiring time delay		Ref. 2, p. 473
Time of flight	Below 10	Very high	0.1 Mev	Insensitive	Source must be pulsed	[12a]

included. The uncertainty introduced by the differentiation is so great that such detectors are practical only in very special applications. Furthermore such spectrometers are very sensitive to γ-rays, but on the other hand they have very high detecting efficiencies. The high detecting efficiency is, however, a misleading characteristic, because the differentiation results in the loss of most of the information.

In summary, for neutron energies below 1 Mev, observations of He^3 disintegrations offer the best way of measuring neutron energies, but this method can be used only if the spectrum does not contain neutrons of energies above 1 Mev. Coincident proportional counters are useful in the energy region below 1 Mev, but the detecting efficiency is rather low. The device which has been employed most frequently and successfully in this energy range is the hydrogen-filled cloud chamber. Its disadvantage is the necessity of evaluating individual tracks.

If the neutron source can be pulsed, time-of-flight techniques are most promising for measurements at not too high neutron energies. These techniques may be used up to energies of the order of 10 Mev. At energies above 1 Mev a thin hydrogenous radiator and counter telescope offer a straightforward method of spectrum analysis, but the efficiency is very low at the lower energies and several radiators and counter fillings may have to be used if the neutron spectrum covers a wide energy range. Other coincidence techniques give higher detecting efficiencies than the thin radiator with telescope.

15. Measurement of total source strength. An important problem in neutron physics is the measurement of the total number of neutrons emitted by a source per unit time. The most generally applicable method for the measurement of source strength was suggested by AMALDI and FERMI[1] and involves the slowing down of the neutrons to thermal or epithermal energies. In most cases one merely wishes to compare a source of unknown strength with a calibrated standard. A much more difficult problem is the absolute calibration of the standard.

For the comparison of neutron sources by slowing down the following methods are used [9b]:

(a) Absorption of neutrons in Mn (or I) in a water tank;

(b) Comparison of thermal flux at one point in graphite;

(c) Integration of the slowing-down density at an epithermal resonance using a graphite moderator;

(d) Integration of thermal neutron density in water.

In method (a) a neutron source is placed into a large tank containing a solution of $MnSO_4$ (or CaI_2) in water. Most of the neutrons are absorbed, after they are slowed down, in Mn and H although a small fraction of the fast neutrons may be absorbed in S or O. The total Mn activity is proportional to the source strength. If the solution is stirred after irradiation, the activity of a sample of the solution will likewise be proportional to the source strength. The activity is independent of the angular distribution of the neutrons from the source, a fact which is important when the strength of an accelerated particle source is measured. Furthermore, the activity is independent of the energy of the primary neutrons provided the tank is large enough that the leakage of neutrons from the bath is small. To determine the activity a thin-walled Geiger counter is dipped into a sample of the solution. For measurements on very weak sources MnO_2 may be precipitated out of a large sample of the solution and then counted.

For comparisons of the strengths of sources it would be simplest to measure the activity of a foil at one point in the moderator (method b). If water is used as the moderator, measurements at one point in the bath will not yield significant results, because the distribution of thermal neutrons in water is a sensitive function of the spectrum of the primary neutrons. Even different Ra (α) Be sources give appreciably different distributions. The slowing down is very rapid because of the small mass of the proton and the large neutron-proton scattering cross section at low neutron energies. Because of the relatively large absorption cross section of hydrogen at thermal energies (0.3 b) and the large scattering cross section, the neutron will be absorbed in the near neighborhood of the point at which it had its first large energy loss and will not diffuse far through the water. These facts result in a steep decrease of the thermal neutron density in water. A change of 1 mm in the position of a detecting foil may cause a change of 2% in the activity. In addition, this steep fall-off will make the activity sensitive to the dimensions of the source. Consequently measurements at one point in the moderator have to be carried out in a medium in which the diffusion length is greater. For this purpose large columns of graphite have been built. A typical column has a size of $1.5 \text{m} \times 1.5 \text{m} \times 2.8 \text{m}$. It is to be expected that the distribution of thermal neutrons in graphite should not depend strongly on the energy of the primary neutrons. These distributions can be calculated. For sources whose neutron energies are known approximately it is possible to find a distance from the source at which the thermal neutron density is relatively insensitive to the neutron energy. For Ra (α) Be sources this distance is about 75 cm. If two sources are to be compared which produce neutrons of appreciably different energies, measurements are carried out at two distances from the source. The two distances can

---

[1] E. AMALDI and E. FERMI: Phys. Rev. **50**, 899 (1936).

usually be chosen so that one can be sure that the correct value of the ratio of the source strengths lies between the two measured ratios.

Method (b) is very sensitive and can be used for weak sources. Particularly if an efficient BF_3 counter is employed instead of foils, sources which emit only a few thousand neutrons per sec may be measured.

For more precise comparisons of sources which give different spectra it is necessary to integrate the neutron density over the volume of the moderator. Either graphite or water may be used as moderators. Since even in the large graphite column just described many of the thermal neutrons leak out of the sides, it is necessary to measure the slowing-down density rather than the thermal neutron density in a graphite moderator. Usually the slowing-down density at the indium resonance level at 1.46 ev is determined. Indium foils are irradiated at about half a dozen points along the long axis of the column both in cadmium absorbers and without the absorbers. The difference between the two activations, corrected for absorption of resonance neutrons in cadmium and for absorption in indium at other resonances, is integrated over the volume. Integrals obtained with two sources are compared and should be in the ratio of the source strengths.

In a large water moderator thermal neutrons are absorbed before they can escape, and the time rate of absorption of thermal neutrons integrated over the volume will be proportional to the source strength. A detector which has a $1/v$ response to slow neutrons will give an activity proportional to the absorption rate of thermal neutrons. Manganese foils are suitable detectors. Foils may be activated at several distances and the volume integral determined, or the integration may be performed by moving a foil continuously through the water at such a rate that the final activity will be proportional to the volume integral.

An absolute determination of source strength is much more difficult than the comparisons described so far, but both method (a) and method (d) may be modified to yield absolute source strength measurements (for example [13][1]). In the Mn-bath method it is necessary to determine the absolute activity of the Mn[2]. In method (d) the foils may be calibrated in a known thermal flux which in turn is measured with a boron counter. Alternatively the foil activity may be determined directly by the coincidence method which was mentioned in Sect. 12.

These methods for absolute determinations of source strength require the knowledge of the absolute cross section of the detecting reaction such as the activation cross sections of Mn or Au. Since these cross sections are not known very accurately, a method has been developed in which a knowledge of the absolute cross section is not needed. For this purpose the same nuclide is used both as absorber in the moderator and as detector, such as a solution of boric acid in water as a moderator and a BF_3 counter as detector. If, as is true for boron, the cross section of the absorber varies as $1/v$ and is large compared to the absorption cross section of all other nuclides in the moderator, the source strength Q is given by

$$Q = 4\pi \frac{N_B}{n} (1 + \varepsilon) \int_0^\infty I r^2 \, dr, \qquad (15.1)$$

where N_B is the density of boron nuclei in the moderator, n the number of boron nuclei in the counter, and I the number of boron disintegrations occurring per

[1] De Juren, Padgett and Curtiss: J. Res. Nat. Bur. Stand. **55**, 63 (1955). — J. A. de Juren and J. Chin: J. Res. Nat. Bur. Stand. **55**, 311 (1955).
[2] R. D. O'Neal and G. Scharff-Goldhaber: Phys. Rev. **69**, 368 (1946). — F. Alder and P. Huber: Helv. phys. Acta **22**, 368 (1949).

unit time in the counter when it is at a distance r from the source. The measurement of source strength is, therefore, independent of the knowledge of the absorption cross section of boron except for a correction term

$$\varepsilon = \frac{\sigma_H N_H}{\sigma_B N_B},$$

where N_H is the nuclear density of hydrogen in the moderator and σ_H and σ_B are the absorption cross sections of hydrogen and boron respectively.

While the slowing-down technique is the most generally applicable method for measuring source strengths, there are some other methods which for some special sources may be more accurate, i.e., the associated particle and the associated product methods.

In the associated particle method the charged particle is observed which is associated with each neutron when it is produced in a nuclear reaction. Since there is one charged particle for each neutron, counting of the number of charged particles will enable one to determine the number of neutrons produced. In experiments with accelerated particle sources, usually the number of associated particles emitted in a given direction is counted and in this way the number of neutrons emitted in the corresponding direction is measured. It is, of course, possible to determine the angular distribution of the associated particles and, thereby, to determine the total neutron source strength. Determinations of the number of neutrons emitted in a given direction have been carried out for the $D + D$ reaction by counting He^3 particles and for the $D + T$ reaction by counting α-particles. The total source strength of a photo-neutron source in which deuterium is disintegrated by γ-rays from RdTh or Na^{24} has also been measured by the associated particle method[1] (also unpublished results). In this case disintegrations take place in deuterium gas contained in a spherical ionization chamber and photoprotons are counted. In order to obtain greater source strength the deuterium gas is then replaced by heavy water. From the number of disintegrations observed in the deuterium gas the number produced in the heavy water may be deduced.

In the associated-product method the number of product nuclei of a neutron-producing reaction is determined. For example, when neutrons are produced in the reaction $Li^7 + p \rightarrow Be^7 + n$ the amount of Be^7 formed may be measured by observing the radioactivity of Be^7[2]. This has so far not been done as an absolute measurement, but only by comparison with a standard source through a Mn-bath measurement. Another reaction to which the associated product method has been applied is $F^{19} + \alpha \rightarrow Na^{22} + n$. Na^{22} decays to stable Ne^{22}, and the amount of Ne may be measured gas-volumetrically [14].

It has also been proposed that the absolute source strength of a Be photo-neutron source may be measured by extraction of helium gas from beryllium metal. In the reaction $Be^9 + \gamma \rightarrow 2He^4 + n$ two atoms of helium are produced per neutron. It is believed that the amount of helium may be determined volumetrically with an accuracy to yield a neutron source strength to 1% [14][3].

A quite different method for measuring the source strength of a radioactive neutron source was suggested by LITTLER[4]. The source is introduced into a chain-reacting graphite-uranium reactor and the change in the effective repro-

[1] BISHOP, COLLIE, HALBAN, HEDGRAN, SIEGBAHN, DU TOIT and WILSON: Phys. Rev. **80**, 211 (1950).

[2] R. TASCHEK and A. HEMMENDINGER: Phys. Rev. **74**, 373 (1948).

[3] E. GLÜCKAUF and F. A. PANETH: Proc. Roy. Soc. Lond. A **165**, 229 (1938).

[4] D. J. LITTLER: Proc. Phys. Soc. Lond. A **64**, 638 (1951).

duction constant is observed. Instead of the neutron-producing source, a neutron absorber such as sodium or phosphorus is then placed into the pile and its effect on the reproduction constant is measured. An absolute β-count on the activated absorber gives the number of neutrons absorbed. By comparing the effects of the source and of the absorber the number of neutrons emitted from the source may be calculated. The accuracy of the method is about 5%.

Total neutron source strengths have been measured by the slowing-down technique with an error of 2 to 3% [1]. Reducing the uncertainty in this method appreciably, appears to be difficult. On the other hand, the associated particle and associated product methods are likely to be capable of higher accuracy. Preliminary tests indicate that an error of 1% may be achieved with these methods.

16. Measurement of thermal flux. Techniques for determining thermal neutron flux are discussed in Ref. [9], [10], and [13]. There are two essentially different methods for absolute thermal flux determinations. Either a fast neutron source of known strength is placed into a moderator and the flux is calculated, or reactions are counted which are induced in a substance the reaction cross section of which is known.

In the first method a graphite column as described in Sect. 15 serves as a moderator. The spatial distribution of the slowing down density of the neutrons is measured by activation of indium foils covered with cadmium. If capture and escape from the graphite column are negligible for the resonance neutrons, the volume integral of the measured activity is simply related to the strength of the fast neutron source placed into the column so that absolute values of the slowing down density may be determined. From the slowing down density the thermal neutron density may be calculated with an accuracy which is limited primarily by the knowledge of the strength of the fast neutron source. Once the thermal neutron density at any point in the column is known, foils or counters for thermal neutrons may be calibrated in the column. This method has several disadvantages. It requires the availability of a standard neutron source and of a large graphite column. Its accuracy is limited by the accuracy of absolute source strength calibrations. Furthermore, the thermal neutron flux is quite low so that either very intense sources or very thick foils must be used. The most accurately known and stable sources are photo-neutron sources and these are relatively weak. Thick foils, on the other hand, depress the neutron density in their neighborhood. If a foil calibrated in the graphite column is then used to measure neutron flux in another medium, such as in a slow neutron beam in air, corrections have to be applied for this effect.

The second method for measuring thermal neutron density is to count absolutely the number of reactions induced by the neutrons. The reactions most frequently utilized are $B^{10}(n, \alpha)$, radiative capture in Au, and fission in U^{235}. In all these cases the process which is observed has a much higher cross section than all other processes induced by thermal neutrons. In order to determine the reaction cross section the total cross section is measured in a simple transmission experiment which does not necessitate a knowledge of absolute flux. The total cross section contains also the cross section for scattering and, for U^{235}, radiative capture. Since the scattering cross sections are small compared to the reaction cross sections of these nuclides at thermal energies, uncertainties in the scattering cross section do not introduce appreciable errors into the reaction cross sections when the scattering cross section is subtracted from the total cross section. For U^{235} the ratio of capture to fission is also well known for thermal neutrons.

---

[1] J. DE JUREN and J. CHIN: J. Res. Nat. Bur. Stand. **55**, 311 (1955).

The number of boron disintegrations may be counted in a BF_3 counter of known active volume and filled to a known pressure [13] or in a counter containing a boron foil[1]. If radiative capture in gold is used, a gold foil is activated and the activity is counted absolutely by the coincidence technique [13]. Fissions in U^{235} may be counted in a chamber with a U^{235} foil.

For slow neutrons the cross section for $B^{10}(n, \alpha)$ is inversely proportional to the neutron velocity over a wide range of velocities, for $Au(n, \gamma)$ deviations from the $1/v$ behavior set in at lower energies, and the U^{235} fission cross section deviates somewhat from the $1/v$ dependence even at thermal energies. As was pointed out in Sect. 4, a detector whose cross section varies as $1/v$ measures neutron density rather than neutron flux. For a determination of the neutron flux a knowledge of the velocity of the neutrons is needed in addition. Actually for most experiments one is interested in the neutron density as measured with a $1/v$ detector.

The nuclides which are most suitable for absolute flux measurements are not those which from the point of view of half-life and cross section are the most convenient ones to use. Other materials may, however, be calibrated in a standard flux and may then serve as secondary standards. For example Mn whose radiative capture cross section also varies as $1/v$ is a very convenient material for slow neutron flux measurements.

The reaction cross sections of B^{10} and Au are believed to be known with an accuracy of the order of 1%, and it should therefore be possible to measure thermal neutron densities with an accuracy of 1 to 2%. When boron disintegrations are observed, an important source of error may be the uncertainty in the isotopic composition of boron. Even if the isotopic composition of the BF_3 used for filling the counter is known, the purification of the gas may alter the ratio of the two isotopes. Counting of Au^{198} disintegrations by the coincidence method is subject to some uncertainty because the internal conversion coefficient of the γ-rays is not known precisely.

There has so far been little need for absolute flux measurements at epithermal energies. Boron disintegrations could be used for such measurements. In addition, determinations of the slowing down density with cadmium-covered indium foils in a graphite column give information about the flux of resonance neutrons.

17. Measurement of fast flux. The problem of measuring a flux of fast neutrons is discussed in detail in Ref. [3], [13], and [12d].

In the description of methods for measuring total source strength the associated particle method was mentioned as a way of determining the number of neutrons emitted in a given direction from certain accelerated particle sources. Examples of reactions to which this method has been applied for measuring fast neutron flux are $D + D \rightarrow He^3 + n$ and $D + T \rightarrow He^4 + n$. The method is also applicable to the reaction $p + T \rightarrow He^3 + n$ at sufficiently high bombarding energies.

If a neutron of energy T_n is emitted at an angle ϑ with respect to the incident particles, the associated particle of energy T_a is emitted at an angle φ given by

$$\sin \varphi = (\mu T_n / m T_a)^{\frac{1}{2}} \sin \vartheta, \qquad (17.1)$$

where μ and m are the masses of the neutron and of the associated particle, respectively. Since it is usually most convenient to keep the angle φ fixed, measurements of neutron flux at different bombarding energies require either

---

[1] J. A. DE JUREN and H. ROSENWASSER: J. Res. Nat. Bur. Stand. **52**, 93 (1954).

changes of the angle ϑ or a knowledge of the angular distribution of the neutrons at the different bombarding energies.

Two factors which limit the accuracy of the associated particle method for measuring neutron flux are the scattering of neutrons in the material surrounding the target and the background of neutrons originating in places other than the target or which are scattered by the floor, the walls, or the air of the laboratory. The former effect may be minimized by constructing the target assembly of sufficiently thin material. The background may be measured by inserting a shadow bar between the source and detector. If the background is very high as, for example, when deuterons are accelerated in a cyclotron, observation of coincidences between associated particles and neutron induced events permits flux measurements even for backgrounds of 1000 times the intensity of the desired neutrons[1].

The associated particle method should be the most precise method for measuring neutron flux since the associated charged particles can be counted with accurately known efficiency.

All the methods which are useful for measuring total source strength may also be applied to fast neutron flux measurements provided the angular distribution of the neutrons from the source is known or can be measured accurately. For example the associated product method mentioned in Sect. 15 may serve to determine the neutron flux from the $Li^7(p, n) Be^7$ reaction by observation of the radioactivity of Be^7. Likewise slowing-down techniques for measuring total source strength combined with a knowledge of the angular distribution of the neutrons allow measurements of fast neutron flux. While this technique is capable of high accuracy, care must be taken to take into account or avoid spurious absorption effects in the moderator. For water which is the most effective and most widely used moderator, the absorption of neutrons by the $O^{16}(n, \alpha)$ reaction introduces an appreciable uncertainty into the measurement, if neutrons of energies above 3 Mev are produced by the source.

Many measurements of fast neutron flux have been carried out with the energy insensitive "Long Counter" developed by HANSON and McKIBBEN[2]. In analogy to the water bath method, a long boron counter is placed along the axis of a cylinder of paraffin with the axis pointing toward the source. The long counter actually differs materially from a water bath. The number of neutrons reflected from the front surface of the paraffin will vary with energy, since both the hydrogen and carbon cross sections are energy dependent. Because of the decrease of the hydrogen cross section with increasing energy fast neutrons will penetrate deeper into the paraffin before being slowed than slower neutrons. Therefore primary neutrons of lower energies have a higher probability of escape from the front surface and will be detected with lower efficiency. This effect is partly compensated by the fact that high energy neutrons have a larger probability of escape from the cylindrical surface. In order to raise the sensitivity of the counter to slow neutrons several holes are drilled into the front face of the paraffin. The boron counter which was used by HANSON and McKIBBEN is 26 cm long and 1.3 cm in diameter. It is imbedded in a paraffin cylinder, 20 cm in diameter and 26 cm long. The paraffin cylinder is surrounded on all sides except the front surface by a layer of B_2O_3 which is surrounded in turn by a layer of paraffin, 8 cm thick. These outer layers form a shield to make the counter less sensitive to neutrons which do not come from the direction of the source. The inner paraffin cylinder has in its front surface eight holes, 2.5 cm in diameter and 9 cm deep. Measurements carried out by the original authors indicated that the sensitivity

[1] CURTIS, FOWLER and ROSEN: Rev. Sci. Instrum. **20**, 388 (1949).
[2] A. O. HANSON and J. L. McKIBBEN: Phys. Rev. **72**, 673 (1947).

of this counter was constant to 10% from 0.5-Mev to Ra (α) Be neutrons and dropped off to 80% of its maximum sensitivity for thermal neutrons. Subsequent tests[1] showed anomalies of about 5% in the sensitivity at energies corresponding to resonances in the scattering cross section of carbon. On the basis of the same tests it was stated that the sensitivity of the counter was constant to $\pm 1\%$ down to energies of a few hundred kilovolts. Additional tests performed at the same laboratory in which the response of the long counter was compared to that of a counter telescope showed that in addition to anomalies caused by carbon resonances there is a gradual decrease of the sensitivity of the long counter by about 15% in the neutron energy range from 2 to 6 Mev. It was pointed out by W. D. ALLEN[2] that statements regarding the dependence of the sensitivity of the long counter on energy should include a specification of the source-counter distance. Measurements of counting rates for different distances show that the counter behaves like a point detector, but that the effective center of the counter moves from about 8 mm inside the front surface of the paraffin for Sb (γ) Be neutrons to 5 cm for Ra (α) Be neutrons.

For measurements of neutron flux the counting rate of the long counter at the position where the flux is to be measured is compared with the counting rate observed when the unknown source is replaced by a standard source. This method, although very convenient, yields only approximate flux measurements. The accuracy is limited by the knowledge of the strength of the standard source. A more serious source of error is the fact that the sensitivity of the counter is somewhat energy dependent and that the effective center of the counter also depends on energy.

The most generally useful method for fast neutron flux measurements, particularly for monoergic neutrons, is the recoil particle method which was described in Sect. 1. For absolute measurements proton recoils are almost always observed. The manner of detecting the recoils depends primarily on the energy of the neutrons which in turn determines the range of the recoils.

At the lowest neutron energies, at which the recoils have a very short range, a proportional counter filled with hydrogen or a hydrocarbon is the most convenient detector. With such a counter the pulse height distribution of recoil protons is measured. This distribution has to be converted into a recoil energy distribution. For this purpose the ionizing power of the recoils as a function of their energy needs to be known, and in addition information about the range-energy relation for the protons is required for wall effect corrections. When these corrections are determined and applied carefully, the method may yield flux determinations in the neutron energy range from 0.05 to 2 Mev with an accuracy of about 2% [1].

A disadvantage of a proportional counter is its rather large size and relatively complicated construction. This results in the difficulty that the point at which the flux is measured is not very accurately defined and not readily accessible for placing at the same location other nuclides whose properties are to be measured in the same flux. A thin hydrogenous radiator reduces these difficulties although the counting rates will be much lower with the radiator. On the other hand, no wall effect corrections need to be applied when a radiator is used. Thin hydrogenous foils are practical as sources of recoils at neutron energies above 0.5 Mev. At neutron energies above 2 Mev the range of the recoils is sufficiently long that

---

[1] NOBLES, DAY, HENKEL, JARVIS, KUTARNIA, McKIBBEN, PERRY and SMITH: Rev. Sci. Instrum. 25, 334 (1954).

[2] W. D. ALLEN: British Atomic Energy Research Est. Report NP/R 1667.

a counter telescope offers the most satisfactory device for counting the recoils emitted into a known solid angle.

In Sect. 8 experiments were mentioned in which proton recoils were observed in a large organic scintillator. In this way very high detecting efficiencies may be achieved, but the accuracy of flux measurements has not approached that obtained with radiators. Scintillators are sufficiently sensitive to electrons that they can be used only for measurements of a neutron flux in the absence of appreciable γ-radiation.

Recording of recoils in photographic emulsions may be the only method usable for flux measurements in some instances in which neutrons of different energies are present and the space available for the detector is very small. The accuracy is, however, not comparable to other methods.

The homogeneous ionization chamber described in Sect. 6 is another flux measuring device. While this detector eliminates wall effects, it is very sensitive to γ-rays although a compensating chamber may reduce ionization by γ-rays considerably.

In Table 6 the methods for absolute fast neutron flux measurements are summarized. In each case the conditions are indicated under which the method is most useful, *i.e.*, the energy of the neutrons, whether they should be monoergic, and the effect of the presence of γ-rays. Several of the methods require the knowledge of either an absolute neutron interaction cross section or of the strength of a standard neutron source. The neutron-proton cross section is known to around 1% up to 20 Mev so that the requirement of a knowledge of the neutron-proton cross section is usually not the principal limitation to the accuracy of the recoil particle method. The strength of standard neutron sources is known to 2 to 3% and may limit the accuracy of methods depending on a standard source.

In view of the difficulties involved in absolute fast flux measurements it is desirable to employ a reaction of known cross section as a secondary standard, particularly when an accuracy of 5% suffices. The reaction should be easily observable, the element should preferably be suitable for deposition on a foil, and the cross section should vary slowly with neutron energy. Fissionable materials satisfy these requirements especially well. Fission fragments have a short range and high energy and are easily detected in pulse counters. The fissionable nuclide the cross section of which is best known is U^{235} in the energy range from 0.1 to 3 Mev [*12d*]. Corrections have to be applied for absorption of fragments in the foil and for neutrons scattered by surrounding objects. Care must be taken to eliminate the effect of fissions induced by thermal and epi-thermal neutrons. This effect may be avoided by using a threshold fission detector rather than U^{235}. Neutron-induced reactions other than fission may be employed as secondary standards, but at present disintegration cross sections for fast neutrons are not sufficiently well known for any suitable nuclide to achieve good accuracy.

18. Measurement of dosage. For studies of the biological effect of radiation the quantity of interest is the radiation dose. The radiation dose is defined as the amount of energy released per unit mass in the irradiated object [*15*]. Since the biological effect of neutrons, per equal dose, is greater than that of photons, it is necessary that a neutron dosimeter not be sensitive to γ-rays.

Inside an extended object the fast neutron dose varies because of scattering and absorption of neutrons as the neutrons penetrate into the object. Near the surface of the object energy will be deposited by the incident neutrons and by neutrons scattered by other parts of the object. The quantity which one usually attempts to measure is the dose caused by first collisions only. The actual maximum dose will be higher than this because of the effect of scattered neutrons.

31*

Table 6. *Methods for measuring fast neutron flux absolutely.*

Method	Energy range or source for which method is most useful	Requires monoergic neutrons	Detects also γ-rays	Requires knowledge of angular distribution of source	Requires knowledge of absolute neutron cross section	Requires standard source	Accuracy (%)	Reference
Associated particles	D+d; D+T; p+T; γ+D	Yes	No	No, if angles are properly chosen	No	No	2	[13]
Associated product	Li+p	No	No	Yes	No	In principle, no		Ref. 2, p. 478
Slowing-down	Most accurate for neutron energies below 3 Mev	No	No	Yes	Yes, unless comparison with standard source or same material is used for moderator and detector	Standard source greatly simplifies procedure	2−3	[13]
Long counter		No	No	No	No	Yes	10	Ref. 2, p. 481
Recoil particle method. Hydrogen-filled proportional counter	0.05 to 2 Mev	Preferably	Yes	No	Yes	No	2−3	Ref. 1, p. 459 [1]
Recoil particle method. Homogeneous ionization chamber		Yes	Yes, but compensation possible	No	Yes	No	5	[13]
Recoil particle method. Hydrogenous foil in counter or telescope	Above 0.5 Mev	Preferably	No	No	Yes	No	2	[12d]
Recoil particle method. Organic scintillator	Above 1 Mev	Preferably	Yes	No	Yes	No	7	Ref. 5, p. 462
Recoil particle method. Photographic emulsion as radiator and detector	Above 0.5 Mev	No	No	No	Yes	No	35	[16]

Animal tissue in which one wishes to measure radiation dose consists principally of H, O, C, and N. Two different ways of measuring first collision neutron doses have been proposed. Either a mixture or compound which simulates the composition of animal tissue is prepared and the ionization in this material is measured, or a counter is constructed whose counting rate as a function of neutron energy is proportional to the calculated first collision tissue dosage.

For the first approach the homogeneous ionization chamber mentioned in Sect. 6 is an appropriate detector, since the number of ion pairs produced in the gas is very closely proportional to the energy lost by the ionizing particles. Instead of a hydrocarbon a plastic lining of a composition of typical animal tissue is used and the chamber is filled with a gas mixture of the same composition. Such a chamber will, however, be sensitive to γ-rays. It is possible to balance out the effect of γ-rays by using a second chamber which has the same response to photons but a different response to neutrons. Although such twin ionization chambers have been used, they are too cumbersome for survey work, and it is difficult to be sure that both chambers really have the same response to γ-rays and are in the same γ-ray flux.

For these reasons the second approach has been more successful. From the scattering cross sections of the constituents of tissue it is possible to calculate the first collision dose as a function of neutron energy per incident neutron flux. The angular distribution of the neutrons scattered by C, N, and O is at present not well enough known at all energies to make the results of the calculations very reliable, but they are probably sufficiently accurate for most practical applications. An attempt is then made to construct a counter whose neutron energy response matches the calculated variation of dosage with energy. A counter of this type is a methane filled proportional counter which contains two paraffin radiators one of which is covered with an aluminum foil[1]. By adjusting the gas pressure, and the thicknesses and areas of the paraffin and aluminum foils, it is possible to obtain a good fit of the sensitivity to the calculated dosage curve. The counter is insensitive to γ-rays, but, because of the use of foils, its response depends on the direction of incidence of the neutrons.

By modification of the long counter (see Sect. 17) it is possible to change its flat response so that the response is proportional to dosage[2]. A BF_3 counter is surrounded by a cylindrical paraffin moderator, 11 cm thick, and neutrons are incident at right angles to the axis of the cylinder.

When it is desired to measure dosage caused by neutron bursts such as might occur accidentally in operations of chain reacting devices, counters are not satisfactory because of counting losses and saturation effects. It is preferable instead to activate various materials whose radioactivity may be counted after the irradiation. A combination of suitable detecting materials may again be chosen to give a response corresponding to the calculated first collision tissue dose. For neutron spectra extending from thermal to several Mev, the energy range may be divided into five intervals[3]. Activation of Au detects thermal neutrons. Pu^{239} shielded by boron, Np^{237}, U^{238}, and S serve as threshold detectors for fast neutrons. For the fissionable materials the γ-rays from fission products are counted and for $S(n, p)$ P the P^{32} activity is observed. The number of neutrons of energies between successive thresholds may be determined and may then be added with appropriate coefficients to yield the neutron dosage.

[1] HURST, RITCHIE and WILSON: Rev. Sci. Instrum. **22**, 981 (1951).

[2] J. DE PANGHER and W. C. ROESCH: Phys. Rev. **100**, 1793 (1955), and private communication.

[3] HURST, HARTER, HENSLEY, MILLS, SLATER and REINHARDT: Rev. Sci. Instrum. **27**, 153 (1956).

References.

[1] ALLEN, W. D., and A. T. G. FERGUSON: Flux Measurements of Fast Neutrons with a Recoil Counter in the Energy Range of 50—2000 kev. British Atomic Energy Research Establishment Report NP/R 1720, 1955.

[2] BARSCHALL, H. H.: Methods for Measuring Fast Neutron Cross Sections. Rev. Mod. Phys. **24**, 120 (1952).

[3] BARSCHALL, H. H., L. ROSEN, R. F. TASCHEK and J. H. WILLIAMS: Measurement of Fast Neutron Flux. Rev. Mod. Phys. **24**, 1 (1952).

[4] BIRKS, J. B.: Scintillation Counters. London: Pergamon Press 1953.

[5] CORSON, D. R., and R. R. WILSON: Particle and Quantum Counters. Rev. Sci. Instrum. **19**, 207 (1948).

[6] CURRAN, S. C., and J. D. CRAGGS: Counting Tubes. London: Butterworth Scientific Publications 1949.

[7] FOWLER, J. L., and J. E. BROLLEY: Monoenergetic Neutron Techniques in 10—30 Mev Range. Rev. Mod. Phys. **28**, 103 (1956).

[8] FRANZEN, W.: Theory and Use of Pulse Ionization Chambers. Princeton: 1951. U.S. Atomic Energy Report NYO-3003.

[9] GRAVES, A. C., and D. K. FROMAN: Miscellaneous Physical and Chemical Techniques of the Los Alamos Project. New York: McGraw-Hill Book Co. 1952. — (a) Preparation of foils. — (b) Neutron Sources.

[10] HUGHES, D. J.: Pile Neutron Research. Cambridge, Mass.: Addison-Wesley Publishing Co 1953.

[11] HUGHES, D. J., and J. A. HARVEY: Neutron Cross Sections. Washington: U.S. Government Printing Office 1955.

[12] International Conference on Peaceful Uses of Atomic Energy. Geneva 1955. — (a) CRANBERG, L.: Time-of-flight techniques applied to fast neutron measurements. Paper 577. — (b) BOLLINGER, L. M.: Recent advances in neutron detection. Paper 580. — (c) ROSEN, L.: Techniques for measurement of neutron cross sections and energy spectra for sources which are continuous in energy and time. Paper 582. — (d) DIVEN, B. C.: Some techniques for measurement of fast neutron flux. Paper 594.

[13] LARSSON, K. E.: A Systematic Study of Methods to Measure a Neutron Flux Absolutely. Ark. Fys. **7**, 323 (1954); **9**, 293 (1955).

[14] LITTLER, D. J.: Report on Conference on Neutron Source Preparation and Calibration. British Atomic Energy Research Establishment Report NP/R 1577, 1955.

[15] MARINELLI, L. D.: Radiation Dosimetry and Protection. Ann. Rev. Nucl. Sci. **3**, 249 (1953).

[16] ROSEN, L.: Nuclear Emulsion Techniques for the Measurement of Neutron Energy Spectra. Nucleonics **11**, 32 (July 1953) and **11**, 38 (August 1953).

[17] ROSSI, B. B., and H. H. STAUB: Ionization Chambers and Counters. New York: McGraw-Hill Book Co. 1949.

[18] SEGRÈ, E.: Experimental Nuclear Physics, Vol. I. New York: John Wiley & Sons 1953. — (a) Detection Methods by H. STAUB. — (b) Passage of Radiations through Matter by H. BETHE and J. ASHKIN.

[19] SEGRÈ, E.: Experimental Nuclear Physics, Vol. II. New York: John Wiley & Sons 1953. — (a) A Survey of Nuclear Reactions by P. MORRISON. — (b) The Neutron by B. T. FELD.

[20] WILKINSON, D. H.: Ionization Chambers and Counters. Cambridge: The University Press 1950.

High Energy Neutron Detectors.

By

R. T. SIEGEL.

With 16 Figures.

Introduction.

In the past few years, accelerators have produced neutrons of energy exceeding one billion (10^9) electron volts, and the next decade will see artificially produced particles of over twenty Bev. At present[1], however, few neutron experiments have been carried out at energies above several hundred Mev, so the detection of neutrons with energy from 20 to 700 Mev will be the main subject of this report, with few exceptions. For energies below 20 Mev the preceding article on *Detection of Neutrons* [1] may be consulted. Many of the references listed there also make some mention of high energy problems. In addition, the high energy situation up to 1953 is summarized in the paper by WATTENBERG [2].

Part I of this report is concerned with the general considerations which affect neutron detectors for high-energy experiments, i.e., the energy and intensities of available neutron beams and the types of interaction which are important at various energies. Part II discusses the devices which have been developed to make use of the available neutron interactions for purposes of detection. Evolution of these detectors has been closely related to the types of neutron measurements for which they are used, so part II also describes some specific applications of the detectors. In part III the properties of detectors are briefly summarized according to their usefulness in various types of experiments.

I. General considerations.

1. Neutron spectra and intensities. Monoenergetic neutrons of energy up to about 30 Mev have been obtained from charged particle reactions, particularly $D(t, n)$ He^4. Extensive information on the experimental techniques in this energy region, including neutron energies, angular distributions, etc., is given in the review by FOWLER and BROLLEY [3]. At higher energies, neutrons are produced internally in accelerators by bombarding thick (~ 1 in) targets of light nuclei with protons or deuterons. The resulting neutron spectra are far from monoenergetic, being broadened by the betatron oscillations and multiple target traversals of the incident beam, as well as by the momentum distributions within the struck nuclei.

The nature of multiple traversals in a synchrocyclotron beam has been studied by KNOX[2], who measured the Na^{24} activity produced in thin aluminum foils placed over internal targets of various materials and thicknesses in the Berkeley 184″ machine. He concluded that with either 340 Mev protons or 190 Mev deuterons incident, a $\frac{1}{2}$ inch Be target would probably offer the best combination

---

[1] Literature survey for this paper concluded July 1956.
[2] W. J. KNOX: Phys. Rev. **81**, 693 (1951).

of small energy spread and high intensity of the emergent neutrons. Although the differences in magnetic field shape, etc., among accelerators were expected to make his results of limited generality, a $\frac{1}{2}$ or 1-inch berylium target has in fact been most often used to obtain high energy neutron beams. However, the neutron spectra have proven to be somewhat broader than could be accounted for simply by the energy distribution of the internal proton beam.

The increased energy width is attributable to the nature of the interaction in which the neutron is produced. In the case of deuteron stripping, a reasonably sharp spectrum is obtained when the internal momentum of a nucleon in the deuteron is small compared to the momentum of the incident deuteron itself. Thus, the full width at half maximum of the energy distribution of the neutrons proceeding forward after a stripping interaction should be

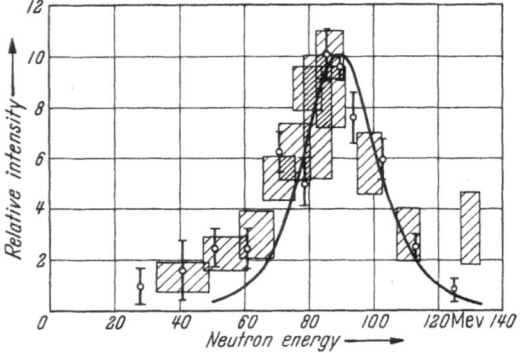

Fig. 1. Neutron spectrum produced by stripping 190 Mev deuterons on a 1.27 cm Be target. Experimental points from J. Hadley, E. Kelly, C. Leith, E. Segrè, C. Wiegand and H. York: Phys. Rev. **75**, 351 (1949). Solid curve from theory of R. Serber: Phys. Rev. **72**, 1008 (1947).

$$\frac{\Delta E}{E} \sim 1.5 \left(\frac{\varepsilon_d}{E_d}\right)^{\frac{1}{2}} \sim \Delta\vartheta$$

where $\varepsilon_d =$ binding energy of deuteron, $E_d =$ kinetic energy of deuteron, and $\Delta\vartheta$ is the angular width of the neutrons about the incident deuteron direction[1]. With 190 Mev deuterons stripped on light nuclei, reasonable agreement with this expression is found (Fig. 1)[2] with an additional low energy tail produced by evaporation neutrons from excited states of the target nuclei. The evaporation effect is more serious if the deuterons impinge upon heavy nuclei, though in this case the stripping portion of the spectrum is enhanced by electric disintegration of the deuteron, which amounts to about one-half the stripping cross-section in uranium[3].

The sharp peaks in the angular and energy distributions from deuteron stripping makes this process a very desirable source of high energy neutrons. According to Knox[4] the forward yield of neutrons (detected by means of a Bi fission chamber) per unit incident particle flux on a Be target is about 40 times greater from 190 Mev deuterons than from 340 Mev protons. However, few high energy machines are built to accelerate deuterons, and proton bombardment is perforce the most commonly used source of high energy neutrons.

Representative neutron spectra from protons bombarding internal accelerator targets are given in Fig. 2. In all cases there is a strong high energy peak, which is displaced donward from the maximum incident proton energy by at least 20 Mev. A large portion of this displacement is undoubtedly due to the actual distribution in energy of the incident protons. Additional displacement is caused by the exclusion principle, which requires that the incident proton retain enough of its energy so that it may occupy a previously empty state in the nucleus. In $_4Be^9$ this requirement is easy to satisfy if the proton merely exchanges with the available loosely bound neutron, so the neutron peak energy lies near the incident

[1] R. Serber: Phys. Rev. **72**, 1008 (1947).
[2] J. Hadley, E. Kelly, C. Leith, E. Segrè, C. Wiegand and H. York: Phys. Rev. **75**, 351 (1949).
[3] W. J. Knox: Phys. Rev. **81**, 687 (1951).
[4] Cf. footnote 2, p. 487.

beam energy. Heavier nuclei generally show neutron peaks displaced further downward, and broadened beyond the Be spectrum by the increased probability of multiple collisions in a single nucleus and the higher thresholds for (p, n) reactions in these nuclei. Because of the broadness of the neutron spectra available at high energies, the energy response of detectors will be of great importance. On the other hand, the energy dependence of neutron interaction cross-sections is less marked at high than at low energies, so that rather wide resolution is acceptable and characteristic of reported experiments.

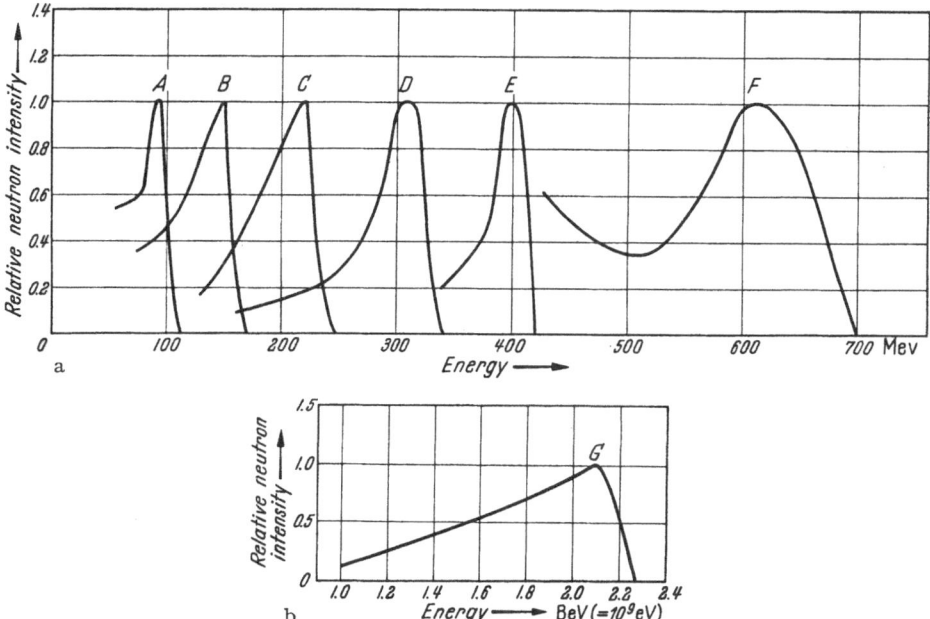

Fig. 2 a and b. Neutron spectra produced by protons of various energies incident on internal accelerator targets. For detailed information see Table 1.

Table 1 gives representative intensities available from high energy accelerator sources, along with some additional information on targets, etc. The contaminants present in these beams differ somewhat from those present at low energies, and have not been very carefully investigated. The most serious are the low energy neutrons which spread the spectrum all the way to zero energy. These can be eliminated by (a) choice of detector (see Sect. 3) or (b) use of a hydrogenous filter, which preferentially removes from the beam neutrons of energy below 150 Mev. Above that energy the filter removes all energies with equal probability because of the almost constant values of neutron total cross-sections (Fig. 3).

High-energy γ-rays from π^0-meson decay near the target also constitute a sizeable contaminant above the threshold for meson production, which is ~145 Mev in Be. For example, the γ-ray content of neutron beams produced by 340 Mev protons on carbon is about 10^{-3} of the neutron flux[1]. However, lead filters can be used to remove the γ-rays without unduly decreasing the neutron flux.

---

[1] Estimate based on data of R. BJORKLUND, W. E. CRANDALL, B. J. MOYER and H. F. YORK, Phys. Rev. **77**, 213 (1950) on γ-ray production, and the neutron production cross-sections of L. SCHECTER, W. E. CRANDALL, G. P. MILLBURN and J. ISE, Phys. Rev. **97**, 184 (1955).

Table 1. *Properties of some high energy neutron beams.*

Spectrum curve in Fig. 2	Cyclotron location	Internal target	Nominal proton energy Mev	Mean neutron energy Mev	Neutron flux (cm⁻² sec⁻¹)	Reference
A	Harvard	$\frac{1}{8}''$ Be	112	90	5×10^4 above 50 Mev	J. A. Hofmann and K. Strauch: Phys. Rev. **90**, 449 (1953).
B	Harwell	0.35'' Be	171	153	2×10^4 above 80 Mev[1]	T. C. Randle, J. M. Cassels, T. G. Pickavance and A.E. Taylor: Phil. Mag. **44**, 425 (1953).
C	Rochester	$\frac{1}{2}''$ Be	245	220	Not quoted	B. K. Nelson, G. Guernsey and G. Mott: Phys. Rev. **88**, 1 (1952).
D	Berkeley	$1\frac{3}{4}''$ LiD	340	307	10^4-10^5 above 200 Mev[1]	J. DePangher: Phys. Rev. **99**, 1447 (1955).
E	Carnegie Tech	1'' Be	425	400	2×10^5 above 365 Mev	A. J. Hartzler and R. T. Siegel: Phys. Rev. **95**, 185 (1954).
F[2]	Academy of Sciences USSR	Be	680	590	2×10^4 above 450 Mev	V. P. Dželepov, B. M. Golovin, U. M. Kazarinov, U. N. Simonov: Paper (II/48) presented at the CERN Symposium 1956.
G	Cosmotron	C	2300	1790	10^3 per cm² per pulse[1]	W. B. Fowler, R. P. Shutt, A. M. Thorndike and W. L. Whittemore: Phys. Rev. **95**, 1026 (1954).

In all the neutron beams described in Table 1 and Fig. 2, the neutrons are observed at 0° (or at most 3°) to the direction of the protons impinging on the internal target. Because of this symmetry about the beam axis, it is expected that none of these beams is polarized, that is, displays a preferred orientation for the neutron spins in any direction perpendicular to the beam. But if neutrons are observed at a larger angle to the incident proton direction, it is possible that a measurable polarization exists, and that information about nuclear interactions can be

Table 2. *Neutron beam polarizations.*

Nominal internal proton energy (Mev)	Angle of emission of neutrons	Effective neutron energy (Mev)	Internal target material	Polarization
170[3]	26°	98	Be	0,085 ± 0.006
230[4]	30°	100−200	Be	0.13 ± 0,03
	30°	100−200	C	0.19 ± 0.02
425[5]	20°	350	C	0.163 ± 0.007
		350	D₂O	0.181 ± 0.016

[1] The flux was not mentioned in the quoted reference, but has been estimated from various data.

[2] This spectrum has a second peak at ∼270 Mev which is thought to be composed of neutrons emitted from collisions of the primary protons in which pions are also produced.

[3] R. G. P. Voss and R. Wilson: Phil. Mag., Ser. VIII **1**, 175 (1956). — P. Hillman, G. H. Stafford and C. Whitehead: Nuovo Cim. **4**, 67 (1956).

[4] A. Roberts, J. Tinlot and E. M. Hafner: Phys. Rev. **95**, 1099 (1954).

[5] R. T. Siegel, A. J. Hartzler and W. A. Love: Phys. Rev. **101**, 838 (1956). *Note added in proof.* G. H. Stafford, S. Tornabene and C. Whitehead: Phys. Rev. **106**, 831 (1957) have found that with 170 Mev protons on Be, C, Cu and W targets it is possible to obtain highly polarized neutron beams at an emission angle of 55°. The measured polarizations of the neutrons (∼75 Mev effective energy) from these four target elements were 0.35 ± 0.05, 0.35 ± 0.06, 0.28 ± 0.09, 0.27 ± 0.06 respectively. (Cf. Ref. 3 above.)

deduced from scattering or production experiments with such an oriented beam [4]. Unfortunately, a polarization of no more than 20% has yet been observed for high energy neutron beams produced in this way, which is to be compared with the almost complete polarization which can be obtained for protons elastically scattered at small angles from light nuclei. Though it is also possible to polarize neutrons by elastic scattering, the resultant intensity would probably be too low for use as a primary beam, so all useful polarized neutron beams have been obtained at a finite (usually ~20°) angle to the protons striking an internal cyclotron target. The spectra of such neutron beams do not exhibit peaks so sharply defined as those in Fig. 2, so for the cases listed in Table 2, effective neutron energy was determined largely by the experimental apparatus.

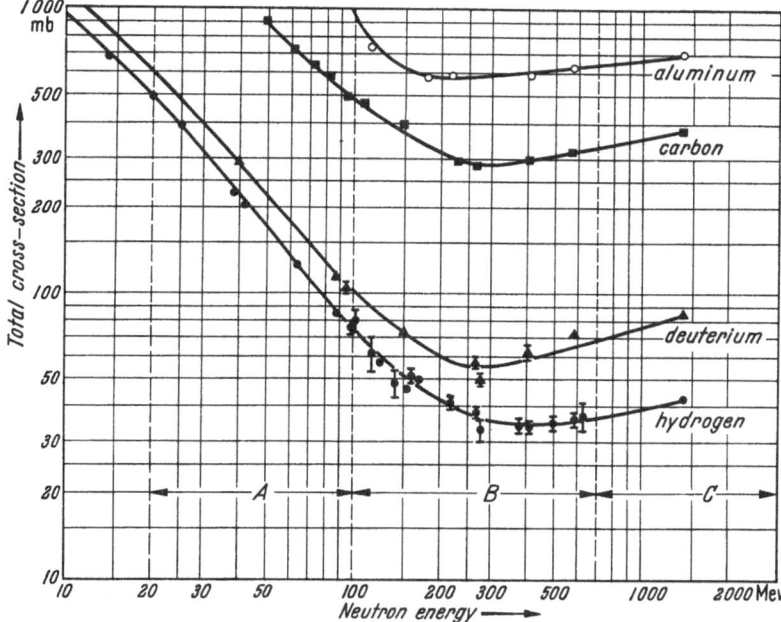

Fig. 3. Total neutron cross-sections vs. energy for some light nuclei. Hydrogen data from Ref. [6], [7] and Ref. 4, p. 499. Other data from various references quoted throughout paper.

2. Energy dependence of detection techniques. The total neutron cross-sections for several nuclei are shown in Fig. 3, where the abscissae are divided into the three indicated bands in order to facilitate a brief analysis of the energy dependence of interactions used for detection.

α) *Region A (20 to 100 Mev)* is one of transition from the types of neutron interaction with nuclei observed at low energies. Here the compound nucleus theory of nuclear reactions is losing its applicability because the mean free path of the incident neutron is now comparable to the size of the struck nucleus. Reactions in nuclei begin to look like cascade processes, with a few successive nucleon-nucleon collisions dissipating all or part of the incoming neutron's energy. If such collisions occur near the nuclear surface, they may result in the emission of a high energy proton or other charged particle which is useful for detection of the primary neutron. On the other hand, loss of an appreciable fraction of its energy by the incident particle often supplies so much energy to the nucleus as a whole that de-excitation can occur through an array of reactions

Thus the simplicity of the excitation reactions usable for detection at lower energy is lost, making the use of activation detectors difficult. In spite of the complicated picture in nuclei, there are a few reactions such as $C^{12}(n, 2n) C^{11}$ and bismuth fission which have thresholds in region A, and have been used as neutron detectors. Others are expected to be studied and used in the future.

In region A the elastic scattering in hydrogen is the best-known and most convenient interaction for detecting neutrons, and the inelastic hydrogen cross-section is still so small that little background is present to mix with the recoil protons. The two-body nature of elastic scattering allows the neutron energy to be computed from a knowledge of the recoil proton angle and energy. The proton energy may be measured by determining range, ionization loss, or by time-of-flight velocity determination, which has been used more often for the incident neutrons themselves. For accurate dE/dx measurements in this energy band, ionization chambers and proportional counters are still useful, though the scintillation counter is in ever-increasing use. Range measurements are perhaps the easiest to perform, and in this region do not suffer from excessive loss of protons by scattering and nuclear absorption in the stopping material. It is also possible to completely stop the recoil protons in a scintillant of manageable size, using pulse-height analysis to measure the energy. Various combinations of the above techniques are often used.

Star production has been studied in Region A (as well as at higher energy) by means of photographic emulsions. It is difficult to measure flux at this energy with emulsions because the recoil protons often leave the thin sensitive layers, an effect which practically eliminates the emulsion as a detector of any neutron reactions besides star production or fission above 50 Mev.

Summarizing region A of Fig. 2, we may say that the recoil proton is dominant for neutron detection, and that only a few other interactions are specifically useful for neutrons of this energy. From Fig. 3 it is apparent that the total cross-sections are dropping steadily towards a low plateau (30 to 40 mb in hydrogen). These interaction magnitudes, combined with the limited intensities available (10^5 cm^{-2} sec^{-1} is representative), place severe statistical limitations on experimental accuracy. It may also be noted that the tendency of experimenters to press towards the highest energies available has resulted in an undeserved lack of attention to this "transition band" of energies.

β) *Region B, 100 to 700 Mev*, has been of great interest because of π-meson production and interactions. The recoil proton method of neutron detection is still preeminent, with a CH_2 radiator and scintillation telescope being most often used to determine range or dE/dx. The loss of protons from scattering and absorption in a recoil telescope absorber is now becoming appreciable ($> 30\%$). Ionization loss is losing its sensitivity as a measure of energy, because the proton velocity is beginning to approach the speed of light. For the same reason, the time-of-flight technique, used often at low energies, has probably reached its limit at a few hundred Mev. On the other hand, recoil protons with $\beta > 0.5$ can be detected with Cerenkov counters, a method not available below 100 Mev. In principle the angle of emission of Cerenkov radiation is a sensitive measure of proton velocity, but in practice this measurement has found limited application because of the small amount of light produced. Of wider usefulness is the threshold type of Cerenkov counter, with an anti-coincidence between two such devices with different thresholds sometimes used to fix a narrow energy band.

Measurement of recoil proton momentum by means of a magnetic field in conjunction with a proton detector is becoming important in Region B, particularly since the protons escape the radiator with little scattering or energy loss.

Thus spectrometers using Geiger or scintillation counter arrays can serve to analyze neutron spectra accurately. The cloud and bubble chambers are their own radiators and serve as self-monitors in many experiments, since a search for rare neutron interactions produces many pictures which contain recoil protons from elastic n-p scattering. In the future, the use of bubble-counting in Glaser chambers will give dE/dx measurements serving as a check on $H\varrho$ determination, and one expects this apparatus to supplant the cloud chamber because of the larger density of matter in the former.

Because of the considerable range and small ionization loss of most secondaries produced by neutrons of energy greater than 100 Mev, ionization chambers and proportional counters are rarely used. Fission chambers have been used to monitor transmission experiments, and also in the interesting neutron-neutron scattering experiment of DZHELEPOV et al.[1] at 300 Mev. Photographic emulsions are limited to investigations of low energy secondaries emitted from stars, fission, or meson production events.

Surveying Region B, it is apparent that the supply of detection mechanisms is fast running out because of relativistic effects and the increasing complexity of the reactions of neutrons with nuclei. For usefulness in detection an interaction must be characterized by simplicity, large cross-section, and if possible maximum sensitivity in the energy region of interest, combined with low sensitivity at other energies. Even the recoil proton method is meeting difficulties which will be accentuated in Region C.

γ) *In this highest energy band* (C) there is practically no counter information available, except for total cross-section measurements made at Brookhaven. Several studies of n-p interactions in cloud chambers have been made, but since the elastic scattering as a function of energy has not been studied at these energies, neutron spectra and intensities cannot be easily measured. Protons emitted from any light nucleus can probably be used in monitoring devices because of the quasi-elastic nature of the interactions inside such nuclei, but the large thickness of absorbers required for proton range measurements will result in the loss of a major portion of these protons. Pressurized gas Cerenkov counters may become useful for proton energy measurements, but until careful studies are made of the interactions of neutrons with protons, the accuracy of quantitative experiments must remain low.

II. Detecting devices.

3. Recoil proton detector-separate radiator. As the most generally used method of neutron detection, the counting of protons from elastic n-p scattering has involved a wide variety of experimental techniques. Of these, a hydrogenous radiator and scintillation or proportional counter telescope to count recoil protons at a small angle to the incident beam are most frequently used and will be discussed first.

Because of their convenience of operation and short resolving time, scintillation counters are now used whenever the recoil proton energy is high enough to penetrate two or three crystals. Proportional counters have nevertheless been used on occasion, notably by BROLLEY et al., at 27 Mev[2] and in several experiments at Harwell[3] in the 150 Mev region.

[1] W. P. DZHELEPOV, B. M. GOLOVIN and V. I. SATAROV: Dokl. Akad. Nauk, SSSR. **99**, 943 (1954).

[2] J. E. BROLLEY, J. H. COON and J. L. FOWLER: Phys. Rev. **82**, 190 (1951).

[3] T. C. RANDLE, J. M. CASSELS, A. E. TAYLOR and T. C. PICKAVANCE: Phil. Mag. **44**, 425 (1953). — A. E. TAYLOR and E. WOOD: Phil. Mag. **44**, 95 (1953).

α) *Relative measurements with recoil proton telescopes.* We consider the measurement of neutron flux for the purpose of determining background counting rates, transmission in a good or poor geometry, energy spectra, or relative neutron angular distributions. The most important property of a "monitor" counter for such applications will be its ability to detect neutrons in the energy range being studied with a constant efficiency which is sufficiently high to insure adequate statistical accuracy. A recoil proton telescope without absorber will often satisfy this condition, but is unfortunately sensitive to all neutrons which produce secondary protons energetic enough to penetrate the last counter of the telescope. An absorber may then be introduced to raise the lower limit of detectable neutron energy, which will considerably reduce the counting rate, since much of the neutron spectrum may be expected to lie at low energies (Fig. 2). Nevertheless, insensitivity to these energies is desirable in order to reduce background and, if the detector is being used to measure relative intensity vs. time of the accelerator beam, to insure that the energy response of the monitor is similar to that of the detector for the process being studied. In this way one can reduce the effects of short-term fluctuations of neutron spectrum shape which are known to occur with cyclotron produced beams[1]. Such instability is probably caused by changes in the internal beam oscillations which affect the proton spectrum incident upon the internal target. It follows that the use of BF_3 or other slow-neutron monitors during a high-energy experiment is somewhat risky unless the cyclotron is known to be stable in neutron beam characteristics.

In Fig. 4 the laboratory differential cross-section for elastic n-p scattering is given as a function of energy for various recoil proton angles[2]. The increasing proportion of exchange scattering at high energies is indicated by the large cross-sections for forward protons. In order to take advantage of this effect, telescopes for counting recoil protons from a hydrogenous radiator are usually placed at 10 to 30° to the neutron beam direction.

If an absorber is used in the telescope, it is often made of heavy metal in order to keep the size of the counter array to a minimum. Although the multiple scattering and absorption of the protons in a high$-Z$ absorber is greater than in light materials, the compactness of a copper or wolfram absorber will permit the last counter of the telescope to subtend a larger solid angle at the radiator and at the center of the absorber, and thus detect more recoil protons than if a low$-Z$ absorber were used.

Though much less widely used than an absorber, a Cerenkov counter can also serve to fix a lower limit on the energy of protons detected in a recoil telescope. Such an arrangement was used by NEDZEL[3], whose telescope consisted of two

---

[1] W. SELOVE, K. STRAUCH and F. TITUS: Phys. Rev. **92**, 724 (1953).

[2] Kinematic relations useful in relating laboratory quantities to their center-of-mass analogues are summarized below for the case of n-p scattering, with the approximation $m_n = m_p = 1$.

E_n = laboratory kinetic energy of incident neutron (velocity βc)

= $\gamma - 1$, where $\gamma = 1/(1-\beta^2)^{\frac{1}{2}}$

E_p = recoil proton lab kinetic energy

Φ, φ = proton laboratory and center-of-mass scattering angle, respectively

Θ, ϑ = neutron laboratory and center-of-mass scattering angle, respectively

$\tan \Phi = (2/\gamma + 1)^{\frac{1}{2}} \cot (\vartheta/z)$

$E_p = \frac{1}{2} E_n (1 + \cos \varphi) = \frac{1}{2} E_n (1 - \cos \vartheta)$

$J = \dfrac{d(\cos \Phi)}{d(\cos \vartheta)} = \dfrac{[(\gamma + 1) - (\gamma - 1) \cos^2 \Phi]^2}{8(\gamma + 1) \cos \Phi}.$

[3] V. A. NEDZEL: Phys. Rev. **94**, 174 (1954).

scintillation counters and a Lucite Cerenkov counter. The minimum detectable proton energy was determined by the index of refraction of Lucite and corresponded to a velocity of $0.67c$; at $10°$ to the neutron beam this would indicate a neutron threshold of 340 Mev. Unfortunately, the telescope exhibited a finite sensitivity below this threshold, which was perhaps caused by high-velocity secondaries (pions) produced in the Lucite. A 3 inch Cu range absorber was

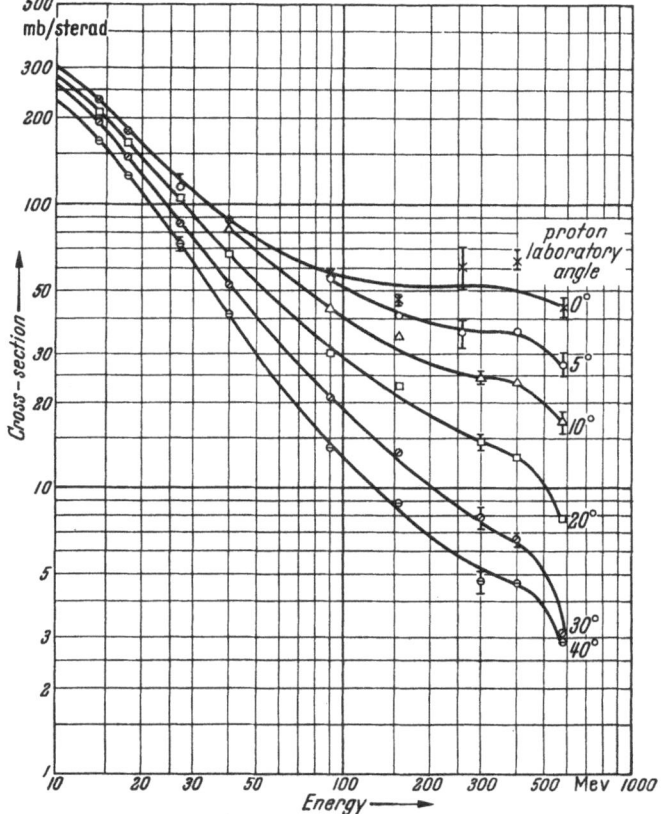

Fig. 4. Laboratory $n\text{-}p$ scattering cross-section vs. energy for several angles. Data from Ref. [6] and Ref. 4, p. 499 was converted to laboratory system of coordinates.

introduced to eliminate this low-energy sensitivity. It would seem that the final arrangement had no advantages over the simple absorber telescope. At higher energies, where range absorbers become excessively thick, it may prove advantageous to use a Cerenkov counter constructed of low density material in order to fix the energy threshold of recoil telescopes.

Thus far, we have been concerned with telescopes placed at small but non-zero angles to the incident neutron beam. In spite of the high exchange ($n\text{-}p$) cross-section, such counter arrays are not often located directly in the primary neutron beam of a cyclotron, because the intensities ($\sim 10^5$ cm^{-2} sec^{-1}) and duty cycles ($<5\%$) of such beams make it quite difficult to cope with the counting rates in this arrangement. The blocking effect is further aggravated by the secondaries produced by the neutrons in the range fixing absorber of the telescope. However, $0°$ telescopes have been used successfully at Harvard[1] in the direct

---

[1] Cf. footnote 1, p. 494.

beam with an apparatus containing five scintillation crystals in order to reduce accidentals.

When the flux is much lower than that mentioned above, it is possible to use a 0° telescope, in which case an anti-coincidence counter or absorber is normally placed ahead of the radiator to prevent detection of charged particles. Zero-degree telescopes have been used in several experiments designed to detect neutrons scattered out of a primary beam, and at Brookhaven to measure transmission cross-sections for Cosmotron neutrons. Three representative detectors of this type are described below:

1. The apparatus used by Chamberlain and Easley[1] to measure neutrons scattered at small angles from protons at 90 and 270 Mev is shown in Fig. 5. Incident protons were absorbed in A_1, any secondary charged particles produced within A_1 being detected by the counter C_1. Protons ejected by neutrons from the CH_2 converter were detected by C_2 and C_3 in coincidence, with charged particle background evaluated by subtraction of $C_1 C_2 C_3$ coincidences. (This is equivalent to placing C_1 in electronic anticoincidence with the $C_2 C_3$ coincidences.) Some systematic error ($\sim 3\%$) was introduced by neutrons emerging from inelastic proton interactions[2] in the copper absorber A_1 and detected by the usual conversion process. The thickness of the converter (1.74 gm/cm² at 90 Mev, 6.53 gm/cm² at 270 Mev) and the angular acceptance of the $C_2 C_3$ combination were such as to yield an efficiency to neutrons of about 0.5%. An unusual feature of this neutron telescope was the shaped absorber A_2 used to fix the minimum energy neutron which produced a proton energetic enough to penetrate counter C_3. The variation in thickness of A_2 compensated for the energy dependence of the recoil proton upon its angle with the incoming neutron. Use of this shaped absorber allowed the acceptance of a rather wide angular spread of protons from the converter without sacrifice of a sharp low-energy cut-off. The uncertainty in "minimum energy to count" was then determined primarily by the thickness of the CH_2 converter and the range straggling of the protons.

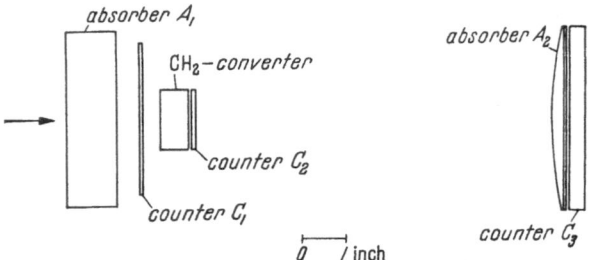

Fig. 5. Neutron detector used for small angle n-p scattering at 90 and 270 Mev by O. Chamberlain and J. M. Easley: Phys. Rev. 94, 208 (1954).

2. A telescope detecting forward recoils was also used in the transmission experiments of Coor et al.[3], at a mean neutron energy of 1.4 Bev, and is shown in Fig. 6. The multiplicity of processes which occur in the radiator at such energies precludes an accurate calculation of the detector efficiency. Although not required for the cross-section measurements, a rough estimate of the absolute neutron flux was therefore obtained from nuclear emulsion data[4], and yielded a value of about 0.1% for the efficiency of the neutron telescope with the 6-inch lead

[1] O. Chamberlain and J. M. Easley: Phys. Rev. 94, 208 (1954). — J. M. Easley: University of California Radiation Laboratory Report UCRL 2693 (1954).

[2] A. J. Kirschbaum: University of California Radiation Laboratory Report UCRL 1967 (1952). — G. Bernardini, E. T. Booth and S. J. Lindenbaum: Phys. Rev. 85, 826 (1952); 88, 1017 (1952).

[3] T. Coor, D. A. Hill, W. F. Hornyak, L. W. Smith and G. Snow: Phys. Rev. 98, 1369 (1955).

[4] Cf. Sect. 8, p. 508.

absorber. With this thickness the energy threshold was about 400 Mev, as determined from the proton energy required to traverse the telescope. (The neutron energy needed to produce a meson energetic enough to pass through the telescope was considerably higher than 400 Mev, because of the energy consumed in creating the meson rest mass.) The authors point out that to further raise the threshold by a significant amount would require an unreasonably thick absorber. For example, a 1 Bev threshold would mean a lead absorber more than three absorption mean free paths in length, which would result in a prohibitive loss of efficiency, particularly in view of the rather small incident flux ($\sim 10^3$/cm^2/pulse). Total and absorption cross-sections of several elements were measured with this detector by moving it along the beam axis so that the half-angle it subtended at the scatterer block varied from about 0.2° to 15° . The "good" geometry (0.2° to 0.5°) determined the total cross-section, and the "poor" geometry (found by observing

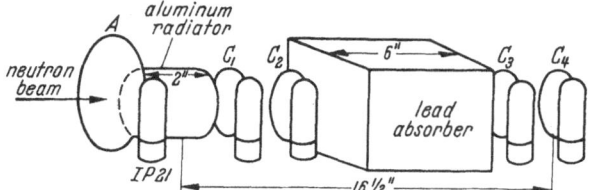

Fig. 6. Neutron detector used for transmission measurements at 1.4 Bev. Coincident counters $C_1 - C_4$ are plastic scintillators 2½ in. in diameter and ⅛ in. thick. Anti-coincidence counter A is plastic scintillator 4 in. in diameter and ⅛ in. thick. From T. Coor, D. A. Hill, W. F. Hornyak, L. W. Smith and G. Snow: Phys. Rev. **98**, 1369 (1955).

a plateau of transmission vs. subtended angle) the absorption. Errors in σ_{tota} were about 2% for the heaviest elements and 4.5% for hydrogen.

3. An interesting variation on the good and poor geometry type of measurement was made by Ball[1] with 300 Mev neutrons. The principle was similar to that used by Coor *et al.* and described above, with a neutron detector consisting of proton absorber, CH_2 converter, and three-crystal telescope (with range absorber) being moved back and forth along the axis of the neutron beam. The geometry was unusual in that the neutrons were collimated not only by a circular hole in the cyclotron shield, but also by a 7'6'' long iron billet placed on the beam axis, so that an annular intensity pattern about 8½'' in mean diameter was obtained at the scatterer position. The detector was placed in the central shadow of the billet in order to take advantage of the large beam area without losing angular resolution. Intensity was thus sufficient for measurement of differential elastic angular distribution, as well as total and inelastic cross-sections to an accuracy of about 5%. It may be noted that effects of possible beam polarization on the elastic scattering measurement were eliminated by the axially symmetric geometry.

In this discussion of relative measurements, attention has been concentrated on recoil telescopes incorporating an absorber to delimit the energy sensitivity of the detector. Many other modifications of the simple radiator plus telescope are possible, and will be described in connection with their use in absolute measurements, since there is no real difference between the detectors used in relative or absolute measurements. However, the latter require quantitative information about the sensitivity vs. energy of the detector, as well as of the neutron beam spectrum.

β) Absolute measurements. Experimental errors may be expected to be considerably larger if absolute measurements of neutron flux are desired, since accurate knowledge of several factors is involved [2].

---

[1] W. P. Ball: University of California Radiation Laboratory Report UCRL-1938 (1952).

The counting rate in a telescope observing recoil protons is in general given by:

$$R = N_n \iint \frac{d\sigma}{d\omega} (\vartheta, E) \, d\omega (\vartheta) \, I(E) \, \varepsilon(E) \, dE . \tag{3.1}$$

The integrals must be taken over the energy (E) spectrum of the beam and the physical dimensions (ϑ) of the apparatus. The factors of the expression above are defined as follows.

1. N_n, the number of hydrogen atoms in the radiator, can be determined to a few tenths of one percent by the accuracy of chemical analysis if a CH_2-C subtraction is used. Usually the CH_2 and C scatterers are machined to thicknesses corresponding to equal stopping powers for protons instead of equal numbers of carbon atoms, in order to reduce the errors arising from uncertainty in scattering of the recoil protons before they emerge from the radiator. If liquid hydrogen is used as radiator, uncertainty in the contraction of the container upon cooling is about one percent, and often two to three percent.

2. $\frac{d\sigma}{d\omega} (\vartheta, E)$ is the differential elastic n-p scattering cross-section. The most accurately measured n-p interaction in this energy range, it still has errors of at least 4% at all angles and energies.

3. $d\omega(\vartheta)$ is an element of solid angle subtended by the telescope, accurately ($<1\%$) calculable from the size of the scintillators and their distance from the radiator.

4. $I(E)$ is the intensity of neutrons vs. energy which one is attempting to measure. Although the relative neutron intensites vs. energy are fairly well known (see Fig. 2), it would obviously be most desirable to detect only a narrow band of energies so that $I(E)$ is effectively constant and may be removed from the integral. Below 100 Mev the rapid variation with energy of $d\sigma/d\omega$ increases the need for narrow-band detection, which is obtained by control of the last factors in Eq. (3.1), namely:

5. $\varepsilon(E) \, \varDelta E$, the product of telescope efficiency and energy width. By means of a range absorber or Cerenkov threshold detector it is possible to make $\varepsilon(E)$ equal to zero for all energies below the peak in the neutron spectrum, and more or less constant above the chosen threshold. This method has been used in the detectors described in Sect. 3α above. In this way the product $I(E) \, \varepsilon(E)$ is made nearly a delta-function, eliminating the integral over energy in Eq. (3.1).

Another approach is to define an energy interval $\varDelta E$ by means of an anti-coincidence counter at the rear of the telescope[1], thus obtaining a narrow detection band $\varDelta E = - (dE/dx) \, \varDelta X$. Here (dE/dx) is the specific ionization loss and $\varDelta X$ is the thickness of the last coincidence counter of the telescope, with a small correction for the depth to which a proton must penetrate in order to produce a detectable scintillation. The method has been extended by other groups at Harvard[2], who used a seven-counter telescope in order to make simultaneous measurements in three energy intervals. This technique is capable of very good energy definition if the last coincidence counter is reasonably thin. Naturally the counting rate will suffer if $\varDelta E$ is made very small, and corrections to $\varepsilon(E)$ for Coulomb scattering and nuclear absorption in the telescope absorber and counters must be made to this as well as to the threshold absorber type of detector. The accuracy with which these corrections can be calculated depends largely on

---

[1] D. Bodansky and N. F. Ramsey: Phys. Rev. **82**, 837 (1951).
[2] J. A. Hofmann and K. Strauch: Phys. Rev. **90**, 449 (1953). — V. Culler and R. W. Waniek: Phys. Rev. **99**, 740 (1955).

experimental results on nuclear absorption and scattering of protons by nuclei[1]. Because of the uncertainties in calculated correction factors, it is customary to measure the proton loss in the telescope experimentally. An external beam of protons of variable energy is usually available, and the transmission $\varepsilon(E)$ of the telescope vs. energy may be measured in such beams[2]. An objection to this calibration method lies in the fact that when the telescope is used as part of a neutron detector, it is not uniformly illuminated by a parallel beam of monoenergetic protons, but by a flux spreading from the radiator and varying in energy over the aperture of the telescope. In order to take account of these effects, it is possible to calibrate a range telescope by using it to count protons from p-p instead of n-p scattering, with the first counters in the telescope serving as a monitor[3]. The transmission through all counters then gives the telescope efficiency under conditions very similar to those of the actual neutron measurement. In one such experiment[3], 3% accuracy was claimed for the absolute efficiency measurements, including corrections for other particles (pions) produced in the p-p collisions used for calibration. At higher energies (> 400 Mev) the number of pions becomes comparable to the number of elastically scattered protons, making it advisable to use the double counter method of detecting both scattered and recoil proton in coincidence during the calibration experiment[4].

The problems arising from loss of protons in the telescope are practically eliminated if (dE/dx) is used to measure the energy of the recoil protons at their full energy, or when they have been only slightly degraded. Such a technique was used by GUERNSEY et al.[5], who analyzed the scintillation pulse heights produced by recoil protons in an anthracene crystal at the rear of their telescope, which did not contain any absorber other than the crystals themselves. By means of a multichannel analyzer, they could study many energy bands simultaneously, a virtue of (dE/dx) measurements which is somewhat clouded by the fluctuations in energy loss[6], which limit pulse height resolution. These fluctuations become even more serious at energies above the 200 Mev available to GUERNSEY et al. because of the increasing flatness of the (dE/dx) vs. E curve, which eventually becomes double valued near minimum ionization and practically horizontal above that. With the technique used by the Rochester group, part of the pulse height spectrum was already double-valued in energy, since some values of energy loss could be incurred either by a low energy proton coming to rest in the crystal, or a high energy proton passing through. This effect can be eliminated by following the (dE/dx) counter with an anticoincidence counter in order to differentiate between the two energy bands.

At the present time absolute neutron flux measurements with proton recoil detectors can be made to an accuracy of 4 to 5%, the principal source of error being in our knowledge of elastic neutron-proton scattering cross-sections. Relative measurements can be made with a precision of one percent or less, depending principally on statistical errors.

4. Recoil proton detector-integral radiator. In this section we include those detectors in which the hydrogenous substance in which recoil protons are produced

[1] Cf. footnote 2, p. 496.

[2] E. KELLY, C. LEITH, E. SEGRÈ and C. WIEGAND: Phys. Rev. **79**, 96 (1950).

[3] A. J. HARTZLER and R. T. SIEGEL: Phys. Rev. **95**, 185 (1954).

[4] V. P. DZHELEPOV, B. M. GOLOVIN, V. M. KAZARINOV and U. N. SIMONOV: Paper No. (II/48) of the CERN Symposium 1956.

[5] G. GUERNSEY, G. MOTT, B. K. NELSON and A. ROBERTS: Rev. Sci. Instrum. **23**, 476 (1952).

[6] K. R. SYMON: Thesis, Harvard University 1949.

is part of the sensitive volume itself, instead of being physically separated from it as in the recoil telescope detector (cf. Sect. 3). Also, we consider only electronic counters, as distinguished from detectors (cloud chambers, etc.), in which the proton identification is visual.

The advantage of the integral radiator-counter lies principally in the high efficiency (10^{-1}—10^{-3}) which can be obtained, as compared with the much lower sensitivity (10^{-3}—10^{-5}) of recoil telescopes. A single counter also is capable of accepting larger incident fluxes than a recoil telescope, which is limited by accidental coincidences to counting rates far below those permissible in organic scintillators of fast decay time ($\leq 10^{-8}$ sec). However, intensities are not often such as to make this last consideration paramount, and the single radiator counter has its own disadvantages. Its rather poor energy resolution and the lack of a flat plateau of counting-rate vs. discriminator level both place severe stability restrictions on the associated electronics. Qualitatively, one may state that the recoil telescope is characterized by good energy resolution and poor efficiency, while the radiator-counter has the opposite properties.

At high energies, the radiator-counter is usually composed of liquid or solid organic scintillant because of its high density compared with gases. The highest energy at which a gas-filled ion chamber has been used is apparently 17.9 Mev, in a "long" propane counter used to monitor an n-p scattering experiment[1]. We may divide the large scintillation detectors into two classes, depending upon whether (a) or not (b) an attempt is made to secure accurate (10 to 20%) energy resolution. Discussion of the characteristics of these two types will be based principally on the analyses by Thresher et al.[2] and Christie et al.[3] of their liquid scintillation counters.

(a) The Oxford group, working at Harwell on problems of small-angle elastic neutron scattering at 105 and 137 Mev, used a volume of terphenyl in benzene 14.0×7.6 cm in area and 44.5 cm long, viewed by four E.M.I. photomultipliers. The last dimension is about 35% of the neutron absorption mean free path at these energies and is large enough so that recoil protons emitted forward by the incident neutrons would in many cases reach the end of their range (18 cm at 150 Mev) before leaving the sensitive volume. By setting the counter bias above protons from quasi-elastic n-p events in carbon, which involve an energy loss of 13 Mev, Thresher et al.[2] were able to calculate the efficiency of their counter with confidence from known n-p cross-sections [6] (see Fig. 4), since the observed pulses were then certain to have arisen from free protons recoiling forward into a cone of about 25° half-angle. Thus the counter served essentially as a directional recoil proton counter with a very long (~ 25 cm) radiator, all of which was used without impairment of the energy resolution. (If such a large radiator were used with a counter telescope, the energy resolution would be seriously affected by the uncertainty in origin of the recoil protons.) The efficiency of this counter was quoted as a few percent.

The excellent agreement between the calculated and observed counting rate vs. pulse height curves is shown in Fig. 7. The calculated curves included the effect of (1) energy dependence of the n-p cross-section, (2) absorption of neutrons in the counter, including inelastic events in carbon, and (3) loss of recoil protons

    [1] A. Galonsky and J. P. Judish: Phys. Rev. **100**, 121 (1955).

    [2] J. J. Thresher, R. G. P. Voss and R. Wilson: Proc. Roy. Soc. Lond. **229**, 492 (1955). — J. J. Thresher, R. G. P. Voss, C. P. van Zyl and R. Wilson: Rev. Sci. Instrum. **26**, 1186 (1955).

    [3] E. R. Christie, B. T. Feld, A. C. Odian, P. C. Stein and A. Wattenberg: Rev. Sci. Instrum. **27**, 127 (1956). — Phys. Rev. **102**, 837 (1956).

through the sides and rear of the counter. The edge effects also altered the angular resolution to an extent which was measured by comparing counting rates in two neutron beams, one larger in area than the detector and the other six times smaller. Application of this counter to elastic neutron scattering experiments also involved the usual corrections for multiple neutron scattering in the hydrogen target, etc. The conservatively estimated errors in the scattering cross-sections observed at various angles totaled 5 to 11%.

Because of the rapid increase in recoil proton range at higher energies, which is accompanied by a reduced probability that protons will reach the end of their range without suffering a nuclear interaction, THRESHER et al. feel that the use of a counter of their design is limited to energies below 150 Mev. Counters longer than theirs (44.5 cm) become cumbersome and difficult to shield from background radiation, so that at higher energies it often becomes advantageous to use the high electron densities of metallic absorbers when defining the proton range, e.g., with a recoil telescope detector.

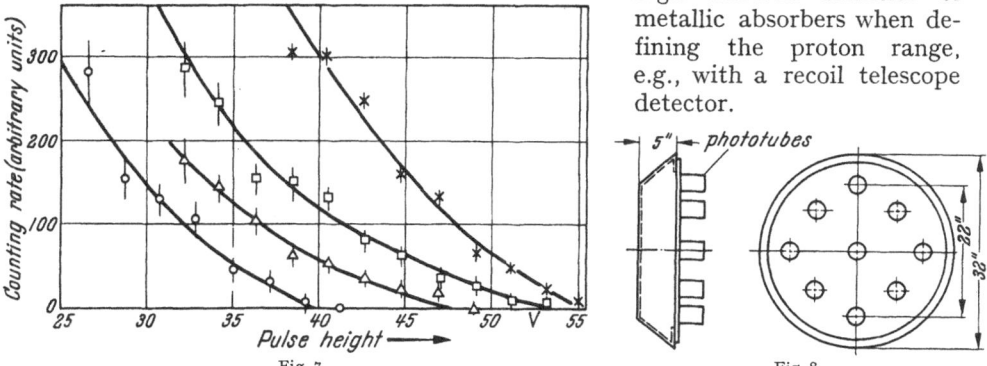

Fig. 7. Pulse-height distributions of neutron counts observed at various angles from carbon and hydrogen scatterers. From J. J. THRESHER, R. G. P. VOSS and R. WILSON: Proc. Roy. Soc. Lond., Ser. A **229**, 492 (1955).

Fig. 8. A large liquid scintillation counter used for neutron detection by E. R. CHRISTIE, B. T. FELD, A. C. ODIAN, P. C. STEIN and A. WATTENBERG: Rev. Sci. Instrum. **27**, 127 (1956).

(b) By counting with a large scintillator not only forward recoil protons but also charged fragments from carbon interactions and large-angle recoil protons, it is possible to obtain even at several hundred Mev, neutron detection efficiencies of 1 to 10% with a sensitive layer about six inches in length. The M.I.T. group[1] has used two such counters, the larger of which is shown in Fig. 8, while the smaller counter was a 4″ diameter by 12″ long volume viewed by a single photomultiplier.

A general expression for the efficiency to neutrons of a thick scintillant is given by these authors:

$$\varepsilon(E_n, B) = \frac{\sigma_H(E_n, B) N_H + \sigma_c(E_n, B) N_c}{\Sigma_t} \left[1 - e^{-\Sigma_t L(E_n, B)}\right] \tag{4.1}$$

where $\varepsilon(E_n, B)$ is the efficiency at bias setting B to neutrons of energy E_n; σ_H and σ_c are the effective cross-sections for producing observable charged particles at E_n and B in hydrogen and carbon respectively; N_H and N_C are the numbers of hydrogen and carbon atoms per cm^3; Σ_t is the reciprocal mean free path; and $L(E_n, B)$ is the effective length of the counter at bias B. This formula is valid as long as multiple scattering effects can be ignored, since it implies that elastic scattering events remove neutrons from the beam.

---

[1] Cf. footnote 3, p. 500.

Some references giving the data required to calculate large-counter efficiency are listed below Fig. 9, which is a set of curves computed for terphenyl-polystyrene plastic scintillators[1], using an arbitrary bias setting B which is assumed to eliminate 10% of the inelastic events in carbon at each neutron energy.

Because of the γ-ray background near the synchrotron, the counter of Fig. 8 was surrounded by a three-inch lead shield, which also served to remove charged particle background. Under the conditions present near a proton accelerator, the charged particles may be removed equally well by a thinner metallic shield (Thresher et al. used 1 to 2 cm of steel at 105 to 140 Mev) or an anticoincidence counter. One effect of the lead wall was to produce secondaries which contributed to the counting rate. The resulting increase in efficiency is difficult to calculate because of the present lack of information on neutron interactions in lead, which emphasizes the need for experimental measurement of efficiencies.

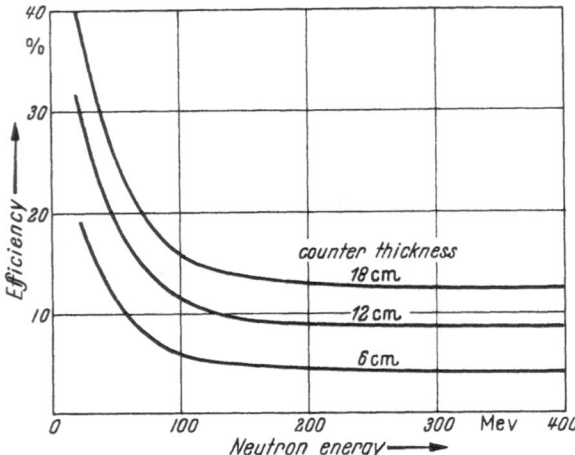

Fig. 9. Efficiency of C_6H_8 neutron counter vs. energy (see discussion in text). Curves calculated from data of Ref. [6]; Ref. 3, p. 494; Ref. 2, p. 496; and D. A. Kellogg: Phys. Rev. 90, 224 (1953).

The counter shown in Fig. 8 was thus calibrated with monoenergetic neutrons emitted at a fixed angle from deuteron photodisintegration, with the γ-ray energy known from the angle and energy of the protons observed in coincidences with the neutrons. A similar arrangement using neutrons scattered from hydrogen may be employed to calibrate neutron counters at a proton accelerator. In this case, the energy of the incident neutrons can be obtained by detecting recoil protons (of energy determined by range, etc.) in coincidence with the neutron counter.

In order to obtain maximum efficiency, it is desirable to set the discriminator bias on a neutron scintillator to the lowest value consistent with noise and extraneous counting rate. Since the counting rate vs. bias curves rise sharply at low settings (Fig. 7), precautions to insure stability of the output pulse height vs. time to within 1% or less are usually necessary. Various automatic gain control devices for this purpose are discussed by Thresher et al.[2].

5. Time-of-flight detectors. In order to detect neutrons within a narrow energy band, it is possible to select a given velocity determined by the neutron flight time over a fixed distance. This method, extensively used for slow neutrons, has been applied by Linlor and Ragent[3] to total cross-section measurements up to an energy of 160 Mev. Their technique consisted of displaying on an oscilloscope the pulses produced by neutrons passing through a scintillator placed 43.7 meters from the internal Be target of the Berkeley cyclotron. The sweep was triggered by the signal generated when a single pulse of deuterons was de-

---

[1] W. A. Love: Private communication.

[2] Cf. footnote 2, p. 500.

[3] W. I. Linlor and B. Ragent: Phys. Rev. **92**, 835 (1953), and University of California Radiation Laboratory Reports UCRL-1952, UCRL-2303, and UCRL-2337.

flected onto the target. In this way the neutron flight time could be measured to $\pm 0.2 \times 10^{-8}$ sec, including the uncertainty in neutron production time.

A Harwell group[1] has also developed instrumentation for applying time-of-flight methods in the 100 Mev region. Their time signals are analyzed by a time sorter circuit, with the uncertainty (20 mμsec) in flight time over a 25 meter path leading to an energy resolution (full width at half-maximum) of 10 Mev for neutrons of 75, and 25 Mev for neutrons of 150 Mev.

These set-ups exhibit two characteristics which one expects of all high energy time-of-flight instrumentation, namely, (a) the first time signal is generated by pulsing the internal accelerator beam, and (b) a fast, high efficiency detector, usually an organic scintillation counter, is placed directly in the neutron beam to give the second time signal. The limitations of this method are summarized by the following relation:

$$\Delta E = - \frac{E^3}{E_0^2} \beta^3 c \frac{\Delta t}{s}.$$

Here $E = T + E_0$ is the total energy of the neutron (T the kinetic energy), $E_0 = m_0 c^2$ its rest energy, βc its velocity, s the flight distance, and $\Delta E = \Delta T$ the uncertainty in energy resulting from the uncertainty Δt in the flight time. It is apparent that ΔE rises rapidly with increasing energy E, in fact as its third power as $\beta \rightarrow 1$. The energy band width is shown as a function of kinetic energy in Fig. 10 for three values of the parameter $\Delta t/s$. The limit of present techniques

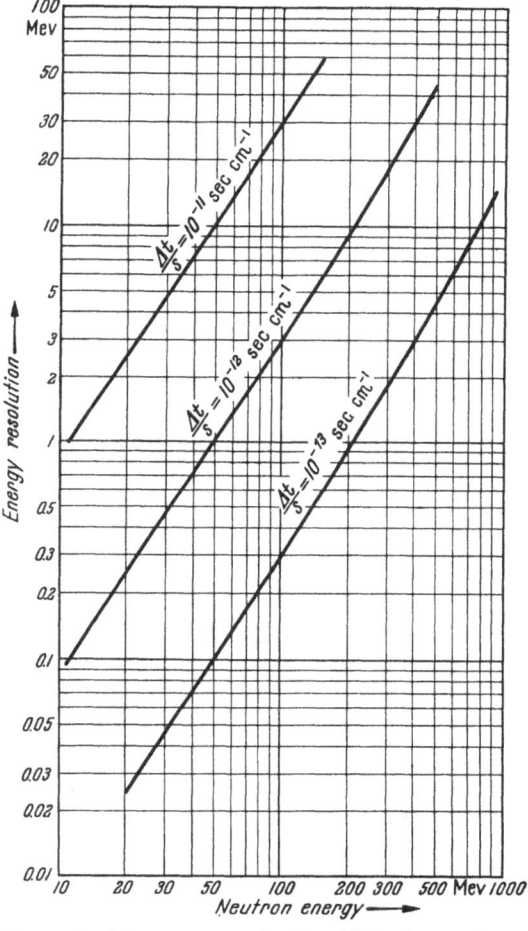

Fig. 10. Resolution vs. energy for time-of-flight detectors (see text for resolution function). Δt = uncertainty in flight time measurement, s = length of flight path.

is approximately at $\Delta t/s = 10^{-13}$ sec cm^{-1}, which corresponds to $\Delta t = 10^{-9}$ sec and $s = 100$ meter. Although it may be possible to reduce Δt by using shorter beam pulses and faster circuitry, the crucial problem is one of neutron intensity, which limits the practical size of flight paths.

The usefulness of the time-of-flight method is greatest in transmission measurements, and its attribute of providing data simultaneously at many energies makes it attractive where β is a reasonable sensitive measure of energy. However, it

[1] J. P. Scanlon, G. H. Stafford and J. J. Thresher, reported by G. H. Stafford in Paper (II/14) of the CERN Symposium 1956.

seems unlikely that this technique will find much application above a few hundred Mev until machines are built with much higher current than is now available.

6. The bismuth fission chamber. A pulse ionization chamber containing fissionable material is often used to count slow or low energy neutrons. Since the discovery that very energetic particles can induce fission even in comparatively light elements[1], it has also been apparent that use of high-threshold fission as a detecting mechanism would result in efficient rejection of much slow-neutron and γ-ray background. A fission chamber should be quite compact and portable (compared to high energy recoil telescopes, for instance) because of the short range of fission fragments.

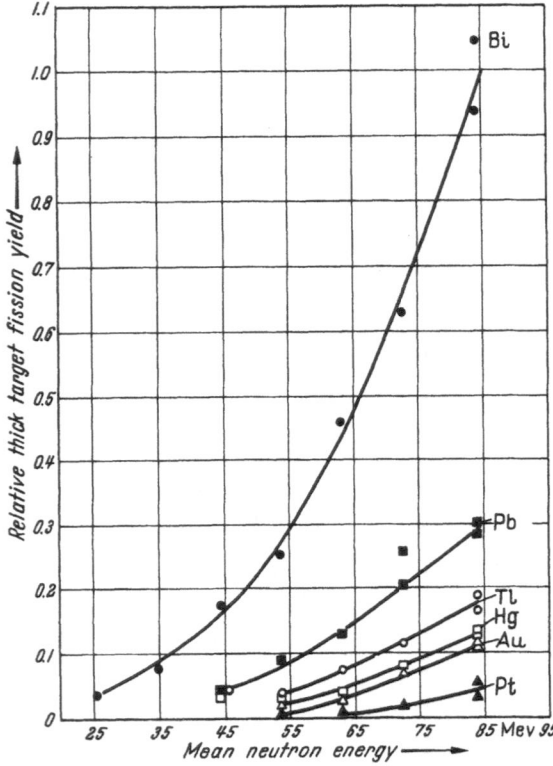

Fig. 11. Relative neutron-induced fission yields for several elements vs. energy. From E. L. Kelly and C. Wiegand: Phys. Rev. 73, 1135 (1948). The mean neutron energy was calculated from the cyclotron target radius. Energy resolution (full width at half-maximum) was calculated as 13.7 Mev at a mean neutron energy of 25 Mev, and 20.9 at 84 Mev.

Yield curves for neutron induced fission were first obtained by Kelly and Wiegand[2] (Fig. 11), and showed that Bi has a yield considerably larger than the other elements investigated. It was therefore chosen as the sensitive element in the first high energy fission chambers[3], and has been the only target used in such detectors.

The various measurements of the bismuth fission excitation function are not notable for consistency. Two experimental methods have been used, one the counting of fission pulses in an ionization chamber containing a Bi coated plate, the other a radiochemical analysis of the fission products. In spite of the tediousness of the second method, and the possibility that some nuclides might be missed in the analysis, it has generally given larger cross-sections by about a factor of two than the ionization chamber technique. A possible explanation for this is given by the work of Wiegand, who noted that a cross-section measured by ionization chamber with 0.3 mg/cm² of Bi on the plates was higher than when a 1 mg/cm² coating was used. Presumably some of the fission fragments cannot escape the thicker layer. Wiegand's fission cross-section value of ∼50 mb for 84 Mev mean energy neutrons incident on the thinner coating is consistent with

[1] M. L. Perlman, R. H. Goeckermann, D. H. Templeton and J. J. Howland: Phys. Rev. **72**, 352 (1947).

[2] E. L. Kelly and C. Wiegand: Phys. Rev. **73**, 1135 (1948).

[3] C. Wiegand: Rev. Sci. Instrum. **19**, 790 (1948).

the radiochemical data of JODRA and SUGARMAN[1]. The data of these authors constitute most of the information used in compiling Fig. 12, which includes measurements to 450 Mev. In this work, protons have been used as the exciting particle.

If Bi fission detectors are used for neutrons, the effective spectra will be changed in shape from those presented in Fig. 2 by the energy dependence of the fission cross-section. This effect is shown in Fig. 13, where DE JUREN[2] has folded the 90 Mev stripping spectrum into the fission excitation curve. The change in shape is not dramatic, but would be if neutrons below 60 Mev formed an appreciable part of the spectrum.

The original design reported by WIEGAND was for a multiplate chamber (570 cm² total area) with 28 parallel Al plates, alternate plates being coated (by evaporation) on both sides with 1 mg/cm² of Bi and held at a high negative potential (field 500 to 700 volt/cm). Such a chamber has an efficiency of ~10⁻⁶ to 90 Mev neutrons. By adjusting the gas (argon with 3% CO_2) pressure in a similar chamber so that the range of fission fragments was equal to the separation between the plates, DE JUREN obtained maximum pulse height from the fission events, while α-particles or spallation fragments would generally have greater range and smaller ionization loss in the gas. Thus he secured a quite respectable plateau of counting rate vs. discriminator

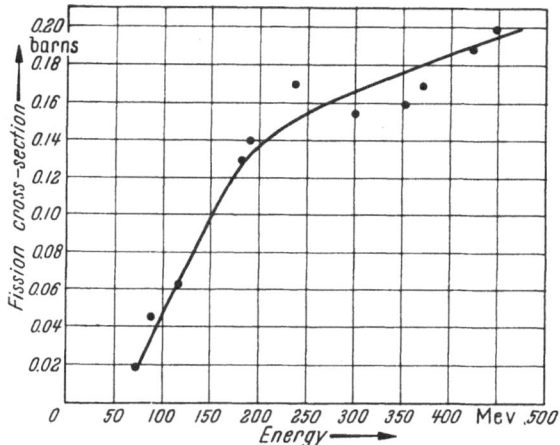

Fig. 12. Bismuth fission excitation function above 75 Mev, for protons incident. From the data of L. S. JODRA and N. SUGARMAN: Phys. Rev. 99, 1470 (1955), corrected by the Al²² (p, 3pn) Na²⁴ results of H. G. HICKS, P. C. STEVENSON and W. E. NERVIK: Phys. Rev. 102, 1390 (1956).

setting in the linear amplifier pulse detector. The fact that α-particles were easily visible above amplifier noise indicates that many more plates could have been used to increase efficiency without introducing excessive capacitance. However, the α-sensitivity is useful as a check of operating characteristics, and a larger chamber of this design would become bulky.

Therefore DE JUREN[3] designed a "giant" chamber with nineteen Al plates of total effective (Bi coated) area 5800 cm². The plates were "paired off" in order to reduce capacitance, i.e., plates at equal distances from the central (uncoated) plate were connected to each other and to a voltage dividing network. Pulses induced on the central plate by electrons collected on any one of the plates were fed into an amplifier, which unfortunately did not produce a flat plateau of counting-rate vs. discriminator bias. This was probably because of the non-uniform deposition of the bismuth (average density 1.4 mg/cm² of Bi as Bi_2O_3) on the plates, a result of the chemical deposition technique used because of the large plate size.

[1] L. S. JODRA and N. SUGARMAN: Phys. Rev. 99, 1470 (1955).
[2] J. DE JUREN: Phys. Rev. 77, 606 (1950).
[3] J. DE JUREN: University of California Radiation Laboratory Report UCRL-1090 (1951).

One of the most interesting uses for a bismuth fission chamber was in the neutron-neutron scattering experiment of Dzhelepov et al.[1]. Here a ring-shaped chamber containing plates 12 and 52 cm in inner and outer diameters respectively was used. The plane of the plates was perpendicular to the neutron beam, which passed through the center of the chamber. By moving the chamber axially along the beam, the angular distribution of neutrons scattered from a long $D_2O - H_2O$ target could be measured.

A promising approach to the problem of obtaining efficient bismuth fission detectors has been made by Furst and Kallmann[2], who succeeded in loading a liquid scintillant with a metal-organic compound containing bismuth, the γ-ray pulse height still remaining 25 to 50% of that from the unloaded scintillant. With their prescription[3], a beam of 100 Mev neutrons can be counted with an efficiency of 10^{-4} by a 10 cm thick scintillant, assuming that the fission events alone are seen. This should provide a detector of simple construction with an efficiency considerably higher than that of even a large multiplate fission chamber.

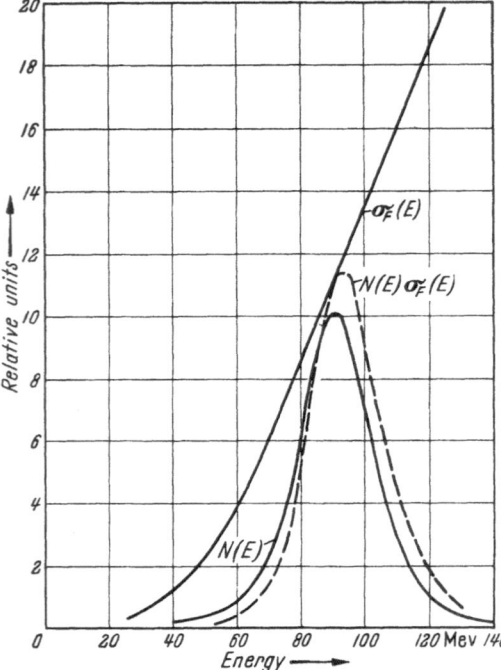

Fig. 13. Effective neutron spectrum from 190 Mev deuteron stripping, when the neutrons are detected with a bismuth fission chamber. $\sigma_f(E)$ — fission cross-section of bismuth for neutrons, $N(E)$ — energy distribution of neutrons, and $N(E) \sigma_f(E)$ — effective spectrum. From J. DeJuren: Phys. Rev. **77**, 606 (1950).

7. Induced radioactivity detectors.

Nuclear reactions induced by neutrons may be useful for detection if the end products are radioactive with a lifetime convenient for counting purposes. In detecting specifically high-energy neutrons, one would prefer to use reactions which have a threshold near the peak of the neutron spectrum, so that low energy neutrons are unable to produce the activity which is to be used as a monitor. Although reactions of the type $(n, \lambda n)$, where $\lambda \gtrsim 2$, would presumably have fairly high thresholds, very little investigation of processes in which more than two neutrons are emitted has been carried out, particularly with neutrons as the exciting particle. However, the results of Kelly and Bell (see Fig. 14) with protons indicate that agreement with statistical theory [4] is good, and similar studies with neutrons might establish the usefulness of several reactions as neutron detectors in the region from 20 to 100 Mev.

The most widely used induced radioactivity detector at high energies has been the $C^{12}(n, 2n) C^{11}$ reaction, with threshold at 20 Mev and 20 min β^+ emission.

[1] V. P. Dzhelepov, B. M. Golovin and V. I. Satarov: Dokl. Akad. Nauk. SSSR. **99**, 943 (1954). — V. P. Dzhelepov, J. M. Kazarinov, B. M. Golovin, V. B. Flizgin and V. I. Satarow: Nuovo Cimento **3**, Suppl. **1**, 61 (1956).

[2] M. Furst and H. Kallmann: Phys. Rev. **97**, 583 (1955).

[3] 3 mg/liter of α-napththylphenyloxazole (α-NPO) in a solution containing by weight 70 xylene, 24% naphthalene, and 6% triphenylbismuthene.

The excitation function is shown in Fig. 15, along with those for the other reactions $Cu^{63}(n, 2n) Cu^{62}$ (10 min, β^+, γ), $Mo^{92}(n, 2n) Mo^{91}$ (15.5 min, β^+), and

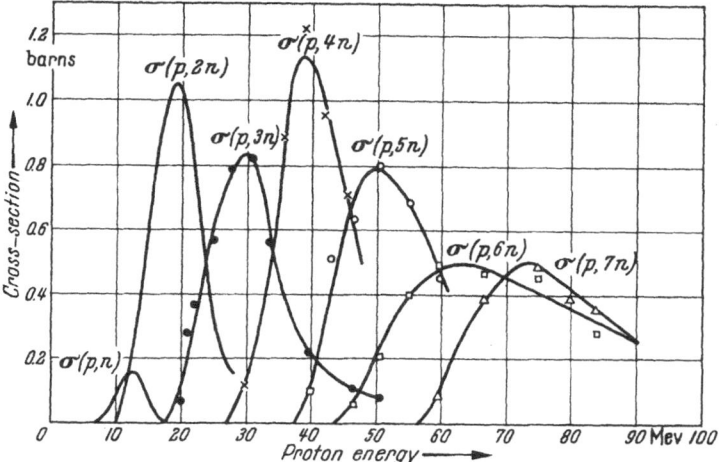

Fig. 14. Excitation functions for $(p, \lambda n)$ reactions in Bi^{209}, where $\lambda = 1 - 7$. From E. L. KELLY: University of California Radiation Laboratory Report UCRL-1044 (1950), and R. E. BELL and H. M. SKARSGARD: Canad. J. Phys. **34**, 745 (1956).

$I^{127}(n, 2n) I^{126}$ (13 days, β^-). Since the accuracy of these measurements is about ten percent in all cases, absolute flux measurements are limited to this precision if these reactions are used as monitors.

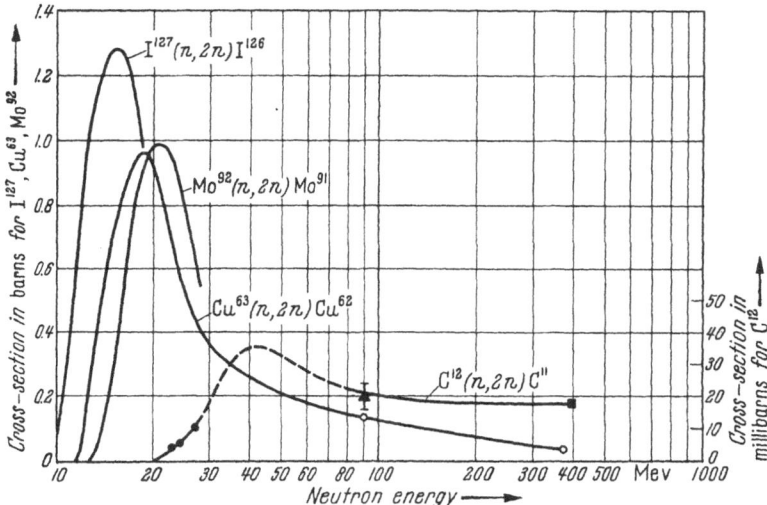

Fig. 15. Excitation functions for some $(n, 2n)$ reactions useful as neutron detectors. Curves below 30 Mev from Ref. [3]. Dashed portion of C^{12} curve estimated from theory of W. HECKROTTE and P. WOLFF: Phys. Rev. **73**, 264 (1948), and experiment of E. O. McMILLAN and R. D. MILLER: Phys. Rev. **73**, 80 (1948).

If one of the reactions of Fig. 15 is used as an absolute flux monitor, some uncertainty will arise from the absolute β-counting which is necessarily a part of the technique. This source of error can be attacked by distributing the material to be activated throughout a scintillating material. The $C^{12}(n, 2n) C^{11}$ and

I^{127} $(n, 2n)$ I^{126} reactions are particularly convenient in this respect[1], since carbon is a constituent of organic scintillants and iodine of NaI (Tl).

Because the absolute cross-sections for producing radioactivity via $(n, 2n)$ reactions are so poorly known, this method has been used mostly in making measurements of relative intensity. An experiment in which the convenience of the activation technique was exploited was that of Bratenahl et al.[2] on the elastic scattering of 84-Mev neutrons from several elements. These authors used carbon detectors at various distances from the neutron beam axis to measure the elastically scattered neutrons, with a copper shield providing protection against charged particles which might excite the C^{11} activity. Absolute differential cross-sections (quoted with an accuracy of $\pm 10\%$) were obtained by placing one of the carbon detectors in the neutron beam as a monitor. Measurements with a proton recoil detector indicated little inelastic scattering, which might have been troublesome since the 20 Mev C^{12} $(n, 2n)$ C^{11} threshold is far below the mean neutron energy. This was not the case in the experiment of Hildebrand and Leith[3], in which the threshold nature of the C^{12} $(n, 2n)$ C^{11} reaction was utilized to give a sharply peaked effective neutron spectrum for use in total cross-section measurements. The primary spectrum (from deuteron stripping) was calculated to have a peak at 35 Mev and a full width at half maximum of 25 Mev. When detected by the C^{11} induced activity this spectrum was modified so as to have an effective peak at ~ 42 Mev with a width of ~ 15 Mev, and was completely free of neutrons below 20 Mev. This is an ideal way in which to make use of threshold detectors.

8. Nuclear emulsions. As previously mentioned, the use of photographic plates as neutron detectors at high energies is made difficult by the small size of the emulsions compared to the ranges of most of the particles produced by neutrons in the plates. In Fig. 16, the ranges in emulsion of various particles may be compared to 600 μ, the sensitive depth of a "thick" plate. Since the determination of a recoil proton's energy depends upon a range and/or a dE/dx measurement, it is apparent that observations of high-energy particles must be confined to trajectories dipping into the emulsion at small angles. Hence neutron spectrum measurements with photographic plates as detectors are usually made with a separate hydrogenous radiator, the emulsion being inclined at a slight angle to the recoil protons, and well shielded from the direct neutron beam. The most extensive measurements of this type have been made with the 14 Mev neutrons from the T(d, n) He^4 reaction[4], and also include measurements of the neutron-proton scattering cross-section[5]. The latter phenomenon has also been studied with plates at 90 Mev[6], with the nuclear emulsion's ability as a detector of low energy protons (1.5 Mev was the lower limit used) of particular value at small neutron scattering angles. Since the low energy protons are heavily scattered by the hydrogenous radiator, the same angular region can be studied more effectively with a neutron detector.

The high-energy neutron interactions which are especially suited to study with emulsions are those in which stars of one or more prongs are produced in

---

[1] J. Sharpe and G. H. Stafford: Proc. Phys. Soc. Lond. A **64**, 211 (1951). — P. S. Baranow and V. I. Goldanskii: Zh. eksp. teor. Fiz. **28**, 621 (1955).

[2] A. Bratenahl, S. Fernbach, R. H. Hildebrand, C. E. Leith and B. J. Moyer: Phys. Rev. **77**, 597 (1950). [For similar uses of the carbon activation see L. Schecter, W. E. Crandall, G. P. Millburn and J. Ise: Phys. Rev. **97**, 184 (1955); and R. D. Miller, D. C. Sewell and R. W. Wright: Phys. Rev. **81**, 374 (1951).]

[3] R. H. Hildebrand and C. E. Leith: Phys. Rev. **80**, 842 (1950).

[4] L. Rosen: Nucleonics **11**, 32, 38 (1953).

[5] J. C. Allred, A. H. Armstrong and L. Rosen: Phys. Rev. **91**, 90 (1953).

[6] R. Wallace: Phys. Rev. **81**, 493 (1951).

nuclei of an emulsion, or nuclei introduced into the emulsion as laminations, by impregnation, or in the form of thin wires. The method of imbedded wires has distinct advantage over the others in that the type of nucleus in which the interaction occurs can be unambiguously determined. A group at Harvard[1] has used Mo and W wires of $14\,\mu$ diameter, as well as nylon filaments, in Ilford G-5 emulsions in order to study neutron interactions in the 50 to 100 Mev range.

Equivalent to the use of nylon filaments is the introduction of gelatin laminations between emulsion layers, a method which allows studies of interactions in light nuclei to be performed without interference from the silver and bromine which form most of the mass of emulsion. BLAU et al.[2] have used gelatin layers of 7 to 8 microns between standard G-5 plates to study energy and angle distribution of prongs initiated by 300 Mev neutrons from the Columbia cyclotron.

An examination of the stars produced by this same neutron beam on the nuclei of unloaded G-5 emulsion has been carried out by BERNARDINI et al.[3], with similar studies at 180 Mev by FISHMAN and PERRY[4], and at 150 Mev by TITTERTON[5]. As remarked by FISHMAN and PERRY, the prong distributions at these different energies are quite similar. Though this fact may be understandable in terms of the detailed shapes of the neutron spectra used and the theory of nuclear reactions [4], it

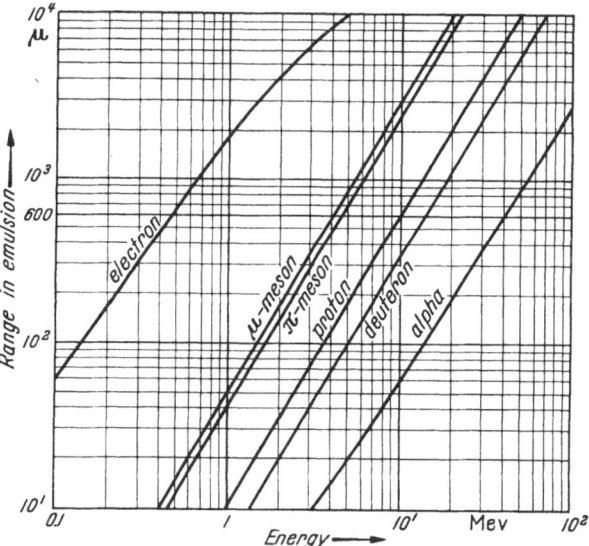

Fig. 16. Range-energy relations in nuclear emulsions (Courtesy Kodak Research Laboratories). All nuclear emulsions have these same curves [A. BEISER: Rev. Mod. Phys. 24, 273 (1952)].

indicates clearly the difficulties involved in trying to go in the other direction, i.e., attempting to measure neutron beam properties by observing stars in nuclear track plates. However, the characteristics of the stars produced in emulsions are at present among the few bits of information known about very high energy neutron interactions. Thus COOR et al.[6] used a prong number distribution for their 1.4 Be beam to estimate the absolute neutron flux above 550 Mev. It appeared to be about 10^3 neutrons per Cosmotron pulse.

The recent development of techniques for obtaining pulsed magnetic fields of several hundreds of kilogauss[7] provide hope that wider application of nuclear

[1] E. G. SILVER and R. W. WANIEK: Rev. Sci. Instrum. 25, 1119 (1954) describe the technique; R. W. WANIEK, L. GARRIDO, R. GOLOSKIE and E. G. SILVER: Phys. Rev. 99, 622 (1955) present some results and give references to other reports.

[2] M. BLAU, A. R. OLIVER and J. E. SMITH: Phys. Rev. 91, 949 (1953).

[3] G. BERNARDINI, E. T. BOOTH and S. J. LINDENBAUM: Phys. Rev. 85, 826 (1952).

[4] H. FISHMAN and A. M. PERRY: Phys. Rev. 86, 167 (1952).

[5] E. W. TITTERTON: Phil. Mag. 42, 109 (1951).

[6] T. COOR, D. A. HILL, W. F. HORNYAK, L. W. SMITH and G. SNOW: Phys. Rev. 98, 1369 (1955).

[7] S. FONER and H. H. KOLM: Bull. Amer. Phys. Soc., Ser. II 1, 298 (1956). — H. P. FURTH and R. W. WANIEK: Rev. Sci. Instrum. 27, 195 (1956).

track plates to high energy problems will occur, since momentum measurements in emulsions by means of such fields will partially compensate for the loss of fast particles from the emulsion. In fact, the nuclear track plate seems to be the only particle detector which can operate effectively in the small volumes to which such high fields are limited at present.

9. The cloud chamber. Considerations similar to those relating to the use of photographic emulsions as neutron detectors also govern the application of the cloud chamber. Both devices are especially suited to investigation of processes which are too complicated for analysis with counter arrays, or which are so little understood that a visual image of the events is necessary for interpretation of data. Among this last class of phenomena are many cosmic ray interactions, e.g., penetrating showers produced by neutral primaries, which have been often studied by a combination of counters and cloud chamber. This points up a major advantage of the cloud chamber compared to the photographic emulsion (and thus far also to the bubble chamber), namely, the former's ability to be triggered by a counter array which preselects events of interest. Also the large sensitive size of a cloud chamber compared to an emulsion is important, as well as a somewhat less tedious problem of scanning and data processing.

Extensive use of the gating property of an expansion cloud chamber has been made in cosmic ray research. Neutral particle-produced showers have been studied in this way by Gregory and Tinlot[1], who triggered a multiplate cloud chamber on events which penetrated a shielded counter telescope placed beneath the chamber but did not set off a counter array placed above the chamber. Rather than characterizing events by their penetrating power only, as would have been the case with a pure counter detector, they were able to analyze the structure and energy involved in each picture. A major difficulty was the uncertainty in shape of the primary neutron spectrum, which in this case was estimated to have a lower limit of 400 Mev, with an average energy of ~ 1 Bev and some neutrons of at least 4 Bev.

The application of cloud chambers to experiments with accelerator-produced neutrons has not required the triggering of expansion cloud chambers. (It is to be noted that random expansion avoids the bias which often enters counter-controlled chamber experiments because of the way in which events are selected.) Since the neutron beam can be pulsed through a cloud chamber by controlling the accelerator, standard technique is to take pictures as rapidly as the chamber can be cycled, or in the case of the diffusion chamber, as often as ion loading of the vapor supply permits. It is usual to admit the neutron beam into the chamber through a thin window, often placed at the end of a long snout in order to reduce the number of tracks resulting from interactions in the chamber walls.

An example of an expansion chamber experiment of this type is the work of DePangher with hydrogen at ten atmospheres[2]. Recoil protons produced by the 300 Mev beam of the Berkeley cyclotron were analyzed in a 21 700 gauss magnetic field, and some inelastic events of various types were also observed. From the proton tracks the incident neutron spectrum was deduced, and the angular distribution of n-p scattering was also measured. The accuracy of n-p elastic scattering data taken with a cloud chamber[3] is usually about 7 to 10%, which compares with 5% or less for most counter experiments. The difference

[1] B. P. Gregory and J. H. Tinlot: Phys. Rev. **81**, 667 (1951).
[2] J. DePangher: Phys. Rev. **99**, 1447 (1955).
[3] C. Y. Chih: Thesis, University of California Radiation Laboratory Report UCRL-2575 (1954) has studied n-p scattering at 90 Mev.

is apparently caused by the tedium of picture analysis, which seems to set a limit of ~2000 tracks on the data analyzed. On the other hand, the cloud chamber data is perhaps less subject to systematic errors.

If the absolute n-p scattering cross-section is known for the neutron energy being studied, the recoil protons seen in a hydrogen-filled cloud chamber can be used to measure the incident flux. This was done, for example, in the experiment of SCHLUTER[1] with 400 Mev neutrons on a 22 atmosphere diffusion chamber, which was concerned with the production of π^0-mesons in hydrogen. He discusses the special precautions necessary in using such a chamber in a neutron experiment, particularly a hydrogeneous filter to suppress neutrons below 100 Mev, careful shielding against ionizing background, and a long (27 sec) period between, pictures to insure complete recovery of the vapor supply.

Many other cyclotron experiments have been performed with the purpose of studying neutron collisions with nuclei of various elements. Gaseous interactions have been observed in D_2, He, O_2, K, among others, and carbon and lead plates have also frequently been used. In these cases the properties of the neutron beam must be known from other evidence, for it is not readily deducable from the pictures themselves.

The work of FOWLER et al., at the Cosmotron[2] illustrates the problems associated with experiments at very high energies. With a hydrogen filled (22 atmosphere) diffusion chamber and a 10 500 gauss field, these authors found it impractical to attempt identification of recoil protons from elastic scattering. The single tracks of such particles were not only difficult to find, but confusion with inelastic events involving π^0 production led to ambiguity. Hence they analyzed the neutron spectrum by measuring three-prong events in which the total energy involved could be estimated. The resultant spectrum, as produced by 2.2 Bev protons on carbon, was shown in Fig. 2. It rises gradually between 1 and 2 Bev, drops to zero at the maximum proton energy, and has a median energy of 1.79 Bev.

The liquid hydrogen bubble chamber promises to become an important tool for nuclear research, and should overcome many of the defects of the cloud chamber and nuclear track plate. With a superimposed magnetic field even chambers of modest size should be application to the problems of high energy neutron detection.

Another device which has been applied to neutron detection is the luminescent chamber. In the original report on this device, ZAVOISKY et al.[3], display some photographs of 380 Mev neutron interactions within a CsI scintillant. However, the development of this device as a particle detector is still in its early stages.

III. Application of detectors to neutron measurements.

10. Relative flux. Most experiments are concerned with measurements of relative intensity only. Total cross-sections in good and poor geometry, angular distributions, and neutron beam intensity vs. time are among the common relative measurements undertaken, and any of the detectors listed in part II can be used for this purpose. The important properties of a detector for such uses

[1] R. A. SCHLUTER: Phys. Rev. **96**, 734 (1954).

[2] W. B. FOWLER, R. P. SHUTT, A. M. THORNDIKE and W. L. WHITTEMORE: Phys. Rev. **95**, 1026 (1954).

[3] E. K. ZAVOISKY, G. E. SMOLKIN, A. G. PLAKHOV and M. M. BUTSLOV: Akad. Nauk. SSSR. **100**, 241 (1955).

are good time stability, high efficiency, and some degree of energy selectivity. Sensitivity to neutrons coming from one direction only is often helpful in reducing background, which may also be reduced by proper collimation of the beam and shielding of the counting apparatus. At high energies the collimation usually takes the form of steel or brass many feet in thickness, with a tapered hole viewing the "point" source of neutrons at the internal target of the accelerator.

Table 3. *Neutron detectors for relative flux measurement above 20 Mev.*

Detector	Useful neutron energy range	Maximum efficiency	Neutron energy sensitivity easily varied	Response to other particles	Remarks
1. Recoil proton telescope with separate radiator					
a) Telescope at 0° to neutron beam	All	10^{-2}	Yes	Low	Directional; very stable operation
b) Telescope at 10° to 30°	All	10^{-3}	Yes	Low	
2. Recoil detector with integral radiator	All	10^{-1}	Yes	Sensitive to γ-rays and protons if not shielded	Poor plateau; non-directional in most shapes
3. Time-of-flight detector	< 200 Mev	10^{-1}	Yes	Low	Accelerator must be pulsed; low neutron intensity results from long flight path
4. Bi fission chamber	> 50 Mev	10^{-4}	No	Efficiency for protons same as for neutrons	Fair plateau; poor time resolution (10^{-6} sec)
5. Induced radioactivity	Depends on reaction used [> 20 Mev for $C^{12}(n,2n)C^{11}$]	10^{-3} for 10 cm C_8H_8 scintillant	No	Protons	
6. Hydrogen-filled cloud chamber	All	Depends on H_2 pressure and size of chamber	No	Low	

The difficulties resulting from γ-ray background near the counters are not so severe as at lower energies, particularly since most high energy neutron detectors are rather insensitive to γ-rays. However, a troublesome background of high energy neutrons and protons is often present. The protons are usually the result of neutron interactions in the outer layers of the collimation and shielding and are therefore not entirely removed by insertion of a protective absorber equal in length to the proton range. Since the mean free path for neutrons above 100 Mev amounts to several inches of lead, counter shielding to remove neutron background often assumes massive proportions, and a cave providing 4π shielding for a counter array is not uncommon.

In Table 3 the properties of neutron detectors are listed according to their usefulness in relative intensity measurements. The proton recoil telescope has also been discussed in this connection in Sect. 3.

11. Absolute flux. For neutrons with energy above 40 Mev, where the asociated particle method (1) cannot be used, practically all absolute flux measurements are made by counting recoil protons from a hydrogenous substance. The most popular arrangement for this purpose is a separate radiator and scintillation telescope, and in Sect. 3 the application of this apparatus to absolute flux determinations was analyzed in detail. A massive scintillation counter offers much higher efficiency than the radiator-telescope combination, but is more difficult to calibrate accurately. It also lacks the broad plateau of counting rate vs. operating conditions (photomultiplier high voltage, discriminator bias, etc.) which is characteristic of a coincidence telescope. Absolute flux may also be measured by counting protons produced in a hydrogen-filled cloud chamber[1], although the method is tedious and of low efficiency.

In Table 4 the various neutron detectors are listed with those properties which are relevant to absolute flux measurements.

Table 4. *Neutron detectors for absolute flux measurement above 20 Mev.*

Detector	Useful neutron energy range	Accuracy %	Response to other particles	Remarks
1. Radiator and recoil proton telescope	All	5	Very low	Limited to accuracy of n-p cross sections
2. Recoil detector with integral radiator	All	10	Moderate	High stability electronics required, limited to n-p accuracy
3. Induced radioactitity via $C^{12}(n, 2n)\,C^{11}$	>20 Mev	10—15	Responds to protons	Limited by knowledge of activation cross-sections
4. Hydrogen-filled cloud chamber	All	10	Low	Limited by n-p accuracy and number of tracks scanned

12. Energy spectra. The analysis of neutron spectra at high energies is always carried out by studying the protons emitted from hydrogen at a small angle to the primary beam. It is necessary for this purpose to know the energy dependence of the elastic n-p scattering cross-section for the angle at which the protons are observed (Fig. 4). The protons may be separated into momentum groups by a magnetic field, as in a spectrometer or cloud chamber, or into energy intervals by a determination of range or specific ionization loss.

BALL[2] has employed a magnetic spectrometer[3] to obtain the spectrum produced by 340 Mev protons on LiD. With the available magnetic field ($2'' \times 12'' \times 30''$ area, 14000 gauss) and the sizes of the acceptance slits and of the counters of the detecting array, it was possible to resolve the energy of the recoil protons at $4°$ to the beam to within ± 13 Mev in the 200 to 300 Mev region. One convenience of the technique was that an entire spectrum could be examined at one time.

The same spectrum used by BALL was also examined by DePangher[4] with a high pressure cloud chamber. (This spectrum is included in Fig. 2.) The accuracy with which the energy of a 300 Mev neutron could be measured with the cloud chamber was ± 27.7 Mev, and the relative peak intensity was determined within 7%. Because of their low efficiency as neutron detectors (due to the small amount

---

[1] Cf. footnote 1, p. 511.

[2] W. P. BALL: University of California Radiation Laboratory Report UCRL-1938 (1952).

[3] J. B. CLADIS, J. HADLEY and W. N. HESS: Phys. Rev. **86**, 110 (1952).

[4] J. DePangher: Phys. Rev. **99**, 1447 (1955). (Cf. p. 510 above.)

of hydrogen present) and the tedium of analysis, cloud chambers are not often used for spectrum studies unless another neutron experiment is being performed, in which case the spectrum is a by-product.

Extensive studies of neutron spectra emitted from various target elements have been carried out with the counter telescope modifications described in Sect. 3. Three experiments are of special note because of the elegance of technique and precision of results:

(a) Hofmann and Strauch[1] examined spectra produced at several angles by 95 Mev protons. The recoil telescope was placed at 20° to the neutron beam, and consisted of seven stilbene scintillation counters electronically connected in such a way that the neutron fluxes within three energy intervals could be analyzed simultaneously. The major factors affecting the uncertainty in neutron energy as measured with this spectrometer were (a) the finite thickness of the crystals, (b) the finite angle of acceptance of the crystals, and (c) multiple scattering in the radiator. Calculation yielded a resolving power (defined as half-width et half-maximum) for this apparatus of 9.8 Mev at 55 Mev neutron energy and 8.3 at 90 Mev.

(b) A Harwell group[2] used a telescope consisting of two proportional counters, a carbon absorber, and an ion chamber for studying neutron spectra produced by 160 Mev protons. The ion chamber was sensitive only to protons near the end of their range, so that the incident neutrons were detected in a narrow energy band. Randle et al. found the half-width of their resolution curves to be 12.5 Mev at a mean neutron energy of 60, and 5.5 at 150 Mev.

(c) Protons of mean energy 244 Mev produced spectra which were analyzed by Nelson et al.[3] by means of proton ionization loss in an anthracene crystal placed at 20° to the neutron beam. The angle chosen was a compromise between the rapid decrease of n-p cross-section with increasing angle and the greater accuracy of (dE/dx) measurements when performed on the lower energy protons emitted at large angles. A pulse height analyzer covered the entire neutron spectrum from 120 to 240 Mev, with proton energy resolution curves of the

Table 5. *High energy neutron spectrometers.*

Spectrometer	Useful energy range	Energy resolution	Effi- ciency	Sensitivity to other particles	References
1. Radiator and proton recoil tele- scope with:					
a) Range absorber	All	7—10% up to 300 Mev	10^{-3}	Low	Ref. 1 and 2, p. 514
b) dE/dx counter	200 Mev	10—15%	10^{-3}	Low	Ref. 3, p. 514
2. Proton magnetic spectrometer with counters	All	8%	10^{-3}	Low	Ref. 2, p. 513
3. Hydrogen-filled cloud chamber	All	Depends on number of tracks scanned	—	Low	Ref. 2, p. 510
4. Time of flight detector	200 Mev	See Fig. 10	High	Low	Ref. 3, p. 502 and Ref. 1, p. 503

[1] J. A. Hofmann and K. Strauch: Phys. Rev. **90**, 449 (1953).

[2] T. C. Randle, J. M. Cassels, T. G. Pickavance and A. E. Taylor: Phil. Mag. **44**, 425 (1953).

[3] B. K. Nelson, G. Guernsey and G. Mott: Phys. Rev. **88**, 1 (1952). (Cf. p. 499, Sect. 3 above.)

crystal having a full width of about 10% at the lowest energy and 15% at the highest.

From these experiments we conclude that it is difficult to reduce the full widths of neutron energy resolution curves much below 10% at energies from 70 to 300 Mev. It is nevertheless possible to know the mean energy of the neutron beam to a much higher accuracy, and the effective energy of the beam can be often located to within a few Mev.

Table V summarizes detector properties as related to energy spectrum determinations.

13. Polarization. Any of the detectors described above in section II may be used with neutron fluxes which are polarized, i.e., have their spins preferentially aligned in one direction. In fact, if one does not desire to measure polarization effects in a particular experiment, it may be advantageous to use those detectors which are insensitive to the degree of polarization of the neutrons, such as recoil proton detectors with integral radiator, induced radioactivity detectors, bismuth fission counters, or time-of-flight detectors. In order to minimize the effects of possible beam polarization it is also common practice to measure angular distributions on both sides of the primary beam, which has the additional desirable effect of reducing errors due to apparatus misalignment.

The polarization of a neutron beam is usually measured with a counter telescope viewing protons emitted from a radiator at a non-zero angle with the incident beam. If the direction of the beam is k_i, and of the secondary protons k_f, then $n = k_f \times k_i$ is the normal to the "scattering plane". If the beam is polarized to a degree P_1 in some direction n' perpendicular to k_i, the intensity of the secondary protons will be proportional to the quantity $I = (1 + P_1 P_2 \, n \cdot n')$, where P_2 is the "analyzing power" of the telescope at the angle $\vartheta = \cos^{-1}(k_f \cdot k_i)$. If P_2 is known, then the azimuthal variation of I may be used to deduce the beam polarization P_1. Usually one measures the intensity factor I on opposite sides of the beam, i.e., where $n \cdot n' = \pm 1$, and the asymmetry $\varepsilon = (I_+ - I_-)/I_+ + I_-$ is then equal to $P_1 P_2$. It is important to choose an angle ϑ at which the magnitude of P_2 is large in order to minimize the error in P_1. In Table 6 the angles at which P_2 is a maximum at various energies are given. Carbon is most often used as a radiator because of its convenience and reasonably large analyzing power.

Table 6. *Maximum analyzing power of carbon and hydrogen scatterers for polarized neutrons vs. energy.*

Effective neutron[1] energy	Scatterer	Proton angle ϑ for maximum P_2 °	Maximum analyzing power P_2 [2]
98 Mev [3]	H	20	-0.09 ± 0.06
		50	$+0.41 \pm 0.11$
	C	20	-0.32 ± 0.09
		45	-0.42 ± 0.25
150 Mev [4]	H	30	-0.15 ± 0.05
		55	$+0.55 \pm 0.15$
310 Mev [5]	H	35	-0.26 ± 0.04
		75	$+0.46 \pm 0.08$
350 Mev [6]	H	25	-0.29 ± 0.03
		65	$+0.25 \pm 0.03$
	C	20	0.16 ± 0.01

---

[1] Determined by the neutron spectrum and absorber thickness used in the proton telescopes.

[2] See text for definition of P_2. Signs for P_2 are in accord with the convention that beam polarization is positive in the direction $n = k_f \times k_i$.

[3] P. HILLMAN and G. H. STAFFORD: Nuovo Cim. 3, 633 (1956).

[4] A. ROBERTS, J. TINLOT and E. M. HAFNER: Phys. Rev. 95, 1099 (1954).

[5] T. J. YPSILANTIS: University of California Radiation Laboratory Report, UCRL-3047 (1955).

[6] R. T. SIEGEL, A. J. HARTZLER and W. A. LOVE: Phys. Rev. 101, 838 (1956).

The measurement of the analyzing power P_2 can be performed by using a neutron beam which has been emitted at the angle ϑ (defined above) from an internal accelerator target constructed of the same material used as radiator for the polarization analyzing telescope. Under certain assumptions [4] the observed asymmetry will in this case be $(P_2)^2$. The sign of the analyzing power P_2 remains indeterminate, but may be deduced from an elastic neutron scattering experiment[1].

A measurement of beam polarization which is comparatively free from theoretical assumptions has been performed by Voss and Wilson[2], who observed the asymmetry of (98 ± 3) Mev neutrons scattered by the Coulomb field of uranium nuclei. In this case the magnitude and sign of P_2 are given by theory[3], affording a direct measure of beam polarization P_1. The detector used by Voss and Wilson was of the integral radiator type described in Sect. 4.

14. Devices for measuring dosage. Since the problems of measuring neutron dosage are somewhat different from those involved in research problems, instruments for this purpose will be discussed briefly in this section. The problem is to determine very low neutron fluxes with counters which are insensitive to other types of radiation and are stable and convenient to operate. The flux densities of interest vary from about 2 to 40 neutrons cm^{-2} sec^{-1}, with currently stated tolerance dose varying from $21 n$ cm^{-2} sec^{-1} at 5 Mev to $33 n$ cm^{-2} sec^{-1} at 200 Mev[4].

Because of the relative constancy of the permissible flux as a function of energy, the use of threshold detectors using the $C^{12}(n, 2n) C^{11}$ reaction (cf. Sect. 7) is appropriate. Cowan and O'Brien[5] have reported that a 7.62 cm diameter by 1 cm thick anthracene crystal irradiated for 30 min, "cooled" (allowed to decay) for 5 min, and counted for 30 min, can be used to detect flux as low as 2.3 cm^{-2} sec^{-1}. Longer irradiations yield useful measurements (presumably to 20 to 30% accuracy) at 5 to 10 $n/cm^2/sec$, with a minimum of 1 $n/cm^2/sec$. The energy spectrum of the incident neutrons (from the Cosmotron) was not reported, but the excitation function of the reaction (Fig. 15) provides an almost uniform response from 50 to 400 Mev. The ease of efficiency calibration (performed with a radioactive β-source) makes this type of detector more desirable for general monitoring purposes than an organic counter used as a recoil proton detector. A large scintillator used as carbon activation detector is also non-directional, which is a distinct advantage compared to a recoil telescope.

The insensitivity of a bismuth fission counter (q.v.) to γ-rays and its low background recommend it for use as a monitoring device. Cowan points out that a 36 liter volume of Kallmann's[6] bismuth loaded scintillant in which fission events are detected with 50% efficiency should give a counting rate of 70 counts/min with an incident flux of 1 $n/cm^2/sec$.

The highly directional response of a proton recoil telescope is undesirable if a general survey is being made, but is helpful in localizing the source of an observed neutron flux. Coombs[7] used a triple telescope with 3-inch paraffin

---

[1] R. T. Siegel: Phys. Rev. **100**, 437 (1955).

[2] R. G. P. Voss and R. Wilson: Phil. Mag. Ser. VIII **1**, 175 (1956).

[3] J. Schwinger: Phys. Rev. **73**, 407 (1948).

[4] Fixed by National Committee on Radiation Protection Subcommittee on Heavy Particles on January 23, 1953.

[5] F. P. Cowan and J. F. O'Brien: Paper 63 of the Geneva Conference on Peaceful Uses of Atomic Energy, 1955.

[6] M. Furst and H. Kallmann: Phys. Rev. **97**, 583 (1955).

[7] W. F. Coombs: U.S. Atomic Energy Commission Document NYO-7592 (1956).

converter during a thorough study of the neutron hazard at the Rochester 130″ synchrocyclotron. With an efficiency of about 1.5×10^{-3} for neutrons above 50 Mev and 5 in. square effective area, the telescope was capable of detecting 10^{-1} n/cm²/sec with an acceptable counting rate.

General references.

[1] BARSCHALL, H. H.: Detection of Neutrons (in this volume). — Summarizes techniques used below ~15 Mev.

[2] WATTENBERG, A.: The Standardization of Neutron Measurements. Ann. Rev. Nucl. Sci. **3**, 119 (1953).

[3] FOWLER, J. L., and J. E. BROLLEY: Monoenergetic Neutron Techniques in 10—30 Mev Range. Rev. Mod. Phys. **28**, 103 (1956).

[4] WOLFENSTEIN, L.: Polarization of Fast Nucleons. Ann. Rev. Nucl. Sci. **6**, 43 (1956).

[5] a) MORRISON, P.: A survey of nuclear reactions.
  b) FIELD, B. T.: The neutron. — Articles in Experimental Nuclear Physics (Editor: E. SEGRÈ), Vol. II. New York: John Wiley & Sons 1953.

[6] HESS, W. N.: A Summary of High-Energy Nucleon-Nucleon Cross-section-Data. University of California Radiation Laboratory Report. UCRL-4639 (January 1956).

[7] CHEN, F. F., C. P. LEAVITT and A. M. SHAPIRO: Total p-p and "p-n" Cross-sections at Cosmotron Energies. Phys. Rev. **103**, 211 (1956). — Includes tabulation of total nucleon-nucleon cross-section data from 41 Mev to 2.6 Bev as of February 29, 1956.

Sachverzeichnis.

(Deutsch-Englisch.)

Bei gleicher Schreibweise in beiden Sprachen sind die Stichwörter nur einmal aufgeführt.

Subject Index.

(English-German.)

Where English and German spelling of a word is identical the German version is omitted.